C0-ALU-536

THE GROUNDNUT CROP

World Crop Series

Available

The Grass Crop
The physiological basis of production
M.B. Jones and A. Lazenby

The Tomato Crop
A scientific basis for improvement
J.G. Atherton and J. Rudich

Wheat Breeding
Its scientific basis
F.G.H. Lupton

The Potato Crop
The scientific basis for improvement
P.M. Harris

The Sugar Beet Crop
Science into practice
D.A. Cooke and R.K. Scott

Forthcoming titles

Bananas and Plantains
S. Gowen

Oats
R.W. Welch

THE GROUNDNUT CROP

A scientific basis for improvement

Edited by

J. Smartt

Department of Biology
Southampton University, UK

CHAPMAN & HALL
London · Glasgow · Weinheim · New York · Tokyo · Melbourne · Madras

Agr
QK
495
.L52
G76x
1994

Published by Chapman & Hall, 2–6 Boundary Row, London SE1 8HN, UK

Chapman & Hall, 2–6 Boundary Row, London, SE1 8HN, UK

Blackie Academic & Professional, Wester Cleddens Road, Bishopbriggs, Glasgow G64 2NZ, UK

Chapman & Hall GmbH, Pappelallee 3, 69469 Weinheim, Germany

Chapman & Hall USA, One Penn Plaza, 41st Floor, New York NY 10119, USA

Chapman & Hall Japan, ITP-Japan, Kyowa Building, 3F, 2–2–1 Hirakawa-cho, Chiyoda-ku, Tokyo 102, Japan

Chapman & Hall Australia, Thomas Nelson Australia, 102 Dodds Street, South Melbourne, Victoria 3205, Australia

Chapman & Hall India, R. Seshadri, 32 Second Main Road, CIT East, Madras 600 035, India

First edition 1994
© 1994 Chapman & Hall

Typeset in 10/12pt Times by Florencetype Ltd, Kewstoke, Avon

Printed in Great Britain by St Edmundsbury Press, Bury St Edmunds

ISBN 0 412 40820 1

Apart from any fair dealing for the purposes of research or private study, or criticism or review, as permitted under the UK Copyright Designs and Patents Act, 1988, this publication may not be reproduced, stored, or transmitted, in any form or by any means, without the prior permission in writing of the publishers, or in the case of reprographic reproduction only in accordance with the terms of the licences issued by the Copyright Licensing Agency in the UK, or in accordance with the terms of licences issued by the appropriate Reproduction Rights Organization outside the UK. Enquiries concerning reproduction outside the terms stated here should be sent to the publishers at the London address printed on this page.

The publisher makes no representation, express or implied, with regard to the accuracy of the information contained in this book and cannot accept any legal responsibility or liability for any errors or omissions that may be made.

A catalogue record for this book is available from the British Library

Library of Congress Catalog Card Number: 94-70983

♾ Printed on acid-free text paper, manufactured in accordance with ANSI/NISO Z39 48–1992 (Permanence of Paper)

Contents

Contents

Contributors

J.J.K.B. Asiedu
The African Regional Centre for Technology
BP 2435
Avenue Djily Mbaye
Dakar
Senegal

B.G. Cook
Senior Pasture Agronomist
Queensland Department of Primary Industries
Cnr Cartwright Road and Louisa Street
PO Box 395
GYMPIE
Queensland 4570
Australia

P. Coolbear
Seed Technology Centre
Massey University
Palmerston North
New Zealand

I.C. Crosthwaite
Extension Agronomist
Queensland Department of Primary Industries
Kingaroy
Queensland
Australia

J.G. Davis
Department of Crop and Soil Sciences
University of Georgia
Coastal Plain Experiment Station
Tifton
GA 31793
USA

Contributors

W.J. Florkowski
Associate Professor
The University of Georgia
College of Agricultural and Environmental Sciences
Department of Agricultural and Applied Economics
Georgia Experiment Station
Griffin
GA 30223–1797
USA

G.J. Gascho
Department of Crop and Soil Sciences
University of Georgia
Coastal Plain Experiment Station
Tifton
GA 31793
USA

R.O. Hammons
Hammons Consultancy
1203 Lake Drive
Tifton
Georgia 31794–3834
USA

T.G. Isleib
Department of Crop Science
North Carolina State University
Raleigh
NC 27695–7620
USA

J.I. Keenan
Department of Surgery
Christchurch School of Medicine
University of Otago
New Zealand

K.J. Middleton (deceased)
J. Bjelke Petersen Research Station
Queensland Department of Primary Industries
P O Box 23
Kingaroy 4610
Queensland
Australia

U.R. Murty
Director
National Research Centre for Sorghum
Rajendranagar
Hyderabad
Andhra Pradesh 500 030
India

R.C. Nageswara Rao
Legumes Program
International Crops Research Institute for the Semi-Arid
  Tropics (ICRISAT)
Patancheru
Andhra Pradesh 502 234
India

S.N. Nigam
Principal Groundnut Breeder
International Crops Research Institute for the Semi-Arid
  Tropics (ICRISAT)
Patancheru
Andhra Pradesh 502 324
India

S. Pande
Saudi Agricultural Development Co. (INMA)
P.O. Box 148
Alkamasen
Wadi-Al-Dawasir 1191
Kingdom of Saudi Arabia

V. Ramanatha Rao
International Board for Plant Genetic Resources
Regional Office for Asia, the Pacific
  and Oceania
7th storey, RELC Building
30 Orange Grove Road
Singapore 1025

G.V. Ranga Rao
Legumes Program
International Crops Research Institute for the Semi-Arid
  Tropics (ICRISAT)
Patancheru
Andhra Pradesh 502 324
India

G.P. Savage
Department of Biochemistry and Microbiology
Lincoln University
Canterbury
New Zealand

S.B. Sharma
International Crops Research Institute for the Semi-Arid
   Tropics (ICRISAT)
Patancheru, Andhra Pradesh 502 324
India

C.E. Simpson
Texas Agricultural Experiment Station
Texas A & M University
Stephenville
Texas 76401
USA

A.K. Singh
Genetic Resources Unit
International Crops Research for the Semi-Arid
   Tropics (ICRISAT)
Patancheru
Andhra Pradesh 502 324
India

J. Smartt
School of Biological Sciences
Department of Biology
Biomedical Sciences Building
University of Southampton
Bassett Crescent East
Southampton SO9 3TU
UK

D.H. Smith
International Crops Research Institute for the Semi-Arid
   Tropics (ICRISAT)
Patancheru
Andhra Pradesh 502 324
India

J. Sprent
Department of Biological Sciences
The University
Dundee DD1 4HN
UK

J.A. Wightman
*Legumes Program International Crops Research Institute for the
Semi-Arid Tropics (ICRISAT)*
Patancheru
Andhra Pradesh 502 324
India

G.C. Wright
Queensland Department of Primary Industries
J. Bjelke Petersen Research Station
Kingaroy
Qld 4610
Australia

J.C. Wynne
Head
Department of Crop Science
North Carolina State University
Raleigh
NC 27695–7620
USA

Series foreword

This book is a most welcome addition to the World Crop Series. For although in terms of world figures groundnut annual production (in shell) at 23 506 000 tonnes is second to and well below that of soyabean at 114 011 000 tonnes of grain, it is of special importance in developing countries. Thus about ten times as much of the groundnut crop is produced in developing countries than in developed countries, whereas well over half the production of soyabean is in developed countries (*FAO Yearbook, 1992*). Moreover, the groundnut crop has a special role in sustaining the agricultural systems of the semi-arid tropics where, because of the stressful climates, farming is particularly difficult.

This series recognizes that, because the disciplines contributing to crop improvement are becoming more specialized, it is increasingly difficult for the specialists involved to understand each other's language and perceptions. It is not only a problem for those active in research but is perhaps an even greater difficulty for those who are involved in agricultural development, extension and education to obtain a critical, balanced and up-to-date view of the scientific advances that are contributing to crop improvement.

At the same time it is evident from the way in which research priorities and programmes are being addressed internationally that increasing emphasis is being placed on multidisciplinary projects. Hence, in spite of increased specialization, the need for scientists to understand each other and integrate their efforts is also increasing. The multidisciplinary approach is evident, for example, in the project-based system of organization and management that has been recently introduced in the International Crops Research Institute for the Semi-Arid Tropics (ICRISAT). This is the institute within the system of eighteen International Centres of the Consultative Group for International Agricultural Research which has been given a special responsibility to work on the groundnut crop and also to conserve the world germplasm collection of this species and its wild relatives. It is not surprising, therefore, that several of the authors of this book are ICRISAT scientists. But interest in the crop is not restricted to the semi-arid tropics; and while the major production regions are India, China and Africa (in that order) it is also important elsewhere, especially in the USA. Scientific expertise in the crop extends even wider and the authors who were invited to contribute to this book by the editor reflect the considerable international interest in groundnuts.

I believe the result of this combined effort will become an essential source of ideas and information for anyone who takes more than a passing interest in this crop. While making no scientific compromises, the language is meant to be comprehensible to anyone with some understanding of biological and agricultural sciences. It should therefore be a considerable aid to the multidisciplinary approaches that are now a feature of agricultural research and development.

E.H. Roberts
University of Reading

Preface

The event which threw the 'Groundnut Crop' into sharpest relief was undoubtedly the monumental fiasco of the East African Groundnut Scheme, conceived in 1946. What had been an interesting botanical curiosity and the source of useful foods, edible oils and feeding stuffs became the butt of innumerable comedians and to be in any way associated with the crop in the 1950s was to be a sitting target for wits and humorists of every stripe. To be informed within 24 hours of arrival in Northern Rhodesia that I was to devote my attention and energies to this crop made me feel that I had drawn the shortest of short straws. It was some while before gloom and despondency lifted but by the end of my first crop season I realized that in fact I had not been given the metaphorical poisoned chalice but something that was to become an abiding interest, which has remained fresh over a period of 40 years.

Three events in that first season (1954–5) launched me on a course which would over the years give me enormous job satisfaction. Stephen Hoyle of the Nyasaland Department of Agriculture (a groundnut enthusiast if ever there was one) made available to me his own variety collection. Early in 1955 I received from Queensland, in response to a request for groundnut material, several selections of a Bolivian groundnut landrace named Mani Pintar which were to prove extremely interesting. The final gift of the fairy godmother was a copy of *The Peanut – the Unpredictable Legume*, which became my groundnut bible during my spell in Africa, as it did for many others in the groundnut (or peanut) world. It was so successful and highly regarded that its successor *Peanuts – Culture and Uses* did not appear until 1973 – a lapse of 22 years. In the meantime, however, two useful if modest publications were produced in 1967: the PANS (Pest Articles and News Summaries) Manual No. 2, *Pest control in GROUNDNUTS* (published by the Overseas Development Agency in the UK) and *Groundnut Production/Groundnut Research 1954–61* published in Lusaka by the Government Printer, Zambia.

Advances in production technology and research continued at an accelerating pace and in 1982 the American Peanut Research and Education Association (now Society) published *Peanut Science and Technology*, a successor to its earlier volume, *Peanuts – Culture and Uses*. This organization has since 1974 published its own journal *Peanut Science* which provides a useful vehicle for publication on any aspect of investigation related to the crop. Prior to that date, APREA had published its own untitled journal for a number of years. In addition, the Proceedings of its annual meetings are invaluable sources of information – going back to the early 1960s if one also includes those of its predecessor, the Peanut Improvement Working Group.

The first major work on the groundnut to emerge from the developing world was published in 1988 by the Indian Council of Agricultural Research with a similar format and organization of material to the APREA publications. This very valuable and substantial piece of work is concerned primarily with the Indian context while the corresponding APREA publications address the North American situation. Thus there appears to be scope for a broader approach to the crop, incorporating experience with the groundnut crop in the rest of the world. This was the motivating force behind the present volume, which attempts where possible to break new ground with the presentation of authoritative reviews such as that of Professor Janet Sprent as well as broad geographic coverage of other topics such as breeding and the control of pests and disease.

Any editor setting about the task of assembling a cast of authors does so with a set of ideals in mind, well knowing whom to recruit. The process is beset with trials and disappointments and this volume has had its fair share or more of these. It is with the deepest regret that the death of a major contributor, Keith Middleton, must be recorded. The chapter on Diseases had been fully drafted by the time of his death and was completed by his colleagues at ICRISAT. I count myself privileged and honoured to have met him in the year before his untimely death. He will be sorely missed.

The current depressed state of the world economy has not been without its effects on contributors who had agreed in good faith to produce chapters but in the event were unable to do so. An individual threatened by the horrors of 'rationalization' may not be in the best condition to discharge obligations gladly accepted in happier times. A somewhat fuller coverage had been anticipated but in fairness to contributors who had met the deadlines the decision was taken to proceed, with the editor making an attempt to fill as far as possible the lacunae left by the contributions which failed to materialize.

Acknowledgement must be given to ICRISAT's work on the groundnut crop, guided by Dr J.G. Ryan, its present director, and his predecessor Dr L.D. Swindale. The encouragement given by Dr Y.L. Nene as Deputy Director General is acknowledged gratefully. The experience of editing this work has been very rewarding and has given me a pretext for making contact with those whose work I have long admired but whom otherwise I might never have approached.

J. Smartt
Southampton, October, 1993

CHAPTER 1

Groundnut production and trade

W.J. Florkowski

This chapter focuses on major issues and implications of groundnut production, processing, distribution and consumption, discussion of yields and area harvested, world prices and trade illustrates strong and weak links in the world groundnut sector.

Revenue generated from groundnut sales amounts to billions of dollars annually and sales for domestic use are important to regional and national economies. Groundnut is a high value product and an important source of cash to farmers in less developed countries, allowing the purchase of other foods and industrial goods.

Groundnut oil is a major vegetable oil in many parts of the world and its sale to domestic and foreign consumers generates revenue. Its production by mills crushing locally grown groundnut provides jobs and income for numerous workers. By-products, such as groundnut cake and meal, directly or indirectly improve nutrition.

Export sales of groundnut, groundnut oil and cake can increase the income of groundnut farmers and processors. Groundnut exporters include a number of countries in Africa, Asia and the Americas but the three major exporters in the 1980s were the USA, Argentina and China. An increase in yield and in the area harvested led to increased production and exports from China and Argentina. Many countries with increasing yields concentrate on satisfying a strong domestic demand but are likely to export in years of world shortage, taking advantage of a high price.

1.1 PRODUCTION, AREA HARVESTED AND YIELD

Several factors determine changes in yield, area harvested and level of production. Applied technology, better management and the weather directly influence yield. The area under crop is determined by groundnut price, expected profits from groundnut relative to other crops, and the

The Groundnut Crop: A scientific basis for improvement. Edited by J. Smartt. Published in 1994 by Chapman & Hall, London. ISBN 0 412 408201.

inputs available. Both policy and weather influence yield and area harvested. Policy is a factor that represents a broad spectrum of macroecono-
mic variables. Foreign exchange needs frequently lead to a government
decision to increase production of an exportable commodity, as was the
case in China. Food needs of growing populations could accelerate adoption of higher yielding cultivars.

In the following sections, discussion of the groundnut industry on each
continent centres on selected producing countries which illustrate changes
in the groundnut sector of specific regions. Trends in area harvested and
production among countries reveal the changing importance of the domestic groundnut sector.

For several countries, continuous information on the industry is not
available for the entire period (1960–1990). Reasons for gaps in the
statistics include creation of sovereign states, social unrest and military
conflicts. For the purposes of this presentation, however, the available
information is adequate.

1.1.1 Groundnut production

Groundnut production is important to several countries with large populations, in which groundnut plays a role as a food crop. The volume of
production reflects the amount of resources allocated to groundnut and its
importance to a particular economy. Table 1.1 compares the average
output for selected countries over three decades (1960s–1980s), and Table
1.2 highlights changes in production levels on a percentage basis. Some of
the more obvious trends are as follows.

(a) North America

In Mexico, production decreased in the 1970s but recovered in the 1980s
with an increase of 22%. In the USA, a production increase of 62%
between the 1960s and the 1970s reflected technological gains.

(b) South America

In the large groundnut countries, production has been decreasing. The
opposite trend is true in many smaller countries. Brazil was the largest
producer in South America in the mid 1960s and the 1970s but its output
decreased by 53% during the 1980s in comparison with the 1970s. Since
1986, Argentina has been consistently the largest producer in South
America: production increased by 23% between the 1960s and the 1970s
but decreased by 30% during the 1980s. The overall decrease in production
in South America was the result of a smaller area being sown to groundnut
in Brazil and Argentina.

(c) Africa

In Africa, countries that reported decreased groundnut production out-numbered those with an increase in the 1980s. The average for individual countries varied widely over the three decades but the continent's total increased slightly, having declined during the 1970s.

In West Africa, there was a decrease in major producing countries during the 1980s over the 1960s, although output was fairly stable for the region as a whole and it is noticeable that the Ivory Coast nearly doubled its output in the 1980s, while in Benin production increased by 134% between the 1960s and the 1980s.

In eastern Africa, overall production increased over the period 1960–1989. In the Sudan, the increase was by 136% during the 1970s, but Uganda's average output in the 1980s dropped to 37% below its 1960s level.

In central Africa, Zaire's output increased by 89% over the three decades. In southern Africa, the trend was for a decrease in production of groundnut.

(d) Asia

All major producers in Asia increased their average output over the three decades. In south-west Asia, India's production climbed steadily over the period, increasing by 25%. In south-east Asia (excluding Vietnam), production increased by 176% between the 1960s and 1980s. In Myanmar, the average output rose steadily, increasing by 16% during the 1970s and a further 36% during the 1980s. Indonesia more than doubled its average output.

East Asia's output grew mainly because of China, which recorded a 90% increase in average annual groundnut production between the 1970s and the 1980s. However, output fell in Taiwan and did so more dramatically in Japan, where it declined by 63% between the 1960s and the 1980s.

(e) Europe

The groundnut is of marginal importance to European farmers. The growing conditions in most European countries are unsuitable for its production. In several countries where it is grown, there may be a comparative advantage in production of other crops. Relatively high yields have not compensated for the small area and, as a result, the production of groundnuts has been low in most of Europe, seldom exceeding 10 000 MT in any one country. Turkey has been the largest producer: in the 1960s, it averaged 25 400 MT annually, increasing this level by 77% in the 1970s and another 34% in the 1980s.

MT = tonnes.

TABLE 1.1   *Average groundnut production in selected countries (1000 tonnes)*

Country	1960–1969	1970–1979	Change 1970s/1960s	1980–1989	Change 1980s/1970s
North America					
Mexico	89.7	59.5	−30.2	72.9	13.4
USA	992.9	1606.8	613.9	1669.7	62.9
South America					
Argentina	325.8	402.1	76.3	281.8	−120.3
Brazil	588.1	582.8	−5.3	272.6	−310.2
Colombia	NA	2.6	NA	4.4	2.2
Ecuador	10.9	10.3	−0.6	9.8	−0.5
Paraguay	13.6	17.5	3.9	36.0	18.5
Uruguay	4.8	2.6	−2.2	1.2	−1.4
Venezuela	2.4	19.6	17.2	15.4	−4.2
Africa					
West Africa	285.5	209.0	−76.5	184.1	−24.9
Benin	24.0	49.9	25.9	56.2	6.3
Burkina Faso	114.1	118.0	3.9	123.7	5.7
Cameroon	120.5	123.5	3.0	185.6	62.1
Ghana	59.6	74.8	15.2	63.7	−9.1
Gambia	100.3	123.1	22.8	114.9	−8.2
Ivory Coast	29.1	44.3	15.2	80.5	36.2
Mali	112.0	124.2	12.2	99.8	−24.4
Niger	235.1	138.0	−97.1	69.3	−68.7
Nigeria	1402.8	607.8	−795.0	489.5	−118.3
Senegal	926.9	875.6	−51.3	712.9	162.7
Togo	15.8	20.0	4.2	28.5	8.5
East Africa	102.6	189.6	87.0	165.3	−24.6
Burundi	7.0	25.0	18.0	57.8	32.8
Egypt	41.5	32.6	−8.9	31.7	−0.9

1.1.2 Area harvested

The average area of groundnut harvested varied considerably from country to country and from continent to continent between 1960 and 1989 (Tables 1.3 and 1.4). In some countries, this area declined following the use of new, higher yielding cultivars which resulted in maintained or increased production on a reduced area. The allocation of land to groundnut cultivation is influenced by changes in relative market price and agricultural policies. In some countries, such as Nigeria, a rapidly growing population increased the demand for food and this was an important reason for withdrawing land from groundnut production. In other countries, including major exporters such as Argentina, South Africa and Brazil, producers

TABLE 1.1 *cont.*

Country	1960–1969	1970–1979	Change 1970s/1960s	1980–1989	Change 1980s/1970s
Sudan	263.2	621.2	358.0	565.3	−55.9
Tanzania	21.3	58.9	37.6	57.6	−1.3
Uganda	179.9	210.5	30.6	114.3	−96.2
Central Africa	179.2	200.8	21.6	265.3	64.5
Malawi	170.6	158.9	−11.7	176.0	17.1
Zaire	187.7	242.6	54.9	354.5	111.9
Southern Africa	112.6	144.1	31.5	94.5	−49.6
Madagascar	33.6	44.0	10.4	33.2	−10.82
Mozambique	66.8	113.4	46.6	69.0	−44.4
South Africa	251.1	301.2	50.1	179.2	−122.0
Zimbabwe	99.0	117.6	18.6	96.6	−21.0
Asia					
South-west Asia	2509.3	2926.5	417.2	3118.1	191.6
India	4951.1	5782.8	831.7	6168.8	386.0
Pakistan	67.4	70.2	2.8	67.3	−2.9
South-east Asia	608.7	669.7	61.0	1221.9	552.2
Indonesia	375.4	504.4	129.0	780.7	276.3
Myanmar	358.8	416.3	57.5	566.7	150.4
Thailand	132.9	210.4	77.5	151.6	−58.8
Vietnam	NA	88.2	NA	151.4	63.2
East Asia	289.0	377.0	88.0	412.6	35.6
China	2185.6	2488.7	303.1	4733.9	2245.2
Japan	135.7	87.8	−47.9	50.4	−37.4
Korea	3.5	7.8	4.3	19.4	11.6
Taiwan	110.1	94.4	−15.7	83.9	−10.5
Europe					
Turkey	25.4	45.0	24.6	60.2	15.2

reacted to improved production technology, changes in relative price, agricultural policies and the weather by reallocating land to other crops.

(a) North America

Mexico and the USA are the two major producers on the continent. The average area harvested in Mexico decreased by 41% during the 1970s, dropping to a minimum of 25 000 ha in 1975, but in the 1980s an initially slow increase in the area harvested accelerated and the average for the decade exceeded that of the 1970s by 27%.

The USA harvested, on average, a fairly stable area between 1960 and 1989. The government groundnut programme limited the extension of this

TABLE 1.2 *Production tonnage increases and decreases in selected countries (based on FAO and USDA statistics)*

Average change (%)	1960s:1970s		1970s:1980s	
	Increases	Decreases	Increases	Decreases
1–10	Pakistan Cameroon Burkina Faso	(Brazil) (Ecuador) (Malawi) (Senegal)	USA India Sudan Burkina Faso	(Pakistan) (Egypt) (Tanzania) (Gambia)
11–25	Argentina Paraguay Myanmar China India Ghana Mali Uganda Zimbabwe South Africa Gambia	(Taiwan) (Egypt)	Mexico Benin Malawi	(Venezuela) (Ghana) (Mali) (Nigeria) (Zimbabwe) (Senegal) (Madagascar) (Taiwan)
26–50	Indonesia Zaire Madagascar Togo	(Mexico) (Uruguay) (Japan) (Niger)	Colombia Myanmar Zaire Togo	(Argentina) (Thailand) (Japan) (Cameroon) (Uganda) (Mozambique) (South Africa)
51 and more	USA Venezuela Thailand Sudan Ivory Coast Benin Burundi Mozambique Tanzania	(Nigeria)	Paraguay Indonesia Vietnam China Ivory Coast Burundi	(Brazil) (Uruguay) (Niger)

area; however, an expansion of the groundnut export market and changes in the government programme led to a slight increase in area harvested over time. In the 1980s, it exceeded the 1960s area by 4%.

(b) South America

The area harvested in Argentina varied widely between 1960 and 1989: it expanded to a record 428 000 ha in 1978 and the average area during the

1970s was 19% greater than in the 1960s. However, the introduction of improved cultivars led to a 39% decrease in that area between 1980 and 1989. In Brazil, the average area harvested decreased by 17% in the 1970s and continued to decrease steadily in the 1980s. In 1990, Brazilian farmers planted only 89 000 ha with groundnut, an area 87% less than the record 694 000 ha in 1967.

Paraguay expanded its area under groundnut production in the mid 1960s. During the late 1960s and the early 1970s the area figure stagnated and in some years declined but it increased again between the mid 1970s and 1990 so that between the 1960s and the 1980s it had grown by 94%. Several other South American countries cultivate groundnut – for example Uruguay and Ecuador – but average area harvested has remained below 20 000 ha.

(c) Africa

The area harvested in Africa increased during the 1970s but the average in the 1980s was 5% less than in the 1960s. The land allocation within individual countries, however, varied in response to each country's specific demands and policies.

In western Africa, area harvested decreased generally between 1960 and 1990. The area harvested in Senegal remains the largest in Africa, despite decreasing by 15% during the 1980s. Nigeria assigned the second largest area to groundnut production in Africa during the three decades but it now ranks third and the average area harvested in the 1980s was only 2% larger than its 1960s average. Cameroon emerged as the country with the third largest area harvested in West Africa: it increased by 90% between the 1960s and the 1980s. In Niger the area harvested decreased by 31% between the 1960s and the 1970s and by 44% between the 1970s and the 1980s. Mali recorded a decrease of 24% in the period 1960–1989, with an especially large drop in the 1970s.

In central Africa, the average area harvested increased over time because of an expansion in groundnut production. Zaire harvested an average area 77% greater in the 1980s than in the 1960s.

In eastern Africa, the area harvested increased in the 1970s and remained stable in the 1980s. In the Sudan, the average became the second largest in Africa in the 1980s, exceeding Nigeria's area by 21%. Uganda decreased its area during the 1970s by 22% compared with the 1960s and by another 32% during the 1980s. Tanzania increased its average area harvested by 117% between the 1960s and the 1980s. In Egypt there was a decrease of 29% during the period 1960–1989.

In southern Africa, the average area harvested in the 1980s was smaller than in the 1960s. The Republic of South Africa devoted a steadily decreasing land area to groundnut production. There was no growth in the

TABLE 1.3   *Average area of groundnuts harvested in selected countries*
(1000 hectares)

Country	1960–1969	1970–1979	Change 1970s/1960s	1980–1989	Change 1980s/1970s
North America					
Mexico	69	41	−28	52	11
USA	576	605	29	597	−8
South America					
Argentina	284	339	55	173	−166
Brazil	482	399	−83	186	−213
Ecuador	11	12	1	10	−2
Colombia	NA	2	NA	3	1
Paraguay	17	20	3	33	13
Uruguay	6	3	−3	2	−1
Venezuela	3	16	13	11	−5
Africa					
West Africa	264	291	27	245	−46
Benin	67	74	7	81	7
Burkina Faso	226	173	−53	203	30
Cameroon	166	199	33	315	116
Ghana	44	99	55	105	6
Gambia	145	148	3	99	−49
Ivory Coast	50	51	1	87	36
Mali	161	129	−32	123	−6
Niger	342	235	−107	132	−103
Nigeria	599	953	354	609	−344
Senegal	1063	1106	43	906	−200
Togo	40	31	−9	39	8
East Africa	129	217	88	208	−9
Burundi	9	18	9	50	32
Egypt	21	17	−4	15	−2

average area harvested in Zimbabwe between the 1960s and the 1970s but
it expanded by 9% in the 1980s.

(d) Asia

Asian countries reported the world's largest area harvested of groundnut.
For the continent as a whole, the rapid expansion in average area harvested during the 1980s surpassed the 1960s area by 47%. The area in India
is larger than that of any other country and is increasing slowly: it averages
more than three times that of China.

In south-east Asia, the area harvested is substantial. Myanmar is the

TABLE 1.3 *cont.*

Country	1960–1969	1970–1979	Change 1970s/1960s	1980–1989	Change 1980s/1970s
Sudan	320	769	449	739	−30
Tanzania	47	86	39	102	18
Uganda	250	196	−54	133	−63
Central Africa	258	254	−4	374	120
Malawi	239	232	−7	258	26
Zaire	277	276	−1	489	213
Southern Africa	160	176	16	138	−38
Madagascar	40	60	20	35	−25
Mozambique	109	200	91	107	−93
South Africa	319	273	−46	223	−50
Zimbabwe	170	170	0	186	16
Asia					
South-west Asia	3506	3604	98	3664	60
India	6967	7159	192	7270	111
Pakistan	45	48	3	57	9
South-east Asia	522	554	32	696	142
Indonesia	380	421	41	579	158
Myanmar	583	651	68	595	−56
Thailand	97	126	29	121	−5
Vietnam	NA	84	NA	121	37
East Asia	353	321	−32	354	33
China	1919	2092	173	2647	555
Japan	63	45	−18	28	−17
Korea	4	8	4	15	7
Taiwan	100	69	−31	93	24
Europe					
Turkey	11	20	9	24	4

Note Reported figures have been rounded to the nearest thousand.

third largest producer of groundnuts in Asia and its area harvested increased by 12% between the 1960s and the 1970s, but in the 1980s it was only 2% above the level of the 1960s. Indonesia increased its average area harvested by 52% between the 1960s and the 1980s and now ranks seventh in the world. The increase in Vietnam between the 1970s and the 1980s was an impressive 44%. Thailand expanded its area harvested during the 1970s but it decreased slightly during the 1980s.

East Asia includes countries with an expanding groundnut sector – especially China – and countries where groundnut production becomes less important, such as Japan. China rapidly expanded its area harvested between 1960 and 1989: the average increased by 9% between the 1960s

TABLE 1.4 *Increases and decreases in area of groundnut harvested in selected countries*

Average change (%)	1960s:1970s		1970s:1980s	
	Increases	Decreases	Increases	Decreases
1–10	USA Ecuador Pakistan China India Ivory Coast Gambia Senegal	(Malawi)	India Guyana Zimbabwe Benin	(USA) (Myanmar) (Thailand) (Sudan) (Mali)
11–25	Argentina Paraguay Myanmar Indonesia Cameroon Benin	(Brazil) (Mali) (Uganda) (Egypt) (Burkina Faso) (Togo) (South Africa)	Pakistan Indonesia Burkina Faso Malawi Tanzania	(Ecuador) (Mozambique) (South Africa) (Senegal) (Egypt)
26–50		(Mexico) (Uruguay) (Taiwan)	Mexico Colombia Taiwan	(Argentina) (Uruguay) (Venezuela)
51 and more	Venezuela Sudan Ghana Mozambique Tanzania		Paraguay Vietnam Cameroon Ivory Coast Zaire	(Brazil)

and the 1970s and by 27% in the 1980s. Japan consistently planted a reduced area: the average in the 1980s was 56% smaller than in the 1960s. Korea is a small producer but has been steadily expanding its area under groundnut cultivation.

(e) Europe

Areas harvested in individual European countries remained below 10 000 ha. Greece, Spain and Italy devoted 3000–5000 ha to groundnut production. In the early 1960s, Yugoslavia grew 21 000 ha of the crop but the area decreased towards the end of the decade.

Turkey consistently increased its groundnut area: the average increased by 82% in the 1970s and the expansion continued in the 1980s.

1.1.3 Groundnut yield

Significant changes in groundnut yield occurred between 1960 and 1989 (Tables 1.5 and 1.6). Increased yields, which reflect the use of improved technology and management, are often necessary to increase exports.

Adoption of improved cultivars adapted to local growing conditions led to relative increases in yields in China and Argentina during the 1980s. Increased yields in many other countries signal the potential of cultivars outside the traditional groundnut-producing countries, though smaller yield gains have been reported in countries located in climatic zones which are drastically different from those of the leading groundnut-producing regions.

(a) North America

Groundnut yields in Mexico and the USA reflect differences in quality of cultivars and production technology. Compared with the 1960s, the average yield in Mexico increased by 13% in the 1970s but then decreased by 5% in the 1980s. In contrast, the USA average yield consistently increased, in relative terms, by 62% between the 1960s and the 1980s.

(b) South America

Between the 1960s and the 1980s, average yield in Brazil increased by 20%. Brazil experienced periods of declining yield in the late 1960s and the first half of the 1980s, but the average has increased steadily overall.

Changes in yield were more dramatic in Argentina. The average yields were smaller than in Brazil during the 1960s and 1970s, but larger in the 1980s when Argentina exceeded its 1970s yield by 41%.

Yield in Paraguay and other South American countries remained considerably lower than in Argentina and Brazil throughout the three decades. However, the average yield in Paraguay increased by 38% and the rate of yield increase was especially high during the 1980s.

(c) Africa

A number of groundnut producers in Africa achieved record yields in the 1960s and 1970s but experienced a decline during the 1980s. Among countries reporting this pattern were Benin, Burkina Faso, Burundi, Mali, Tanzania, Uganda, Zaire and the Republic of South Africa.

In western Africa, the overall average yields remained stable between 1960 and 1990 but fluctuated in individual countries. The average Gambian yields have been consistently above 500 kg/ha since 1983. Senegal reported considerably lower and more variable yields than Gambia. Yields in Nigeria, a large producer of groundnuts, exceeded its average yield of the 1970s by 32% in the 1980s. In Ghana and Cameroon average yields

TABLE 1.5 *Average groundnut yields in selected countries (kg/ha)*

Country	1960–1969	1970–1979	Change 1970s/1960s	1980–1989	Change 1980s/1970s
North America					
Mexico	1329	1497	168	1420	−77
USA	1724	2652	928	2792	140
South America					
Argentina	1154	1174	20	1660	486
Brazil	1231	1383	152	1481	98
Ecuador	985	956	−29	925	−31
Colombia	NA	1600	NA	1302	−298
Paraguay	822	869	47	1137	−268
Uruguay	735	1058	323	550	−508
Venezuela	987	1235	248	1501	266
Africa					
West Africa	743	757	14	752	−5
Benin	379	751	372	694	−57
Burkina Faso	510	651	141	601	−50
Cameroon	716	620	−96	502	−118
Gambia	706	855	149	1165	310
Ghana	1708	761	−947	559	−202
Ivory Coast	596	872	276	916	44
Mali	792	1124	332	856	−268
Niger	724	540	−184	508	−32
Nigeria	NA	619	NA	817	198
Senegal	876	781	−95	804	23
Togo	423	756	333	846	90
East Africa	977	1278	301	1092	−186
Burundi	696	1496	800	1073	−423
Egypt	2004	1920	−84	2201	281

declined between 1960 and 1989 – in Ghana by 67% and in Cameroon by 30%. The average yield in the Ivory Coast climbed by 54% over the same period. In Benin and Burkina Faso the average 1970s yields were higher than in the 1960s or the 1980s; Niger reached its highest average yield in the 1960s and Mali in the 1970s.

In eastern Africa average yields vary widely. Egypt's, the highest in Africa, increased by 10% between the 1960s and the 1980s. Uganda's yield increased by 72% during the 1970s compared with the 1960s but the average then decreased by 31%. Tanzania reported a rapid increase of average yield during the 1970s but, again, these gains were eroded during the 1980s and decreased by 37%. In relative terms, average yield in Mozambique decreased by 62% between 1960 and 1989. The Sudan also

TABLE 1.5 *cont.*

Country	1960–1969	1970–1979	Change 1970s/1960s	1980–1989	Change 1980s/1970s
Sudan	835	817	−18	753	−64
Tanzania	574	910	336	572	−338
Uganda	724	1245	521	860	−385
Central Africa	699	794	95	711	−83
Malawi	715	684	−31	681	−3
Zaire	682	903	221	742	−161
Southern Africa	866	792	−74	678	−114
Madagascar	875	845	−30	960	115
Mozambique	1073	534	−539	413	−121
South Africa	796	1106	310	801	−305
Zimbabwe	718	682	−36	536	−146
Asia					
South-west Asia	1050	1139	89	1015	−124
India	712	807	95	841	34
Pakistan	1388	1470	82	1188	−282
South-east Asia	997	1199	202	1136	−63
Indonesia	988	1185	197	1506	321
Myanmar	620	644	24	950	306
Thailand	1382	1671	289	1255	−416
Vietnam	NA	1297	NA	833	−464
East Asia	1465	1505	40	1543	38
China	1140	1195	55	1760	565
Japan	2148	1943	−205	1773	−170
Taiwan	1107	1376	269	1092	−284
Europe					
Turkey	2332	2289	−43	2324	35

suffered a slow but steady decline of 10% in average yield between the 1960s and the 1980s.

In central Africa, average yields increased but then decreased, leaving them at a slightly higher level in the 1980s than in the 1960s. In southern Africa, the average yield declined steadily between 1960 and 1989. The Republic of South Africa, a major African producer, experienced an increase in average yield during the 1970s but the average during the 1980s was 28% lower than in the 1970s. Zimbabwe's yield declined by 21% in the 1980s compared with the 1960s.

(d) Asia

In south-west Asia, India's average yield increased by 13% between the 1960s and the 1970s and by 4% in the 1980s.

TABLE 1.6 *Yield increases and decreases in selected countries in comparison to average yields in the period 1960–1969 (based on FAO and USDA statistics)*

Average change (%)	1960s:1970s		1970s:1980s	
	Increases	Decreases	Increases	Decreases
1–10	Argentina	(Japan)	USA	(Mexico)
	Paraguay	(Egypt)	Brazil	(Ecuador)
	Myanmar	(Sudan)	India	(Japan)
	China	(Malawi)	Ivory Coast	(Benin)
	Pakistan	(Zimbabwe)	Senegal	(Burkina Faso)
				(Cameroon)
				(Niger)
				(Sudan)
11–25	Mexico		Venezuela	(Colombia)
	Brazil	(Senegal)	Egypt	(Pakistan)
	Venezuela	(Niger)	Togo	(Taiwan)
	Taiwan			(Thailand)
	Thailand			(Zaire)
	India			(Mali)
	Indonesia			(Zimbabwe)
	Gambia			(Mozambique)
26–50	Uruguay		Argentina	(Uruguay)
	Burkina Faso	(Cameroon)	Paraguay	(Vietnam)
	Ivory Coast	(Mozambique)	Myanmar	(Ghana)
	Mali		China	(South Africa)
	Zaire		Indonesia	(Tanzania)
	South Africa		Nigeria	(Uganda)
			Gambia	
51 and more	USA	(Ghana)		
	Benin	(Nigeria)		
	Tanzania			
	Uganda			

Yields in eastern Asia increased in the 1970s but decreased slightly in the 1980s. However, Myanmar's average yield in the 1980s was 48% higher than in the 1970s. Yields in Thailand peaked between 1975 and 1977 but declined below these record levels: the average yield in the 1980s was 9% lower than in the 1960s.

Yield in Indonesia increased over time. The average increased by 20% in the 1970s and was 52% higher in the 1980s than in the 1960s. Vietnam is expanding its groundnut production but its average yield in the 1980s was 36% below that of the 1970s.

In south-east Asia, yields in Taiwain and Japan were on average lower in the 1980s than in the 1960s. Yields in China remained stagnant during the 1960s and 1970s but then increased, reaching a record 2036 kg/ha in 1987, and decreased again in 1989–1990. The average yield in the 1980s was 48% higher than in the 1970s. The average yield in Korea has been increasing rapidly but its groundnut sector remains small.

(e) Europe

Competition from other crops and a highly developed agriculture in Europe contribute to average yields which are among the highest in the world. The exception is the former USSR but groundnut is produced largely in non-European regions of the Commonwealth of Independent States.

Among small European producers, Italy attains high average yields which frequently exceed 2 000 kg/ha. Spain and Greece have also achieved high yields but those of Yugoslavia and Bulgaria have been highly variable.

The largest groundnut producer in Europe is Turkey, whose average yield in the 1960s was 2 332 kg/ha. This decreased by 2% in the 1970s and in the 1980s it was 0.3% below the 1960s level.

1.2 STOCKS AND PRICES

The level of world groundnut stocks influences both supply and prices in the groundnut trade. A record level of 944 000 MT was achieved in 1976/77 as the result of accumulation of large stocks in the USA, India and Argentina. An increased demand for oil-bearing crops in the preceding year was probably a major factor in this accumulation. Large domestic stocks in Argentina and the USA led to large exports in the following years.

Between 1971/72 and 1990/91, the USA consistently maintained the largest stocks in the world. These ranged from 178 000 MT in 1972/73 to 646 000 MT in 1985/86 but declined in the late 1980s. India recorded its highest stocks (450 000 MT) in 1974/75, following the world food shortage of the early 1970s, but the levels have varied widely from year to year. Since 1986/87, India has reported no groundnut in stock. Argentina reported large stocks between 1975/76 and 1977/78 but levels decreased during the 1980s.

The combined stocks of other countries have decreased from a record level of 615 000 MT in 1974/75; during the 1980s, they exceeded 400 000 MT in 1982/83 only. In 1990/91, the groundnut inventory was 169 000 MT in these other countries of the world – excluding the European Community (EC) countries, which maintained stocks ranging from none to 58 000 MT between 1971/72 and 1990/91.

Stock levels and price influence domestic consumption and the level of exports. Changes in price in major producing and exporting countries change export levels. During the 1960s, groundnut prices were the lowest in African countries: pricing was particularly competitive in Senegal, Gambia and the Sudan and relatively low in the Republic of South Africa, Malaysia and Nigeria, while prices in China, Argentina and the USA were amongst the world's highest.

Prices remained low in African countries in the early 1970s but gradually increased until those in the Sudan and the Republic of South Africa were among the highest in the world by the late 1970s. Between the late 1970s and the early 1980s, the USA, Argentina and China were more competitive in relation to African exporters.

In the USA, prices were supported under a government programme. The early 1980s brought a change in the programme through the introduction of a two-tier price system. Groundnut sold at a higher price for quota; production above the quota level, selling at a lower price, was called 'additionals'. The price of additionals responded to the expected prices of groundnut oil, meal, other vegetable oils and world groundnut demand. The two-tier price system lowered the price of exported groundnut. The price on the world market became less influenced by the US government programme.

Table 1.7 shows export prices for major exporting countries from 1960 to 1987. The USA and Argentina became more competitive in the 1970s, as did China in the late 1970s when it began its large-scale exports. During the 1980s, the prices of many exporters became less attractive to buyers, given the quality and the intended use.

International groundnut price provides information to producers, brokers, and consumers. The price for each type of groundnut reflects supply and demand conditions and the quality attributes of traded cultivars.

The supply situation changes frequently because harvests in the northern and southern hemispheres occur at different times of the year. A major factor that influences the supply is the weather. Weather-related supply changes cause price variability and alter quantities exported by specific countries. Quality also influences price, but crop quality information is less readily available. Groundnut quality is determined during trading and processing.

Prices in different parts of the world interact with each other, reflecting the many uses of groundnut. Groundnuts from different sources are reasonable substitutes, especially for a similar use. Some specialized end users, however, prefer groundnut with specific quality attributes and pay higher prices to satisfy their needs.

Timely information concerning current and expected price on the international market affects planting decisions, stocks and quantities of applied inputs. A European-based company publishes prices of the three major groundnut exporters: the USA, China, and Argentina. Reports include

TABLE 1.7 *Groundnut export prices for major exporters (US dollars per ton) (based on FAO statistics)*

Year	US	CN	AG	SD	SAF	GB	ID	BZ	THL	SEN	VT	MAL
1960	235	247[a]	257	192	195	185	242	NA	188	173	NA	NA
1961	255	247[a]	NA	186	188	166	276	220	183	172	NA	NA
1962	318	233[a]	257	163	188	160	254	185	184	173	NA	NA
1963	239	235[a]	NA	159	184	150	233	168	198	173	NA	NA
1964	228	234[a]	225	173	186	154	273	190	197	173	NA	198
1965	251	232[a]	200	155	219	199	205	222	202	172	NA	288
1966	251	221[a]	174	210	208	163	96	251	196	175	NA	243
1967	218	233[a]	211	178	183	154	208	230	305	174	NA	185
1968	230	233[a]	NA	161	191	133	213	228	295	137	NA	185
1969	358	269[a]	333	230	197	185	280	222	309	165	NA	196
1970	382	275[a]	393	246	197	194	293	229	307	188	NA	226[a]
1971	292	297[a]	296	229	214	233	287	247	317	197	NA	244
1972	261	319[a]	303	242	207	223	278	244	401	269	NA	251
1973	435	447[a]	481	275	275	267	470	359	470	387	NA	268
1974	600	640	464	548	334	463	507	557	633	573	NA	300
1975	622	680	516	482	332	493	552	606[a]	655[a]	530	NA	291
1976	685	560	507	396	297	453	753	634	682	380	242	473
1977	693	642	839	577	397	550	969	849	723	477	290	531
1978	733	710	1000	641	418[a]	529	1408	853	702	602	429	815
1979	741	701	526	715	581	569	743	857	864	948	614	798
1980	731	728	628	518	947	409	805	793	1380	807	620	767
1981	943	1047	942	1037	792	670	1192	1241	1195	1116	625	693
1982	864	702	559	385	614	374	889	863	835	408	603	612
1983	784	590	440	796	692	389	912	905	1042	435	589	625
1984	762	713	700	833	750	535	947	1025	760	450	652	658
1985	676	679	450	763	644	490	652	845	619	619	553	382
1986	727	588	502	905	619	637	571	893	566	527	419	453
1987	787	686	435	612	671	606	762	930	479	412	385	323

[a] Estimate

NA = not available

US = United States, CN = China, AG = Argentina, SD = Sudan, SAF = Republic of South Africa, GB = Gambia, ID = Indonesia, BZ = Brazil, THL = Thailand, SEN = Senegal, VT = Vietnam, MAL = Malawi.

prices for the US Runner 40/50s, Argentine Runners, and Chinese Hsujis. Quarterly price quotations allow traders and producers to adjust expectations and make appropriate decisions.

1.3 INTERNATIONAL TRADE

The strong world demand for groundnut oil on the one hand and groundnut products on the other determines the economic significance of the crop. This dichotomy of demand has led to the evolution of two markets. Groundnuts traded as an oil-bearing crop compete with other oil-bearing crops such as palm oil, soybean, rapeseed, etc. – groundnuts as a source of oil are important in many less developed countries. Used as a snack, groundnuts compete with other snack foods – for example, tree nuts. In the form of butter, groundnuts compete with other spreads. Groundnut snacks and peanut butter are consumed largely in developed industrial economies.

Quality characteristics of groundnuts as snack foods or as kernels for oil reflect the two distinct uses of groundnut that influence the volume traded, price, and quality attributes. Groundnuts intended for different uses follow separate geographical trade patterns and use specialized marketing channels. The needs of end-users communicated through market price influence the selection of cultivars and levels of production. A separate market exists for groundnut meal and cake (by-products from groundnuts crushed for oil), which compete with other feed protein supplements and are purchased primarily by countries with large commercial livestock production.

1.3.1 International trade in groundnuts

The volume of world trade increased slowly during the period 1971/72–1990/91. During the 1970s, the volume traded on the world market ranged between 950 000 MT and 1 200 000 MT – except in the 1975/76 season, which was characterized by severe drought, especially in Europe. This drought led to increased demand for imported oil-bearing crops. During the 1980s, the volume traded edged upwards, although at an erratic pace, and ranged from 1 000 000 MT to 1 300 000 MT. In the years 1985/86 and 1990/91 the volume traded was 1 365 000 MT and 1 339 000 MT, respectively.

The geographical composition of exporters changed between 1971/72 and 1990/91. The USA exported the largest volume during the period 1971/72–1974/75. After a short crop in 1975, Europe looked for alternative sources of oil and meal and imported groundnuts as a substitute for other oil-bearing commodities which, in normal crop years, were also less expensive. Simultaneously, according to the existing programme in the USA, the price support for groundnuts increased to compensate for increased production costs, resulting in the highest price of any major exporting country. This encouraged importers to search for other suppliers: the Sudan exported 293 000 MT in 1976; India exported 239 000 MT, while Senegal reported record exports of 173 000 MT. Countries which surpassed the USA in exports in 1976 were those with lower prices: the price of Sudanese groundnut was 42% lower and Indian 30% lower.

Before 1980, other major exporters included the Republic of South Africa, the Sudan and Senegal. Since 1980, both China and Argentina have surpassed the Republic of South Africa and the Sudan in export volumes. A short 1980 crop in the USA led to difficulties in meeting importer's demands and the USA itself imported 82 000 MT in 1980/81, leaving net exports of only 46 000 MT.

The shortage in the USA groundnut supply coincided with changes in economic and agricultural policies in China. China's groundnut exports increased following the change in regulating domestic agricultural production and the implementation of economic incentives for farmers. China exported 305 000 MT in 1980/81; the figure declined between 1981/82 and 1984/85 but China has since exported the largest quantity of groundnut in the world, except for 1986/87 and 1989/90.

Argentina had a history of groundnut exports dating back to World War II. Despite the deteriorating terms of trade during the period 1935–1946, Argentina expanded its exports in 1940–1945 (Mundlak *et al.*,1989). For years its groundnut exports were negligible but they were increased following record production in 1976/77 and 1978/79 (a result of increased yields and a near-record area harvested). The need to generate foreign exchange caused a steady increase in the volume exported and the quality of Argentine groundnut satisfied buyers' requirements. Between 1985/86 and 1989/90, volume exceeded 100 000 MT; in 1990/91, exports from Argentina reached a record 215 000 MT. Furthermore, domestic economic policies maintained a competitive edge and Argentina's groundnut price between 1984 and 1990 was the lowest among major exporting countries.

The USA, China and Argentina will continue to share the bulk of the world export market in the foreseeable future. Argentina can easily expand the area harvested, while the USA and China can alter allocation of land to groundnut production, depending on domestic needs and opportunity costs.

Changes in quality and consumption have influenced shifts in origins of groundnut exports. Groundnuts, intended for use as a source of edible oil, have become over time a food item eaten as kernels or paste. This change called for different quality attributes. Crushing qualities remained important to oil producers but appearance, flavour and texture became more important for snack foods and peanut butter.

The composition of the import market remained stable but a few countries, such as Indonesia, Korea, and Malaysia, increased imports after 1979. Major importers included Canada, western European countries (12 EC nations and Switzerland, Sweden and Norway), Japan and the former USSR. Total imports by Canada and Japan remained largely unchanged, while the USSR imports increased steadily. Since reaching a peak in 1976, western European countries have reduced total imports to about 50% of 1976 levels.

The three major competitors in the western European market are the USA, Argentina and China. The USA market share in western Europe increased from 14% in 1973 to 37% in 1986, with a peak of 54% in 1979.

Western European countries absorbed 60% of the USA exports. Argentina and China increased their shares from 7% and 8% in 1982 to 20% and 21% in 1986, respectively. The Republic of South Africa has also been a significant exporter to the western European market but has lost more than half of its market share.

China and the USA have shared equally about 85% of the market in Japan, with some annual variation in quantities exported. The USA shipped about 10% of its groundnut exports to Japan. Argentina also exported small quantities to Japan. The USA dominated the Canadian market for years: from 1973 to 1986, Canada alone bought more than 20% of US groundnut exports each year, except in 1980. Chinese exports are increasing steadily and, while the market share of Chinese groundnut in Canada accounted for only 5% in 1982, it increased to 11% in 1984 and exceeded 15% in 1986. The major sources for Indonesia are China and Vietnam, with an increasing importance for the latter. The opening of Indonesia's market to Vietnam groundnuts coincided with improved political relationships between the two countries.

A relatively strong dependence on exports characterizes the USA's position in the world market. The USA is not the largest groundnut producer in the world but is consistently the largest exporter. The position of Argentina is even more sensitive to world markets because the groundnut is largely an export commodity: the domestic market is small and Argentina exports groundnut to earn foreign currency. China, the second largest producer, uses most of its crop for domestic consumption; exports represent only a fraction of its total production and can be expanded.

1.3.2 International trade in groundnut oil

Groundnut oil is an important vegetable oil in some regions where other oil-bearing crops are not grown; in other regions, it is a substitute for other vegetable oils. In some cultures, consumption habits call specifically for the use of groundnut oil in cooking.

World exports of groundnut oil increased sharply between 1960 and 1968, reaching 510 000 MT, but the trade decreased in the late 1960s. African exports dominated the world oil market and the varying volume of groundnut oil on that market resulted from the variable supply coming from Nigeria and Senegal. Senegal was the largest exporter in the 1960s, followed by Nigeria and then Argentina.

During the 1970s, the most significant change was a steady decline of Nigerian oil shipments. Towards the end of the decade Nigeria became a net importer of groundnut oil. Senegal remained the largest groundnut oil exporter but at highly variable rates during the 1970s. Brazil and China emerged as new oil exporters in the 1970s: Brazil slowly expanded its exports and maintained them through the decade, while China's oil exports were especially high during the first half of the 1970s and at the end of the decade. Argentina and the Republic of South Africa continued to export oil during the 1970s.

In the 1980s exports from Senegal, Argentina and Brazil decreased and the Sudan began to export significant quantities of groundnut oil.

Major buyers include the EC members (purchasing oil from Argentina and Africa) and the countries of Asia where the groundnut oil trade is largely intra-continental.

1.3.3 International trade in groundnut cake and meal

The trade in groundnut cake and meal followed a different pattern from the trade in shelled nuts or oil. European countries were the major buyers of meal in the first half of the 1960s; Japan and the Soviet Union also purchased significant quantities of cake and meal.

In the first half of the 1960s, India was the largest cake and meal supplier, its share frequently exceeding 50% of the total world cake and meal exports. Argentina and Senegal were other major exporters. In the second half of the decade, India and Senegal continued to export large quantities of groundnut cake and meal but Argentina reduced its exports while Brazil rapidly increased its share of the world cake and meal market.

During the 1970s, India and Senegal continued to dominate cake and meal exports; the importance of exports from Argentina decreased and the Sudan's exports slowly expanded. After peaking in the early 1970s, Brazilian exports declined. Myanmar exported fairly large quantities of cake and meal in the 1960s but its share decreased by the mid 1970s.

The international trade decreased during the second half of the 1970s and the presence of aflatoxin in cake exported from Senegal may have contributed to a decrease in exports to Europe. A decrease in trade did not change the relative importance of India's and Senegal's cake and meal exports. The Sudan emerged as a large cake and meal supplier towards the end of the decade; Myanmar decreased its exports while exports from Argentina and Brazil were highly variable. Europe remained the primary destination for cake and meal exports.

In the 1980s, the trade continued at the 1970s level, according to the geographical pattern established in earlier decades. The exception was the year 1985, when exports reached a record low level of 385 000 MT.

1.4 CONCLUDING COMMENTS

This overview of the world groundnut yields, area harvested and production indicates considerable regional differences. Groundnuts are cultivated in many countries but their relative economic importance varies. As in the case of many other cultures, production is highly developed in countries with well-functioning agriculture. Differences in applied technology and management are major factors contributing to regional differences in yield and production. Marketing infrastructure (including transportation, storage, quality control, advertising and promotion) is an important factor influencing consumption of groundnut and groundnut products.

The quality–price relationship determined by end users is likely to increase in importance, and is influenced by both biological and economic factors. Quality attributes preferred by buyers can be met by cultivar development. The increasing importance of health concerns resulted in the expansion of canola production which has become the largest oil crop in the world (Fletcher *et al.*, 1992). Re-examination of peanut oil properties is a desirable long-term goal. Protecting quality after harvest and the timely supply of adequate volume are essential for the smooth operation of regional and international markets.

Opportunities and constraints on the world market will be generated primarily by variable production in different countries and regions. Droughts, sudden disease outbreaks or insect attacks can occur with little warning, limiting time for adjusting market strategy. Canada, Japan and western Europe will remain the largest buyers. Large but variable quantities have been exported to Indonesia. Expanding markets include countries with well-developed industry and service sectors, such as those of the Pacific Rim. New markets include the central and east European countries where economic reforms are being implemented. Although groundnut products can become easily available, they remain expensive.

The former USSR represents a special case. It was steadily increasing its imports, their volume being comparable in size to Canada's. Recent political and economic changes and the emergence of new independent states will bring changes in lifestyle and consumption which may help exports. Some of the new states may increase their own production after the system of central planning disintegrates and privatization of the economy progresses.

ACKNOWLEDGEMENTS

The author would like to thank D. Cummins, S. M. Fletcher and J. C. Purcell for their helpful comments on earlier drafts of this chapter.

REFERENCES

Fletcher, S.M., Zhang, P. and Carley, D.H. (1992) *Peanuts: Production, Utilization and Trade in the 1980s*. Department of Agricultural and Applied Economics, College of Agriculture, University of Georgia, FS-91-32.

Food and Agricultural Organization *Production Yearbook*, United Nations Organization, various issues.

Food and Agricultural Organization *Trade Yearbook*, United Nations Organization, various issues.

Mundlak, Y., Cavallo, D. and Domenech, R. (1989) *Agriculture and economic growth in Argentina, 1913–1984*, IFPRI, Research Report 76.

US Department of Agriculture *World Oilseed Situation and Market Highlights*, Foreign Agriculture Circular, various issues.

CHAPTER 2

The origin and history
of the groundnut

R.O. Hammons

2.1 INTRODUCTION

Groundnut is a native New World crop. Early explorers found it cultivated extensively in both Mesoamerica and South America. Remnant
pericarp (fruit hull) tissue recovered from archaeological sites in Peru
dates its purposeful agricultural use there at approximately 3900–3750
years before the present (YBP). No one is certain how much earlier
domestication occurred but it probably first took place in the valleys of
the Paraná and Paraguay river systems in the Gran Chaco area of South
America.

Today, groundnut is an important oil, food and forage crop generally
distributed in tropical, subtropical and warm temperate zones. The exact
origin of the principal cultigen, *Arachis hypogaea* L., remains a subject of
scientific inquiry. Early Spanish, Portuguese, Dutch, German and other
explorers found Indians cultivating the crop on many islands in the
Antilles, on the northeast and east coasts of Brazil, in all the warm
regions of the Río de la Plata basin (Paraguay, Bolivia, northern
Argentina, extreme southwest Brazil), extensively in Peru and sparsely in
Mexico.

This chapter relates the history of the crop as revealed in art, artifacts
and archaeological remains from prehistoric sites; it traces early post-
Columbian accounts of the plant or fruit in contemporary natural historical
narratives of the sixteenth and seventeenth centuries and reviews current
information concerning the crop's centre of origin and geographic dispersion. It also lists linguistic affinities in common names and notes the impact
of the European vegetable oil industry on production expansion in West
Africa.

The Groundnut Crop: A scientific basis for improvement. Edited by J. Smartt. Published in
1994 by Chapman & Hall, London ISBN 0 412 408201.

2.2 PREHISTORIC GROUNDNUT AGRICULTURE

Peruvian civilization apparently began along the eastern slopes of the Andes near the tropical Amazonian lowlands, where many wild relatives of domesticated crops occur (Burger, 1989). However, archaeological evaluation of agricultural sites in that area is difficult because of climatic conditions. The coast of Peru at about 11° S is an extremely dry desert due to the cold off-shore ocean currents and no natural vegetation exists (Cohen, 1977). Along these shores are the oldest known examples of monumental architecture in the New World. The native peoples reshaped their environment to create distinctive Andean agricultural systems. About 4200 YBP the coastal society began building small scale irrigation systems (Burger, 1989). By 3200 YBP the demographic, technological and socio-economic foundations for central Andean civilization were in place.

The finding of well-preserved groundnut fruits in terracotta jars in the prehistoric burial sites at Ancón on the Peruvian coast north of Lima conclusively demonstrated the New World origin of the cultivated crop (Squier, 1877). Specimens recovered there date to the early Ancón culture, approximately 2740–2490 YBP (Towle, 1961).

Further north near Trujillo, archaeologists found funerary vases decorated with replicas of groundnut pods sculptured in relief, some of which used moulds made from actual pods. An earthenware pan recovered from a grave at the fishing village of Chimbote, Peru, had groundnut fruits painted upon the handle.

From about 1900 to 1300 YBP the agricultural peoples of the Mochica culture grew irrigated crops such as maize, beans, squash, peppers, potatoes, manioc and groundnut amid the searing desert of coastal Peru. The Moche empire extended along a 136 km swath of coastal Peru. Recently, near the village of Sipán, 7.5 km west of Pampa Grande in the heart of the Lambayesque valley, Walter Alva (1988) unearthed the richest documented burial ground of a pre-Hispanic ruler, finding treasures of unprecedented artistic magnificence. From above the 'royal' tomb they recovered from clandestine looters a pair of gold groundnut pods three times larger than natural size. The artisans faithfully reproduced the ridges and indentations that are present on one major cultivar grown in the area even today. Two unusual necklaces exhibited quality in attention to detail and care in execution: half of each necklace was of silver and the other half gold, fashioned into groundnut pods. A third necklace displayed a gold groundnut pod but its alternating lapis and gold beads reveal an extensive Moche trade network, with gold originating in the eastern Andes and the lapis from Chile.

The best dates yet established for the occurrence of groundnut in Peru are Bird's (1948) finding of its remains at Huaca Prieta near latitude 8° S on the coastline of the Chicama valley. Bird placed the appearance of groundnut there prior to maize and probably contemporaneous with warty squash. The precedence of groundnut to maize here is unique for Peruvian sites. Neither groundnut nor warty squash plant remains appeared in the preceramic refuse, suggesting their introduction concurrent with the first pottery. Carbon dating of the pottery, and thus for groundnut, is for the beginning of the ceramic period and ranges from 3500 to 3200 YBP. However, Cohen (1977) reports groundnut remains from the Gaviota Complex (preceramic period 6) in the Ancón-Chillón region of the coast of Peru with a possible age of 3750–3900 YBP. Lima bean, jack bean, sweet potato, peppers and squashes occurred with groundnut in this period, but not maize or the common bean that later became staples of Peruvian agriculture.

The usual cultivar of *A. hypogaea* found in the coastal sites of Peru has the long, slender, ribbed pod and sharp, recurved beak typical of subspecies *hypogaea* var. *hirsuta* (Krapovickas, 1968). Its similarity in external morphology, pod size and seed number to a major phenotype sold today in markets of coastal Peru reflects its antiquity in cultivation in ancient Peru. Towle (1961) dates remains of a smaller podded type that she recovered from a site at Supe, Peru, to the Early Ancón period. The modest reticulation and lack of dorsal humps on its pods suggest selection under domestication. The presence of groundnut, manioc and chilli pepper – domesticated east of the Andes – in early ceramic cultures to the west shows prolonged and effective contact between residents of the arid coast of Peru and those of the Gran Pantanal region.

The discovery of ethnobotanical samples of cultivated groundnut in a Coxcatlan cave in the Tehuacán valley of Mexico established groundnut as a cultigen on both American continents many centuries before their discovery by the Europeans. These date to about 1900 YBP. Although present in later phases (1200–450 YBP), they were never abundant (MacNeish, 1965; Smith, 1967). The surviving Aztec codices lack known illustrations of the groundnut. Current evidence suggests that groundnut was an introduced and unimportant crop in Mexico.

Phytomorphic representations of groundnut have yet to be found in Brazil, Bolivia or the Antilles, where the climate is less favourable for the preservation of archaeological plant remains. Thus, for the present, the deductions drawn from archaeological data must be supplemented and extended by evidence from natural historians and other disciplines.

2.3 POST-COLUMBIAN HISTORICAL NARRATIVES

2.3.1 Sixteenth century natural histories

The first written notice of the groundnut in the chronicles and natural histories of the sixteenth century is that of El Capitan Gonzalo Fernández, subsequently known as de Oviedo y Valdés, the first historian of the New

World. Ferdinand and Isabella sent Oviedo to Santo Domingo in 1513, where he soon became governor of Hispaniola (now Haiti/Dominican Republic) and royal historiographer of the Indies. In 1525 he sent Emperor Charles V his *Sumario Historia*, printed in Toledo two years later (Oviedo, 1527), and in 1535 he began publishing his *Historia General de las Indias* (Oviedo, 1535, 1547), a task finally completed three centuries later by the Spanish Royal Academy of History (Oviedo, 1851).

Oviedo (1527) describes groundnut (*maní*) as a very abundant, ordinary food crop sown and harvested in the gardens and fields of the Indians on Hispaniola and other islands. Its fruit was about the size of a pine nut with the shell. The Indians considered groundnut a healthy food and consumption was high.

Three centuries later the completed *Cronica* reported that the fruit grew underground, that pulling upon the branches uprooted the plant for harvest and that the seeds eaten raw or roasted were very tasty (Oviedo, 1851). These statements are not in the several sixteenth century editions (Hammons, 1982).

Although the specific geographic area where farmers domesticated groundnut is unquestionably South America, Oviedo's 1527 narrative also established distribution and adaptation of the crop as far as the West Indies, and its widespread and ordinary usage there before the discovery of the Americas by Europeans.

Ulrich Schmidt of Straubing, a German adventurer who spent 20 years in the Rio de la Plata basin, subsequently wrote a historical account of the conquests by the emperor's forces there. Schmidt (1567) encountered groundnut in 1542 when his expedition up the Río Paraguay from Asunción met the Cheriguanos (or Guaycurús) Indians who bartered maize, manioc and groundnut with the soldiers. Later, in the area inhabited by the Mepenes (or Abipones) he recorded a flourishing agriculture with maize, manioc, potatoes, sweet potatoes and groundnut in abundant supply.

The first delibrate introductions of groundnut to Europe probably went unrecorded. The New World presented a bountiful wealth of domesticated seed, root and tree crops, including maize, the amaranths, four species of bean, warty and summer squash, groundnut, pumpkin, gourds, tomato, chilli pepper, potato, manioc, sweet potato, the New World yam, arrowroot, cacao (*Theobroma*), papaya, pineapple, the cultivated tobaccos and cotton, *inter alia*. From the time of Columbus' first voyage, explorers collected useful and exotic American plants for introduction to Europe. One may speculate on the introduction of groundnut – plentiful in the Antilles – to Europe early in the sixteenth century. However, extensive reviews of early literature by writers of groundnut history (Dubard, 1906; Higgins, 1951; Krapovickas, 1968; Gillier and Silvestre, 1969; Hammons, 1973, 1982) do not document such an occurrence.

The writer who first gave Europe an account of the more useful plants discovered by the Spanish adventurers in the New World was Dr. Nicholas Monardes, a physician of Seville, who published a small treatise on medicinal plants (Monardes, 1569). Other Spanish editions appeared in 1571 and 1580, and the work passed through many editions in many languages, including an English version in 1577 by Frampton. Better known than the original Spanish or the vernacular translations is the excellent Latin version included in the *Exoticorum* of Clusius (1605), which spread this knowledge throughout the world of learned men.

Monardes never visited America but derived his information and specimens from the navigators and explorers who frequently arrived in Seville. He saw only the nameless underground fruit, which he described as having neither root nor plant! He wrote that the fruit was:

> half a finger round and had a small kernel [*sic*] that parted into two halves . . . It may be eaten green (raw) or dry, but the best way is to toast it.

Since the samples came to Monardes from the sandy banks ('coasts') of the Rio Marañón, he postulated that this area was the only part of the 'Indias' where the fruit grew. As a cultural note, he observed that both the Indians and the Spanish in Peru held the groundnut in great esteem.

The second writer to publish a book upon American natural history was Jean de Léry. A Calvinist missionary with the Huguenot colony founded by the Chevalier de Villegagnon in 1555 on a small island in the bay of Río de Janeiro (now Guanabara bay), Léry remained in Brazil less than five years. In 1578, he published the account of his voyage to the land of Brazil, describing groundnut (*manobí*) as 'a fruit the savages' grew in the soil like truffles. He wrote that it had a hard pod containing grey-brown seeds with size and taste similar to the hazelnut. Although Léry said he had eaten the fruit often, he could not say whether the plant had leaves or seed.

At about the same time as Monardes thought groundnut grew only along the river Marañón, the Portuguese naturalist Gabriel Soares de Souza (who lived in Brazil from 1570 to 1587) believed that the plant occurred only in that country. Soares (1587) described the groundnut plant, its cultivation and the common native custom of using smoke to cure the crop. His detailed account of production practices is still of interest: the seed was planted 'a hand's breadth apart' in February, in humid soil prepared solely by Indian or halfbreed females who, in May, each harvested what they had planted. The plant's leaves resembled those of Spanish beans, and the branches trailed along the ground. Soares mistakenly placed the fruit pods on the root ends. The pods resembled pine nuts but had a thicker shell and contained three or four seeds each. The seed, when eaten raw, had the taste of chickpea. In-shell roasting was the usual method of preparation but Soares considered that the shelled, toasted product had a better taste. Portuguese women in Brazil made several confections from mixtures of sugar and groundnut.

Following the conquest of the Aztecs, the conquistadors sent to the emperor many reports of the natural resources in Mexico. These documents, in the Spanish archives, are generally unavailable to researchers, so that access to information concerning possible distribution and use of groundnut in Aztec agriculture is limited. Friar B. de Sahagún (1820–30), compiler of an encyclopedia in the Náhuatl language (1558–1566), mentioned a folk-medicine use of *tlalcacautl* (Náhuatl for groundnut) but did not list it among the principal food plants of central Mexico. Nor is it among records of the tributes Montezuma extracted from tribes conquered by the Aztecs. The compound name *tlalcacautl* is cited as evidence of its late arrival in Mexico. Krapovickas (1968) suggested that the introduction was probably a subspecies *hypogaea* type grown in the Antilles. Despite recent archaeological evidence for the relative antiquity of cultivation, the absence of other *Arachis* species in Mexico is substantive evidence that the groundnut is not native to the area, nor was it domesticated there.

2.3.2 Selected seventeenth century narratives

Early in the seventeenth century, descriptions and illustrations of groundnut appeared regularly in the European literature and botanical gardens grew the plant as a curiosity. Many natural historians were annotators, compilers, copiers, editors, illustrators and translators who systematized the observations of others but rarely saw the plants (or animals) whose descriptions and figures they placed in their folios. In contrast are the field studies of four naturalists: Marcgrave de Liebstad (1648, 1658), Cobo (1653), Tertre (1654) and Labat (1697, 1742), whose descriptions and figures came from material they observed and collected in natural habitats. An English translation of Monardes' book appeared in 1577 and editions in several other languages quickly followed. The first figure of the beaked pods of a Brazilian cultigen with two or three seed cavities is that of Jan de Laet, naturalist, editor and managing director of the Dutch West Indies Company. The company's ship captains brought him plant collections from many parts of the New World. However, Laet's description (1625, 1640) follows Léry's text (1578).

After the Dutch wrested control of the northeastern Brazilian coast from the Spanish Empire in 1630, Governor-General J. Maurice von Nassau-Siegen commissioned a scientific expedition (1638–1641) of Pernambuco by his personal physician Willem Piso and his friend George Marcgrave of Liebstad. A naturalist, astronomer and geographer, Marcgrave collected the plants reported in their books. After Marcgrave's death in 1640, Jan de Laet became his literary executor and edited and published the Marcgrave–Piso notes and figures. Their *Natural History of Brazil* includes Marcgrave's eight books (1648, 1658). Marcgrave described the groundnut (*mandubi*) plant in some detail, illustrating the two-seeded fruits, quadrifoliate leaves with opposite leaflets, and flowers in the axillary position, but he erroneously placed the fruits growing on the roots. In a second edition (1658) issued by Piso, Marcgrave's description reappears but two pods of a different botanical variety (apparently copied from an edition of Laet's 1625 book) were added to the illustration. Cobo (1653), who also thought the plant produced fruit on the roots as in sweet potato, gave essentially the same description.

The first French botanist in the Antilles, the priest J.B. du Tertre, described the groundnut plant brought to him by the Carib Indians in Guadeloupe (Tertre, 1654). In the location, configuration and morphology of organs, however, Tertre's figure bears a striking likeness to that of Marcgrave (1648), and Tertre's three-segmented pod is that figured in Marcgrave's revised work (1658).

Labat, another priest, who lived in the French Antilles for 12 years (1693–1705), gave a remarkable description of groundnut in his *New Voyage to the Isles of America*, written in 1697:

> the fruits came from a plant hardly a foot tall, a creeper with many slender stems . . . leaves resembling sweet clover, and nasturtium-coloured flowers. . . . The short life of the flowers is due to their shrivelling in the sun. The fruit is found in the earth . . . attached by filaments to hairs that the roots put out [*sic*] from stems distributed on the surface of the earth, where they enter and produce pods.

Pods were 2.7–4.0 cm in length and 0.9–1.35 cm in diameter. Labat (1697, 1742) described shell reticulation, seed shape and number per pod, and the testae, and he recorded different culinary uses of the seed. He found some indigestibility from eating the seed raw and claimed that 'roasted groundnut stimulates the appetite and thirst'.

In view of Labat's comment concerning groundnut's persistence of volunteering in fields after harvesting, the type he saw may have possessed appreciable fresh-seed dormancy. Labat's figure (1697, 1742) reverses the illustration in du Tertre (1654) and both appear identical with the branch and opened pod figured in Marcgrave (1658).

British naturalists came late to the West Indies. Sir Hans Sloane visited Jamaica, Barbados and St. Kitts in 1687–88. Sloane's catalogue (1696) describes groundnut and, more importantly for this review, he cited at least 14 authors who recorded groundnut in or from the Americas (Hammons, 1982). His frequent references to the plant show the prevalence and wide use of groundnut as a food crop in the islands he visited.

By the beginning of the eighteenth century, at least 20 botanical or historical works describing the groundnut were in print and in wide circulation among historians. Besides the universal Latin editions, translations appeared in most of the major languages of western Europe. The main exception is the important work of Soares de Souza, written in 1587 but not published until 1825. All these authors knew that groundnut was native to the Americas.

2.4 INDIAN VERNACULAR NAMES

The distribution and variety of local names of the groundnut are important in the study of its origin. The Spanish introduced the name *maní* from the West Indies and substituted it for other Indian names in Spanish America, except Mexico. There they accepted the alternative Náhuatl word *cacahuate*, modified it and took it to Spain as *cacahuete*. The Portuguese *amendoim* stems from many cognate names still used in Brazil in the late twentieth century.

Cobo (1653) listed Indian vernacular names to document groundnut's diffusion in pre-discovery America: 'This root [*sic*] is called *maní* in the language of Hispaniola, Mexicans call it *cacahuate*, and the Peruvian Indians call it *inchic* in the Quichua language and *chocapa* in Aymara.'

Marcgrave (1648) used the Brazilian Indian *mandubi*, and, citing Monardes (1574), '*anchic* of Peru, the same is called *maní* in Spanish.' Tertre (1654) coined the French name *pistache* because of the seed shape and its taste similar to the European pistachio. Sloane (1696) compiled eleven common name synonyms: *manobí, mandovy, munduvi, anchic, ibimaní, maní, ynchic, pistache, mandues*, earthnuts or pindalls.

Krapovickas (1968) successfully associated the latin names of many earlier authors with the subspecific variation within *Arachis hypogaea* and then associated the ethnic, linguistic and geographic centres of origin of the cultivated groundnut. He also (1968, 1973) correlated Amerindian names with groundnut diffusion throughout South America. In the Tupi-Guaraní region, he found numerous variants of *manduvi*. Vernacular names showed little linguistic affinity in the remainder of the continent. The greatest diversity of Indian names occurs on the eastern slopes of the Andes, where Bolivia and Peru join and where *manduvi* variants are interspersed with other names of Arawak affinity.

The Arawaks inhabited a vast area extending from the Caribbean to the heart of South America as far as the Bolivian border with the Chaco (Hammons, 1982). The presumed centre of origin of *A. hypogaea* is the region where Arawak linguistic influences predominate and Krapovickas (1968) presented a tenable hypothesis of Arawak responsibility for the spread of groundnut from such a centre to the Caribbean basin.

2.5 ORIGIN OF GROUNDNUT

2.5.1 Geographical origin of genus *Arachis*

The genus *Arachis* L. comprises a large and diverse group of diploid and tetraploid taxa native only to South America. There *Arachis* ranges geographically from the equator near the mouth of the Amazon to 34° S latitude on the northern bank of the Río de la Plata in Uruguay, and

extends westward from the Atlantic to the Paraná and the eastern foothills
of the Andes (35–66° W longitude). The northern boundary is usually
marked by the southern extent of the Amazonian rainforest, except where
Arachis spp. may occur where the forest is penetrated by the more open
vegetation of the *cerrado* breaks. Gregory *et al.* (1980) theorize that the
inferred centre of distribution for the genus is the Brazilian Planalto Ellipse
and that the only plausible natural agent of distribution is the downward
flow of soil and water. Biological dispersal mechanisms are very poor, with
the annual dispersion rate limited to a radius varying from several centimetres to a few metres.

Both Krapovickas (1968, 1973) and Gregory *et al.* (1980) postulated a
Planalto profile from Corumbá to Joazeiro, Brazil, as the centre from
which distribution of *Arachis* arose. Half of the eight sections into which
the genus naturally divides botanically occur there.

Although 50–70 species may exist in the genus, only 23 species, one
variety and one hybrid are validly published (Stalker, 1990). All are
indigenous to the area east of the Andes and south of the Amazon
rainforest.

2.5.2 Origin of *Arachis hypogaea*

Linnaeus (1753) described the first *Arachis* in his *Species Plantarium*. Five
species, each from different sections, are now cultivated. *A. hypogaea* L.,
the principal cultigen, and *A. villosulicarpa* Hoehne are grown for their
edible seeds and were improved by the indigenous peoples of South
America. Recently, forms of *A. glabrata* Benth., *A. repens* Handro, and
A. pintoi Krap. et Greg. *nom. nud.* were purposely adapted for grazing use
and released. *A. repens* is also in use as the principal ground cover (lawn)
at the Museum of Natural History in São Paulo and at the Clube Praia das
Cigarras resort, 7 km west of San Sebastiao, Brazil (R. Hammons, unpublished field party notes, 2 June 1968). The low genetic affinity of these five
species shows their independent domestication. This is especially interesting for *A. hypogaea* and *A. villosulicarpa*, both cultivated for their qualities
for human nutrition: *A. hypogaea* is widely spread throughout the world
but *A. villosulicarpa* has limited use only by the Indians of the Rondonia
area of the Mato Grosso, Brazil (Krapovickas, 1968). (Conversely, one
may argue that *A. hypogaea* is the only domesticant, whereas the others
are cultivated, i.e. they can survive in the wild.) *Arachis hypogaea*
($2n = 4x = 40$) is a member of section *Arachis nom. nud.*, along with
its tetraploid progenitor, *A. monticola* Krap. et Rig. (Stalker, 1990).
Identification of the diploid ancestors is inconclusive. Krapovickas (1968)
proposed southern Bolivia and northwestern Argentina for the origin of *A.
hypogaea*. The area is an important centre of diversity for subsp. *hypogaea*.
Although *A. hypogaea* is a largely self-pollinating, annual, herbaceous
legume, earlier work of Krapovickas and Rigoni (1960) showed that the

great variability of groundnut in the Guaraní region (the basins of the Paraná–Paraguay river systems) resulted from a hybrid swarm of the original prototypes with natural crossing occurring freely. An extension of this hypothesis, with different prototypes, to the rest of South America gave rise, according to Krapovickas (1968), to the patterns of variability that he and Gregory *et al.* (1973, 1980) delineated as distinct gene centres.

The eastern foothills of the Andes constitute a wide range of ecologically distinct environments. In this proposed centre of origin, where small-scale cultivation is practised, Krapovickas (1968) cited the diversity of uses as further evidence of the antiquity of groundnut. Immature to fully ripe seeds are eaten raw or cooked. They may be boiled, broiled, roasted, toasted, fried or ground into a paste for mixing with other food. Soups may occasionally contain boiled whole young pods. The common refreshing non-alcoholic drink, *chica de maní*, is popular there and soap is made from oil pressed from groundnut seed. (Note: in western Brazil, *chica* is an alcoholic drink.)

A Bolivian origin is also supported by the wide range in seed and pod morphology documented there, for example, by Cardenas (1969). However, no one is certain of the exact origin of groundnut. All the linguistic, cytological, genetic, morphological, biochemical and geographic data indicate that the eastern foothills of the Andes is the region where *A. hypogaea* originated. Information from the large reservoir of groundnut landraces collected over a vast area of South America, beginning in 1959, should provide new insight on genetic homologies, biochemical affinities, original habitat, domestication and dispersal.

2.6 DISPERSAL

Prior to the early 1500s, groundnut was unknown outside the Americas. Worldwide distribution of at least two distinct forms – a two-seeded Brazilian and a three-seeded Peruvian – took place comparatively soon after the discovery of the New World (Dubard, 1906). Many authorities credit the Portuguese with enriching African agriculture by introducing groundnut there from Brazil, carrying it subsequently to the Malabar coast of south-western India, and possibly to other lands. However, I found no documentation that they did intentionally introduce the seed to those lands.

Gibbons *et al.* (1972) observed that cultivar clusters in Africa of subsp. *fastigiata* var. *vulgaris* represented both the Guaraní region and the region of the eastern slopes of the Andes in Bolivia and parts of western Brazil (Gregory *et al.* 1973). As the latter authors note:

> It is also fairly plain that the peanuts of the eastern slopes of the Andes must have reached Africa from Portuguese boats plying the

Amazon and not from the immediate interior back of the northeast coast of Brazil.

The Peruvian type *A. hypogaea* var. *hirsuta* went to the western Pacific, to China, Indonesia (Java), and to Madagascar. Dubard (1906) documented the concurrence in morphology and configuration of random pod samples from the latter three places and between these and the 'humpbacked' material found in tombs at Ancón, Peru. Their most plausible path was up the west coast from Peru to Mexico, thence across the Pacific as an item of trade on ship crossings that were regularly scheduled between Acapulco and Manilla (the Philippines) for the 250 years prior to 1815 (Krapovickas,1968).

In the African and Asian lands where it was carried, the groundnut readapted for environmental and specialized agricultural requirements. Precisely how and when groundnut entered the United States is unknown. Indeed, there may be instances of its being introduced indirectly from the European farming tradition into colonial America. Nor can one rule out introduction directly from its native South America and secondary Mesoamerican agriculture. Any such technological borrowing could have been mediated by contacts between landowners, the inspection of fields and botanical gardens, or the reports of travellers. Probable locations for the transfer of seed and husbandry information could be the Iberian peninsula, southern France or Italy. The diffusion of technology from the Old World back to the New may always remain partly conjecture. The anonymity of the event should not devalue the achievement.

Burkhill (1901) cited Clusius (1605) as saying that:

> slavers took as food for their captives on the voyage from the Guinea Coast of Lisbon, roots of the sweet potato, '*besides certain nuts*', which Sloane identifies as fruits of *Arachis*.

But, as Burkhill stressed, Clusius' observation places Sloane's identification in question (Hammons, 1982).

A small-podded genotype with a spreading habit of growth was the earliest form successfully introduced into commerce in the south-eastern United States. It is a long-season groundnut, possibly from Africa (Higgins, 1951), and its pod and seed morphology, aspect and branching pattern agree with phenotypes described and illustrated in the West Indies by Tertre (1654) and Labat (1697, 1742). Thus, direct introduction from the Caribbean cannot be excluded (Hammons, 1982).

The Guaraní region of north-eastern Argentina, Paraguay and south-western Brazil is the centre of variation for the spanish type (subsp. *fastigiata* var. *vulgaris*) and it was distributed from this region. According to Krapovickas (1968), F.L. Gilli and G. Xuarez documented its introduction into Europe: seed from Brazil came in 1784 to Don José Campos in Lisbon, who sent a portion on to Rome. Tabares de Ulloa (1799) spread the type in Valencia. From there, Lucien Bonaparte carried it to the south of France in 1801 (Burkhill, 1901; André, 1932). The Spanish, who cultivated the crop both for cooking oil and for use in preparing a chocolate-covered confection, extracted the first oil in Europe (Dubec, 1822).

The spanish type is a small two-seeded form adapted to adverse environmental conditions; it has a shorter growing cycle and lacks appreciable fresh-seed dormancy. T.B. Rowland successfully introduced it from Malaga, Spain, in 1871 to Norfolk, Virginia (USA), where he distributed seed without cost among planters (Anon., 1918b).

Gregory *et al.* (1980) associated the virginia type (subsp. *hypogaea* var. *hypogaea*) with both the Bolivian and Amazonian geographical regions, but its origin needs further clarification. Extensive secondary variations occurred in Africa (Gibbons *et al.* 1972). McClenny (1935) has placed its cultivation in Virginia as early as 1844; other records suggest that its introduction followed the pioneer work of Rowland. The jumbo virginia of United States commerce may be from a chance hybrid (Anon., 1918a).

Recently, Williams (1991b) presented ethnobotanical evidence for the Bolivian origin of the valencia type (subsp. *fastigiata* var. *fastigiata*), which Krapovickas (1968) postulated spread throughout the world from Paraguay and central Brazil. Williams (1991a, b) investigated the farming practices of indigenous growers, cultivating six distinct landraces along the Río Beni using age-old agricultural procedures. Groundnut is cropped on exposed riverine sandbars during the low-water season. A selected portion of the harvest is replanted in upland gardens as a seed crop. The cropping system exerts strong selection pressure for the subspecific valencia characters of earliness, sequential flowering, loss of seed dormancy, and strong pegs. Bees are frequent visitors to the flowers. Such data, with the archaeological and historical evidence, and the diversity of valencia landraces, further support the notion that ancient people living in this region also developed the valencia botanical type (Williams, 1991a, b).

Although Dubard (1906) described the fruit, Beattie (1911) apparently chose the name 'valencia' to designate an introduction into the United States from Valencia, Spain. Soon 'valencia' became a generic name for biotypes with similar pod configuration and a unique arrangement of vegetative and reproductive branches.

'Waspada', with approximately 50% earlier maturity, was the first cultivar introduced to achieve a specific agricultural goal: a shorter growing season. Brought to Java in 1875, the cultivar eventually replaced the common form that matured in eight to nine months (Holle, 1877).

2.7 ETHNOGRAPHY

Ethnographic investigations among the major Indian tribes of South America document the widespread cultivation and use of groundnut as a food crop throughout much of the continent. The studies also provide indirect evidence supporting its domestication long before the Spanish conquest. In the seven volumes of the *Handbook of South American Indians*, Steward and his collaborators (1943–1959) traced groundnut dispersion through records of food plants sown and harvested by native peoples. The following information summarizes the major findings.

The Spaniards found more than 40 new food plants – including groundnut – for pre-Columbian civilization in the fertile highland basins and coastal valleys of the central Andes. Mètraux (1942) reported that the Spaniards who penetrated eastern Bolivia with Gonzalo de Solis Holguin, Governor of Santa Cruz de la Sierra, in 1617 and again in 1624 were amazed at the large plantations where the Mojo (*c.* 15° S and 65° W) and Bauré tribes grew groundnut, preferably along sandy river beaches of the Mamoré's tributaries in the upper Amazon river system. Some 375 years later, Williams (1991a, b) provided the first detailed ethnobotanical examination of this agricultural practice, in areas to the west in a diverse rainforest environment on the Río Beni.

Sixteenth century Indians in Paraguay grew groundnut as a main crop for trade in the markets of Asunción. Throughout the vast tropical forest many indigenous aborigine tribes grew groundnut as a staple crop on farms in natural or manmade clearings. In central Brazil, large gardens guaranteed an economy of abundance for the Tapirape tribe (*c.* 10° S, 52° W). Here, as Soares (1587) observed for the east coastal area, women planted and harvested the crop. In the Caribbean basin, the Arawak tribe grew groundnut with the aid of irrigation (Stewart, 1943–1959). Irrigation was always practised where groundnut was grown in the arid climate of coastal Peru (Cohen, 1977; Burger, 1989).

2.8 EARLY INDUSTRIAL DEVELOPMENTS

Tabares de Ulloa (1799), later bishop of Valencia, invented the first machine to shell groundnut. A description of this device appeared in an 1805 supplemental issue of Rozier's *Traité Général d'Agriculture*.

2.8.1 Beginning of the groundnut oil industry

Worldwide shortage of oil in Europe during early decades of the nineteenth century led to industrial development of the groundnut oil industry. Conflicting claims, that fail to differentiate between samples imported for experimental trial and shipments to mills for crushing, cloud the question of credit for initiating the commercialization of groundnut production in Africa for export, and when this occurred. Once trade began, however, exports increased at a rapid rate. For example, Gambian exports to Britain went from 213 baskets in 1834 to 47 tons in 1835, and jumped to thousands of tons by the early 1840s (Brooks, 1975).

America recorded its first imports from Gambia in 1835 and American purchases dominated the Gambian market from 1837 to 1841. The American interest in the West African groundnut differed from that of the Europeans: the latter desired the oil product, wheras Americans relished the roasted groundnut (Brooks, 1975).

French industrialists also entered the groundnut trade during the 1830s. Flückiger and Hanbury (1879) credit Jaubert, a French colonist and trader at Gorée near Cape Verde, with first suggesting commercialization by sending an oil sample to Marseilles, France, in 1833. He initiated the industry with a shipment of 722 kg to Marseilles in 1840, when France reduced the tariff on groundnut (Brooks, 1975). Following the import of a large tonnage from Cape Verde in 1848, groundnut became a most important raw material for the French oil industry for more than a century (Schlossstein, 1918). In the first decade of the twentieth century, France led the world in tonnage of groundnut crushed for oil, followed in volume by Germany, Holland, Austria and then England (Schlossstein, 1918).

In the United States, initial expansions in land area cropped and in production came in response to the increased need for oil for various uses in times of shortages due to war and other causes. Following the Civil War, the crop area doubled and trebled between 1865 and 1870. Other spurts coincided with the Spanish–American War and World Wars I and II. Moreover, as the boll weevil destroyed the cotton crop in the southern states early in this century, farmers turned to groundnut for economic relief.

The first groundnut pod polisher, invented by T.B. Rowland in the 1870s, was operated by draught animal (mule) power. It embodied mechanical principles used for all subsequent polishers (Anon., 1918b).

2.8.2 Peanut butter

Until the late 1890s consumption of groundnut in America was mostly roasted-in-the-shell. However, in 1894, Dr. John Harvey Kellogg, MD, an eminent health reform advocate and director of the Sanitarium [sic] in Battle Creek, Michigan, developed and introduced peanut butter as a nutritious, easily digested health food for patients (Powell, 1956). Kellogg, who observed a strictly vegetarian diet, obtained the first two patents for making peanut butter (US Pat.No. 580 787, issued 13 April 1897; US Pat.No. 604 493, issued 24 May 1898). To promote his health food ideas, Kellogg elected not to enforce his patent rights and several brands of peanut butter were on the market by 1899.

A former employee of the Sanitarium, Joseph Lambert, began to manufacture and market three machines – a roaster, a blancher and a hand nut-grinding mill – for making peanut butter in 1896 (Grohens, 1920).

2.8.3 Subsequent development

The commercialization of groundnut food products expanded rapidly thereafter. Scientists developed new cultivars and cultural practices, and improved agricultural and industrial machinery became available for all phases of groundnut production, processing and manufacturing.

Periodically, groundnut oil is proposed for use as a substitute or extender for diesel fuel. When Rudolf Diesel demonstrated the engine that bears his name at the Paris exposition of 1890, it was powered entirely with groundnut oil (Nitske and Wilson, 1965). This practice has not been feasible subsequently, both economically and mechanically (Hammons, 1981).

Without the primitive plant selectors who fashioned the groundnut into useful sources of oil and protein, we would not have this crop whose fruit can be eaten raw or cooked and whose seed can be stored for future use and for seed. Who were these early selectors? Anthropologists and ethnographers say that they probably were women – women who gathered the seed and tended the crop, as Soares de Souza (1587) and Williams (1991a) have noted. As Burton (1981) comments, they lacked a knowledge of genetics and reproductive behaviour in plants, of mutations and metabolic pathways, of pathogens or pests and their control; they had neither written language, libraries nor computers. But they had their plants, though undoubtedly a much more restricted germplasm base than groundnut breeders would wish, as Williams (1991a, b) among others, has illustrated. They knew their plant material from living with it: they were motivated because their lives depended on their own success. And they had time on their hands.

What did they do? They used a rudimentary but primary activity of all plant breeding procedures – selection – to choose the material best adapted for their goal: food for survival. They increased the yield, perhaps many times, from the basic stocks inherited from their predecessors. As the crop dispersed over the millennia from the centres of origin, new selectors sought or preserved changes – for environmental adaptation, for resistance to pests and diseases, and for more nutritious qualities – essential for genetic diversity under continued domestication.

Thus, by the time of the European encounter with the Americas, *Arachis hypogaea* already existed in many landraces, distributed and under cultivation throughout much of South and Central America, and the Antilles and other islands of the Caribbean basin. The full extent of its dispersion may soon became known as scholars see the narrative reports kept in the royal Spanish archives. Recent breakthroughs in decoding the complex of hieroglyphs forming the written system of the Mayas may reveal a broader use of groundnut in agriculture and the diet in Mexico than that recorded by Sahagún in the Náhuatl language in 1558–1566, but suppressed until 1820.

Additional archaeological evidence could come as Alva (1988) completes the inventory of Moche culture material from the tomb near Sipàn,

Peru, and further Moche or Classic Mayan records will likely be revealed by the renewed interest in the prehistory of native peoples in Mesoamerica.

The groundnut, by the very nature of its underground fruit placement, was not among the colourful New World crops that excited the sixteenth century European chef; moreover, except in Spain, it was ill-adapted to Europe's agricultural requirements. Although the plant soon became a curiosity in Europe's botanical gardens, it was poorly understood long after Linnaeus (1753) formalized its name, as Gregory *et al.* (1973) have well described.

The food historian has yet to document fully the importance and diversity of groundnut in the diet of native peoples prior to their encounter with Europeans. The groundnut was overshadowed by other novel foods: maize, potatoes, green beans, pineapple and tomato. Even the capsicum peppers made their way more quickly into world cuisine than the nutritious groundnut.

The Quincentennial in 1992 of the encounter between the two worlds may yet focus renewed attention on the groundnut – one of the 13 crops that stand between humans and starvation.

REFERENCES

Alva, W. (1988) Discovering the New World's richest unlooted tomb. *National Geographic Magazine*, **174**, 510–549.

André, E. (1932) L'arachide. *Comptes Rendus de l'Academie d'Agriculture*, **18**, 552–561.

Anon. (1918a) The romance of the peanut. *The Peanut Promoter*, **1**,(2), 48–49. (Probable author: O.C. Lightner, editor.)

Anon. (1918b) Father of the peanut industry is dead. *The Peanut Promoter*, **1**,(8), 24–29. (Probable author: O.C. Lightner, editor.)

Beattie, W.R. (1911) The peanut. *Farmer's Bulletin* **431**, US Department of Agriculture, Washington.

Bird, J.B. (1948) America's oldest farmers. *Natural History* (New York), **57**, 296–303; 334–335.

Brooks, G.E. (1975) Peanuts and colonialism: Consequences of the commercialization of peanuts in West Africa, 1830–70. *Journal of African History*, **16**, 29–54.

Burger, R.L. (1989) Long before the Inca. *Natural History*, **98** (Feb), 66–73.

Burkhill, I.H. (1901) Groundnut or pea-nut (*Arachis hypogaea* Linn.). *Bulletin Miscellaneous Information*, Kew, **178–180**, 175–200.

Burton, G.W. (1981) Meeting human needs through plant breeding: Past progress and prospects for the future, in *Plant Breeding II* (ed. K. Frey),The Iowa State University Press, Ames, Iowa, pp. 433–465.

Cardenas, M. (1969) *Manual de plantas economicas de Bolivia*. Imprenta Icthus, Cochabamba, Bolivia. *Maní* pp. 130–136.

Clusius, C. (L'Ecluse, C. de) (1605) Atrebatis, . . . Exoticorum libri decum, . . . Historiæ . . . Describitur. Antwerp, Ex *Officina Plantiniana Ralphelengic*, lib. 2, cap. 29, p. 57, fig. 5; cap. 60, p. 344.

Cobo, B. (1653) *Historia del Nuevo Mundo*, republished 1890 (ed. J. de la Espada), Sociedad de Bibliofilos, Andaluces, Sevilla, Spain. Vol.1, cap. 12, *maní*, pp. 359–360.

Cohen, M.N. (1977) Population pressure and the origins of agriculture: An archeological example from the coast of Peru, in *Origins of Agriculture* (ed. C.A. Reed), Mouton, The Hague. pp. 135–177.

Dubard, Monsieur (1906) De l'origine de l'arachide, *Muséum National d'Histoire Naturelle* (Paris), **5**, 340–344.

Dubec, Monsieur (1822) Sur la pistasche de terre (*Arachis hypogaea*). Cited from extract by B. Lagrauge, *Journal de Pharmacie*, **8**, 231–235.

Flückiger, F.A. and Hanbury, D. (1879) *Pharmacographia, A History of the Principal Drugs of Vegetable Origin met with in Great Britian and British India*, 2nd edn, Macmillan & Co., London. Oleum Arachis, pp. 186–188.

Gibbons, R.W., Bunting, A.H. and Smartt, J. (1972) The classification of varieties of groundnut (*Arachis hypogaea* L). *Euphytica* **21**, 78–85.

Gillier, P. and Silvestre, P. (1969) *L'arachide*, G.-P. Maisonneure et Larose, Paris.

Gregory, W.C., Gregory, M.P., Krapovickas, A. *et al.* (1973) Structures and genetic resources of peanuts, in *Peanuts: Culture and Uses*, (ed. C.T. Wilson), American Peanut Research and Education Association, Stillwater, Oklahoma, pp. 47–133.

Gregory, W.C., Krapovickas, A. and Gregory, M.P. (1980) Structures, variation, evolution and classification in *Arachis*, in *Advances in Legume Science*, eds R.J. Summerfield and A.H. Bunting), Royal Botanic Gardens, Kew, pp. 469–481.

Grohens, A.P. (1920) Peanut butter history and development of the peanut butter industry, *The Peanut Promoter* **3** (8), 65, 67, 69, 71, 78.

Hammons, R.O. (1973) Early history and origin of the peanut, in *Peanuts: Culture and Uses*, (ed. C.T. Wilson), American Peanut Research and Education Association, Stillwater, Oklahoma, pp. 17–45.

Hammons, R.O. (1981) Peanut varieties: Potential for fuel oil. *American Peanut Research and Education Society Proceedings*, **13**,(1), 12–20.

Hammons, R.O. (1982) Origin and early history of the peanut, in *Peanut Science and Technology*, (eds H.E. Pattee and C.T. Young), American Peanut Research and Education Society, Yoakum, Texas, pp. 1–20.

Higgins, B.B. (1951) Origin and early history of the peanut, in *The Peanut – the Unpredictable Legume*, National Fertilizer Association, Washington, DC, pp. 18–27.

Holle, K.F. (1877) Verzamelingstaat van den uitslag der proeven, genomen met een nieuwe soort van katjang tanah, . . . *Tydschrif voor Nyverheid en Landbouw in Nederlandsch Indie*, **21**, 360–361.

Krapovickas, A. (1968) Origen, variabilidad y difusion del maní (*Arachis hypogaea*). *Actas y Memorias Congress International Americanistas* (Buenos Aires), **2**, 517–534. English version (1969): The origin, variability and spread of the groundnut (*Arachis hypogaea*), [translation by J. Smartt] in *The Domestication and Exploitation of Plants and Animals*, (eds P.J. Ucko and I.S. Falk), Gerald Duckworth, London, pp. 427–441.

Krapovickas, A. (1973) Evolution of the genus *Arachis*, in Agricultural Genetics: Selected Topics, (ed. R. Moav), National Council for Research and Development, Jerusalem, pp. 135–151.

Krapovickas, A., and Rigoni, V.A. (1960) La nomenclature de las sub-species y variedades de *Arachis hypogaea* L. *Revista de Investigaciones Agricola*, **14**,(2), 197–228.

Labat, J.B. (1697, 1742) *Nouveaux Voyage aux Isles de l'Amerique*, (1742) new edn, (ed. Ch. J.B. Delespine), Paris, Vol. 4, pp. 365–69.

Laet, J. de (1625, 1633) *Nieuvve Wereldt, ofte Beschrijvinghe van West-Indien*, I. Elzevier, Leyden, p. 446. (1630) 2nd Dutch edn, *Beschrijvinghe van West-Indiæ*, Elzeviers, Leyden, p. 510; (1633) Latin edn, enlarged, *Novus Orbis seu Descriptionis Indiæ Occidentalis*, cum privilegio, Elzevirios, Leyden. Batavorum, lib. 18, cap. 11, p. 568; (1640) French edn, *L'histoire du Nouveau Monde* . . ., B. and A. Elseuiers, Leyden, chap. 11, p. 503.

Léry, J. de (1578) *Historie d'un Voyage faiet en la Terre du Brésil autrement dite Amerique*, A. Chuppin, Rochelle, 1st edn, manobi, p. 215. 1960, 3rd edn, Martins, São Paulo, manobi, p. 162.

Linnaeus, C. (Linné, C. von) (1753) *Species Plantarium*, Laurentii Salviae, Holmiae 2, 741.

MacNeish, R.S. (1965) The origins of American agriculture. *Antiquity*, **39**, 87–94.

Marcgrave de Liebstad, G. (1648) Mandubi, in Piso, W., et Margravi, G., *Historia naturalis Brasiliae* (ed. J. de Laet), L. Elzevirium, Amsterdam, 2 vols; Vol. l, p. 37.

Marcgrave de Liebstad, G. (1658) Mandubi, in Pisonis, G., *Medici Amstelaedamensis de Indiae utriusque Historia naturali et Medicae*, (2nd edn), F. Hackim, Leyden and L. Elzevirium, Amsterdam, lib. 4, cap. 64, p. 256.

McClenny, W.E. (1935) *History of the peanut*, The Commercial Press, Suffolk, Virginia, 22 pp.

Mètraux, A. (1942) The native tribes of eastern Bolivia and western Matto Grosso. *Bulletin* **134**, Bureau American Ethnology, Smithsonian Institution, Washington.

Monardes, N. (1569, 1574, 1577) *Historia Medicinal de las Cosas que se traen de Nuestras Indias Occidentales que siruen en Medicinia*. H. Diaz, Seville. English translation from 1574 edn, J. Frampton, 1577, Wm. Norton, London. Reprinted, 1925, Constable & Company, London, 2 vols; vol. 2, p. 14. Latin edn by Clusius, *De Simplicibus Medicamentios*, Antwerp, 1605.

Nitske, W.R. and Wilson, C.M. (1965) *Rudolf Diesel – Pioneer of the Age of Power*, University of Oklahoma Press, Norman, Oklahoma.

Oviedo y Valdés, G.F. de [also known as Fernandez] (1527) *Sumario de la natural y General Historia de las Indias*. Toledo, con privilegio imperial.

Oviedo y Valdés, GF. de (1535, 1547) *Primera parte de la Historia General de las Indias*. Seville, con privilegio imperial, lib. 7, cap. 5, p. 74, del maní. Reissued (1547) as part of the *Cronica*.

Oviedo y Valdés, G.F. de (1851) *Cronica de las Indias*, Salamanca, lib. 23, cap. 12, p. 193. (Completed work issued by the Spanish Royal Academy of History in 1851.)

Powell, H.B. (1956) *The Original has this Signature – W.K. Kellogg*, Prentice-Hall, Inc., Englewood Cliffs, New Jersey.

Sahagún, B. de (1820–30) *Historia General de las cosas de Neuva España*. Republished 1956, A.M. Garibay-K., Editorial Porrua, S.A., Mexico, D.F., 4 vols; v. 3, *tlalcacuatl*, p. 173. Original Ms in Náhuatl, 1558–1566.

Schlossstein, H. (1918) The peanut situation as affected by world events. *The Peanut Promoter*, **1**, (4), 32–33.

Schmidt, U. (Faber, U.) (1567) *Weltbuchs von Newen erfundnen Landtschafften: Warhafftige . . .*, Original German edn 2 vols., Frankfort am Main. Translation (1891) from the original German text: *The Conquest of the River Plate (1535–1555), Part I. Voyage of Ulrich Schmidt to the Rivers La Plata and Paraguai*, (ed. L.L. Dominguez), Hakluyt Society Publication, vol. 81, London, pp. 25, 37, 40–41, 63.

Sloane, H. (1696) *Catalogus Plantarium que in Insula Jamaica, etc.*, D. Brown, London, pp. 72–73, 221.

Smith, C.E., Jr (1967) Plant remains, in *The Prehistory of the Tehaucan Valley* (ed. D.S. Byers), University of Texas, Austin, Texas, vol. 1, pp. 220–225.

Soares de Souza, G. (1587, 1825, 1851) *Tratado Descriptivo do Brasil em 1587*, first published 1825, in Colleção de Noticias para a Historia e Geographia das Nacões Ultramarinas, que vivem nos Dominios Portuguezes, Academie de Sciencias, Lisbon. Republished 1851, separate issue, *Revista trimonsal do Instituto Historico e Geographico do Brasil*, **14**, 1–423. Amendoim, cap. 47, pp. 173–176.

Squier, E.G. (1877) *Peru: Incidents of Travel and Exploration in the Land of the Incas*, Macmillan, New York and London.

Stalker, H.T. (1990) A morphological appraisal of wild species in section *Arachis* of peanuts. *Peanut Science*, **17**, 117–122.

Stewart, J.H. (ed.) (1943–1959) *Handbook of South American Indians*, Bulletin 143, Bureau of American Ethnology, 7 vols, Smithsonian Institution, Washington.

Tabares de Ulloa, F. (1799) *del Cacahuate ó Maní de América, Semananario de Agricultura y Artes*, **5** (123), 289–294.

Tertre, J.B. du (1654) *Histoire Générale des Isles . . . et autres dans l'Amérique*, Paris. (1667–71) rev. edn, 4 vols, *Histoire Générale des Antilles Habitées les François*, T. Jolly, Paris, v. 2, p. 121 and plate.

Towle, M.A. (1961) *The Ethnobotany of Pre-Columbian Peru*, Aldine Publishing Company, Chicago.

Williams, D.E. (1991a) *Peanuts and peanut farmers of the Rio Beni: Traditional crop genetic resource management in the Bolivian Amazon*, PhD dissertation, City University of New York, New York.

Williams, D.E. (1991b) Ethnobotanical evidence for the Bolivian origin of Valencia peanut, (Abstract), *American Peanut Research and Education Society, Proceedings*, 23, p. 22.

CHAPTER 3

Botany – morphology and anatomy

V. Ramanatha Rao and U.R. Murty

3.1 INTRODUCTION

Groundnut (*Arachis hypogaea* L.) is one of the world's major food legume crops. It originated in South America, where the genus *Arachis* is widely distributed. The natural distribution of all the *Arachis* species is confined to Argentina, Bolivia, Brazil, Paraguay and Uruguay (Krapovickas, 1973; Krapovickas and Rigoni, 1957; Ramanatha Rao, 1987). *A. hypogaea* is not known to occur in the wild state. Groundnut is presently cultivated in over 80 countries from 40° N to 40° S in tropical and warm temperate regions of the world. The origin and evolution of the genus as well as of the cultivated species are fairly well known but more evidence is required to make explicit conclusions. Though considerable information on morphology, cytology and genetics has been accumulated along with information on the distribution of the species in the genus, the taxonomic treatment is still incomplete. One of the major reasons for this is the discovery of additional 'species' after every collecting mission to centres of diversity of the genus (Valls *et al.*, 1985). The morphology, development and anatomy of *Arachis* presented problems to groundnut workers.

This chapter attempts to review these topics and to clarify certain ambiguities. However, most of the discussion is on the cultivated groundnut rather than on its wild relatives. Information on origin, evolution and taxonomy is limited to the extent that it is relative to morphology and development.

3.2 DESCRIPTION OF THE GENUS

Arachis is a perennial or annual legume with three or four leaflets, stipulate leaves, papilionate flowers, tubular hypanthium and subterranean fruits. The 'peg', which is an expanded intercalary meristem at the base of the

The Groundnut Crop: A scientific basis for improvement. Edited by J. Smartt. Published in 1994 by Chapman & Hall, London. ISBN 0 412 408201.

basal ovule (Gregory *et al.*, 1973), is unique to the genus. The expansion of the intercalary meristem results in a lomentiform carpel of one to five segments, each containing a single seed with two very large cotyledons and a straight embryo. A general description of *Arachis* is as follows (after Bentham and Hooker, 1862):

> Seeds with thick and fleshy cotyledons; short radicle growing into a well-developed taproot; plants low suberect herbs, often prostrate and even creeping; leaves abruptly bipinnate, tetrafoliolate leaves with two pairs of opposite leaflets, rarely trifoliolate, exstipellate, stipules adnate to the petiole at the base. Flowers, crowded in simple or compound monopodium, look like dense axillary spikes, sessile in the leaf axil, or very shortly pedicellate. Calyx lobes five, often dentate, calyx tube filiform, lobes membranaceous, the four upper ones connate, the lower one slender, separate. Petals and stamens inserted at the apex of the tube, standard oblong, obovate to suborbicular, wings oblong, free, keel incurved, prostrate. All stamens connate to form a closed staminal tube, stamens 10, usually one absent, anthers alternate, elongate, subbasifixed, the alternate ones versatile. Ovary subsessile towards the base of the calyx tube, usually aerial, occasionally subterranean, two- to three-ovuled. When the flower withers and falls away, the ovary shows a stalk, which elongates and becomes reflexed and rigid and the ovary is continuous with the same, acute at the apex; style long, filiform with a minute terminal stigma; the pod ripens inside the soil, oblong, thick, indehiscent, subtorulose, articulate or non-articulate. Seeds one to three, irregularly ovoid, rich in oil and protein.

3.3 DESCRIPTION OF THE SPECIES

In this section, brief morphological descriptions of the valid *Arachis* wild species are given. The authors feel that these descriptions serve as useful guides to *Arachis* species because a taxonomic monograph of the genus is still lacking. Much of what is given below is an English translation of the original descriptions given by the respective authors (the protologues) and is adapted from Ramanatha Rao (1988). Where appropriate, some additional information is also provided. Most of the measurements given are averages.

Besides the valid names, a few *nomina nuda* or *nomina inedita* are used in the literature. Some of these are *A. chacoense* (= *A. chacoensis*), *A. correntina*, *A. cardenasii*, *A. pintoi*, *A. stenosperma* (= *A. stenocarpa*), *A. sylvestris*, *A. ipaensis*, *A. spegazzini*, *A. oteroi*, *A. lignosa*, *A. duranensis* and *A. macedoi* (Ramanatha Rao, 1988; Resslar, 1980). These are taxonomically invalid and should not be used; if used they should accompany the collector numbers. The 22 species described so far are listed in

TABLE 3.1   *Valid* Arachis *epithets with citation and botanical assignment*

Species	Author citation	Section/Series
A. batizocoi Krap. et Greg.	Krapovickas *et al.*, 1974	A1
A. villosa Benth.	Bentham, 1841	A2
A. diogoi Hoehne	Hoehne, 1919	A2
A. helodes Mart. ex Krap. et Greg.	Krapovickas and Rigoni, 1957	A2
A. hypogaea L.	Linnaeus, 1753	A3
A. nambyquarae Hoehne	Hoehne, 1922	A3
A. monticola Krap. et Rig.	Krapovickas and Rigoni, 1957	A3
A. tuberosa Benth.	Bentham, 1841	E1
A. guaranitica Chod. et Hassl.	Chodat and Hassler, 1904	E1
A. paraguariensis Chod. et Hassl.	Chodat and Hassler, 1904	E2
A. benthamii Handro	Handro, 1958	E2
A. martii Handro	Handro, 1958	E2
A. rigonii Krap. et Greg.	Krapovickas and Gregory, 1960	E3
A. repens Handro	Handro, 1958	C
A. burkartii Handro	Handro, 1958	R1
A. glabrata Benth.	Bentham, 1841	R2
A. hagenbeckii Harms	Kuntze, 1898	R2
A. prostrata Benth.	Bentham, 1841	EX
A. marginata Gard.	Gardner, 1842	EX
A. villosulicarpa Hoehne	Hoehne, 1944	EX
A. lutescens Krap. et Rig.	Krapovickas and Rigoni, 1957	EX
A. pusilla Benth.	Bentham, 1841	T

Table 3.1. Though some workers include *A. angustifolia* (Chod. et Hassl.) Killip (in Hoehne, 1940) as one of the described species (Smartt, 1990), the validity of the name as given in the literature is questionable. *A. X batizogaea* Krap. et Fern., an experimental hybrid origin from a cross between *A. hypogaea* and *A. batizocoi* (Krapovickas *et al.*, 1974) is also not included. *A. correntina* (Burk) Krap. et Greg. *nom. nud.* is *A. villosa* var. *correntina* Burkart, but needs validation according to the International Code of Botanical Nomenclature. It has been suggested that *A. namby-quarae* should be regarded as a variety of *A. hypogaea* (John *et al.*, 1954) but its description is not given here because it is simply a form of *A. hypogaea* (Smartt and Stalker, 1982) and not a distinct species.

Arachis batizocoi Krap. et Greg.

Annual, prostrate herb; erect main stem, up to 15 cm, hairy; tetrafoliolate; outer surface of stipules hairy, 2.5 cm × 1 cm, mostly adnate to the petiole; petiole 5 cm long on main stem, 4 cm on branches, hairy; leaflets elliptic on main stem, 3 cm × 2 cm, orbicular on branches, 2.5 cm in diameter, hairy underneath, ciliate margins, tip acute on main stem, rounded, obtuse or mucronate on branches, coriaceous to brittle; corolla yellow, generally with pink or purple blush on the back of the standard; hypanthium long; pods articulate, 10 mm × 5 mm, with prominent beak, seeds light tan, 8 mm × 4 mm.

Arachis villosa Benth.

Perennial, prostrate; stems sometimes angular, villous, 5–14 cm tall; little branching, primaries 40–100 cm or longer; tetrafoliolate; stipules elongate, falcate or lanceolate, 1–2 cm long, rigid, very hairy, shortly adnate; petioles 2–3 cm, hairy; rachis 3–7 mm; leaflets broadly ovate, rigid, mucronate, emarginate, short dense hairs above, sparsely hairy below, 0.8–1.5 cm × 0.5–1.0 cm, generally apical leaflets longer than basal; thread-like hypanthium, 2–4 cm, hairy; corolla glabrous; calyx upper lip tri- or bilabiate, 6–7 mm, lower lip wide, entire or toothed; standard yellow or yellow-orange, 12–15 mm in diameter; ovary with two ovules; pegs up to 20 cm, two or three articles of pod separated by isthmus which rots after seeds mature, isthmus 2 cm, articles 11–16 mm × 7–10 mm, slightly ridged, beak prominent; seeds reddish brown or tan, 10 mm × 5 mm.

Arachis diogoi Hoehne

Perennial; prostrate; stem angular, well-branched, dense interwoven white hairs on stem, branches, leaves, stipules; internodes 4–6 cm long; tetrafoliolate; stipules adnate to the petiole up to 6 mm, long-acuminate, striate, 2 cm long; petioles 4–6 cm long; leaflets narrow-oblong to linear-oblong, tapering to base and apex, base subrounded, apex acute, glabrous above, thinly hairy below and on margins, 3–4 cm × 0.7–1.0 cm; flowers one to four per axil, hypanthium 4–7 cm long, thin, hairy; calyx bilabiate, lobes 10 mm long, upper lip minutely three-toothed, lower lip entire, narrow-acute; standard yellow or orange-yellow, suborbicular, 12 mm in diameter, apex emarginate, base contracted in short claws, reflexed above; wings with rounded tip, subfalcate, dilatate above the middle, base clawed, eared above the claw; keel narrow, base clawed, eared above, falcate, tip rostrate; stamens alternately short and long; anthers narrow up to 2.5 mm long; pegs 3–8 cm long, pericarp slightly reticulated, pods mostly uniarticulate, articles oblong, very slightly beaked.

Arachis helodes Mart. ex Krap. et Rig.

Perennial; taproot without rhizomes or stolons; stem highly branched, prostrate; tetrafoliolate; stipules falcate, adnate to half of their length, free part lanceolate–acuminate; petioles short; leaflets 0.5–1.1 cm × 0.4–0.8 cm, obovate–orbicular, emarginate, tip barely mucronate, margins slightly thickened with few hairs, glabrous above and glabrous or subglabrous below; hypanthium thread-like, 3–6 cm long with scattered hairs; calyx bilabiate, upper lip irregularly four-toothed, lower lip entire, linear–acuminate, falcate, 7 mm long; standard lemon-yellow or yellow, orbicular. 10 mm in diameter; wing conspicuously eared; keel beaked, curved; anthers eight, style filiform, longer than stamens, hairy at the tip, ovules two; peg 5 cm long; articles 8–9 mm × 5–6 mm, pericarp thin, fragile, hardly reticulate.

Arachis monticola Krap. et Rig.

Annual; taproot with profuse lateral roots; stem erect, radially symmetrical. 30 cm tall, lateral branches distichous, creeping, hairy; stipules subfalcate, adnate to the petiole for one third of their length; leaflets oblong–obovate; flowers up to five per axil; hypanthium thread-like, hairy, 5 cm long; calyx bilabiate, upper lip irregularly four-toothed, lower lip entire. 10 mm long; standard yellow, suborbicular, 15 mm × 17 mm, keel falcate, 12 mm long; anthers eight of which four are oblong, basifixed and four small, globose and dorsifixed, style filiform, top hairy; ovules two; pegs 10 cm long; pods biarticulate, articles 13 mm × 8 mm, isthmus 2–7 cm long, pericarp reticulate, beaked; seeds light tan, 10 mm × 5 mm.

   A. monticola is regarded as subspecies of *A. hypogaea*, because of the high level of cross-compatibility and marked similarities between the two (Smartt, 1964). It is very difficult to maintain this species in a pure form due to the ease with which natural crossing occurs with *A. hypogaea*. The fact that both alternate and sequential branching forms of *A. monticola* occur (Gibbons, 1966) is a further confirmation of such a conclusion. *A. monticola* could be regarded as either a subspecies or a botanical variety of cultivated groundnut (Ramanatha Rao, 1988) but this needs to be validated.

Arachis tuberosa Benth.

Perennial; roots with oblong woody tubers extending downward; stems simple, very little branching, erect, rarely prostrate, softly adpressed, villous, 20–30 cm high, angular, slightly sinuous; trifoliolate; stipules rigid, striate, 2–3 cm long, adnate to the petiole from two thirds of their length to almost to the tip, margins ciliate; petioles short; leaflets oblong, coriaceous to rigid, marginate, reticulate, tip obtuse, base tapering, glabrous, 3.5–4.0 cm × 1.0–1.3 cm; inflorescence subterranean, concentrated around the crown of the stem or rarely at the base of branches; ovary underground but hypanthium exposes flowers; hypanthium thread-like, erect, hairy, 5–7 cm long; standard orange-yellow; pods mostly uniarticulate, 12 mm × 6 mm, beak prominent, pericarp very fragile, slightly reticulated; seeds light tan, 8 mm × 5 mm, tend to lose viability when dry.

Arachis guaranitica Chod. et Hassl.

Perennial; obconical root; erect stem, stems originating from the root tubers also erect; trifoliolate; closed leaf sheath forming tube covering internodes; stipules long and congested, adnate to petiole, 3–6 cm long, tip 3–7 mm free; leaflets glabrous, linear–lanceolate, 5.0–9.5 × 1.5–3.0 cm, marginal vein thick and midrib grooved, coriaceous, densely branched striations; inflorescence almost entirely basal and arise in clusters around the crown, flowers with subterranean ovary, flowers exposed by 6 cm long hypanthium, hirsute; calyx limb bilabiate, upper lip four-toothed, hairy, 5 mm × 2.5 mm, lower lip entire, hairy and 7 mm wide; standard 18 mm wide; wing emarginate; keel pointed, rostrate; stamens nine, five anthers globose and four elongate; style thin, tip attenuate, hairy; pegs horizontal, may grow up to 1 m at a depth of 5 cm; pods articulate, 10 mm × 5 mm, with slight beak, seeds light tan, 8 mm × 4 mm, tend to lose viability when dry; related to *A. tuberosa*, differs by having glabrous stems and narrow leaflets.

Arachis paraguariensis Chod. et Hassl.
Root narrow, elongate, tuberous; stem erect, 20–30 cm tall; the basal leaflets without lamina or only with two leaflets, rest tetrafoliolate; stipules 15–20 mm long, partially adnate to petiole, free part 10–13 mm long, linear, softly hairy; petiole 12–25 mm long, woolly; leaflets oblong–lanceolate, slightly mucronate, marginal vein thickened, midrib with 8–12 secondary veins regularly spread, glabrous above and light green hairy below, margin woolly, 10–24 × 6–9 mm; flowers mostly subterranean; hypanthium 14–30 mm long, densely hairy; calyx bilabiate, upper lip four-toothed, hairy, 5 mm × 2.5 mm, lower lip entire, hairy, 7 mm wide; standard orange, 16–18 mm wide; wing emarginate; keel pointed, rostrate; stamens nine, five anthers globose, four elongate; pegs horizontal, may grow up to 80–90 cm, pods articulate, 8–11 mm × 4–6 mm, with slight beak, seeds light tan, 5–8 mm × 3–4 mm.

Arachis benthamii Handro
Perennial; roots partially tuberous when old; stem erect, 40 cm or more, decumbent, simple or branched, terete, hirsute; tetrafoliolate; stipules narrow–lanceolate, acuminate, veined–striate, hairy outside, glabrous or sparsely hairy inside, margin ciliate, 25–32 mm long, 8–10 mm adnate to the petiole; petiole 18–22 mm, grooved, very hairy; rachis 5–6 mm long, grooved, hairy; leaflets, subcoriaceous, ciliate, basal pair normally obovate or obovate–oblong, apex obtuse, rounded or slightly emarginate, mucronate, base obtuse, usually hairy above, 32–37 mm × 17–21 mm; hypanthium filiform, 6 cm long; calyx deeply bilabiate, upper lip entire, hairy outside, hairs setaceous, glabrous inside; standard orange, glabrous; pod uni- or biarticulate, isthmus thread-like, 5 cm long, articles of pod oblong or obovate, beaked, woolly, hairs later fall off, 8–11 mm × 4–6 mm, pericarp thin, reticulate; seeds reddish.

Arachis martii Handro
Perennial; stem prostrate, branched, hirsute, 10–17 cm long; tetrafoliolate; stipules linear–lanceolate, veined–striate, hairy outside, margin cili-

ate, reaching to the tip of rachis or beyond, 12–15 mm long, 4–5 mm adnate to the petiole; petiole 8–10 mm long, hirsute, grooved, rachis 2.5–3.5 mm long; leaflets subsessile, membranaceous, ciliate, basal pair elliptic, apical pair obovate or suborbicular, tip obtuse or rounded or slightly emarginate, slightly mucronate, base obtuse, puberulous above, adpressed hairy below, 7–12 cm × 4–9 mm; hypanthium thread-like, hairy, 4 cm long; calyx deeply bilabiate, upper lip four-toothed, hairy outside, hairs setaceous; standard glabrous, yellow.

The above description is from a herbarium specimen only. Seed and germplasm of this species are not available. Several collecting missions to the recorded area of distribution of this species, including one by the authors, were in vain.

Arachis rigonii Krap. et Greg.

Perennial, herbaceous; no rhizomes or stolons, lateral roots horizontal; stems cylindrical, short, 4 cm, glabrous or subglabrous; well-branched, creeping; tetrafoliolate; stipules falcate, adnate portion 13–15 mm long and 3 mm wide with rigid spines on outer surface, rest 10–12 mm free, 2.5 mm wide at the base, marked with longitudinal veins, hairy margin; petiole 6 cm, grooved, rachis 10–15 mm long, grooved; leaflets obovate, apical leaflets 50 mm × 32 mm, basal 45 mm × 28 mm, glabrous above hairy below, mucronate; flowers four or five per axil; hypanthium thread-like, 4–7 cm long, subglabrous; calyx bilabiate, upper lip irregularly three- or four-toothed, lower lip entire, falcate, standard orange or yellow with orange markings, 12–14 mm × 15–17 mm; wings yellow, 8 mm × 6 mm, tongued; keel 6–7 mm, beaked, curved, subfalcate; pegs 5–25 cm, hairy at the base; fruits biarticulate, 11–13 mm × 6–7 mm, isthmus 1–4 cm long, pericarp slightly reticulated, covered with hairs, prominently beaked; seeds 9 mm × 4 mm, light tan.

Arachis repens Handro

Perennial; taproot obtriangulate, many secondary roots without thickenings; stem well-branched, often hollow; stem, branches, outer surface of stipules, petiole, rachis, lower leaf surface and outside of calyx covered with fine adpressed hairs; stems and branches prostrate or ascending in shady places, terete, adventitious roots at nodes; internodes 3–5 cm long, with adpressed hairs; stipules veined–striate, 15–20 mm long, free portion as long as or shorter, lanceolate or falcate, glabrous or sometimes with few hairs, adpressed at the base, ciliate margin; petiole 2–4 cm, grooved; rachis 3–7 mm long; leaflets membranaceous, elliptic–oblong, or obovate–oblong, tip obtuse or rounded, slightly mucronate, base obtuse, glabrous above, 20–35 mm × 8–12 mm; hypanthium thread-like, hirsute with short hairs, 8.5–11.0 cm long; calyx deeply bilabiate, 7 mm long, upper lip three-toothed, lower lip two-toothed, setaceous outside; standard yellow,

8–13 mm × 16–17 mm; wings and keel yellow; fruit biarticulate, peg 5 cm long, isthmus 1.5–3.5 cm long, articles 8–13 mm × 5–6 mm.

Arachis prostrata Benth.

Perennial; taproot and lateral roots form lomentiform tuberoids, 9 mm × 5 mm; main stem thin, erect, becomes prostrate later on, 5–12 cm long, covered almost totally by stipules; centrally branched, lateral branches procumbent, 90 cm long; internodes on main stem short, 5 mm long, villous, on laterals longer; tetrafoliolate; on main stem stipules rigid, striate, villous at the base, margins ciliate, hairs 1.5–2.0 mm long, adnate portion 8 mm long, free portion 10–12 mm and 2 mm wide at the base; rachis 5 mm long; leaflets commonly obovate, or oblong, 19–20 mm × 11–12 mm; on prostrate laterals stipules shortly adnate, 3–4 mm and free portion 5–9 mm long and 2.0–2.5 mm wide at the base; petioles 3–5 mm long, grooved and villous; leaflets obovate, 9–13 mm × 7–9 mm, mucronate, margins hairy, glabrous or slightly hairy above, densely hairy below, upper surface shining; flowers larger on laterals than on main stem; hypanthium 2–4 cm, densely hairy; calyx bilabiate, villous, upper lip 4 mm long, lower lip entire; standard orange with purple markings on the back (dorsal) face, 8 mm × 9–10 mm; wings yellow, 5 mm long; peg 1–5 cm long, aerial portion purple and hairy, underground portion with adventitious roots, isthmus short, fruits biarticulate, rarely form a single cavity with two seeds, 5–8 mm × 4–5 mm, covered with hair; seeds 4–6 mm × 3–4 mm, tan or reddish brown.

Arachis marginata Gard.

Perennial; taproot with many laterals with numerous tuberiform thickenings, 10–25 cm × 5–7 cm; stem erect, 10–20 cm, hardly branched; long and silky hairs cover the whole plant except the upper leaf surface; internodes 12–30 mm long, quadrangular, villous; tetrafoliolate; adnate portion of stipules 7 mm long and 5 mm wide, villous and free portion 13 mm long, glabrous; petioles of basal leaves short, 2 mm long, and apical leaves 3 cm long, grooved; rachis 7–12 mm long, grooved; leaflets oblong–obovate, apical leaflets 40–44 mm × 26–30 mm, basal leaflets 32–36 mm × 20–25 mm, upper surface shining, smooth, glabrous with one line of long hairs on the midrib; lower surface with thickened veins, reticulate, with long adpressed hairs, hairs rigid and denser on the midrib, margin thickened, especially on lower surface, setose, sometimes with short hairs; flowers small, covered by stipules on basal nodes; hypanthium thread-like, 40 mm long, villous; standard yellow with purple lines on the back, lines less distinct on front face; pegs long, up to 60 cm, mostly horizontal; fruits articulate, articles 14–16 mm × 5–8 mm, highly reticulated, strongly beaked; seeds reddish, 7–10 mm × 4–6 mm.

Arachis villosulicarpa Hoehne

Perennial; long thickened taproot; main stem erect, up to 60 cm, angular, densely whitish villous, internodes 2–4 cm long; well-branched, branches prostrate up to 1 m or decumbent; tetrafoliolate; stipules lanceolate, acuminate, lower 10–17 mm adnate to petiole, upper 20–30 mm free, 3–9 mm wide at the base, fine linear margins, sparsely hairy; petiole 19–50 mm long, grooved, abruptly reflexed below; rachis 5–13 mm long, grooved, sparsely hairy; leaflets elliptic or oblong–elliptic, apical leaflets 20–50 mm × 6–22 mm, basal leaflets 16–40 mm × 5–15 mm, marginal veins slightly thickened, thinly puberulous or glabrous, shiny upper surface, ciliate margin; flowers axillary, on main stem covered by stipules, one to many in axils; hypanthium slender, hairy, 2–5 cm long; calyx bilabiate, upper lip entire or two-toothed, lower lip thin, linear, 5–6 mm long, hairy outside; standard suborbicular, orange with purple markings on the back, reflexed, 10–12 mm × 12–14 mm; wings oblong, orange with yellow blush at the tip; keel incurved, beaked; fruit biarticulate, pegs 6–26 cm, pod exterior densely villous or covered with tomentose hairs, 20–25 mm × 7–9 mm; seeds oblong, yellow-purplish, 16–23 mm × 6–7 mm.

A. villosulicarpa is considered as a cultigen in the genus Arachis, cultivated by a small group of Amerindians in north-western Mato Grosso state of Brazil. This species has not been found in the wild (Valls et al., 1985). Interspecific hybrids with A. hypogaea have been reported (Raman, 1976; Sundaram, 1985), however similar attempts at hybridization elsewhere have failed (Smartt, 1990).

Arachis lutescens Krap. et Rig.

Perennial without rhizomes or stolons; root with thick and elongated tubers about 10 cm × 5 cm; most of the aerial parts covered with hairs except the upper leaf surface; stem erect, 3–10 cm tall, laterals procumbent, 60 cm long; internodes 10–25 mm long; tetrafoliolate, rarely with a supernumerary apical leaflet; on main stem adnate portion of stipule subfalcate, 6–8 mm × 2 mm, free portion linear–lanceolate, 10–16 mm long; petioles on main stem longer, 15–45 mm long, shorter on branches, 2–6 mm long; leaflets on main stem oblong, sometimes obovate, apical leaflets 15–17 mm × 9–12 mm, basal leaflets 21–23 mm × 8–10 mm; obovate or oblong on branches, apical leaflets 8–18 mm × 6–11 mm, and basal leaflets 15–16 mm × 9–11 mm, upper leaf surface glabrous, shining, lower surface hairy, hairs 2 mm long, sometimes adpressed; hypanthium threadlike, villous, 2–4 cm long; calyx villous, bilabiate, upper lip irregularly three-toothed, lower lip entire, linear–acuminate, subfalcate, 6 mm long; standard orange with central yellow portion, suborbicular, 10 mm × 11 mm; wings 6 mm long, yellow with orange blush in front; keel falcate; fruits biarticulate, pegs 2–5 cm long, aerial portion pubescent, underground portion without roots, articles of pods 6–9 mm × 5–6 mm, apical article generally longer, pods densely hairy, isthmus short, 1–3 mm long; seeds 5 mm × 3.5 mm.

Arachis burkartii Handro

Perennial; rhizomatous, rhizomes 35 cm × 2 cm; stem short, erect, simple with little branching, branches short; stipules narrow–lanceolate, long–acuminate, striate, hairy outside, ciliate, 15 mm or longer, one third adnate to the petiole; petioles long, 7 cm, terete, slightly grooved above, hairy; rachis 4–8 mm long; leaflets rigid, coriaceous when dry, ciliate, subsessile, elliptic or obovate or obovate–lanceolate or obovate–oblong, apex acute, mucronate; base obtuse or rounded, glabrous above, adpressed hairy below, thick-nerved, 2–41 cm × 10–12 mm; hypanthium thread-like, hairy, 6–9 cm long; calyx deeply bilabiate, upper lip entire, outside hairy; hairs setaceous, sparsely mixed; calyx hairy, bilabiate, upper lip four-toothed, lower entire and acuminate; standard suborbicular, orange; wings oblong, obtuse; pods 10 mm long, pericarp slightly reticulated; seed tan, tip elongated or rounded.

Arachis glabrata Benth.

Perennial; rhizomatous, prostrate or ascending; stems and rhizomes thick, rhizomes subterranean, long, branched; stem semi-woody, short, well-branched; branches prostrate, subglabrous; stipules narrow, acuminate, 20 mm × 2 mm; petiole 3–6 cm long, sparsely hairy, hairs long; tetrafoliolate, leaflets oblong–elliptic, 22–30 mm × 10–18 mm, tip acute or sometimes obtuse, glabrous above, rarely hairy below, margins may be serrulate; numerous flowers but few fruits; standard yellow or orange-yellow, 15–20 mm wide; hypanthium 3–6 cm long, hairy; calyx hairy, bilabiate, upper lip four-toothed, lower entire and acuminate; standard suborbicular; wings oblong, obtuse; pods 10 mm long, pericarp slightly reticulated, seed tan, tip elongated or rounded.

Wide variation for morphological features occurs between accessions of *A. glabrata* collected so far. Variation in plant size, growth habit, leaf canopy, leaflet shape, size and hairiness and flower colour have been observed. Seed set is rare though not uncommon in the area of distribution.

Arachis hagenbeckii Harms.

Perennial, central taproot with rhizomes; rhizomes long, 10–35 cm, well-branched, 1.5–5 mm thick with adventitious roots; stems prostrate, subglabrous; tetrafoliolate; stipules 30 mm × 4–7 mm; petiole 4–8 mm long, glabrous or sparsely puberulous, at the point of attachment to the leaflet densely hairy; leaflets oblong, elliptic–lanceolate, narrowly oblong or obovate–oblong, base obtuse or rounded, tip obtuse to subacute, slightly mucronate, glabrous on both surfaces, subcoriaceous, marginate, 30 mm × 8–15 mm; long hypanthium, subglabrous or puberulous, 5–7 cm long; calyx falcate, bilabiate, upper lip irregularly four-toothed, lower lip entire, up to 10 mm long, sparsely hairy outside; standard suborbicular, 12–15 mm wide, yellow; keel hyaline, falcate; style hairy at the tip; pegs 6 cm long; fruits rare, generally uniarticulate, ovoid, acuminate, 16 mm × 8 mm.

Arachis pusilla **Benth.**

Perennial (annual?); taproot well-branched; main stem short, 3–8 cm tall, erect, angular, densely hairy; basal branches thin, procumbent or decumbent, subglabrous; tetrafoliolate; on main stem adnate portion of stipules 10–12 mm long, free portion 20–23 mm, minutely puberulous at the margin, hairy at the base, petioles 50–60 mm long, rachis 17–19 mm; on lateral branches adnate portion of stipules 5–7 mm, free portion 10–17 mm, petioles 25–40 mm long, rachis 8–15 mm long; leaflets ovate or obovate on main stem; apical leaflets 29 mm × 17 mm, basal 28 mm × 16 mm, leaflets on branches oblong–obovate, apical leaflets 17–26 mm × 9–17 mm, basal leaflets 15–26 mm × 14–17 mm, membranaceous, acute, glabrous or adpressed pilose, mucronate; hypanthium thread-like, 37–43 mm long, slightly pubescent; calyx bilabiate, upper lip irregularly three- or four-toothed, lower lip entire, hairy outside, 5 mm long; standard orange, crescent yellow, suborbicular, 10 mm in diameter, wings lemon yellow, 6–7 mm long; keel hyaline, falcate; fruit 2–3 articulate, articles 10–12 mm × 6–7 mm, covered with short hairs, pegs almost horizontal, 30–72 cm long, inserted basally to pod, isthmus 20–37 cm long; seeds 8 mm × 4 mm, light tan.

The name *A. pusilla* for the above description may be incorrect (Krapovickas, 1988) but it is given here because there is as yet (1994) no authentic renaming of the species.

3.4 INFRASPECIFIC DIFFERENTIATION AND MORPHOLOGY OF *ARACHIS HYPOGAEA*

As in the case of interspecific taxonomy of the genus *Arachis*, subspecific classification of *A. hypogaea* has been considered by various workers (for a review see Ramanatha Rao, 1987; 1988). Most of the earlier classifications were based on growth habit, presence or absence of dormancy and maturity. The later attempts included characters such as branching pattern and location of reproductive branches. The classification by Gregory *et al.* (1951) was a comprehensive study in which groundnut was divided into two large botanical groups, virginia and spanish-valencia, on the basis of branching pattern as described by Richter (1899). The presence or absence of reproductive axes (inflorescence) on the main stem and the arrangement of reproductive (R) and vegetative (V) axes on the primary laterals were the most important criteria. The main axis was denoted as **n**, and the primary, secondary, and tertiary lateral branches **n+1**, **n+2**, and **n+3**, respectively.

The virginia group is characterized by the absence of **R** axes on the main stem. Further, alternating pairs of **V** and **R** axes are borne on the cotyledonary lateral and other **n+1** branches. This system is termed the 'alternate branching pattern'. The first two branches on the **n+1** lateral are always vegetative. The alternate branching pattern is repeated in higher order branches, with some irregularity, especially on the distal nodes. The spanish–valencia group is characterized by the presence of **R** axes in a continuous series on successive nodes of lateral branches, on which the first branch is always reproductive. This system is called the 'sequential branching pattern'. **R** axes are also borne directly on the main axis at higher nodes. Most **n+2** and **n+3** nodes are reproductive. Generally the spanish types show some amount of **n+2** branching in an irregular fashion. Valencia types may not have any **n+2**s, if present they will be in a sequence, distal to the 5th to 8th nodes of the **n+1** branch.

A few other systems of classifications were also attempted based on morphological characters, with different degrees of importance given to characters such as growth habit and branching pattern (Bhavanisankar Rao and Raman, 1960; John *et al.*, 1954; Seshadri, 1962; Varisai Muhammad *et al.*, 1973a; Varisai Muhammad *et al.*, 1973c). All these later systems are deficient because either the germplasm considered for classification was limited or the systems did not take into account the classical taxonomic nomenclature (Dubard, 1906; Harz, 1885; Waldron, 1919). Krapovickas studied extensive collections and proposed a system of subspecific classification of *A. hypogaea* (Krapovickas, 1969; Krapovickas, 1973; Krapovickas and Rigoni, 1960). In this system, subspecies were found to be associated with the five geographic regions: Guaranian, Bolivian, Peruvian, Amazonian (Rondonia and north-west Mato Grosso) and the region of Goiás and Minas Gerais. The number of regions was extended to six, to include north-east Brazil (Gregory and Gregory, 1976). This comprehensive subspecific classification is summarized below, along with a brief description of each class:

Arachis hypogaea L. (Linnaeus, 1753)
Subsp. *hypogaea* (Krapovickas and Rigoni, 1960)
Habit procumbent, decumbent or erect; branching alternate; inflorescence simple and never borne directly on the main axis, first branch on the cotyledonary lateral always vegetative; two or two to four seeds per pod; pod beak prominent, slight or absent; pod constriction prominent, slight or absent; pod very large (20 mm) or small (<10 mm); generally tan seed coat colour but red, white, purple and variegated forms exist; seed dormancy usually present; foliage dark green.

   Var. *hypogaea* (= *A. africana* Lour; type 'Braziliano' Dubard, 1906; type 'Virginia' (Gregory *et al.*, 1951)); Bolivia and Amazonia. Habit procumbent, decumbent or erect; main axis in procumbent forms short (not exceeding 40–50 cm); stem usually not very hairy; usually two-seeded; pod beak generally not very prominent, medium–late maturing.

   Var. *hirsuta* Kohler (= *A. asiatica* Lour; type 'peruano' (Dubard, 1906 in part)); Peru. Habit procumbent; main axis may exceed 1 m; stem fairly hairy; pods strongly beaked and ridged, with two to four seeds; very late maturing.

Subsp. *fastigiata* Waldron (Waldron, 1919)
Habit erect or decumbent; branching sequential; inflorescence simple or compound, always present on main axis; first branches on cotyledonary laterals reproductive; seed dormancy usually absent; foliage usually lighter in colour than in subsp. *hypogaea*

   Var. *fastigiata* (= type 'peruano' (Dubard, 1906 in part)); type 'valencia' (Gregory *et al.*, 1951); Guarania, Goiás, Minas Gerais, Peru, and north-east Brazil. Vegetative branches on primaries absent or regularly placed at the distal nodes; inflorescence usually simple; pods with two or two to four (rarely five) seeds; beak absent, slight or prominent; size medium to small; testa colour tan, red, white, yellow, purple or variegated.

   Var. *vulgaris* Harz (= type 'spanish' (Gregory *et al.*, 1951)); Guarania, Goiás, Minas Gerais, and north-east Brazil. Vegetative branches occasional and irregularly placed; inflorescence compound; pods usually with 2 seeds; beak present or absent; size medium to small; testa colour tan, red, white, or purple.

   There have been some attempts to link the morpho-agronomic classification (Bunting, 1955; 1958; extended by Smartt, 1961) with the taxonomic treatment (Krapovickas, 1969; Krapovickas and Rigoni, 1960) by some workers (Gibbons *et al.*, 1972; Varisai Muhammad *et al.*, 1973a, 1973c). These attempts were based on studies of limited amount of germplasm available to the authors. However, much larger geographically and genetically more diverse collections cannot be explained by any of the above systems. With the many intermediate forms that are now available, any classification will be very difficult and may not be of great practical value except locally.

3.5 MORPHOLOGY AND DEVELOPMENT

The morphological description and developmental aspects of groundnut were very confused for a long time. The geocarpic fruit, the complex branching pattern and the highly condensed (telescoped) nature of the reproductive axis (inflorescence) have been mainly responsible for such a confusion. Many of the earlier workers were unable to associate the underground fruits to the plant or to the aerial flower. Marggraff (1648) illustrated the fruits as growing on the roots. The first accurate description of flowers was published by Poiteau (1806). However, it was only in 1950 that Smith published for the first time a clear and correct account of the aerial flowers and subterranean fruits of the groundnut (Smith, 1950). A brief, updated description was published by Hammons (1981).

In the following account the developmental aspects are combined with morphological description. This is because morphology at any point in time is best described in relation to the developmental stage.

3.5.1 Seed

Seeds show large variation for their size, shape and colour. The testa or seed coat is thin and papery. Generally the seed coat constitutes three unicellular layers: the epidermis or sclerenchyma, the middle parenchyma and the inner parenchyma. These layers represent the integuments of the maturing ovule and are maternal in origin (Glueck *et al.*, 1979). The surface as well as the transverse sections of testa show a great deal of diversity (Zambettakis and Bockellee-Morvan, 1976). Groundnut cultivars can be grouped according to the size of the wax layer, the joining of the epidermal cells, thickness of cell walls and presence of cracks in the epidermal layer. The seeds of wild species basically resemble the cultivated groundnut, except that they are much smaller in size.

(a) Size

Seed size is an important economic character. It is fairly stable for any given cultivar and is highly diagnostic in nature. Seed lengths ranging from 7 to 21 mm and seed diameters from 5 to 13 mm have been observed (Ramanatha Rao, 1988; Retamal *et al.*, 1990). Seed size, together with the seed mass, has been used extensively in agronomic classification of groundnut. Larger seed types are preferred for confectionery purposes, while most of the oil types have medium to small seeds. Seed size in wild *Arachis* is much smaller and length ranges between 8 and 18 mm (16–23 mm in *A. villosulicarpa*) and diameter between 4 and 7 mm.

(b) Mass

Seed mass (weight) is an important economic as well as diagnostic character. Depending on the material studied and site of evaluation, various ranges have been reported: 0.2–1.0 g (Seshadri, 1962); 0.17–1.24 g (Ramanatha Rao, 1988); and 0.54–2.38 g (Retamal *et al.*, 1990). In general, cultivars belonging to var. *hypogaea* tend to have larger and heavier seeds; those belonging to var. *fastigiata* have smaller and lighter seeds; and the wild species have much lower seed mass.

(c) Colour

Colour of the seed coat or testa is an important diagnostic character. It is also an important market trait. Broadly, groundnuts can be classified into those possessing non-variegated testa (one solid colour) and those having variegated testa (more than one colour). Generally, seed colour deepens with storage in shell or as shelled seed over a period of time (Bunting, 1955) so that observations immediately after harvest or a long time after harvest can be unreliable. Different solid colours such as white, rose, flesh, wine, red, light purple and dark purple can easily be recognized. Within each class, the intensity may vary depending on maturity, environment, genotype or the interaction between genotype and environment. Variegated testa colour has been associated with splitting or rupture of the outer epidermis due to differential growth rates of testa and embryo (Ashri and Yona, 1965; Stokes and Hull, 1930; Yona, 1964) and inhibition of full development of the outer epidermal layer of testa in some regions (Branch and Hammons, 1979). However, these theories can explain only the variegation when white is involved. The exact genetic or physiological mechanisms for variegation of testa are yet to be fully understood. In some cases a white spot may appear on the seed coat opposite to the micropylar end (Srivastava, 1968).

The various solid colours observed in cultivars include off-white, yellow, pale tan, light tan, dark tan, rose, grey-orange, light red, red, dark red, light purple and dark purple. Among the variegated types, two components can be recognized: the major (primary) colour and the minor (secondary) colour superimposed on the major colour. The four observed colour combinations are: red and white, purple and white, light tan and dark tan, tan and purple. In the case of combinations involving white, it can be almost pure white or off-white. Similarly, different shades of purple and red occur. The minor colour may vary in its degree from thin streaks to large blotches. Because of the subjective nature of this character, use of guides such as the Royal Horticultural Society (RHS) Colour Chart is recommended (IBPGR/ICRISAT, 1981). Seed colour is best observed at least two weeks after harvest (not later than six weeks) and after thorough drying.

Wild *Arachis* species show much less variation. The main colours that have been recorded in different species are: yellowish-tan, light tan, dark tan and reddish brown.

(d) Primordia in seed

Each seed consists of two massive cotyledons, upper stem axis and young leaf primordia (epicotyl), hypocotyl and primary root. Unlike other seeds of Papilinoideae, the embryo of the groundnut is straight. It contains all the primordial leaves and the above-ground parts that appear during the first two weeks of growth. The epicotyl consists of three buds, one terminal with four leaf primordia and two cotyledonary laterals with one or two leaf primordia (Gregory et al., 1951; Maeda, 1972). Some instances of poly-embryony have been recorded (Patel and Narayana, 1935; Raman and Nagarajan, 1958; Smith, 1950). The speculation about the presence of flower primordia in seed (Schwabe, 1971) could not be confirmed (Star-itsky, 1973).

3.5.2. Seedling

The morphology, growth and development of the groundnut seedling was studied by a number of workers (Badami, 1933, 1935; Bouffil, 1947; Richter, 1899; Yarbrough, 1949, 1957a, 1957b). The radicle consists of about half hypocotyl and half primary root during the first few hours of germination, depending on the depth of planting. In moist soils at 27 °C, the primary root emerges in 24–36 hr. The primary root may grow from 0.5 to 4.0 cm in four days. Lateral roots generally appear after the second day, and as many as 100 laterals could be formed in five days (Yarbrough, 1949).

3.5.3 Root system

Groundnut is a herbaceous annual with a taproot and a fairly well-developed root system (Figure 3.1). However, the root and hypocotyl may be modified to form tubers (section Erectoides) or tuberoids (sections Erectoides and Extranervosae). Diameter of the primary root may vary from a few millimetres in annual species to 8 cm in perennial species (Gregory et al., 1980). Adventitious rooting at nodes of branches that come in contact with soil is common, and is more frequent in species belonging to section Caulorhizae. Most of the root system is generally concentrated at a depth of 5–35 cm (Narasinga Rao, 1936). The root spread is confined to a radius 12–14 cm. The spreading types usually have a more vigorous root system than the bunch types (Mohammad and Khanna, 1932; Seshadri et al., 1958).

(a) Taproot

On the second day after germination the taproot appears with a large root cap. It elongates rapidly and grows almost vertically downwards, diverting only to overcome obstacles. The taproot can grow to a length of 50–55 cm in ill-drained clayey soils (Badami, 1935), or up to 90–130 cm (Bruner, 1932). The root elongation depends on the cultivar, soil and available

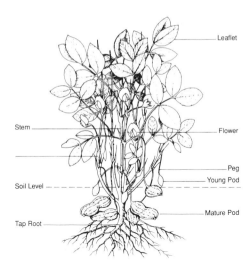

Figure 3.1 Groundnut plant.

moisture (Intorzato and Tella, 1960; Seshadri *et al.*, 1958; Yarbrough, 1949).

(b) Lateral roots

Lateral roots appear on the third day after seed germination in the upper region of the taproot. As many as 100–120 may be produced by the fifth day which grow to a length of 15–20 cm (Yarbrough, 1949). The lateral roots are basically similar to taproots. The main differences are the lack of central pith and being diarch in contrast to the tetrarch primary root (Gregory *et al.*, 1973).

(c) Hypocotyl

The hypocotyl (the portion of the stem which lies between the primary root and the cotyledons) is succulent early on and grows up to 10 mm in diameter. The region between the hypocotyl and the primary root is bounded by the intact epidermis of the hypocotyl (Yarbrough, 1949). This collar region is swollen due to the enlargement of the cortex. It elongates to bring the cotyledons to the soil surface and its length depends mainly on the depth of planting (Bouffil, 1947). The emergence is complete in about 10 days after sowing and the hypocotyl grows to 1–11 cm (Bouffil, 1947; Seshadri, 1962; Yarbrough, 1949). It loses its fleshy nature, becomes fibrous and changes from white to creamy-white to brown, becoming indistinguishable from the taproot by about four weeks after germination. Later it may produce many roots.

(d) Root hairs

The outer layers of the primary root generally slough off as root growth proceeds (Petit, 1895) causing loss of the structural base for root hairs (Richter, 1899). However, under certain conditions formation of root hairs, as clusters or rosettes in the junction of root axils, has been observed (Badami, 1935; Reed, 1924; Waldron, 1919). These hairs were less fragile than those found in other legumes and ranged from 2 mm to 4 mm in length (Allen and Allen, 1940). Some of these hairs were septate (Chandler, 1978). These structures have often been referred to as root hairs, a misnomer in the strict botanical sense.

Though root nodules are not plant structures in a botanical sense, their close association with the plant makes them appear so. Nodules first appear when plants are about 15 days old. A group of nitrogen-fixing bacteria belonging to the genus *Bradyrhizobium* is responsible for the formation of nodules. Though groundnut does not possess real root hairs, an association between the tufts of hairs in root axils and nodulation has been found (Nambiar *et al.*, 1983).

3.5.4 Stem

The stem is generally angular, pubescent and solid with a large central pith in the early stages (section 3.6.2). As the plant grows, stems tend to become hollow and cylindrical and shed most of the hair, especially on lower internodes. There is no indication of woody development in *A. hypogaea* but some wild species can develop woody stems (section 3.3). The main stem develops from the terminal bud of the epicotyl flanked by two opposite cotyledonary laterals. Though main stem height depends on genotype, it is influenced by environment to a considerable degree and ranges from 12 cm to 65 cm (Ramanatha Rao, 1988). In wild *Arachis*, the main stem tends to be much shorter, ranging from 12 cm to 35 cm, with a few exceptions. Stem thickness is highly variable, although Seshadri (1962) reported that generally the bunch types have thicker internodes, short and highly condensed at the base and longer at the higher nodes. The basal stem diameter could be as much as 8 cm in some of the wild species.

(a) Colour

Anthocyanin pigments in the epidermal cells of the stem can give different shades of colour. Stem colour is determined by the absence or presence and intensity of pigmentation. The common colours observed are purple, pink, dark red, light red or green (absence of anthocyanin) and numerous shades of purple, pink and red. The colour development is influenced significantly by exposure to sunlight and recording typical colours is generally difficult. Hence, the stem colour could be recorded as present or absent, classifying the cultivars into two broad groups (IBPGR/ICRISAT, 1981).

(b) Hairiness

Groundnut has been described as a glabrous to hirsute herb, indicating the extent of variability for this character. A range of two to four grades of hairiness were recognized (Hayes, 1933; John et al., 1954; Patel et al., 1936; Patil, 1965; Varisai Muhammad et al., 1973c). Because of the difficulty in defining various grades of hairiness, it is pragmatic to recognize only two grades: scarce and abundant. Generally the upper internodes should be observed because at lower levels the hairs are not persistent. The hairs are arranged in regular rows on the stem. It is possible, by careful observation, to classify the arrangement into the number of rows of hairs (two, four, six and irregular).

Usually the following three types of hairs have been observed on plant parts:

- Long hairs (up to 3 mm), are generally septate and uniseriate and distributed irregularly. Occasionally the cell walls of the top three or four cells disintegrate, giving the appearance of a single long cell. The outer walls may sometimes be thickened.
- Short hairs (<1 mm), generally occur densely along with the long hairs.
- Glandular hairs (trichomes, spines) are long and bristle-like, green with chlorophyll and multicellular with a bulbous base.

(c) Growth habit

The groundnut plant has a distinct main stem and a variable number of lateral branches. The carriage of laterals determines the growth habit of the plant. Two distinct forms of growth habit – spreading (runner, trailing, procumbent and prostrate) and erect (upright, erect bunch and bunch) – have long been recognized and have provided the basis for both agronomic and taxonomic classifications.

There has been no agreement on classification of growth habit in groundnut (for a review, see Ramanatha Rao, 1988). This is further complicated by the complexity of inheritance of this character (Hammons, 1973; Wynne and Coffelt, 1982). A large number of intermediate forms occur because of cytoplasmic and genetic factors that interact together and also interact with light environments for the expression of growth habit (Ashri and Goldin, 1963; Ziv et al., 1973). Recognizing bunch (erect) and runner (trailing) forms as the major forms, it was suggested that the cytoplasmic–genic interactions that determine growth habit act on the biosynthetic pathways, producing phytohormones and their inhibitors (Ashri, 1976; Resslar and Emery, 1978).

The above discussion suggests that growth habit is not a discrete character, and its inheritance is complex. Several loci interacting among themselves and with the cytoplasm may produce various grades. However, for a given homozygous accession, its growth habit can be defined as stable. In the earlier classifications, the runner and spreading bunch forms were associated with alternate branching (subsp. *hypogaea*) and the erect habit with sequential branching (subsp. *fastigiata*) (Krapovickas, 1969; Krapovickas, 1973). However, if one considers the enormous variation that exists in the current groundnut genetic resources, such association is no longer very clear-cut. Hence, the need to reclassify the growth-habit types is recognized. Growth habit needs to be described purely on the basis of observations on plants grown at a wide spacing in a given environment. Nevertheless, it must be noted that a vast majority of runner and spreading bunch forms show alternate branching, while the erect forms are generally sequentially branched. The major collections would include, along with landraces, a number of accessions derived through breeding and also accessions from South America. A certain amount of introgression might have occurred at subspecific level in the centre of diversity. In such germplasm the following growth habits could be observed which are based only on the carriage of plant and the position of primary branches in relation to the main stem (IBPGR/ICRISAT, 1981) (Figure 3.2):

1. Procumbent 1: The main stem is erect and may vary in height. The lateral branches are prostrate.
2. Procumbent 2: This is similar to Procumbent 1, but the main stem has

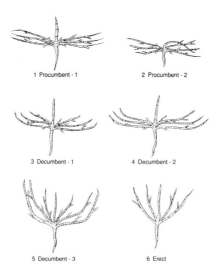

Figure 3.2   Growth habit.

a tendency to bend and continue to trail on the ground. This type is generally encountered in wild *Arachis*.

The common runner (= spreading, trailing, creeping or prostrate) habit group includes the above two forms.

3. Decumbent 1: The main stem is distinct. The laterals, which normally trail on the ground, tend to become upright at distal nodes.
4. Decumbent 2: The main stem is distinct. The laterals tend to be almost upright at the median nodes.
5. Decumbent 3: The main stem becomes indistinct as most of the lateral branches are almost upright and give the plant a bushy appearance.

   The spreading bunch (= semi-spreading, bunch runner, and runner bunch) group generally includes the above three forms.
6. Erect: the main stem is indistinguishable from the laterals. The laterals are at an acute angle to the main stem. This form includes the erect bunch or bunch groups in the earlier classifications.

(d) Branching pattern

The arrangement of reproductive (**R**) axes on the main stem (**n**) and **R** and **V** axes on the primary lateral branches (**n+1**) is the basis for classifying the branching pattern. Groundnuts were classified into two botanical groups based on the branching pattern (Gregory *et al.*, 1951), which were named as alternate branching and sequential branching (Bunting, 1955) (section 3.4). Such a system was confirmed and used in classifying *A. hypogaea* along with other correlated characteristics (Krapovickas, 1969). The cultivars of var. *hypogaea* produce numerous secondary (**n+2**) and tertiary (**n+3**) branches, while subsp. *fastigiata* produces a limited number of **n+2**s and rarely **n+3** branches.

Describing groundnuts as alternate and sequential forms had been found adequate until recently. The order of occurrence of vegetative and reproductive nodes on the basal primary branch of *A. hypogaea* subsp. *hypogaea* var. *hypogaea* was studied. Wide variation was observed in vegetative and reproductive node numbers and it was suggested that the sequence in runner types could be a result of introgression. Because of the large variation, it was felt that branching pattern might not be a dependable attribute to classify groundnuts (Bhagat *et al.*, 1987). However, more studies of this nature, over different environments, would be required to draw such definitive conclusions on such an important character. It is also essential to note that branching pattern should be observed on the cotyledonary lateral, which may be difficult to identify in well-grown runner types.

Intermediate types from infraspecific crosses have been observed and there is recognition of the serious problem that such forms can cause in taxonomic treatment, due to the increased frequency of introgressed forms, arising from artificial or natural hybridization between the two

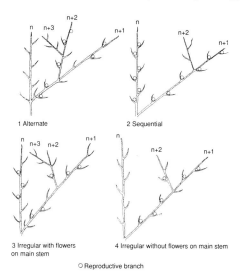

Figure 3.3 Branching pattern.

subspecies (Maeda, 1973) . This is indeed so, as the assembly of a large number of cultivars includes many variants that cannot be assigned to either group (Ramanatha Rao, 1988). To account for such types it was proposed to have two more groups: irregular branching-1 (this is similar to the alternate pattern but **R** axes occur on the main stem also) and irregular branching-2 (similar to the sequential pattern but with no **R** axes on the main stem) (Figure 3.3).

3.5.5 Leaf

The leaves that first appear are from the primordia in the seed and new ones are formed much later. Leaves are tetrafoliolate except in species belonging to the section Trifoliolatae (= series Trifoliolatae of the section Erectoides), which have three leaflets. Occasionally small and abnormal leaflets may appear. The leaves, which are paripinnate, are borne spirally in a 2/5 phyllotaxy and their arrangement on the main stem and higher-order branches is distichous. Leaflets are opposite, subsessile, elliptic (variable) and shortly mucronate with entire ciliate margin. Stipules are prominent, linear and adnate to some length to the petiole and become free at the pulvinus. Foliaceous stipules have been reported (Mouli and Kale, 1982). The leaflets are borne on a slender, grooved and jointed rachis. The leaf exhibits nyctiotropic (sleep) movements, where the adaxial surfaces of the leaflets come together and the petiole droops during each night. Considerable variation exists in leaf characters such as foliage colour, shape, hairiness and size.

Stomata are present on both the leaf surfaces (Smartt, 1976b). Stomatal frequency and distribution on leaf epidermis were studied and a higher stomatal count was recorded on the upper surface than on the lower (Lin *et al.*, 1986).

(a) Colour

Leaf colour is an important character to differentiate the subspecies in groundnut. *A. hypogaea* subsp. *hypogaea* has dark green foliage in contrast to the light green of subsp. *fastigiata*. Leaf colour recordings are highly subjective: differentiating subtle shades always poses problems and the intensity of light at the time of observation can also create difficulties. The use of a guide such as the RHS Colour Chart has been recommended (section 3.5.1 (c)). Some of the wild species, especially those belonging to section Extranervosae and a few cultivars belonging to var. *fastigiata*, often have shiny upper leaf surfaces.

(b) Shape

The shape of leaflets varies from oblong to lanceolate. The variation for leaflet shape is much greater in wild *Arachis* than in the cultivated species. Linear–lanceolate or narrow–lanceolate leaflets have been recorded (Bhide and Desale, 1970). Various shapes observed include cuneate, obcuneate, elliptic and lanceolate. However, variation exists for leaflet shape even within a plant. This variation may be between the leaflets on basal nodes and the ones on upper nodes and also between the apical leaflets and the basal leaflets of the same leaf. In most of the wild species, the shape of the leaflets on the main stem leaves may differ from that of the leaflets on lateral branches. For standardization purposes, the shape is best recorded from the fully expanded apical leaflet of the third leaf on the main stem and also the shape of the apical leaflet of the fifth leaf on primary lateral in the case of wild species (IBPGR/ICRISAT, 1981).

(c) Size

Leaflet size is variable among cultivars, within a cultivar, or even on the same plant, depending on the position of the leaf on the plant. The apical pair is generally larger than the basal. Leaflets on the main stem are slightly larger than those on the branches in both cultivated and wild species. Size is influenced to some extent by environment. For standardization purposes, the size of the apical leaflet of the third leaf on the main stem is considered and it is a diagnostic character. Subsp. *fastigiata* usually has larger leaflets than subsp. *hypogaea*. The size of a leaflet varies from 24 mm × 8 mm to 86 mm × 41 mm. In wild species, the range is much larger: 15 mm × 8 mm to 95 mm × 30 mm (Ramanatha Rao, 1988).

(c) Hairiness

The leaflets are pubescent, as are the stems, mainly on the abaxial surface and on the margins. Generally midribs are also hairy, often more densely than the leaf surface. As described earlier (section 3.5.4 (b)), all the three hair types – long, short and glandular – can be found on the petiole, rachis and leaflet. There is significant variation in hair distribution, shape, size and number in *A. hypogaea*. In wild species, stem hairiness (type and distribution) has taxonomic significance. The length, shape and number of hairs appear to be associated with resistance to leaf hoppers (Campbell *et al.*, 1976). Five different grades of hairiness can be distinguished: almost glabrous, sparse and short, sparse and long, profuse and short, and profuse and long. Differences in grades have been noticed between young and mature leaves.

3.5.6 Inflorescence

The inflorescence, which is a reduced simple or compound monopodium, appears as a cluster of flowers in the leaf axil (Figure 3.4). In some wild species it appears as a cluster around the crown of the plant, with subterranean ovaries and the hypanthium exposing other flower parts. The inflorescence, consisting of up to three (but occasionally more) flowers, is spike-like and always occurs in the axils of cataphylls or foliage leaves. The reproductive (**R**) and vegetative (**V**) axes appear to occur at the same nodes because of the short internodes below the first cataphyll branch (Norden, 1980). The inflorescence is a small replica of the vegetative branch. In the axils of scale-like leaves, either a very short peduncle bearing a single flower (simple inflorescence) or a peduncle bearing secondary branches (compound inflorescence) is produced. In var. *hypogaea* the inflorescence expands very little during maturation. In var. *vulgaris* it expands moderately, while in var. *fastigiata* the inflorescence internodes may elongate to form a conspicuous branch that may terminate with a pair of small leaflets. This branch may also bear inflorescence and may be confused with normal **V** axes. But **R** axes can be identified by their numerous scale leaves at the basal nodes. The length of the reproductive axes is cultivar-dependent and may exceed 10 cm in some cases.

3.5.7 Flowers and flowering

The typical papilionoid corolla is inserted on the top of the hypanthium and surrounds the staminal column (Figure 3.5). Flowers are enclosed between two bracts, one of which is simple, subtending a short peduncle, and the other bifid, subtending the pedicel. The flower is sessile but appears stalked after the growth of a tubular hypanthium just before anthesis. The calyx is five-toothed, one of which is free and anterior to the keel. The length of the posterior lobe in relation to the standard petal varies from half the height of the standard to three-fourths. The other four are fused up to near their tips at the back of the standard and sometimes the calyx is considered to be two-toothed. The colour of the standard petal is generally orange, with a basal central area, the crescent, which may be marked with a different colour. The wings are generally yellow, occasionally marked with a diffuse band of deeper yellow, orange or brick-red at the tip, generally referred to as 'blush'. The keel is almost hyaline; it is faintly yellow and closely encloses the staminal column. Generally, the keel and staminal column are curved at right angles (reflexed) in the mid-section. Under normal conditions the flower stalk (actually the hypanthium) is about 4–5 cm long; the standard is 10–22 mm wide, and 8–16 mm long.

The stamens are 10, monadelphous, with the staminal column surrounding the ovary. Two are usually sterile. Of the remaining eight, four are globose and uniloculate alternating with four oblong anthers, of which three are biloculate and the one opposite the standard is uniloculate (Smith, 1950). However, it has been observed that the oblong anthers become biloculate and globose uniloculate when the anthers reach maturity (Xi, 1991). The filaments are fused for two thirds of their length and form acute angles with their fused bases, because at the point of separation the filaments are sharply bent. Three basic morphological groups were identified in pollen: spheroidal, prolate and columnar-spheroidal. Differences among cultivars in shape, size, exine sculpturing and poration of pollen grains may be useful for varietal classification and taxonomic studies (Pen *et al.*, 1987). The pistil consists of a single ovary surrounded by the base of the hypanthium. The ovary is about 1.5 mm long

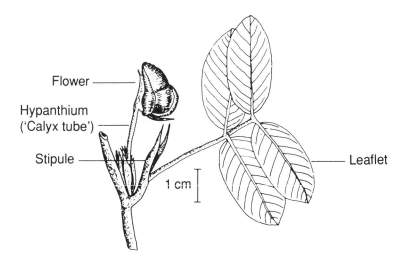

Figure 3.4 Inflorescence.

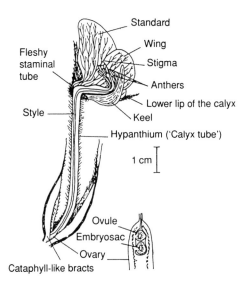

Figure 3.5   Flower – a longitudinal section (after Smith, 1950).

and 0.5 mm wide, and has two to four, sometimes five and rarely six ovules. The long filiform style has two sharp bends and is covered with upward-slanting hairs near the top on the surface facing the standard (Gregory *et al.*, 1951; Smartt, 1976a; Umen, 1933). The stigma is club-shaped or clavate, usually at anther level or protruding slightly above.

(a) Standard petal colour

The most common colour of the standard is orange but other colours such as white, lemon-yellow, yellow, yellow-orange and brick-red (garnet, russet-brown, burnt-orange, amber) also occur. The colour of the flower (or, to be precise, the standard petal) is distinct and is cultivar-dependent. The colour of the crescent portion is considered separately. As with any colour characteristics, observation is quite subjective and can be difficult. The use of a colour chart is recommended. Wings and keel are generally yellow or orange-yellow in colour. In some accessions a darker orange or purple blush could be found on wings.

(b) Standard crescent

The inside of the standard petal has distinct markings at the base from which lines of different intensities radiate and this is usually referred to as the crescent. The intensity of colour of the crescent ranges from complete absence to a prominent pattern. Variation is also observed in the colour of these markings between cultivars. Five gradations in the colour of the crescent were reported, from the faintest (almost absent) to the most prominent deep purple colour (Hayes, 1933; John *et al.*, 1954). Usually three expressions involving the standard crescent can be observed: presence or absence, compactness of the radiating lines and brightness of the colour. The species belonging to the section Extranervosae have the markings on the back of the standard. In species belonging to section Ambinervosae, the markings appear on both surfaces of the flower. *A. batizocoi* has a pink or purple blush on the back of the standard.

(c) Flowering

Groundnut is indeterminate in growth to a degree. Depending on genotype, environment and temperature, flowering starts at about 25 days after emergence. As the daily mean temperature rises from 20 °C to 30 °C, the number of days required for the first flowering is reduced from 38 to 25 in subsp. *hypogaea* and from 35 to 24 in subsp. *fastigiata* (Ono, 1979). The most prolific flowering occurs between 5 and 11 weeks after planting, depending on the duration of the cultivar and the season, with a high degree of first-formed flowers producing mature fruits.

Usually four or five stages of flowering can be distinguished. Very few flowers are produced at stage I, followed by rapid flowering in stage II. A peak is reached at stage III, followed by a decline in the number of flowers produced at stage IV. In some cases there can be two peaks of flowering. Very few flowers that open after the first and/or second peaks produce mature fruits, unless early flowers are prevented from functioning normally due to some environmental stress (Bear and Bailey, 1973; Mohammad and Khanna, 1932). Groundnut produces more flowers than the plant can sustain to develop into pods. About 40% of the flowers fail to develop from the outset, while another 40% produce only pegs (Smith, 1954). Less than 20% of the flowers produce mature fruit (Donovan, 1963) under best conditions and sometimes less than 15% (Lim and Hamdan, 1984). Genotypes which flower early show greater synchrony, and those which produce most of the flowers during the first two weeks of the flowering period produce greater numbers of pods (Sastry *et al.*, 1985). Removal of some flowers every day can prolong flowering (Bear and Bailey, 1973: Seshadri, 1962).

On a given day, usually only one flower in an inflorescence reaches anthesis, except in var. *vulgaris* in which two or more flowers may reach anthesis on the same day. Intervals between successive flowers on an inflorescence may vary from one to several days. A day before anthesis the flower bud is 6–10 mm long. During the day the hypanthium elongates slowly and the buds grow to 10–20 mm. The elongation is faster during the night and at the time of anthesis the buds are about 50–70 mm long (Smith, 1950). Buds generally open at the beginning of the light period; it may be delayed in cold or wet weather. The dehiscence of anthers takes place 7–8 hours before the flower opens or sometimes much later (Bolhuis *et al.*, 1965). Though the stigma was reported to be receptive from 24 hours before to 12 hours after the opening of the flower (Hassan and Srivastava, 1966), generally it is receptive only a few hours before anthesis (Sastri and Moss, 1982). On warm, sunny days most of the floral parts may wither within 5–6 hours after flowering. Only the ovary and base of the style remain turgid after the day of anthesis. Flowering generally occurs during the early hours in the morning with some delay on cold/wet/dull days. Generally, anther dehiscence occurs in the bud about one hour before flowering. The pollen tube grows at the rate of about 1 cm per hour, resulting in fertilization 5–6 hours after pollination (Lim and Gumpil, 1984).

3.5.8 Pollination and fertilization

Normally the stigma protrudes above the anthers and is receptive before anthesis. As the stigma is enclosed in the keel and surrounded by the staminal column, self-pollination is most common with a high frequency of cleistogamy (Murty *et al.*, 1980). Rarely, the stigma and anthers are exerted from the keel and, as the stigma is receptive, some amount of outcrossing may occur. Outcrossing ranged from less than 1.0% to 3.9%, depending on the season, genotype and location, and several bee species were found to be the pollen vectors (Culp *et al.*, 1968; Gibbons and Tattersfield, 1969; Hammons, 1963, 1964; Leuck and Hammons, 1969; Srinivasalu and Chandrasekharan, 1958). Fertilization is complete in about 6 hours after pollination, and within 5–6 hours the flower may wither.

3.5.9 The peg

Generally fertilization is completed before midday. After fertilization, the flower droops, the corolla closes, the calyx tube bends and the flower withers. The ovary at the base of the calyx tube starts growing actively within a week by the activation of the intercalary meristem present below the ovary. The green ovary turns purplish from the tip downwards. The developing ovary pierces through the floral parts to reveal an elongating peg. The peg or carpophore is a stalk-like structure that carries the fertilized ovules at its tip. Its growth is positively geotropic until it enters the soil to some depth (up to 5–7 cm). The tip then becomes diageotropic or almost horizontal. Only then does it begin to develop into a fruit. The peg usually withers away if it fails to contact and penetrate the soil after it extends to about 15 cm. In humid conditions, however, some var. *fastigiata* cultivars can form aerial pods. These are underdeveloped, green and small. True breeding aerial podding types have also been reported (Prasad, 1985) and the physiology of such aerial podding, producing normal viable seeds, needs further investigation.

The unique nature of the peg led to several misconceptions about its homology and it has been described as a gynophore, a carpophore, or an apetalous flower (Brennan, 1969; Jacobs, 1947). The peg has been defined as the young fruit during the stalk-like phase of its development (Smith, 1950). The numerous minute plastids that develop after fertilization in the epidermal walls of the peg were found to be responsible for its positive geotropic movement (Patel and Narayana, 1935). The elongation of the peg ceases as a result of auxin formed in its distal portion which moves with strict basipetal polarity through the peg tissue and the intercalary meristem (Jacobs, 1947, 1951a, 1951b). Sometimes, the peg or carpophore has been incorrectly named gynophore (Smartt, 1976a). Bharathi and Murty (1980) suggested that the peg is a highly reduced gynophore and is discernible as a meristem at the base of the ovary. However, considering that the intercalary meristem basal to the ovary elongates and produces articulated pods in most of the wild *Arachis* species, the peg is a carpophore (Ramanatha Rao, 1988).

(a) Colour

The colour of the above-ground portion of the pegs varies from shades of pink to purple, in contrast to the normal green; and such colours are due to anthocyanin pigments. Different peg colours such as light purple, dark purple and green (no pigmentation) have been reported, as has a loose association of the green colour of pegs with white testa colour (Varisai Muhammad *et al.*, 1973c). Though the development of pigment in the peg is cultivar-dependent, it is greatly influenced by exposure to sunlight and grades of pink or purple can occur on the same plant. Pegs can be broadly classified as pigmented or green but the portion in the soil is generally white (colourless).

(b) Hairiness

The developing pegs have minute white hairs that give a downy appearance (Seshadri, 1962). These hairs are deciduous and are shed as pegs mature. Above-ground parts of pegs may have stomata, lenticels and multicellular trichomes whereas below-ground parts may bear unicellular structures resembling root hairs which can reach very high density and may be up to 0.75 mm long. Similar structures may be found on the developing pod, but at later stages these hairs degenerate and large lenticels may then be present on pegs (Webb and Hansen, 1989). The species belonging to section Extranervosae produce roots on the underground portion of the peg.

(c) Size

The depth at which the pods are produced is determined by the extent of the peg's penetration into the soil. Though significant cultivar differences have been reported (Seshadri *et al.*, 1955), variation within the same cultivar for the size of the peg is also very great. The length of the peg was reported to be closely related to the differences in branching pattern (Ono, 1979).

In wild *Arachis* species, the length of the peg is highly variable and can be up to a metre in some species. In the context of distribution, the length of the peg is important to produce the fruit as far away as possible from the mother plant. A long peg, together with a long isthmus, can place the seeds at a considerable distance from the mother plant. The orientation of the peg in the wild species – horizontal or vertical – is an important taxonomic criterion.

The thickness of the peg varies from about 1 mm to 2 mm. An assessment of the thickness of the pegs in the different types showed that the cultivars of subsp. *fastigiata* had thicker pegs than those of subsp. *hypogaea* (Seshadri, 1962). The strength of the peg is an agronomic attribute governing the ease of harvesting of the crop.

3.5.10 The fruit

The fruit, referred to as a pod, is a slightly modified lomentiform indehiscent carpel. The cultivar differences in the proportion of pegs that penetrate the soil developing into pods were found to be significant (Seshadri, 1962; Singh *et al.*, 1981). The mature pod normally contains up to four seeds. Occasionally five or even six seeds per pod have been recorded. Single-seeded pods may be produced when all the ovules except the proximal abort (Smartt, 1976a). The pod size may range up to 8.0 cm × 2.7 cm. The fruit consists of valves, structurally dehiscent but functionally indehiscent. Under pressure, pods split along the longitudinal suture. This is the normal line of dehiscence and the mechanical tissue of the mesocarp is interrupted along this line (Richter, 1899). The shell consists of an outer spongy layer, a middle fibrous and woody layer, and an internal layer which turns thin and papery upon maturation. Mechanical tissue on the dry pod gives it a reticulate pattern with longitudinal ridges (up to 10) and less prominent transverse ridges.

(a) Pod development

The ovule grows gradually up to flowering and the growth continues after fertilization to some extent (Figure 3.6). The embryo, dormant during peg elongation, begins to develop three to four days after commencement of pod development (Murty *et al.*, 1980). The basal ovule develops first. The peg becomes diageotropic and almost horizontal when it reaches its maximum depth in the soil and ceases to grow, and pod development starts. The enlargement of the pod proceeds from base to apex. The pod expands rapidly underground by the growth of a large parenchymatous tissue (endocarp) lying between the ovules and shell layers (Schenk, 1961). The pod grows to its maximum size in about three weeks after the penetration of the peg into the soil. The endocarp recedes as the ovules grow and disappears completely by the time the seeds mature. During this period the inner face of the shell becomes increasingly darker in colour, associated with an increased tannin content, and becomes very dark brown on maturation (Gregory *et al.*, 1973). It takes about 60 days from the time of fertilization to full maturity (Patel *et al.*, 1936). The size of the mature pod and seeds is influenced by genotype, soil and method and time of cultivation (Shibuya, 1935). The normal podding zone is located 4–7 cm below the soil surface. The optimum soil temperature in the podding zone is 31–33°C (Ono, 1979). Higher or lower temperatures prolong the duration of pod development. Lower soil temperatures (around 23°C) increase the number of pods and pod weight but the filling period will be longer, thus increasing the number of days to maturity (Dreyer *et al.*, 1981).

(b) Number of seeds per pod

The number of seeds per pod may range from one to five or even six. Single-seeded pods may occur in almost all cultivars but single-seededness

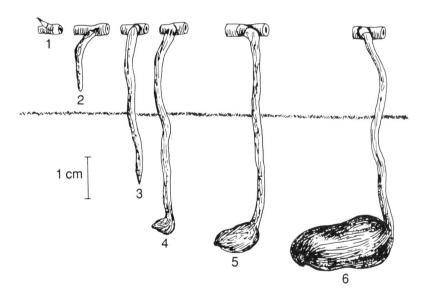

Figure 3.6 Pod development stages: 1. ovary after fertilization; 2. aerial peg, 5–7 days; 3. peg penetrating soil, 18–22 days; 4. pod enlargement, 14–21 days; 5. early stage in pod development; 6. immature pod. (After Smith, 1950.)

is not genetic (Bunting, 1955) and it is usually a result of abortion of all but the proximal ovules (Smartt, 1976a). Three-seeded pods are found in both *hypogaea* and *fastigiata* subspecies. The number of seeds per pod is basically cultivar-specific, though it is influenced to some extent by season and other factors (Seshadri, 1962). Three- or four-seeded pods may be more common in bunch types than in semi-spreading and spreading cultivars, although two-seeded pods are the most common (Varisai Muhammad *et al.*, 1973c). Cultivars belonging to var. *fastigiata* possess predominantly three- or four-seeded pods. The number of seeds per pod has been extensively used for cultivar classification (Gibbons *et al.*, 1972; Varisai Muhammad *et al* , 1973c).

Most of the wild species, in contrast to the cultivated, produce articulated pods. Generally they are biarticulate, though they may appear to be uniarticulate since the isthmus between the articles quickly degenerates with maturity of the proximal article.

(c) Beak

The tip of the indehiscent fruit may end in an appendage called the beak. The prominence or presence or absence of the beak is dependent on the cultivar and is a good diagnostic trait. Five grades can be used to classify groundnuts based on this character: absent, slight, moderate, prominent and very prominent. There can be some variation between the pods of the same cultivar and the observation may be somewhat subjective; nevertheless, it is a very useful character. The shape of pod beak also varies but is very difficult to define.

(d) Constriction

The constriction of the pod is an important character as it affects the developing seed. Non-constricted pods generally have seeds with flattened ends and are in contact with other seeds in the pod. Such a contact sometimes results in embryo damage (Coffelt and Hammons, 1974). Seeds from unconstricted pods tend to split during shelling and so cultivars with no constriction are generally undesirable. On the other hand, pods with deep constrictions tend to carry soil on them. They also break during harvesting and shelling, reducing the market value. Most commercial cultivars have some constriction between each seed; a few do not, especially the cultivars belonging to subsp. *fastigiata*. Under abnormal conditions (such as aerial fruiting), elongated pods with a filiform isthmus may be produced by a cultivar that is normally only slightly constricted (Smartt, 1976a). In several wild species a filiform isthmus is characteristic, which is developmentally complete, unlike the peg.

Pod constriction shows considerable variation both within and between cultivars. However, the range of variation within a cultivar is not great enough to prevent visual classification. Groundnut cultivars can easily be graded into five groups based on the extent of pod constriction: none, slight, moderate, deep and very deep.

(e) Reticulation

The mechanical tissue of the shell gives a reticulated or ridged appearance to the pod. Usually the longitudinal ribbing is prominent but vertical ridges may be prominent in some cultivars. Reticulation (venation, ribbing, ridging) on the shell is an obvious visual characteristic. Typical var. *fastigiata* cultivars may be without any reticulation; however, some Peruvian forms show very prominent ribbing. As with constriction, reticulation also contributes to some extent to the cleanliness of pods at the time of harvest. Five grades can be distinguished: smooth, slight, moderate, prominent and very prominent.

(f) Size

Though the size of the pod is influenced to a some extent by soil and other environmental conditions, it is an excellent diagnostic character. A range of 8–20 mm in diameter has been reported and groundnuts have been divided into five classes, depending on the mean length and mean pod weight (John *et al.*, 1954; Varisai Muhammad *et al.*, 1973b). For a much larger collection, a range of 11–83 mm for pod length and 9–27 mm for pod diameter has been recorded (Ramanatha Rao, 1988). Ranges of 18–58 mm for pod length, 9–19 mm and 7–39 mg for pod weight have also been recorded (Retamal *et al.*, 1990). Pod length was found to be the most useful diagnostic character in cultivar identification, in conjunction with the other morphological characters (Salma, 1985).

   In the case of the wild *Arachis*, the pod is usually articulated, the articles being separated by the isthmus. The length of the isthmus is species-dependant and varies from less than 1 cm to 30–35 cm. In the strict botanical sense, the pod consists of the articles and the isthmus, but for diagnostic purposes the size of a single article (preferably the apical) is used.

3.6 ANATOMY

Groundnut yields are comparatively low, especially in the developing countries. While the reasons for this are many, information on anatomy of the various organs will help in understanding the physiological processes governing higher yields. This information is likely to have direct bearing in formulating research programmes in the disciplines of plant breeding, agronomy, pathology, entomology and physiology. Efficient plant types can be visualized in terms of disease, insect and drought stress, photosynthetic efficiency and harvest index if more precise information on the anatomy of various vegetative and reproductive organs is available. Detailed accounts of groundnut anatomy can be obtained from many works (Badami, 1935; Bharathi, 1981; Bharathi and Murty, 1984; Murty, 1988; Petit, 1895; Richter, 1899; Suryakumari, 1984; Suryakumari *et al.*, 1983, 1984, 1989a, 1989b; Waldron, 1919; Yarbrough, 1949, 1957, 1957b).

3.6.1 Root

As groundnut roots typically lack an intact epidermis because of the lack of specific epidermal initials, root hairs of the usual type rarely appear. Cells originating in the lateral root cap region and in the outer cortex form the surface layers of the root, the outermost of which is split and sloughs off. Experiments have clearly established an active absorptive function of solutes for this region (Richter, 1899; Yarbrough, 1949). Root hairs are not entirely absent and are seen under special conditions and where humidity is high and also more on lateral roots than on the primary root (Badami, 1935; Waldron, 1919, Yarbrough, 1949). The primary root has a tetrarch vascular cylinder (Figure 3.7(c)). Meta and protoxylem tissues are differentiated as in other dicotyledons. The large pith portion of the primary root breaks down with age. During the early stages of seedling growth, the cortex and pith contain abundant food stores. The root and hypocotyl develop active cork and vascular cambium.

Lateral roots are formed by the activity of the endodermis and pericycle which actively contribute cells to the new root apex (Petit, 1895; Richter, 1899). Lateral roots arise opposite the primary xylem plates. A four-ranked arrangement of lateral roots is thus developed on the primary root. The lateral roots are diarch and do not have a pith (Figure 3.7(d)). Thus they differ from the primary roots of other dicotyledonous roots (Yarbrough, 1949).

3.6.2 Stem

The description of the anatomy of *Arachis* stem can be found in several earlier works (Reed, 1924; Waldron, 1919; Yarbrough, 1957b). The structure of the stem in general conforms to that of the typical dicotyledon type. The epidermis of the stem consists of small cells with a thick cuticle. The hypodermis is formed of angular collenchyma. The cortex consists of two parts: the first part is chlorenchyma, and the inner part consists of larger celled and thin-walled parenchyma (Yarbrough, 1957b). The vascular

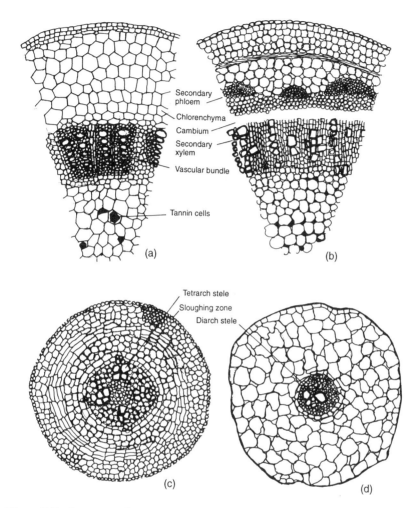

Figure 3.7 Structure of root and stem. Transections of (a) primary stem; (b) stem (secondary structure); (c) primary root; (d) lateral root with diarch stele and no pith.

bundles are endarch and vary from 20 to 40. Five or six of them are larger than the rest. Tannin cells occur in the stem (Figure 3.7(a)).

During the secondary growth of the stem, many pith cells break down and form a cavity. An active fascicular and an interfascicular cambium form considerable secondary tissue. Periderm develops in outer cortical cells. Most cells of the cork cambium collapse before periderm development (Figure 3.7(b)). Both the main axis and the cotyledonary laterals are obconical in form when the primary tissues only are considered. The number of fascicular bundles is greater in the upper internodes than in the lower ones. The diameter of the pith is greater likewise (Yarbrough, 1957b).

3.6.3 Leaf

Routine information on leaf structure in *Arachis* was provided in earlier studies (Reed, 1924; Yarbrough, 1957a). Studies on possible relationships between leaf structure and resistance to leaf diseases were also made by some workers (D'Cruz and Upadhyaya, 1961; Hemmingway, 1957; Suryakumari *et al.*, 1989a). A similar study was made by Pallas to find out the reason for the abnormally higher conductance from the adaxial side of the leaf than from the abaxial (Pallas, 1980). Detailed studies of leaf structure in cultivated and wild species, interspecific hybrids and experimental polyploids are also available (Suryakumari, 1984; Suryakumari *et al.*, 1983, 1989a).

(a) Cultivated varieties

Stomata
Stomata are present on both leaf surfaces with a higher frequency on the adaxial surface, with few exceptions. Stomata are mostly of the paracytic type but occasionally anisocytic, anomocytic and hemiparacytic types also occur. Their frequency varies between 183 mm^2 to 302 mm^2. Stomata vary in length from 16–18 μm (Suryakumari, 1984).

Epidermal cells
The epidermal cells are polygonal with straight anticlinal walls. In some cultivars, the adaxial cells are larger than the abaxial cells.

Epidermal hairs
The epidermal hairs are non-glandular and uniseriate, with a thick basal cell. The body of the hair consists of a short subterminal cell and a long terminal cell. The hairs range in length from 0.02 to 0.92 mm. Epidermal hairs occur on both surfaces of the leaves (Suryakumari, 1984).

Structure of lamina
Both the upper and lower epidermis are uniseriate with a thin cuticle. The mesophyll is divisible into the spongy and palisade tissues. The palisade cells occur beneath the upper epidermis only, loosely arranged in two to four vertical layers. The poorly developed spongy parenchyma is represented by a uniseriate layer. A layer of water-storage cells occurs beneath the lower epidermis and the spongy layer. Tannin cells project inwards from the cells of the upper epidermis. The thickness of the lamina

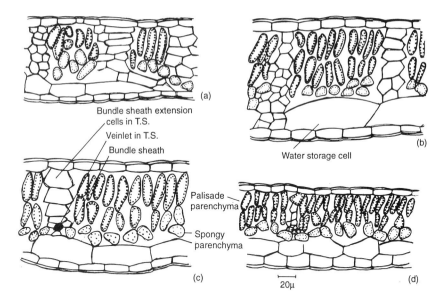

Figure 3.8 Leaf transections: (a) subsp. *fastigiata*; (b) subsp. *hypogaea*; (c) *A. batizocoi*; (d) *A. pusilla*.

varies from 100–250 μm and the vertical extent of palisade tissue ranges from 40–120 μm (Suryakumari *et al.*, 1983).

(b) Wild species

Wild *Arachis* species differ from the cultivated species in stomatal length, frequency of hairs and magnitude of tissue ratio (i.e. the ratio of all other tissues of the leaf to palisade height). *A. batizocoi* differs from the other species in section Arachis in the number of stomata and hairs, length of hairs, thickness of lamina, extent of palisade, palisade cell height and palisade cell breadth. The two subspecies of *Arachis hypogaea* differ from each other in the frequency of stomata, lamina thickness and palisade parameters (Suryakumari *et al.*, 1983) (Figure 3.8(a)–(d)).

(c) Interspecific hybrids

Due to increased use of wild *Arachis* species in groundnut improvement, especially for the incorporation of disease resistance, more and more stable interspecific hybrids are becoming available. Interspecific hybrids at triploid, tetraploid and hexaploid level show a gradual increase in leaf thickness, palisade thickness and diameter with the increase in the ploidy level. However, these characters deviate at the tetraploid level, the order being 4x > 3x < 5x < 6x.

(d) Polyploids

Diploid wild species of *Arachis* exhibit higher values than the tetraploid species for the following parameters: frequency and length of hairs; frequency of stomata; leaf thickness; and thickness and diameter of palisade. In *Arachis*, features such as decreased size of epidermal cells, high frequency of epidermal hairs, strongly developed palisade and higher percentage of water storage tissue were found to be associated with drought resistance (Suryakumari, 1984).

3.6.4 Reproductive organs

Detailed accounts of microsporogenesis, megasporogenesis and reproduction in groundnut have been described by several workers (Badami, 1935; Banerji, 1938; Reed, 1924; Shibuya, 1935; Smith, 1956a, 1956b). Similar studies of wild species have also been reported recently (Bharathi and Murty, 1984; Suryakumari *et al.*, 1989a).

(a) Cultivated varieties

For the most part, embryological features of *A. hypogaea* are of the dicotyledonous type. In the following account, only the salient features are given and the detailed routine descriptions are avoided.

Microsporogenesis and male gametophyte
The androecium consists of eight functional stamens and two sterile filaments. In the functional stamens, four have oblong anthers, of which three are tetrasporangiate and one bisporangiate (Figure 3.9(j)). The other four stamens have bisporangiate globose anthers. A very young anther in transverse section shows an epidermis enclosing a homogeneous mass of cells (Figure 3.9 (a)–(c)). The development of anther wall corresponds to the dicotyledonous (basic) type (Davis, 1966; Xi, 1991). In a mature anther, four wall layers – epidermis, endothecium, middle layer and tapetum – are distinct (Figure 3.9 (d)–(h)). Cells of the endothecium become radially elongated and develop typical fibrous thickenings at the time of meiosis in the pollen mother cells. The middle layer is ephemeral and disappears during the maturation of pollen (Figure 3.9 (h)–(i)). In the initial states, the tapetum is uniseriate with uninucleate cells which become bi- or tetranucleate with a vacuolated cytoplasm. Sometimes these nuclei fuse to form polyploid nuclei (Figure 3.9 (k)–(n)). The tapetal cells are glandular in nature. They degenerate before the dehiscence of the anthers (Figure 3.9(o)).

The primary sporogenous cells divide mitotically to increase the number of microspore mother cells which undergo normal meiosis. Occurrence of cytomictic channels between microsporocytes, correlated with meiotic

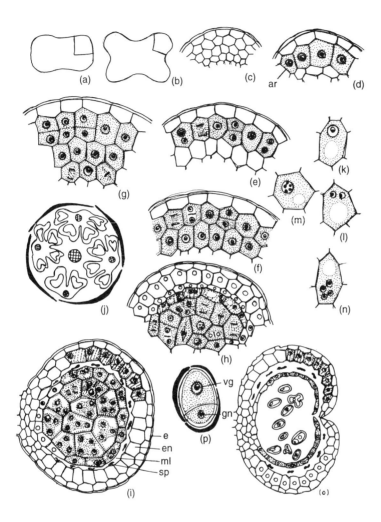

Figure 3.9 Microsporogenesis in *Arachis hypogaea* and male gametophyte (ar, archesporial cell; e, epidermis; en, endothecium; gn, generative nucellus; ml, middle layer; sp, sporogenous tissue; vg, vegetative nucellus): (a), (b) transverse sections of anthers at successive stages of development; (c) a portion of young anthers marked in (a) showing a homogeneous mass of parenchymatous cells; (d) a portion of (b) enlarged to show hypodermal archesporial cells; (e), (f), (g) anthers showing different stages of development of wall layers; (h), (i) mature anthers showing meiosis in pollen mother cells, the disintegrating middle layer and fibrous endothecium; (j) one bisporangiate and three tetrasporangiate anthers; (k), (l), (m), (n) tapetal cells showing one, two and four nuclei and also polyploid nucleus; (o) another showing disintegrated middle layer and tapetum, and fully developed fibrous bands in endothecium; (p) tricolpate 2-celled pollen grain. (After Murty, 1988.)

synchrony, has been reported (Xi, 1991). Cytokinesis is both successive and simultaneous, divisions resulting in tetrahedral, T-shaped or isobilateral tetrads (Figure 3.9(i)). The microspores soon separate from each other and each secretes its own wall,which differentiates into an outer thick exine and inner thin intine.

The uninucleate microspore with centrally situated nucleus increases in size. Due to the vacuolation of cytoplasm, the nucleus migrates to the periphery and divides into a large vegetative cell and a small densely cytoplasmic generative cell. The pollen grains are tricolpate and bicelled at anthesis (Figure 3.9 (p)).

Megasporogenesis and megagametophyte
The ovules are hemi-anatropous, crassinucellate and bitegmic. The micropyle is formed by both the integuments. The archesporium in the ovule is sub-hypodermal and is represented by a single cell. It directly functions as the megaspore mother cell (MMC) (Figure 3.10 (a)–(c)). The MMC increases in size and divides meiotically to produce a linear, occasionally T-shaped tetrad of megaspores (Figure 3.10 (d), (e)). The chalazal megaspore undergoes three mitotic divisions and gives rise to an eight-nucleate polygonum type embryo sac (Figure 3.10 (f)–(k)). The mature embryo sac is seven-celled with a three-celled egg apparatus formed of two synergids and an egg cell. The synergids are pyriform in shape and have filiform apparatus. The two polar nuclei in the central cell fuse to form a secondary nucleus during fertilization. The antipodal cells at the chalazal end are ephemeral and disintegrate by the time the embryo sac is organized (Figure 3.10 (k)).

The nucellus disintegrates by the time the embryo sac reaches maturity. However, some nucellar cells in the chalazal region become thick-walled and form a hypostase, and at the micropylar end form an operculum. After the disintegration of the nucellus, the cells of the inner integuments adjacent to the embryo sac become radially elongated and densely cytoplasmic and form an integumentary tapetum or endothelium.

Fertilization
Fertilization in groundnut is porogamous. After pollination, one of the synergids degenerates. The pollen tube enters the embryo sac through this degenerating synergid soon after pollination (Murty and Rao, 1979). Triple fusion precedes syngamy (Figure 3.11). Fertilization takes place 12–18 hours after anthesis.

Endosperm
Endosperm is *ab initio* nuclear. A fertilized embryo sac contains abundant starch, which is a conspicuous feature during its early development. The starch grains are mostly digested when the endosperm has 8–16 nuclei. Free nuclear divisions in the endosperm continue until the embryo reaches

the early dicotyledonous stage. A higher number of endosperm nuclei (about 5000) occurs at this stage (Figure 3.12 (a)–(f)).

Wall formation progresses centripetally. Cell wall formation in the endosperm first takes place around the embryo and then gradually extends (Figure 3.12 (g)). The entire endosperm does not become cellular and the nuclei at the narrow chalazal end continue to lie embedded in a common

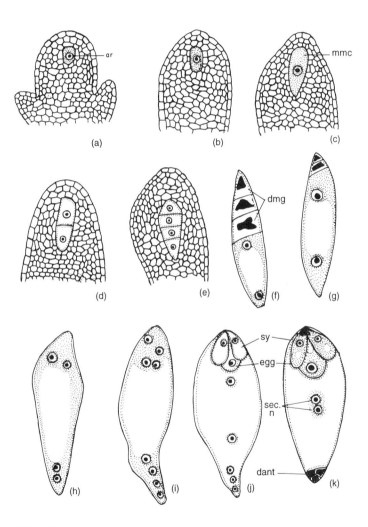

Figure 3.10 Megasporogenesis and mature embryo sac development in *A. hypogaea* (ar, archesporial cell; dant, degenerating antipodal; mmc, megaspore mother cell; sec.n, secondary nucleus; sy, synergid): (a) longitudinal section of an ovule showing subhypodermal archesporial cell; (b), (c) enlarged megaspore mother cell; (d) diad of megaspores; (e) megaspore tetrad; (f)–(j) successive stages of megasporogenesis; (k) mature embryo sac.

cytoplasm. The coenocytic structure thus formed functions as an haustorium. During the development of the embryo, the endosperm is absorbed gradually and in the mature seed there is no endosperm.

Embryogeny
The embryogeny conforms to the Soland type of Johansen (Johansen, 1950). The first division of the zygote is transverse, resulting in an apical cell and a basal cell (Figure 3.12 (c)). These cells again divide transversely

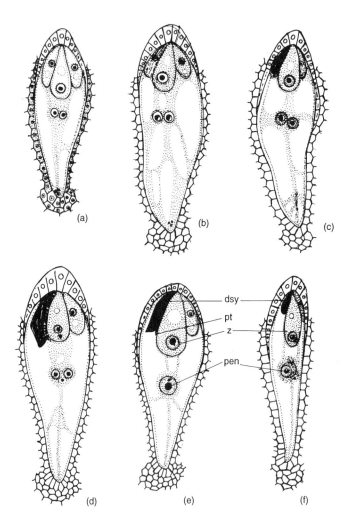

Figure 3.11 Fertilization in *A. hypogaea* (dsy, degenerated synergid; pen, primary endosperm nucleus; pt, pollen tube; z, zygote): (a)–(f) successive stages in syngamy and triple fusion.

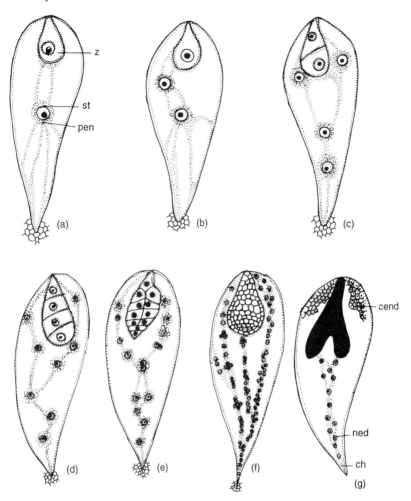

Figure 3.12 Endosperm and embryo development in *A. hypogaea* (cend, cellular endosperm; ch, chalazal end; ned, nuclear endosperm; pen, primary endosperm nucleus; z, zygote): (a) embryo sac in which syngamy and triple fusion are completed; (b)–(e) successive stages in the development of the embryo and endosperm; (f) embryo sac showing globular embryo and endosperm with *c*. 3000 nuclei; (g) embryo sac showing dicotyledonous embryo and cellular endosperm at micropylar end.

and longitudinally and form a quadrant and then an octant embryo. At this stage the embryo-forming region is clearly distinguishable from the suspensor region by its smaller cells and denser cytoplasm. The embryo forming region mostly divides by oblique walls and the basal cell, after two transverse divisions, mostly divides by oblique walls and gives rise to a massive suspensor. Localized divisions in apical meristems at two points at the distal end of the globular embryo result in a bilobed heart-shaped embryo. The later stages are marked by progressive increase in cells resulting in an embryo with a massive radicle, two cotyledons and a massive hypocotyledonary region (Figure 3.12 (g)). The embryo is straight and consists of a robust radicle, a hypocotyledonary axis and an epicotyl made up of three incipient shoot axes bearing nine or more differentiated leaves.

Seed coat
Initially the outer and inner integuments are two-layered. Periclinal divisions take place in both the integuments, making them three-layered. Further periclinal divisions occur in the derivative layers of the integument only when the outer integument attains 11–13 layers. The outermost layer of cells becomes columnar with thick walls, constituting the outer palisade (Figure 3.13 (a)–(d)).

Starch grains appear in the cells of the outer five to six layers. The cells towards the inner integument do not contain starch grains and are comparatively small. During the later stages, the inner integument gradually disintegrates (Figure 3.13 (e)) and is completely obliterated in the mature seed. Further divisions in the outer integument continue and in the mature seed it is represented by 23–27 layers. The mature seed coat consists of (1) a uniseriate outer palisade of columnar cells, (2) a middle layer of parenchymatous cells, 20–25 cells thick containing starch grains and (3) a single layer of isodiametric cells which represents the innermost layer of the outer integument (Figure 3.13 (f)) (Suryakumari, 1984).

(b) Wild species

Information on the reproductive behaviour of some of the wild species, including diploids and tetraploids belonging to the section Arachis ('A. chacoense' Krap. et Greg. (nom. nud., PI 276235, 2n = 20); A. villosa Benth. 2n = 20; 'A. duranensis' Krap. et Greg. (nom. nud., PI 219823, 2n = 20); and A. monticola Krap. et Rig. 2n = 40)), the section Rhizomatosae (A. hagenbeckii Harms. 2n = 40; and A. glabrata Benth. 2n = 40) and two undescribed tetraploid Arachis species (2n = 40), has been provided in previous reports (Bharathi and Murty, 1984; Suryakumari et al., 1989a).

No significant differences occur in the developmental morphology of the species belonging to the different sections of the genus. This is true especially of microsporogenesis and male gametophyte development and of embryo sac development. However, minor differences occur. For example, during megasporogenesis, the archesporial cell is hypodermal in origin in the wild species. It divides into a primary parietal cell and a primary sporogenous cell. The periclinal and anticlinal divisions of the primary parietal cell produces a three-layered parietal tissue and the primary sporogenous cell functions as megaspore mother cell (Figure 3.14

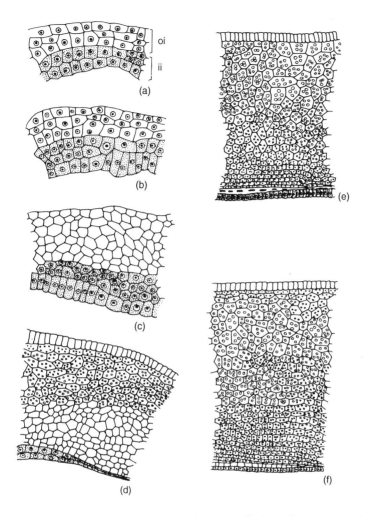

Figure 3.13 Anatomy of seed coat in *A. hypogaea* (ii, inner integument; oi, outer integument).

(a)–(h)). Except in an undescribed tetraploid species, in which a triad was observed, all other species have a linear or T-shaped megaspore tetrad. Fertilization occurs more slowly in the wild species, since fusion of male and female nuclei seems to be a slow process and the two nuclei are detectable in close proximity to each other. The time taken for fertilization varies from 6 to 48 hours. The number of free endosperm nuclei is much less than in *A. hypogaea* at the dicotyledonous stage of the embryo.

In the seed-sterile species *A. hagenbeckii* Harms. the primary endosperm nucleus divides only once. Thereafter the two daughter nuclei

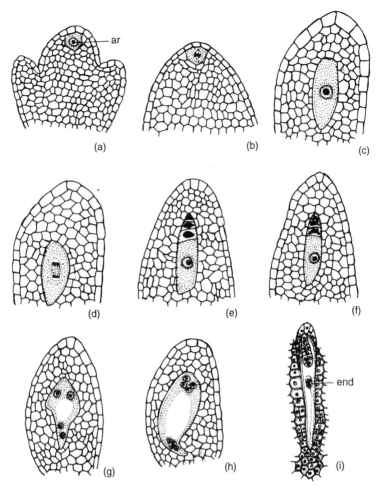

Figure 3.14 Megasporogenesis and megagametophyte in wild species (ar, arche-sporium; end, endothelium): (a), (b) hypodermal archesporium; (c)–(h) successive stages in the development of embryo sac; (i) degenerating fertilized embryo sac, showing crushing of the embryo sac.

degenerate. No further development takes place. The embryo does not develop beyond the four-celled stage. The endothelium becomes hyper-plastic and crushes the embryo sac (Figure 3.14 (f)).

4.7 CONCLUDING REMARKS

The above morphological and anatomical descriptions of the groundnut mainly consider the typical forms. However, several variants have been reported from artificial or natural mutations or through hybridization (Branch *et al.*, 1982; Gopani and Vaishnani, 1970; Hammons, 1964; Seshadri and Seshu, 1956). The morphological observations have been mostly limited to the cultivated species. With the increased use of wild species germplasm in studying and improving groundnut crops, there is a need for extension of such description to wild *Arachis* and for the study of the genus as a whole.

Large gaps still exist in our knowledge of the genus as a whole and more investigations are necessary. Further anatomical studies, including at least a few samples of all 70 or so species, need to be carried out. These may not only have taxonomical significance but may also help to identify any useful association of morphological characters with anatomical features. Such a comprehensive study of the genus as whole, which is presently lacking, will be useful in gaining a better understanding of how wild species can be used to improve the commercial groundnut crop.

REFERENCES

Allen, O.N. and Allen, A.K. (1940) Response of the peanut plant to inoculation with rhizobia, with special reference to morphological development of the nodules. *Botanical Gazette*, **16**, 305–314.

Ashri, A. (1976) Plasmon divergence in peanuts (*Arachis hypogaea*): A third plasmon and locus affecting growth habit. *Theoretical and Applied Genetics*, **48**, 17–21.

Ashri, A. and Goldin, E. (1963) Genic and cytoplasmic interactions in peanuts. *Proceeding of the 11th International Congress of Genetics*, **1**, 203.

Ashri, A. and Yona, J. (1965) Nature and inheritance of seed coat splitting in peanuts (*Arachis hypogaea* L.). *Israel Journal of Agricultural Research*, **15**, 109.

Badami, V.K. (1933) Botany of groundnut – Part I. *Journal of the Mysore Agricultural and Experimental Union*, **14**, 188–194.

Badami, V.K. (1935) Botany of groundnut – Part II. *Journal of the Mysore Agricultural and Experimental Union*, **15**, 59–70.

Banerji, I. (1938) A note on the embryology of groundnut (*Arachis hypogaea*). *Journal of the Bombay Natural History Society*, **40**, 539–543.

Bear, J.E. and Bailey, W.K. (1973), Earliness of flower opening and potential for pod development in peanut, *Arachis hypogaea* L. *Journal of the American Peanut Research and Education Association*, **5**, 26–31.

Bentham, G. (1841) On the structure and affinities of *Arachis and Voandzeia*. *Transactions of the Linnean Society of London*, **18**, 155–162.

Bentham, G. and Hooker, J.D. (1862) *Genera Plantarum*, Hookerian Herbarium,. Kew, London.

Bhagat, N.R., Lalwani, H.B. and Reddy, P.S. (1987) Variation in the node sequence in var. *hypogaea* collection of cultivated peanuts. *Canadian Journal of Botany*, **65**, 1245–1248.

Bharathi, M. (1981) *Cytogenetic studies in the genus Arachis L.* Ph.D., Osmania University, Andhra Pradesh, India.

Bharathi, M. and Murty, U.R. (1980) The botany of fruit stalk in groundnut *Arachis hypogaea*. *Current Science*, **49**, 948.

Bharathi, M. and Murty, U.R. (1984) Comparative embryology of wild and cultivated species of *Arachis*. *Phytomorphology*, **34**, 58–56.

Bhavanisankar Rao, M. and Raman, V.S. (1960) Studies in the genus *Arachis*. VIII. Note on classification of groundnut. *Indian Oilseeds Journal*, **4**, 15–16.

Bhide, M.E. and Desale, S.C. (1970) A small leaf mutant in groundnut. *Poona Agricultural College Magazine*, pp. 113–114.

Bolhuis, G.G., Frinking, H.D., Leeuwaugh, I. *et al.* (1965) Observations on the opening of flowers, dehiscence of anthers and growth of pollen tubes in *Arachis hypogaea*. *Netherlands Journal of Agricultural Science*, **7**, 138–140.

Bouffil, F. (1947) Biologie, écologie et sélection de l'arachide au Senegal. *Bulletin Scientifique de Direction de l'Agriculture de l'Élevage et des Forêts*, **1**, 1–112.

Branch, W.D. and Hammons, R.O. (1979) Inheritance of testa color variegation in peanut. *Crop Science*, **19**, 786–788.

Branch, W.D., Hammons, R.O. and Kirby, J.S. (1982) Inheritance of white-stem peanut. *Journal of Heredity*, **73**, 301–302.

Brennan, J.R. (1969) The peanut gynophore. *The Biologist*, **51**, 71–82.

Bruner, W.E. (1932) Root development of cotton, peanut and tobacco in central Oklahoma. *Proceedings of the Oklahoma Academy of Science*, **12**, 20–37.

Bunting, A.H. (1955) A classification of cultivated groundnut. *Empire Journal of Experimental Agriculture*, **23**, 158–170.

Bunting, A.H. (1958) A further note on the classification of cultivated groundnuts. *Empire Journal of Experimental Agriculture*, **26**, 254–248.

Campbell, W.V., Emery, D.A. and Wynne, J.C. (1976) Resistance of peanuts to the potato leaf hopper. *Peanut Science*, **3**, 40–43.

Chandler, M.R. (1978) Some observations on the infection of *Arachis hypogaea* L. by *Rhizobium*. *Journal of Experimental Botany*, **29**, 749–755.

Chodat, R. and Hassler, E. (1904) Enumeration des plantes recoltées au Paraguay. *Bulletin de l'Herbier Boissier, Series 2*, **4**, 884–887.

Coffelt, T.A. and Hammons, R.O. (1974) Inheritance of pod constriction in peanuts. *Journal of Heredity*, **65**, 94–96.

Culp, T.W., Bailey, W.K and Hammons, R.O. (1968) Natural hybridization of peanuts, *Arachis hypogaea* L., in Virginia. *Crop Science*, **8**, 108–111.

D'Cruz, R. and Upadhyaya, B.R. (1961) Stem and leaf anatomy of *Arachis*. *Indian Oilseeds Journal*, **60**, 121–122.

Davis, G.L. (1966) *Systematic Embryology of Angiosperms*, John Wiley, New York.

Donovan, P.A. (1963) Groundnut investigations at Matapos Research station. *Rhodesia Agricultural Journal*, **60**, 121–122.

Dreyer, J., Duncan, W.G. and McCloud, D.E. (1981) Fruit temperature, growth rates and yield of peanuts. *Crop Science*, **21**, 686–688.

Dubard, M. (1906) De l'origine de l'arachide. *Bulletin du Muséum d'Histoire Naturelle*, **5**, 340–344.

Gardner, G. (1842) Gardinerianae. Leguminosae B, in *Icones Plantarum* (vol. 1 Pt. 2 Plate 500). (ed. J.D. Hooker).

Gibbons, R.W. (1966) The branching habit of *Arachis monticola*. *Rhodesia, Zambia and Malawi Journal of Agricultural Research*, **4**, 9–11.

Gibbons, R.W., Bunting, A.H. and Smartt, J. (1972) The classification of varieties of groundnut, *Arachis hypogaea* L. *Euphytica*, **21**, 78–85.

Gibbons, R.W. and Tattersfield, J.R. (1969) Out-crossing trials with groundnuts, *Arachis hypogaea* L. *Rhodesia Journal of Agricultural Research*, **7**, 71–75.

Glueck, J.A., Clark, L.E. and Smith, O.D. (1979) Testa comparisons of four peanut cultivars. *Crop Science*, **17**, 777–782.

Gopani, D.D. and Vaishnani, N.L. (1970) Mutant form of groundnut. *Indian Journal of Agricultural Sciences*, **40**, 431–437.

Gregory, W.C. and Gregory, M.P. (1976) Groundnut, *Arachis hypogaea*, in *Evolution of Crop Plants* (ed. N.W. Simmonds), Longman, London, pp. 151–154.

Gregory, W.C., Smith, B.W. and Yarbrough, J.A. (1951) Morphology, genetics and breeding, in *The Peanut – The Unpredictable Legume*, National Fertilizer Association, Washington, DC, pp. 28–85.

Gregory, W.C., Gregory, M.P., Krapovickas, A. *et al.* (1973) Structure and genetic resources of peanuts, in *Peanuts – Culture and Uses*, American Peanut Research and Education Association, Stillwater, Oklahoma, pp. 47–133.

Gregory, W.C., Krapovickas, A. and Gregory, M.P. (1980) Structure, variation, and evolution and classification in *Arachis*, in *Advances in Legume Science* (eds R.J. Summerfield and A.H. Bunting), Royal Botanical Gardens, Kew, London, pp 469–481.

Hammons, R.O. (1963) Artificial cross pollination of the peanut with bee-collected pollen. *Crop Science*, **3**, 562–563.

Hammons, R.O (1964) Krinkle, a dominant leaf marker in the peanut *Arachis hypogaea* L. *Crop Science*, **4**, 22–24.

Hammons, R.O. (1973) Early history and origin of the peanuts, in *Peanuts – Culture and Uses*, American Peanut Research and Education Association, Stillwater, Oklahoma, pp. 17–46.

Hammons, R.O. (1981) *Arachis hypogaea* L., in *Handbook of Legumes of World Economic Importance* (ed. J.A. Duke), Plenum Press, New York, pp. 19–22.

Handro, O. (1958) Especies novas de *Arachis* L. *Arquivos de Botanica do Estado de S. Paulo*, **3(4)**, 177–81.

Harz, C.D. (1885) *Landwirthschaftliches Samenkundliches Handbuch für Botaniker, Landwirthe, Gaertner, Droguisten, Hygieniker*, T. Parey, Berlin.

Hassan, M.A. and Srivastava, D.P. (1966) Floral biology and pod development of peanut studied in India. *Journal of Indian Botanical Society*, **45**, 92–102.

Hayes, R.T. (1933) The classification of groundnut varieties with preliminary note on the inheritance of some characters. *Tropical Agriculture, Trinidad*, **10**, 318–327.

Hemmingway, J.S. (1957) The resistance of groundnuts to *Cercospora* leafspots. *Empire Journal of Experimental Agriculture*, **25**, 60–68.

Hoehne, F.C. (1919) Leguminosas. *Botanica*, **45(8)**, 71–72.

Hoehne, F.C. (1922) Leguminosas. *Botanica*, **74(12)**, 21–22.

Hoehne, F.C. (1940) Arachis. *Flora Brasílica*, **25(122)**, 1–20.

Hoehne, F.C. (1944) Duas novas especies de Leguminosas do Brasil. *Archivos de Botanica do Estado de São Paulo*, **2(1)**, 15–16.

IBPGR/ICRISAT (1981) *Groundnut Descriptors*, IBPGR Descriptor Series, IBPGR, Rome and ICRISAT, Patancheru.

Intorzato, R. and Tella, R. (1960) Sistema radicular do amendoim. *Bragantia*, **19**, 119–123.

Jacobs, W.P. (1947) The development of the gynophore of the peanut plant, *Arachis hypogaea* L. 1. The distribution of mitoses, the region of greatest elongation and the maintenance of vascular continuity in the intercalary meristem. *American Journal of Botany*, **34**, 361–370.

Jacobs, W.P. (1951a) Auxin relationships in an intercalary meristem: Further studies on the gynophore of *Arachis hypogaea* L. *American Journal of Botany*, **38**, 307–310.

Jacobs, W.P. (1951b) The growth of peanut plants at various diurnal and nocturnal temperatures. *Science*, **114**, 205–206.

Johansen, D.A. (1950) *Plant Embryology*. Waltham, Massachussets.

John, C.M., Venkatanarayana, G. and Seshadri, C.R. (1954) Varieties and forms of groundnut *Arachis hypogaea* Linn. Their classification and economic characters. *Indian Journal of Agricultural Sciences*, **24**, 159–193.

Krapovickas, A. (1969) Origin, variability and distribution of groundnut (*Arachis hypogaea*), in *The Domestication and Exploitation of Plants and Animals*, (eds P.J. Ucko and G. Dimbleby), Duckworth, London, pp. 427–441.

Krapovickas, A. (1973) Evolution of the genus *Arachis*, in *Agricultural Genetics – Selected Topics*, (ed. R. Moav), National Council for Research and Development, Jerusalem, pp. 135–151.

Krapovickas, A. (1988) personal communication.

Krapovickas, A., Fernandez, A. and Seeligmann, P. (1974) Recuperacion de la fertilidad en un hybrido interspecifico esteril de *Arachis*, Leguminosae. *Bonplandia*, **3**, 129–142.

Krapovickas, A. and Gregory, W.C. (1960) *Arachis rigonii – Nueva especie silvestre de maní. Revista de Investigaciones Agricolas*, **14(2)**, 157–160.

Krapovickas, A. and Rigoni, V.A. (1957) Nuevas especies de *Arachis*: vinculados al problema del origen del maní. *Darwiniana*, **17**, 431–455.

Krapovickas, A. and Rigoni, V.A. (1960) La nomenclatura de las subspecies y variadades de *Arachis hypogaea* L. *Revista de Investigaciones Agricolas*, **14**, 197–228.

Kuntze, O. (1898) *Revisio Generum Plantarum*. Delau & Co., London.

Leuck, D.B. and Hammons, R.O. (1969) Occurrence of atypical flowers and some associated bees, Apoidea in the peanut, *Arachis hypogaea* L. *Agronomy Journal*, **61**, 958–960.

Lim, E.S. and Gumpil, J.S. (1984) The flowering, pollination and hybridization of groundnuts (*Arachis hypogaea* L.). *Pertanika*, **7**, 61–66.

Lim, E.S. and Hamdan, O. (1984) The reproductive characters of four varieties of groundnuts (*Arachis hypogaea* L.). *Pertanika*, **7**, 25–31.

Lin, Z.F., Li, S.S. and Lin, G.S. (1986) The distribution of stomata and photosynthetic pathway in leaves. *Acta Botanica Sinica*, **28**, 387–395.

Linnaeus, C. (1753) Species Plantarum. *Holmiae*, **2**, 741.

Maeda, K. (1972) Growth analysis on the plant type in peanut varieties, *Arachis hypogaea* L. IV. Relationship between the varietal difference of the progress of leaf emergence on the main stem during pre-flowering period and the degree of morphological differentiation of leaf primordia in the embryo. *Proceedings of the Crop Science Society of Japan*, **41**, 179–186.

Maeda, K. (1973) The recent botanical taxonomic system of peanut cultivars. *Japanese Agriculture Research Quarterly*, **7**, 228–232.

Marggraf, G. (1648) Georgo Marcgravi, historia rerum naturalium Brasiliae, in *Historia Naturalis Brasiliae* vol. (cited from Gregory *et al.*, 1973), (eds W. Piso and G. Marggraf), Liber Primus.

Mohammad, A.A. and Khanna, K.L. (1932) Studies on germination and growth in groundnut. *Agriculture and Live-Stock in India*, **3**, 91–115.

Mouli, C. and Kale, D.M. (1982) An early maturing groundnut with foliaceous stipule marker. *Current Science*, **51**, 132–134.

Murty, U.R. (1988) Anatomy, in *Groundnut* (ed. P.S. Reddy), Indian Council of Agricultural Research, New Delhi, pp. 65–76.

Murty, U.R. and Rao, N.G.P. (1979) A rapid method for the study of fertilization in groundnut, *Arachis hypogaea* L. *Current Science*, **48**, 492–493.

Murty, U.R., Rao, N.G.P., Kirti, P.B. and Bharati, M. (1980) Fertilization in groundnut, *Arachis hypogaea* L. *Oléagineux*, **36**, 73–76.

Nambiar, P.T.C., Nigam, S.N., Dart, P.J. and Gibbons, R.W. (1983) Absence of root hairs in non-nodulating groundnut, *Arachis hypogaea* L. *Journal of Experimental Botany*, **34**, 484–488.

Narasinga Rao, V. (1936) A note on preliminary physiological observations on the groundnut. In *Proceedings of the Association of Economic Biologists*, Coimbatore, Tamil Nadu, India, pp. 1–6.

Norden, A.J. (1980) Peanut in *Hybridization of Crop Plants*, (eds W.R. Fehr and H H. Hadley), American Society of Agronomy – Crop Science Society America, Madison, pp. 443–456.

Ono, Y. (1979) Flowering and fruiting of peanut plants. *Japanese Agriculture Research Quarterly*, **3**, 226–229.

Pallas, J.E. (1980) An apparent anomaly in peanut leaf conductance. *Plant Physiology*, **65**, 848–851.

Patel, J.S., John, C.M. and Seshadri, C.R. (1936) The inheritance of characters in the groundnut *Arachis hypogaea*. *Proceedings of the Indian Academy of Sciences*, **3**, 214–233.

Patel, J.S. and Narayana, G.V. (1935) A rare instance of polyembryony in *Arachis hypogaea*. *Current Science*, **4**, 32–33.

Patil, V.H. (1965) *Genetic studies in groundnut Arachis hypogaea L.*, M.Sc. Thesis, Poona University, Poona.

Pen, S.Y., Zhuang, W.J. and Huang, J.H. (1987) Lens and electron microscopic scanning on the pollen morphology of different types and varieties of peanut *Arachis hypogaea* L. *Journal of Fujian Agricultural College*, **16**, 313–319.

Petit, A.S. (1895) *Arachis hypogaea* L. *Memoirs of the Torrey Botanical Club*, **4**, 275–296.

Poiteau, M. (1806) Observations sur l'*Arachis hypogaea*. *Academie des Sciences (Paris), Memoires Sciences, Mathématiques et Physiques*, **1**, 455–462.

Prasad, M.V.R. (1985) Aerial podding in groundnut (*Arachis hypogaea* L.). *Indian Journal of Genetics and Plant Breeding*, **45**, 89–91.

Raman, V.S. (1976) *Cytogenetics and Breeding in Arachis*, Today and Tomorrow Printers and Publishers, New Delhi.

Raman, V.S. and Nagarajan, S.S. (1958) Studies in the genus *Arachis*. III. Polyembryony in the groundnut *Arachis hypogaea* Linn. *Indian Oilseeds Journal*, **2**, 74.

Ramanatha Rao, V. (1987). Origin, distribution and taxonomy of *Arachis* and sources of resistance to groundnut rust (*Puccinia arachidis* Speg), in *Groundnut rust disease: Proceedings of the International Group Discussion Meeting, 24–27 September 1984*, ICRISAT, Patancheru, India, pp. 3–15.

Ramanatha Rao, V. (1988) Botany, in *Groundnut* (ed. P.S. Reddy), Indian Council of Agricultural Research, New Delhi, pp. 24–64.

Reed, E.L. (1924) Anatomy, embryology and ecology of *Arachis hypogaea*. *Botanical Gazette*, **78**, 289–310.

Resslar, P.M. (1980) A review of the nomenclature of the genus *Arachis*. *Euphytica*, **29**, 813–817.

Resslar, P.M. and Emery, D.A. (1978) Inheritance of growth habit in peanuts: cytoplasmic or maternal modifications. *Journal of Heredity*, **69**, 101–106.

Retamal, N., Lopez-Vences and Duran, J.M. (1990) Seed morphology of 75 genotypes of peanut *Arachis hypogaea* L. grown in Spain. *Plant Genetic Resources Newsletter*, **80**, 1–4.

Richter, C.G. (1899) Beitrage zur Biologie von *Arachis hypogaea*, in *Inaug. Diss. Kgl. Bot. Gart. Breslau* (ed. A. Schrieber), Breslau, pp. 39.

Salma, I. (1985) Identification of fifteen groundnut cultivars based on their morphological traits. *MARDI Research Bulletin*, **13**, 285–289.

Sastri, D.C. and Moss, J.P. (1982) Effects of growth regulators on incompatible crosses in the genus *Arachis* L. *Journal of Experimental Botany*, **33**, 1293–1301.

Sastry, K S.K., Chari, M., Prasad, T.G. *et al.* (1985) Flowering pattern and pod development in bunch types of groundnut: is there a relationship between synchrony in flowering and pod development? *Indian Journal of Plant Physiology*, **28**, 64–71.

Schenk, R.V. (1961) Development of the peanut fruit. *Georgia Agricultural Experiment Station Technical Bulletin*, **22**, 5–53.

Schwabe, W.W. (1971) Physiology of vegetative reproduction and flowering in groundnut, in *Physiology VI* (ed. A.F.C. Steward), Academic Press, New York.

Seshadri, C.R. (1962) *Groundnut*. The Indian Central Oilseeds Committee, Hyderabad.

Seshadri, C.R., Bhavanisankara Rao, M. and Srinivasalu, N. (1955) Studies on the gynophore in the groundnut, in *Proceedings of the 6th Scientific Workers Conference, Department of Agriculture, Tamil Nadu*, Department of Agriculture, Coimbatore, Tamil Nadu.

Seshadri, C.R., Bhavanisankara Rao, M. and Varisai Muhammad, S. (1958) Studies on root development in groundnut. *Indian Journal of Agricultural Sciences*, **28**, 211–215.

Seshadri, C.R. and Seshu, G. (1956) A mutant form of groundnut. *Madras Agricultural Journal*, **43**, 199–200.

Shibuya, T. (1935) Morphological and physiological studies on the fructification of peanut

(*Arachis hypogaea* L.). *Memoirs of the Faculty of Science and Agriculture, Taihoku Imperial University*, **17**, 1–120.

Singh, H., Yadav, A.K., Yadava, T.P. and Chabra, M.L. (1981) Genetic variability and heritability for morpho-physiological attributes in groundnut. *Indian Journal of Agricultural Sciences*, **52**, 432–434.

Smartt, J. (1961) Groundnut varieties of northern Rhodesia and their classification. *Empire Journal of Experimental Agriculture*, **29**, 153–158.

Smartt, J. (1964) Interspecific hybridization in relation to peanut improvement. In *Proceedings of the 3rd National Peanut Research Conference*, Alabama, pp. 53–56.

Smartt, J. (1976a) Comparative evolution of pulse crops. *Euphytica*, **25**, 139–143.

Smartt, J. (1976b) *Tropical Pulses*, Longman, London.

Smartt, J. (1990) *Grain Legumes: Evolution and Genetic Resources*, Cambridge University Press, Cambridge.

Smartt, J. and Stalker, H.T. (1982) Speciaton and cytogenetics in *Arachis*, in *Peanut Science and Technology*, APRES, Yoakum, USA, pp. 21–49.

Smith, B.W. (1950) *Arachis hypogaea* L. Aerial flower and subterranean fruit. *American Journal of Botany*, **37**, 802–815.

Smith, B.W. (1954) *Arachis hypogaea*, reproductive efficiency. *American Journal of Botany*, **41**, 607–616.

Smith, B.W. (1956a) *Arachis hypogaea* L. Embryogeny and the effect of peg elongation upon embryo and endosperm growth. *American Journal of Botany*, **43**, 233–240.

Smith, B.W. (1956b) *Arachis hypogaea* L. Normal megasporogenesis and syngamy with occasional single fertilization. *American Journal of Botany*, **43**, 81–90.

Srinivasalu, N. and Chandrasekharan, N.R. (1958) A note on natural crossing in groundnut, *Arachis hypogaea* Linn. *Science and Culture*, **23**, 650.

Srivastava, A.N. (1968) *Classification and inheritance studies in groundnut Arachis hypogaea Linn*. PhD Thesis, Agra University, Uttar Pradesh.

Staritsky, G. (1973) Seed morphology and early development of the groundnut (*Arachis hypogaea* L). *Acta Botanica Neerlandica*, **22**, 373–379.

Stokes, W.E. and Hull, F.H. (1930) Peanut breeding. *Journal of the American Society of Agronomy*, **22**, 1004–1009.

Sundaram, N. (1985) Studies on cytogenetics of *Arachis* at Regional Research Station, Vridhachalam, Tamil Nadu, India, in *Proceedings of the International Workshop on Cytogenetics of Arachis, 31 Oct–2 Nov 1983*, ICRISAT, Patancheru, pp. 137–139.

Suryakumari, D. (1984) *Leaf anatomical and embryological studies in the genus Arachis L*. PhD Thesis, Andhra University, Waltair, India.

Suryakumari, D., Seshavatharam, V. and Murty, U.R. (1983) Comparative leaf anatomy of the wild species and cultivated varieties of the genus *Arachis*. *Oléagineux*, **38**, 27–40.

Suryakumari, D., Seshavatharam, V. and Murty, U.R. (1984) Association of rust resistance with number of tannin sacs in groundnut. *Current Science*, **53**, 604–606.

Suryakumari, D., Seshavatharam, V. and Murty, U.R. (1989a) Leaf anatomical features of some interspecific hybrids and polyploids in the genus *Arachis* L. *Journal of Oilseeds Research*, **6**, 75–84.

Suryakumari, D., Seshavatharam, V. and Murty, U.R. (1989b) Studies on the embryology of the genus *Arachis* L. *Journal of Oilseeds Research*, **6**, 85–91.

Umen, D.P. (1933) *Biology of Peanut Flowering*, Lenin Academy of Agricultural Science, Krasnodar, English translation by Amerind Publisher, New Delhi, India.

Valls, J.F.M., Ramanatha Rao, V., Simpson, C.E. and Krapovickas, A. (1985) Current status of collection and conservation of South American groundnut germplasm with emphasis on wild species of *Arachis*, in *Proceedings of an International Workshop on Cytogenetics of Arachis*, ICRISAT, Patancheru, India, pp. 15–35.

Varisai Muhammad, S., Ramanathan, T. and Ramachandran, M. (1973a) Classification of *Arachis hypogaea* L. var. *fastigiata*. *Madras Agricultural Journal*, **60**, 1399–1402.

Varisai Muhammad, S., Ramanathan, T. and Ramachandran, M. (1973b) Classification of *Arachis hypogaea* L. var. *procumbens*. *Madras Agricultural Journal*, **60**, 1403–1408.

Varisai Muhammad, S., Ramanathan, T. and Ramachandran, M. (1973c) Variation in morphological characters of *Arachis hypogaea* L. *Madras Agricultural Journal*, **60**, 1373–1379.

Waldron, R. A. (1919), The peanut *Arachis hypogaea*, its history, histology and utility. *Penn University Botany Lab Contributions*, **4**, 301–338.

Webb, A.J. and Hansen, A.P. (1989) Histological changes of the peanut (*Arachis hypogaea*) gynophore and fruit surface during development, and their potential significance for nutrient uptake. *Annals of Botany*, **64**, 351–57.

Wynne, J.C. and Coffelt, T.A. (1982) Genetics of *Arachis hypogaea* L., in *Peanut Science and Technology*, APRES, Yoakum, USA, pp. 50–94.

Xi, X.-Y. (1991) Development and structure of pollen and embryosac in peanut (*Arachis hypogaea* L.). *Botanical Gazette*, **152(2)**, 164–172.

Yarbrough, J.A. (1949) *Arachis hypogaea*. The seedling, its cotyledons, hypocotyl and roots. *American Journal of Botany*, **36**, 758–772.

Yarbrough, J.A. (1957a) *Arachis hypogaea*. The seedling, its epicotyl and foliar organs. *American Journal of Botany*, **44**, 19–30.

Yarbrough, S.J. (1957b) *Arachis hypogaea*. The form and structure of the stem. *American Journal of Botany*, **44**, 31–36.

Yona, J. (1964). *The nature and inheritance of seed coat splitting and color in peanuts Arachis hypogaea L.* M.S. Thesis, Faculty of Agriculture, Hebrew University Jerusalem, Israel.

Zambettakis, C. and Bockellee-Morvan, A. (1976) Research on structure of the groundnut seed coat and its influence on the penetration of *Aspergillus flavus*. *Oléagineux*, **31**, 219–228.

Ziv, M., Halevy, A.H. and Ashri, A. (1973) Phytohormones and light regulation of growth habit in peanuts (*Arachis hypogaea* L.). *Plant Physiology*, **14**, 727–735.

CHAPTER 4

Biosystematics and genetic resources

A.K. Singh and C.E. Simpson

4.1 INTRODUCTION

Biosystematics determines taxonomic status from experimental evidence of the genetic diversity that arises within plant groups as a result of evolution. It thereby assists applied biologists to utilize this diversity more effectively. Without adequate analysis and assimilation of biosystematic information, appropriate use of such diversity (by any method and in any field) would be difficult and unpredictable. It is assumed that the evolution which brings about differentiation between and within groups of organisms has a built-in mechanism of isolation, the action of which is reflected in sound systems of classification. Biosystematics has to uncover this through appropriate investigations and provide unequivocal taxonomic nomenclature to the taxa which comprise the existing diversity. Given that organisms each carry built-in information on the pattern in which their development is ordered, it is the primary function of taxonomy and biosystematics to utilize this information and produce systems of classification and hypotheses to elucidate these patterns. The better such information, the more useful it will be in answering biological questions, particularly in relation to differentiation and the development of divergence.

In *Arachis*, though interest in the use of genetic resources from both cultivated *Arachis hypogaea* L. and wild *Arachis* species has increased and extensive studies have been made of evolutionary relationships between species of sect. *Arachis* and the cultivated groundnut (Smartt *et al.*, 1978a, 1978b; Singh, 1988; Singh *et al.*, 1991), taxonomic descriptions of all the taxa of this genus are still not available, and confusion exists with regard to the usage of botanical names and phylogenetic relationships among the taxa (Resslar, 1980).

Groundnut genetic resources are available in the form of the naturally evolved landraces of cultivated groundnut (*A. hypogaea*) in various centres of diversity, breeding lines or material developed in different groundnut

The Groundnut Crop: A scientific basis for improvement. Edited by J. Smartt. Published in 1994 by Chapman & Hall, London. ISBN 0 412 408201.

producing areas, and also the 70–80 wild *Arachis* species that are native in South America. Together these present a satisfactory range of genetic diversity. However, there may still be much more genetic variability in centres of diversity that could be lost with the further expansion of improved cultivars and the spread of modern civilization to remote areas and peoples.

Significant progress has been made in characterization and evaluation of groundnut germplasm. This has led to the identification of a number of lines with desirable sources of resistance to both biotic and abiotic hazards (Amin *et al.*, 1985; Mehan, 1989; Subrahmanyam *et al.*, 1990). But use of the germplasm in genetic improvement of *A. hypogaea* is still in its infancy, restricting progress in broadening the genetic base of the cultigen. In this chapter we review the progress and present the current state of biosystematics in the genus *Arachis* and of the available groundnut genetic resources, highlighting promising areas for future research.

4.2 NOMENCLATURE

4.2.1 Subgeneric

The genus *Arachis* belongs to the Papilionoid legumes (Leguminosae, Papilionoideae or Fabaceae), tribe *Aeschynomeneae*, and subtribe *Stylosanthinae*. Based on morphological similarities, cross-compatibility relationships and pollen fertility of interspecific hybrids (where available), Krapovickas (1969, 1973), IBPGR (1990) and Gregory *et al.* (1973, 1980) proposed a subgeneric classification dividing the genus into sections and series (Table 4.1). However, none of these classifications were validly published according to the rules of the International Code of Botanical Nomenclature (Resslar, 1980). Therefore, all the sectional epithets which have been proposed to date must be considered *nomina nuda* and have no status under the code. Any of the proposed systems may be used equally but the classification suggested by Gregory *et al.* (1980) has been used most commonly, despite its non-validity, and has served well for defining broad cross-compatibility relationships between taxa within the genus and delimiting taxonomic boundaries. We follow the Gregory *et al.* (1980) plan as modified by Krapovickas (1990) in the course of our discussion in this chapter.

4.2.2 Specific and subspecific

The first *Arachis* species to be described was the groundnut itself (*A. hypogaea*), by Linnaeus (1753). Krapovickas and Rigoni (1960) and Krapovickas (1969) divided this species into two subspecies, each of which has two botanical varieties (Table 4.2). Gregory *et al.* (1980) suggested that

TABLE 4.1 *Taxonomic subdivisions of genus* Arachis *(Krapovickas, 1969, 1973, 1990; Gregory* et al., *1980) and proposed genomes*

Sections	Series	Genome	$2n=$
Arachis	1. *Annuae*	A, B, D	20
	2. *Perennes*	A	20
	3. *Amphiploides*	AB	40
Erectoides	1. *Trifoliolatae*	E_1	20
	2. *Tetrafoliolatae*	E_2	20
Procumbensae	–	P	20
Caulorhizae	–	C	20
Rhizomatosae	1. *Prorhizomatosae*	R	20
	2. *Eurhizomatosae*	2R	40
Extranervosae	–	Ex	20
Ambinervosae	–	AM	20
Triseminalae	–	T	20

the taxon, *A. nambyquarae* Hoehne, be considered only a variety of subspecies *hypogaea*. In 1841 Bentham described some wild species including *A. villosa*, *A. tuberosa*, *A. glabrata* and *A. pusilla*. Taxonomic revision of the genus was attempted by Chevalier (1933, 1934, 1936), Hoehne (1940) and Hermann (1954). Gregory and Gregory (1979) have estimated that the genus consists of about 40–70 species. However, at present, only 23 species, one variety (*A. nambyquarae*) and a hybrid (*A. batizogaea* Krap. et Fern., which was thought to have originated from a cross between *A. hypogaea* and *A. batizocoi*) have been validly described and named (Table 4.3). In addition, there are 12 invalidly named species which regularly appear in the literature, such as *A. cardenasii*, *A. duranensis* and *A. chacoense* (corrected spelling = *chacoensis*). According to Resslar (1980), they should be designated by placing the abbreviation '*nom.nud.*' after the author citation. In the case *A. villosa* var. *correntina* (one of the validly described varieties of *A. villosa*, which sometimes is also referred to as *A. correntina*) biosystematic effort has solved the nomenclatural problem, showing that morphologically and also in cross-compatibility patterns *A. villosa* and *A. correntina* are two distinct species.

A revision of the taxonomy of the genus is being prepared by Krapovickas and Gregory to cope with the large number of new and distinct accessions which have been collected in recent years. This monograph should alleviate problems which have been compounded by recent germplasm collections. The non-availability of published descriptions of most of the *Arachis* species has resulted in assigning affinities wrongly, especially by those not familiar with the genus, and others who continue to add to the list of invalid names.

TABLE 4.2   *Classification of groundnut (*Arachis hypogaea*)*

Subspecies	Variety	Botanical type	Branching pattern	Growth habit	Seed/pod
hypogaea	*hypogaea*	virginia	alternate	prostrate to erect	2–3
	hirsuta	peruvian runner	alternate	prostrate	2–4
fastigiata	*fastigiata*	valencia	sequential	erect	3–5
	vulgaris	spanish	sequential	erect	2

4.3 SYSTEMATICS

Systematic studies have been little appreciated by funding authorities, who have been reluctant to support them because they were considered to be basic taxonomic research and to have little practical value. These misleading perceptions are now changing because of the valuable contributions such studies are making to wide hybridization in crop improvement. It has become clear that crops like wheat and maize, which are well understood biosystematically, have progressed more in their genetic improvement than those crops which lack such information. Biosystematic information, particularly on phylogeny and cross-compatibility relationships between a cultigen and its wild relatives, has greatly assisted breeders and cytogeneticists involved in exploitation of interspecific variability for genetic improvement of cultivated species.

4.3.1 Classical

With the advent of new techniques our ability to understand the taxonomic and phylogenetic relationships, even at the molecular level, has increased and data are now available from several lines of investigation for integrated analysis to produce a sound taxonomic system. However, a classical approach, based on morphological similarities and dissimilarities, is still primary and paramount. The genus *Arachis* is morphologically quite distinct from its close relatives because of its unique reproductive system leading to production of geocarpic fruit. On the basis of similarities in morphological features of leaves, leaflets, inflorescence and anthers, *Arachis* has been placed in the tribe *Aeschynomeneae* and subtribe *Stylosanthinae* together with relatives such as *Stylosanthes* (Taubert, 1884). The genus has been further divided into sections and series by Krapovickas (1969, 1973), Gregory *et al.* (1973, 1980) and Krapovickas (1990), based on morphological features of the root system, orientation of peg, venation on the standard petal and size of flowers. Cross-compatibility relationships have been used to support many conclusions in this work (Table 4.1).

TABLE 4.3 *List of validly described species of the genus Arachis (sub-generic classification after Gregory et al., 1973; sectional modification by Krapovickas, 1990)*

Species	Series	Section
1. A. batizocoi Krap. et Greg.	Annuae Krap et Greg. nom. nud.	Arachis nom. nud
2. A. glandulifera Stalker	"	"
3. A. helodes Martius ex Krap. et Rig.	Perennes Krap. et Greg. nom. nud	Arachis nom. nud.
4. A. villosa Benth.	"	"
5. A. villosa var. correntina Burkart (A. correntina Krap. et Greg. nom. nud.)	"	"
6. A. diogoi Hoehne	"	"
7. A. hypogaea L.	Amphiploides Krap. et Greg. nom. nud.	"
8. A. monticola Krap. et Rig	"	"
9. A. batizogaea Krap. et Fern.	"	
10. A. repens Handro	–	Caulorhizae Krap. et Greg. nom. nud.
11. A. guaranitica Chod. et Hassl.	Trifoliolatae Krap. et Greg. nom. nud.	Erectoides Krap. et Greg. nom. nud.
12. A. tuberosa Benth.	"	"
13. A. benthamii Handro	Tetrafoliolatae Krap. et Greg. nom. nud.	
14. A. martii Handro	"	"
15. A. paraguariensis Chod. et Hassl.	"	"
16. A. rigonii Krap. et Greg.	–	Procumbensae Krap. nom. nud.
17. A. marginata Gardner	–	Extranervosae Krap. et Greg. nom. nud.
18. A. lutescens Krap. et Rig.	–	"
19. A. prostrata Benth.	–	"
20. A. villosulicarpa Hoehne	–	
21. A. burkartii Handro	Prorhizomatosae Krap. et Greg. nom. nud.	Rhizomatosae Krap. et Greg. nom. nud.
22. A. glabrata Benth.	Eurhizomatosae Krap et Greg. nom. nud.	
23. A. hagenbeckii Harms	"	"
24. A. angustifolia (Chod. et Hassl.) Killip	"	"
25. A. pusilla Benth.	–	Extranervosae Krap. et Greg. nom. nud.

The description of 23 described species of the genus *Arachis* has been considered by Gregory *et al*. (1980) as inadequate for developing a sound morphological basis of species recognition. He pointed to limitations in the herbarium specimens deposited in the major herbaria of the world, and to the undue emphasis given to vegetative aerial plant parts which have strong resemblances, even among taxa that are only distantly related; he said that insufficient attention had been given to reproductive and subterranean parts that are more stable, distinct and useful in differentiating valid taxonomic entities. For example, the strong morphological resemblance, particularly in leaf shape, between *A. hagenbeckii*, *A. chacoensis* and some accessions of *Erectoides* and *Arachis* has led to erroneous identification of one *Erectoides* accession as *A. diogoi*. Now that *A. diogoi* has been re-collected, even the novice can correctly identify the mistakenly identified *Erectoides* if provided with complete herbarium specimens for purposes of comparison.

4.3.2 Biosystematics

Biosystematics is concerned with the evolution of isolating mechanisms that have led to differentiation and genetic isolation of populations; hence it is a valuable tool when combined with classical data in delimitation of species and their recognition. It is easy to separate two taxa in the case of complete isolation but this is more difficult and subjective when isolation has been only partial. From an applied point of view, the information based on the concept of biological species – i.e. that populations which cross freely and are capable of free gene-exchange belong to a single biological species and those which do not are outside it – is very important. It defines quite satisfactorily the ease and extent of possible genetic introgression and hence is of great applied value in strategic crop improvement research. Because *Arachis* is predominantly self-pollinated, geographic isolation and other extrinsic isolating mechanisms are expected to be more important initially in differentiation and evolution of species than intrinsic genetic isolation. The following biosystematic studies have been made to promote understanding of the species and their differentiation.

(a) Cytotaxonomy

Ghimpu (1930) and Kawakami (1930) were the first to determine the chromosome number ($2n = 40$) for *A. hypogaea*. Later, Husted (1933, 1936) identified a pair of small chromosomes and a pair of chromosomes with secondary constriction and a satellite, as markers in the genomic complement of *A. hypogaea*, designating them 'A' and 'B' respectively. Smartt (1965), Smartt *et al*. (1978a, 1978b), Stalker and Dalmacio (1981), and Singh and Moss (1982) observed many diploid species of sect. *Arachis* with such an 'A' chromosome and *A . batizocoi* without such a pair. Smartt *et al.* (1978a) first designated them as the A and B genomes respectively. Singh and Moss (1982) confirmed statistically the distinction of two genomes, and formed two clusters – one represented solely by *A. batizocoi* and the other by the remaining taxa (of those analysed) of sect. *Arachis*. The larger group was further subdivided into three sub-groups: *A. cardenasii*; *A. duranensis* and *A. spegazzinii*; and *A. villosa* and *A. correntina* (Singh and Moss, 1982, 1984a). In taxa with $2n = 4x = 40$, *A. monticola* and *A. hypogaea* were found to be karyotypically very similar. Stalker (1991) reported another species, *A. glandulifera* Stalker (*A. spinaclava* Krap et Greg.), in sect. *Arachis* with a highly asymmetrical karyotype, distinctly different from A and B genomes; he designated this as the 'D' genome. Somatic complements of many more taxa from other sections have been studied, although not in great detail, i.e. *A. glabrata* $2n = 4x = 40$ (Gregory, 1946), several other accessions of Rhizomatosae (Singh, 1985b) and a number of accessions with $2n = 20$ from sect. *Erectoides* (Singh *et. al.*, 1990). In these accessions, very little karyotypic similarity with the species of sect. *Arachis* has been observed. In tetraploid accessions of *Rhizomatosae*, several chromosomes were observed in quadruplicate, indicating their autotetraploid nature (Singh, 1985b). These observations indicate that structural differentiation of chromosomes at the diploid level and two independent courses of polyploidization – autopolyploidization in section *Rhizomatosae*, and amphidiploidization in section *Arachis* have been involved in differentiation and evolution of species in the genus.

Meiotic studies in *Arachis* species have revealed that chromosome pairing in most diploid species is regular, forming 10 bivalents (Smartt *et al.*, 1978a: Resslar and Gregory, 1979; Singh and Moss, 1982). Consistent quadrivalent associations were observed only in tetraploid ($2n = 4x = 40$) species of sect. *Rhizomatosae*, supporting their autotetraploid nature. Such associations could also form as a result of fixation of a structural heterozygote in rhizomatous clones of these accessions. A few quadrivalent associations have also been observed in some accessions of *A. hypogaea* ($2n = 4x = 40$).

(b) Cytogenetical

The degree of genetic and genomic isolation as a result of differentiation at individual loci, through rearrangement of chromosome segments or polyploidy (both auto- and allo-), could result in cross-incompatibilities, either partial or complete. Cytogenetical analysis of hybridization products, the study of meiotic pairing, and pollen and pod fertility in hybrids may provide real insight into the probable mode of speciation and evolution on the one hand, and the possibilities of gene exchange for use by the groundnut breeder on the other. The first viable interspecific hybrid reported in the genus *Arachis* was between *A. hypogaea* ($2n = 40$) and *A.*

villosa var. *correntina* ($2n = 20$) and was produced by Krapovickas and Rigoni (1951). Since then, the cultivated *A. hypogaea* has been crossed with as many as 37 species (Simpson, unpublished data) and crossing between different species from both the same section and different sections has resulted in production of more than 250 successful hybrids (Kumar *et al.*, 1957; Smartt and Gregory, 1967; Gregory and Gregory, 1979; Singh and Moss, 1984a, 1984b; Singh, 1985b; Pompeu, 1977; Simpson, 1991a). Such results have generally corroborated the cross-compatibility relationships established by Gregory and Gregory (1979), who showed that, despite ploidy differences, it is easier to produce intrasectional than intersectional hybrids. They found that intrasectional hybrid pollen fertility at the same ploidy level averaged around 30.2% in sect. *Arachis*; 0.2% in sect. *Extranervosae* and 86.8% in sect. *Caulorhizae*. Intersectional hybrids were completely female-sterile with very low or no pollen fertility; some failed to flower at all.

Sect. *Arachis*, to which *A. hypogaea* belongs, contains a number of diploid wild species reported to be resistant to several diseases of groundnut, such as rust, early and late leaf spots, and rootknot nematodes (Abdou *et al.*, 1974; Subrahmanyam *et al.*, 1980, 1983; Nelson *et al.*, 1989). They are therefore important potential sources of genes for breeding for disease resistance in *A. hypogaea*. Concerted biosystematic efforts have been made to understand phylogenetic relationships among these species of sect. *Arachis* (Smartt and Gregory, 1967; Stalker and Wynne, 1979; Resslar and Gregory, 1979). Singh and Moss (1984a) performed a comprehensive genome analysis in sect. *Arachis* and reported that hybrids between the diploid species they studied, except those with *A. batizocoi*, had near normal bivalent frequencies (9.1–9.8) with moderate to high pollen fertility (60–91%). Hybrids between *A. batizocoi* and other species had low bivalent frequency (5.2–6.9) and very low pollen fertility (3–7%). These results confirmed genetic separation of *A. batizocoi* (B genome) from the other set of species of sect. *Arachis* (A genome). Formation of 8–9 bivalents in some of the pollen mother cells (PMCs) of AB hybrids also suggested that both genomes have a similar basic gene complement. Chromosome pairing in triploid hybrids between *A. hypogaea* and the diploid wild species corroborated the fact that the two genomes present in *A. hypogaea* are common to the genomes found in diploid wild species belonging to sect. *Arachis*. Furthermore, it indicated that *A. batizocoi* is the closest diploid relative of *A. hypogaea*, probably more so to subspecies *fastigiata* than to subspecies *hypogaea*, with homology to one set of the genome (10 chromosomes) of *A. hypogaea*. The second set is homologous to the other group of diploid species containing an A genome, forming an average of 9 bivalents in their triploid hybrids involving *A. hypogaea*. The tetraploid, *A. monticola*, crossed freely with *A. hypogaea* and showed normal pairing of chromosomes in the hybrids, indicating that, biologically, *A. monticola* may just be the wild form of *A. hypogaea*.

In continuation of the above biosystematic studies, Singh (1986a, 1986b, 1988) further studied chromosome pairing, pollen fertility and pod fertility in hybrids between cultivated tetraploid *A. hypogaea* (AA and BB) and both synthetic autotetraploids and amphidiploids from AA and BB genome diploid species of sect. *Arachis*. The objective was to verify the above inferences and to identify the most probable ancestors of *A. hypogaea*. Formation of higher mean bivalent associations in hybrids between *A. hypogaea* X AA BB amphidiploids compared with *A. hypogaea* X AA AA amphidiploids supported the earlier inferences that, most probably, two sympatric related diploid species – one with AA and the other with a BB genomic constitution – hybridized and produced *A. hypogaea* through amphidiploidization. *Arachis batizocoi* and *A. duranensis* were considered to be the most probable donors of B and A genomes respectively. There was no complete pairing (i.e. 20 II) in these crosses to the amphidiploids, since both diploid ancestors and the tetraploid cultivated species have undergone a long process of evolution, leading to their further genetic divergence. However, it is possible that the actual progenitor species of *A. hypogaea* have yet to be collected, or may be extinct.

Very few intersectional hybrids have been produced between the diploid species of sect. *Arachis* and sect. *Erectoides*, with and without embryo rescue to overcome postzygotic incompatibility barriers (Gregory and Gregory, 1979; Singh, 1989). However, a high level of sterility in these hybrids has restricted the studies on chromosome pairing and pollen fertility. Attempts to cross synthetic tetraploids of sect. *Arachis* and amphidiploid or natural tetraploids of sect. *Erectoides* and sect. *Rhizomatosae* respectively have virtually failed. Singh (1988) observed that crossing at the tetraploid level was more difficult than at the diploid level. Gregory and Gregory (personal communication) produced several hybrids between diploid sect. *Arachis* annual types and complex amphidiploid hybrids within sect. *Erectoides*. Stalker (1981) studied meiosis in hybrids between two diploid ($2n = 20$) species, *A. duranensis* and *A. stenocarpa* ($= A. stenosperma$) of sect. *Arachis* and an amphidiploid (*A. rigonii* × *A.* sp. GKP 9841 $= 4x$) of sect. *Erectoides* and observed a high frequency of bivalents indicating chromosomal homologies between the members of sect. *Arachis* and sect. *Erectoides*. Stalker (1981) also analysed chromosomal homologies in some other complex hybrids produced by Gregory and Gregory (1979) between sect. *Erectoides* and sect. *Rhizomatosae* which suggested distant homologies between the chromosomes of the two sections, and common ancestry.

The cytogenetical data in *Arachis* indicate that two chromosome series, $2n = 2x = 20$ and $2n = 4x = 40$, are present in the genus, though most species are diploid with $2n = 2x = 20$. Polyploidy probably arose twice in the genus, independently, once in sect. *Rhizomatosae* through autotetraploidization and again in sect. *Arachis* through amphidiploidization. Cultivated tetraploid *A. hypogaea* (AA BB) could have originated via domestication of the wild tetraploid species *A. monticola*, which most probably originated through amphidiploidization of an F_1 hybrid between a pair of species containing AA and BB genomes. The autotetraploid genome of sect. *Rhizomatosae* has some homology with the A genome of sect. *Arachis* and the E genome of sect. *Erectoides*, and so there are substantially common chromosomes in all three of these sections.

(c) Chemotaxonomy

Seeligman and Krapovickas (Krapovickas 1973, Krapovickas *et al.* 1974) were probably the first to look at the chemical variability in the genus by chromatographic analysis of flavonoids in the leaves of *Arachis* species. They detected more than 20 compounds, and chemical variation coincided well with the centre of morphological variation of the genus (Mato Grosso). There was no variation west of the meridian of 51°, which can be considered as the axis of the area of distribution of the genus, as it runs parallel to the Paraguay and Uruguay rivers. The presence of common flavonoid chromatographic spots in *A. hypogaea*, *A. monticola* and diploid wild species of sect. *Arachis* corroborated their close phylogenetic relationship.

Several workers have studied both seed protein and isozyme profiles through polyacrylamide gel electrophoresis (PAGE) and immunochemical methods (Neucere and Cherry, 1975: Cherry, 1975; Klozová *et al.*, 1983a, 1983b; Krishna and Mitra, 1988; Singh *et al.*, 1991). Tombs and Lowe (1967) noted that one of the major storage proteins, arachin, is polymorphic and they identified three forms of it. The gross fraction of protein profile in the taxa of section *Arachis* did not differ much in overall mobility patterns, and expressed considerable similarity even considering the differences in ploidy levels (Singh *et al.*, 1991). This corroborates the inferences of Klozová *et al.* (1983b), based on similarities of patterns resolved by immunochemical methods (immunoelectrophoresis and double diffusion), that protein patterns in *Arachis* have been relatively conservative in evolution. Singh *et al.* (1991) have further indicated that the variation in protein profiles of accessions of the same species is very low, particularly in a small sampling of *A. hypogaea* cultivars. If this pattern holds true in analyses of a broad spectrum of *A. hypogaea*, protein profiles may not be of much practical value in differentiation and identification of cultivars, but protein profiles can be of great value in differentiation and broad classification of the genus and for species identification. These studies support interspecific relationships consistent with the taxonomic scheme developed by Krapovickas and Gregory (Gregory *et al.*, 1980: Krapovickas, 1990), particularly for the breakdown of the genus into sections. They also support species relationships, and the probable ancestry of *A. hypogaea* deduced from cytogenetical evidence (Singh, 1988).

Preliminary cytophotometric studies on amount of cDNA performed by

Resslar (1980) recorded 4.92–5.98 pg DNA per cell for diploid species and nearly double that for tetraploids. Annual diploids have 1 pg less DNA than perennial diploid species, and tetraploid taxa belonging to two *A. hypogaea* subspecies and *A. monticola* show some small but consistent differences. Restriction Fragment Length Polymorphism (RFLP) studies carried out in the USA (Kochert *et al.*, 1991) have shown only a very low level of variation in a small sample of tetraploids that included *A. hypogaea* cultivars from the USA and wild *A. monticola*. However, the wild diploid species from sect. *Arachis* show considerable variation in their RFLP bands, though the RFLP patterns of tetraploids were more complex than those of diploids. Nevertheless, the two constituent genomes of amphidiploids could usually be distinguished. They suggested *A. ipaensis*, *A. duranensis* and *A. spegazzinii* to be the most closely related diploid progenitors of the tetraploid *A. monticola* and *A. hypogaea*, rather than *A. batizocoi*, as indicated by cross-compatibility data cytogenetic studies and protein profiles. This could be explained by assumption of certain structural changes occurring at the microevolutionary level within the genomes of *A. hypogaea*, and the genetic divergence these have caused in the molecular sequences resolved by the probe and restriction enzyme combinations used by Kochert *et al.* (1991). This is probably one of the limitations of RFLP techniques, where early erroneous conclusions are possible, based on the homology between fractions of genomes, before a broad array of material has been analysed using all possible combinations. One needs to use caution in drawing such conclusions on homologies before evaluating all possible combinations of probe and restriction endonucleases to reveal most of the genomic sequences. Critical investigations are still needed to resolve such contradictions.

The studies above indicate that useful variation exists for chemical characters in the genus. However, it is clear that the potential value may lie in resolving the problems of classification between broad groups, which may help in establishing species relationships and their delimitation, and in assigning affinities to new taxa. There is obviously considerable scope for genetic and molecular characterization of various genomes in the genus using these biochemical techniques, and for using the information in further classification and developing appropriate technologies for their exploitation. Limited variation detected in *A. hypogaea* may be misleading and the situation could well change following analysis of a broader spectrum of the available germplasm. Hence, the application of biochemical techniques in groundnut is still to be developed to enable its utilization in (1) resolving polymorphism among cultivars of groundnut, (2) fingerprinting of cultivars and (3) identification of molecular markers associated with desirable traits, which can be used in indirect screening of germplasm for specific traits to aid in the production of improved new varieties.

From the above biosystematic studies, it is possible to establish tentatively a series of genomes which parallel the *Arachis* sections. Some of the taxonomic entities are monotypic, such as sect. *Triseminale* with *Arachis* sp. GKP 12881, and 12922 and series *Prorhizomatosae* of sect. *Rhizomatosae* with *A. burkartii*, while sect. *Caulorhizae* is only represented by a pair of taxa. This provides a very narrow level of variability for making inferences; nevertheless, the strong genetic barriers observed between these taxonomic entities are sufficient to recognize the twelve genomes shown in Table 4.1. These correspond to the section and series as initially proposed by Gregory and Gregory (1979) and expanded by Krapovickas (1990).

The sect. *Erectoides* comprises only diploid species which are classified into two distinct series corresponding to two sub-genomes . In *Rhizomatosae*, based on compatibility, it is proposed that tetraploid *Eurhizomatous* species have two genomes, one with homologies to sect. *Erectoides* and the other to sect. *Arachis* (Krapovickas, 1973; Gregory and Gregory, 1979). However, the *Rhizomatosae* are assumed to be primitive and it is difficult to see *Arachis* evolving from *Rhizomatosae* (if the section is indeed primitive), as diploid rhizomatous *A. burkartii* is genetically and physically isolated from other *Arachis* species. Singh (1985b), based on karyomorphology (chromosomes in quadruplicate) and meiotic behaviour (with quadrivalents), has indicated that tetraploid rhizomatous species appear to be autotetraploids. Hence, it is more likely that genomic homologies reflected between sections *Erectoides*, *Arachis* and *Rhizomatosae* on the basis of cross-compatibilities may be due to a pivotal genome common to all three sections that have evolved in parallel, evolving sect. *Arachis* and sect. *Erectoides* through genetic and chromosomal alteration on the one side, and sect. *Eurhizomatosae* via polyploidy on the other. This appears plausible because several annual species of sect. *Arachis* do cross with the species of tetrafoliolate *Erectoides* (Gregory and Gregory, 1979; Singh, 1989) although only two crosses were successful without embryo rescue techniques. In sect. *Arachis*, the genome analysis does not conform with series classification. Series *Annuae* consists of species with the B genome, sub-genomes of genome A, and another recently discovered D genome, while series *Perennes* consists of only A genome species. The series Amphiploides combines both A and B genomes. These genomic classifications work best in the light of biosystematic evidence collected to date and are subject to modification in case evidence from further collection favours a different interpretation.

4.4 PHYLOGENY

Phylogenetically the genus *Arachis* has a number of characters in common with *Stylosanthes*; more so than with any other related genus, although some characters are not common to all species. Thus it is plausible to suppose that *Arachis* evolved from a form such as *Stylosanthes*. Similarly within the genus *Arachis*, based on morphological relationships, sect. *Erectoides*, series *Trifoliolatae* with its trifoliolate leaves and *Extranervosae* with its syncolpated pollen have been considered to be the most primitive. *A. guaranitica* of series *Trifoliolatae* of *Erectoides* has basal flowers, grouped in a cluster at the crown of the plant. They appear to have a subterranean ovary, and a hypanthium 5 cm in length, which exposes the androecium and stigma, protected by the keel petal, for pollination. All *Trifoliolatae* and one species of the *Extranervosae* also have fruit produced from the crown of the plant. In series *Tetrafoliolatae* all species have flowers above the cotyledonary node (termed aerial for simplicity) in addition to those with basal ovaries with proximity to the soil for fruiting through decumbent branches. In other sections, such as *Ambinervosae*, *Caulorhizae*, *Rhizomatosae* and *Arachis*, the flowers are normally aerial on procumbent branches. But the trifoliolate characteristic itself may not be primitive as the condition in *Tetrafoliolatae* is closest to that of *Zornia* in the *Aeschynomeneae*. The peculiar fruiting habit may also not be primitive. In addition, populations of *A. tuberosa* (a member of the *Trifoliolatae*) have been collected with tetrafoliolate leaves; therefore the trifoliolate form may be a simple genetic mutant, advanced and specialized in nature and not primitive at all. However, the *Extranervosae* are clearly differentiated from the rest of the genus by their unique pollen type, which is supposed to be very primitive. Moreover, species of this section have developed strong genetic barriers and only cross with species of sect. *Ambinervosae*.

The fact that some species of the sect. *Ambinervosae* can be crossed with some species of sect. *Extranervosae* and tetrafoliolate *Erectoides* can be considered as indicative of affinity of the *Ambinervosae* between some of the ancient sections and those more recently evolved.

Arachis, *Rhizomatosae* and *Caulorhizae* are the other derived sections. The latter two have developed systems of vegetative reproduction. The *Rhizomatosae*, which have tetraploid taxa, are thought to be the most recent (Krapovickas, 1973). On account of crossing affinities, Krapovickas (1973) and Gregory and Gregory (1979) presumed that some rhizomatous diploids might have crossed with a diploid *Erectoides* followed by doubling of the chromosome complement to give rise to the centrally located and broadly distributed tetraploid ($2n = 4x = 40$) Eurhizomatous species. This phylogenetic pattern is in agreement with the postulated evolution of flowering and fructification types (Krapovickas, 1973). However, cytogenetic evidence does not fully support this hypothesis and to date no hybrids have been produced between *Prorhizomatosae* and *Erectoides*. The location of *Prorhizomatosae* and *Erectoides* is also problematic. The distance between the closest collection sites of these two groups does not indicate that they were close enough to hybridize by natural means to be credible progenitors of *Eurhizomatosae*. Further biosystematic investigations are needed to resolve phylogenetic relationships between these sections.

The derived sections do not necessarily have a wide distribution. For example, that of the section *Extranervosae* is only exceeded by sects. *Eurhizomatosae* and *Arachis* in its geographic area. According to Krapovickas (1973), it is reasonable to assume that the formation of aerial flowers on prostrate branches is more recent than that of the flowers grouped on the crown. The production of subterranean fruits and pegs originating above ground is probably an ancestral character. The geographical distribution of these characters supports the contention that prostrate plants have been the more effective in spreading to new areas.

The cultivated *A. hypogaea*, which is of primary importance, is phylogenetically very close to the tetraploid species *A. monticola* and to diploid annual species such as *A. duranensis*, *A. spegazzinii* (a form of *A. duranensis*) and *A. batizocoi* on the basis of morphological affinities, in conjunction with cytological, genetical and biochemical affinities. They are also found in the same region of northwest Argentina and southern Bolivia along with primitive *A. hypogaea* (Valls *et al.*, 1985) and have been considered an immediate ancestor of *A. hypogaea*. Tetraploid *A. monticola* differs from *A. hypogaea* in several characters, such as biarticulated fruits (a character it shares with other species of the genus *Arachis*), a much smaller fruit size, pegs as much as three times as long as those of *A. hypogaea*, very weak peg–pod attachment and being able to survive without human intervention.

4.5 GENETIC RESOURCES

The information that has been obtained by biosystematists in the genus *Arachis* and in *A. hypogaea* is of considerable value for breeders involved in improvement of the cultivated groundnut. Studies on cross-compatibility are of particular importance as they have indicated the taxa that are most likely to be accessible to the breeders for improvement of the cultigen. Besides germplasm of *A. hypogaea*, the cross-compatible species of sect. *Arachis* (which is the most recently evolved and evolving, section) provide another reservoir of easily accessible genetic resources. Therefore the pool of landraces and cultivars of cultivated groundnut which have evolved under different levels of natural and human-induced selection pressure, along with the wild species of sect. *Arachis* and other sections of the genus, together provide a wealth of material for genetic modification and improvement of *A. hypogaea*.

The distribution pattern of wild species and the origin and spread of the cultivated groundnut throughout South America and later to other countries is not totally understood. The peculiar distribution patterns of some taxa and the apparent wide gaps between some collections raise many questions regarding probable evolutionary pathways. Incomplete sampling of the total of potentially available material and the apparent absence of important intermediate species and landraces (via extinction, introgression or simply absence of collection) are factors that limit knowledge of local evolution. The extremes of phenotypic variation within the genus appear to be represented in our collections, but additional collections continue to expand the bulk of what is considered extreme, while new collections from extreme areas of the range of the genus continue to make our knowledge more comprehensive.

The genus extends over more than 2.6 million km^2 of the South American continent, from north-east Brazil to north-west Argentina, and from the south coast of Uruguay to north-west Mato Grosso south of the Amazon; and from the base of the Andes to the Atlantic (Hammons, 1982). Many of the areas where additional taxa probably exist have not been accessible. The extensive distribution of the genus, coupled with the lack of adequate transportation networks and the harshness of the environment, make the task more difficult in the light of financial and political considerations. A similar situation exists for collections from areas of early introductions in Africa and Asia, which may still have useful genetic diversity but which may be lost with further expansion of improved cultivars and modern technology into remote areas.

Significant progress has been made in the evaluation of genetic resources, leading to identification of a number of lines from cultivated *A. hypogaea* and wild *Arachis* species with desirable traits and sources of resistance to both biotic and abiotic hazards that reduce groundnut production worldwide (Amin *et al.*, 1985; Mehan, 1989; Subrahmanyam *et al.*, 1990). However, utilization of these genetic resources is restricted by lack of information on reaction to other contemporary stresses, illustrating their usefulness, and the detailed knowledge regarding the gene(s) mediating them (it is essential to select appropriate breeding strategies for their effective utilization).

4.6 GENE POOLS

The above sources of genetic diversity have been classified into four gene pools (Wynne and Halward, 1989; Smartt, 1990). The first order or primary gene pool consists of taxa belonging to the cultivated species; the secondary gene pool consists of *A. monticola*. Arguments have been presented that, based on crossability and fertility of hybrids between them, *A. monticola* and *A. hypogaea* should be considered as part of the same gene pool of a single biological species. Crossability studies have also indicated that sect. *Procumbensae* (which is essentially sympatric with sect. *Arachis* throughout its distribution) produces normal pegs and pods when cross pollinated with sect. *Arachis* but the embryo aborts at an early stage of development, indicating evolution of postzygotic barriers to hybridization during the course of evolution

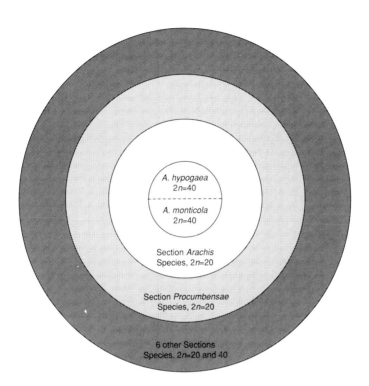

Figure 4.1 Gene pools of genus *Arachis*.

(Singh, 1989). Therefore, in all probability *Procumbensae* has co-evolved with series *Perennes* of sect. *Arachis* and must share genes with *A. hypogaea*.

On the basis of these observations, the five or four gene pool theory fits the genus *Arachis* very well (Figure 4.1). These pools may be described as follows:

1. The primary gene pool consists of landraces of the cultivated groundnut from the primary centre of origin and diversity in South America, and cultivars and breeding lines developed in various groundnut-growing countries around the globe. It may include *A. monticola* found in north-west Argentina which has otherwise been considered as a secondary gene pool.
2. The secondary/tertiary gene pool consists of diploid species that are

cross-compatible with *A. hypogaea*, despite ploidy differences, and are placed in sect. *Arachis* with *A. hypogaea* and *A. monticola*.
3. The tertiary/fourth gene pool consists of sect. *Procumbensae*, which in all probability co-evolved with series *perennes* of sect. *Arachis* and can share genes with *A. hypogaea* on overcoming postzygotic barriers.
4. A fourth/fifth gene pool contains the rest of the *Arachis* species that are cross-incompatible or weakly cross-compatible to sect. *Arachis* and are classified into six other sections.

The chance of exploiting the third and fourth order gene pools will depend upon advances in biotechnological techniques of genetic trans-formation and somatic hybridization. The process will be time-consuming and expensive. The most accessible gene pools will be the primary and secondary ones and their actual breeding value is likely to be more productive and somewhat more predictable. The characters of wild *Arachis* species which have most immediate attraction to groundnut workers con-cern resistance to diseases and pests, for which the variation within the primary and secondary gene pools of *A. hypogaea* and *A. monticola* is very limited.

4.7 CENTRES OF ORIGIN

The centre of origin of the genus *Arachis* was most probably in central Brazil (Gregory *et al.*, 1980). The geocarpic habit of the plant suggests that long-distance dispersal has been along streams and rivers of the South American continent. According to Gregory *et al.* (1973, 1980), the most ancient species are found in higher elevations. Their immediate descendent relatives occupy the next lower erosion surface and their distantly evolved descendants occupy a still lower and more recent eroded surface. To support this view, they found unique associations of botanical features with highland conditions. Further, as seeds moved to lower elevations they became isolated in major river valleys: thus different sections of the genus evolved independently in a parallel fashion. Sect. *Caulorhizae* was isolated and evolved in valleys of the Jequitinhonha and São Francisco rivers, *Triseminalae* in the São Francisco, *Ambinervosae* in the drainage system of north-east Brazil, and series *Prorhizomatosae* in the Uruguay River valley. Sect. *Arachis* probably evolved in the basin of the Paraguay or in the upper reaches of the middle Amazon, which is invaded by the upper reaches of the Paraguay. In both, the captured species are somewhat confined in the uppermost part of the drainage systems, though they are widely distributed in their original location. Species in the sect. *Erectoides* occur in the extensive dissection of the Central Brazilian Planalto by both the Paraná and branches of the Paraguay rivers. *Eurhizomatosae* occur sympatrically with *Erectoides* but extend beyond the Paraguay and Paraná, south of

Corrientes, east into São Paulo and north of Rosário Oeste. However, the belief that species in different sections are geographically isolated has changed, since considerable overlaps in distribution between the members of sections *Arachis*, *Extranervosae* and *Rhizomatosae* have been recorded (Simpson, 1984; Valls *et al.*, 1985) and major sectional groups of the genus have been found with widespread distributions.

Cultivated *A. hypogaea* most probably originated in the region of southern Bolivia and northern Argentina, since *A. monticola*, the only wild tetraploid species that crosses with *A. hypogaea*, is found in this area. Hence this is the region of the presumed centre of origin of cultivated groundnuts (Krapovickas, 1969). Most of the diploid sect. *Arachis* species that may have given rise to the segmental allotetraploid (*A. monticola*/ *A. hypogaea*) – such as *A. batizocoi* (Krapovickas *et al.*, 1974; Smartt *et al.*, 1978a, 1978b; Singh and Moss, 1984a; Singh, 1986b, 1988), *A. duranensis* (Seetharam *et al.*, 1973; Gregory and Gregory, 1976; Singh, 1986b, 1988) and *A. spegazzinii* (a form of *A. duranensis*; Kochert *et al.*, 1991) – are also distributed here. These observations support the conclusion that this region of southern Bolivia and northern Argentina was the centre of origin and primary centre of diversity for *A. hypogaea*. This does not however take into account recent archaeological evidence which could significantly alter present perceptions.

4.8 CENTRES OF DIVERSITY

The genus *Arachis* is naturally restricted to Argentina, Bolivia, Brazil, Paraguay and Uruguay in South America, although some species such as *A. repens*, *A. glabrata* and *A. pintoi* are now cultivated elsewhere. *Arachis* species are found from the mouth of the Amazon (0°) south across the São Francisco and the Jequitinhonha, and into the mild temperate zone to 34° S on the shores of the south Atlantic in southern Uruguay. Specimens have been found west to the Paraná River, south of Corrientes, and from the Gran Chaco region of Argentina. The north-west limits of the genus appear irregular but information may be incomplete because of lack of accessibility to certain areas. The southern and eastern boundaries of the multi-galleried Amazon Selva along a curved line south and south-westwards from the mouth of the Amazon through Goiás, Pará, and Central Mato Grosso to the north Beni in Bolivia form a tentative northern boundary for *Arachis* species. From there it proceeds south along the foothills of the Andes to Salta (Figure 4.2).

Arachis grows from sea level to 650 m above sea level on the Planalto, from southern Mato Grosso to southern Goiás, and to 1450 m a.s.l. near Jujuy. It is found among vegetation types from broken forest to open grassland. Its species grow submerged, among stones bathed with water, in

dry gravel, and in flood plain alluvium; they are found from semi-arid
locations to regions that receive an average of more than 2000 mm rain
annually, and in regions subject to great flood or intense droughts. They
grow most commonly in friable or somewhat sandy soil but also on soils
that vary from almost pure sand to reduced humic clays, sands of various
alluvia, and dark red humic latosols to pure lateritic caps and gravel.
Adaptation of wild *Arachis* species to such diverse conditions led to the

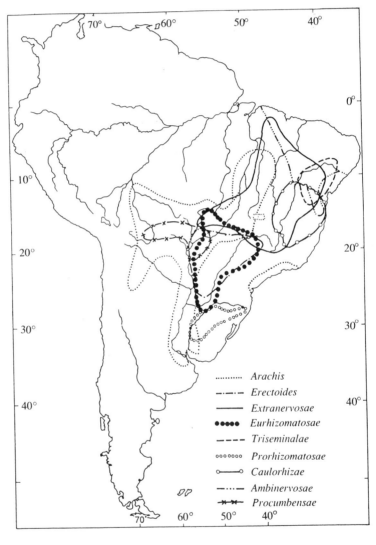

Figure 4.2 Geographical distribution of genus *Arachis*.

development of greater genetic variability and more combinations of characters than those presented by the cultivated species. The above conditions probably contributed to the development of geocarpy and tuberiform roots of various forms, but the most important factor was probably the harshness of the dry season. This and the need to escape dry-season fires (an almost inevitable annual occurrence) contributed heavily to tuberiform roots and geocarpy. Under a helping human hand, the tuberiform roots were selected against in the cultigen but the geocarpic fruit was retained.

The highest number of species occurs in Brazil, where all eight sections are found and to which country four sections of the genus *Arachis* are restricted. Bolivia has the second highest number of species, followed by Paraguay, Argentina and Uruguay. Most species occurring in Brazil are restricted to the west central region but there is a group of endemic species in the semi-arid north-east. Further differentiation in patterns of genetic variability in different sections occurred as a result of their adoption of different ecological niches where they were caught within a series of land uplifts during their movements downstream in the associated drainage systems.

Sect. *Arachis* occurs from northern Goiás and southern Pará, south and west to the base of the last-erosion surface exposed in the Pantanal (100–175 m a.s.l.), along the Paraguay across north-central Bolivia to the foothills of the Andes, where it was caught up in the Pleistocene uplifts and where distinct new species mark the drainage systems, unlike the occurrence of entire sections that characterize the older Miocene system. Genetic isolation is not as strongly marked between most species of sect. *Arachis* compared with those of other sections.

The probable region of origin of the cultivated groundnut (discussed above) presents a wide range of distinct environments and there is evidence that natural hybridization and the establishment of recombinants have occurred in these areas.

Archaeological evidence suggests that groundnut has been cultivated for over 3500 years. It was probably first domesticated in northern Argentina and eastern Bolivia. It was subsequently introduced to Africa, India and the Far East by the Portuguese; and from the west coast of South America to the western Pacific to Indonesia and China by the Spaniards in the early sixteenth century. It is possible that the groundnut may have travelled to China well before the time of Columbus but additional evidence is needed to support this idea. There were introductions to East Africa from Asian countries. By the middle of the sixteenth century the groundnut was introduced to North America and to other parts of the world. Since this period of introduction numerous forms have evolved (Hammons, 1982). The introductions from Bolivia to Africa and Asia are more recent.

In South America, where the greatest amount of genetic diversity is found, Krapovickas (1969) and Gregory and Gregory (1976) recognized six gene centres for cultivated groundnuts (Figure 4.3). These centres are:

1. the Guarani region
2. Goiás and Minas Gerais (Brazil)
3. Rondônia and north-west Mato Grosso (Brazil)
4. The eastern foothills of the Andes in Bolivia
5. Peru
6. North-eastern Brazil.

It is now apparent that at least one more primitive centre exists in Ecuador, where a distinct group of landraces is grown at present.

These centres of diversity contain a high level of genetic variation due to introgressive hybridization followed by human selection, resulting in production of typical hybrid swarms, which might have been produced by crossing between types with divergent characters. However, most can still be linked with specific subspecies and botanical varieties. The Guarani region is dominated by erect valencia types (subsp. *fastigiata*); subsp. *hypogaea* is very rare here. There is little evidence of introgression between the two subspecies in this region, but hybrid swarms of intermediate races between two botanical varieties of subsp. *fastigiata* do exist, and both var. *fastigiata* valencia, Pôrto Alegre, and var. *vulgaris* Negrito, spanish, are identified. It is possible that the spread of valencia types to other parts of the world has occurred from Paraguay or central Brazil but a more likely point of embarkation is from the north-east coast (i.e. the Amazon). The Guarani region is also the centre of variation for the variety *vulgaris*, and spanish types were probably also disseminated from this region (Krapovickas, 1969; Gregory and Gregory, 1976).

The second region of Goiás and Minas Gerais has a varietal pattern distinct from the Guarani, but still is dominated by erect groundnuts (subsp. *fastigiata*), with very few examples of subsp. *hypogaea*. Cultivated races belong to both varieties, *fastigiata* and *vulgaris*, without much indication of introgression. Rondônia, the third region of diversity, is typically represented by the nambyquarae type of subsp. *hypogaea* and the Mato Grosso by a yellow testa, erect type and *A. villosulicarpa*. The fourth region in the eastern foothills of the Andes is a great centre of variability of subsp. *hypogaea* var. *hypogaea* with few races of var. *fastigiata*. In Bolivia there are indications of introgression between the two subspecies, and 'Overo' and 'Cruceno' types of groundnut are probably the product of such introgression (Krapovickas, 1969; Gregory and Gregory, 1976).

From Peru, three distinct types of groundnut are collected, one of which is like that found in pre-Columbian tombs: it has fruit with prominent constrictions, veins and beak and belongs to subsp. *hypogaea*. This is called the 'chinese' type in the USA. A second type, with similar fruit characters but belonging to subsp. *fastigiata* var. *fastigiata*, has also been found in this region. These two together were called the 'peruvian' type by Dubard (1906). A third type belongs to subsp. *fastigiata* var. *fastigiata* but has smooth pods with three to five seeds and almost no beak.

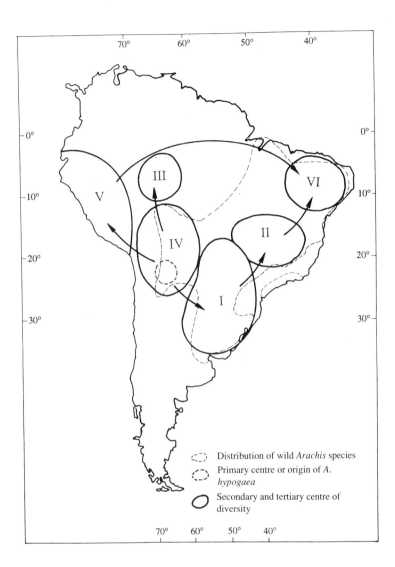

Figure 4.3 Centres of origin and diversity of *Arachis hypogaea* in South America (adapted from Gregory and Gregory, 1976). I *A. hypogaea fastigiata* var. *fastigiata* and var. *vulgaris*; II *A. hypogaea fastigiata* var. *fastigiata*; III *A. hypogaea hypogaea* var. *hypogaea*; IV *A. hypogaea hypogaea* var. *hypogaea*, *A. hypogaea fastigiata* var. *fastigiata*; V *A. hypogaea hypogaea* var. *hypogaea* and var. *hirsuta*, *A. hypogaea fastigiata* var. *fastigiata*; VI tertiary *A. hypogaea fastigiata* and var. *vulgaris*.

Williams (1991) studied the region of the north Beni of Bolivia/Peru. He collected some extraordinary types which appear to be intermediate between subsp. *hypogaea* and subsp. *fastigiata*.

The sixth centre, north-east Brazil, is regarded as a tertiary centre with all types. The Ecuadorian type is similar to the var. *fastigiata* types of Peru but morphologically distinct and might even be considered intermediate between vars. *fastigiata* and *hypogaea*.

Africa has been regarded as another important tertiary centre of diversity. Introductions from Brazil probably became established concomitantly with the slave trade in West Africa, while the east coast probably received material which had come from the west coast of South America via the Philippines, China and India. More recently Smartt (1990) suggested significant introductions from Bolivia, India and China can also be considered as sources of much variability for cultivated groundnut at various times.

4.9 COLLECTIONS

Collection and assembly of groundnut genetic resources has been extensive. It received much impetus from the establishment of the International Board for Plant Genetic Resources (IBPGR) in 1974, and the inclusion of groundnut in the mandate of the International Crops Research Institute for the Semi-Arid Tropics (ICRISAT) in 1976. Before this, extensive groundnut collections were maintained only at the Southern Regional Plant Introduction Station, USA, at Agricultural Research Stations of the USA, and in countries including Brazil, Argentina, Senegal, Israel, Taiwan, India, Nigeria, Malawi and Zimbabwe.

With the establishment of the Genetic Resources Unit at ICRISAT, various collections could be assembled at a central place for use by the international community. The world collection was also enhanced through specific collection expeditions to various groundnut-growing regions of the world and to centres of diversity in South America (Figure 4.4). In these expeditions various international agencies, such as IBPGR, and national programmes, such as those of countries of exploration, and the United States Department of Agriculture (USDA) have collaborated very closely (Simpson, 1982, 1984), leading to the establishment of the world's largest repository of groundnut germplasm at ICRISAT and in the USA under the USDA. ICRISAT and the USDA now maintain around 14 000 and 7545 accessions respectively, and provide basic genetic stocks to the international scientific community for further improvement of the groundnut. They include both cultivated and wild *Arachis* species. The most complete, accessible catalogues of collections are also the ones maintained at USDA and ICRISAT with information on 7545 and 12 160 accessions respectively. Many of the collections held at different places are duplicated through exchange of material, and the total number of different genetic types is less than that represented by the above numbers.

Systematic collection expeditions in South America were started by Archer in 1936 (Valls *et al.*, 1985). Besides the collection of landraces of cultivated groundnut, efforts were made in these expeditions to collect distinct *Arachis* species (Simpson, 1982). The expedition undertaken in 1959 resulted in the collection of six old and five new *Arachis* species. From 1961 to 1967, five new and three old species were collected. A species belonging to sect. *Ambinervosae* was collected for the first time in 1967. A 1968 expedition resulted in collection of one old, one new and several undescribed and unnamed species from sect. *Erectoides*. Many more new wild *Arachis* species were collected in expeditions from 1976 to 1979; most of these were from Bolivia and the edge of the Pantanal in Brazil. The 1976 expedition collected typical *A. diogoi* and re-collected *A. lutescens*. In 1980–81 a species referred to in some literature as *A. sylvestris* was collected for the first time. New *Ambinervosae* species were discovered in Brazil along with several new species of sect. *Arachis* and sect. *Erectoides* (*Trifoliolatae*), and sect. *Procumbensae* in west and central Brazil. The re-collection of *A. tuberosa*, *A. lutescens* and *A. prostrata* occurred in 1982. *A. marginata* was re-collected with additional unnamed species of sect. *Extranervosae* and annual species of sect. *Arachis* penetrating into north-east Brazil, further extending our knowledge of the area of their natural occurrence. In 1983, two new species from sect. *Extranervosae* and sect. *Erectoides* (*Tetrafoliolatae*) were collected. Each expedition provided additional recorded locations of both wild and cultivated groundnut, as a result of which the map of natural occurrence for the different sections was greatly modified from the earlier one presented by Gregory *et al.* (1980) (see Simpson, 1984). Figure 4.2 illustrates the current perception of the distribution of *Arachis* species. The ranges of several sections, such as *Erectoides* and *Ambinervosae*, were expanded due to widespread distribution of *A. sylvestris* from the *Ambinervosae*, and due to discovery of a *Tetrafoliolatae* species east of the previously known range and of new species of sect. *Procumbensae* in east Bolivia and north of the Pantanal (Simpson, 1984).

The area of sects. *Ambinervosae* and *Triseminalae* also showed significant expansion south to the valley of the Jequitinhonha, overlapping with *Caulorhizae* and *Extranervosae*. The monotypic series *Prorhizomatosae* was also extended to the east coast of Brazil.

Table 4.4 lists the number of accessions available at some of the major centres. The collections are non-discriminatory in the sense that both landraces and materials developed by breeders and/or released cultivars were collected, since any present-day genotype could contain genes that may be of use in the future. These collections can be classified into:

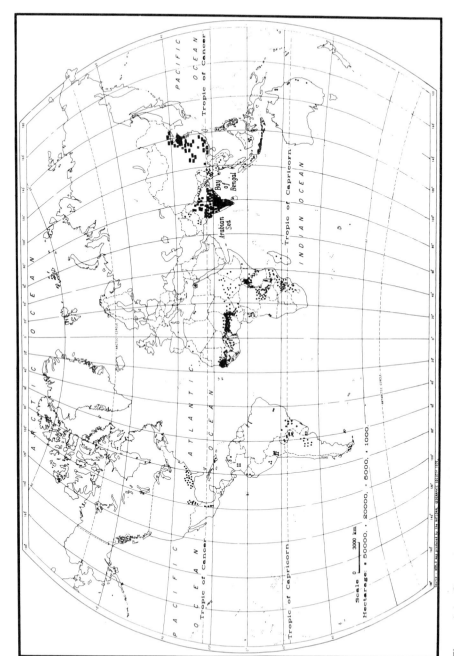

Figure 4.4 Groundnut-growing regions in the world.

TABLE 4.4   *Status of* Arachis *germplasm at different gene banks*

Resource	ICRISAT	Griffin Georgia	Texas A & M	NCSU	Campinas[b]
Accessions	7817	7545	3403		
Landraces	5146	–[a]	–	–	
Named cultivars	230	–	75	–	
Breeding lines	4553	–	11 498	–	
Genetic stocks	131	–	–	–	
Wild *Arachis* spp.	195	498	798	294	596
Taxonomic representation:					
Section					
Arachis	12 044	–	15 204	147	137 (wild)
Erectoides	21	–	68	32	68
Caulorhizae	2	–	34	2	60
Rhizomatosae	90	–	208	76	71
Extranervosae	2	–	63	5	178
Triseminalae	1	–	7	2	7
Procumbensae	0	–	30	–	29
Ambinervosae	0	–	27	2	46
Interspecific derivatives	165	261	705	–	–
Unknown	1651	–	–	28	–

[a] Information not available
[b] Instituto Agronomico, Campinas, Brazil.

- Accessions: world collection assembled from other agencies with an accession number.
- Landraces: non-uniform material (unselected) collected from farmers' fields or purchased in local markets.
- Breeding line: material developed by breeders but not released as cultivars.
- Genetic stocks: genotype identified by special features or sources of resistance to biotic and/or abiotic stresses.
- Wild species: *Arachis* species belonging to one of the eight sections of genus *Arachis*.

The assembly of such large collections is an excellent example of co-operation between various international, national and state agencies. However, there are several gaps in these collections of both wild and cultivated species, as mentioned previously. According to earlier priorities, major areas have been explored, and now countries and regions with limited groundnut cultivation but which may still hold extensive variability

TABLE 4.5   *Priorities for future collections*

Country	Areas to be covered	Justification
Argentina	1. North and north-west	To collect *A. monticola*
	2. Corrientes and Entre Rios	To collect *A. villosa*
Brazil	1. Rios Tapajós and Xingu	Unexplored
	2. North of Ilha do Bananal	Unexplored
	3. The valley of Rio Parnaíba	To collect sect. *Arachis* and *A. hypogaea*
	4. Rio Paraná to coast of Brazil and Rio de Janeiro to Uruguay	To collect sect. *Arachis* and *A. hypogaea*
	5. City of Campina Grande	To collect *A. martii*
	6. Northern part of Mato Grasso	To collect *A. villosulicarpa* and *A. hypogaea*
	7. North-west part of Goiás, Maranhão, northern Piauí state	To collect additional germplasm of *Arachis* and *A. hypogaea*
	8. Ceará, Rio Grande do Norte and Paraíba states	Landraces of *A. hypogaea*
	9. South-eastern Amazon region	"
	10. Rondônia and Acre States	"
Bolivia	1. Northeast Bolivia along the valleys of Mamore and Guaporé	To collect sect. *Arachis* and *Procumbensae*
	2. Bañados del Izozog	
	3. Eastern Bolivia	
Paraguay	1. East of Paraguay river	To collect lost species
	2. West of Paraguay river and Rio Paraguay valley	To collect sect. *Arachis*
	3. North Chaco of Paraguay	To collect sect. *Arachis*
Uruguay	Along Uruguay river	To collect *A. villosa* and landraces of *A. hypogaea*

should be explored to make the collection more comprehensive. It is important to do this before germplasm is lost due to drastic changes in agricultural patterns. This is especially true for South America, which has both wild and cultivated groundnut and where natural habitats are increasingly being disturbed. Table 4.5 presents priorities for future collections, partially based on an IBPGR paper by Simpson (1990).

4.10 GENETIC VARIABILITY

The wide range of genetic diversity in the cultivated and wild *Arachis* species contradicts the earlier apprehensions of lack of variability.

TABLE 4.5 *cont.*

Country	Areas to be covered	Justification
Peru	1. South and southwest	To collect *fastigiata* type
	2. Central and east	To collect *fastigiata* type
	3. North and north-east	To collect *fastigiata* and *hirsuta* types
Ecuador	Western half	To collect Zaruma types
Colombia	West and north	Unexplored and unrepresented
Mexico	Whole country	Landraces and *hirsuta* types
Caribbean islands	Whole countries	Unexplored
China	Unexplored	Landraces and *hirsuta* type
Kampuchea	Whole country	Unexplored and unrepresented
Laos	Whole country	Unexplored and unrepresented
Thailand	Whole country	Underexplored
Vietnam	1. North and northeast	Landraces and *hirsuta* type
	2. Southern state	Landraces
Bangladesh	Whole country	Unexplored and unrepresented
Nepal	Interior regions	Landraces
Central African Republic	Whole country	Landraces for humid tropics
Gabon	Whole country	Early introduction zone
Angola	Whole country	"
Republic Malagasay	Whole country	Isolated
Cameroon	Whole country	To re-collect lost landraces
Namibia	Coastal regions	Early introduction zone
Senegal	Whole country	To re-collect and preserve lost variation

Together they present a considerable amount of diversity. At ICRISAT, and at most other major repositories, a multidisciplinary approach is adopted for characterization of groundnut germplasm, conducted by a team of scientists with expertise in different disciplines who follow a common groundnut descriptor list developed by IBPGR/ICRISAT (1981) and revised subsequently. For these characterizations, emphasis has been given to characters considered to be highly heritable. The different morphoagronomic characters recorded are listed in Table 4.6. These descriptors have also been applied to the 2000 most recent collections from South America (Simpson *et al.* 1992). Other centres have applied these and other

TABLE 4.6 *List of characters recorded*

I		Morphological characterization
	1.	Life form
	2.	Growth habit
	3.	Branching pattern
	4.	Stem pigmentation
	5.	Stem hairiness
	6.	Peg colour
	7.	Standard petal colour
	8.	Standard petal markings colour
	9.	Leaf colour
	10.	Leaflet length
	11.	Leaflet width
	12.	Leaflet shape
	13.	Leaflet hairiness
	14.	Number of seeds per pod
	15.	Pod beak
	16.	Pod constriction
	17.	Pod reticulation
	18.	Pod length
	19.	Pod width
	20.	Seed colour, primary
	21.	Seed colour, secondary
	22.	Seed colour, type of variegation
	23.	Seed length
	24.	Seed width
	25.	Seed mass
II		Preliminary evaluation
	26.	Days to emergence
	27.	Shelling percentage
	28.	Days to maturity
III		Further evaluation
	29.	Oil content
	30.	Protein content
	31.	Reaction to abiotic stresses.
	32.	Reaction to biotic stresses.

descriptor lists to large numbers of germplasm accessions. For example, Pietrarelli *et al.* (1985a, 1985b) evaluated all the germplasm ever collected in Peru and Bolivia; and Howard *et al.* (1985) applied IBPGR/ICRISAT descriptors to a collection of South American germplasm lines. Further evaluation of germplasm for additional descriptors such as reaction to diseases, insect pests, drought and nutritional components has also been carried out at ICRISAT and other locations.

The available germplasm (at ICRISAT as well as other centres) has exhibited a wide range of variation for morphoagronomic characters. Growth habit in wild *Arachis* species and the cultigen have striking similarities: a relatively short main axis and four or more horizontal branches arising from the axil of the cotyledon and the first pair of foliage leaves. Further branching is basically of two types – alternate and sequential. In the alternate type, pairs of axillary vegetative branches are produced alternating with one or two axillary inflorescences; in the sequential type, either a distal sequence of inflorescences or an unbroken sequence of axillary inflorescences is present. Various combinations of the two also occur frequently. *A. monticola* has both types (Gibbons, 1966) and so has *A. hypogaea*. As much as one third of the germplasm collection of *A. hypogaea* is a mixture of alternate and sequential, either within or between plants.

In growth habit, cultigens can be of the procumbent runner type with short (generally) or long main axis and laterals growing horizontally to various lengths. Other genotypes may have decumbent laterals which have an ascending tendency or are erect with shortened internodes in both alternate and sequential branching patterns. These have been designated as D1, D2, D3 and E, according to the angle between main axis and branches. Spanish and valencia are mostly erect while virginias are generally spreading. Valencia and spanish genotypes have few secondary and tertiary branches, while virginias have many such branches. A consistent difference between the two subspecies and botanical types is that spanish and valencia types have inflorescences on the main axis, while pure virginias generally do not. However, recent extensive hybridization has resulted in several intermediate types.

Leaves of virginia types are generally darker green than those of spanish or valencia. In diploid species also, the annuals have lighter green leaves than the perennials, though there is great variation in the depth of colour. Mutations, recombinations and their segregants have evolved many more types of leaf shape and colours.

The range of flower colour in *Arachis* is not great. Flowers are usually orange or yellow, with or without marking, but certain accessions have white and even brick red flowers (the latter only in *A. hypogaea*). Pods in wild *Arachis* species are mostly lomentiform. The isthmus between the seeds can vary from a few millimetres to greater than 10 cm (Gregory *et al.*, 1980). The cultigens have great variability for pod characteristics due to different selection pressures for this commercially important trait. Genotypes differ from deep to almost no pod constrictions, and from prominently reticulated to essentially smooth pod surfaces. Some have a distinct beak while others have no beak at all, and pod length varies from 1 to 9 cm. Similarly, there is considerable variation in seed size. Wild *Arachis* species, *A. cardenasii* and *A. correntina*, have very small seed (but by no means the smallest), weighing about 20–25 mg each. Some of the smallest seeded genotypes of *A. hypogaea* may approach this level, while the largest seed may weigh as much as 1.3 g each. Cultigens also show great variability for testa colour, varying from white to tan, purple, dark red and black, with different types of variegation or self colours.

This range of variability recorded for 12 160 accessions at ICRISAT is summarized in Table 4.7. Further screening of groundnut germplasm to assess the potential – particularly their reaction to different biotic and abiotic stresses – has been carried out worldwide, resulting in identification of a large number of accessions with desirable traits and sources of resistance to many stress factors. The results of such screening at ICRISAT and other major centres are summarized in Table 4.8.

Wild *Arachis* species have shown great variability in both perennial and annual life forms, sequential or alternate branching patterns, and open or compact prostrate habit. The reaction of wild species to various groundnut pests and diseases has shown greater variation than is present in cultivated groundnut and they provide a large number of sources for introgression of disease resistance. Species such as *A. glabrata*, *A. hagenbeckii*, *A. repens* (Gibbons and Bailey, 1967), *A. chacoensis*, *A. villosulicarpa*, *A. sp.* GKP 10596 (Abdou *et al.*, 1974; Foster *et al.*, 1981; Kolawale, 1976) and *A. stenocarpa* (= *A. stenosperma*) (Sharief *et al.*, 1978) were reported to be resistant to early leaf spot. *A. cardenasii* and several other taxa from section *Arachis* (Kolawale, 1976), *Caulorhizae*, *Ambinervosae* and *Rhizomatosae* (Abdou *et al.*, 1974; Subrahmanyam *et al.*, 1985b) were found resistant to late leaf spot. Many accessions of *Arachis* species have been found resistant to rust.

Many wild *Arachis* species have been identified as resistant to viruses. *A. glabrata* and *A. repens* were reported to be symptomless carriers of groundnut rosette virus (Gibbons, 1969). *A. pusilla* (GKP 12922 probably not authentic *A. pusilla*, better designated *Arachis* sp.), *A. correntina* (GKP 9530) and *A. cardenasii* (GKP 10017) showed field resistance to tomato spotted wilt virus, while *A. chacoensis* (GKP 10602) showed no infection either by grafting or by thrips inoculation (Subrahmanyam *et al.*, 1985a). Similarly Herbert and Stalker (1981) found a high level of resistance to peanut stunt virus in species of sect. *Arachis*, *A. duranensis* (K 7988), *A. villosa* (B 22585) and *A. villosa* var. *correntina* (Manfredi 8); *Caulorhizae*; *Erectoides*; and *Rhizomatosae*. For peanut mottle virus, Demski and Sowell (1981) identified six accessions of *Rhizomatosae* as immune, and Subrahmanyam *et al.* (1985) reported no infection after mechanical or air-brush inoculation in GK 12922, *A. chacoensis* (GKP 10602), *A. cardenasii* (GKP 10017) and *A. correntina* (GKP 9530). *A. monticola* (GKPBSSc 30062) was found resistant to cylindrocladium black rot, (Fitzner *et al.*, 1985). Castillo *et al.* (1973) reported resistance to northern rootknot nematodes in three accessions of *Rhizomatosae*, and Nelson *et al.* (1989, 1990) found 55 accessions of 22 species of *Arachis* with resistance to rootknot nematode (*Meloidogyne arenaria*).

A large number of *Arachis* species accessions have also been screened for resistance to insects such as thrips, leaf-hoppers and corn earworms

TABLE 4.7 *Range of variation in cultivated groundnut observed at ICRISAT, Patancheru*

Character	Minimum	Maximum	Intermediate(s)
Life form	Annual	–	–
Growth habit	Erect	Procumbent	Decumbent
Branching pattern	Sequential	Alternate	Irregular
Stem pigmentation	Absent	Present	–
Stem hairiness	Glabrous	Woolly	Hairy, very hairy
Reproductive branch length	>1 cm	10 cm	Continuous
Number of flowers/ inflorescence	1	5	2, 3, 4
Peg colour	Absent	Present	–
Standard petal colour	Yellow	Garnet	Lemon yellow, light orange, orange, dark orange
Standard petal markings	Yellow	Garnet	Lemon yellow, light orange, orange, dark orange
Leaf colour	Yellowish green	Dark green	Light green, green, bottle green
Leaflet length	17 mm	94 mm	Continuous
Leaflet width	7 mm	52 mm	Continuous
Leaflet L/W ratio	1	6	Continuous
Leaflet shape	Cuneate	Lanceolate	Obcuneate, elliptic
Hairiness of leaflet	Subglabrous	Profuse and long	Scarce and short, scarce and long, profuse and short
Number of seeds/pod	1	5	2, 3, 4
Pod beak	Absent	Very prominent	Slight, moderate, prominent
Pod constriction	Absent	Very deep	Slight, moderate, deep
Pod reticulation	Smooth	Prominent	Slight, moderate
Pod length	14 mm	65 mm	Continuous
Pod width	7 mm	20 mm	Continuous
Seed colour pattern	One	Variegated	–
Seed colour	Off white	Dark purple	Yellow, shades of tan, rose, shades of red, grey-orange, shades of purple
Seed length	4 mm	23 mm	Continuous
Seed width	5 mm	13 mm	Continuous
100-seed weight	14 g	136 g	Continuous
Days to emergence	4	18	Continuous
Days to 50% flowering	17	54	Continuous
Days to maturity	75	>155	Continuous
Fresh seed dormancy	0 days	>66 days	Continuous
Oil content	31.8%	55.0%	Continuous
Protein content	15.5	34.2	Continuous

TABLE 4.8 *Results of screening of* Arachis *germplasm at ICRISAT Center and Texas A & M*

Specific trait	ICRISAT		Texas A & M	
	Number screened	Number identified	Number screened	Number identified
Disease resistance:				
Early leaf spot	–	–	2500	28 (2)
Late leaf spot	9400	76 (26)	2500	40 (3)
Rust	9400	141 (35)	1500	12
TSWV	7400	23 (6)	–	–
PMV	1800	2 (2)	–	–
Aspergillus flavus	582	17 (4)	–	–
Pod rot	3222	24 (6)	–	–
Sclerotinia blight	–	–	4100	1 (1)
Wet blotch	–	–	50	35 (4)
Pest resistance:				
Thrips	5000	14 (7)	–	–
Jassids	6500	30 (7)	–	–
Termites	520	20 (6)	–	–
Aphids	300	4 (1)	–	–
Leafminer	930	18 (6)	–	–
Rootknot nematodes	–	–	116	55 (3)
Multiple resistance	9400	85 (45)	2500	12 (3)
Abiotic stresses/nutrition:				
Drought	742	38 (8)	–	–
N Fixation	342	4 (2)	–	–
High oil	8868	44 (10)	–	–
High protein	8868	51	–	–

Numbers in parentheses indicate number commonly used in breeding programmes.

(Stalker and Campbell, 1983), army worm (Lynch *et al.*, 1981), lesser cornstalk borers (Stalker *et al.*, 1984) and spider mites (Leuck and Hammons, 1968; Johnson *et al.*, 1977). High levels of resistance have been identified in a number of accessions (Stalker and Moss, 1987).

Wild *Arachis* species have also been studied for nutritional quality and some species, such as *A. villosulicarpa*, have been reported to have high (1.44–1.66%) tryptophan contents (Amaya *et al.*, 1977). Recently *A.* sp. (KSSc 36008), *A. paraguariensis*, *A. rigonii* and *A. appressipila* have been found with more than 60% oil content, while *A.* sp. GK 12922, *A.* sp. VKR 6110 and *A.* sp. GK 30126 have high linoleic acid levels (Jambunathan *et al.*, 1991).

4.11 CONSERVATION OF GERMPLASM

Most groundnut germplasm has been conserved in the form of either pods or seed except for some accessions (most of which belong to sect. *Rhizomatosae*) which produce very few or no seed. These non-seed producers are being conserved as plants. The major repositories of the world collection of groundnut germplasm are genebanks at ICRISAT; in the USA at the Southern Regional Plant Introduction Station in Georgia, the North Carolina State University at Raleigh and the Texas Agricultural Experiment Station at Stephenville; in Brazil at Empresa Brasileira De Pesquisa Agropecuaria (EMBRAPA)/Centro Nacional de Recursos Geneticos (CENARGEN), Brasilia and Instituto Agronomico, Campinas; and in Argentina at Instituto Botanica del Nordeste (IBONE), Corrientes and Instituto Nacional de Technologia Agropecuaria (INTA), Manfredi.

It is established that the two key factors influencing the seed's viability during storage are temperature and moisture. Duration of viability is mainly influenced by seed moisture content. Control and optimization of these factors can dramatically improve the longevity of seeds in storage. At ICRISAT, and probably in most other gene banks, the seeds are dried to 5–7% moisture content and are stored in controlled-environment chambers with low temperature and humidity, following the guidelines of the International Board for Plant Genetic Resources (IBPGR). Many locations have three storage regimes:

1. Short-term chambers which are maintained at a temperature around 18 °C and relative humidity (RH) of 30–50%. They hold freshly harvested material before it is transferred to chambers with lower temperature and humidity.
2. Medium-term chambers which are maintained at 4 °C and 20–30% RH and conserve almost all the accessions in the form of pods; made up principally of the working and/or active collections.
3. Long-term chambers which are maintained at −20 °C with or without control of humidity. Almost all the germplasm will eventually be stored in these chambers as shelled seeds in closed containers (aluminium pouches at ICRISAT) to comprise the base collection.

At ICRISAT the present emphasis is to bring all the accessions into long-term chambers with a recommended amount of seeds. This will minimize chances of genetic drift and loss of seed viability during conservation. During transfer of germplasm to different storage conditions, utmost care has to be taken to avoid any chances of genetic contamination by mechanical mixing. Pathologists and virologists sample the materials to ensure that the genebanks are stocked only with pathogen-free seeds. Large scale multiplications, which would reduce the number of multiplication generations, can further assist in reducing chances of genetic drift or mutations during conservation; however, low seed multiplication rate and the large seed size of the groundnut are serious constraints.

The preservation of wild *Arachis* species presents special problems, particularly in the case of accessions that produce few, if any, seeds and these only after long periods. Rhizomatous accessions produce very few seeds and are also quite heterozygous. Conservation of these seeds may not represent the total variability present in the parental clones and/or their populations. These accessions are being conserved as plants, which are propagated vegetatively. Conservation of the wild *Arachis* species, even by conventional methods, demands tremendous effort. Because of their perennial nature, they can pose problems of disease and pest build-up if maintained in a glasshouse. In the field, they have to be isolated to avoid outcrossing and mechanical mixing because of their extensive vegetative growth. Therefore most of these accessions in the USA and at ICRISAT are maintained in containers, i.e. pots, baskets or concrete rings. Excess vegetative growth necessitates regular maintenance of these stocks through frequent pruning and repotting.

It is advisable to maintain these stocks of vegetatively propagated species in duplicate and under diverse conditions so that they can adapt and can be suitably conserved without the danger of being lost. This practice is quite successful at present at ICRISAT, the Texas A & M University Agricultural Experiment Station at Stephenville, the Crop Science Department of North Carolina State University, USDA at Griffin, Georgia, EMBRAPA/CENARGEN in Brazil and IBONE and INTA in Argentina. This international team effort has resulted in successful germplasm conservation with free exchange of material and information limited only by the quarantine restrictions of individual countries and by financial resources.

The research on chemical, physical and environmental factors affecting the viability of seed during storage has been very limited as has the utilization of high technology *in vitro* techniques for conservation of germplasm. There are possibilities of overcoming many constraints, particularly in the case of wild *Arachis* species, by using *in vitro* techniques for conservation micropropagation and germplasm distribution. The vegetatively propagated rhizomatous species and many other species that produce very few seeds might be stored as shoots and embryo axes *in vitro* or in cryopreservation. However, before this can be attempted it will be necessary to establish protocols for regeneration of plants from shoots and excised embryo axes. Recently, a protocol was established at ICRISAT for germination of embryo axes excised from healthy mature seeds (Adib *et al.*, 1991) in an attempt to overcome the problem of large seed and pod size, which restricts conservation of large numbers of seeds with the full range of variability. The *in vitro* techniques could also help in overcoming low seed multiplication through micropropagation, but suitable regeneration protocols from various explants causing no genetic instability and culture-induced variation are needed to make it operative on a regular basis. Once this is achieved, the technique could be used in distribution of germplasm, particularly of those wild *Arachis* species which produce very few seeds. Strengthening of research in these areas could make utilization of groundnut germplasm more effective and help fill some of the existing gaps in genebanks.

4.12 UTILIZATION

Breeding activities around the world have made significant progress in the last two decades but exploitation of available genetic variability is still limited. Breeders prefer their known sources of previously identified resistances and are often reluctant to use those recently identified because of lack of information on their reaction to other biotic and abiotic hazards, centre of origin, and whether they have the same components of resistance governed by the same genes/alleles. Poor agronomic potential of these sources is another possible factor contributing to this situation. Hesitation in utilizing new germplasm sources in resistance breeding is dangerous, as the old sources may rest on a very narrow genetic base. Epiphytotics of any pathogen may wipe out crops, causing considerable economic loss and subsequent scarcity of planting materials. Interspecific variability has not been fully exploited, though it has provided encouraging results (Stalker *et al.*, 1979; Singh, 1985b; Singh and Gibbons, 1986; Singh, 1989; ICRISAT, 1986; Simpson, 1991a).

   Progress in breeding for resistance to early and late leaf spot and rust has been significant. Southern runner, a high-yielding cultivar resistant to late leaf spot has been released in the USA. Resistance breeding at ICRISAT has developed high-yielding cultivars ICG(FDRS)4 and ICG(FDRS)10 with strong resistance to rust and low level resistance/tolerance to late leaf spot. Indian national programmes have also released several other lines. Using several wild *Arachis* species from section *Arachis*, ICRISAT breeders have been able to transfer disease resistance into *A. hypogaea*. A large number of *A. hypogaea*-like interspecific derivatives with high yield and resistance to rust and/or late leaf spot have been developed involving *A. cardenasii, A. batizocoi, A. duranensis, A. stenocarpa* (HLK-410, = *A. stenosperma*) and *A. chacoensis* (ICRISAT, 1985; Singh and Gibbons, 1986). Recently, ICRISAT scientists have also identified an interspecific line, 259-2, originating from crosses with *A. cardenasii*, with multiple resistance to rust and to early and late leaf spots (Singh, 1988; ICRISAT, 1990). Resistance to groundnut rosette virus has been identified from an interspecific derivative involving *A. chacoensis* (ICRISAT, 1992). Simpson (1991a) has reported a successful pathway for introgressing leaf spot resistance from *A. cardenasii* and *A. chacoensis* through *A. batizocoi*.

Introgression of nematode resistance is reported by Simpson (1991a) and by Starr *et al.* (1990).

Progress in breeding for resistance to soil-borne fungi, aflatoxin development and virus diseases such as tomato spotted wilt virus (bud necrosis) has been very difficult and slow. Smith *et al.* (1991) have released a cultivar, Tamspan 90, with high levels of resistance to pythium pod rot and sclerotinia blight. The research efforts in these directions have not been adequate but there may also be a dearth of qualitative sources of resistance for most of these diseases.

4.13 CONCLUSIONS AND PERSPECTIVES

Significant progress has been made in biosystematics of section *Arachis* of the genus *Arachis*, leading to identification of cross-compatibility relationships, genomes and phylogenetic relationships between the diploid species of section *Arachis* and the cultivated tetraploid species *A. hypogaea*. This information is of vital importance and has led to the establishment of several breeding strategies for exploitation of genetic resources available in these species. Several species have already contributed useful material. However, progress in taxonomy and biosystematics in the genus *Arachis* as a whole, particularly in relation to the seven other sections of the genus, has been very slow. There is an urgent need to provide an appropriate taxonomic and biosystematic treatment of the genus to avoid perpetuating confusion and to facilitate free communication and better exchange of information and material. This would help towards effective utilization of the vast genetic reservoir offered by the wild *Arachis* species for improving the agronomic and pathogen resistance status of *A. hypogaea*.

A. *hypogaea* is a segmental allopolyploid which originated from the diploid species of sect. *Arachis* with AA and BB genomes very similar to those present in *A. hypogaea*. The wild species of sect. *Arachis* hybridize with *A. hypogaea*, and introgression of desirable genes from them is possible through appropriate ploidy and genomic manipulations which increase effective meiotic recombination.

With regard to groundnut genetic resources, there are positive trends in collection characterization, conservation and utilization. Considerable progress has been made and large numbers of accessions are being conserved in various state, national and international genebanks. However, there is still much potential diversity and there is the urgent need to collect additional germplasm from centres of origin and diversity in South America. Also, areas of early introduction in Africa and Asia are in danger of extinction and need attention. Efforts on characterization and evaluation of germplasm need to be accelerated. Information on preliminary characteristics and genetic potential is essential for utilization otherwise these collections will continue to be underused. Special attention needs to be given to the already identified sources of desirable traits for a thorough evaluation to provide a complete picture of their merits and demerits, so that they can be utilized effectively and with confidence by breeders.

Significant progress has been made in genetic improvement of the groundnut but greater advances can be expected, particularly in the field of breeding for disease resistance. Many sources of resistance to biotic and abiotic stresses have been identified within the cultivated *A. hypogaea* but the materials often lack desired quality characteristics. Efforts to identify or increase variability combined with high quality characters are needed. Considerable attention has been given to evaluation of wild *Arachis* species, which have been found to possess high levels of resistance or immunity to many diseases such as rust, early and late leaf spots, peanut stunt virus, tomato spotted wilt virus, nematodes and other pests. Some of the cross-compatible species within sect. *Arachis* have been exploited successfully and stable 40-chromosome hybrid derivatives have been obtained in several interspecific crosses. The high level of resistance of these derivatives indicate that wild *Arachis* species can contribute much to improvement of the cultivated groundnut. Species outside sect. *Arachis* have not been successfully used because of differences in ploidy level, cross-incompatibility and genomic incompatibility. Methodologies of embryo rescue have not been productive and hence attention should be given to techniques dealing with manipulation at the cellular and molecular level, e.g. protoplast fusion and genetic transformation. However, establishment of highly productive regeneration protocols are essential prerequisites. These will open new vistas for exploitation of not only wild species of genus *Arachis* for introgression of desirable genes but also of genes beyond *Arachis*, present elsewhere in the plant kingdom.

REFERENCES

Abdou, Y.A.M., Gregory, W.C. and Cooper, W.E. (1974) Sources and nature of resistance to *Cercospora arachidicola* and *Cercosporidium personatum* in *Arachis* spp. *Peanut Science* 1, 6–11.

Adib, Sultana, Singh, A.K., Jana, M.K. *et al.* (1991) In-vitro germination of excised embryo of groundnut, in *Groundnut – a global perspective: Proceedings of an International Workshop*, 25–29 Nov. 1991, ICRISAT, Patancheru, India, p. 419.

Amaya, F.J., Young, C.T. and Hammons, R.D. (1977) The tryptophan content in the US commercial and South American wild genotypes of the genus *Arachis*. A survey. *Oléagineux* 32, 225–229.

Amin, P.W., Singh, K.N., Dwivedi, S.L. *et al.* (1985) Thrips (*Frankliniella schultzei* Trybom) and termite (*Odontotermes sp.*) in groundnut (*Arachis hypogaea*). *Peanut Science* 12, 58–60.

Castillo, M.B., Morrison, T.S., Russell, C.L. *et al.* (1973) Resistance to *Meliodogyne hapla* in peanut. *Journal of Nematology* 5, 281–285.

Cherry, J.P. (1975) Comparative studies of seed protein and enzyme of species and collection of *Arachis* by gel electrophoresis. *Peanut Science* 2, 57–65.

Chevalier, A. (1933) Monographie de l'Arachide. *Revue de Botanique Appliquée et d'Agriculture Tropicale* 13, 689–789.

Chevalier, A. (1934) Monographie de l'Arachide. *Revue de Botanique Appliquée et d'Agriculture Tropicale* 14, 565–632, 709–755, 833–864.

Chevalier, A. (1936) Monographie de l'Arachide. *Revue de Botanique Appliquée et d'Agriculture Tropicale* **15**, 637–871.

Demski, J.W. and Sowell, G., Jr (1981) Resistance to Peanut Mottle Virus in *Arachis* spp. *Peanut Science* **8**, 43–44.

Dubard, M. (1906) De l'origine de l'arachide. Bull. Mus. Hist. Nat. Paris. *Journal of Nematology* **5**, 340–344.

Fitzner, M.S., Alderman, S.C. and Stalker, H.T. (1985) Greenhouse evaluation of cultivated and wild peanut species for resistance to Cylindrocladium black rot. *Proceedings of American Peanut Research and Education Society*, **17**, 28 (Abstr).

Foster, D.J., Stalker, H.T., Wynne, J.C. *et al.* (1981) Resistance of *Arachis hypogaea* L. and wild relatives to *Cercospora arachidicola* Hori. *Oléagineux* **36**, 139–143.

Ghimpu, V. (1930) Recherches cytologiques sur les genes. *Hordeum, Acacia, Medicago, Vitis et Quercus. Archives d'Anatomie Microscopique* **26**, 136–234.

Gibbons, R.W. (1966) The branching habit of *Arachis monticola. Rhodesia, Zambia, and Malawi Journal of Agriculture Research* **4**, 9–11.

Gibbons, R.W. (1969) Groundnut rosette research in Malawi, in *Third African Cereals Conference Zambia and Malawi.* Mimeo Report. pp. 1–8.

Gibbons, R.W. and Bailey, B.E. (1967) Resistance to *Cercospora arachidicola* in some species of *Arachis. Rhodesia, Zambia, and Malawi Journal of Agriculture Research* **5**, 57.

Gregory, W.C. and Gregory, M.P. (1976) Groundnuts, in *Evolution of crop plants,* (ed. N.W. Simmonds), Longman Group Ltd., London, pp. 151–154.

Gregory, W.C. and Gregory, M.P. (1979) Exotic germplasm of *Arachis* L., interspecific hybrids. *Journal of Heredity* **70**, 185–193.

Gregory, W.C., Gregory, M.P., Krapovickas, A. *et al.* (1973) Structure and genetic resources of peanuts, in *Peanut Culture and Uses,* (ed. C.T. Wilson), American Peanut Research Education Association, Inc., Stillwater, Oklahoma, pp. 47–133.

Gregory, W.C., Krapovickas, A. and Gregory, M.P. (1980) Structure, Variation, Evolution and Classification in *Arachis,* in *Advances in Legume Sciences,* (eds R.J. Summerfield and A.H. Bunting) Royal Botanical Gardens, Kew, pp. 469–481.

Hammons, R.O. (1982) Origin and early history of peanut, in *Peanut Science and Technology,* (eds H.E. Pattee and C.T. Young), American Peanut Research and Education Society, Yoakum, pp. 1–20.

Harlan, J.R. and de Wet, J.M.J. (1971) Toward a rational classification of cultivated plants. *Taxon* **20**: 509–517.

Herbert, T.T. and Stalker, H.T. (1981) Resistance to peanut stunt virus in cultivated and wild *Arachis* species. *Peanut Science* **8**, 48–52.

Hermann, F.J. (1954) A synopsis of genus *Arachis. Agriculture Monograph USDA* 19, pp. 26.

Hoehne, F.C. (1940) *Leguminosas-Papilionadas.* Genero *Arachis. Flora Brasilica* **25(2)**, 122, 1–20.

Howard, E.R., Higgins, D.L., Thomas, G.D. *et al.* (1985) IBPGR/ICRISAT Minimum Descriptors of *Arachis hypogaea* L. Collections II. The variability within certain characters. *Proceedings American Peanut Research and Education Society.* (Abstract) Vol. 17: 26.

Husted, L. (1933) Cytological studies of the peanut *Arachis* I. Chromosome number and morphology. *Cytologia* **5**, 109–117.

Husted, L. (1936) Cytological studies of the peanut *Arachis* II. Chromosome number, morphology and behaviour and their application to the origin of cultivated form. *Cytologia* **7**, 396–423.

IBPGR (1990) International Crop Network Series. 2. *Report of workshop on genetic resources of wild Arachis species. Including Preliminary Descriptors for Arachis (IBPGR/ICRISAT).* International Board for Plant Genetic Resources. p. 37.

IBPGR/ICRISAT (1981) *Groundnut Descriptors,* International Board for Plant Genetic Resources, Rome, Italy, pp. 1–23.

ICRISAT (1985) *Annual Report 1984,* ICRISAT, Patancheru, India.

ICRISAT (1986) *Annual Report 1985,* ICRISAT, Patancheru, India p. 250.

ICRISAT (1990) *Annual Report 1989,* ICRISAT, Patancheru, India.

ICRISAT (1991) *Annual Report 1990,* ICRISAT, Patancheru, India.

Jambunathan, R., Singh, A.K., Gurtu, S. *et al.* (1991) Assessment of seed quality traits in wild *Arachis* species, in *Groundnut – a global perspective: Proceedings of an International Workshop*, 25–29 Nov. 1991, ICRISAT, Patancheru, India, p. 412.

Johnson, D.R., Wynne, J.C. and Campbell, W.V. (1977) Resistance of wild species of *Arachis* to the two-spotted spider mite, *Tetranychus urticae*. *Peanut Science* **4**, 9–11.

Kawakami, J. (1930) Chromosome number on *Leguminosae*. *Botanical Magazine* (Tokyo) **44**, 319–328.

Klozová, E., Turkova, V., Smartt, J. *et al.* (1983a) Immunological characterization of seed protein of some species of the genus *Arachis* L. *Biologia Plantarum* **25**, 201–208.

Klozová, E., Suachulova, J., Smartt, J. *et al.* (1983b) The comparison of seed protein patterns within genus *Arachis* by polyacrylamide gel electrophoresis. *Biologia Plantarum* **25**, 266–273.

Kochert, G., Halward, T., Branch, W.D. *et al.* (1991). RFLP variability in peanut (*Arachis hypogaea* L.) cultivars and wild species. *Theoretical and Applied Genetics* **81**, 565–570.

Krapovickas, A. (1969) The origin, variability, and spread of the groundnut (*Arachis hypogaea*), in *The Domestication and Exploitation of Plant and Animals*, (eds R.J. Ucko and C.W. Dimbleby), Duckworth, London, pp. 427–440.

Krapovickas, A. (1973) Evolution of the genus *Arachis*, in *Agricultural Genetics, Selected Topics* (ed. R. Moav), National Council for Research and Development, Jerusalem, pp. 35–157.

Krapovickas, A. (1990) Classification of *Arachis*, in *IBPGR 1990. International Crop Network Series. 2. Report of workshop on genetic resources of wild Arachis species. Including Preliminary Descriptors for Arachis* (IBPGR/ICRISAT). International Board for Plant Genetic Resources, Rome, p. 37.

Krapovickas, A., and Rigoni, V.A. (1951) Estudios citologicas en el genero *Arachis*. *Revista de Investigaciones Agricolos* (Buenos Aires) **5**, 289–293.

Krapovickas, A., and Rigoni, V.A. (1960) La nomenclatura de las subspecies y variedadas de *Arachis hypogaea* L. *Revista de Investigaciones Agricolos* **14**, 197–228.

Krapovickas, A., Fernandez, A. and Seeligmann, P. (1974) Recuperation de la fertilidad en un hibrido interspecifico esteril de *Arachis (Leguminosae)*. *Bonplandia* **3**, 129–142.

Krishna, T.G. and Mitra, R. (1988) The probable genome donors to *Arachis hypogaea* L. based on *Arachis* seed storage protein. *Euphytica* **37**, 47–52.

Kumar, L.S.S., D Cruz, R. and Oke, J.G. (1957) A synthetic allohexaploid in *Arachis*. *Current Science* **26**, 121–122.

Leuck, D.B. and Hammons, R.O. (1968) Resistance of wild peanut plants to mite *Tetranychum humidellus*. *Journal of Economic Entomology* **61**, 687–688.

Linnaeus, C. Von (1753) Species plantarum. Laurentii Salviae, *Holmiae*

Lynch, R.E., Branch, W.D. and Gasner, J.W. (1981) Resistance of *Arachis* species to the Fall Armyworm, *Spodepterpa Frugiperda*. *Peanut Science* **8**, 106–109.

Mehan, V.K. (1989) Screening groundnut for resistance to seed invasion by *Aspergillus flavus* and aflatoxin production. *Aflatoxin contamination in groundnuts: Proceedings of the International Workshop*, 6–9 Oct. 1987, ICRISAT, Patancheru, India, pp. 323–345.

Melouk, H.A. and Banks, D.J. (1978) A method of screening peanut genotype for resistance to Cercospora leaf spot. *Peanut Science* **5**, 112–114.

Nelson, S.C., Simpson, C.E. and Starr, J.L. (1989). Resistance to *Meloidogyne arenaria* in *Arachis spp.* germplasm. Supplement to the *Journal of Applied Nematology (Annals of Applied Nematology)* **21**, 654–660.

Nelson, S.C., Starr, J.L. and Simpson, C.E. (1990). Expression of resistance to *Meloidogyne arenaria* in *Arachis batizocoi* and *A. cardenasii*. *Journal of Nematology* **22**, 423–425.

Neucere, N.J. and Cherry, J.P. (1975) An immunochemical survey of proteins in species of *Arachis*. *Peanut Science* **25**, 66–92.

Pietrarelli, J.R., Krapovickas, A., Vanni, R.O. *et al.* (1985a) Los manies cultivados de Peru (The cultivated peanuts of Peru). *XX Jornadas Argentinas de Botanica*, Salta. Sept. 16–20, 1985. (Abstract.)

Pietrarelli, J.R., Krapovickas, A., Vanni, R.O. *et al.* (1985b) Los manis cultivados en Bolivia. (The cultivated peanuts in Bolivia). *XX Jornadas Argentinas de Botanica*, Salta. Sept. 16–20, 1985. (Abstract.)

Pompeu, A.S., (1977) Cruzamentos entre *Arachis hypogaea* e as especies *A. villosa* var.

correntina, A. diogoi e *A. villosulicarpa. Ciencia e Cultura* **29**, 319–321.

Resslar, P.M. (1980) A review of nomenclature of the genus *Arachis* L. *Euphytica* **29**, 813–817.

Resslar, P.M. and Gregory, W.C. (1979). A cytological study of three diploid species of the genus *Arachis* L. *Journal of Heredity* **70**, 13–16.

Resslar, P.M., Stucky, J.M. and Miksche, J.P. (1981) Cytophotometric determination of the amount of DNA in *Arachis* L. Sect. *Arachis* (Leguminosae). *American Journal of Botany* **68**, 149–153.

Seetharam, A., Nayar, K.M.D., Sree Kantaradhya R. *et al.* (1973) Cytological studies on interspecific hybrids of *Arachis hypogaea* × *A. duranensis. Cytologia* **38**, 277–280.

Sharief, Y., Rawlings, J.O. and Gregory, W.C. (1978) Estimates of leafspot resistance in three interspecific hybrids of *Arachis. Euphytica* **27**, 741–751.

Simpson, C.E. (1982) Collection of *Arachis* germplasm (1976–1982). *Plant Genetic Resources Newsletter* **52**, 10–12.

Simpson, C.E. (1984) Plant Exploration: Planning, Organization, and Implementation with Special Emphasis on *Arachis,* in *Conservation of Crop Germplasm – An International Perspective,* (ed. W.L. Brown, T.T. Chang *et al.*), Crop Science Society of America Special Publication, No. 8, pp. 1–20.

Simpson, C.E. (1990) Collecting wild *Arachis* in South America – Past and Future, in *IBPGR 1990, International Crop Network Series, 2, Report of workshop in genetic resources of wild Arachis species. Including Preliminary Descriptors for Arachis (IBPGR/ICRISAT).* International Board for Plant Genetic Resources, Rome, p. 10–37.

Simpson, C.E. (1991a) Pathways for introgression of pest resistance into *Arachis hypogaea* L. *Peanut Science* **18**, 22–26.

Simpson, C.E. (1991b) Global collaborations find and conserve the irreplaceable genetic resources of wild peanut in South America. *Diversity* **7**, 59–61.

Simpson, C.E., Higgins, D.L., Thomas, G.D. *et al.* (1992) *Catalog of Passport data and minimum descriptors of Arachis hypogaea L. germplasm collected in South America, 1977–1986, MP-1737.* Texas Agricultural Experiment Station, Texas A & M University System, College Station, Texas, p. 244.

Singh, A.K. (1985a) Genetic introgression from compatible wild species into cultivated groundnut, *Proceedings of International Workshop on Cytogenetics of Arachis,* 31 Oct.–2 Nov. 1983, ICRISAT, Patancheru, India, pp. 107–117.

Singh, A.K. (1985b) Cytogenetic analysis of wild species of *Arachis. Project Report (1978–82).* ICRISAT, Patancheru, India. p. 71.

Singh, A.K. (1986a) Utilization of wild relatives in the genetic improvement of *Arachis hypogaea* L. 7. Autotetraploid production and prospects in interspecific breeding. *Theoretical and Applied Genetics* **72**, 164–169.

Singh, A.K. (1986b) Utilization of wild relatives in genetic improvement of *Arachis hypogaea* L. 8. Synthetic amphidiploids and their importance in interspecific breeding. *Theoretical and Applied Genetics* **72**, 433–439.

Singh, A.K. (1988) Putative genome donors of *Arachis hypogaea* (Fabaceae) evidence from crosses with synthetic amphidiploids. *Plant Systematics and Evolution* **160**, 143–151.

Singh, A.K. (1989) Exploitation of *Arachis* species for improvement of cultivated groundnut. *Progress Report 1988.* ICRISAT, Patancheru, India. p. 73.

Singh, A.K. and Gibbons, R.W. (1986) Wild species in crop improvement: Groundnut – a case study, in *Advances in genetics and crop improvement,* (eds P.K. Gupta and J.R. Bahl), Rastogi Publication, Meerut, p. pp. 297–308.

Singh, A.K. and Moss, J.P. (1982) Utilization of wild relatives in genetic improvement of *Arachis hypogaea* L. 2. Chromosome complements of species of section *Arachis. Theoretical and Applied Genetics* **61**, 305–314.

Singh, A.K. and Moss, J.P. (1984a) Utilization of wild relatives in genetic improvement of *Arachis hypogaea* L. 5. Genome analysis in section *Arachis* and its implication in gene transfer. *Theoretical and Applied Genetics* **68**, 350–364.

Singh, A.K. and Moss, J.P. (1984b) Utilization of wild relatives in genetic improvement of *Arachis hypogaea* L. 6. Fertility in triploids. Cytological basis and breeding implications. *Peanut Science* **11**, 17–21.

Singh, A.K., Sivaramakrishnan, S., Mengesha, M.H. *et al.* (1991) Phylogenetic relations in section *Arachis* based on seed protein profile. *Theoretical and Applied Genetics* **82**, 593–597.

Singh, A.K., Venkateshwar, A., Moss, J.P. *et al.* (1990) Chromosome number and karyomorphology of some new accessions of genus *Arachis*. *International Arachis Newsletter* **8**, 11–14.

Smartt, J. (1965) Cross-compatibility relationships between cultivated peanut *Arachis hypogaea*. L. and other species of genus *Arachis*. PhD Thesis, North Carolina State University, Raleigh.

Smartt, J. (1990) Grain Legumes, in *Evolution and Genetic Resources*, Cambridge University Press, Cambridge, pp. 30–84.

Smartt, J. and Gregory, W.C. (1967) Interspecific cross-compatibility between the cultivated peanut *Arachis hypogaea* L. and other members of the genus Arachis. *Oléagineaux* **22**, 455–459.

Smartt, J., Gregory, W.C. and Gregory, M.P. (1978a) The genomes of *Arachis hypogaea* 1. Cytogenetic studies of putative genome donors. *Euphytica* **27**, 665–675.

Smartt, J., Gregory, W.C. and Gregory, M.P. (1987b) The genomes of *Arachis hypogaea* 2. The implications in interspecific breeding. *Euphytica* **25**, 677–680.

Smith, O.D., Simpson, C.E., Grichar, W.J. and Melouk, H.A. (1991) Registration of 'Tamspan 90' peanut. *Crop Science* **31**, 1711.

Stalker, H.T. (1981) Intersectional hybrids in the genus *Arachis* between section *Erectoides* and *Arachis*. *Crop Science* **21**, 359–362.

Stalker, H.T. (1991) A new species in section *Arachis* of peanut with a D genome. *American Journal of Botany* **78**, 630–637.

Stalker, H.T. and Moss, J.P. (1987) Speciation, cytogenetics and utilization of *Arachis* species. *Advances in Agronomy* **41**, 1–40.

Stalker, H.T. and Dalmacio, R.D. (1981) Chromosomes of *Arachis* species, Section *Arachis* (*Leguminosae*). *Journal of Heredity* **72**, 403–408.

Stalker, H.T. and Campbell, W.V. (1983) Resistance of wild species of peanut to an insect complex. *Peanut Science* **10**, 30–33.

Stalker, H.T. and Wynne, J.C. (1979) Cytology of interspecific hybrids in the section *Arachis* of peanuts. *Peanut Science* **6**, 110–114.

Stalker, H.T., Wynne, J.C. and Company, M. (1979) Variation in progenies of an *Arachis hypogaea* × diploid wild species hybrid. *Euphytica* **28**, 675–684.

Stalker, H.T., Campbell, W.V. and Wynne, J.C. (1984) Evaluation of cultivated and wild peanut species for resistance to the lesser cornstalk borer (Lepidoptera: Pyralidae) *Journal of Ecological Entomology* **77**, 53–57.

Starr, J.L., Schuster, G.L. and Simpson, C.E. (1990) Characterization of the resistance to *Meloidogyne arenaria* in an interspecific *Arachis* spp. hybrid. *Peanut Science* **17**, 106–108.

Subrahmanyam, P., Ghanekar, A.M., Nolt, B. *et al.* (1985a) Resistance to groundnut diseases in wild *Arachis* species, in *Proceedings of International Workshop on Cytogenetics of Arachis*, 31 Oct.–2 Nov. 1983, ICRISAT, Patancheru, India, pp. 49–55.

Subrahmanyam, P., Moss, J.P. and Rao, V.R. (1983) Resistance to peanut rust in wild *Arachis* species. *Plant Disease* **67**, 209–212.

Subrahmanyam, P., Moss, J.P., and McDonald, D. *et al.* (1985b) Resistance to leafspot caused by *Cercosporidium personatum* in wild *Arachis* species. *Plant Disease* **69**, 951–954.

Subrahmanyam, P., McDonald, D., Redd, L.J. *et al.* (1990) Resistance to rust and late leafspot of groundnut at ICRISAT Center. Problem and Progress. *Proceedings of Fourth Regional Groundnut Workshop for South Africa*, 19–23 Mar. 1990, Arusha, Tanzania. ICRISAT, Patancheru, India, pp. 85–92.

Taubert, P. (1884) *Leguminosae*, in *Die naturlichen Pflanzen familian*, (eds A. Engler and R. Prantl) Teil. Abt. 3: 70–388 (*Arachis*). Verlag von Wilhelm Engelman, Leipzig.

Tombs, M.D. and Lowe, M. (1967) A determination of the sub-units of arachin by osmometry. *Biochemical Journal* **181**, 181–187.

Valls, J.F.M., Rao, V.R., Simpson, C.E. *et al.* (1985) Current status of collection and conservation of South American groundnut germplasm with emphasis on wild species of *Arachis*. *Proceedings of an International Workshop on Cytogenetics of Arachis*, 31 Oct.–2 Nov. 1983, ICRISAT, Patancheru, India, pp. 15–39.

Williams, E.D. (1991) Exploration of Amazonian Bolivia yields rare peanut landraces. *Diversity* **5**, 12–13.

Wynne, J.C. and Halward, T. (1989) Cytogenetics and genetics of *Arachis*, in *Critical Reviews in Plant Science*, (ed. B.V. Conger) CRC Press, Boca Raton, pp. 189–220.

CHAPTER 5

Reproductive biology and development

Peter Coolbear

The possibility that any plant – never mind a major commercial crop – might possess a reproductive strategy whereby fertilized ovaries bury themselves in the soil before seed development can properly commence is, to say the least, unusual. It is understandable, therefore, that the reproductive botany of the groundnut (peanut) remained a source of surprise and confusion well into the middle of this century and that it was not until then that a truly definitive description of this process was published (Smith, 1950). Forty years later, this geocarpic habit still raises a series of intriguing and fundamental questions for seed scientists.

The aim of this chapter is to review the major events in this process from flower development onwards. Special emphasis has been placed on factors affecting seed yield and quality, taking the quality aspects of the discussion through to seed storage and the particular problems of groundnut which result from the fact that 50% of the weight of the seed may be oil reserve.

5.1 FLOWER DEVELOPMENT, POLLINATION AND FERTILIZATION

Groundnuts may flower as early as 3 weeks after planting. Van Rossem and Bolhuis (1954) found floral initiation in the cotyledon axillary buds at a very early stage in the seedling growth of a spanish cultivar, Schwarz 21. Typical flowering patterns for both a virginia and a spanish cultivar are shown in Figure 5.1. In this study, flowering occurred over a period of about 11 weeks, with peak production occurring at around 8 weeks after sowing in the spanish type and after 10 or 11 weeks in the more prolifically flowering virginia type. In both cultivars flowering was clearly cyclical and this pattern was apparently independent of both daylength and environ-

The Groundnut Crop: A scientific basis for improvement. Edited by J. Smartt. Published in 1994 by Chapman & Hall, London. ISBN 0 412 408201.

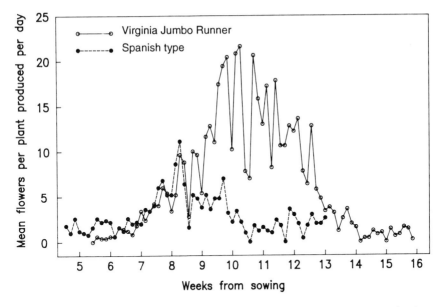

Figure 5.1 Flowering pattern in groundnut. Typical patterns of flower production in two cultivars grown in North Carolina, USA. (Redrawn from Smith, 1954.)

ment *per se*. Bagnall and King (1991 a,b) have recently demonstrated for a range of cultivars that while the onset of flowering is independent of daylength (except under low light and temperature conditions when short days are promotory), the number of flowers produced is greatly enhanced by short days. Flowering rates are also a function of plant size and are promoted by increasing mean temperatures up to about 30 °C and/or increased light intensity. On the other hand, they may be reduced by intraplant competition with developing fruits and environmental stresses of various kinds, in particular a shortage of nutrients (Smith, 1954).

The compressed spical inflorescences of *Arachis hypogaea* comprise at least three flowers but it is unusual for more than one to elongate at the same time, although the second and third flower may then expand on successive days in warm weather. In cool or wet conditions this interval may be greatly increased. Opening usually takes place in the early morning; on the day prior to this, the sessile 6–10 mm flower bud will double in length, after which the rate of expansion growth of the hypanthium more than trebles overnight. At anthesis the flower may be up to 70 mm in length from the base of the hypanthium, a 20 mm standard being borne on a 50 mm hypanthial tube (Figure 3.5). During this time the style elongates in pace with the hypanthium and remains close to the dehiscing anthers. Self-pollination normally occurs either just before or at the same time as flower opening, although it should be noted that some wild species of groundnut, (e.g. *A. lignosa*) may require insect pollination (Banks, 1990).

Generally, in *A. hypogaea*, groundnut ovaries contain between two and five ovules, most commonly two (Figure 5.2(a)). However, development of only a single seed in the fruit is frequent, as is the production of 'pops' – empty pods. In certain conditions, losses of potential yield can be more than 90% (Smith, 1946). The reasons for this will be explored later.

Sporogenesis and gametogenesis occur 2 days prior to anthesis when bud length is around 5 mm (Smith, 1956a). There is general agreement in the literature that the antipodal cells degenerate several hours before fertilization takes place (Smith, 1956a, Periasamy and Sampoornam, 1984; Pattee and Mohapatra, 1987). This is in marked contrast to many other legumes where these cells persist for some time (usually days) after fertilization and are thought to play a key role in initial endosperm development, possibly by providing a readily available supply of specific nutrients or endogenous plant growth regulators as they break down (e.g. Chen and Gibson, 1971). A well established feature of the embryo sac is that it contains a large quantity of starch grains which will provide the nutrients for early endosperm growth (Reed, 1924; Smith, 1956b). Details of embryo sac anatomy are shown in Figure 5.2(b).

Considering the distance it has to grow, pollen tube extension is rapid. Germination of pollen is promoted by red light, auxin, gibberellic acid (GA) and many of the phenols present as glycoside derivatives in the stigma and style. Postgerminative pollen tube growth requires both transcriptional and translational activity and is promoted by ethylene and red light, although GA can substitute for the phytochrome response in darkness (Malik and Chhabra, 1978). Smith (1956a) estimated that, on average, a pollen tube took about 9 h to reach the ovary and fertilization occurred 10–18 h after pollination. Later authors dispute his view that the synergids disintegrate prior to this event. Periasamy and Sampoornam (1984) suggest that the cellular synergids are essential to this process and that pollen tube penetration actually occurs through the vacuole of one of them, although Pattee and Mohapatra (1987) were unable to confirm this. In Smith's (1956a) study less than 7% of ovules failed to be fertilized, although there was a small additional percentage where there was successful fusion of only one of the two pollen nuclei. Malik and Chhabra (1978), however, cite very much lower fertilization percentages in India. In their survey, successful pollination is quoted as being between 5% and 59% in spreading types and 22–68% in bunch forms with extracted pollen having very poor viability. It is unclear whether this difference between India and the USA is a function of local varieties or environment.

5.2 FRUIT DEVELOPMENT

After fertilization the flower withers rapidly and over a period of several days the fruit develops as a positively geotropic stalk-like structure, the

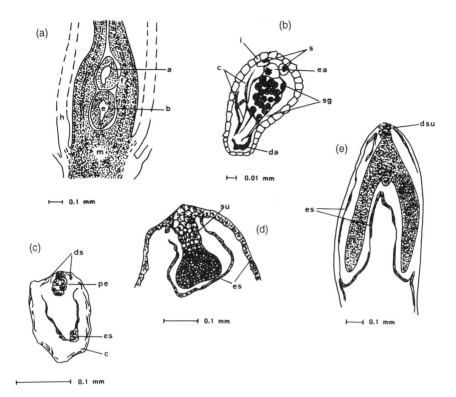

Figure 5.2 Embryology of groundnut (a, apical ovule; b, basal ovule; c, cytoplasmic strands; da, degenerating antipodals; ds, degenerating synergids; dsu, degenerating suspensor; e, embryo sac; ea, egg apparatus; f, funiculus of apical ovule; h, hypanthium; i, integumentary tapetum; m, meristem; pe, pro-embryo; s, synergids; sg, starch grains; su, suspensor): (a) an ovary at anthesis; (b) embryo sac at anthesis, with plentiful starch grains enclosed in compartments defined by cytoplasmic strands; (c) four-cell stage in embryogenesis; (d) heart-shaped embryo stage; (e) embryo after cotyledon initiation. The endosperm tissue remains poorly differentiated, parts of it remaining acellular throughout development. ((a)–(d) are interpretations of plates in Pattee and Mohapatra, 1987; (e) in Periasamy and Sampoornam, 1984.)

'peg', which eventually penetrates the soil. The peg is often described, even in recent literature (e.g. Ziv 1981; Amoroso and Amoroso, 1988), as a gynophore. As Smith (1950) points out, this is technically incorrect because the ovary is sessile: there is no ovary stalk to elongate. The cells which elongate comprise the basal tissue of the ovary itself and thus a more appropriate term for the peg is a carpophore (Smartt, 1976).

Growth of the carpophore is intimately associated with embryo development. Peg extension is slow at first, taking 5–6 days to penetrate the bracts

(Smith, 1950). As the fruit reaches 3–4 mm in length it becomes positively geotropic and starts to grow towards the soil (Periasamy and Sampoornam, 1984), the rate of elongation increasing rapidly between 5 and 10 days after fertilization (Figure 5.3). By the time this rapid elongation starts, the embryo, which is now in the 8–12 cell stage, becomes quiescent (Smith, 1956b). Reduction in embryo growth may simply be a function of nutrient deficiency due to the high demand of the rapidly growing carpophore, or there may be more subtle control via pH changes within the ovary apex (Schenk, 1961). Zamski and Ziv (1976) have demonstrated that initial proembryo development is an essential prerequisite for peg elongation and suggest that auxin and GA produced by the proembryo promote its growth. Ziv (1981) provides evidence which indicates that carpophore growth is also mediated by phytochrome. The carpophore is actively photosynthetic, containing two or three layers of chlorenchyma with abundant stomata on the endodermis, these later being replaced by lenticels as secondary thickening occurs.

Growth usually continues until penetration of the soil occurs 8–14 days after fertilization (Schenk, 1961). Zamski and Ziv (1976) found that, in the virginia runner type of groundnut, light inhibited pod formation and pegs failing to reach soil might grow up to 20 cm before withering and dying. They were also able to demonstrate that the mechanical stimulus of the peg entering the soil is important for successful pod development, stimulating ovary wall thickening and the diageotropic orientation of the pod due to rapid cell enlargement on the dorsal side of the carpel near the basal ovule. Once pod development starts, seed development resumes. If an ovule is unfertilized there is generally little pod development around that ovule (Periasamy and Sampoornam, 1984). Typically, basal ovules resume growth 1–2 days before apical ones and endosperm growth recommences 1–2 days before embryo growth (Smith, 1956b; Pattee and Mohapatra, 1987) (Figure 5.3).

Depth of pod placement depends on soil structure (due to both its mechanical properties and light penetration), but usually pods are placed about 5 cm below the soil surface. Periasamy and Sampoornam (1984) noted that the presence of light may not be completely inhibitory to pod development in all cultivars. In the erect valencia groundnut some aerial pod development does occur and viable, small seeds may be produced. This is unusual and in general the successful development of high quality seed is dependent upon soil penetration by the peg.

Schenk (1961) undertook a detailed study of the underground development of whole fruits in both a virginia bunch type and a spanish type. Fresh weights of whole pods increased very rapidly during the first 2 weeks of subterranean growth and they attained maximum size after 3 weeks. Subsequently there was little increase in total fresh weight, although it fluctuated with soil moisture status. Dry weights increased more gradually, reaching a maximum after 8 weeks in virginia bunch (Figure 5.4) and after

6 weeks in the faster developing spanish type. Respiration rates were highest during the period of maximum pod growth (Figure 5.4). After 2–3 weeks there was very little change in shell dry weights, although fresh weights decreased rapidly as tissues collapsed and structural changes occurred as the shells hardened. Hull colour changes were a useful indicator of pod maturity (section 5.4.2). The seeds themselves tended to develop more slowly, following a typical pattern which will be discussed in detail in the next section. In this study, spanish pods were ready for lifting after about 7 weeks of underground development, and virginia bunch pods after 11 weeks. Times vary depending on cultivar and production conditions but, in general, runner types take about 180 days from planting to

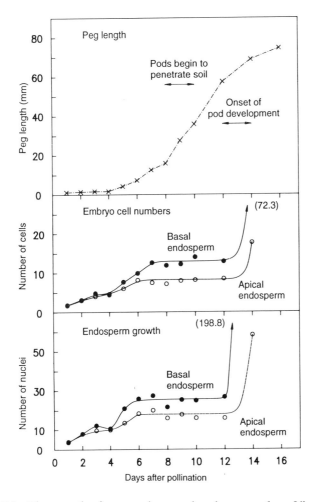

Figure 5.3 The growth of peg, embryo and endosperm of cv. Virginia Runner during peg elongation and soil penetration. (Data from Smith, 1956b.)

maturity in the US, compared with 130–135 days for virginia bunch and around 120 days for spanish types (Woodroof, 1973).

At digging, pods may still have a moisture content of 35–55% and are usually left to dry in the field until kernel moistures are down to 18–25%, when they are brought in and artificially dried for storage. Late digging and prolonged windrowing before combining both result in potential loss of yield as pods become detached from the carpophore (Schenk, 1961; Woodroof, 1973).

5.3 SEED DEVELOPMENT

In most orthodox seeds, whether endospermic or non-endospermic, there are three clearly defined phases of development (e.g. Hyde *et al.*, 1959). A phase of rapid growth with concomitant increases in both fresh and dry weight (Stage I) is followed by a period of food reserve accumulation, where the rate of dry weight increase is often constant but seed moisture contents begin to fall (Stage II). The seed then loses water in the third, ripening stage, usually with little change in dry weight (Stage III).

These three stages summarize the different morphological and physiological processes which occur during development and provide useful reference points for subsequent discussion. During Stage I, growth is

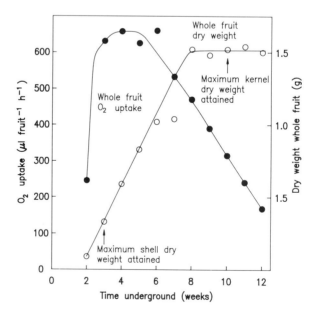

Figure 5.4 Oxygen uptake and dry weight increases by whole fruits of virginia bunch groundnuts during development. (Redrawn from Schenk, 1961.)

mainly by cell division. At the end of this stage all the key structures of the seed are present, even if not fully differentiated. Growth in Stage II is thus by cell enlargement as seed reserves are laid down. The loss of moisture content reflects the increasing proportion of insoluble food reserves in storage tissue. At the end of this period, maximum seed dry weight is attained. This point has been variously referred to as morphological (Anderson, 1944), functional (Grabe, 1956) or physiological maturity (Shaw and Loomis, 1950). It is perhaps unfortunate that the last of these terms is the one in general use, as the physiological development of the seed may be far from complete at this time. Many seeds require the stimulus of further drying to become fully germinable (e.g. Kermode *et al.*, 1986). It is generally accepted, however, that a seed's maximum vigour potential (i.e. its potential to produce a strong and healthy seedling under variously stressful field conditions) is in place at this time.

Because of the obvious difficulties of working with an indeterminate crop with the added complication of fruits developing underground, few workers have characterized the growth of groundnut kernels in these terms. A study of this type, using tagged carpophores just before they entered the soil, was undertaken by Pickett (1950) but he estimated that there were variations in physiological age within samples of up to 7 days. Data shown in Figure 5.5 were obtained by Aldana *et al.* (1972) by harvesting only fruits produced by the second internode of the first lateral reproductive branch of field grown plants of cv. NC-2 (a virginia bunch variety). Although data collection did not commence until 4 weeks after the buried ovary had become diageotropic (at which stage the endosperm has been fully absorbed), it is possible to identify the general pattern of seed development outlined above. For these seeds, stage I was completed 8–9 weeks after flowering when the increase in dry weight became linear. Maximum dry weight was achieved at 13–14 weeks. At 14 weeks after flowering, these fruits were considered ready for lifting, at a time when Stage III of seed development had barely begun. On inverting, seed moisture contents are usually higher than 35% (Woodroof, 1973) and Stage III is completed during windrowing. Depending on cultivar and local environment, Stages I and II of seed development may be completed 8–14 weeks after flowering.

Although the above description summarizes the broad pattern of seed development, it should be remembered that each component – seed coat, endosperm and embryo – are genetically different and each follows its own programme. The details of the developmental pattern for each system are discussed individually below.

5.3.1 Embryology

After fertilization the contents of the zygote become much denser and the cell enlarges. In common with most species, groundnut cell division lags behind that of the endosperm, the first mitosis not occurring until the beginning of the second day after fertilization (Smith, 1956b). The basal cell undergoes a second mitosis within 24h but another day may elapse before the terminal cell divides. However, as mentioned previously, after 5 days the quiescent period for both embryo and endosperm growth occurs with rapid extension growth of the peg. Smith (1956b) found growth resuming 10–12 days after pollination (Figure 5.3). Having observed the contraction of the embryo sac during the quiescent phase, Pattee and Mohapatra (1987) concluded that the system was under considerable metabolic stress at this time. This suggestion tallies with the extensive losses in potential yield which occur during pegging (section 5.6.1).

The suspensor of the developing groundnut seed is typically large. Early growth of this organ is characterized as much by cell enlargement as by cell division, while the cells derived from the terminal cell will continue to divide without much expansion to produce a tight ball. After 3–4 weeks, the undifferentiated globular embryo may consist of 500–1000 cells. At this time the suspensor is likely to have around one tenth of this number of

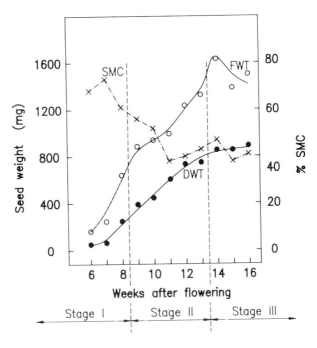

Figure 5.5 Changes in seed weight (less testa) and seed moisture content (SMC) during groundnut seed development. Material for this study was harvested from fruits produced at the second internode of the first reproductive lateral in var. NC-2. The characteristic stages of seed development are indicated on the graph. (Calculated from Aldana *et al.*, 1972.)

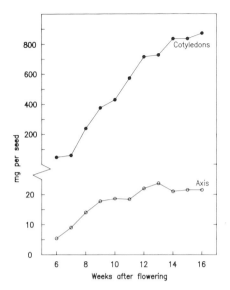

Figure 5.6 Dry weight changes of the embryo axis and cotyledons during seed development in groundnut var. NC-2. (Redrawn from Aldana *et al.*, 1972.)

cells; in contrast, there may now be 5000 nuclei in the acellular endosperm (Smith, 1956b). From this stage onwards the embryo will successively become heart-shaped (Figure 5.2(d)) and differentiate well defined cotyledonary outgrowths which arise at opposite sides of the proembryo (Figure 5.2(e)). Being constrained by the embryo sac and the developing seed coat, the cotyledons curve round in such a way that they come together in parallel. Between them lies the plumule, which is a complex meristematic region showing not only a terminal bud but also two lateral buds in the axes of the cotyledons and several well developed leaf primordia (Reed, 1924; Gelmond, 1971a; Periasamy and Sampoornam, 1984). The radicle and hypocotyl are discernible from the suspensor region only at a late stage (Reed, 1924) and protrude from the base of the cotyledons, making the seed highly susceptible to mechanical damage (Gelmond, 1971a). Figure 5.6 shows how weight changes in axis and cotyledons differ as development proceeds. Most of the dry weight of the axis has been attained by the end of Stage I, while the bulk of cotyledon growth is in Stage II, reflecting the storage function of this tissue. Schenk (1961) noted that seeds of virginia bunch groundnuts were capable of germination about 6 weeks after the end of Stage I.

5.3.2 Endosperm development

The triploid endosperm nucleus begins to divide within a few hours of fertilization (Smith, 1956b) and the two daughter nuclei migrate to opposite poles of the embryo sac. Each nucleus takes half the starch grains with them (Periasamy and Sampoornam, 1984). Despite early descriptions, subsequent endosperm development is entirely coenocytic. After 4 days the endosperm reaches the 16 nuclei stage, by which time the starch grains of the embryo sac had disappeared in the virginia and valencia cultivars described by Smith (1956b) and Periasamy and Sampoornam (1984), but they have recently been observed to persist much longer in the Indian cultivar NS-5 (Suryakumari *et al*, 1989).

As the embryo sac elongates, the endosperm remains as strands of largely peripheral multinucleate cytoplasm, although another film of endospermic cytoplasm surrounds the embryo (Figure 5.2(c)–(e)). Cellularization of the endosperm does not begin until 5 or more weeks from fertilization, when the embryo has begun to develop its cotyledonary outgrowths. However, there still seems to be debate in the literature concerning the extent of the cellularization process. Periasamy and Sampoornam (1984) consider that cellularization only occurs at the chalazal end of the embryo sac, but both Prakash (1960) and Suryakumari *et al.* (1989) report endosperm cellularization right to the suspensor. Periasamy and Sampoornam (1984) suggest the possibility of confusion of endosperm tissue with the inner layers of the developing seed coat and the light and scanning electron micrographs (SEMs) of Pattee and Mohapatra (1987) would appear to bear out this suggestion. Nevertheless some SEM pictures from these authors seem to indicate some cell wall formation near the embryo, although they themselves refrain from speculation, recognizing the problems of fixation and shrinkage in this fragile tissue. It is quite likely, too, that there is some genotypic variation between cultivars. For instance, Suryakumari *et al* (1989) described the appearance of endosperm nodules at an early stage of embryo development. These were discrete structures within the endosperm containing four to five nuclei. There appears to be no mention of this type of structure in other descriptions.

Whatever the case, it seems clear that cellularization of the endosperm is incomplete and, in common with many other legumes, this tissue is wholly absorbed as the cotyledons enlarge in Stage II of development.

5.3.3 Seed coat

Although the coat of the mature dry seed is thin and papery, it may comprise as many as 25 layers of collapsed parenchymatous cells. There is a considerable divergence of opinion about the origin of these layers. Suryakumari *et al.* (1989) suggest that they are entirely derived from the outer integument; however, Periasamy and Sampoornam (1984) suggest that the major part of the seed coat is chalazal in origin. Unfortunately the photographic evidence for this is inconclusive in their paper while none is presented by Suryakumari *et al.* (1989). While this type of development has been recognized in other families, it does not seem to have been reported previously in the *Leguminosae* (Corner, 1976).

Schenk (1961) demonstrated that seed coats attain maximum dry weight by the end of Stage I of development. However, their moisture contents and metabolic activity are still high. During Stage II skins begin to darken and tannin contents increase considerably as dry weights start to decrease. About 8–10 weeks after fertilization, they lose moisture and collapse. Interestingly the progress of seed coat development in the virginia bunch type followed a very similar time course to the spanish type, despite the fact that kernel development was so much more rapid in the latter cultivar.

There is an extensive network of eight to ten bundles of vascular tissue through the seed coat (Periasamy and Sampoornam, 1984). This network is crucial in supplying nutrients to the developing embryo during Stage II of development. Schenk (1961) notes that the vascular strands in the skins of virginia bunch types persist after the rest of the testa tissue has collapsed around the middle of Stage II. There are, of course, no direct vascular connections to the embryo, nutrients being unloaded into the apoplast surrounding the developing cotyledons.

5.4 THE BIOCHEMISTRY OF SEED DEVELOPMENT

Figure 5.7 shows the patterns of DNA, RNA and protein accumulation in cotyledons and axes of developing kernels described by Aldana *et al.* (1972). DNA levels reached a maximum in both tissues after 8 weeks of development in these seeds, confirming that embryo growth by cell division was essentially complete at the end of Stage I of seed development. The loss of RNA from the cotyledons between weeks 10 and 13 (Figure 5.7(b)) remains unexplained. Levels of ribonuclease, although variable, were high at all stages of development. The observation of an increase in RNA levels after the seed was ready for harvesting was complemented in later work by Pattee *et al.* (1981), who found a dramatic fall in cotyledonary ribose as seed matured. It is possible that these late changes may represent the onset of germinative activity in an over-mature seed, although actual radicle emergence is likely to be inhibited by dormancy in most cases (section 5.5).

As shown in Figure 5.6, the cotyledonary reserves accumulate during Stage II of development. The mature seed comprises approximately 30% protein, around 50% lipid and 12% starch (Bewley and Black, 1978). Pickett (1950) determined the time course of accumulation of these components in virginia bunch type (Figure 5.8), although his data may be confounded by high variation in maturity classes at each harvest.

Seed storage protein is rich in arginine, aspartate, glutamate and glycine, but methionine levels are low (Yatsu and Jacks, 1972). There are particularly high levels of free glutamate during development, while the high arginine levels found in immature kernels rapidly decrease as the seed

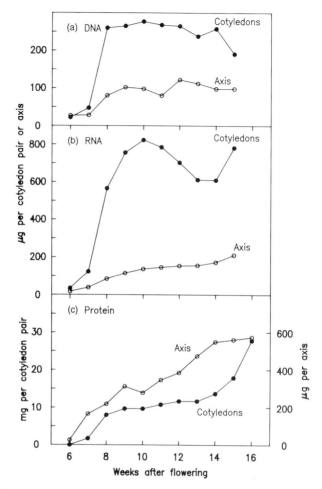

Figure 5.7 Accumulation of DNA, RNA and protein in axes and cotyledons during development of groundnut kernels, var. NC-2. Note change of scale for different tissues in graph (c). (Redrawn from Aldana *et al.*, 1972.)

approaches maturity (Mason *et al.*, 1969; Young, 1973). Major storage proteins comprise a 12–14 S legumin type glycoprotein (molecular weight ≈350 kD), two or more vicilins (mwt between 140 and 190 kD) and small amounts of a 2S globulin (mwt ≈20 kD) (Derbyshire *et al.*, 1974).

The pattern of protein accumulation found by Aldana *et al.* (1972) in a similar type (Figure 5.7(c)) represents a study from more homogeneous material than that used by Pickett (1950) and shows some interesting differences from the pattern shown in Figure 5.8. Data from the study by Aldana *et al.*, (1972) show a triphasic pattern of protein accumulation in the cotyledons. This is similar to that found in maize endosperm by Ingle *et*

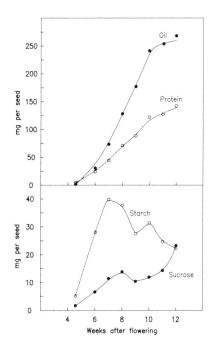

Figure 5.8 Changes in the chemical composition of whole kernels during the development of virginia bunch type groundnuts. (Redrawn from Pickett, 1950.)

al. (1965) where a lag phase in net protein accumulation during early Stage II resulted from the interconversion (via breakdown and reassembly) of proteins no longer required for cell division to storage proteins. If this figure is compared with the changes in dry weight occurring in the same seeds (Figure 5.5), it can be seen that extensive protein accumulation is still occurring in early Stage III of seed development, i.e. after the seed has reached maximum dry weight.

The unusually high lipid content of mature groundnut kernels is made up largely of triglycerides whose major fatty acid components are 18:1, oleic (≈50%) and 18:2, linoleic (≈24%) acids, but 16:0 through to 24:0 saturated fatty acids are also present in significant amounts, as is 20:1 unsaturated fatty acid (Worthington *et al.*, 1972; Bewley and Black, 1978). Variation between genotypes is considerable: for example, in their survey of 82 lines, Worthington *et al.* (1972) found oleic acid contents varied between 36% and 69% and linoleic acid between 14% and 40%. Linolenic acid (18:3) is usually present in only small amounts (between 0.04% and 0.5% – Pattee *et al.*, 1970 and Wallerstein *et al.*, 1989, respectively). Environmental factors during seed production may also affect the oil content of seeds (section 5.7.1).

5.4.1 The metabolism of seed development

Little work seems to have been done on the cellular biochemistry of the seed during development. Both protein and lipid are stored in membrane-bound bodies within the cotyledons. It is likely that the site of both lipid accumulation and storage protein synthesis is the endoplasmic reticulum, with dictyosomes being responsible for the accumulation of storage protein in vacuoles. However, the details of these processes have yet to be worked out (Bewley and Black, 1978). Yatsu and Jacks (1972) made a good case suggesting that the lipid-containing spherosomes in groundnut and other seeds are bounded by half unit membranes. More recently, Cao and Huang (1986) have demonstrated that diacylglycerol acyltransferase (an enzyme which would be required for triacylglyceride assembly in the lipid bodies) is localized in the rough endoplasmic reticulum of maize and probably other seeds. This suggests that, during lipid body development, triacylglycerides are inserted into the middle of the lipid bilayer of endoplasmic reticulum membranes.

Most studies on groundnut seed development have, understandably, focused on the flavour producing compounds in the kernel. Unfortunately, these data have often been obtained from mixed samples of pods of varying maturity and cannot be related precisely to stages of development post-anthesis. Thus Mason et al. (1969) demonstrated that, while many free amino acids remained relatively constant during development, there was a slow but steady increase in phenylalanine as seeds matured and, as already mentioned, there were dramatic decreases in arginine levels such that the amount in the kernel had fallen approximately ten-fold from its levels in early development. Similarly, they demonstrated that sucrose levels fell during development, then rose again as seeds finally matured, confirming earlier findings by Pickett, 1950 (Figure 5.7(b)). Sucrose is by far the most predominant soluble carbohydrate, comprising about 80% of the total at maturity, with stachyose forming another 16% (Pattee et al., 1981). Starch levels peak around the end of Stage I of seed development and then begin to fall (Figure 5.7(b)). A high proportion of the starch during early development is present in the seed coat (Schenk, 1961; Suryakumari et al. 1989).

Research on volatiles present in developing groundnut seeds by Pattee et al. (1970) indicate that the maximum metabolic activity of seeds occurs during Stage II of seed development and is probably related to larger seed size while moisture contents are still adequate, rather than high levels of metabolic activity per se. Schenk's (1961) data for pod respiration show that this reaches a maximum around the beginning of Stage II of seed development (Figure 5.4). During active lipid synthesis the respiratory quotient of developing kernels approaches two, falling to unity as this process stops. One of the most noteworthy aspects of the study by Pattee et al. (1970) was the high level of lipoxygenase activity detected from the ninth week of development onwards: the activity of this enzyme might be expected to reduce the levels of polyunsaturated fatty acids accumulated by the maturing seed. Levels of conserved lipoxygenase in dried kernels may be important in the subsequent storage behaviour of seeds (section 5.8.2).

5.4.2 Maturity indices based on the process of development

A major problem in groundnut production is determining the appropriate time to harvest. Inevitably, because of the plants' extended flowering pattern, there will be a wide range of maturity classes within the crop. Immature kernels tend to produce off-flavours, especially when cured at high temperatures (e.g. Kramer et al., 1963). Various attempts have been made to develop rapid methods of determining kernel maturity, some of which are based on biochemical changes occurring during development. These include spectrophotometric analysis of whole seeds to screen for the high xanthophyll levels found in immature cotyledons (Kramer et al., 1963); assays of volatiles produced by seeds, the levels of which decrease as seeds mature (Pattee et al., 1970); density measurements of cured kernels which decrease as they mature, due to the fact that tissues of mature seeds are less likely to collapse during heating (Miller and Burns, 1971); or screening for the high arginine levels found in immature seeds (Young, 1973). However, the most useful method seems to be based on measuring the darkening of internal hull tissues during maturation (e.g. Miller and Burns, 1971). It has recently been shown that this browning is due to flavonoid metabolism in the maturing pod, especially an increase in the pigment luteolin. As such changes are not normally affected by variations in environment during seed production, this simple method of maturity determination is more reliable than most (Daigle et al., 1988).

5.5 SEED DORMANCY

Dormancy may occur in many varieties of groundnuts and is found in both bunch and runner types (Toole et al., 1964). In general the presence of dormancy in groundnut is not a practical problem as it will usually wear off in storage before the next planting season. However, lack of dormancy in some lines can cause vivipary in areas when rains delay harvest, e.g. Maharashtra in India (Bhapkar et al., 1986) and parts of Africa (Smartt, 1976).

Various authors (e.g. review by Bhapkar et al., 1986) have demonstrated that the coat-imposed component of groundnut seed dormancy involves water soluble germination inhibitors and, in some cases, resistance to initial water uptake by the dry seed. Sengupta (1989) has suggested that the inability to leach phenolic compounds rapidly on imbibition may be one cause of seed dormancy. Despite the assertions of Bhapkar *et al.* (1986), groundnut dormancy mechanisms do not reside entirely in the seed coat: there is also a component of 'true', embryo-based dormancy. Toole *et al.* (1964) clearly demonstrated that, whilst seed coat removal improves germinability of many maturity classes, ethylene may be required to obtain full germination of either mature intact seeds or immature decoated ones. In a single lot varietal comparison, they demonstrated that there may be considerable differences in dormancy between varieties, but it is equally clear that different lots of the same variety may show differing dormancy characteristics depending on production conditions. For example, dormant seeds may lose dormancy if digging is delayed and they are allowed to remain in the soil after maturity. As a general rule, immature seeds of a dormant line have deeper dormancy than mature ones, while basal seeds are more dormant than apical seeds in the same maturity class.

In a series of papers, Ketring and Morgan (1970, 1971, 1972) have investigated the embryo component of groundnut seed dormancy in detail. They have shown that the evolution of ethylene by the embryo is an essential prerequisite for germination and have estimated that groundnut seeds must be able to produce 2–3 nl/g fresh weight in order to break dormancy, the required threshold ethylene concentration for basal seeds often being twice that of apical seeds. The germination promoting effects of GA or, more potently, cytokinin are via their promotion of ethylene production. Surprisingly, auxin, known to promote ethylene production in vegetative tissue, seems to be ineffective in groundnut seeds. Applied abscisic acid (ABA) maintains the dormancy of groundnut by inhibiting ethylene production, although its effect can be competed out by cytokinin (Ketring and Morgan, 1972). It is well established in other species that one of the major options available to plants to prevent seed germination during development is endogenous ABA (e.g. Kermode *et al.*, 1986), although this has not been confirmed in groundnut *in vivo* – all Ketring and Morgan's evidence was gained from studies with exogenously applied ABA. Seeds with little or no dormancy can thus be expected to have put their ethylene synthesis metabolism in place during development and also to possess either low levels of ABA or reduced sensitivity to this plant growth regulator.

Sharma and Sengupta (1988) have shown that a non-dormant groundnut line possessed higher levels of α-amylase and protease during seed development than a dormant cultivar. While it is unlikely that this hydrolase activity (almost certainly sequestered in the mature seed) is directly related to dormancy control, these results are interesting in that they may represent other effects of ABA in the dormant line.

TABLE 5.1 *Abortion during seed development in cv. Virginia Jumbo Runner, plants grown in adequate calcium (data reworked from Smith, 1954)*

	Number of ovules or seeds	Losses at each stage (% total)
Available ovules[a]	4774	–
Fertilized ovules	4454	6.7
Ovules placed below the soil surface	2907	25.2
Ovules resuming development after quiescent phase	1093	45.2
Seeds completing development	544	11.5

[a] Based on the number of flowers observed, assuming two ovules per flower.

5.6 LOSSES OF POTENTIAL YIELD

Actual seed yields in groundnut may be as low as 10% of potential yield based on the numbers of flowers produced. Accordingly, a discussion of seed development would be incomplete without some consideration of the reasons for this loss. Much more detailed discussions of groundnut agronomy will be presented in later chapters.

5.6.1 Losses of potential yield prior to underground pod development

As has already been described, low pollination efficiency may be a major limiting factor for seed production in some parts of the world, but even when most ovules become fertilized, considerable intraplant competition means that abortion can occur throughout development. Other stages at which potential yield was lost in Smith's (1954) study are summarized in Table 5.1. Only 68% of flowers produced pegs, most of the rest remaining as dormant fertilized ovules which may persist for several weeks before withering away. However, if these ovules resume growth unusually high seed losses occur. Pegging will then take place and development will proceed as normal. Bunting and Elston (1980) noted that in some of the older cultivars less than 20% of flowers may produce pegs: carpophore elongation on the first flower of the inflorescence inhibits the development of fruits of other fertilized flowers on the spike. Similarly, late inflorescences have little chance of producing successful fruit unless earlier ones are damaged in some way. In their detailed study over several cultivars, Bagnall and King (1991b) showed that short days greatly increased the proportion of pegs successfully setting pods.

In Smith's (1954) study, only one third of the pegs produced penetrated the soil and commenced pod development (Table 5.1). Underwood *et al.* (1971) showed that the penetration force of the extending carpophore was only equivalent to 3–4 g, so that pegs have difficulty penetrating capped soils. Normally, pegs produced by early flowers have the highest chance of successful pod production, although if soil conditions are initially poor but improve later during the flowering period it will be later flowers that eventually produce successful fruit.

5.6.2 Calcium nutrition, water stress and pod development

Abortion continues throughout fruit and seed development. Many enlarging pods fail to reach maturity, while many ovules within these pods do not set seed. Calcium nutrition in the pod region is a crucial factor in pod and seed abortion, but calcium deficiency symptoms are often difficult to detect above ground (Smartt, 1976).

Smith (1954) found that the proportion of empty segments in developed pods increased from 18% to 47% under low calcium conditions. Skelton and Shear (1971) used $^{45}Ca^{++}$ labelling techniques to demonstrate that calcium ions move extremely slowly to fruits developing in the soil. Under normal soil moisture conditions there is little if any water movement to developing fruits via the xylem and Ca^{++} is almost immobile in phloem. Thus any calcium required by the developing seed has to be absorbed directly from the soil solution by the fruit itself. Skelton and Shear found that an increased proportion of pegs failed to set fruit on penetrating soils deficient in calcium: 80% of all one-segment pods aborted, but all two-segment pods succeeded in developing one seed. As with early flowers being most successful at fruit production, this seems to be another demonstration of a finely tuned competitive hierarchy within the groundnut plant.

Kvien *et al.*, (1988) have shown that the surface area of the pod is the key factor determining calcium uptake by developing fruits, but other factors such as days to maturity or pod thickness are also important. Genotypes with thick, dense hulls tend to accumulate Ca^{++} at the expense of the developing seed. Lauter and Meiri (1990) have shown that the low salt tolerance of groundnuts is probably related to the inhibition of calcium ion uptake by developing pods via competition from Na^+ for calcium exchange sites.

Moisture around the pod zone is crucial for successful seed development in most cultivars, but in many regions the top few centimetres of soil are liable to dry out in the later weeks of the growing season. In a comparative study, Wright (1989) demonstrated that water deficits in the top 8 cm of soil caused a reduction in the seed yield of two pot-grown spanish type cultivars even when deeper roots were adequately watered. Interestingly, the yield components of each cultivar were affected somewhat differently:

one showed a decrease in the proportion of pegs producing successful pods, the other showed reduced yield due to a decrease in the average number of seeds per pod. In neither case was seed weight affected. In both cultivars, water stress in the pod zone resulted in the seeds' calcium accumulation being reduced by 50%, as was the case in a third cultivar, a virginia type which showed no decrease in yield as a result of a water stressed pod zone. Wright ascribes this finding to the fact that the calcium levels in the seeds of the third cultivar never became limiting: seeds from these stressed plants had calcium levels comparable with the two spanish type controls. Whether calcium is the sole limiting factor under these conditions requires further confirmation: unexpectedly, the vegetative growth of the virginia type cultivar was much poorer than in either of the spanish types.

5.6.3 Plant population effects

In general, a planting population of at least 100 000 plants per hectare is recommended for groundnuts (Smartt, 1976). Many workers (e.g. Mozingo and Coffelt, 1984) have shown the benefit of increased plant populations, but results have been variable and clearly depend on other factors. Kvien and Bergmark (1987) identified some of the important parameters involved. They showed that, at 30 000 plants/ha, population density was the key limiting factor and yield improvements of nearly 30% could be effected by increasing the population eight-fold. Plants at high density tended to increase stem growth at the expense of assimilate partitioning to reproductive tissue. At high populations, yield was sensitive to planting date: a delay in planting by 5 weeks reduced yield by as much as 27%. At low densities plants were better able to compensate for late sowing. Water stress reduced yield in all cases, despite the fact that plants showed some compensation for the stress by increasing harvest index.

As previously emphasized, a key problem with the groundnut crop is the range of pod maturities encountered at harvest. This in turn may be reflected in changes in seed size distribution. In general, as population density increases, numbers of seeds in the larger size grades tend to increase (Figure 5.9). The reason for this is that high interplant competition at high densities tends to suppress the development of later reproductive growth and, typically, earlier flowers are more successful at setting seed (Kvien and Bergmark, 1987). Sung and Chen (1990) have shown that, while cotyledon cell numbers are relatively constant, cell expansion rates tend to be much faster in early formed pods. Nevertheless in Kvien and Bergmark's (1987) study, between 64% and 69% of pods failed to reach maturity in early sowings at high density, irrespective of field location. Very immature pods will not be picked up during machine harvesting, while slightly more advanced ones will contribute small seed to the harvest.

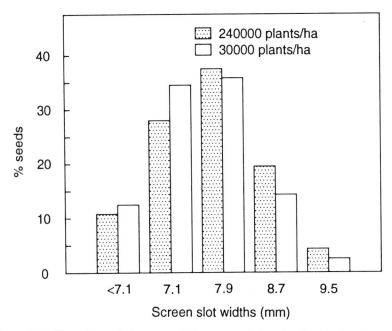

Figure 5.9 The effect of plant population on seed size distribution in Florunner peanut. Data averaged over two planting sites × two planting dates. (Calculated from Kvien and Bergmark, 1987.)

5.7 PRODUCTION CONDITIONS AND SEED QUALITY

The next two sections of this review deal with factors affecting seed quality in groundnut, with special emphasis on quality factors relating to planting value. Section 5.7 reviews the effects of production conditions, while in section 5.8 some consideration is given to the impact of storage on seed quality.

While there is quite a large body of literature on the effects of environmental conditions during development on seed yield in a range of crops, this is less often extended to seed quality in terms of either composition or planting value. Many of these effects may be mediated by seed size and maturity status – for example, low soil moisture and high temperatures are factors which may singly or in combination reduce the duration of the reserve accumulation stage and result in smaller seeds. Alternatively, cooler, wetter conditions may considerably delay seed maturity. While the buffering properties of the soil may mean that some of these effects may be less evident in groundnuts than in the cereals, it is also likely that some important effects are overlooked owing to the indeterminate growth habit of the crop, which results in a wide range of maturity and size classes in any control crop grown under favourable conditions. Seed size *per se* may not always be related to maturity in physiological terms, but it is clear that small groundnut seed have reduced food reserves and produce smaller seedlings which often have lower survival rates. These effects may persist until yield (Sivasubramanian and Ramakrishnan, 1974; Knauft *et al.*, 1990).

This section of the review attempts to collate some general ideas on this problem under the headings of key environmental factors.

5.7.1 Temperature

Low temperatures can depress seed yield by reducing flower numbers (Chang *et al.*, 1985) and generally decrease rates of subsequent pod and seed development, while high temperatures may inhibit fertilization and/or embryo growth. Night temperatures near zero can severely impair the planting value of windrowed groundnuts, and this loss of quality is probably caused by damage to cell and mitochondrial membranes, resulting in increased leakage from dry seeds and higher levels of anaerobic respiration as they begin germination (Singleton and Pattee, 1989). Again, there is an interaction with seed maturity: small seeds respond much more severely to chilling injury than larger ones. On the other hand, in the growing crop, lower night temperatures may in some genotypes allow recovery from the adverse effects of supra-optimal daytime temperatures (Smartt, 1976).

Qualitative changes in seed food reserves as a consequence of differing temperature regimes during seed development have been noted by many authors. Several of these effects can be explained by seed size and/or maturity effects, as was found by McMeans *et al.* (1990) who showed that sucrose concentrations in developing seeds of cv. Florunner tended to be depressed by higher soil temperatures. As Figure 5.8 shows, increases in sucrose levels are a relatively late event in seed development and high temperatures are likely to curtail Stage II early. Similarly, it is not surprising that immature seeds or seeds grown under cooler conditions tend to have lower levels of oil and protein reserves (Bovi, 1983; Nagaraj *et al.*, 1989). What is more interesting is that lower temperatures during seed development can affect oil quality, favouring high proportions of polyunsaturated fatty acids, especially linoleic acid, at the expense of oleic acid levels (Bovi, 1983). Another temperature related effect was noted by Chang *et al.* (1985), who reported that low day–night alternations in temperature favoured high oil yields, while higher temperature variation favoured kernel protein contents. The physiological basis of this is not known.

5.7.2 Water stress

It is clear from various studies that high humidity is essential for successful groundnut seed development. This is evidenced by the fact that pods

TABLE 5.2 *The effects of water stress during production on subsequent seed quality of cv. Florigiant, (values are averages from four seasons' experiments; adapted from Pallas* et al., *1977)*

Stress conditions[a]	Irrigation water used (cm)	Sound mature kernels[b](%)	TSW (g)	Germination (%)[c]
⩾0.02 Mpa	53	66	900	92
⩾0.2 Mpa	46	61	940	89
⩾1.5 Mpa	34	55	790	82
Plants wilted overnight	23	43	640	69[d]

[a]Water potential at 30 cm
[b]Kernel diameter ⩾ 6.36 mm
[c]Of sound, mature kernels
[d]Adjusted for missing value for one season
TSW, thousand seed weight

usually develop several centimetres below the soil surface, where surface drying effects are avoided (Smartt, 1976). Water stress will reduce groundnut yields but there is also good evidence that it will reduce the quality of seeds produced. Pallas *et al.* (1977) demonstrated that the cultivar Florigiant was susceptible to water potentials of −1.5 MPa or greater. These production conditions cause decreases in percentages of sound mature kernels, seed weight and germinability (Table 5.2). Comparative studies with Florunner and Tifspan showed that these smaller seeded cultivars were less susceptible. Recent work by Ketring (1991) using other cultivars has confirmed these results, showing that water deficits typically result in the production of seeds with reduced early seedling growth rates. Earlier work by this group (Ketring *et al.*, 1978) indicated that the development of ethylene synthesis capacity (section 5.5) was generally associated with seed vigour and that this may be impaired by water stress conditions during seed development.

The chemical composition of groundnut kernels may be affected by water stress during seed production, but there may also be major interactions between genotype and the timing of the water deficit. Ross and Kvien (1989) looked at changes in sugars and phenolics in kernels subjected to water stress during the growth of the parent plants. Typically, mid-season drought caused increases in fructose, glucose and phenolic compounds in mature seeds, but the effects on sucrose contents were highly variable. Four out of eight cultivars evaluated showed increased kernel sucrose contents as a result of water stress at some stage of development. One line showed significant decreases in seed sucrose levels when water-stressed 50–80 days after sowing, while the cultivar Tifton 8 showed wide variations in sucrose contents, as compared with seeds from unstressed controls, with around 40% increases when stressed at 20–50 days, but a decrease of equal magnitude when parent plants were stressed at 50–80 days.

5.7.3 Mineral nutrition

Specific effects of various mineral element deficiencies during production on subsequent seed quality have been documented by several authors. The importance of adequate calcium nutrition in preventing abortion has already been mentioned, although adequate Ca^{++} may also be important to maintain the planting value of set seed. Cox et al. (1976) found that a seed calcium content of 420 ppm or higher was necessary for high germinability in cv. Florigiant. Ca^{++} concentrations below this value resulted in decreased germination, falling to almost half the control value when the calcium content was reduced to 200 ppm. Maeda et al. (1986) noted that liming caused a reduction in seed size in groundnuts produced in Brazil (presumably due to pH effects, cf. Smartt, 1976) but that these seeds had improved germinability and vigour.

Interactions between calcium and boron deficiency in groundnut are well known. Cox and Reid (1964) and Harris and Brolmann (1966) showed that blackening of plumules was caused by calcium deficiency whilst cotyledonary damage, especially hollow heart, was largely a function of boron deficiency, although lack of Ca^{++} may also contribute to discolouration of the cotyledons (Table 5.3). Harris and Brolmann (1966) demonstrated that plumule damage due to Ca^{++} deficiency was caused by disruption of vascular tissue at the base of the plumule.

TABLE 5.3 *Influence of calcium and boron nutrition during seed development on embryo damage in groundnuts (data from Cox and Reid, 1964)*

	Damaged kernels (%)			
Site of damage	0.029% Ca^{++}		0.039% Ca^{++}	
	8 ppm B	12 ppm B	8 ppm B	12 ppm B
Plumules	34	30	9	7
Cotyledons	14	8	6	1

Boron deficiency may or may not affect seed yield, depending on cultivar and other cultural factors (Rerkasem et al., 1988). For instance, there may be significant interactions with other mineral elements besides calcium. Patil et al. (1987) reported that combined applications of boron and phosphate were much more effective than either element alone in improving seed numbers, seed weights, protein and oil contents from groundnuts grown in boron deficient soils in India.

5.7.4 Other management factors

Plant spacing has been shown to affect the maturation pattern of some cultivars of groundnut. For example, Mozingo and Coffelt (1984) found that increased interplant competition within rows resulted in plants setting fewer late pods, which allowed an increased proportion of large kernels without an overall reduction in yield. However, extensions of this kind of work into attempts to manipulate groundnut seed yield or quality via chemical manipulation, e.g. the anti-auxin Kylar (Wynne *et al.*, 1974), seem to have been largely unsuccessful. Worthington and Smith (1974) have shown that foliar applications of chemicals to the parent crop can affect the nutritional value of groundnut kernels: Kylar tended to decrease the proportion of linoleic acid, while the fungicide benomyl had the opposite effect.

5.8 QUALITY FACTORS AND THE PROCESS OF SEED DETERIORATION

5.8.1 General principles of groundnut storage

It is well known that groundnut seeds are more difficult to store than most other seed crops, with the exception, perhaps, of soybean (Barton, 1961). Any losses in seed viability will be preceded by considerable losses in planting value at earlier stages of storage. This poor storability is attributed to the high oil content of these seeds (e.g. Gelmond, 1971b) and many studies have focused on the relationship between oil composition and storability of groundnut kernels and of groundnut oil itself. This topic will be discussed in section 5.8.3, but it is important to preface it with a short general discussion on the principles of seed storage and then to review present ideas about the mechanisms involved in seed deterioration.

The basic principles of seed storage are well known. Essentially, the key points are to process seed carefully, clean it well and then keep seed moisture content and storage temperature as low as possible. In standard commercial practice, target seed moisture contents (SMCs) are around 8% for unshelled seeds or 6% for shelled kernels, which are usually more difficult to store (Woodroof, 1973; Shewfelt and Young, 1977; Navarro *et al.*, 1989). Shelled kernels are normally traded in the US food industry at an SMC of around 7.5% (Woodroof, 1973) but seeds intended for planting should be handled more carefully. It should also be recognized that relative humidity (RH), rather than seed moisture content *per se*, is the key criterion for many physiological events and microfloral associations during seed storage. Groundnut seeds, having such a high oil content, equilibrate at much lower SMCs for a given RH than do cereal grains (Halloin, 1986). Standard practice in the US is to store the crop in shells under controlled

RH in cool stores. If financial constraints on storage facilities prevail, more attention should be given to keeping seed moisture and store RH low than reducing the storage temperature.

Several empirical studies have been carried out on the effects on seed germinability of different storage conditions. In comparing different accounts in the literature, it should be remembered that there may be differences in the storage behaviour of different cultivars, the quality of the particular seed lot studied, the method of evaluation of germinability subsequent to storage and/or possible contamination of the lot by storage microflora. Recent work includes that by Navarro *et al.* (1989), who suggest that seed moisture content should not exceed 8% if shelled seeds are to be maintained at 90% germination for 6 months at 15 °C, but that a 1% reduction in seed moisture will more than compensate for a 10 °C rise in storage temperature. In earlier studies, Gelmond (1971b) had shown that seeds stored at 4 °C and 30% RH (expected seed moisture content ≈4.5%) still retained good planting value after three years' storage. Going back further in the literature, Barton (1961) recommend storage SMCs as low as 3% when storage temperatures exceed 26 °C: however, reducing the SMC of unshelled kernels below 6% makes seed very prone to mechanical damage, especially loss of the thin testa on subsequent processing (Gelmond, 1971b).

Vertucci and Roos (1990) attempted to locate applied research of this type within an appropriate theoretical framework based on the physiological status of the seed in store. They calculated that, at an SMC of around 5%, the viscosity of water within the peanut kernel should be so high as effectively to inhibit any deleterious chemical reactions in an aqueous system. In the same study, they showed that reducing SMCs below 4% caused the rate of deterioration of groundnut kernels to begin to increase again, presumably due to the loss of bound water which is essential to the structural integrity of major macromolecules. Accordingly, it can be concluded that a moisture content of 4–5% is the ideal safe level for shelled kernels but, as indicated above, maintaining SMCs as low as this may be inappropriate for short-term bulk seed storage, especially as kernels are liable to skin at this level. However, such information is invaluable for the practice of the long-term seed storage required for genetic conservation.

5.8.2 Mechanisms of deterioration in stored groundnut

Seed deterioration is a complex matrix of processes involving both oxidative and hydrolytic damage. While many facets of cellular structure and metabolism may be affected by the deterioration process, key events seem to be either loss of membrane integrity or genetic damage. One fundamental problem in the study of the physiology of deterioration is the difficulty of separating secondary deteriorative events (the consequences of earlier damage) from primary damage-initiating reactions. The relationship between these events may vary according to cultivar, the prestorage history of the seed lots and, in particular, the nature of the storage conditions themselves (e.g. Priestley, 1986). Sometimes, too, interpretation of experimental data may not be as straightforward as it appears. For example, observations of increased leachate from deteriorating peanuts during early imbibition (e.g. Nautiyal *et al.*, 1988) are usually interpreted to be an indication of membrane deterioration, but they may equally be due to increases in solutes available to leak out of rehydrating/reorganizing cells. Pattee *et al.* (1981), for instance, reported that sucrose and stachyose levels increased dramatically in stored immature seed but not in mature kernels. Further complications arise when it is recognized that seeds are capable of some degree of self-repair at higher moisture contents and also have a considerable array of detoxification and antioxidant systems.

Vertucci and Roos (1990) have demonstrated that oxidative phosphorylation activity becomes negligible in groundnut at ≈16% SMC. It is unreasonable, therefore, to assume that much repair activity will occur in groundnuts at moisture contents below this level. However, it is essential to store them at much lower moisture contents than this because of high rates of physiological deterioration and the proliferation of seed storage fungi, especially *Aspergillus* and *Penicillium* species. There may be significant growth of *Aspergillus* when the SMC of groundnut is as low as 9% (Halloin, 1986). Although tannins present in the testa may inhibit the germination of fungal spores, and other phenolic leachates also have fungistatic properties (Sanders and Mixon, 1978; Kalaichelvan and Mahadevan, 1987), these defence mechanisms are of limited value in prolonged storage at raised seed moistures and their effectiveness may vary considerably with cultivar and the condition of the seedlot. The basis of good seed storage, therefore, is to attempt to slow down the various deteriorative events by keeping moisture contents as low as possible. Reducing storage temperature will, of course, have similar effects.

It is generally accepted that polyunsaturated fatty acids are more prone to oxidative damage than other stored lipids within seeds (e.g. Harman and Mattick, 1976; Bewley, 1986). In the presence of oxygen, free radicals (highly reactive molecular groups with an unpaired electron) can attack fatty acid chains containing more than one double bond to form lipid peroxides in a potentially devastating chain reaction sequence (Wilson and McDonald, 1986; Benson, 1990). Both storage and membrane lipids may be affected in this way, with damage to cell membranes being particularly critical in seed deterioration, especially as lipid peroxides may also attack membrane proteins. Lipid peroxidation may occur spontaneously, initiated by the action of light or metal ions (autoxidation), or it may be enzymically driven by the lipoxygenase enzyme. There is considerable controversy, however, on the relative importance of free radical damage in seed deterioration compared with other types of damage (e.g. Wilson and McDonald, 1986). For instance, free radicals are likely to be quenched by free water in the cell, while natural antioxidants and detoxification systems are also present.

Based on observations and research done by the food industry, it is often assumed that autoxidation is a key deteriorative event in the dry storage of intact groundnut kernels as well as in their products. However, very few biochemical evaluations of deterioration mechanisms in kernels appear to have been published and these seem inconsistent with each other. From the composition of volatiles released from kernels stored at 6% SMC and ambient temperatures, Pattee *et al.* (1971) concluded that autoxidation reactions were of minor importance and that both lipoxygenase and other enzyme activities were necessary to produce the range of compounds detected. While they legitimately argued that, even at 6% SMC, there may be localized areas of significantly higher moisture content to permit activity of these enzymes, it is not possible to demonstrate the *in vivo* activity of these enzymes unequivocally.

Pearce and Abdel Samad (1980) approached the problem by looking at substrate depletion in deteriorating groundnut stored under either good or adverse conditions. In no case were they able to demonstrate any evidence of lipid peroxidation damage, despite the fact that there was considerable depletion of phospholipid and loss of membrane integrity – events which preceded losses of germinability. They concluded that hydrolytic damage, rather than autoxidation or lipoxygenase-driven reactions, were the most important events in deterioration. In contrast, Chen and Fu (1986) reported that ageing both increased lipid peroxide levels and caused a decrease in natural antioxidants (e.g. glutathione, ascorbate and superoxide dismutase). Of particular interest is their finding that treatment of kernels with ascorbate or glutathione solutions improved seed vigour. Clearly we need to know much more about the specific mechanisms of deterioration in groundnut seeds.

5.8.3 Oil content and storability of groundnuts

Despite the uncertainties outlined in the previous section, it does appear that there is a correlation between polyunsaturated fatty acid levels and storability of groundnuts. As most groundnut cultivars have only trace amounts of polyunsaturated fatty acids other than linoleic acid (section 5.4), levels of the latter compound in relation to the other major fatty acid component, oleic acid, are often used as an index of storability, viz the $C_{18:1}:C_{18:2}$ ratio or oil stability index. Low values of this index are generally associated with reduced storability of both intact kernels and groundnut products (Shewfelt and Young, 1977).

In general the oil stability index seems much higher for virginia runner types than for virginia bunch, valencia or spanish peanuts (Raheja *et al.*, 1987; Nagaraj *et al.*, 1989). Table 5.4 shows the data range obtained by the former group of authors in their comprehensive study. Moore and Knauft

TABLE 5.4 *The range of oil contents and stability indices of three types of groundnuts determined in a survey of 82 cultivars (data from Raheja et al., 1987)*

	Virginia runner	Virginia bunch	Spanish bunch
Oleic acid (as % total oil content)[a]	50.6–54.7	37.6–49.1	36.3–49.7
Linoleic acid (as % total oil content)	29.7–33.4	33.8–46.7	30.0–44.3
Oil stability index[b]	1.53–1.84	0.82–1.45	0.83–1.53
Iodine values[c]	99.2–100.9	101.5–114.2	93.7–111.2

[a]Total oil contents of the three types were very similar and around 50% dry weight
[b]Oil stability index is the $C_{18:1}:C_{18:2}$ ratio
[c]Iodine values are an indication of the proportion of double bonds in the oil or fat sample, but are also affected by the proportions of free fatty acids present (Cocks and van Rede, 1966)

(1989) have identified two recessive genes in groundnut which independently influence oleic acid levels in kernels. The expression of either gene can raise the proportion of oleic acid in peanut oil to around 70%, which is likely to improve the storability of both extracted oil and intact kernels but impairs the nutritional quality of the product.

Apart from genetic variation, factors during seed production can also affect oil stability. Young *et al.* (1972) noted that mature kernels tended to be higher in 18:0 and 18:1, but lower in 18:2 fatty acids. Similarly, Mozingo *et al.* (1988) reported that larger kernels tended to have higher $C_{18:1}:C_{18:2}$ ratios. This group was also able to show that growing season and field location both affected this parameter, but relevant climatic data were not supplied. Bovi (1983), however, observed that the $C_{18:1}:C_{18:2}$ ratio was reduced in kernels growing in cooler locations.

As might be concluded from the previous sections, it is a dangerous assumption to infer that oil stability is entirely a function of $C_{18:1}:C_{18:2}$ ratios. In a major study, Worthington *et al.* (1972) examined the correlation between the time taken for expressed oil to become rancid (the autoxidation induction period) and the lipid composition of that oil. Late maturing types were generally higher in oleic and lower in linoleic acid levels, but it can be seen from some of their data (shown in Table 5.5) that the importance of oleic:linoleic acid ratios varies considerably with growing season. In 1967 and 1968 it accounted for a high proportion of the variation in oil stability, but in 1965 correlations were very poor. Clearly other factors are involved – no doubt levels of natural antioxidants are important, but what else is involved and the precise nature of interactions with environmental conditions during seed production are still a matter of debate, as is the exact relationship between this type of data and the situation in whole kernels.

TABLE 5.5 *Mean oleic and linoleic acid compositions over 82 groundnut genotypes in three different production years and the correlation between the autoxidation induction period (AIP) and the $C_{18:1}:C_{18:2}$ oil stability index. (Standard errors of individual means are shown in brackets. Data from Worthington et al., 1972)*

	Growing season		
	1965	1967	1968
Oleic acid (% total oil content)	50.1 ($\pm$ 10.5)	49.1 ($\pm$ 9.03)	50.3 ($\pm$ 10.09)
Linoleic acid (% total oil content)	24.5 ($\pm$ 8.20)	30.1 ($\pm$ 7.21)	30.4 ($\pm$ 8.33)
Correlation between AIP and $C_{18:1}:C_{18:2}$ ratio (R^2 value)	10%	73%	53%

R^2, coefficient of determination

5.9 SOME FINAL CONCLUSIONS

From the foregoing discussion, it is clear that we have a reasonably comprehensive picture of the morphological and compositional changes occurring during seed development in groundnuts. We are also beginning to build up a body of knowledge on how to manipulate oil content and storability by a combination of breeding programmes and appropriate agronomy. There are, however, two key areas where there are major gaps in our knowledge. Firstly, we know almost nothing about how seed development in groundnut is controlled. We know very little about the role played by endogenous plant growth regulators in the reproductive biology of this species and virtually nothing about how they mediate intraplant competition and losses of potential yield. This is certainly an area ripe for future research. The second area for future work is to improve our understanding of how pre- and post-harvest environmental factors affect seed quality, particularly in terms of planting value. In this instance there is scope for both careful empirical research and some basic physiological studies, especially on seed deterioration in store.

REFERENCES

Aldana, A.B., Fites, R.C. and Pattee, H.E. (1972) Changes in nucleic acids, protein and ribonuclease activity during maturation of peanut seeds. *Plant and Cell Physiology*, **13**, 515–21.

Amoroso, V.B. and Amoroso, C.B. (1988) Gynophore and pod development in *Arachis hypogaea* L. *Central Mindanao University Journal of Science*, **1**, 2–11.

Anderson, J.C. (1944) The effect of nitrogen fertilisation on the gross morphology of timothy (*Phleum pratense* L.). *Journal of the American Society of Agronomy*, **36**, 584–7.

Bagnall, D.J. and King, R.W. (1991a) Responses of peanut (*Arachis hypogaea*) to temperature, photoperiod and irradiance. 1. Effect on flowering. *Field Crops Research*, **26**, 263–77.

Bagnall, D.J. and King, R.W. (1991b) Responses of peanut (*Arachis hypogaea*) to temperature, photoperiod and irradiance. 2. Effect on peg and pod development. *Field Crops Research*, **26**, 279–93.

Banks, D.J. (1990) Hand-tripped flowers promoted seed production in *Arachis lignosa*, a wild peanut. *Peanut Science*, **17**, 22–4.

Barton, L.V. (1961) *Seed Preservation and Longevity*, Leonard Hill Ltd., London.

Benson, E.A. (1990) *Free Radical Damage in Stored Plant Germplasm*, IBPGR, Rome.

Bewley, J.D. and Black, M. (1978) *Physiology and Biochemistry of Seeds in Relation to Germination, Volume 1, Development, Germination and Growth*, Springer-Verlag, Berlin.

Bewley, J.D. (1986) Membrane changes in seeds as related to germination and the perturbations resulting from deterioration in storage, in *Physiology of Seed Deterioration*, CSSA Special Publication No. 11, Crop Science Society of America, Inc., Madison, pp. 27–45.

Bhapkar, D.G., Patil, P.S. and Patil, V.A. (1986) Dormancy in groundnut – a review. *Journal of Maharastra Agricultural Universities*, **11**, 68–71.

Bovi, M.C.A. (1983) Genotypic and environmental effects on fatty acid composition, iodine value and oil content of peanut (*Arachis hypogaea* L.). *Dissertation Abstracts International B*, **44**, 406.

Bunting, A.H. and Elston, J. (1980) Ecophysiology of growth and adaptation in the groundnut – an essay on structure, partition and adaptation, in *Advances in Legume Science*, (eds R.J. Summerfield and A.H. Bunting), Royal Botanic Gardens, Kew, pp. 495–500.

Cao, Y. and Huang, A.H.C. (1986) Diacylglycerol acyltransferase in maturing oil seeds of maize and other species. *Plant Physiology*, **82**, 813–20.

Chang, H.H., Wang, Y.F. and Hsieh, F.H. (1985) [Effect of day/night temperature on seed yield and oil and protein contents of groundnuts]. *Memoirs of the College of Agriculture, National Taiwan University*, **25**, 91–7.

Chen, C.-C. and Gibson, P.B. (1971) Seed development following the mating of *Trifolium repens* x *T. uniflorum*. *Crop Science*, **11**, 667–72.

Chen, G.Y. and Fu, J.R. (1986) [Deterioration of groundnut seeds and peroxidation]. *Acta Scientarum Naturalium Universitatis Sunyat Seni*, **3**, 69–75.

Cocks, L.V. and van Rede, C. (1966) *Laboratory Handbook for Oil and Fat Analysis*, Academic Press, London, pp. 109–13.

Corner, E.J.H. (1976) *The Seeds of Dicotyledons, Volume 1*, Cambridge University Press, Cambridge.

Cox, F.R. and Reid, P.H. (1964) Calcium–boron nutrition as related to concealed damage in peanuts. *Agronomy Journal*, **56**, 173–6.

Cox, F.R., Sullivan, G.A. and Martin, C.K. (1976) Effect of calcium and irrigation treatments on peanut yield, grade and seed quality. *Peanut Science*, **3**, 81–85.

Daigle, D.J., Conkerton, E.J., Sanders, T.H. and Mixon, A.C. (1988) Peanut hull flavonoids – their relationship with peanut maturity. *Journal of Agricultural and Food Chemistry*, **36**, 1179–81.

Derbyshire, E., Wright, D.J. and Boulter, D. (1974) Legumin and vicilin, storage proteins of legume seeds. *Phytochemistry*, **15**, 3–24.

Gelmond, H.G. (1971a) Growth and development of the peanut plant (*Arachis hypogaea*) in relation to seedling evaluation in the germination test. *Proceedings of the International Seed Testing Association*, **36**, 121–30.

Gelmond, H.G. (1971b) Moisture content and storage of peanut seed (*Arachis hypogaea* L.). *Proceedings of the International Seed Testing Association*, **36**, 159–71.

Grabe, D.F. (1956) Maturity in smooth bromegrass. *Agronomy Journal*, **48**, 253–6.

Halloin, J.M. (1986) Microorganisms and seed deterioration, in *Physiology of Seed*

Deterioration, CSSA Special Publication No. 11, Crop Science Society of America, Inc., Madison, pp. 89–99.

Harman, G.E. and Mattick, L.R. (1976) Association of lipid oxidation with seed ageing and death. *Nature*, **260**, 323–4.

Harris, H.C. and Brolmann, J.B. (1966) Comparison of calcium and boron deficiencies of the peanut. II. Seed quality in relation to histology and viability. *Agronomy Journal*, **58**, 578–82.

Hyde, E.O.C., McLeavey, M.A. and Harris, G.S. (1959) Seed development in ryegrass and red and white clover. *New Zealand Journal of Agricultural Research*, **2**, 947–952.

Ingle, J., Bietz, D. and Hageman, R.H. (1965) Changes in composition during development and maturation of maize seeds. *Plant Physiology*, **40**, 835–9.

Kalaichelvan, P.T. and Mahadevan, A. (1987) Groundnut prohibitins. I. Prohibitins in groundnut seed leachate. *Indian Journal of Plant Pathology*, **5**, 59–62.

Kermode, A.R., Bewley, J.D., Dasgupta, J. and Misra, S. (1986) The transition from seed development to germination: a key role for desiccation? *HortScience*, **21**, 1113–8.

Ketring, D.L. (1991) Physiology of oil seeds. IX. Effects of water deficit on peanut seed quality. *Crop Science*, **31**, 459–63.

Ketring, D.L. and Morgan, P.W. (1970) Physiology of oil seeds. I. Regulation of dormancy in Virginia-type peanut seeds. *Plant Physiology*, **45**, 268–72.

Ketring, D.L. and Morgan, P.W. (1971) Physiology of oil seeds. II. Dormancy release in Virginia-type peanut seeds by plant growth regulators. *Plant Physiology*, **47**, 488–92.

Ketring, D.L. and Morgan, P.W. (1972) Physiology of oil seeds. IV. Role of endogenous ethylene and inhibitory regulators during natural and induced after-ripening of dormant Virginia-type peanut seeds. *Plant Physiology*, **50**, 382–7.

Ketring, D.L., Simpson, C.E. and Smith, O.D. (1978) Physiology of oil seeds. VII. Growing season and location effects on seedling vigour and ethylene production by seeds of three peanut cultivars. *Crop Science*, **18**, 409–13.

Knauft, D.A., Gorbet, D.W. and Wood, H.C. (1990) The influence of seed size on the agronomic performance of a small-seeded Spanish peanut line. *Proceedings – Soil and Crop Science Society of Florida*, **49**, 135–8.

Kramer, H.A., Gates, J.E., Demaree, K.D. and Sidwell, A.P. (1963) Spectrophotometric investigations on peanuts with particular reference to estimation of maturity. *Food Technology*, **17**, 1044–6.

Kvien, C.S. and Bergmark, C.L. (1987) Growth and development of the Florunner peanut cultivar as influenced by population, planting date and water availability. *Peanut Science*, **14**, 11–16.

Kvien, C.S., Branch, W.D., Sumner, M.E. and Csinos, A.S. (1988) Pod characteristics influencing calcium concentrations in the seed and hull of peanut. *Crop Science*, **28**, 666–71.

Lauter, D.J. and Meiri, A. (1990) Peanut pod development in pegging and rooting zones salinised with sodium chloride. *Crop Science*, **30**, 660–4.

Maeda, J.A., Do Lago, A.A. and De Tella, R. (1986) [Effect of liming and NPK fertilizer application on the quality of groundnut seeds]. *Pesquisa Agropecuária Brasileira*, **21**, 941–4.

Malik, C.P. and Chhabra, N. (1978) Groundnut pollination and its control mechanism, in *Physiology of Sexual Reproduction in Flowering Plants*, (eds) C.P. Malik, A.K. Srivastava, N.C. Bhattacharya, R. Singh), Kalyani Publishers, New Delhi, pp. 31–38.

Mason, M.E., Newell, J.A., Johnson *et al.* (1969) Non-volatile flavour components of peanuts. *Journal of Agricultural and Food Chemistry*, **17**, 728–32.

McMeans, J.L., Sanders, T.H., Wood, B.W. and Blankenship, P.D. (1990) Soil temperature effects on free carbohydrate concentrations in peanut (*Arachis hypogaea* L.) seed. *Peanut Science*, **17**, 31–5.

Miller, O.H. and Burns, E.E. (1971) Internal colour of Spanish peanut hulls as an index of kernel maturity. *Journal of Food Science*, **36**, 669–70.

Moore, K.M. and Knauft, D.A. (1989) The inheritance of high oleic acid in peanut. *Journal of Heredity*, **80**, 252–3.

Mozingo, R.W. and Coffelt, T.A. (1984) Row pattern and seeding rate effects on value of Virginia-type peanut. *Agronomy Journal*, **76**, 460–2.

Mozingo, R.W., Coffelt, T.A. and Wynne, J.C. (1988) Market grade effects on fatty acid composition of five peanut cultivars. *Agronomy Journal*, **80**, 73–5.

Nagaraj, G., Chauhan, S. and Ravinada, V. (1989) Peanut composition and oil quality as influenced by genotype and harvest stages. *Journal of the Oil Technologists' Association of India*, **21**, 60–3.

Nautiyal, P.C., Vasantha, S., Sureja, S.K. and Thakkar, A.N. (1988) Physiological and biochemical attributes associated with the loss of seed viability and vigour in groundnut (*Arachis hypogaea* L.). *Oléagineux*, **43**, 459–63.

Navarro, S., Donahaye, E., Kleinerman, R. and Hakam, H. (1989) The influence of temperature and moisture content on the germination of peanut seeds. *Peanut Science*, **16**, 6–9.

Pallas, J.E., Stansell, J.R. and Bruce, R.R. (1977) Peanut seed germination as related to soil water regime during pod development. *Agronomy Journal*, **69**, 381–3.

Patil, G.D., Patil, M.D., Patil, N.D. and Adsule, R.N. (1987) Effects of boronated super-phosphate, single superphosphate and borax on yield and quality of groundnut. *Journal of Maharashtra Agricultural Universities*, **12**, 168–70.

Pattee, H.E. and Mohapatra, S.C. (1987) Anatomical changes during ontogeny of peanut (*Arachis hypogaea* L.) fruit: mature megagametophyte through heart-shaped embryo. *Botanical Gazette*, **148**, 156–64.

Pattee, H.E., Singleton, J.A., Johns, E.B. and Mullen, B.C. (1970) Changes in the volatile profile of peanuts and their relationship to enzyme activity during maturation. *Journal of Agricultural and Food Chemistry*, **18**, 353–6.

Pattee, H.E., Singleton, J.A. and Johns, E.B. (1971) Effects of storage time and conditions on peanut volatiles. *Journal of Agricultural and Food Chemistry*, **19**, 134–7.

Pattee, H.E., Young, C.T. and Giesbrecht, F.G. (1981) Seed size and storage effects on carbohydrates of peanuts. *Journal of Agricultural and Food Chemistry*, **29**, 800–2.

Pearce, R.S. and Abdel Samad, I.M. (1980) Changes in the fatty acid content of polar lipids during ageing of seeds of peanut (*Arachis hypogaea* L.). *Journal of Experimental Botany*, **31**, 1283–90.

Periasamy, K. and Sampoornam, C. (1984) The morphology and anatomy of ovule and fruit development in *Arachis hypogaea* L. *Annals of Botany*, **53**, 399–411.

Pickett, T.A. (1950) Composition of developing peanut seed. *Plant Physiology*, **25**, 20–4.

Prakash, S. (1960) The endosperm of *Arachis hypogaea* Linn. *Phytomorphology*, **10**, 60–64.

Priestley, D.A. (1986) *Seed Ageing*, Comstock Publishing, Cornell University Press, New York.

Raheja, R.K., Batta, S.K., Ahuja, K.L. *et al* (1987) Comparison of oil content and fatty acid composition of peanut genotypes differing in growth habit. *Plant Foods for Human Nutrition*, **37**, 103–8.

Reed, E.L. (1924) Anatomy, embryology and ecology of *Arachis hypogaea*. *Botanical Gazette*, **78**, 289–310.

Rerkasem, B., Netsangtip, R., Bell, R.W. *et al.* (1988) Comparative species responses to boron on a typic tropaqualf in northern Thailand. *Plant and Soil*, **106**, 15–21.

Ross, L.F. and Kvien, C.S. (1989) The effect of drought stress on peanut seed composition. I. Soluble carbohydrates, tartaric acid and phenolics. *Oléagineux*, **44**, 295–301.

Sanders, T.H. and Mixon, A.C. (1978) Effect of peanut tannins on percent seed colonization and in vitro growth by *Aspergillus parasiticus*. *Mycopathologia*, **66**, 169–173.

Schenk, R.U. (1961) Development of the peanut fruit. *Georgia Agricultural Experiment Stations Technical Bulletin N.S.* **22**.

Sengupta, U.K. (1989) Changes in phenolic compounds during seed development and germination of groundnut cultivars. *Seed Research*, **17**, 36–42.

Sharma, A. and Sengupta, U.K. (1988) Changes in α-amylase and protease activity during seed development in dormant and non-dormant groundnut (*Arachis hypogaea*) cultivars. *Indian Journal of Experimental Biology*, **26**, 732–3.

Shaw, R.H. and Loomis, W.E. (1950) Basis for the prediction of corn yields. *Plant Physiology*, **25**, 225–44.

Shewfelt, A.L. and Young, C.T. (1977) Storage stability of peanut based foods: a review. *Journal Food Science*, **42**,1148–52.

Singleton, J.A. and Pattee, H.E. (1989) Effect of chilling injury on windrowed peanuts. *Peanut Science*, **16**, 51–4.

Sivasubramanian, S. and Ramakrishnan, V. (1974) Effect of seed size on seedling vigour in groundnut. *Seed Science and Technology*, **2**, 435–47.

Skelton, B.J. and Shear, G.M. (1971) Calcium translocation in the peanut. *Agronomy Journal*, **63**, 409–412.

Smartt, J. (1976) *Tropical Pulses*, Longman, London.

Smith, B.W. (1946) Macrosporogenesis and embryogeny in *Arachis hypogaea* L. as related to seed failure. *American Journal of Botany*, **33**, 826.

Smith, B.W. (1950) *Arachis hypogaea*. Aerial flower and subterranean fruit. *American Journal of Botany*, **37**, 802–15.

Smith, B.W. (1954) *Arachis hypogaea*. Reproductive efficiency. *American Journal of Botany*, **41**, 607–16.

Smith, B.W. (1956a) *Arachis hypogaea*. Normal megasporogenesis and syngamy with occasional single fertilization. *American Journal of Botany*, **43**, 81–9.

Smith, B.W. (1956b) *Arachis hypogaea*. Embryogeny and the effect of peg elongation upon embryo and endosperm growth. *American Journal of Botany*, **43**, 233–40.

Sung, F.J.M. and Chen, J.J. (1990) Cotyledon cells and seed growth relationships in CO_2-enriched peanuts. *Peanut Science*, **17**, 4–6.

Suryakumari, D., Seshavatharam, V. and Murthy, V.R. (1989) Studies on the embryology of the genus *Arachis* L. *Journal of Oilseeds Research*, **6**, 85–91.

Toole, V.K., Bailey, W.K. and Toole, E.H. (1964) Factors influencing dormancy of peanut seeds. *Plant Physiology*, **39**, 822–32.

Underwood, C.V., Taylor, H.M. and Hoveland, C.S. (1971) Soil physical factors affecting peanut pod development. *Agronomy Journal*, **63**, 953–4.

Van Rossem, A. and Bolhuis, G.G. (1954) Some observations on the generative development of the peanut. *Netherlands Journal of Agricultural Science*, **2**, 302–3.

Vertucci, C.W. and Roos, E.E. (1990) Theoretical basis of protocols for seed storage. *Plant Physiology*, **94**, 1019–1023.

Wallerstein, I.S., Merin, U. and Rozenthal, I. (1989) Comparison of kernels of three Virginia-type peanut cultivars. *Lebensmittel-Wissenschaft und Technologie*, **22**, 179–181.

Wilson, D.O. and McDonald, M.B. (1986) The lipid peroxidation model of seed ageing. *Seed Science and Technology*, **14**, 269–300.

Woodroof, J.G. (1973) *Peanuts: Production, Processing, Products*, 2nd edn, The AVI Publishing Co., Westport, Connecticut.

Worthington, R.E., Hammons, R.O. and Allison, J.R. (1972) Varietal differences and seasonal effects on fatty acid composition and stability of oil from 82 peanut genotypes. *Journal of Agricultural and Food Chemistry*, **20**, 727–30.

Worthington, R.E. and Smith, D.H. (1974) Modification of peanut oil fatty acid composition by foliar applications of 2,2-dimethyl succinohydrazide (Kylar). *Journal of Agricultural and Food Chemistry*, **22**, 507–8.

Wright, G.C. (1989) Effect of pod zone moisture content on reproductive growth in three cultivars of peanuts (*Arachis hypogaea*). *Plant and Soil*, **116**, 111–4.

Wynne, J.C., Baker, W.R. and Rice, R.W. (1974) Effects of spacing and a growth regulator, Kylar, on size and yield of fruit of Virginia-type peanut cultivars. *Agronomy Journal*, **66**, 192–4.

Yatsu, L.Y. and Jacks, T.J. (1972) Spherosome membranes, half unit-membranes. *Plant Physiology*, **49**, 937–43.

Young, C.T., Mason, M.E., Matlock, R.S. and Waller, G.R. (1972) Effect of maturity on the fatty acid composition of eight varieties of peanuts grown at Perkins, Oklahoma, in 1968. *Journal of the American Oil Chemists' Society*, **49**, 314–7.

Young, C.T. (1973) Automated colorimetric measurement of free arginine in peanuts as a means to evaluate maturity and flavour. *Journal of Agricultural and Food Chemistry*, **21**, 556–8.

Zamski, E. and Ziv, M. (1976) Pod formation and its geotropic orientation in the peanut, *Arachis hypogaea* L., in relation to light and mechanical stimulus. *Annals of Botany*, **40**, 631–6.

Ziv, M. (1981) Photomorphogenesis of the gynophore, pod and embryo in peanut, *Arachis hypogaea* L. *Annals of Botany*, **48**, 353–9.

CHAPTER 6

The composition and nutritive value of groundnut kernels

G.P. Savage and J.I. Keenan

6.1 INTRODUCTION

Groundnuts were originally considered to be food for animals; then they were used as food for slaves. They have now become an important source of protein in many developing countries. A summary of the important food uses of groundnuts and the fortified foods that use groundnuts or groundnut protein has been given by Singh and Singh (1991).

The use of groundnuts by the US armed forces as peanut butter and as snacks has boosted their popularity in western countries as well. It should be remembered that groundnuts are primarily grown as a source of edible vegetable oil; they are the fourth most important source of edible oil and the third most important source of vegetable protein.

6.2 PROXIMATE COMPOSITION

The proximate composition of groundnut kernels and a number of products made from them are presented in Table 6.1, which contains a summary of the accepted values that can be found in the literature. (Where significant variations from these values occur, they will be discussed in the appropriate section.) The type of processing alters the gross composition significantly, from full fat flour which contains essentially the same amounts of the major classes of nutrients as the raw kernels, to protein isolates which contain more than 90% crude protein ($N*6.25$) on a dry weight basis as well as traces of other components.

The Groundnut Crop: A scientific basis for improvement. Edited by J. Smartt. Published in 1994 by Chapman & Hall, London. ISBN 0 412 408201.

TABLE 6.1 *Proximate composition of groundnuts and groundnut products (Natarajan, 1980)*

	Kernels	Full fat flour	Defatted flour	Protein concentrate	Protein isolate
Moisture	5.0	2.5	7.5	4.5	3.4
Protein	30.0	27.4	57.0	60.0	95.0
Fat	48.0	44.4	0.6	8.0	0.5
Carbohydrates	15.5	21.5	30.0	18.5	–
Crude fibre	3.0	2.3	4.6	4.0	–
Ash	2.0	2.0	4.6	2.1	0.5

6.3 OIL AND FATTY ACIDS

Groundnuts are valued for their high quality oil content. About two thirds of the world production of groundnuts is utilized as an edible oil, making it one of the world's leading oil seed crops. Most of the oil is found in the cotyledons which comprise approximately 72.4% of the kernel (Fedeli *et al.*, 1968; Woodroof, 1983). Groundnuts are also consumed whole as snacks, and in confectioneries and peanut butter. In this form, they are usually roasted before use. This accentuates the flavour which contributes to consumer acceptance. Studies have shown that the oil content and composition undergo very little, if any, change as a result of roasting (Hoffpauir, 1953; Iverson *et al.*, 1963; Sekhon *et al.*, 1970; Khalil and Chughtai, 1983).

6.3.1 Quantitative changes in oil content and fatty acid composition

Studies have shown considerable compositional and quantitative changes in the lipids and fatty acids of groundnut seeds during the growth period to maturity, i.e., 15–80 days. The oil percentage increases directly in proportion to the increase in dry seed weight, which is most rapid between 7 and 8 weeks, reaching a maximum at 11 weeks (Sanders, 1980a). From 7 weeks onward, lipid synthesis replaces starch as the dominant reserve accumulation mechanism (Pickett, 1950; Abdel Rahman, 1982b). The maturity of the seed also has an important effect on the composition of the oil (Pickett, 1950). The free fatty acid levels decrease (from approximately 4.5% to 0.7%), offset by triglyceride formation. The oleic acid (18:1) levels of the triglycerides and compound lipids increase rapidly. Shibahara *et al.* (1977) report a corresponding decrease in the proportions of linoleic (18:2) and palmitic acid (16:0). Abdel Rahman (1982a,b) and Sanders (1980b) found the linoleic acid (18:2) level to be changeable. The greatest change apart from oleic acid was in the levels of behenic acid (Sanders, 1980b).

These changes can be important, as oil stock groundnuts are usually those that are rejected or diverted from edible channels. Rejection can be for a number of reasons, including small kernel size (often an indication of kernel immaturity). Weathered or insect infected nuts (a sign of inadequate storage) are also used (Woodroof, 1969). Whiston *et al.* (1959) showed decreased oil content and increased free fatty acid levels in kernels (especially those that were damaged) stored for 24 months.

6.3.2 Analysis of kernel content

Given the changing nature of the fatty acids in immature kernels, it is essential to select mature kernels for analysis in order to obtain highly reproducible results. Typically they are smooth, with little or no wrinkling of the testa surface. They must also have a dark-coloured interior pericarp surface with a thin, faded testa (Worthington *et al.*, 1972; Young and Waller, 1972). The internal distribution of the oil varies and this can be significant when only part of the kernel is to be sampled. The tip opposite the germ contains the most unsaturated oil; the centre of the kernel has been reported as having the lowest iodine value, indicative of more saturated fatty acids (Young and Waller, 1972). Ideally one quarter to one half of the kernel is taken from the end opposite to the germ end.

Oil can be extracted by mechanical or solvent means from the kernel. A silver-plated Carver press can be used to expel the oil. Alternatively, solvent extraction (with petroleum ether or chloroform-methanol) can be carried out in a Soxhlet apparatus for 18 hours. Chloroform-methanol extraction always extracts the maximum amount of lipid (Varma and Kumar, 1968; Sanders, 1980). A comparative study by Wallerstein *et al.* (1989) did not show any significant differences between Soxhlet extraction and the modified Gerber method, which involves sulphuric acid digestion of the seed pulp and phase separation of the oil. Other methods include a rapid butyrometric method (Shukla *et al.*, 1980) or oil analysis by nuclear magnetic resonance, a non-destructive and rapid analytical method which would be of use in breeding programmes (Jambunathan *et al.*, 1985).

Individual fatty acid content is determined after the preparation of methyl esters from the oil samples by transmethylation with 3% sulphuric acid in methanol. Analysis is then by gas liquid chromatography (Worthington and Holley, 1967).

6.3.3 Total oil content

Groundnuts can be divided into three distinctive types on the basis of plant characteristics and habits of growth: spanish (bunch), virginia (spreading bunch) and runner. Data from several sources (Hoffpauir, 1953; Hopkins and Chisholm, 1953; Iverson *et al.*, 1963; Sekhon *et al.*, 1972a; Sekhon *et al.*, 1973; Woodroof, 1983; Taira, 1985; Wallerstein *et al.*, 1989) show that the oil content of these three varieties differs to some extent, as follows:

Spanish	42.0–53.8% of dry weight
Virginia	45.0–58.6% of dry weight
Runner	41.23–53.6% of dry weight

In a study covering eight countries, Taira (1985) reported a significant geographical effect on the oil content of a spanish cultivar which ranged from 46.9% to 58.6%.

Other authors have reported oil content without specifying the groundnut type. These range from 42.3–56.5% (Crawford and Hilditch, 1950; Iverson *et al.*, 1963; Rao *et al.*, 1965; Oke, 1967; Sekhon *et al.*, 1970; Sekhon *et al.*, 1972b; Badami *et al.*, 1980; Abdel Rahman, 1982a; Khalil and Chughtai, 1983) while Roberson *et al.* (1966) reported 44.4–54.4% oil content in 30 commercial peanut butter samples.

Because two thirds of the groundnuts cultivated worldwide are grown for their oil content, a level of more than 50% oil is desirable. Plant breeding techniques are now being used to improve the quantity (and quality) of oil in cultivars.

6.3.4. Oil composition

The quality of the oil is important from the point of view of human nutrition as well as the stability of the oil during storage.

(a) Saponifiable fraction

The saponifiable fraction of groundnut oil consists mainly of triglycerides of long-chain fatty acids. The phospholipids lecithin and cephalin also occur but are thought to be less extractable from cured seed (Sanders, 1980b). They are also thought to settle out of crude oil. Alkaline hydrolysis of oil or fat gives a 'saponification number' which indicates the nature of the fatty acids present, since those with longer carbon chains liberate less acid for each gram of oil hydrolysed. The saponification value is only of interest if the groundnut oil is going to be used for industrial purposes as it has no nutritional significance. Saponification numbers reported for groundnut oil range from 190 (spanish type) to 298.8 (Crawford and Hilditch, 1950; Hopkins and Chisholm, 1953; Sekhon *et al.*, 1970).

Fatty acid composition
Eight fatty acids account for more than 98% of the total fatty acid composition of groundnut oil (Worthington *et al.*, 1972; Woodroof, 1983). Palmitic, oleic and linoleic account for more than 80% of total fatty acids after 30 days growth (Shibahara *et al.*, 1977). They eventually contribute approximately 90% of total fatty acids at kernel maturity (Young and Waller, 1972; Sekhon *et al.*, 1972a,b) but varying ratios of these three fatty acids are found within the three major types (Table 6.2). The virginia type, which is later maturing, and the runner type are generally higher in

monounsaturated fatty acids (oleic acid). The earlier maturing spanish type has a higher linoleic acid (i.e. polyunsaturated) content but total saturated fatty acids are also higher (Woodroof, 1983; Worthington *et al.*, 1972). Virginia cultivars have been reported to contain higher levels of linoleic than oleic acid (Sekhon *et al.*, 1973), which is not observed in spanish cultivars (Sekhon *et al.*, 1972a).

Strains grown under similar climatic and agronomic conditions still show varietal differences which have been attributed to genetic causes (Sekhon *et al.*, 1972b). This makes manipulation of the oil content and composition through plant breeding techniques a practical objective (Sekhon *et al.*, 1972a). An alternative has been the development of mutant breeding using gamma-irradiation (35–75 kR dose range) (Sharma *et al.*, 1981). While having no effect on the overall oil content, the process produced mutants with a higher oleic and lower linoleic acid content than the parent strain. The two mutants showed a substantial increase in oleic acid, with a corresponding decrease in palmitic acid, a ratio which is nutritionally more desirable. The other fatty acids showed minor variation.

In a study of 30 commercial peanut butters, Roberson *et al.* (1966) found comparable composition and fatty acid levels, with the small differences between various brands probably due to the variety of groundnuts used or the amount of added hydrogenated fat. Little difference was found between smooth, medium and chunky blends of the butters.

Differences in fatty acid composition
It has been reported that the composition of groundnut oil is influenced by environmental and seasonal effects in addition to varietal differences, genotypic variation and maturity (Young *et al.*, 1974c; de Bertorelli, 1976;

TABLE 6.2   *Fatty acid composition of groundnut oil (% of total fatty acid)*

Fatty acid type	Spanish	Virginia	Runner	Unspecified
16:0	8.2–18.4	6.0–13.5	8.6–13.3	6.0–20.12
16:1ω9	0.1–2.3	0–1.1	0.1–0.2	0.1–1.47
18:0	1.1–6.2	1.0–4.9	2.0–3.0	0.8–5.3
18:1ω9	36.4–53.1	36.7–61.1	40.9–60.4	35.7–71.5
18:2ω6	20.5–40.3	21.6–48.3	21.0–35.2	13.0–41.1
18:3ω6	0.1–0.3	0.1–0.5	0.1–0.2	0–1.5
20:0	1.0–2.1	1.1–1.8	1.0–1.5	0.9–3.5
20:1ω9	0.7–1.4	1.1–1.8	0.9–1.9	0.6–2.0
22:0	3.0–4.4	2.7–3.7	2.8–3.9	1.3–5.1
24:0	0.9–2.0	1.5–2.0	1.2–2.6	0.6–5.9

Based on data from Hoffpauir, 1953; Hopkins and Chisholm, 1953; Iverson *et al.*, 1963; Worthington and Holley, 1967; Sekhon *et al.*, 1972a; Sekhon *et al.*, 1972b; Worthington *et al.*, 1972; Sekhon *et al.*, 1973; de Bertorelli, 1976; Badami *et al.*, 1980; Kritchevsky *et al.*, 1981; Abdel Rahman, 1982a; Nouvelot *et al.*, 1983; Woodroof, 1983; Taira, 1985; Wallerstein *et al.*, 1989.

Taira, 1985). These effects include annual climatic fluctuations, variable soil conditions and agricultural practices. The environment (especially temperature) during seed formation can modify both the oil content and the fatty acid distribution pattern in a given cultivar (Worthington *et al.*, 1972; Liu *et al.*, 1984). Taira (1985) showed that the daily mean temperature during ripening correlated negatively with the linoleic acid content. Young *et al.* (1974c) reported several significant first and second order interactions with the three major fatty acids involving variety, location and, most noticeably, soil moisture conditions. Inanga *et al.* (1990) studied the effect of calcium levels in the soil during the growing period. They found that, while calcium deficiency can lower phospholipid content, it has no effect on the fatty acid composition of the triglyceride.

Linolenic acid
One characteristic of groundnut oil is its low linolenic acid (18:3) content. Studies have shown that authentic groundnut oil contains considerably less than 1% linolenic acid (0.02–0.3%) and that this value is not greatly influenced by geographical distribution (Worthington and Holley, 1967; Worthington *et al.*, 1972; Worthington, 1977; Taira, 1985). In such small amounts it is able to exert an effect on the oil flavour without contributing to oxidative rancidity. Ideally it can also be used for the identification of authentic groundnut oil and the detection of adulteration. However, the acceptance of a value of less than 1.0% linolenic acid would still permit the adulteration of groundnut oil with 10% or more soya bean oil as the linolenic acid content of the mixture would still be less than 1% (Worthington, 1977).

Several authors do not report linolenic acid levels, while others such as Iverson *et al.* (1963) were unable to detect it. The methodology used to detect the fatty acids is important. Ideally GLC (gas liquid chromatography) liquid phases with polarity characteristics that allow elution of C18 fatty acid methyl esters prior to the elution of the C20 series should be used; otherwise there may be difficulties in distinguishing between linolenic acid, eicosanoic acid (20:1) and, less frequently, arachidonic acid (20:0). Hoffpauir (1953) reported an arachidonic acid range of 2.4–4.0%. The usually reported range is 0.9–3.5%. This would suggest misidentification of linolenic and eicosanoic acids as arachidonic acid. Similarly, Badami *et al.* (1980) reported that one cultivar contained 1.46% linolenic acid but made no mention of eicosanoic or behenic acids.

(b) Unsaponifiable fraction

The unsaponifiable fraction of groundnut oil consists of a mixture of classes of compounds including hydrocarbons, tocopherols, sterol esters, free sterols, glycolipids and others. This fraction has been reported as ranging from 0.7% to 1.6% (Crawford and Hilditch, 1950; Hoffpauir, 1953; Hopkins and Chisholm, 1953; Fedeli *et al.*, 1968; Badami *et al.*, 1980), which indicates that individual compounds within a class are usually present in only trace amounts.

Phytosterols account for 0.16–0.25% of the total oil content and are the main component of this fraction (Hoffpauir, 1953; Worthington and Hitchcock, 1984; Kritchevsky *et al.*, 1981). The main components of the free sterol fraction are β-sitosterol, campesterol and stigmasterol. Steryl esters also account for a portion of this fraction (Fedeli *et al.*, 1968). The characteristic odour and flavour of groundnut oil is attributed to the presence of extremely minute quantities (1.8 g/ton) of higher hydrocarbons ($C_{15}H_{30}$ and $C_{19}H_{38}$). Squalene is the major component of this fraction (Fedeli *et al.*, 1968; Worthington and Hitchcock, 1984). The tocopherols in this fraction have an important role in oil stability (discussed later in this chapter). They have another important role, shown by their effect on other vitamins – mainly vitamin A.

6.3.5. Oil quality

Vegetable oils are promoted as being healthy on the basis of their varying degrees of unsaturation. This can be measured by the number of grams of iodine absorbed across the double bonds by 100 g of oil. Reported iodine values for groundnut oil range from 84.5 to 103.0 (Crawford and Hilditch, 1950; Hoffpauir, 1953; Hopkins and Chisholm, 1953; Fedeli *et al.*, 1968; Sekhon *et al.*, 1970; Worthington *et al.*, 1972; Khattab *et al.*, 1974; Badami *et al.*, 1980). The iodine value only gives an indication of the total amount of unsaturated fatty acids in an oil. It is more important to determine the amounts and ratios of the predominant fatty acids.

A number of tests can be used to determine the quality and freshness of groundnut oil. The oil from mature, freshly ripened groundnuts should contain less than 1% free fatty acids. Therefore the presence of higher levels in fresh oil may be an indication of tissue damage and the subsequent degradation of lipids by lipolytic enzymes. This usually occurs during storage (Crawford and Hilditch, 1950; Sanders, 1980a). The free fatty acid content of groundnut oil (as % oleic acid) has been reported to range from 0.1% to 7.2% (Crawford and Hilditch, 1950; Hoffpauir, 1953; Rao *et al.*, 1965a; Sekhon *et al.*, 1970; Sanders, 1980; Kritchevsky *et al.*, 1981). Phillips and Singleton (1978) noted a problem with the hydrolysis of fatty acid triglycerides when measuring free fatty acids, which could account for the higher values. Their method avoided this problem. Similarly, when exposed to air, triglycerides containing highly unsaturated fatty acids tend to undergo a complex process of autoxidation, resulting in a number of products that are responsible for the off-taste characteristic of rancidity. Fresh oils should have a negligible peroxide value, e.g. 0.68 mmol/kg (Khattab *et al.*, 1974).

Whether oil is stored on the shelf for long periods or submitted to accelerated oxidation tests by exposure to 100 °C in air, an increase in free fatty acid content and peroxide value will occur (Varma and Kumar, 1968; Arya *et al.*, 1972; Wallerstein *et al.*, 1989).

The degree of unsaturation of groundnut oil is due almost entirely to the oleic and linoleic acid contents and may be expressed as a ratio of these acids, i.e. oleic/linoleic (or 18:1/18:2) ratio. These two fatty acids can account for up to 80% of the total fatty acids (Young and Waller, 1972; Abdel Rahman, 1982a; Woodroof, 1983). A high linoleic acid content, with its two unsaturated bonds, is undesirable because it decreases the shelf life of groundnut products (Sekhon *et al.*, 1972a,b). Therefore a higher ratio of oleic to linoleic acid in groundnut oil and other groundnut products is considered an indicator of a more stable product. However, in nutritional terms, a higher linoleic acid content is more desirable because of its hypocholesterolaemic effect (Young and Waller, 1972; Taira, 1985). Oleic/linoleic ratios, calculated from the literature, range from 0.76 to 5.5 (Hoffpauir, 1953; Worthington and Holley, 1967; Sekhon *et al.*, 1972a; Sekhon *et al.*, 1972b; Sekhon *et al.*, 1973; Iverson *et al.*, 1963; Badami *et al.*, 1980; Kritchevsky *et al.*, 1981; Abdel Rahman, 1982a; Taira, 1985; Wallerstein *et al.*, 1989).

Plant breeding can produce a groundnut oil with a desirable level of unsaturation (Sekhon *et al.*, 1973). A rapid microanalytical method for determining the oleic/linoleic ratio by GLC has been developed. It uses only a small portion of kernel, which means the remainder can be planted for reproduction (Young and Waller, 1972). However, Worthington *et al.* (1972) conducted a study which indicated that yearly or seasonal variations in environmental conditions also have an unknown but very pronounced effect on oil stability, largely unrelated to the levels of linoleic acid in the seed.

The tocopherols, which are powerful antioxidants, also contribute to the stability of groundnut oil and other groundnut products containing oil. It is possible that antioxidants and/or synergists other than tocopherols are present in the crude oil and that these compounds are removed in refining (Worthington *et al.*, 1972). Shibahara *et al.* (1977) found the amounts of tocopherols increased in direct correlation with the accumulation of un-saturated fatty acids of triglycerides in the seeds from 40 days to maturity. They normally range from 0.3 to 0.5 g/kg (Hoffpauir, 1953). No β-tocopherols are found in groundnut oil but α-0.18 to 0.3 g/kg, γ-0.18 to 0.22 g/kg and δ-tocopherols have been identified The α-form is found predominantly at the early stage of seed development, whereas the γ-form is found at maturity.

The degradative changes that take place in the unsaponifiable fraction after oxidation can inhibit the activity of pancreatic lipase on the oil. The antioxidant properties and sterol content decrease markedly in groundnut oil (Ramamurti and Banerjee, 1950) on storage. This could contribute to the depressed growth seen in experimental animals fed rancid groundnut oil.

Another factor involved in oil quality is related to its use as a cooking medium. Refined groundnut oil contains mostly monomeric material, no dimers and only a very small quantity of polymers. As the oil is heated up to 260 °C, the monomers decrease, the polymers increase and the dimers appear in increasing amounts with temperature and time. Similarly the percentage decrease in the rate of *in vitro* digestibility falls to 52% at 260 °C, attributable to the formation of compounds which inhibit lipolysis. Peroxide values and free fatty acid values increase (Khattab *et al.*, 1974) while linoleic acid is reported to decrease by as much as 95% after 6 hours heating at 260 °C. Iodine values also decrease. Different values would suggest the quantity of oil and surface area in contact with the air during heating are relevant to the degree of deterioration (Arya *et al.*, 1972; Vidyasagar *et al.*, 1974; Khattab *et al.*, 1974). In contrast, Kokatnur *et al.* (1964) reported a negligible increase in peroxide value after heating groundnut oil for 8 hours at 200 °C.

6.3.6 Groundnut oil and human health

Groundnut oil, despite its unsaturated fatty acid components, has been found to be extremely atherogenic to rhesus monkeys (Vesselinovitch *et al.*, 1974), rabbits (Kritchevsky *et al.*, 1971 Kritchevsky *et al.*, 1973; Kritchevsky *et al.*, 1984) and rats. The effect of groundnut oil in atherogenic diets (i.e. cholesterol-containing) resembles that of a more saturated fat but closer examination of the lesions shows that those due to saturated fat are primarily lipid-rich 'foam' cells with little collagen deposition. The typical lesions caused by groundnut oil are due to prominent intimal cell proliferation associated with a particularly high collagen content, and present as raised and sometimes very thick fibrotic areas (Kritchevsky *et al.*, 1971; Ehrhart and Holderbaum, 1980). Rhesus monkeys developed progressive atherosclerosis with many of the morphological features of human atherosclerosis (Vesselinovitch *et al.*, 1974).

It has been postulated that arachidonic and behenic acids may contribute towards the atherogenicity of groundnut oil. However, the removal of these long-chain saturated fatty acids (which comprise 5–6% of groundnut oil) in a specially formulated test oil of similar iodine value (105) to groundnut oil did not prevent the formation of milder atherosclerotic lesions (Kritchevsky *et al.*, 1971). Therefore while atherogenicity may be correlated with iodine value, some aspect of triglyceride structure must also be important. Kritchevsky *et al.* (1981) showed atherogenic differences between groundnut oil from different geographic locations which would suggest structural or compositional changes. Pancreatic lipase specifically hydrolyses fats esterified to the 1 and 3 position of triglycerides, which means lipid absorbed by the intestinal wall consists of

2–monoglyceride and free fatty acids. It has been shown that the triglyceride structure of groundnut oil and the much less atherogenic randomized groundnut oil differ in that the former contains more triglycerides with linoleic acid (18:2) at the 2 position. While this alone would not explain its atherogenicity over corn oil or the specially formulated test oil, there is the possibility that more monounsaturated fatty acids in the 1,3 positions may play an important role. Bezard and Sawadogo (1983) reported poor absorption into tissue triacylglycerols of long-chain saturated fatty acids in comparison with monounsaturated oleic acid.

Randomized groundnut oil has a fatty acid composition identical to groundnut oil but it has a different triglyceride structure and considerably reduced atherogenicity. It has been shown to enhance cholesterogenesis *in vivo*, which suggests it presents as being more unsaturated, despite having the same iodine value as groundnut oil (Kritchevsky *et al.*, 1982). Therefore the resulting micelle formed from the 2-monoglyceride, free fatty acids and bile salts may differ in structure and composition, which may in turn affect the structure of circulating lipoproteins (Kritchevsky *et al.*, 1973; Kritchevsky *et al.*, 1984).

The ingestion of groundnut oil by chicks is accompanied by only mild hypercholesterolaemia, even when fed with large amounts of cholesterol (Vesselinovitch *et al.*, 1974; Kritchevsky *et al.*, 1971). This is in contrast to chicks fed cholesterol with hydrogenated groundnut oil. They showed a massive increase in plasma cholesterol (Banerjee *et al.*, 1965). An increase in the proportion of β-lipoprotein as LDL (low density lipoprotein) has also been noted (Kritchevsky *et al.*, 1976; Vesselinovitch *et al.*, 1974).

It appears that the chemical nature of the lipid deposited in the aortic wall may at least partly determine the quantity and quality of the atherosclerotic lesions. While the influence of dietary fat on pre-established lesions appears to be principally a function of the level of unsaturation (Kritchevsky *et al.*, 1978), animals fed a cholesterol-free diet with the addition of groundnut oil still showed the characteristic lesions, although they contained relatively little lipid (Kritchevsky *et al.*, 1971; Vesselinovitch *et al.*, 1974; Kritchevsky *et al.*, 1976). The fact that the characteristic lesions were produced, even in the absence of cholesterol, also strongly suggests a role for groundnut oil triglycerides.

Studies of rapeseed (mustard) oil, which has a monounsaturated fatty acid content similar to groundnut oil, have shown it to have a similar or increased atherogenic effect, especially when the oils are fed at high concentrations in the diet for an extended period of time (Flanzy, 1979; Cluzan *et al.*, 1979; Sen and Gupta, 1979; Ray *et al.*, 1980).

6.3.7 Absorption of fat

Levine and Silvis (1980) noted that the total fat in groundnuts was not always efficiently absorbed in the gastrointestinal tract. Visual analysis of stool samples taken from people fed whole groundnuts (76 g/day) showed that portions of the nuts remained intact and were therefore not available for lipid digestion (or any other nutrient). In these diets, 17.8% of the 80 g of fat fed per day was found in the faeces. Simple grinding of whole groundnuts (peanut butter) resulted in far less malabsorption of the fat in the groundnuts (7.0% remained undigested). As expected, when the fat was consumed as groundnut oil most was absorbed during its passage through the tract (only 4.5% remained in the faeces).

6.4 PROTEIN

As more groundnuts are grown to meet the worldwide demand for oil, greater amounts of groundnut protein are becoming available. This protein is showing increasing potential as a food source, especially in the developing world where the lack of adequate protein can be a very serious dietary problem. Research is leading to increased utilization of the protein, often as part of processed foods.

Protein makes up 12.0–36.4% of the groundnut kernel (Hoffpauir, 1953; Sekhon et al., 1970; Derise et al., 1974; Lusas, 1979; Woodroof, 1983). The percentage increases during ripening, with the seeds from older plants showing a higher capacity to accumulate protein than seeds from young plants. The maximum increase occurs at 9 weeks; it then becomes approximately stable (Abdel Rahman, 1982b; Basha, 1991).

The storage proteins arachin and conarachin account for over 63% and 33% of the total protein content of groundnuts. The ratio of these globulins ranges from 2:1 to 4:1, depending on the fractionation technique used.

Conarachin, found in the cytoplasm, has a basic nitrogen content of 6.5%, which is very high for seed globulins. It has a chemical score of 68–82% because of low levels of threonine (Thomson, 1952; Hoffpauir, 1953; Woodroof, 1983). Conarachin is a heterogeneous protein complex with a high molecular weight. More recently it has been divided into three subgroups, using DEAE cellulose chromatography. A major subfraction α-conarachin has also been identified, and it is thought that this may consist of two compounds whose molecular weights range from 140 000 to 295 000, depending on the ionic strength and pH of the buffer in which they are dissolved.

Arachin, the second major storage protein, has a molecular weight of 330 000 and is localized in the protein bodies (aleurone grains) of the seeds. It is much less soluble than conarachin and has a much lower chemical score, due to limiting amounts of cystine and methionine. Polymorphism of arachin has been reported (Tombs, 1965).

The lysine and methionine contents of conarachin I and II are significantly higher than in arachin. A significant improvement in the overall nutritive value of groundnut protein would be obtained if the proportion of conarachin I and II, which is approximately 33% of the total protein content, was increased at the expense of the arachin content. This is especially important as it is quite clear that methionine and lysine are the first and second limiting amino acids in groundnut protein.

A methionine-rich protein (MRP), consisting of six polypeptides with molecular weights between 15.5 and 20 kD, has also been identified in groundnut seed. Studies have shown considerable intervarietal variation in the MRP composition, which shows the potential for improving the nutritional value of groundnuts either by conventional plant breeding or by genetic engineering (Basha, 1991). High performance liquid chromatography has been developed as a rapid alternative to gel filtration in detecting qualitative and quantitative differences in groundnut seed protein composition (Basha 1988).

6.4.1 Allergic reactions to groundnut protein

In some countries, groundnuts account for a significant proportion of the dietary protein for both adults and children. Unfortunately groundnut protein can cause immediate hypersensitivity reactions such as angioedema and asthma in some individuals. Adverse reactions can range from abdominal discomfort to anaphylatic shock. Among foods that produce such allergic reactions groundnut is considered second only to milk and eggs in the total numbers of people affected.

Most individuals who are allergic to groundnuts avoid these problems by eliminating groundnuts and groundnut products from their diets. In the past this has not been a problem as groundnuts are usually easily recognized in food. More recently identification has become a little more difficult as groundnut flour is now added to a wider range of formulated and processed foods.

Nordlee *et al.* (1991), using an *in vitro* radioallergosorbent test (RAST), were able to show that groundnut oil, groundnut hull flour and a product made from acid hydrolysis of groundnuts were not allergenic. Peanut butter and products that had been processed to remove oil or heat treated (roasted) were still allergenic. The remarkable heat stability of groundnut allergens has been investigated by Neucere *et al.* (1969). Roasting of groundnuts (145 °C for 1 hour) produces little overall change in allergenicity as measured by RAST.

Barnett *et al.* (1983) were able to show that allergenicity is spread through both crude arachin and conarachin fractions of groundnut protein and that this activity was spread through the various molecular groupings they were able to separate. Sachs *et al.* (1981) were able to isolate and partially characterize a major groundnut allergen (Peanut-I) but they were unable to suggest which major fraction of the storage protein it constituted. They were able to show that Peanut-I was an acidic glycoprotein with non-identical subunits. Peanut I appeared to be the most potent allergen they isolated from groundnuts but not all of the allergenic activity in groundnut extracts could be accounted for by this fraction.

The reason why groundnut protein appears to be so allergenic is due to its remarkable thermal stability (Barnett *et al.*, 1983) and its low protein digestibility which means that it will remain in contact with the mucosa of the small intestine for a longer time than a more easily digestible protein. It is strongly recommended that individuals hypersensitive to groundnut should avoid all products which may contain significant amounts of groundnut protein.

6.4.2 Amino acid composition

The amino acid composition of groundnuts is given in Table 6.3 with the FAO reference protein for comparison While most of the amino acids are found as part of protein and peptides, non-protein or free amino acids (FAA) also occur. These are thought to react with glucose and fructose, produced by the hydrolysis of sucrose during the browning process, to produce the typical roasted groundnut flavour, colour and aroma (Woodroof, 1983; Young *et al.*, 1974a). Specifically these are glutamic acid, glutamine, asparagine, phenylalanine, aspartic acid and histidine. The ratio of these amino acids alters with seed size, i.e. the different response to roasting depends on the maturity of the seed (Young *et al.*, 1974a). Threonine and serine levels also increase with maturity (Young *et al.*, 1974b). Marshall *et al.* (1989) have developed a rapid method for the determination of free amino acids in whole groundnuts.

High levels of free arginine, tyrosine, lysine and threonine have been shown to affect the flavour adversely. Arginine and proline are found in higher concentrations in immature kernels. Arginine levels decrease as the kernels mature (Woodroof, 1983) but proline levels are more variable and may be indicative of disease or other adverse conditions (Young *et al.*, 1974a).

The presence of two non-protein amino acids (γ-methyleneglutamine and γ-methyleneglutamic acid) in mature kernels has been confirmed (Young *et al.*, 1974b). Lee *et al.* (1965) have also reported an imino acid derivative, N-methylhydroxyproline, in groundnut flour. At least six acidic peptides have been reported in addition to the glutamic and aspartic acids (Sundar *et al.*, 1976). The role of these peptides in the seed has not been established.

Table 6.3 shows that groundnuts are a reasonable source of essential amino acids. Hoffpauir (1953) reported only lysine and methionine to be deficient. Other authors suggest that isoleucine is also deficient (Evans and Bandemer, 1967; Abdel Rahman, 1982a). Khalil and Chughtai (1983)

TABLE 6.3 *Amino acid composition of groundnut protein (g/100 g kernels)*

	Groundnuts	FAO[a]
Essential		
Isoleucine	1.86–4.3	4.2
Leucine	6.12–7.0	4.8
Lysine	3.0–4.27	4.2
Methionine	0.79–1.6	2.2
Methionine + Cystine	0.80–3.3	4.2
Phenylalanine	4.60–5.4	2.8
Threonine	2.43–2.9	2.6
Tryptophan	0.60–2.0	1.4
Valine	2.55–4.5	4.2
Non-essential		
Alanine	3.4–4.2	
Arginine	10.6–11.84	
Aspartic acid	11.56–14.1	
Cystine	1.3–2.42	
Glutamic acid	19.2–22.46	
Glycine	5.6–6.77	
Histidine	2.1–2.51	
Proline	4.3–6.36	
Serine	4.9–5.30	
Tyrosine	3.6–4.40	

[a]FAO pattern of amino acid requirements (Evans and Bandemer, 1967)

Based on data from: Hoffpauir, 1953; Evans and Bandemer, 1967; Chopra and Bhatia, 1970; Lusas, 1979; Alid *et al.*, 1981; Sharma *et al.*, 1981; Abdel Rahman, 1982a.

found only leucine, phenylalanine + tyrosine to exceed the reference protein, in contrast to Sharma *et al.* (1981) who reported only phenylalanine. Like many other legumes, groundnut protein is also limited by the availability of some essential amino acids. McOsker (1962) reported lysine, methionine and threonine to be equally limiting while others report lysine followed by threonine and methionine + cystine to be first limiting in groundnut protein (Sharma *et al.*, 1981; Khalil and Chughtai, 1983). Methionine alone (Cheema and Ranhotra, 1967; Evans and Bandemer, 1967) or in combination with cystine (Evans and Bandemer, 1967) has also been reported as the first limiting amino acid.

Most amino acids in proteins can be satisfactorily determined by the use of an automatic amino acid analyser after acid hydrolysis of the protein. Unfortunately tryptophan is destroyed during acid hydrolysis. A collaborative study conducted by Westgarth and Williams (1974) showed that the method devised by Miller (1967) using alkaline hydrolysis gave consistent results for tryptophan on a standard groundnut sample, mean 0.98 g/16 g N. This value increased to a mean of 1.00 g/16 g N (range 0.91–1.07) when the recovery of added tryptophan was considered. Amaya *et al.* (1977) carried out an extensive survey of the tryptophan content of ground-nuts grown commercially in the USA and reported that the tryptophan content ranged from 1.05 to 1.41 g/16 g N. They acknowledged that the tryptophan content of groundnuts was not nutritionally significant in the light of the low levels of lysine and methionine. They noted an interesting correlation between the tryptophan and total protein content among the cultivated genotypes investigated. A wider variation in this ratio was found when some wild genotypes were considered.

There appear to be considerable varietal differences in the total amino acid content reported by various authors. These could, in part, be accounted for by differing seed sizes and maturity as well as varying methods of analysis. Other factors include location and environmental conditions during growth, curing and storage. An example of the wide variation of amino acid contents of a number of different groundnut varieties in a single study is given by Heinis *et al.* (1975). Variation in amino acids levels between different cultivars of groundnuts should pro-vide the opportunity for further genetic improvement. Glutamic acid, aspartic acid and arginine account for approximately 45% of the total amino acids; therefore a significant reduction of these three should pro-duce significant increases in other amino acids (Young *et al.*, 1973).

As lysine and methionine are the two main limiting essential amino acids, various authors have determined the variables that may affect their levels. Chopra and Sidhu (1965) attributed genetic similarity to the small variation in total nitrogen, lysine and methionine levels found in their study. In a further study Chopra and Sidhu (1967) showed a significant difference ($P < 0.01$) in methionine levels between different spreading and erect cultivars. Lysine levels were less significantly variable ($P < 0.05$). Chopra and Bhatia (1970) showed no significant difference in the levels of either of these amino acids due to habit or location of growth, but they noted some intervarietal differences. These were more marked for lysine (2.67–4.27%) than methionine (0.79–0.96%).

In contrast, a study of three cultivars (virginia, red and white skinned spanish) by Conkerton and Ory (1976) showed no significant difference in lysine or methionine levels (3.1–3.3% and 0.9–1.1% respectively). This is supported by Young (1980) who studied the three types, spanish, virginia and runner. He reported a significant location effect for all the amino acids but variety had no effect on methionine, lysine or arginine levels. Looking at cultivars of the spanish type, Young *et al.* (1974a) showed intervarietal, location and environmental (irrigation) differences to be considerable for a number of amino acids while lysine and methionine levels were unaffected. Similarly, habit and location of growth showed no significant influence on methionine levels in a study by Cheema and Ranhotra (1967). These variables seem to play a more significant role in the levels of the amino acids related to organoleptic qualities rather than nutritive quality (Young *et al.*, 1974a). However, Young *et al.* (1973) investigated the amino acid levels of defatted groundnut meal produced from 16 different varieties of groundnuts with widely differing protein content. They found nearly two-fold variations in the limiting essential amino acids (lysine, isoleucine, methionine, threonine and valine) – large enough to be significant for breeding purposes. From this it could be interpreted that varietal differences are linked to total nitrogen levels.

Amaya *et al.* (1977) investigated the correlation between tryptophan levels and total protein as a means of indicating the abundance of other essential amino acids. They found the former tends to remain constant while the total protein varies substantially with different environmental conditions. Sharma *et al.* (1981) demonstrated a marked change in protein quality in mutants produced by gamma-irradiation. They showed lower sulphur amino acids levels and a higher tryptophan content compared to the parent.

6.4.3 Processing of groundnuts and their products

Often processing only involves roasting prior to consumption as a snack food or grinding into peanut butter. Some authors report that the overall protein content of groundnuts appears to be little affected by roasting (Sekhon *et al.*, 1970; Woodroof, 1983) while Derise *et al.* (1974) and Khalil and Chughtai (1983) reported a slight increase, probably because of an overall decrease in weight due to loss of moisture and non-nitrogenous volatiles. In many groundnut producing countries, the protein-rich meal that remains after oil extraction is used as animal feed or fertilizer. This is now being recognized as a valuable supplement for human diets. Further processing provides flour, protein concentrates and isolates.

Groundnut flour can range from full-fat to defatted, with the protein level increasing as the fat level falls. The protein content of defatted groundnuts is reported to range from 43.8% to 68.6% (Howe *et al.*, 1965; Cheema and Ranhotra, 1967; Fetuga *et al.*, 1973; Ayres *et al.*, 1974; Bookwalter *et al.*, 1978, 1979; Ory and Conkerton, 1983; Khalil *et al.*, 1983b; Alid *et al.*, 1981; Ghuman *et al.*, 1990). Screw pressing or pre-pressing followed by solvent extraction are the two main commercial methods for producing the defatted product. The latter then requires the solvent to be stripped from the meal by steam heating, which also removes the raw groundnut flavour (Natarajan, 1980). As with other processes, the heating stage is critical; severe heat can affect the availability and utilization of the essential amino acids. Ayres *et al.* (1974) found the nitrogen solubility index fell from 92% in defatted raw groundnuts to 59% for pre-pressed solvent extracted flour.

Further processing of the meal to produce protein concentrates and isolates has the advantage of removing the insoluble and partly indigestible carbohydrates which constitute approximately one third of the raw material. Groundnut protein concentrate (PPC), containing 65–70% protein, can be produced by wet-milling, which gives simultaneous recovery of oil and protein (Natarajan, 1980). Khan *et al.* (1975) reported levels of 54.3–56.1%. Alternatively concentrates can be prepared by leaching meals with dilute acids and aqueous alcohols to give a product with higher protein content, considerably lower trypsin inhibitor content and negligible aflatoxin levels. The nutritional quality of these concentrates remain unaffected but the protein solubility is decreased, owing to denaturation that occurs during extraction (Nagaraj and Subramanian, 1974).

Groundnut protein isolates are the most refined form of groundnut proteins, going one step further than the concentrate process and removing water-insoluble polysaccharides, water-soluble sugars and other minor constituents (Anantharaman *et al.*, 1969; Bhatia *et al.*, 1960; Natarajan, 1980). Depending on the method of isolation, they contain 90% (wet-milling) to 95% (solvent extracted meal or screw press) protein.

The overall effect of processing groundnuts to produce flour, concentrate and isolates is to increase markedly the crude protein and consequently the essential amino acid content by the removal of the carbohydrate fractions (Table 6.4).

Fermentation of groundnut kernels or defatted groundnut flour (DPF) supposedly enhances the digestibility by breaking down the complex

TABLE 6.4   *Essential amino acids levels in groundnut products (g/100 g)*

	Flour	Concentrate	Isolate
Essential			
Isoleucine	1.64–3.4	3.5–4.3	3.6
Leucine	3.64–6.79	6.7–7.09	6.6
Lysine	2.70–4.55	2.9–3.9	3.0
Methionine	0.51–1.21	0.9–1.1	1.0
Methionine + Cystine	1.8	2.05	
Phenylalanine	2.69–5.54	5.5–5.6	5.6
Phenylalanine + Tyrosine	4.9	9.94	
Threonine	1.46–3.4	2.5–3.04	2.5
Tryptophan	0.55–1.0	0.9–1.24	1.0
Valine	2.18–4.1	4.27–4.7	4.4
Non-essential			
Cystine	1.97–3.31	1.14–1.4	1.4
Tyrosine	3.36–3.96	4.1–4.34	4.3

Based on data from: Young *et al.*, 1973; Fetuga *et al.*, 1973; Ayres *et al.*, 1974; Khan *et al.*, 1975; Conkerton and Ory, 1976; Bookwalter *et al.*, 1978; Lusas, 1979; Alid *et al.*, 1981.

protein structures to peptides of varying lengths and free amino-acids, which in turn increases the nutritive value (Coelho and Bhat, 1959; Rao and Rao, 1972; Quinn *et al.*, 1975; Bhavanishankar *et al.*, 1987). Alternatively, proteolytic enzymes can be used to modify and extract protein from groundnut flour and press cake (Natarajan, 1980). The product is known as hydrolysed vegetable protein. Texturized proteins have also been developed from DPF (Alid *et al.*, 1981).

6.4.4 Availability of amino acids

The key to amino acid availability in groundnuts and their products lies in the processing they undergo before consumption. Raw groundnuts contain a number of antinutritive factors including trypsin inhibitors, haemagglutinins, goitrogens, saponins and phytic acid. While the level of trypsin inhibitor found in groundnuts is only approximately 20% of that found in raw beans, it is still enough to cause pancreatic hypertrophy in rats receiving 15% protein intake from groundnuts (Lusas, 1979). Groundnuts are unusual in that, while most of the protein is concentrated in the outer cotyledon layer, the antitrypsin activity is concentrated in the inner part of the cotyledon (Zimmermann *et al.*, 1967). Mild heat treatment, an inherent part of most of the processes, is sufficient to destroy the trypsin inhibitor and haemagglutinin in the seeds.

Mild dry heating increases available lysine and therefore protein efficiency ratio (PER) values but continued high-temperature roasting decreases the protein solubility and protein quality. Roasting has been reported to decrease the measurable free amino acid content by 77.8% (dry), 64.6% (microwave) and 52% (oil) (Lusas, 1979). Aspartic acid, serine, glutamic acid, proline and phenylalanine levels increase while threonine, tyrosine, lysine, arginine, methionine, cystine and tryptophan levels all fall, with lysine, methionine and threonine suffering maximum loss (in that order) (Khalil and Chughtai, 1983). McOsker (1962) also found these three to form the limiting amino acid sequence in roasted groundnuts but in a slightly different order. He found roasting at 170 °C for 36 minutes resulted in the destruction of 15% lysine, 11% threonine and 10% methionine. Similarly Woodroof (1983) reported a decrease in lysine levels from 3.4% to 1.9% for meal prepared by cooking at 250 °F for 2 hours.

Anantharaman and Carpenter (1971) looked closely at the effects of heat and found that higher temperature, longer heating time and higher moisture content all resulted in reduced values for each amino acid investigated. Autoclaving at 107 °C for 30 minutes (similar to heating in commercial groundnut meal production) has little effect on cystine and lysine levels. More drastic heating (121 °C for 4 hours) reduces FDBP-available lysine by nearly 40% and total cystine by nearly 30% while appearing to have little effect on available methionine and threonine levels. The reduction in ϵ-NH$_2$ lysine groups is attributed to binding to hydrolysed sucrose. Linkages are formed which are not hydrolysed by digestive enzymes.

Quinn *et al.* (1975) found that fungal fermentation of DPF gave a slight decrease in arginine and proline levels while glycine and alanine seemed to be slightly increased. They also noted slight increases in lysine and methionine levels. Bhavanishankar *et al.* (1987) noted certain fungal strains have higher proteolytic activity, resulting in increased free lysine and methionine levels. In contrast Sugimura *et al.* (1965) found considerably increased levels of free amino acids, especially proline and glutamic acid, while arginine levels decreased. The samples were autoclaved in each of these studies prior to fermentation.

The available lysine content of groundnuts is high (2.69 g/16 g N) compared with the levels found in cereals (Johri *et al.*, 1988) but on long-term storage (up to 24 months) the levels can fall considerably, especially at high temperatures (40 °C). Similar falls in available lysine occurred even when they were stored in a vacuum.

6.4.5 Protein quality

The nutritional quality of protein depends on the amount and kind of essential amino acids which become available during digestion. If any one of the essential amino acids is deficient, it becomes the limiting factor in the utilization of that protein. Completeness of digestion and absorption of the essential amino acids is also important (Watts *et al.*, 1959). The nutritional value has been assessed for a number of different groundnut products, e.g. meal, flour or defatted oil cake. The use of heat in the processing of these products can reduce the solubility of the protein as well as destroying some of the sulphur-containing amino acids (Ladell and Phillips, 1959). This variable alone may account for some of the variability of results that have been reported.

In vivo biological evaluation to determine the biological value (BV) (Watts *et al.*, 1959), protein efficiency ratios (PER) (Howe *et al.*, 1965; Nagaraj and Subramanian, 1974; Ghuman *et al.*, 1990) and net protein utilization (NPU) (Khalil and Chughtai, 1983) are commonly employed to evaluate protein quality. A summary of some typical values for groundnut and groundnut products are shown in Table 6.5. Alternatively, microbiological techniques can be used for the rapid testing of protein quality. These measure the growth of the organism in question (usually *Tetrahymena pyriformis* W) to give the relative nutritive value (Ghuman *et al.*, 1990).

Because groundnuts are deficient in some of the essential amino acids (notably lysine and methionine), their relative nutritive value is decreased in comparison with a reference protein. They have a chemical score of between 58 and 70 (Khalil and Chughtai, 1983). Cheema and Ranhotra (1967) used the chemical score method to assess the biological value of

TABLE 6.5　*Comparisons of protein quality for raw and processed groundnuts*

	Raw	Roasted/ boiled defatted
True digestibility	89–93	
Biological value	32–75	
Nutritive value		47.2–54.2
Protein efficiency ratio	1.5–1.8	1.45–1.76
Net protein utilization		34–44
Available lysine	2.83%	

Based on data from: Hoffpauir, 1953; Clegg, 1960; Howe *et al.*, 1965; Evans and Bandemer, 1967; Nagaraj and Subramanian, 1974; Lusas, 1979; Woodroof, 1983; Alid *et al.*, 1981; Khalil and Chughtai, 1983; Ghuman *et al.*, 1990.

defatted groundnuts which ranged from 51.3 to 58.0. Biological value is more often determined by incorporating the protein (usually at 10%) in a test animal's diet. The reported BV for raw groundnuts is 50–75% of the reference proteins and the PER is 1.5–1.8 (Lusas, 1979). Studies have indicated that the digestibility and utilization of groundnut protein from kernels, shown by nitrogen balance, is poor (Watts *et al.*, 1959; Hoffpauir, 1953; Edwards *et al.*, 1966; Woodroof, 1983). This is probably due in part to the presence of a trypsin inhibitor and fibre. Khalil and Chughtai (1983) compared NPU from several strains after roasting and found the differences in nutritional quality (34 to 44) were minimized by the higher amounts of lysine in the strains containing the lower amounts of total protein.

Coelho and Bhat (1959) compared the digestibility, BV and PER of raw vs roasted/boiled kernels after fermentation. While the levels of digestibility were not significantly different, the BV increased from 56% to 75.5% and 76.4% respectively after boiling and fermentation. Roasting appeared to be the least effective treatment, resulting in a BV of 69.6%. In contrast the PER values for all groups were comparable (1.14–1.38) which suggests that, while the fermentation of groundnuts may help in an improved utilization of their proteins, it contributes nothing towards digestibility or growth. Bhavanishankar *et al.* (1987) found a significant difference in the PER of two different strains of fungi (1.36 and 1.75) and this suggests that one important factor in fermentation is the fungal strain used.

Defatted groundnut flour (DPF) shows little nutritive difference compared with raw groundnuts, with PERs ranging from 1.45 to 1.76 (Nagaraj and Subramanian, 1974; Howe *et al.*, 1965; Alid *et al.*, 1981; Ghuman *et al.*, 1990). Mild heating of the flour has little effect, but autoclaving for 4 hours at 121 °C reduced its protein value from 60% to 53.9% (Ananthara-man and Carpenter, 1971). Ghuman *et al.* (1990) report that the chemical score for S-amino acids and lysine ranges from 37% to 50% and from 47% to 55% respectively, with a significant and positive correlation between PER and methionine levels. They also found highly significant correlations between conarachin values and the values of the two limiting amino acids, methionine and lysine. This suggests that cultivars with a high conarachin content are better for breeding purposes. Evans and Bandemer (1967) report a protein nutritive value (weight gain × 100/weight gain of rats fed casein) ranging from 32% to 49%. A similar assay utilizing *Tetrahymena pyriformis* W found a range from 45.6% to 54.2% (Ghuman *et al.*, 1990). This result was significantly linked to the chemical score and PER and can be used as a quick method to measure protein quality.

The limited availability of lysine and methionine are linked to decreased BV and PER compared with the WHO/FAO reference protein. Authors differ in their views as to the effects of supplementing the groundnut protein with the deficient essential amino acids. These differences may in part be accounted for by the varying amounts added. Howe *et al.* (1965) reported lysine, methionine and threonine to be equally limiting in a diet containing 10% DPF. They used 0.4% threonine, 0.2% lysine and 0.2% methionine to give a PER (2.22) close to that of casein (2.5). Alid *et al.* (1981) reduced the threonine to 0.3% (which still exceeded the WHO/FAO reference pattern for essential amino acids), with a resulting PER of 2.18. McOsker (1962) showed that supplementation of roasted groundnuts with at least 0.31% L-lysine, 0.19% DL-threonine and 0.21% DL-methionine showed feed efficiency equal to or better than that of a 15% casein diet.

Evans and Bandemer (1967) showed no effect with supplementation of 10% groundnut protein with 0.3% methionine, 0.1% isoleucine and 0.1% lysine. They postulated the need for threonine as well. In another experi-ment, they measured bacterial growth, this time increasing the methionine to 3.3%, isoleucine to 0.7% and the lysine to 0.5%. The protein nutritive value increased from 26% to 81%. Adding 0.2% valine and 0.2% threo-nine caused this to decrease to 45%. Joseph *et al.* (1974) reported that the addition of 1.5% L-lysine and 1% DL-methionine to a diet containing groundnut protein at a dietary level of 10% increased the PER from 1.65 to 2.07 after 4 weeks. Addition of 1% threonine further increased this to 2.6. The supplementation of an extrusion texturized protein with 0.3% DL-threonine, 0.2% L-lysine and 0.2% DL-methionine gave a similar PER (2.18) to the similarly supplemented starting product (DPF), which suggests that there is no important change in the protein quality during this process (Alid *et al.*, 1981).

Quinn *et al.* (1975) found that feeding fermented groundnut flour (at 15%), supplemented with 0.4% methionine, 0.25% lysine and 0.13% tryptophan, made no difference to the PER (2.37–2.44) in comparison with supplemented, non-fermented flour (2.5–2.53). It would be interesting to know the PER values for unsupplemented flour. Anantharaman and Carpenter (1971) found supplementation of both unheated and mildly heated DPF improved the PER, the order of limitation being methionine > lysine > threonine. Severe heating of the flour changed this order to lysine > cystine > methionine > threonine.

It was thought by some that raising the dietary level of groundnut protein might increase its biological usefulness. Lusas (1979) reported rats fed 16.7–20% groundnut protein grew as well as those fed 12–24% casein; however, Murphy and Dunn (1950) showed a diet containing 19% groundnut protein needed to be supplemented with 1.6% lysine plus 0.58% methionine to be adequate for lactation and reproduction. Lysine supplementation alone was insufficient.

It is important to note that treatment of the groundnut meal or flour to reduce aflatoxin contamination can markedly reduce the protein quality, reflected in lower PER values (Ihekoronye, 1987). However, it is still suitable for supplementing high carbohydrate, cereal-based foods.

6.4.6 Supplementation of diet

Protein quality is often assessed on the 'completeness' of the single protein, rather than its supplementary value in a mixture from a variety of plant and animal sources. More recently the use of groundnut meal is becoming more recognized, not only as a dietary supplement for children on protein-poor, cereal-based diets in economically underdeveloped countries, but also as an effective treatment for children with protein malnutrition, i.e. kwashiorkor (Rao et al., 1965a; Joseph et al., 1960). The nutritional value can be improved by combining it with an animal protein such as skim-milk powder or plant proteins which complement it, or by supplementing it with the deficient essential amino acids (Waldo and Goddard, 1950; Sanchez et al., 1972; Joseph et al., 1974). Rao et al. (1965) showed that, while the protein content remained the same (49.2), supplementing 50% groundnut flour with 50% full-fat soya flour and 1% each of L-lysine and DL-methionine increased the PER to 2.8 and NPU to 72.3.

Protein-rich groundnut flour can be used to supplement corn or wheat flour for the production of a more nutritious bread, a staple food throughout the world (Khalil et al., 1983a; Khalil et al., 1983b). Supplementing wheat flour with 30% groundnut flour gave a 100% increase in protein content of the bread. However, the best result in terms of PER, NPU and organoleptic evaluation was achieved when the bread was supplemented with 20% groundnut flour (Khalil et al., 1983a). Ory and Conkerton (1983) found that the use of more than 12.5% protein flour required dough conditioners to prevent a decrease in loaf volume during baking.

While supplementation of bread with protein flour increases the overall protein content and generally results in a more positive nitrogen balance (due to a better balance of the essential amino acids), increased fibre content can lead to decreased digestibility of the product (Khalil *et al.*, 1983b). Therefore while the PER might increase in the supplemented product, it may still fall a long way short of the reference protein. Khalil *et al.* (1983a) showed the PER of bread increased from 1.21 to a maximum 1.61 (supplemented with 20% groundnut flour) while the NPU increased from 62% to 80% at this level. Substitution of one half of the groundnut flour with either soy or chick-pea flour has been shown to improve the nutritive quality (Bookwalter *et al.*, 1978; Khalil *et al.*, 1983b). Bread made with groundnut protein concentrate showed superior flavour, taste and crumb colour but concentrations above 10% reduced loaf volume considerably (Khan *et al.*, 1975). Available lysine does increase with supplementation, but it is also lost during cooking (Bhat and Vivian, 1980). Bookwalter *et al.* (1978), using 15% defatted groundnut flour, increased the PER from 1.4 to 2.0 with the addition of 0.18% L-lysine. No further improvement was noted with increased lysine levels (Bookwalter *et al.*, 1979). Alternatively, the addition of 20% skim-milk powder substantially increases the PER (Waldo and Goddard, 1950; Joseph *et al.*, 1974). However, if these dietary supplements are further processed by cooking (e.g. as biscuits made with groundnut flour, wheat flour and skim-milk powder) the available lysine content is reduced, which in turn lowers the value of the protein (Clegg, 1960).

Children with kwashiorkor fed experimental blends of two thirds groundnut protein isolate combined with one third dry skim-milk (to give an average protein content of 69%) or 50% isolate with 50% casein plus 2% lysine and 1% methionine (to give an average protein of 83.6%) gave a similar clinical response to those fed skim-milk (Webb *et al.*, 1964). The isolate contained approximately 90% protein. Shurpalekar and Bretschneider (1969) found only 22% of isolate was required to make biscuits with a protein content of 28.8%, suitable for the prevention and treatment of malnutrition. These biscuits had good organoleptic and keeping qualities in comparison with biscuits made with defatted groundnut flour, which were reported to be hard (Ory and Conkerton, 1983). Alternatively, the protein isolate can be combined with lentil flour, raw sugar and hydrogenated groundnut oil, to be eaten as a sweet (Dumm *et al.*, 1966).

Groundnut protein, once used only as animal feed after the extraction of oil from kernels, is gaining credibility as a readily available source of protein to meet world demand. While invariably deficient in the essential amino acids lysine, methionine and threonine, the protein quality is of a sufficient standard to increase protein efficiency ratios when added to a staple cereal diet. Processing of the meal to remove carbohydrates and fibre increases the digestibility and protein content of the final product. Supplementation of the deficient amino acids, with either another plant protein or the deficient amino acids, can provide a food capable of meeting dietary requirements even in the malnourished.

The quality of groundnut protein would also be improved if the levels of the three main limiting amino acids were increased in the seed protein. If plant breeders could identify cultivars that consistently produce higher levels of any (or all) of these three, regardless of location or environmental effects, the resulting protein would have even more value.

6.4.7 The effect of storage on protein quality

Yannai and Zimmermann (1970) noted that storage of groundnut meal for 6–18 months at relative humidities of 40% and 60% resulted in a slight drop in the nutritive value of the meal. The PER of the meal increased on storage to 24 months and at some storage temperatures was greater (3.19) than the PER value of the freshly prepared meal (2.84). It is interesting to note that the trypsin inhibitor activity of the unheated groundnut meal decreased significantly during this storage trial. The samples stored at the highest temperatures showed greater reductions. It is quite possible that the increases in nutritive value observed by Yannai and Zimmermann (1970) were a direct result of trypsin inhibitor degradation. It should be noted that the successful storage of this meal at adverse temperatures and relative humidities for up to two years was a direct result of the efficient removal of oil from the dehulled seeds at the commencement of the experiment.

6.5 CARBOHYDRATE CONTENT

Carbohydrates in groundnut kernels consist of water-soluble carbo-hydrates (monosaccharides, disaccharides) and oligosaccharides (including starch, raffinose and stachyose). These together comprise the total avail-able carbohydrate fraction. A summary of typical values is shown in Table 6.6. Two reviews of groundnut carbohydrates with good summaries of the non-starchy polysaccharide fraction have been published by Tharanathan *et al.* (1975, 1979).

Carbohydrate levels are high in newly germinated groundnut kernels but the levels begin to fall from an average of 46% at 5 weeks as the maturing seed begins to store oil (Abdel Rahman, 1982b). Mature kernels are reported to contain 9.5–19.0% total available carbohydrate as both soluble and insoluble carbohydrate (Rao *et al.*, 1965a; Oke, 1967; Abdel Rahman, 1982b; Woodroof, 1983). Water-soluble carbohydrates comprise 3.0 to 8.9% of the available carbohydrate (Sekhon *et al.*, 1970; Wallerstein *et al.*, 1989; Rodriguez *et al.*, 1989). Seasonal variation plays a role in the levels of both soluble and insoluble carbohydrates. Rodriguez *et al.* (1989) analysed soluble carbohydrate levels from a crop on two successive years. The first

TABLE 6.6   *Carbohydrates in groundnut kernels and flour (%)*

	Kernels	Flour
Soluble sugars		
Monosaccharides	0.1–0.3	0.13
Sucrose	1.9–5.2	7.70
Total water-soluble carbohydrates	3.0–8.92	
Oligosaccharides		
Raffinose	0.14	
Stachyose	0.71	
Starch	6.7–10.0	
Total available carbohydrates	9.5–18.6	22.1–32.85
Insoluble carbohydrates		
Total insoluble carbohydrates	12.2–15.67	
Crude fibre	1.2–4.3	2.7–5.76
Neutral-detergent fibre	8.3–11.2	8.4
Acid-detergent fibre	4.3–5.9	
Lignin	1.1–3.3	

Based on data from: Anantharaman *et al.*, 1959; Ladell and Phillips, 1959; Howe *el al.*, 1965; Rao *et al.*, 1965a; Oke, 1967; Daghir *et al.*, 1969; Yannai and Zimmermann, 1970; Ayres *et al.*, 1974; Bookwalter *et al.*, 1979; Alid *et al.*, 1981; Woodroof, 1983; Ory and Conkerton, 1983; Wallerstein *et al.*, 1989; Rodriguez *et al.*, 1989.

year ranged from 7.14% to 8.92%; the following year's crop ranged from 5.49% to 7.82%. Analysis of insoluble carbohydrates showed little difference between the two crops (12.79% to 15.67% compared with 12.18% to 15.43%).

The amount of total available carbohydrate falls slightly if the groundnuts are skinned, whereas boiling can account for a loss of 4% (Woodroof, 1983). Water-soluble carbohydrates comprise 3.0 to 6.17% of the available carbohydrate (Sekhon *et al.*, 1970; Wallerstein *et al.*, 1989). Roasting the kernels causes an increase in available carbohydrate, probably due to the loss of volatiles and water. It also results in the hydrolysis of sucrose, the predominant water-soluble carbohydrate, to glucose and fructose. These then react with free amino acids (Maillard reaction) to give the characteristic flavour of roasted groundnuts (Young *et al.*, 1974b). This is supported by Sekhon *et al.* (1970) who reported an increase in reducing sugars, offset by a decrease in non-reducing sugars and total water-soluble carbohydrate during roasting. Differing soluble carbohydrate levels, due to varietal/ environmental differences or level of maturity, may all contribute to variation in roast groundnut kernel characteristics (Rodriguez *et al.*, 1989).

Carbohydrate levels in groundnut products depend on the method of processing used (Table 6.6). Groundnut flour, made from defatted groundnuts, contains protein and the soluble and insoluble carbohydrates. Groundnut protein concentrate also contains some insoluble carbohydrates, whereas groundnut isolates should only contain protein (Natarajan, 1980).

The total carbohydrate level in groundnut flour is reported to range between 22.11 and 32.85% (Anantharaman et al., 1959; Daghir et al., 1969; Yannai and Zimmermann, 1970; Fetuga et al., 1973; Ayres et al., 1974; Conkerton and Ory, 1976; Bhat and Vivian, 1980; Alid et al., 1981; Carew et al., 1988; Woodroof, 1983).Sometimes reported as nitrogen-free extractives (NFE), it is usually estimated by the difference in proximate analysis, which can account for the wide range in reported values. For example, Rao et al. (1965) found 18.5% carbohydrate in a 1:1 blend of groundnut and soya flour, compared with 28.9% found by Shurpalekar and Bretschneider (1969). The other major difference between the two was in moisture levels – 7.2% for the former compared with 2.9% for the latter. Ostrowski et al. (1971) reported groundnut concentrate with 24.1% nitrogen-free extractives, a similar level to that found in groundnut flour, whereas Khan et al. (1975) gave a figure of 18.5%.

The starch levels in groundnut flour inhibit the formation of extrusion-texturized protein. They tend to gelatinize upon extrusion and disrupt the protein matrix necessary for texture formation (Alid et al., 1981). This can be controlled by the addition of surfactants.

Whole-wheat flour has a carbohydrate content which ranges from 72.1% to 81.2%. The addition of groundnut flour to wheat flour decreases the total carbohydrate levels while increasing protein content (Bhat and Vivian, 1980; Khalil et al., 1983b; Ory and Conkerton, 1983). In contrast, products such as muffins made from 100% groundnut flour had a carbohydrate level of 23.2% (Ory and Conkerton, 1983). Tharanathan et al. (1976) analysed the soluble carbohydrate composition of defatted groundnut flour. Glucose was the most predominant (2.9%), followed by fructose (2.2%) and sucrose (0.9%). Raffinose and stachyose levels were low (0.94% and 0.20% respectively). Since they are both poorly digested and responsible for flatulence in humans, low levels make the groundnut flour more acceptable as a food supplement than some other plant proteins, such as soybean flour which can contain higher levels of these indigestible oligosaccharides.

The fermentation of groundnut meal initially causes a slight fall in carbohydrate levels (after 18 hours), which then increases again. It is thought that the increase may be due to degradation of the complex carbohydrates, such as raffinose and stachyose (Bhavanishankar et al., 1987). Quinn et al. (1975) used a 4-day fermentation period and found the carbohydrate levels fell from 35.1–35.8% to 29.3–33.2%. The variation was attributed to the different strains of fungi used for the fermentation, some of which hydrolyse stachyose and/or raffinose better than others. The preparation of a miso-like product, which involves two separate and distinct fermentations, resulted in an increase in reducing sugars from 4.34% to 8.57% after 20 days fermentation (Rao and Rao, 1972).

6.5.1 Fibre

Plant fibre is generally resistant to hydrolysis by human digestive enzymes and remains after food has passed through the small intestine. As such it is referred to as 'unavailable carbohydrate', even though it undergoes some bacterial degradation in the colon. The fibre content of a food can lead to decreased digestibility of the proteins, therefore reducing amino acid availability. Groundnuts contain fibre which consists principally of the carbohydrates cellulose, hemicellulose and pectin. This fraction also contains smaller amounts of mucilages, gums and lignin. Wallerstein et al. (1989) report that the lignin content of groundnuts ranges from 1.1% to 3.3% for three cultivars. It should be noted, however, that lignin is not a polysaccharide – it is just found associated with the fibre fraction. The terms crude fibre, acid-detergent fibre and neutral detergent fibre denote different methods to estimate different fractions present in food. While the estimation of crude fibre seems to be widely used, it is not a true indication of total dietary fibre (Ory and Conkerton, 1983).

Crude fibre levels in raw groundnuts are reported to range from 1.2% to 5.0% (Rao et al., 1965a; Oke, 1967; Derise et al., 1974; Woodroof, 1983). This figure increases slightly for acid-detergent fibre (4.3% to 5.9%), and significantly for neutral-detergent fibre (8.3% to 11.2%) (Wallerstein et al., 1989). Fibre content decreases slightly with boiling or removal of skins, whereas roasting leads to a slight increase (Woodroof, 1983; Derise et al., 1974).

The reported levels of crude fibre for groundnut flour are similar: 2.70–5.76% (Anantharaman et al., 1959; Ladell and Phillips, 1959; Howe et al., 1965; Woodroof, 1969; Natarajan, 1980; Daghir et al., 1969; Yannai and Zimmermann, 1970; Fetuga et al., 1973; Ayres et al., 1974; Bookwalter et al., 1979 Alid et al., 1981). In contrast to this, Carew et al. (1988) give a figure of 10.6% for crude fibre in groundnut meal – almost double the amount – which may be attributable to the different methods used to measure this fraction. Crude fibre in groundnut concentrate ranges from 4.0% to 6.3% (Ostrowski et al., 1971; Khan et al., 1975).

The fibre content of a flour is increased if it is made from unskinned groundnuts. Ory and Conkerton (1983) calculated the neutral-detergent fibre (NDF) level of groundnuts and skins to be 8.4%, whereas the NDF of the skins alone ranged from 35.1% to 41.1%. As such, the skins contribute to the total fibre content of the flour. This could be exploited in developed countries where levels of dietary fibre are often low. The addition of 12.5% groundnut flour in breads increases the dietary fibre content from 0.3% (wheat) to 0.4–0.6% (Ory and Conkerton, 1983). Kahn et al. (1975) attained slightly higher levels with 15% DPF (0.7–0.8%) and 15% groundnut protein concentrate (0.9%). Muffins, made from 100% groundnut flour, attained fibre levels of 2.2% (Ory and Conkerton, 1983).

Fermentation of groundnut meal for up to 22 hours showed no change in the crude fibre content of any of the samples (Bhavanishankar *et al.*, 1987). Quinn *et al.* (1975) reported a slight increase in crude fibre content of the fermented samples, perhaps due to the loss of volatiles during fermentation (this process took 4 days at 28 °C).

A compound not classified as fibre but often found associated with it in foods is phytic acid. This is present as phytin (the mixed calcium, magnesium and potassium salt of phytic acid) in groundnut flour at a concentration of 3.2%. Phytic acid is able not only to bind calcium in the diet and reduce availability but also to bind other useful elements such as zinc and magnesium. Groundnut kernels contain 3.3% of the total phosphorus as phytic acid phosphorus, so that this is also unavailable (Oke, 1967).

6.6 VITAMINS

Raw groundnuts are known to be an excellent source of certain vitamins, especially E, K and the B group. The groundnut is one of the richest sources of thiamin (B_1) in plants. Dougherty and Cobb (1970a,b) noted that 90% of the total thiamine content of groundnuts was contained in the testa, principally in the cell wall fraction. They also noted that approximately 11% of this migrated into the cotyledons when they were dried. Chaturvedi and Geervani (1986) noted that groundnuts were also an excellent source of niacin which ranged from 16.7 to 21.7 mg/100 g DM basis for a number of Indian cultivars. They observed that boiling resulted in a small loss of niacin, while roasting and frying resulted in a small increase due to loss of moisture. In human experiments they observed that 91.7% of the niacin in boiled groundnuts was available, falling to 87.7% for roasted groundnuts. Chaturvedi and Geervani (1986) suggest that the niacin content of groundnuts makes a significant contribution to the diets of people living in dryland areas. This is important as diets in these areas contain limited amounts of the essential amino acid tryptophan; niacin, however, in adequate amounts can spare tryptophan for protein synthesis.

Rao *et al.* (1965b) noted that the tocopherol content of Indian groundnut oil was 0.93 mg/g oil and it was made up of 36% α and 64% γ tocopherol. The total amounts of tocopherol found in groundnut oil were similar to those found in other locally produced oils and to imported soya bean oil.

The colour of groundnut oil is in part due to the carotenoid content. It is well known that there is a marked reduction in the colour of groundnut oil as the seed matures. Pattee *et al.* (1969) showed that this reduction in colour was due to a rapid increase in the oil content of the seed which diluted the carotenoid content. Pattee and Purcell (1967) showed that a sample of oil extracted from immature virginia groundnuts contained 60 μg β-carotene and 138 μg lutein/litre of oil. The total concentration of carotenoids from mature groundnuts was less than 1 μg/litre of oil.

6.7 MINERAL COMPOSITION

The total ash content of groundnuts and products derived from groundnuts is shown in Table 6.7. The rise in the total ash content of defatted groundnut flour is expected as fat is removed from the kernels; in the same way a reduction in the total ash content of protein concentrates or isolates is to be expected as these are easily removed in the washing process. Abdel Rahman (1982b) noted that the total ash content ranged from 1.95% to 2.10% in the growing seed in the developmental period from 5 to 12 weeks.

The mineral content of raw groundnuts and some of their products is shown in Table 6.8. In general, groundnuts provide a wide range of mineral elements to supplement human dietary requirements. They have the advantage that no element is found in excessive amounts in the kernel. It should be noted, however, that a considerable amount of minerals is lost during processing in the manufacture of protein concentrates.

Derise *et al.* (1974) showed that, for three different cultivars grown under similar conditions in the USA, the mineral element content was very similar for a range of elements. They also showed, for the same range of mineral elements, that roasting the groundnuts led to a rise in mineral content as volatiles were lost (except in the case of sodium where a small fall was measured). The range of chromium in groundnuts is 3–41 mg/kg in the kernels. This amount is nutritionally significant, considering that a chromium intake of 50–200 µg/day has been tentatively recommended for adults. Groundnuts also supply significant levels of copper in the diet: 100 g of roasted kernels would supply all the daily needs of copper (Khalil and Chughtai, 1983).

Akrida-Demertzi *et al.* (1985) did note that the level of lead in raw fat extracted groundnut flour was 5.3 µmg/kg for groundnuts grown in a remote coastal area. This level rose to 27.6 µg/kg in groundnuts grown in an inland area of Greece noted for higher levels of air pollution (from

TABLE 6.7 *Total ash content of groundnuts and groundnut products (g/100 g DM)*

	Kernels	Roasted	Partially defatted flour	Defatted flour	Protein concentrate	Isolate
Ash	1.8–3.1	1.7–8.9	4.2–5.3	3.0–4.8	2.1	0.5
Silica free ash		0.1–0.4				

Based on data from: Hoffpauir, 1953; Ladell and Phillips, 1959; Howe *et al.*, 1965; Rao *et al.*, 1965a; Sekhon *et al.*, 1970; Ayres *et al.*, 1974; Derise *et al.*, 1974; Quinn *et al.*, 1975; Conkerten and Ory, 1976; Bookwater *et al.*, 1979; Natarajan, 1980; Alid *et al.*, 1981; Abdel Rahman, 1982a; Woodroof, 1983; Khalil and Chunghtai, 1983; Wallerstein *et al.*, 1989.

TABLE 6.8   *Mineral content of groundnuts and groundnut products (mg/100 g DM)*

	Raw	Roasted	Defatted flour	Isolate	Butter
Ca	44.0–87.8	55.2–91.0	92–200	40–100	46.7
Mg	157.0–200.0	174.3–196.0	10–343	10–70	148
P	137.0–470.3	288.2–538.2	80–100	–	188
Na	5.8–66.0	4.2–8.5	1.4–33.3	0.3–0.5	388
K	618.2–890.0	643.5–734.1	1150–1450	5–2430	682
Fe	1.2–2.0	1.3–2.1	1.5–10.9	1.9–6.7	1.6
Cu	0.74–1.60	1.3–1.7	0.6–5.2	2.1–14.6	0.7
Zn	3.3–6.2	4.4–6.7	4.7–7.0	3.0–18.2	2.9
Mn	1.7–19.0	1.9–2.2	3.1–6.1	1.6–3.4	1.4
Cl	5.8–10.1	5.7–6.4			5.8
Cr	0.2–0.4	0.3–0.5			0.26
Co	–				0.11
Al	2.3	–			
Mo	0.07				
B	1.8–3.2				

Based on data from Oke, 1967; Derise *et al.*, 1974; Galvaro *et al.*, 1976; Morris *et al.*, 1978; Lusas, 1979; Khalil and Chughtai, 1983; Salazar and Young, 1984a,b.

leaded petrol). The lead levels fell only slightly on roasting. It is interesting to note that, although some minerals are extracted into the groundnut oil fraction, the levels are quite low except for cobalt and lead which appear to be concentrated into the oil fraction (Akrida-Demertzi *et al.*, 1985). The boron content of raw groundnuts ranges from 1.8 to 3.2 mg/100 g and as boron deficiency has been associated with cotyledon damage (hollow heart) (Salazar and Young, 1984a) it is unlikely to fall below this range. It is interesting to note that boron has also been found to modify the relative concentration of the basic volatile aroma constituents of roasted groundnuts (Salazar and Young, 1984a).

6.8 OTHER COMPONENTS

A number of compounds can affect the typical groundnut flavour, nuttiness, sweetness and bitterness. Many of these can be altered by variety, growing conditions, methods of harvesting and storing and processing. The bitter flavour in groundnuts is due in part to the saponins, which are found principally in the embryo but also in the cotyledons under certain field conditions (Dieckert and Morris, 1958). Four different saponins could be separated using glass paper chromatography. Maga and Lorenz (1974) report that the total amount of free phenolic acids in defatted groundnut flour was 267 mg/kg. Of the 27 different phenolic acids identified, four were found to occur in amounts which could contribute to the astringency of the flour.

The red skins, which represent 2.0–3.5% of the kernels, contain tannins and related pigments that cause an undesirable colour in the protein unless removed during initial processing. Stansbury *et al.* (1950) found that the red skins contained about 7% tannin along with thiamine. These tannins may also be responsible for some of the bitterness associated with raw groundnuts.

6.8.1 Antinutritive factors

Many legumes contain a range of toxic factors which can adversely affect the nutritive value of the seed. These toxic factors are a group of unrelated chemical compounds with varying effects on metabolic processes. A consistent feature is that the effect of all these factors can be reduced by traditional processing and cooking methods and they are not considered to affect seriously the utilization of cooked groundnut seeds. The effect they have when consumed as raw seeds are probably minor when compared with other legume seeds.

An example of a toxic factor which is found in insignificant amounts in groundnuts is the report by Oke (1967) that oxalic acid occurs in Nigerian groundnuts at a level of 4.0 g/kg. This is comparable to many other grain legume seeds but is low when compared with some other fruits and vegetables.

6.8.2 Trypsin inhibitors

It is widely reported that raw legume seeds contain trypsin inhibitors and groundnuts are no exception. Tur-Sinai *et al.* (1972) extracted an inhibitor from defatted groundnuts and showed that it had activity towards trypsin and chymotrypsin (10480 TIU and 4793 CTIU respectively in 370 g of defatted groundnut flour). They went on to show that the inhibitor lacked the amino acids tryptophan, methionine, isoleucine and cysteine (i.e. it contained no sulphur bridges). They suggested that the inhibitor which they had extracted from raw groundnuts was different to that observed in soya beans. Soya bean inhibitors have a double headed complex which can bind trypsin and chymotrypsin at each end; in contrast, groundnut inhibitor active sites are very close to each other and may even coincide. The overall effect of trypsin inhibitors is to cause a general delay in protein digestion which lowers the overall digestibility of the groundnut protein. Moist heat treatments are the most effective way of reducing their activity. Perkins and Toledo (1982) showed that almost complete destruction could be achieved in 45 minutes at 120 °C; at 100 °C, 98% destruction was only achieved after cooking for 180 minutes. These results confirm the observations of Tur-Sinai *et al.* (1972) that the trypsin inhibitor was stable when boiled for 15 minutes at pH 8.

Mir and Hill (1979) showed that groundnut meals prepared by extracting the fat using petroleum ether at 38–46 °C for 48 hours contained trypsin inhibitor levels which were about 11% of those found in similarly treated soya bean meal.

Sitren *et al.* (1985) showed that the trypsin inhibitor activity of raw groundnut flour was higher than a similarly processed soya bean flour (144.7 vs 112.1 μg lectin/g flour); moist heat (121 °C 15psi for 20 minutes) was most effective at reducing the effect of the trypsin inhibitor. Dry heat (177 °C for 20 minutes) resulted in only 76% (groundnut) and 61% (soya bean) inhibition of the trypsin inhibitor. In a subsequent feeding experiment it was interesting to note that, after 4 weeks, the rats eating the raw and dry-heated groundnut-based diets (which would contain high levels of trypsin inhibitor) were consuming as much diet as the rats consuming the moist heated groundnut diet. The rats fed the raw and dry heated soya bean diet had considerably reduced intakes when compared to the moist heat treatment diet. This experiment suggests that rats at least can adapt to the trypsin inhibitor found in groundnuts and also demonstrates that it is not readily destroyed during heat processing.

6.8.3 Haemagglutinins

A non-specific haemagglutinin has been isolated in skin-free groundnut kernels by Dechary *et al.* (1970), who noted that the amino acid composition of purified groundnut haemagglutinin was very similar to the non-specific haemagglutinin concanavalin. The low levels of haemagglutinins found in groundnuts and the fact that they are readily inactivated by heat suggest that haemagglutinins are not a significant problem in groundnuts.

6.8.4 Goitrogens

Groundnuts have been reported to produce goitrogenic effects when fed at high levels to animals (Srinivasan *et al.*, 1957). A phenolic glycoside is thought to be responsible and is reported to be present in the testa, which is commonly removed in most cooking processes. It was noted in these experiments that the goiterogenic effect was counteracted by the addition of iodine in the experimental diets. In iodine deficient areas, goitrogens may have an effect, when large amounts of groundnuts are consumed in the diet but to date no details of adverse effects have been published.

6.9 CONCLUSIONS

Traditionally, groundnuts have been considered as a source of high quality vegetable oil; however, the annual production of protein has reached very significant proportions. The groundnut is now the world's fourth most important source of edible oil and third most important source of vegetable protein. It makes a significant contribution to the diets of many people in developing nations. The manufacture of many food products consisting of various cereals fortified with a concentrated protein source from groundnuts has done much to alleviate malnutrition in many countries. The range of products that can be made from groundnuts is perhaps the reason why the consumption of this product has increased around the world. In part the special taste and aroma that can be achieved on cooking, especially when roasted, is an important feature of this food's acceptance in an extremely wide range of cultures.

Groundnuts characteristically contain high levels of oil and protein and low levels of carbohydrates and ash. In other ways groundnuts are similar to many other grain legumes. Groundnut protein is deficient in some essential amino acids (methionine and lysine) but overall the digestibility of the protein is superior to many other plant protein sources. Groundnuts do contain a number of antinutritive factors but these tend to be found at lower levels and appear to be less active than in other grain legumes. All of these factors can be reduced by traditional processing and cooking methods and they are not considered to affect seriously the utilization of nutrients in cooked groundnut seeds.

Considerable potential still exists to improve the nutritive value of groundnuts by plant breeding and by developing improved processing techniques. The potential of groundnuts to produce large amounts of protein should be encouraged as this vegetable protein production is very efficient when compared with alternative animal production systems.

ACKNOWLEDGEMENTS

The assistance of the staff of the Lincoln University Library to obtain many of the references used in this work is gratefully acknowledged.

REFERENCES

Abdel Rahman, A.-H.Y. (1982a) Compositional study on some Egyptian peanut varieties. *Rivista Italiana Delle Sostanze Grasse*, **59**(6), 287–288.

Abdel Rahman, A.-H.Y. (1982b) Changes in chemical composition of peanut during development and ripening. *Rivista Italiana Delle Sostanze Grasse*, **59**(6), 285–286.

Akrida-Demertzi, K., Tzouwara-Karayanni, S.M. and Voudouris, E. (1985) Differences in concentration of essential and toxic elements in peanuts and peanut oil. *Food Chemistry*, **16**(2), 133–139.

Alid, G., Yanez, E., Aguilera, J.M. *et al.* (1981) Nutritive value of an extrusion-texturized peanut protein. *Journal of Food Science*, **46**(3), 948–949.

Amaya, F.J., Young, C.T. and Hammons, R.O. (1977) Tryptophan content of the US commercial and some South American wild genotypes of the genus *Arachis*, A survey. *Oléagineux*, **32**(5), 225–229.

Anantharaman, K. and Carpenter, K.J. (1969) Effects of heat processing on the nutritional value of groundnut products. 1. Protein quality of groundnut cotyledons for rats. *Journal of the Science of Food and Agriculture*, **20**, 703–708.

Anantharaman, K. and Carpenter, K.J. (1971) Effects of heat processing on the nutritional value of groundnut product. 2. Individual amino acids. *Journal of Science and Food Agriculture*, **22**(8), 412–418.

Anantharaman, N., Subramanian, N., Bhatia, D.S. and Subrahmanyan, V. (1959) Processing of groundnut cake for edible protein isolate. *Indian Oilseeds Journal*, **3**, 85–90.

Arya, S. S., Vidyasagar, K. and Nath, H. (1972) Storage properties and acceptability trials on refined groundnut oil as a substitute for hydrogenated oil. *The Indian Journal of Nutrition and Dietetics*, **9**, 145–152.

Ayres, J.L., Branscomb, L.L. and Rogers, G.M. (1974) Processing of edible peanut flour and grits. *Journal of the American Oil Chemists' Society*, **51**(4), 133–136.

Badami, R.C., Shivamurthy, S.C., Joshi, M.S. and Patil, K.B. (1980) Yield characters and chemical examination of five varieties of hybrid bunch type peanuts with respect to their parents. *Fette Seifen anstrichmittel*, **82**(10), 400–401.

Banerjee, S., Rao, P.N. and Ghosh, S.K. (1965) Biochemical and histochemical changes in aorta of chicks fed vegetable oils and cholesterol. *Proceedings of the Society for Experimental Biology and Medicine*, **119**, 1081–1086.

Barnett, D., Baldo, B.A. and Howden, M.E.H. (1983) Multiplicity of allergens in peanuts. *Journal of Allergy and Clinical Immunology*, **72**(1), 61–68.

Basha, S.M. (1988) Resolution of peanut seed proteins by high-performance liquid chromatography. *Journal of Agricultural and Food Chemistry*, **36**(4), 778–781.

Basha, S.M. (1991) Deposition pattern of methionine-rich protein in peanuts. *Journal of Agricultural and Food Chemistry*, **39**(1), 88–91.

Ben-Gera, I. and Zimmerman, G. (1972) Changes in the nitrogenous constituents of staple foods and feeds during storage. *Journal of Food Science and Technology*, **9**(3), 113–118.

Bertorelli, L.O. de (1976) Effectos varietales y ambientales sobre la composicion en acidos grasos del aceite de maní. *Revista Facultad de Agronomica Alcance (Maracay)*, **9**(1), 37–52.

Bezard, J. and Sawadogo, K.A. (1983) Structure glyceridique du tissu adipeux périrénal de rats soumis à un régime à base d'huile d'arachide. *Reproduction and Nutritional Development*, **23**(1), 65–80.

Bhat, C.M. and Vivian, V.M. (1980) Effect of incorporation of soy, peanut and cotton seed flours on the acceptability and protein quality of chapatis. *Journal of Food Science and Technology*, **17**(4), 168–171.

Bhatia, D.S., Kalbag, S.S., Subramanian, N. *et al.* (1960) Production of protein isolate from groundnut. *Proceedings of the Symposium on Proteins*, Mysore, 260–265.

Bhavanishankar, T.N., Rajashekaran, T. and Sreenivasa Murthy, V. (1987) Tempeh-like product by groundnut fermentation. *Food Microbiology*, **4**, 121–125.

Bookwalter, G.N., Warner, K., Anderson, R.A. and Bagley, E.B. (1978) Cornmeal/peanut flour blends and their characteristics. *Journal of Food Science*, **43**, 1116–1120.

Bookwalter, G.N., Warner, K., Anderson, R.A. and Bagley, E.B. (1979) Peanut-fortified food blends. *Journal of Food Science*, **44**, 820–825.

Bucker, E.R., Jnr, Mitchell, J.H., Jnr and Johnson, M.G. (1979) Lactic fermentation of peanut milk. *Journal of Food Science*, **44**, 1534–1538.

Buxton, J., Grundy, H.M., Wilson, D.C. and Jamison, D.G. (1954) The absence of anti-thyroid properties for rats in two types of groundnut oil in common use in Nigeria. *British Journal of Nutrition*, **8**, 170–172.

Cama, H.R. and Morton, R.A. (1950) Changes occurring in the proteins as a result of processing groundnuts under selected industrial conditions. 2. Nutritional changes. *British Journal of Nutrition*, **4**, 297–316.

Carew, S.N., Olomu, J.M. and Offiong, S.A. (1988) Amino acid supplementation of ground-nut meal protein in broiler diets. *Tropical Agriculture (Trinidad)*, **65**(4), 329–332.

Carpenter, K.J. and Ellinger, G.M. (1951) The effect of processing methods on the avail-ability of lysine in groundnut meals. *Biochemistry Journal*, **48**, liii–liv.

Chandra, S., Prasad, D.A. and Krishna, N. (1985) Effect of sodium hydroxide treatment and/or extrusion cooking on the nutritive value of peanut hulls. *Animal Feed Science and Technology*, **12**, 187–194.

Chaturvedi, A. and Geervani, P. (1986) Bioavalibility of niacin from processed groundnuts. *Journal of Nutrition Science and Vitaminology*, **32**, 327–334.

Cheema, P.S. and Ranhotra, G.S. (1967) Evaluation of groundnut (*Arachis hypogaea*) varieties for protein quality. *Journal of Nutrition and Dietetics*, **4**, 93–95.

Cherry, J.P. (1977) Potential sources of peanut seed proteins and oil in the genus *Arachis*. *Journal of Agricultural and Food Chemistry*, **25**(1), 186–193.

Chopra, A.K. and Bhatia, I.S. (1970) The influence of habit of growth and environment on the protein quality of groundnut. *Journal of Research*, **7**, 69–74.

Chopra, A.K. and Sidhu, G.S. (1965) Lysine and methionine contents of some strains of groundnut of the Punjab. *Journal of Research*, **2**, 49–53.

Chopra, A.K. and Sidhu, G.S. (1967) Nutritive value of groundnut. 1. Amino acid compo-sition of different varieties of groundnut grown in the Punjab. *British Journal of Nutrition*, **21**, 519–525.

Clegg, K.M. (1960) The availability of lysine in groundnut biscuits used in the treatment of kwashiorkor. *British Journal of Nutrition*, **14**, 325–329.

Cluzan, R., Suschetet, M., Rocquelin, G. and Levillain, R. (1979) A. Recherches anatomo-phologiques chez le rat ingerant différentes doses d'huile d'arachide ou d'huile de colza à faible teneur en acide erucique (Huile de colza Primor). 5a. Étude histologique du myocarde. *Annales de Biologie Animale Biochimie Biophysique*, **19**, 497–500.

Coelho, D.B. and Bhat, J.V. (1959) Studies on groundnut fermentation. 1. Effect of fermentation on the nutritive value of the product. *Journal of the Indian Institute of Science*, **41**, 1–8.

Conkerton, E.J. and Ory, R.L. (1976) Peanut proteins as food supplements, a compositional study of selected Virginia and Spanish peanuts. *Journal of the American Oil Chemists' Society*, **53**(12), 754–756.

Crawford, R.V. and Hilditch, T.P. (1950) The component fatty acids and glycerides of groundnut oil. *Journal of the Science of Food and Agriculture*, **1**, 372–379.

Daghir, N.J., Ayyash, B. and Pellett, P.L. (1969) Evaluation of groundnut meal protein for poultry. *Journal of the Science of Food and Agriculture*, **20**, 349–354.

Dechary, J.M., Leonard, G.L. and Corkern, S. (1970) Purification and properties of a nonspecific hemagglutinin from the peanut (*Arachis hypogaea*). *Lloydia*, **33**, 270–274.

Derise, N.L., Lau, H.A. and Ritchie, S.J. (1974) Yield, proximate composition and mineral element content of three cultivars of raw and roasted peanuts. *Journal of Food Science*, **39**, 264–266.

Dieckert, J.W. and Morris, N.J. (1958) Bitter principles of the peanut. Isolation, general properties, and distribution in the seed. *Journal of Agricultural and Food Chemistry*, **6**, 930–933.

Dougherty, R. and Cobb, W.Y. (1970a) Localisation of thiamine within the cotyledon of dormant groundnut (*Arachis hypogaea* L.) *Journal of the Science of Food and Agriculture*, **21**, 411–415.

Dougherty, R.H. and Cobb, W.Y. (1970b) Characterisation of thiamine in the raw peanut (*Arachis hypogaea* L.) *Journal of Agricultural Food Chemistry*, **18**, 921–925.

Dumm, M.E., Rao, B.R.H., Jesudian, G. and Benjamin, V. (1966) Supplemented groundnut protein isolated in pre-school children. *Journal of Nutrition and Dietetics*, **3**, 111–116.

Edwards, C.H., Thompson, E.S. and Tyson, M.H. (1966) Nitrogen balances and growth of rats fed vegetable-protein diets. *Journal of the American Dietetic Association*, **48**, 38–44.

Ehrhart, L.A. and Holderbaum, D. (1980) Aortic collagen, elastin and non-fibrous protein synthesis in rabbits fed cholesterol and peanut oil. *Atherosclerosis*, **37**(3), 423–432.

Evans, R.J. and Bandemer, S L (1967) Nutritive values of some oilseed proteins. *Cereal Chemistry*, **44**(5), 417–426.

Fedeli, E., Favini, G., Camurati, F. and Jacini, G. (1968) Regional differences of lipid composition in morphologically distinct fatty tissues. 3. Peanut seeds. *Journal of the American Oil Chemists' Society*, **45**, 676–679.

Fetuga, B.L., Babatunde, G.M. and Oyenuga, V.A. (1973) Protein quality of some Nigerian feedstuffs. 1. Chemical assay of nutrients and amino acid composition. *Journal of the Science of Food and Agriculture*, **24**(12), 1505–1514.

Fisher, H. (1965) Further studies on the limiting amino-acids in differently processed ground-nut meals. *Journal of the Science of Food and Agriculture*, **16**, 390–393.

Flanzy, J. (1979) Comparison of the physiological effects of rapeseed and groundnut oil. *Annales de Biologie Animale Biochimie Biophysique*, **19**(2b), 467–552.

Galvaro, L.C.A., Lopez, A. and Williams, H.L. (1976) Essential mineral elements in peanuts and peanut butter. *Journal of Food Science*, **41**, 1305–1307.

Ghuman, P.K., Mann, S.K. and Hira, C.K. (1990) Evaluation of protein quality of peanut (*Arachis hypogaea*) cultivars using *Tetrahymena pyriformis*. *Journal of the Science of Food and Agriculture*, **52**, 137–139.

Heinis, J.L., Pastor, J. and Campbell, E.B. (1975). Amino acids in 96 peanut varieties. *Proceedings of the American Peanut Research and Education Association*, 12–17.

Hoffpauir, C.L. (1953) Peanut composition. Relation to processing and utilization. *Agricultural and Food Chemistry*, **1**(10), 668–671.

Hopkins, C.Y. and Chisholm, M.J. (1953). Some fatty acids of peanut, hickory and acorn oils. *Canadian Journal of Chemistry*, **31**, 1173–1181.

Howe, E.E., Gilfillan, E.W. and Milner, M. (1965) Amino acid supplementation of protein concentrates as related to the world protein supply. *American Journal of Clinical Nutrition*, **16**, 321–326.

Ihekoronye, A.I. (1987) Nutritional quality of acid-precipitated protein concentrate from the Nigerian 'Red Skin' groundnut. *Journal of the Science of Food and Agriculture*, **38**(1), 49–55.

Inanaga, S., Arima, E. and Nishihara, T. (1990) Effect of Ca on composition of fat body of peanut seed. *Plant and Soil*, **122**, 91–96.

Iverson, J.L., Firestone, D. and Horwitz, W. (1963) Fatty acid composition of oil from roasted and unroasted peanuts by GLC. *Journal of the Association of Official Agricultural Chemists*, **46**, 718–725.

Jambunathan, R., Raju, S.M. and Barde, S.P. (1985) Analysis of oil content of groundnuts by nuclear magnetic resonance spectrometry. *Journal of the Science of Food and Agriculture*, **36**(3), 162–166.

Johri, T.S., Agrawal, R. and Sadagopan, V.R. (1988) Available lysine and methionine contents of some proteinous feedstuffs. *Indian Journal of Animal Nutrition*, **5**, 228–229.

Joseph, K., Rao, M.N., Swaminatham, M. *et al.* (1960) The nutritive value of protein blends similar to FAO reference protein pattern in amino acid composition. *Proceedings of the Symposium on Proteins*. Mysore, India.

Khalil, J.K., Ahmad, B., Ahmad, I. and Hussain, T. (1983a) Nitrogen balance in human subjects as influenced by corn bread supplemented with peanut and chickpea flours. *Pakistan Journal of Scientific and Industrial Research*, **26**(2), 83–86.

Khalil, J.K., Ahmad, I. and Iqbal, P. (1983b) Nutritional and organoleptic evaluation of wheat bread supplemented with peanut flour. *Pakistan Journal of Scientific and Industrial Research*, **26**(2), 87–90.

Khalil, J.L. and Chughtai, M.I.D. (1983) Chemical composition and nutritional quality of five peanut cultivars grown in Pakistan. *Plant Foods for Human Nutrition*, **33**, 63–70.

Khan, M.N., Rhee, K.C., Rooney, L.W. and Cater, C.M. (1975) Bread baking properties of aqueous processed peanut protein concentrates. *Journal of Food Science*, **40**, 580–583.

Khattab, A.H., El Tinay, A.H., Khalifa, H.A. and Mirghani, S. (1974) Stability of perox-idised oils and fat to high temperature heating. *Journal of the Science of Food and Agriculture*, **25**(6), 689–696.

Kokatnur, M.G., Ambekar, S.Y., Rao, D.S. and Asan, M.S. (1964) Effect of heat oxidized peanut oil on rat growth. *Indian Journal of Biochemistry*, **1**, 106–108.

Kritchevsky, D., Tepper, S.A., Vesselinovitch, D. and Wissler, R.W. (1971) Cholesterol vehicle in experimental atherosclerosis. II. Peanut Oil. *Atherosclerosis*, **14**(1), 53–64.

Kritchevsky, D., Tepper, S.A., Vesselinovitch, D. and Wissler, R.W. (1973) Cholesterol vehicle in experimental atherosclerosis. 13. Randomized peanut oil. *Atherosclerosis*, **17**, 225–243.

Kritchevsky, D., Tepper, S.A., Kim, H.K. *et al.* (1976) Experimental atherosclerosis in rabbits fed cholesterol-free diets. 5. Comparison of peanut, corn, butter, and coconut oils. *Experimental and Molecular Pathology*, **24**, 375–391.

Kritchevsky, D., Tepper, S.A. and Story, J.A. (1978) Cholesterol vehicle in experimental atherosclerosis. *Atherosclerosis*, **31**, 365–370.

Kritchevsky, D., Tepper, S.A., Scott, D.A. *et al.* (1981) Cholesterol vehicle in experimental atherosclerosis. *Atherosclerosis*, **38**, 291–299.

Kritchevsky, D., Davidson, L.M., Weight, M. *et al.* (1982) Influence of native and random-ized peanut oil on lipid metabolism and aortic sudanophilia in the Vervet monkey. *Atherosclerosis*, **42**, 53–58.

Kritchevsky, D., Tepper, S.A., Klurfeld, D.M. *et al.* (1984) Experimental atherosclerosis in rabbits fed cholesterol-free diets. 12. Comparison of peanut and olive oils. *Atherosclerosis*, **50**, 253–259.

Ladell, W.S.S. and Phillips, P.G. (1959) Groundnut flour as a protein source in the Nigerian diet. *Journal of Tropical Medicine and Hygiene*, **62**, 229–237.

Lee, L.S., Morris, N.J. and Frampton, V.L. (1965) Cyclic amino acid derivative from peanut flour. *Journal of Agricultural and Food Chemistry*, **13**, 309–311.

Levine, A.S. and Silvis, S.E. (1980) Absorption of whole peanuts, peanut oil and peanut butter. *The New England Journal of Medicine*, **303**(16), 917–918.

Liu, S.Y., Chan, K.L. and Yang, J.H. (1984) Variation in oil content of the main oil crops of Taiwan. *Fette Seifen Anstrichmittel*, **86**(12), 466–468.

Lusas, E.W. (1979) Food uses of peanut protein. *Journal of the American Oil Chemists' Society*, **56**(3), 425–430.

Maga, J.A. and Lorenz, K. (1974) Gas-liquid chromatography separation of the free phenolic acid fractions in various oilseed protein sources. *Journal of the Science of Food and Agriculture*, **25**, 797–802.

Marshall, H.F. Jr, Shaffer, G.P. and Conkerton, E.J. (1989) Free amino acid determination of whole peanut seeds. *Analytical Biochemistry*, **180**, 264–268.

McOsker, D.E. (1962) The limiting amino acid sequence in raw and roasted peanut protein. *Journal of Nutrition*, **76**, 453–459.

Miller, E.L. (1967). Determination of tryptophan content of feedstuffs with particular reference to cereals. *Journal of Science of Food and Agriculture*, **18**, 381–386.

Mir, Z. and Hill, D.C. (1979) Nutritional value of peanut meals from Ontario grown peanuts. *Journal of the Institute of Canadian Science and Technologic Aliment*, **12**(2), 56–60.

Morris, N.M., Conkerton, E.J., Piccolo, B. and Ory, R.L. (1978) Retention of minerals in protein isolates prepared from peanut flour. *Journal of Agricultural and Food Chemistry*, **25**(5), 1028–1031.

Murphy, E.A. and Dunn, M.S. (1950). Nutritional value of peanut protein. *Food Research*, **15**, 498–510.

Nagaraj, H.K. and Subramanian, N. (1974) Studies on groundnut protein concentrates prepared by alcohol and acid washing of the defatted flour. *Journal of Food Science and Technology*, **11**(2), 54–57.

Natarajan, K.R. (1980) Peanut protein ingredients, preparation, properties and food uses. *Advances in Food Research*, **26**, 215–273.

Neucere, N.J., Ory, R.L. and Carney, W.B. (1969) Effect of roasting on the stability of peanut proteins. *Journal of Agricultural Food Chemistry*, **17**, 25–28.

Nordlee, J.A., Taylor, S.L., Jones, R.T. and Yunginger, J.W. (1991) Allergenicity of various peanut products as determined by RAST inhibition. *Journal of Allergy and Clinical Immunology*, **68**(5), 376–382.

Nouvelot, A., Bourre, J.M., Sezille, G. *et al.* (1983) Changes in the fatty acid patterns of brain phospholipids during development of rats fed peanut or rapeseed oil, taking into account differences between milk and maternal food. *Annals of Nutrition and Metabolism*, **27**, 173–181.

Oke, O.L. (1967) Chemical studies on some Nigerian pulses. *West African Journal of Biological and Applied Chemistry*, **9**, 52–55.

Ory, R.L. and Conkerton, E.J. (1983) Supplementation of bakery items with high protein peanut flour. *Journal of the American Oil Chemists' Society*, **60**(5), 986–989.

Ostrowski, H., Jones, A.S. and Cadenhead, A. (1971). Nitrogen metabolism of the pig. 3. Utilisation of protein from different sources. *Journal of the Science of Food and Agriculture*, **22**(1), 34–37.

Pattee, H.E. and Purcell, A.E. (1967) Carotenoid pigments of peanut oil. *Journal of the American Oil Chemists' Society*, **44**, 328–330.

Pattee, H.E., Purcell, A.E. and Johns, E.B. (1969) Changes in carotenoid and oil content during maturation of peanut seeds. *Journal of the American Oil Chemists' Society*, **46**, 629–631.

Perkins, D. and Toledo, R.T. (1982) Effect of heat treatment for trypsin inhibitor inactivation on physical and functional properties of peanut protein. *Journal of Food Science*, **47**, 917–922.

Phillips, R.J. and Singleton, B. (1978) The determination of specific free fatty acids in peanut oil by gas chromatography. *Journal of the American Oil Chemists' Society*, **55**(2), 225–227.

Pickett, T.A. (1950) Composition of developing peanut seed. *Plant Physiology*, **25**, 210–224.

Quinn, M.R., Beuchat, L.R., Miller, J. *et al.* (1975) Fungal fermentation of peanut flour, effects on chemical composition and nutritive value. *Journal of Food Science*, **40**, 470–474.

Ramamurti, K. and Banerjee, B.N. (1950) Studies on Indian edible oils. Unsaponifiable matter. *Indian Journal of Medical Research*, **38**, 377–383.

Rao, G.R. (1974) Effect of heat on the proteins of groundnut and Bengal gram *(Cicer arietinum)*. *The Indian Journal of Nutrition and Dietetics*, **11**, 268–275.

Rao, M.N., Rajagopalan, R., Swaminathan, M. and Parpia, H.A.B. (1965) Studies on a vegetable protein mixture based on peanut and soya flours. *Journal of the American Oil Chemists' Society*, **42**, 658–661.

Rao, M.N. and Rao, T.N.R. (1972) Development of predigested protein-rich food based on oilseed meals and pulses. *Journal of Food Science and Technology*, **9**(2), 57–62.

Rao, S.K., Rao, S.D.T. and Murti, K.S. (1965a) Compositional studies on Indian groundnuts – III. *Indian Oilseeds Journal*, **9**, 5–13.

Rao, M.K.G., Rao, S.V. and Achaya, K.T. (1965b) Separation and estimation of tocopherols in vegetable oils by TLC. *Journal of the Science of Food and Agriculture*, **16**, 121–124.

Ray, S., Talukder, G., Sengupta, K.P. and Chatterjee, G.C. (1980) Studies on fatty acid composition of serum lipids in rats fed mustard oil or groundnut oil containing diets. *Indian Journal of Experimental Biology*, **18**, 1509–1511.

Roberson, S., Marion, J.E. and Woodroof, J.G. (1966) Composition of commercial peanut butters. *Journal of the American Dietetic Association*, **49**, 208–210.

Rodriguez, M.M., Basha, S.M. and Sanders, T.M. (1989) Maturity and roasting of peanuts as related to precursors of roasted flavour. *Journal of Agricultural and Food Chemistry*, **37**(3), 760–765.

Sachs, M.I., Jones, R.T. and Yunginger, J.W. (1981) Isolation and partial characterization of a major peanut allergen. *Journal of Allergy and Clinical Immunology*, **67**(1), 27–34.

Salazar, A.J. and Young, C.T. (1984a) An azomethine H automated method for boron determination in peanuts. *Journal of Food Science*, **49**(1), 72–74.

Salazar, A.J. and Young, C.T. (1984b) An automated methylthymol blue method for calcium determination in peanuts. *Journal of Food Science*, **49**(1), 209–211.

Sanchez, A., Fuller, A.B., Yahiku, P.Y. and Baldwin, M.V. (1972) Supplementary value of black-eyed peas, peanuts and egusi seed on the typical West African diet of plant origin. *Nutrition Reports International*, **6**(3), 171–179.

Sanders, T.H. (1980a) Effects of variety and maturity on lipid class composition of peanut oil. *Journal of the American Oil Chemists' Society*, **57**(1), 8–11.

Sanders, T.H. (1980b) Fatty acid composition of lipid classes in oils from peanuts differing in variety and maturity. *Journal of the American Oil Chemists Society*, **57**(1), 12–15.

Sekhon, K.S., Ahuja, K.L. and Sandhu, R.S. (1970) Chemical composition of raw and roasted peanuts. *Indian Journal of Nutrition and Dietetics*, **7**, 243–246.

Sekhon, K.S., Ahuja, K.L., Sandhu, R.S. and Bhatia, I.S. (1972a) Variability in fatty acid composition in peanut. 1. Bunch group. *Journal of the Science of Food and Agriculture*, **23**(8), 919–924.

Sekhon, K.S., Ahuja, K.L. and Jaswal, S.V. (1972b) Fatty acid composition of the Punjab peanuts. *The Indian Journal of Nutrition and Dietetics*, **9**, 78–79.

Sekhon, K.S., Ahuja, K.L., Jaswal, S. and Bhatia, I.S. (1973) Variability in fatty acid composition in peanut. 2. Spreading group. *Journal of the Science of Food and Agriculture*, **24**(8), 957–960.

Sen, A. and Gupta, K.P.S. (1979) Effect of feeding common edible oils to rats on the lipid profile of heart tissue. *Indian Journal of Experimental Biology*, **17**, 1277–1279.

Sharma, N.D., Mehta, S.L., Patil, S.H. and Eggum, B.O. (1981) Oil and protein quality of groundnut mutants. *Plant Foods for Human Nutrition*, **31**, 85–90.

Shibahara, A., Fukumizu, M., Yamashoji, S. *et al.* (1977) Changes in the compositions of lipids, fatty acids and tocopherols in peanut seeds during maturation. *Journal of the Agricultural Chemical Society of Japan*, **51**(10), 575–581.

Shukla, G.B., Brahmachari, G.B., Sharma, C.K. and Murthi, T.N. (1980) A butyrometric method for rapid determination of the oil content of groundnut seeds. *Journal of Food Science and Technology*, **17**(5), 242–244.

Shurpalekar, S.R. and Bretschneider, F. (1969) High protein biscuits containing groundnut protein isolate. *Plant Foods for Human Nutrition*, **1**, 247–251.

Sitren, H.S., Ahmed, E.M. and George, D.E. (1985) *In vivo* and *in vitro* assessment of antinutritional factors in peanut and soy. *Journal of Food Science*, **50**, 418–422.

Singh, B. and Singh, U. (1991) Peanut as a source of protein for human foods. *Plant Foods for Human Nutrition*, **41**, 165–177.

Springhall, J.A. (1967) The use of peanut plant meal and bean seed meal in chicken starter rations. *Papua and New Guinea Agricultural Journal*, **19**, 112–114.

Srinivasan, V., Moudgal, N.R. and Sarma, P.S. (1957) Studies on goitrogenic agents in food. *Journal of Nutrition*, **61**, 87–95.

Stansbury, M.T., Field, E.T. and Guthrie, J.D. (1950) The tannin and related pigments in the red skins (testa) of peanut kernels. *Journal of the American Oil Chemists' Society*, **27**, 317–321.

Sugimura, K., Suh, K.B., Rao, T.N.R. *et al.* (1965) Studies on amino acid contents of foods (Report 2): amino acid contents (total and free) of low salt fermented products of groundnut and soybean. *Journal of the Japanese Society of Food and Nutrition*, **18**, 70–72.

Sundar, R.S., Hariharan, K. and Rao, D.R. (1976) Major acidic peptides of *Arachis hypogea* (Peanut). *Lebensmittel-Wissenchaft und Technologie*, **22**, 180–182.

Taira, H. (1985) Oil content and fatty acid composition of peanuts imported in Japan. *Journal of the American Oil Chemists' Society*, **62**(4), 699–702.

Tharanathan, R.N., Wankhede, D.B. and Rao, M.R.R. (1976) A Research Note. Mono and oligosaccharide composition of groundnut (*Arachis hypogaea*). *Journal of Food Science*, **41**, 715–716.

Tharanathan, R.N., Wankhede, D.B. and Rao, M.R.R. (1979) Groundnut carbohydrates – a review. *Journal of the Science of Food and Agriculture*, **30**(11), 1077–1084.

Tharanathan, R.N., Wankhede, D.B. and Rao, M.R.R. (1975) Carbohydrate composition of groundnuts (*Arachis hypogaea*). *Journal of the Science of Food and Agriculture*, **26**(6), 749–754.

Thomson, R.H.K. (1952) The proteins of *Arachis hypogaea* and fibre formation. *Biochemistry Journal*, **51**, 118–123.

Tombs, M.P. (1965) An electrophoretic investigation of groundnut protein, the structure of arachins A and B. *Biochemistry Journal*, **96**, 119–133.

Tur-Sinai, A., Birk, Y., Gertler, A. and Rigbi, M. (1972) A basic trypsin- and chymotrypsin-inhibitor from groundnuts (*Arachis hypogaea*). *Biochimica et Biophysica Acta*, **263**, 666–672.

Varma, N.R. and Kumar, A.N. (1968) Effect of heat on the yield and physico-chemical characteristics of oil and protein from groundnut. *Journal of Food Science and Technology*, **5**, 69–70.

Vesselinovitch, D., Getz, G.S., Hughes, R.H. and Wissler, R.W. (1974) Atherosclerosis in the rhesus monkey fed three food fats. *Atherosclerosis*, **20**, 303–321.

Vidyasagar, K., Arya, S.S., Premavalli, K.S. *et al.* (1974) Chemical and nutritive changes in refined groundnut oil during deep fat frying. *Journal of Food Science and Technology*, **11**(2), 73–75.

Waldo, M.M. and Goddard, V.R. (1950) A study of the protein value of soy and peanut flours in stock diets for rats. *Journal of Home Economics*, **42**(2), 112–115.

Wallerstein, I.S., Merin, U. and Rosenthal, I. (1989) Comparison of kernels of three Virginia-type peanut cultivars. *Lebensmittel-Wissenchaft und Technologie*, **22**, 179–181.

Watts, J.H., Booker, L.K., McAffe, J.W. *et al.* (1959) Biological availability of essential amino acids, to human subjects. 1. Whole egg, pork muscle and peanut butter. *Journal of Nutrition*, **67**, 483–496.

Webb, J.K.G., John, T.J., Begum, A. *et al.* (1964) Peanut protein and milk protein blends in the treatment of kwashiorkor. *American Journal of Clinical Nutrition*, **14**, 331–341.

Westgarth, D.R. and Williams, A.P. (1974) A collaborative study on the determination of tryptophan in feedingstuffs. *Journal of the Science of Food and Agriculture*, **25**(5), 571–575.

Whiston, A., Hay, R.J. and Raymond, W.D. (1959) The effect of storage for different periods of time in Nigeria on the quality of groundnuts with special reference to their value for the manufacture of protein fibres. *Tropical Science*, **1**, 149–181.

Woodroof, J.G. (1969) Composition and use of peanuts in the diet. *World Review of Nutrition and Dietetics*, **11**, 142–169.

Woodroof, J.G. (1983) *Peanuts Production, processing, products*, 3rd edn, Avi Publishing Company, Inc., Westport, Connecticut.

Worthington, R.E. and Holley, K.T. (1967) The linolenic acid content of peanut oil. *Journal of the American Oil Chemists' Society*, **44**, 515–516.

Worthington, R.E. (1977) The linolenic acid content of peanut oil. *Journal of the American Oil Chemists' Society*, **54**(4), 167–169.

Worthington, R.E., Hammons, R.O. and Allison, J.R. (1972) Varietal differences and seasonal effects on fatty acid composition and stability of oil from 82 peanut genotypes. *Journal of Agricultural Food Chemistry*, **20**(3), 727–730.

Worthington, R.E. and Hitchcock, H.L. (1984) A method for the separation of seed oil steryl esters and free sterols, application to peanut and corn oils. *Journal of the American Oil Chemists' Society*, **61**(6), 1085–1088.

Yannai, S. and Zimmermann, G. (1970) Influence of controlled storage of some staple foods on their protein nutritive value in lysine limited diets. *Journal of Food Science and Technology*, **7**, 190–196.

Young, C.T. (1980) Amino acid composition of three commercial peanut varieties. *Journal of Food Science*, **45**(4), 1086–1087.

Young, C.T. and Waller, G.K. (1972) Rapid oleic/linoleic microanalytical procedure for peanuts. *Journal of Agricultural and Food Chemistry*, **20**(6), 1116–1118.

Young, C.T., Waller, G.R. and Hammons, R.O. (1973) Variations in total amino acid content of peanut meal. *Journal of the American Oil Chemists' Society*, **50**(12), 521–523.

Young, C.T., Waller, G.R., Matlock, R.S. *et al.* (1974a) Some environmental factors affecting free amino acid composition in six varieties of peanuts. *Journal of the American Oil Chemists' Society*, **51**(6), 265.

Young, C.T., Matlock, R.S., Mason, M.E. and Waller, G.R. (1974b) Effect of harvest date and maturity upon free amino acid levels in three varieties of peanuts. *Journal of the American Oil Chemists' Society*, **51**(6), 269–273.

Young, C.T., Worthington, R.E., Hammons, R.O. *et al.* (1974c) Fatty acid composition of Spanish peanut oils as influenced by planting location, soil moisture conditions, variety and season. *Journal of the American Oil Chemists' Society*, **51**(7), 312–315.

Zimmermann, G., Weissmann, S. and Yannai, S. (1967) The distribution of protein, lysine and methionine, and antitryptic activity in the cotyledons of some leguminous seeds. *Journal of Food Science*, **32**, 129–130.

CHAPTER 7
Mineral nutrition

G.J. Gascho and J.G. Davis

7.1 INTRODUCTION

High yields of quality groundnuts require good nutrition. Sixteen elements [(carbon (C), hydrogen (H), oxygen (O), nitrogen (N), phosphorus (P), potassium (K), calcium (Ca), magnesium (Mg), sulphur (S), zinc (Zn), manganese (Mn), iron (Fe), copper (Cu), boron (B), molybdenum (Mo), and chlorine (Cl)] are considered essential for plants although not all have been proved essential for any particular species. Of the sixteen, C, H and O are supplied in the atmosphere or in the air space of the soil and are called non-mineral nutrients. They are usually assumed to be plentiful, even though O may be limiting due to excessive water in soil pores. Therefore neither fertilizer nor soil amendments are applied for the expressed purpose of supplying these three elements. In addition, nickel (Ni) and cobalt (Co) are essential for some legumes and sodium (Na), silicon (Si), selenium (Se), and aluminum (Al) have been shown to be beneficial in some cases. Quantities of nutrients removed by groundnut pods and vines are presented in Table 7.1.

The groundnut is unique in one aspect of its nutrition: the seed develops via nutrients it gathers directly from the soil rather than those transported from roots to shoots and back to the seeds. This unique aspect has required

TABLE 7.1 *Nutrient uptake/removal in groundnuts (kg/ha) (Gascho, 1992)*

Plant part	Yield	N	P	K	Ca	Mg	S
Pods	3 t/ha	120	11	18	13	9	7
Vines	5 t/ha	72	11	48	64	16	8
	Total	192	22	66	77	25	15

The Groundnut Crop: A scientific basis for improvement. Edited by J. Smartt. Published in 1994 by Chapman & Hall, London. ISBN 0 412 408201.

much research and it guides the applications of nutrients, especially calcium, for greatest yield, quality and seed germination.

Groundnut mineral nutrition research has been carried out for several decades (Harris, 1959). Previous reviews of the subject have covered the available knowledge up to 1982 (York and Colwell, 1951; Gillier and Silvestre, 1969; Reid and Cox, 1973; Cox et al., 1982). However, in the decade following the last of these reviews, additional advances have been made in our understanding of mineral nutrition of the groundnut. This understanding has been applied to enhance yield and quality of the crop. In this chapter, emphasis will be placed on important principles of nutrition and particularly to advances made in the past decade. Each required element is presented individually.

7.2 CALCIUM (Ca)

Calcium is often considered the essential element most commonly deficient for groundnut in non-calcareous soils. The consequences of Ca deficiency are blackened plumules, high incidences of pod rot and unfilled pods (termed 'pops') resulting in low yield and substandard grade. Groundnuts produced in Ca deficient conditions will exhibit much poorer germination than those produced with an adequate Ca supply. The effects of poor Ca nutrition are elaborated on later.

7.2.1 Needs related to uptake

For the groundnut, the exceptional requirement for Ca is, for the most part, not in order to grow a healthy plant but for developing a properly filled pod with a high quality seed. Bell et al. (1989) evaluated the external Ca requirements of six tropical legumes in flowing solution culture and found that the Ca requirements of red spanish groundnuts for growth and nodulation were not great in relation to pigeon pea, guar, soybean and cowpea. Guar and pigeon pea required a concentration of $> 50 \mu M$ Ca for nodule formation while groundnuts had nodules at $2 \mu M$ Ca. Maximum growth of groundnut was found at $50 \mu M$, soybean at $100 \mu M$ and cowpea at $2500 \mu M$ Ca. Even though Ca requirements for growth and nodulation of the groundnut are not great, the exceptional needs of Ca for groundnut seed maturation and quality are stressed below.

Barber (1984) stated:

> Calcium appears to have its greatest significance in providing the appropriate balance for levels of other nutrients within the plant. Dependent on the ambient Ca concentration near the absorbing surface mass flow and diffusion can be the dominant mechanisms of Ca absorption by roots.

For most plants, Ca is oversupplied to the root surface by mass flow and only a small proportion of the Ca is absorbed and subsequently translocated via the xylem to leaves in response to the flow of the transpiration stream. Much of the Ca is deposited in the leaf as Ca oxalate and then it is not re-translocated. This is a different mechanism from that employed in root uptake of several other ions. In the root, the uptake of many ions is termed 'active', indicating that the roots select the required nutrient elements from the soil solution. Uptake of Ca is fairly 'passive', i.e. roots take up the elements as they are presented in the soil solution without a selection mechanism. Therefore, the amount of Ca taken up by the plant is dependent both on its concentration in the soil solution and on the amount of water moving into the plant (Mengel and Kirkby, 1982). Because Ca is not easily translocated within the plant, Ca needed during reproductive growth must be directly supplied by uptake.

Classic research with the groundnut has shown that Ca is transported upward in the plant from root uptake via the xylem, but little or no Ca is transported from the leaves downward through the pegs to the developing pod via the phloem (Bledsoe *et al.*, 1949; Wiersum, 1951; Mizuno, 1959). The Ca absorption problem for the developing fruit is compounded by the fact that there is little water movement upward from the peg to the plant tops in order to provide a gradient to carry needed Ca into the developing fruit via mass flow (Wiersum, 1951; Beringer and Taha, 1976; Wolt and Adams, 1979).

Calcium for seed development must be absorbed by the peg via passive uptake by diffusion (Sumner *et al.*, 1988). Therefore, it is clear that a relatively high concentration of Ca is needed in soil in the pod development zone in order to avert Ca deficiency. Calcium in that zone must be replenished throughout peg and seed development periods in order to maintain a gradient of Ca^{2+} toward and into the pod. The developmental period extends from 20–80 days following the entrance of the peg into the soil. Mizuno (1959) determined that 92% of the Ca was taken up by the pod during that time period and 69% was taken up between day 20 and day 30. Smal *et al.* (1989) found that withholding Ca from the pod zone during the first 30 days following initial pegging resulted in the smallest seeds and the least seed dry weight. They determined that a continuous Ca supply of 3.75 μM Ca solution resulted in greatest seed size and weight, but suggested that the 30-day period following pegging is most critical.

For soils with great Ca supplying power and rapid replenishment of soil solution Ca, no deficiency is expected. However, the groundnut is often grown on sandy soils which possess limited ability to supply Ca to replenish the soil solution. This problem is compounded because such soils are often relatively droughty and the groundnut, being relatively drought-tolerant in comparison with most plants grown in semi-humid regions, is often grown in localities with limited rainfall during some portion of the period of peg and seed development. Limited soil moisture at this time worsens the situation in droughty soils: added Ca compounds will not dissolve without moisture, so that there is a lack of Ca solution; there is also a lack of a diffusion gradient so that Ca^{2+} cannot move toward the pod. Therefore, Ca deficiency results.

Effects of drought on Ca uptake by developing fruits have been reviewed by Boote et al (1982). They emphasized studies showing poor uptake in drought periods and the enhancement of uptake with adequate water in the fruiting zone. Groundnuts showing Ca deficiency symptoms during periods of drought also had lower Ca concentrations in seeds and hulls than those showing no Ca deficiency symptoms. Alva et al. (1991) determined soil solution Ca following gypsum application and incubation in the laboratory. They imposed four drying cycles over a period of 70 days in two soils that represent the textural extremes of the production industry in Georgia. Following the drying cycles, the soils were returned to field capacity moisture and soil solution Ca was determined. Soil solution Ca increased with increasing soil moisture in both soils for the first 14 days following gypsum application. Subsequently, soil solution Ca decreased for all moisture regimes in the Bonifay sand (grossarenic Plinthic Kandiudult), but this was true only in the driest regime for the Greenville sandy loam (thermic Rhodic Kandiudult). The results indicate that soil solution Ca deficiency is especially likely in soils such as the Bonifay sand due to low moisture retention. Calcium nutrition for the developing nut is intimately associated with soil moisture. For this reason it appears simplistic to cite a critical soil Ca test level for groundnuts. The true 'critical level' will depend on the moisture during nut development and will change, both geographically and temporally, even for a given cultivar.

Concentrations of other cations in the soil, particularly K and Mg, can affect Ca uptake and thereby affect yield and quality. Several reports of the deleterious effects of high soil K and seed K on groundnuts have been published (Bolhuis and Stubbs, 1955; Hallock and Allison, 1980; Alva et al., 1989a, 1991; Lynd and Ansman, 1989). Top growth is depressed with the addition of K alone, but increased by the addition of K + Ca (Lynd and Ansman, 1989). Using the Mehlich 1 (double acid) extractant, Alva et al. (1989a) determined that an optimum Ca/K ratio in the pegging zone topsoil is near 10:1.

7.2.2 Genotypes and Ca nutrition

In general, greater soil concentrations of Ca are required for larger seeded than for smaller seeded cultivars (Walker et al., 1976; Walker and Keisling, 1978). However, Cox et al. (1982) cited several exceptions. Gaines et al. (1989) determined both runner (cv. Florunner) and virginia type (cv. NC-7 and Early Bunch) responses to gypsum application relative to Mehlich 1 soil Ca. Yield response in the Florunners was limited to experiments with

Mineral nutrition

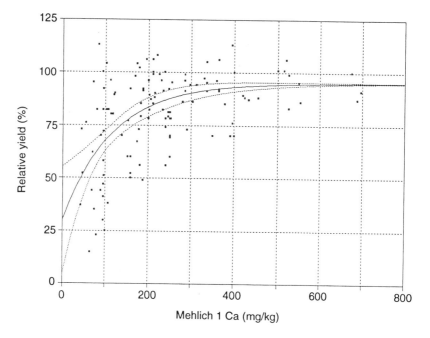

Figure 7.1 Relative pod yield of virginia type receiving no calcium additions to those receiving calcium as affected by Mehlich 1 extractable calcium in soil prior to addition of any calcium. Dotted lines are 95% confidence levels. Data from 122 experiments conducted in Alabama, Georgia, North Carolina and Virginia.

Mehlich 1 soil Ca less than 270 mg/kg, while the virginia type responded to all levels of soil Ca available in the experiments (Gaines *et al.*, 1989). Data from experiments conducted in Alabama, Georgia, North Carolina and Virginia have been combined to show the effect of Mehlich 1 soil Ca on relative yield of the virginia type (Hodges *et al.*, 1993). Maximum yield was attained at about 525 mg Mehlich 1 Ca/kg (Figure 7.1). Likewise, experiments conducted in Alabama and Georgia were combined to indicate that maximum yield of the runner type was attained at a Mehlich 1 Ca test of approximately 225 mg Ca/kg (Figure 7.2).

No good evidence exists to show that the internal Ca requirements of groundnuts are affected by seed size. Adams *et al.* (1993) found that the Ca concentrations in seed which produced maximum germination ranged from 381 to 414 mg/kg for four small-seeded runner cultivars. Those values are close to the 420 mg/kg which was found to give maximum germination of Florigiant, a larger-seeded cultivar, in an earlier study in North Carolina (Cox *et al.*, 1976). Gascho *et al.* (1992) correlated germination with seed Ca to find that the large-seeded virginia type (cv. GK-3) had an internal Ca requirement of approximately 600 mg/kg, which is similar to the internal requirement for maximum germination found for the small-seeded runner type (cv. Florunner) and considerably greater than those reported by Adams *et al.* (1993) or Cox *et al.* (1976). At least for the present, the need for greater ambient Ca concentrations in soil for the large-seeded cultivars is explained by acceptance of the view that diffusion from soil solution is the mechanism by which Ca enters the pod. Sumner *et al.* (1988) evaluated eight genotypes with widely varying pod and maturity characteristics over two seasons and four drought stress periods applied at different times during the crop and found strong relationships between Ca concentration in the hull and surface area of the pod. A larger-seeded pod has a lesser ratio of surface area to weight than a smaller-seeded pod, so that it is less efficient in diffusion and requires a greater concentration of soil solution Ca and/or greater soil moisture in order to provide adequate Ca to the pod. Acceptance of diffusion as the mechanism by which Ca moves to the developing peg explains the observed need for greater Ca applications for the large-seeded cultivars.

Variables other than the ratio of seed size to surface are involved in external Ca needs among cultivars. Kvien *et al.* (1988) used stepwise regression to select five genotypic characteristics which significantly influence seed and hull Ca concentrations:

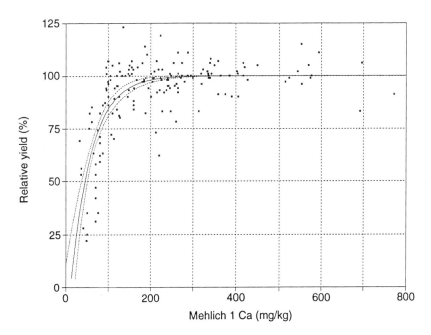

Figure 7.2 Relative pod yield of runner type receiving no calcium additions to those receiving calcium as affected by Mehlich 1 extractable calcium in soil prior to addition of any calcium. Dotted lines are 95% confidence levels. Data from 168 experiments conducted in Alabama and Georgia.

- days required to mature a pod;
- specific hull weight;
- pod surface area;
- hull thickness;
- pod volume.

Cox *et al.* (1982) suggest that Ca efficiency may be increased by breeding the Ca-efficient characteristics of small-seeded runner types into the large-seeded cultivars. Recent literature does not reveal much effort in this regard. In an attempt to identify groundnuts suitable for culture on marginal soils and for low-input agriculture, Branch and Gascho (1985) evaluated 24 cultivars of four USA market types (runner, virginia, spanish and valencia) for their tolerance of low soil fertility. They found that the standard south-eastern USA runner-type cultivar, Florunner, gave the greatest pod and total sound mature kernel yield but the kernels had a great incidence of concealed damage, indicative of Ca deficiency. The pre-eminence of the Florunner cultivar may be due, at least in part, to its Ca efficiency – defined as its ability to yield in a low ambient Ca environment. However, improvement of nutrient efficiency does not appear to be a major objective with breeders in their efforts to improve yield, grade and disease resistance and to decrease aflatoxin concentrations. Recent evaluations of the effects of supplemental Ca application on yield, grade and seed quality of promising runner-type cultivars (Adams *et al.*, 1993) suggest that these cultivars may require different ambient soil solution Ca from that for Florunner. In their study, Ca concentrations in seeds averaged by cultivar across 14 sites and for plots receiving and not receiving gypsum were 360 mg/kg for Florunner, 356 mg/kg for Sunrunner, 345 mg/kg for GK-7 and 301 mg/kg for Southern Runner. These results indicate that no promising runner-type cultivar is more efficient than Florunner in Ca absorption. The relatively new Southern Runner cultivar was less efficient, especially in cases where no gypsum was applied: seed Ca concentration for Florunner was 319 mg/kg, and for Southern Runner 235 mg/kg.

7.2.3 Symptoms of Ca deficiency

The important consequences of Ca deficiency occur in the reproductive stages of development, but some indication of insufficiency for quality seed production may be evident in the vegetative stages of growth. Symptoms summarized by Cox *et al.* (1982) from the literature include:

- more abundant foliage that remains green later in the season;
- tendency for a greater number of flowers, many of which may be infertile;

- localized pitted areas on the lower surface of fully developed leaves that can later develop into brown chlorotic spots, which may have halos on their perimeters, coalesce and cause senescence of the leaves;
- death of root tips and terminal buds (only in extreme deficiency).

Death of seedlings due to Ca deficiency occurred in a sandy field when the Mehlich 1 soil Ca level in the upper 15 cm was 6–21 mg/kg (Gascho *et al.*, unpublished data). Pod production, grade, germination of the seed produced and seedling survival were reduced greatly by and are the final results of Ca deficiency. The primary reason for these results is the high proportion of pods with aborted and shrivelled fruits ('pops') and darkened plumules ('black heart'). These deleterious effects are strongly related to concentrations of Ca in the soil as well as in seed and hull (Adams and Hartzog, 1980; Hallock and Allison, 1980; Walker and Csinos, 1980).

Germination of seed produced in low Ca soils is greatly reduced. Harris and Brolmann (1966c) determined 100% germination and 98% seedling survival when seed Ca was >800 mg/kg and 23% germination with no seedling survival for seeds with <300 mg/kg.

Garren (1964), Hallock and Garren (1968) and Walker and Csinos (1980) provide data indicating that pod rot (caused by several organisms, but especially *Pythium myriotylum* Drechs. and *Rhizoctonia solani* Kuehn.) is related to low Ca supply, based on experiments where pod rot incidence was reduced by application of gypsum. Hallock and Garren (1968) and Csinos and Gaines (1986) suggested that the balance among Ca, K and Mg was important, with high K or Mg supply increasing the incidence of pod rot. The reported effect of Ca on pod rot has been disputed by Filonow *et al.* (1988), who infected soil with *Pythium myriotylum* Drechs. and *Rhizoctonia solani* Kuehn. and found that increasing Ca applications had no effect on pod rot incidence. Their work supports the effect of the organisms on the aetiology of pod rot, but argues against Ca deficiency in the hull as a primary cause of the problem. Recent work in Georgia supports the relationship between Ca and pod rot. Gascho *et al.* (1993) conducted field experiments in soils with a Mehlich 1 soil Ca of less than 200 mg/kg and found that pod rot averaged 8–9% in control plots of both the small-seeded Florunner and the larger-seeded GK-3. Pod rot incidence was decreased significantly by the application of Ca.

Calcium supply may also be related to aflatoxin infection. Davidson *et al.* (1983) and Wilson *et al.* (1989) determined that Ca applications on sandy soils with a low Ca status decreased infection by *Aspergillus flavus*, which is sometimes associated with high levels of aflatoxin production.

7.2.4 Diagnosis of Ca needs

Soil analysis is the most useful diagnostic tool for determining the need for supplemental Ca fertilization. Samples taken from the surface 15–20 cm prior to planting the crop are often used. The choice of critical Ca level varies with the soil extractant and the type of groundnut grown. Adams and Hartzog (1980) determined that 125 mg/kg of Mehlich 1 extractable Ca was the critical value for cv. Florunner in Alabama. Hodges *et al.* (1993) consolidated data from the literature for 168 experiments conducted in Alabama and Georgia and determined that 95% of maximum yield was obtained when the Mehlich 1 extractable Ca was 200 mg/kg (Figure 7.2). In practice, supplemental Ca applications for runner groundnuts are normally recommended for soil tests less than 150 (Alabama) to 250 mg Ca/kg (Georgia) to help ensure adequate Ca (Cope *et al.*, 1981; Plank, 1989a).

For the larger-seeded virginia type, soil test Ca is less valuable since supplemental Ca is normally recommended regardless of the soil test level (Plank, 1989a). Gaines *et al.* (1989) indicated 95% of maximum yield was attained for virginia type at a Mehlich 1 test reading of approximately 775 mg Ca/kg. Parker *et al.* (unpublished) combined data from the literature for 88 experiments conducted with large-seeded virginia types in Alabama, Georgia, North Carolina and Virginia. They determined that 98% of maximum yield was obtained when Mehlich 1 Ca was 525 mg/kg. No doubt there is a soil Ca level above which no yield or quality responses to applied Ca would be expected for large-seeded cultivars but some additional data, particularly in relation to grade, is needed before revising present gypsum recommendations for the larger groundnut. For all types, a particular soil test value should not be chosen as a 'critical level', as that terminology implies precision greater than possible in farmers' fields. The level of soil test above which a response is not expected will vary with many genotypic characteristics but also due to the soil moisture available to transport Ca to the developing fruit.

An ideal soil test would evaluate Ca concentration in the soil solution just prior to bloom. Applications of a slowly soluble Ca source (such as gypsum) could be made to supply Ca if needed. Attempts have been made to measure concentrations directly in the soil solution or to use weak extractants that mimic the soil solution without solubilization of normally soil-solution-insoluble Ca. Smal *et al.* (1988) evaluated 0.01 N $NaNO_3$ in comparison with a stronger extractant (Mehlich 1) which is widely used to measure cations in the acid soils of the south-eastern USA. They concluded that yield and grade parameters were better related to Ca levels produced by the $NaNO_3$ extractant than Ca by the Mehlich 1 solvent. Such tests with weak extractants should be valuable for measuring soil Ca in the pegging zone, especially following the application of limestone prior to planting as they should not solubilize limestone particles to the same degree as the stronger extractant.

However, later studies have re-emphasized that Mehlich 1 is the superior extractant (Alva *et al.*, 1991). However, it may give unreasonably high Ca levels in cases where limestone has been applied recently and where sampling may encounter limestone particles which can be dissolved by

Mehlich 1. In some cases, Ca extracted by $NaNO_3$ was less correlated with peanut Ca needs than was the standard extractant for soil Ca (Mehlich 1) in experiments conducted in Georgia. Alva *et al.* (1991) found that the $NaNO_3$ extractant did not predict a need for supplemental Ca on a very sandy site, but Mehlich 1 extraction accurately predicted a response. Adequate soil solution Ca (evaluated by a weak extractant) for such a soil is no guarantee that adequate Ca will be available when required by the groundnut. Mehlich 1 extracts are more reliable for predicting Ca needs than are 0.01 N $NaNO_3$ extracts on soils with low ability to hold Ca. The limitations of the weak extractants probably occur because such tests make no allowance for or evaluation of the Ca which may become available during the developmental period in which Ca is in constant demand. That time can vary with cultivar and locality, but can be for a period of up to 60 days in the peanut belt of the south-eastern United States. Need for supplemental Ca application to soils which have little ability to supply Ca from reserves can be grossly underestimated by soil solution Ca or by weak extractants.

If limestone is applied by the preplant incorporated method (section 7.2.5), the post-planting soil test will be of less value as there will probably be some lime particles in the pod development zone that may be dissolved by the strong extractant, raising the test to a very high value. However, those particles may not be dissolved in the pod developmental period, in time to benefit the current crop.

Foliar analysis has been used to a limited extent. Due to the unique method of uptake of the developing peg discussed above, it is easy to understand why it has been difficult to obtain satisfactory correlations between leaf nutrient concentrations, yield and seed quality parameters. Nicholaides and Cox (1970) predicted a Ca deficiency when Ca in the tops of 9-week-old plants is less than 1.2%. Others have determined different critical levels varying quite widely (Cox and Reid, 1964). Determination of foliage Ca concentrations is diagnostic for gross deficiencies in growth of the plant, but not as a tool that should be relied upon to ensure a good harvest. Sufficiency levels in leaves are given in Table 7.2.

7.2.5 Correction of Ca deficiency

Getting soluble Ca into the pegging zone and maintaining a supply of that element to replenish depletion by the crop and losses to leaching can be especially difficult in a sandy soil with a poor Ca supply. Very soluble Ca sources such as $CaCl_2$ are generally expensive and short-lived due to leaching losses. Gypsum (calcium sulphate) is much less soluble, but its application on the soil surface at first bloom is generally appropriate for supplying Ca over the 45–60 day period in which the peanut has a great requirement (Sridhar *et al.* 1985).

In addition to mined gypsum, by-product and phosphogypsum (Alva *et al.* 1989b, Gascho and Alva, 1990) are available in many places at lower cost. When soil Ca is less than the chosen threshold level, gypsum is either broadcast or banded over the groundnut row at first bloom. Most products have a range of particle sizes that solubilize over time to provide Ca. Gypsum has been a very successful source of Ca; however, its application can add considerably to production costs. Dependent on source and recommended rate of application, the applied costs may range from about $40 to $80/ha.

Limestone is an important source of Ca for crops grown in the acidic soils of south-eastern USA and other locales. The main response to limestone is due to its supply of Ca, whether applied for that purpose or to increase soil pH. Adams and Hartzog (1980) conducted 16 field experiments with runner-type groundnuts in Alabama with the conclusion that: 'Lime appeared to do little more than serve as a source of calcium. Spring-applied lime provided all of the Ca needed for maximum yield and grade when it was properly incorporated into the pegging zone.' Since good farmers will lime their soils for their chosen crop rotation, they should attempt to apply it in a manner that will supply Ca to their crop and thereby reduce or eliminate the expense of a gypsum application. Adams and Hartzog (1980) have shown that, when limestone is applied to the soil prior to planting and incorporated to a shallow depth of 5–12 cm, adequate

TABLE 7.2 *Sufficiency levels of nutrients in groundnut leaf dry matter (Gillier and Silvestre 1969; Plank, 1989b)*

(a) Macronutrients (%)

Plant part	Time	N	P	K	Ca	Mg	S
7th leaf	40 days after planting	3.3–3.9	0.15–0.25	1.0–1.5	2.0	0.3	0.19–0.25
Upper mature leaf	Bloom	3.0–4.5	0.20–0.50	1.7–3.0	1.25–2.0	0.3–0.8	0.20–0.35

(b) Micronutrients (mg/kg)

Plant part	Time	Mn	Fe	Zn	Cu	B	Mo	Al
Upper mature leaves	Bloom	20–350	50–300	20–60[†]	5–20	20–60	0.1–5.0	<200

[†]Ca:Zn ratio <50:1.

TABLE 7.3   *Effect of limestone placement[1] on soil pH and Ca on groundnut yield and grade*

Placement	Soil		Pod yield	SMK[2]
	pH	Ca		
		mg/kg	kg/ha	%
Turned under	5.7	32	2285	66
PPI	6.4	155	4426	73
No lime	5.7	30	2455	66

[1] 2.2 tons dolomite limestone/ha, either turned under with a mouldboard plough, or applied on soil surface following turning and incorporated approximately 8 cm (PPI), or no lime applied.
[2] SMK, sound mature kernels

Ca is normally available for runner-type groundnuts as judged by the fact that responses to gypsum applied at bloom were rare following such applications. Gascho and Hodges (1991) confirmed these results in Georgia, but found that limestone incorporated before planting, applied at rates adequate to increase pH to the recommended value, did not supply adequate Ca for the larger seeded virginia type. On the other hand, Bell (1985a),working with Virginia Bunch on the irrigated sands of the Ord River area of Australia, determined that shallow incorporation of 1000 kg/ha of limestone (90% $CaCO_3$) before planting produced maximum yields of sound mature kernels. In his study, additional applications of gypsum at flowering had no significant effects on yield.

Much of the inconsistency in reported results may be due to the prevailing moisture conditions. A wet growing season will result in good solution Ca regardless of whether the Ca source is gypsum or limestone. If conditions are extremely wet on a deep sand, there is the possibility that Ca may be leached to a depth below the pegging zone. Application of gypsum at bloom followed by dry soil conditions can result in poor solution of Ca, while heavy rainfall on such a soil could result in Ca leaching.

Limestone placement is critical in evaluating its usefulness to the crop. Most limestone applied in agricultural fields is incorporated much deeper than the pegging zone either by harrows or ploughs. Nearly all groundnut fields in the US are turned by mouldboard ploughing in order to bury trash and provide a loose bed for nut development and removal. The modern mouldboard plough essentially inverts the top 25–30 cm of soil. If limestone is applied prior to ploughing, it will remain far below the zone of nut development and not provide Ca needed by the groundnut. Gascho and Hodges (1991, Table 7.3) found no benefit to runner-type cultivars in preploughing application and benefits equal to gypsum application for limestone incorporated before planting.

Unfortunately, the practice of shallow incorporation of limestone has been adopted only slowly in many growing areas of the world. In some cases, limestone is not available or is too expensive, and no Ca is applied. In other cases, limestone is ploughed under or otherwise incorporated too deeply into the soil or gypsum is considered the only appropriate source of Ca.

Dolomite is the limestone of choice when soils are deficient in Mg but, for the groundnut, the soil test must be very low to derive a benefit from Mg application. Adams and Hartzog (1980) determined that dolomite was equal to calcite as a Ca source (consideration must be given to the lower Ca concentration in the dolomite) for all cases tested in Alabama except one where Mehlich 1 Mg was 4 mg/kg. Gascho and Hodges (1991) also found dolomite to be superior, on sandy soils in Georgia, only when soil Mg was very low. However, most rotational crops have a greater Mg requirement than groundnut, and their needs must also be considered in a liming programme.

7.3 NITROGEN (N)

Approximately 190 kg/ha of N is removed with a 3 ton/ha groundnut pod crop when the vines are also removed (Gascho, 1992). The groundnut is a legume and under most conditions enough N is fixed through symbiotic relations with *Bradyrhizobium* spp. (Chapter 8) to avoid deficiency throughout the plant's life-cycle, including the production of a good yield with high quality. During reproductive stages, N is continually mobilized from leaves to the developing fruit (Kvien *et al.*, 1986) and under some conditions N deficiency can occur. For that reason, some N is commonly applied to groundnuts in most growing areas of the world (Gascho, 1992). However, N is not recommended in the USA, except in cases where symbiotic fixation is low or expected to be inefficient. In China, 30 kg N/ha is commonly top-dressed following an initial application of organic manure and PK fertilizer. In India, 10–25 kg N/ha is commonly applied; in Senegal the recommendation is approximately 12 kg N/ha.

Conditions that affect N supply to the groundnut include type, variety, presence of inoculum, crop rotation, soil type, soil moisture and temperature. Cox *et al.* (1982) summarized N fertilization studies prior to 1982 by stating:

> There seem to be a number of conditions conducive to obtaining a response from fertilizer N. The sequentially branched botanical varieties are more responsive than the alternate branched. In India, responses to N seem more likely if P is also low, but this did not hold true for more acid soils in South America. Nitrogen response could not be shown if other factors were more limiting. Substantial amounts of N were not ordinarily needed unless the site was exceedingly low in N or no effective rhizobia were present.

Studies since 1982 have not appreciably changed any of the previous interpretations of the needs of groundnut for fertilizer N. Several studies still lead to differing conclusions, due to variations in the factors given above. Walker *et al.* (1982), Hiltbold *et al* (1983) and Cope *et al.* (1984) found no responses to N application for the runner type in Georgia and Alabama. In further studies, Walker *et al.* (1984) found that yields of non-nodulating groundnuts and Florunner and Tifrun cultivars were linearly increased with the number of foliar applications of N at 13.5 kg/ha beginning 28 days after emergence. Patel *et al.* (1988) found that N application increased yield in only one of five years of their study in India. Mali *et al.* (1988) found increased N uptake and yield due to application of 20 kg/ha at planting in India, while Lal and Saran (1988) also measured increases in pods/plant and in oil and protein contents.

Inoculation of seed with a correct rhizobium strain seldom increases growth and yield by alleviation of N deficiency. Cox *et al.* (1982) stated:

> Native rhizobia are abundant and apparently able to fix adequate N at current yield levels. Both host plants and rhizobia strains may be selected to improve N uptake, so a means of improvement is available if needed. Neither management practices nor most soil conditions have been shown to affect N fixation greatly, but the rate may be decreased substantially by adverse climatic conditions.

In 1983, J.C. Wynne and associates in North Carolina began publication of several studies on N fixation. In the initial experiment, Ball *et al.* (1983) found no increased plant weight or yield in either spanish or virginia types due to inoculation with rhizobium, while application of N fertilizer increased both plant weight and yield of both groundnut types but reduced N-fixing rates. From their data they suggested that there is no need for inoculation when the soil contains a high population of the proper strain of rhizobium, but either the groundnut–rhizobia symbiosis does not fix enough N for maximum yield or fixed N is not used as efficiently as fertilizer N. Arrendell *et al.* (1985) used host selection in an attempt to increase symbiotic N fixation and found a significant correlation of nitrogenase activity with shoot weight, but lower correlations between nitrogenase and yield. They concluded that selection for greater N-fixing activity should be possible and should result in indirect selection for improved yield. However, in a later study, Arrendell *et al.* (1988) were unable to generate any variability in N-fixing ability among F2 selections. Finally, Phillips *et al.* (1989) determined the presence of non-additive genetic effects in two crosses and concluded that early generation selection for N fixation factors would be ineffective. Thus it appears that little progress has been made in improving N fixation by selection to date.

Nitrogen application has been considered a control measure for cylindrocladium black rot (CBR) in some instances. Black *et al.* (1984) found that N decreased CBR severity for nodulating groundnuts when inoculum density was low but increased CBR severity in non-nodulating lines.

Nitrogen concentrations in leaves decrease with plant age and vary with cultivar, making leaf N a difficult diagnostic tool for determination of need. Analysis of the seventh leaf 40 days after planting should be in the sufficiency range of 3.3–3.9% (Gillier and Silvestre, 1969; Plank, 1989b).

7.4 PHOSPHORUS (P)

The groundnut is grown on P deficient soils in many areas of the world. On a global scale, P may be the most deficient element. However, the deficiency is primarily limited to areas which have never been fertilized with P, where P fertilizers are not available or their cost is prohibitive. Phosphorus deficiency in groundnuts is well documented in several areas (Cox *et al.*, 1982; Chauhan *et al.*, 1988; Bell, 1985b; Mali *et al.* 1988; Survanvesh and Morrill, 1986; Dwivedi *et al.* 1987) but is unusual in countries such as the USA, where P fertilizers are commonly used (Cope *et al.* 1984; Walker *et al.* 1982). In general, P deficiency in groundnuts can be corrected easily by application of P fertilizers since the crop is grown on sandy soils with low amounts of clays – soils in which P fixation is not a common problem. Fertilization generally not only provides enough P for the crop but also increases available P in such soils since the plant requirement and removal are low (Table 7.1) and little P is lost by leaching, even in sandy soils. An exception is made for groundnut culture on calcareous sands, where P fixation can lead to P deficiency in spite of fertilization.

Soil analysis is considered the best method to determine P fertilizer needs. Effectiveness of extractants varies between locations and care must be exercised in comparisons of 'available soil P' values in the literature. The predominant extractants, proposed critical levels for groundnut and responses obtained in groundnut soils throughout the world are discussed by Cox *et al.* (1982). These will not be discussed here.

Soil P levels required for groundnut are often lower than those required for other crops (Cope *et al.* 1984). A summary of data from India, Bolivia, South Africa, Guyana, Colombia, Israel, Uganda, China, Australia and the USA, with a number of extractants, suggest very low critical levels of approximately 10 mg P/kg. A review of recent research emphasizes the low P requirements of runner varieties. In Alabama, Hartzog and Adams (1988a,b) found no correlation between Mehlich 1 extractable P and either yield or grade of runner groundnuts from fertilization in 39 experimental sites where soil P ranged from 1 to 45 mg/kg. Cope *et al.* (1984) found no response to P fertilizer in long term experiments even when Mehlich 1 P was reduced to 11 mg/kg by continuous cropping without fertilization. The critical level for Mehlich 1 P in the south-eastern runner-type peanut belt is approximately 10 mg/kg (Mitchell and Adams, 1993).

Bell (1985b) found responses to applied P at very low levels of 0.5 M $NaHCO_3$ extractable P on sands in Australia. He determined that the critical level for $NaHCO_3$ soil P was 7.3–7.9 mg/kg for Virginia Bunch. Chauhan *et al.* (1988) also had some responses, but only at very low levels of soil P (6.15 mg/kg, unknown extractant). Since 1982, the Mehlich 3 extractant has been introduced and is now in use in North Carolina and is under consideration in other south-eastern states currently using the Mehlich 1 extractant (formerly called Double Acid). Mehlich 3 commonly extracts 1.5 to 2 times more P than Mehlich 1 (Gascho *et al.*, 1990). Therefore critical levels are considered to be proportionally greater for Mehlich 3. The corresponding critical level for Mehlich 3 is 20–24 mg/dm³ (Cox, unpublished data). In areas where P fertilizers are readily available and relatively inexpensive, maintenance of such low levels is nearly assured by P applications to rotational crops, and P fertilization of the groundnut crop is generally not needed.

For areas where P fertilizers are not plentiful, or where P is fixed due to highly calcareous conditions and responses to P have been noted, some lines may be developed which are tolerant of low soil P. Dwivedi *et al.* (1987) conducted field experiments on a P deficient calcareous soil and noted variation in dry matter accumulation and P accumulation in plant parts for one adapted line in comparison with five unadapted lines. From their results, they proposed a model for high-yielding P deficiency resistant plant type: $R = (Ay \times 1.32) \times (An \times 1.09)$, where R is the resistance value for P deficiency or low P supply, A is an adapted cultivar, y is yield, n is P uptake on a P deficient soil, and 1.32 and 1.09 are constants. The authors claim that the model will predict the resistance of lines or cultivars to P deficiency.

Some researchers have considered foliar applications of P in cases where soil applications are fixed by great concentrations of Fe, Al, or free phase $CaCO_3$. Survanvesh and Morrill (1986) made foliar applications of P in the greenhouse to find that it was effective in increasing growth and plant P concentrations when the supply of P to the roots was inadequate, but not when root availability was adequate. In field experiments, Walker *et al.* (1982) found no response to foliar P when adequate preplant fertilizer was applied on a non-fixing soil.

Plant P analysis has also been useful, to an extent limited by the fact that effective applications of P cannot usually be made following a diagnostic tissue test. Therefore plant analysis is only helpful in knowing what to do when groundnuts are next planted in a given field. Care must be taken in the interpretation of plant analysis results since concentrations vary significantly and by plant part and age. Bhan (1977) noted that leaf P declined from 0.35% to 0.15% between 30 and 120 days following planting. Fertilization with P, which increased yields, only increased P concentration by about 0.05% throughout that period. Partitioning of uptake indicates that only 10% of the total is taken up in the vegetative phase while 39%

and 51% are taken up in the reproductive and maturation stages, respectively (Longanathan and Krishamoorthy, 1977). Foster (1980) set a critical P level of 0.29% P for leaves at flowering. Others use a range of sufficiency levels and a compilation of these (Gascho, 1992; Gillier and Silvestre, 1969; Plank, 1989b) indicates a wide variance. Reported ranges are 0.15–0.25% for the seventh leaf 40 days after planting and 0.25–0.50% for upper mature leaves at bloom.

7.5 POTASSIUM (K)

Removal of K from soil by groundnuts is considered to be low relative to soybeans or bermudagrass when the vines are returned (Table 7.1). Removal of the vines as hay quadruples K removal. These data should be considered in planning a cropping system but may not be of great importance when it is grown in a rotation with several other crops, since groundnuts will not respond to direct K fertilization unless soil available K is low (Scarsbrook and Cope, 1956; Walker et al., 1979). Reports of responses to K application are not consistent.

Scarsbrook and Cope (1956) reported an average pod yield response of 170 kg/ha to K fertilization when the Mehlich 1 soil K test was rated 'low' by the soil K interpretation in Alabama. They found no response when K was rated 'high'.

However, the world literature reports rather mixed responses to applied K and, in many cases, does not provide good supporting evidence or information on soil K levels. Piggot (1960) established that K application was necessary for groundnuts in Sierra Leone. Potassium is rarely needed in India but is added as insurance (Kernick, 1961). Goldsworthy and Heathcote (1963) conducted trials in northern Nigeria and found no response to K application. Later, Heathcote (1972) found that, under continuous cropping, groundnut responded to K in the third and fourth years but not in the first two years. This was consistent with the idea that groundnut removes relatively small amounts of K and will respond to K once soil K is reduced to low concentrations. Haggin and Koyumjisky (1966) recorded significant responses to K in only two of 24 fields in Israel. Oil and protein percentages as well as yield of tropical groundnut crops responded positively to 15 kg K/ha in two experiments with initial soil K levels of 0.28 and 0.40 meq K/100 g as neutral normal ammonium acetate (Kayode, 1987). However, yield responded negatively to greater rates of K application in Kayode's study.

Hartzog and Adams (1973a) conducted 34 on-farm K fertilization trials in Alabama without yield response and concluded that direct fertilization of groundnuts was not a good practice, but that K fertilization should be made to the crops grown in rotation with groundnuts. In Ca and K experiments on sandy soils in the coastal plain of Georgia, Walker et al.

(1979) found a decrease in yield with increased K fertilization rates of 0, 112 and 224 kg/ha, which resulted in Mehlich 1 soil test K levels of 20, 25 and 91 mg K/ha, respectively. On long-term fertility plots, Cope *et al.* (1984) found modest but significant response to 19 kg/ha of fertilizer K over a period of 30 years. Mehlich 1 soil K test remained fairly constant at 44 mg/kg during the 30-year period. They concluded that the relative response of multiple crops to soil (and fertilizer) K levels was: cotton > grain sorghum > corn > soybean > wheat and groundnuts.

Potassium levels in the soil and applied K should be considered in relation to levels of other cations, especially Ca, as they compete for uptake by the developing pods. This subject is covered in section 7.2.1.

Indications of optimal soil K levels for yield and quality vary according to the extractant used, which in turn is determined by region and soil type. For the Mehlich 1 extractant used in the runner-type peanut belt of the southern coastal plain of the USA, the identified critical level is 11–13 mg K/kg (Walker *et al.*, 1979; Mitchell and Adams, 1993). The minimum sufficiency level used in K recommendation is 20 mg K/kg in order to provide 'insurance' of adequate available K, bearing in mind sampling and analytical errors (Mitchell and Adams, 1993). Cox (unpublished studies) evaluated the Mehlich 3 extractant for use in the virginia-type peanut belt of North Carolina and determined a critical level between 39 and 47 mg K/l. However, correlations of yield response to surface soil K levels are not high in the soils of the peanut belt of the south-east USA because much of the K not utilized by crops accumulates in the subsoil. Some of this K is accessible to the deep roots of the groundnut. Therefore a low soil test from the top soil does not take into account the K available in the subsoil, and can underestimate K availability and recommend K fertilization when there is scant chance of response. Recognition of the unresponsiveness of the groundnut has recently directed the states in the south-eastern runner-type peanut belt of the USA to reduce application of K by recognizing a lower critical level in the Mehlich 1 soil test K of 20 mg/kg (Mitchell and Adams, 1993).

Leaf K has not been used to any large extent as a guide to fertilization of the groundnut since the timing of such a diagnostic tool is generally too late for making any correction. Also, as reviewed above, the problem may be high rather than low leaf K, thus leaving little that may be done to make a correction. The exception is to add a soluble Ca source, such as gypsum, when the K/Ca ratio is high. However, leaf K concentration can serve as a diagnostic for future groundnut fertilization in fields with a similar history. Since the review of Cox *et al.* (1982), Walker *et al.* (1989) have determined that, for maximum yield of runner-type groundnuts, the minimum sufficiency levels in youngest mature leaflet samples at approximately 100 days following planting should be 1.0% K. However, normal diagnostic tissue sampling should occur much earlier (Table 7.2).

7.6 MAGNESIUM (Mg)

Little response has been recorded for application of Mg to groundnuts. Cox *et al.* (1982) stated: 'The literature is almost barren of reports that peanut yields increased by Mg fertilization.' Responses only occur at very low soil test levels and they are most likely on deep, excessively drained sands. Since 1982, Hartzog and Adams (1988b) have reported no response to $MgSO_4$ on a McLaurin loamy sand with a Mehlich 1 Mg test of 3.5 mg/kg. Walker *et al.* (1989) found no response to Mg on a Fuquay sand with an initial Mehlich 1 Mg test of 7 mg/kg, but application of 67 kg Mg/ha increased yield by an average of 15% over the three years of the study on a Lakeland sand with an initial soil test of 4 mg Mg/kg. In the latter study, both K and Mg were deficient and application of both increased yield by 69% over the no K and no Mg control. It is very likely that the deep rooting pattern of the groundnut allows the plant to forage deeply for Mg as well as for other nutrients. As mentioned for K, some soils retain greater concentrations of Mg in the subsoil than in the surface horizon. Such is the case for the Fuquay sand while the Lakeland soil has essentially no colloidal matter in the rooting zone. The findings of Walker *et al.* (1989) therefore seem consistent with the subsoil retention explanation for the lack of response to fertilizer Mg in soils that have had dolomitic lime applied in the past, but have low soil test Mg just prior to groundnut. Schmidt and Cox (1992) did not find a yield response that could be attributed to Mg fertilization in North Carolina soils with soil tests as low as 0.02 cmol/l (approximately equivalent to 5 mg/kg). Their soil tests were with Mehlich 3, which removes about the same concentration of Mg as Mehlich 1 (Gascho *et al.*, 1990; Hanlon and Johnson, 1984).

Soil analysis may be useful for prediction of Mg needs if the whole of the rooting depth is sampled. However, such samples seem impractical and surface samples are probably not reliable. Walker *et al.* (1989) suggested a Mehlich 1 extractable level of 11 mg/kg in the topsoil. For the present, the sufficiency level is set at 15 mg/kg in the runner peanut belt of Alabama, Florida and Georgia (Hodges *et al.*, 1993). Schmidt and Cox (1992) indicated that sufficient leaf Mg was attained when Mehlich 3 Mg in the soil was as low as 0.06 cmol/l (approximately equivalent to 15 mg/kg). At this low concentration, Mg was only 3% of the cation exchange capacity of the soil. The actual critical level has not been attained in most studies and is often complicated by the groundnut plant's ability to forage deeply in soil horizons which may supplement the surface horizon in supplying Mg.

Use of leaf Mg concentrations for the groundnut suffers from all the drawbacks listed for K above, including variation with plant age. Walker *et al.* (1989) suggest a critical concentration of 0.2% in the recently mature leaflets of runner-type varieties sampled at approximately 100 days after planting. Schmidt and Cox (1992) found no Mg deficiency with leaf concentrations as low as 0.15%. As for soil analysis, the lower limits of the sufficiency range have not been reached in enough studies to provide a firm critical level (Table 7.2).

Magnesium may also interfere with Ca uptake by the developing pod; however, published evidence is scanty. Gascho *et al.* (1992) found that germination percentage of seed is somewhat better related to a ratio of Ca/(Ca+K+Mg) in the nut than to Ca concentration.

7.7 SULPHUR (S)

Even though sulphur is a required nutrient for groundnut, reports of responses to S applications are scarce in recent literature and are often confounded by the fact that elemental S application not only supplies S but also decreases soil pH, thereby changing the availability of other nutrients. Also, application of S as gypsum supplies Ca, an element of paramount importance in the groundnut (section 7.2). Recent reports from India indicate some yield responses on calcareous soils to S applications as ammonium sulphate, single superphosphate, elemental S and gypsum (Maliwal and Tank, 1988; Bahl *et al.*, 1986; Sahu and Singh, 1987; Hago and Salama, 1987). Bahl *et al.* (1986) found that application of both S and Zn resulted in a synergistic effect on increased yield while the main effect in the studies of Hago and Salama (1987) was to increase chlorophyll a and b in an area where groundnuts were chlorotic due to Fe deficiency. These results are attributed to decreased soil pH due to S application. Cox *et al.* (1982) provided a review of reports of S responses in groundnut growing areas of the world and discussed the complications of the reports due to interactions with pH, P and micronutrients.

In the past, S was supplied to most groundnuts incidentally due to the application of S-containing fertilizers such as single superphosphate. However, most modern fertilizers, such as triple superphosphate, are concentrated in the major nutrients and contain little S; therefore, more S deficiency is expected to appear with time in many crops. Groundnuts are probably unique, as much of the crop receives gypsum to supply Ca and thus adequate S is provided incidentally. In addition, several of the fungicides used in the past contained S, but use of these for groundnuts has now decreased.

Atmospheric S is deposited on soils from the burning of fossil fuels. Such deposits are particularly significant in industrialized areas and are usually sufficient for crop needs. Common levels of S deposits from the atmosphere range from 10 to 20 kg/ha/year. Sulphur concentrations can also be significant in irrigation waters, thus eliminating possibility of any response to application (Cox *et al.* 1982).

Sulphur deficiency is most likely on very sandy soils which possess little anion exchange capacity and will not hold the sulphate anion. The depth of the sand is an important factor, since clay accumulation in deeper horizons within the rooting zone will tend to hold the sulphate anion and will lessen the possibility of S deficiency.

Soil tests for S are not important in groundnuts as critical levels have not been established. Such an effort would probably not be rewarding: the soils with S deficiency problems are those where S is leachable, and so S will be too elusive for any diagnostic soil test.

Several attempts have been made to establish critical levels of S in the groundnut plant. Although the literature is not consistent with regard to a specific S critical level, in general the S concentration should be similar to the P concentration. The S concentration is also related to the N concentration and an N:S ratio of about 15:1 is desirable, based on a balance of S-containing and non-S-containing amino acids (Bockelee-Morvan and Martin, 1966; Lund and Murdock, 1978).

7.8 MICRONUTRIENTS

Micronutrient availability in soils for the groundnut and other plants is related to soil pH as well as soil physical and other chemical characteristics. Most micronutrients are more likely to be deficient at high pH, particularly in calcareous soils. Micronutrients in this category are manganese, zinc, iron and copper. Manganese and zinc may occur at toxicity levels in very acid conditions. In contrast, molybdenum is less available in acidic soils.

The single most important consideration in proper availability of micronutrients for plants is soil pH. Secondly, cation and anion exchange capacity can affect leaching and critical toxic levels in the soil. Thirdly, interactions of nutrients can result in nutrient imbalance (e.g. a low Ca:Zn ratio may induce Zn toxicity). Fourthly, the ambient concentrations of micronutrients in soils and in irrigation waters vary due to soil formation factors and to previous cropping history.

7.8.1 Boron (B)

Boron can be important when present at either deficient or toxic levels. The major problem in groundnut is deficiency on highly weathered sandy soils due to their inability to retain the mobile-borate anion. All groundnut growing states of the USA recommend boron application to the crop. Perry (1971) recommended 0.6 kg B/ha for sandy soils and 1.1 kg/ha for heavy soils, but warned against over application in view of potential B toxicity.

(a) Deficiency

A deficiency most often results in internal nut damage termed 'hollow heart' which greatly reduces the quality and value of the crop. The condition refers to cotyledons which are small and discoloured. Early research in Florida found that B deficiency resulted in hollow heart, compacted branch terminals and cracks on pods (Harris and Gilman, 1957). These symptoms were first described in greenhouse studies and were later evaluated in field studies with B application in several growing areas. Application of 1.9 kg H_3BO_3/ha increased yield and grade in the greenhouse, but no B deficiency was detected in field studies. Harris and Gilman (1957) also noted differences in B response between runner varieties. Harris (1963) described B deficiency in cv. Florigiant groundnuts as hollow heart, with a mosaic pattern on the foliage. Boron deficiency symptoms are similar for runner, spanish and virginia types (Harris, 1965). Harris and Brolmann (1966a) showed that B deficiency symptoms were accentuated by application of complete fertilizer without B. Deficiency affects the flowering pattern by producing a less intense but longer flowering period (Harris and Brolmann, 1966b). Harris and Brolmann (1966c) illustrated the influence of B deficiency on the cotyledons and showed that plumules were discoloured and plumule tips were pointed or poorly developed. Harris (1968) indicated that B should be applied at soil B $\leq$0.2 mg/kg. He also noted that spanish and runner types developed B deficiency symptoms earlier than the virginia type, and that the virginia type recovered more quickly after B application. Shiralipour et al. (1969) determined that B deficiency resulted in increased nitrogen and amino acid contents in groundnut leaves without affecting protein levels.

Research in North Carolina using sand cultures showed that B deficiency resulted in deep green, mottled leaves and terminal death with extensive secondary branching and decreased internode length (Reid and York, 1958). Cox and Reid (1964) found that 0.6 kg B/ha decreased plumule damage in a field study. They also showed that liming increased soil available B, but did not increase B content in groundnut kernels. Stoller (1966) illustrated the influence of long day length and high light intensity on accentuating B deficiency symptoms.

In Virginia, hollow-heart symptoms were noted in 1958 but were not confirmed to be B deficiency until 1965 (Anon, 1965). Research showed that 1.1 kg B/ha decreased damage, but 2.2 kg/ha could be toxic. Hallock (1966) also showed marked decrease in hollow heart by B application, but rates of 1–2 kg B/ha could be phytotoxic. He also stated that B deficiency is more common in sandy, droughty soils. Allison (1980) recommended 0.6 kg B/ha application at early bloom.

In Alabama, Hartzog and Adams (1968) determined that topdressing 1.1 kg B/ha had no effect on yield, and increased grade in only one out of five experiments. Hartzog and Adams (1971) reported that, in eight experiments with hot-water extractable soil B <0.07 mg/kg, hollow heart failed to develop, and yield and grade were unaffected by B fertilization. Hartzog and Adams (1973a) again reported no yield or grade effect of B fertilization, and determined that B could be applied preplant in a herbicide tank

mixture or sprayed on with fungicide without damaging plants. Cope *et al.*
(1984) reported that yields were not affected by application of B (in
combination with Zn, Mn, Cu and Mo).

Snyman (1972) in South Africa and Morrill *et al.* (1977) in Oklahoma
determined that application of fertilizer B nearly eliminated hollow heart.
In some cases outside of the USA, deficiency has also been reported to
reduce yield. In India, Saxena and Mehrotra (1985) found a significant
response in yield to the application of 1 kg B/ha on a loam soil in a trial
where up to 2 kg/ha was applied. In similar trials on a sandy loam they
found yield response to only 0.5 kg B/ha. Yields in an experiment in China
(Zhang *et al.*, 1986) were increased by up to 11.5 % by application of B (as
borax). On the other hand, Blamey *et al.* (1981) in South Africa emphasized the narrow window between deficiency and toxicity. They found
decreased growth and yield by residual and applied B applications on a
sandy loam soil with an initial pH of 3.9 (0.1 N KCl). Annual B applications of 1 kg/ha significantly reduced both kernel and hay yields.
Applications of limestone at rates of up to 2.4 tons/ha in previous crops did
not reduce toxicity problems, suggesting that the increased root growth
with higher pH increased the uptake of BO_3 enough to offset any fixation
of B due to precipitation of $Al(OH)_3$.

In India, Rao *et al.* (1960) reported that 11 kg B/ha did not affect yield.
Gopal and Rao (1972) also found no benefit from B application but gave a
critical level of 25 mg/kg in middle leaves. Other studies showed that: 9 kg
B/ha increased yield and oil content (Ganesan and Sundararajan, 1972);
15 kg/ha borax increased yield (Muthuswamy and Sundararajan, 1973);
6 kg/ha borax or boric acid increased yield and quality (Asokan and Raj,
1974); B application increased oil content (Sankaran *et al.*, 1977); B spray
had no yield effect (Swamy and Reddy, 1983); 2 mg/kg B gave maximum
yield (Golakiya and Patel, 1986); pod yield was highest when Ca:B ratio in
pods was 218–224 (Golakiya, 1989); and 5 kg B/ha increased yield on a soil
with 0.16 mg B/kg (Jadhao *et al.*, 1989). Boron deficiency was also reported
in Malawi on light, sandy soils (Anon, 1972).

Research on spanish-type groundnuts in Oklahoma suggested a critical
level of 30 mg B/kg in young leaves (Chrudimsky, 1970). Hill and Morrill
(1974) found B deficiency in 50% of their field locations but reported that
B application did not affect yield or grade. They stated that hollow heart
was related to soil B (hot-water soluble) <0.15 mg/kg and leaf B of
26–30 mg/kg. Hill and Morrill (1975) found that B application improved
kernel grade, except at high potassium rates. Morrill *et al.* (1977) concluded that soil B ≤0.15 mg/kg and leaf B <30 mg/kg require B fertilization at a rate of 0.6 kg B/ha.

Diagnosis of B nutritional problems is by both soil and plant analysis but
neither has been entirely satisfactory. Reliability of soil B tests has been
considered rather low, due mostly to the fact that few correlation data have
been produced in recent years. The critical values (above which there

should be no deficiency) of 0.05 mg B/kg established by Cox and Reid (1964) for acid soils and 0.2 mg B/kg by Hill and Morrill (1974) for calcareous soils by the hot-water soluble method are the best estimates available. In the south-eastern USA 0.6 kg B/ha is recommended when soil B <0.2 mg/kg.

The critical leaf B concentration, below which spanish-type groundnut is deficient, is in the range 25–30 mg/kg (Morrill *et al.*, 1977). Toxic concentrations reported for spanish and virginia types vary from 60 to 112 mg/kg (Gopal, 1971a, 1971b; Luke, 1969; Blamey and Chapman, 1979). In the south-eastern USA, 0.6 kg B/ha is recommended if leaf B is <30 mg/kg (young leaves). Very little B-related research has been reported in the past decade. There is a need for studies to evaluate diagnostic B concentrations using modern testing methods such as inductively coupled plasma (ICP) and/or direct current plasma (DCP) emission spectroscopy methods, which are less laborious.

Boron deficient conditions in sandy soils can be alleviated by foliar applications. Walker (1967) stated that 0.6 kg B/ha applied as a foliar spray increased yields on Ruston and Tifton soils but not on Greenville soil (a finer textured soil). Application of a total of 0.6 kg B/ha is commonly tank-mixed with and split equally between the first two fungicide applications.

(b) Toxicity

Boron can be toxic to groundnuts; therefore, it should be applied at the recommended rate only. McGill and Bergeaux (1966) warned of exceeding 0.6 kg B/ha, and Stoller (1966) showed that the critical toxicity level for B was 100 mg/kg in shoots. Morrill *et al.* (1977) stated that 1–1.5 kg B/ha caused toxicity and reduced yields. Application of 1 kg B/ha reduced yields in Australia (Blamey *et al.*, 1981). A 10% yield decline was associated with 58 mg B/kg in young leaves. In Israel, B concentration of 0.29 mM in nutrient solution decreased yield (Lauter *et al.*, 1989).

Research in India has shown that B toxicity results in chlorosis of leaf tips which extends marginally and interveinally, followed by marginal necrosis (Harigopal and Rao, 1964). Chlorosis was related to decreased leaf chlorophyll, protein N, and Fe. Boron toxicity decreased yield, and leaching was recommended as an ameliorative method (Harigopal and Rao, 1967). Gopal (1968) stated that 10 mg B/l in solution and 85 mg/kg in leaves were related to toxicity. Gopal (1969) showed that B interfered with the ability of Fe to complex with proteins. However, Fe addition in a pot study made no difference (Gopal, 1970a). Boron content of middle leaves of toxic plants was >1100 mg/kg. Boron also decreased leaf Cu (Gopal, 1970b). Gopal (1971a) found 3 mg B/kg in soil to be toxic, and Gopal and Rao (1972) found 140 mg/kg to be the critical level in middle leaves. Boron toxicity decreased total N and protein N in chlorotic leaves, but increased soluble N and free amino acids (Gopal, 1971b). Boron application at

>6 kg/ha borax gave an adverse effect (Asokan and Raj, 1974), and 5 kg/ha borax resulted in toxic symptoms (Reddy and Patil, 1980). Care should be taken not to overapply B to groundnuts.

7.8.2 Chlorine (Cl)

Chlorine toxicity has been described for soybeans in Georgia (Parker *et al.*, 1983) but has not been found in groundnuts. Chlorine is an essential element for plant production but Cl deficiency has not been described for groundnuts.

Schilling and Hirsch (1974) found no correlation between leaf Cl and groundnut yield in a study in Senegal. Chloride effects on cv. Florunner were studied in the greenhouse and field in Georgia (Parker *et al.*, 1984). Addition of Cl to soil increased Cl concentration in leaves but there was no significant effect on dry matter production (greenhouse) or pod yield (field).

There are no data that would warrant fertilizer Cl recommendations for groundnuts.

7.8.3 Copper (Cu)

Copper is sometimes applied to agricultural crops as a micronutrient but is commonly applied in the form of pesticides, particularly fungicides. Bledsoe and Harris (1947, 1948, 1949) reported that application of 11 kg $CuCl_2$/ha increased the proportion of sound to shrivelled nuts for runner groundnuts. Three years after application, the residual effect of Cu on quality was maintained. Harris (1952) described Cu deficiency symptoms as affecting the bud area in particular, as well as causing small, irregular leaflets with marginal necrosis and mild chlorosis and small yellow-white spots on the foliage. Harris (1952) found that the spanish type is more sensitive to Cu deficiency than the runner type, but that yields for all three cultivars studied (two runners and one spanish) were increased more than 300% by applying 11 kg/ha Cu (as $CuCl_2$) to an Arredondo loamy fine sand (pH 5.7). Copper application also decreased seed shrivelling and increased the percentage of sound mature kernels (SMKs). The residual effect of soil Cu application (11 kg/ha) to oats, wheat, rye or cotton in rotation with groundnuts was found to be equally effective as groundnut foliar applications (0.2 kg $CuCl_2$/ha). However, Harris (1952) concluded that, in general, yields in Florida had not been increased by Cu applications (though yields were increased on the Gainesville experimental farm) and, therefore, Cu application was not recommended.

In research in Georgia, Boswell (1964) stated that no definite pattern was found between Cu application and yields. Gopal (1970a) reported that

B treatment decreased leaf Cu concentration and Cu-protein enzyme activity in groundnuts. Gopal (1975) stated that 10 mg/l B in sand culture decreased leaf Cu concentration by 24%. Katawatin *et al.* (1989) determined that phosphorus fertilization decreased Cu uptake.

7.8.4 Iron (Fe)

Iron deficiency can be a serious problem in calcareous soils but no problems have been reported on acidic soils. Perkins (1964) stated that the total Fe content of most acidic soils in Georgia is greater than 1%; therefore he assumed that Fe is available in Georgia in sufficient amounts for crop production. Iron deficiency results in interveinal chlorosis (starting in the youngest leaves), followed by chlorosis of the entire leaf (whitish-yellow) and brown spots leading to marginal necrosis. Lachover and Ebercon (1972b) showed that yield response to Fe application is related to $CaCO_3$ concentration in the soil. Papastylianou (1989) surveyed 35 groundnut fields in Cyprus and determined that plants were chlorotic in soils with $CaCO_3 > 20$–25% and when Fe content was less than 2.5 kg/ha. Dungarwal *et al.* (1974) applied 500 kg S/ha to a clay loam of pH 8.4 and increased yields by 197% evidently due to reduced pH and increased Fe availability.

Lachover *et al.* (1970) applied an iron chelate (FeEDDHA) to a soil in Israel with pH 7.9 and 15% $CaCO_3$ and measured a 50% increase in pod yield and a 40% increase in hay yield. Lachover and Ebercon (1971) showed that Fe chelate applied to a soil of pH 7.9 and 11% $CaCO_3$ caused leaves to green up and increased yield. Foliar applied Fe chelate increased greenness and yield, but Fe polyflacenoid and Fe acetate increased greenness without improving yield (Hartzook *et al.*, 1971). Yields were increased 359% by application of 10 kg Fe/ha (as FeEDDHA) to a loamy clay with pH 7.9 and 31% $CaCO_3$ (Lachover and Ebercon, 1972a).

Reddy and Patil (1980) applied $FeSO_4$ spray to spanish-type groundnuts grown on an Indian soil with pH 7.5 (2.5% $CaCO_3$ and 9 mg/kg Fe) and measured no yield increase. Hallock (1964) applied Fe chelates to groundnuts grown in Virginia and found no yield effect. Schneider and Anderson (1972) measured yield response to FeEDDHA in Texas.

Patil *et al.* (1979) determined that foliar application of $FeSO_4$ produced greater yields than soil applied $FeSO_4$ on a black clay soil with pH 7.7 (2.5% $CaCO_3$ and 1.26 mg/kg available Fe). They also noted that high phosphorus fertilization was related to chlorosis and diminished yields.

Nitrogen fertilizers did not improve leaf colour in groundnuts with Fe chlorosis and N levels in green and chlorotic leaves were not different (Lachover and Ebercon, 1971). Kafkafi and Neumann (1985) showed that no Fe chlorosis was observed when NH_4-N was 20% or greater of the total soil N. They suggested that NH_4^+ uptake resulted in H^+ efflux from roots which reduced the soil Fe^{3+} to Fe^{2+} which groundnut plants then absorbed.

Kafkafi and Neumann (1985) recommended the use of a nitrification inhibitor in addition to ammonium fertilizers. Iron stress was found to be stronger in nodulated than in non-nodulated plants and high plant Fe levels enhanced N fixation (Terry et al., 1988).

Zaharieva et al. (1988) showed that high soil Mn levels can be related to Fe deficiency. Iron and Mn do not compete for absorption sites; Mn actually inactivates Fe metabolic activity by decreasing the Fe^{2+} concentration in groundnut plants. Zaharieva et al. (1988) also stated that Fe efficient cultivars can overcome Fe chlorosis without additional Fe supply. Hartzook et al. (1974a,b) determined that yield of untreated Fe efficient cultivars was roughly equal to the yield of inefficient cultivars which were fertilized with an Fe chelate.

Iron deficiency cannot be determined by plant analysis since chlorotic leaves may contain more Fe than healthy leaves (Papastylianou, 1989). Rao et al. (1987) stated that total Fe is unsatisfactory as a measure of Fe status in plant tissue, and that Fe status was better assessed from an estimate of ferrous-Fe (by extraction with o-phenanthroline). They determined that the critical level for ferrous-Fe in the youngest fully opened leaf was 6 mg/kg. Rao et al. (1987) showed that chlorotic leaves had lower extractable Fe but higher total Fe, so that Fe deficiency is due to poor utilization of Fe within a leaf, not absorption or translocation problems.

Estimated critical levels are 2.5 mg Fe/kg in soil and 6 mg/kg ferrous-Fe in leaves.

7.8.5 Manganese (Mn)

(a) Deficiency

Manganese deficiency is a problem only on high pH soils. Rich (1956) stated Mn deficiency is a problem for the crop in Virginia. He reported that Mn concentration in the plant was inversely related to soil pH, calcium, and magnesium levels, in a study using 32 Coastal Plain soils. Visual Mn deficiency symptoms were evident when leaf Mn was below 10 mg/kg (Rich, 1956). Manganese deficiency in groundnut has been observed on soils with pH values as low as 5.8. Anderson (1964) reported that research in Georgia showed no yield effect of $MnSO_4$ additions (10–50 kg/ha) to a Tifton loamy sand with pH 6.5, a Norfolk sandy loam or a Greenville clay loam. Hickey et al. (1974) recorded significant yield increase for crops grown on a Lakeland sand (pH 6.3, extractable soil Mn 0.67 mg/kg) due to addition of 40 kg Mn/ha ($MnCl_2$). Foliage Mn levels were also increased from 59 to 155 mg/kg. The recommendation from the 1980 *Virginia Peanut Production Guide* is to apply foliar Mn at a rate of 0.8–1.1 kg/ha in each of up to three applications, when interveinal chlorosis, symptomatic of Mn deficiency, is evident (Allison, 1980).

Hallock (1979) showed that soil Mn treatments did not yield significantly less than spray treatments for virginia-type groundnuts but deficiency symptoms were greater for soil applied treatments. Parker and Walker (1986) studied the interaction of Mn response with soil pH on a Pelham sand. Manganese deficiency occurred on plots with pH levels near 6.8 (Mehlich 1 soil Mn = 3.3 mg/kg) but not in plots with pH levels of 5.2 (soil Mn = 2.1 mg/kg) or 6.0 (soil Mn = 2.5 mg/kg). At pH 6.8, soil application of Mn at 0, 10, 20 and 40 kg/ha resulted in yields of 3.41, 5.40, 5.73 and 6.37 tonnes/ha, respectively. Critical Mn levels in the leaves were 13, 15, 15, 15, 13 and 12 mg/kg at 7, 9, 11, 13, 15 and 17 weeks after planting. Parker and Walker (1986) concluded that maintaining a soil pH near 6.0 was optimal for groundnut production.

Soil Mn applications can be used to prevent Mn deficiency when the soil pH is known to be >6.0. Foliar Mn applications can correct Mn deficiency, diagnosed through foliar symptoms, more rapidly than soil Mn applications.

(b) Toxicity

Manganese toxicity can be a problem in low pH soils. Morris and Pierre (1949) reported that a concentration of 10 mg Mn/l in a nutrient solution reduced growth to 76% of the control and resulted in chlorosis of leaf margins. Plant Mn concentration was 1245 mg/kg. Groundnuts were the least sensitive (of five legumes studied) to Mn toxicity and had the lowest plant Mn concentrations. Boyd (1971) described Mn toxicity symptoms for the crop as interveinal leaf chlorosis followed by marginal leaf necrosis. Boyd also found that soil Mn (NH_4OAc extractable) was well correlated with leaf necrosis. Severe symptoms occurred when soil Mn was greater than 10 mg/kg and foliar Mn was greater than 50 mg/kg.

Benac (1976) found that a nutrient solution with Mn concentration ≥ 20 mg/kg caused stunting of plants. Tissue Mn of plants with leaf necrosis was ≥ 4000 mg/kg, and the highest Mn concentration was found in the leaves. In a nutrient solution test Nambiar and Anjaiah (1989) found that groundnut plants with Mn toxicity symptoms had between 1040 and 3070 mg/kg Mn in the plant tissue. High Mn levels decreased dry matter accumulation and nitrogen uptake.

High Mn levels may magnify Fe deficiency. Zaharieva (1986) found that the $Mn:Fe^{2+}$ ratio in plants was greater than 1:1 in chlorotic plants. Zaharieva et al. (1988) stated that groundnut plants with Mn toxicity (dark brown marginal leaf spotting) had >450 mg/kg Mn in the leaves. Application of Fe as FeEDDHA eliminated leaf spotting. Zaharieva et al. (1988) also suggested that Fe application could induce Mn deficiency.

It is difficult to relate nutrient solution tests to field situations. More research is needed in the area of Mn toxicity in groundnuts.

7.8.6 Molybdenum (Mo)

Molybdenum is essential for nitrogen fixation and is therefore recommended for some legumes (e.g. soybeans, alfalfa). However, it is currently not recommended for groundnuts.

Plot trials in Senegal showed that Mo had a harmful effect on groundnuts (Bouyer and Collot, 1952). Molybdenum deficiency symptoms were not obtained in a sand culture experiment, though slight chlorosis was evident late in the growing season (Reid and York, 1958). Harris (1959) stated that Mo application caused foliage to be a darker green and frequently increased the size of the foliage, but it has never caused a significant increase in yield in research in Florida.

Rao *et al.* (1960) reported that an application of 140 g Mo/ha in India increased pod yield. Walker (1967) found that 0.2 kg Mo/ha increased yield by 224 kg/ha on a Tifton soil but had no effect on yield on a Greenville soil. Welch and Anderson (1962) found that Mo availability was increased by liming and that Mo application increased Mo concentration in leaves, but no deficiency symptoms were evident in areas which received no Mo. They stated that seed Mo concentration is high enough to provide the plant's Mo requirement even in a low Mo soil. Sellschop (1967) stated that Mo deficiency is best corrected by liming. Parker (1964) reported that Mo often improved plant colour but gave a yield response in only one of 15 experiments conducted in Georgia. Boswell *et al.* (1967) showed that yield was not well correlated with leaf or soil Mo content, and Mo addition increased nitrogen content of foliage. However, the yield effect of Mo was inconsistent.

Heinis (1972) determined that Mo fertilization increased N and methionine contents in leaves. Graham (1979) stated that Mo is essential for nodule formation and function and Mo deficiency, which is more common on acid soils, can cause nitrogen deficiency. Kiat (1979) reported that Mo application had no significant influence on yield, nitrogen fixation or Mo concentration in groundnut tissue.

In research in India, it was found that 1 kg/ha NH_4 molybdate increased yield of spanish groundnut (Reddy and Patil, 1980). The soil test level was 0.5 mg/kg available Mo, and pH was 7.5. The authors suggested that this beneficial effect may be due to increased N availability which resulted in increased protein in groundnut kernels. Kene *et al.* (1988) found that Mo increased nodulation and nodule N content for groundnut.

Most of the literature agrees that Mo increases greenness and N content of groundnut leaves, but yield increases due to Mo application are rare. More research is required to determine under what conditions Mo fertilizers may be beneficial. Currently, there is no consistent data to recommend Mo fertilization for groundnuts.

Zinc (Zn)

(a) Deficiency

Zinc deficiency usually occurs under high pH conditions. Carter (1964) summarized Georgia research and showed that, even though zinc fertilization sometimes increased yield and sometimes decreased yield, in general the differences were not significant. Sellschop (1967) stated that Zn insufficiency was less conspicuous in groundnut than in maize, and recommended 16–22 kg Zn/ha where the problem is common. Schneider and Anderson (1972) found that a Zn application of 0.1 kg Zn/ha gave a positive yield response for spanish groundnut, but that 90 kg/ha reduced yield. In a calcareous soil with <0.3 mg/kg soil Zn, applications of $ZnSO_4$ from 10 to 60 kg/ha had no significant yield effect (Lakshminarasimhan et al.,1977). Spraying 0.8% ZnO on a calcareous vertisol increased yield of valencia-type groundnut (Mupawose, 1978). Phosphorus application can show an antagonistic effect on Zn uptake (Chahal and Ahluwalia, 1977). Zinc deficiency is associated with high soil pH and high available P levels (Graham, 1979). However, Patil et al. (1979) applied $ZnSO_4$ to groundnuts with severe chlorosis, which was attributed to high soil pH and heavy phosphorus fertilization, but there was no yield response to either soil or foliar applications.

Reddy and Patil (1980) stated that 0.5 mg/kg Zn in soil and 22 mg/kg Zn in leaves at flowering were the critical levels for Zn deficiency in groundnuts. Rhoads et al. (1989) applied Zn to soil in a greenhouse study and determined that the cultivar Southern Runner was more sensitive to Zn deficiency than Sunrunner. They suggested a critical soil Zn (Mehlich 1) level of 2.5 mg/kg at soil Ca >400 mg/kg.

Bell et al. (1990) described Zn deficiency symptoms in groundnuts as decreased internode length and restricted development of new leaves. They also found that Zn deficient plants accumulated reddish pigments in stems, petioles and leaf veins. They stated that 20 mg/kg in upper stems and leaves and 25 mg/kg in recently matured leaves (at early pegging) had been used previously as critical levels for Zn deficiency diagnosis. They recommended that the blade of the youngest fully expanded leaf be used for diagnosis with 8–10 mg/kg Zn as its critical value. Zinc deficiency is related to high soil pH, high soil Ca and high soil P. Leaf critical level is 20–25 mg/kg, and foliar application is probably the best way to correct Zn deficiency.

(b) Toxicity

Zinc toxicity was first reported by Quintana (1972) who noted that application of 90 kg Zn/ha as $ZnSO_4$ decreased yields and resulted in 67 mg Zn/kg in plant tissue. Keisling et al. (1977) described Zn toxicity symptoms as chlorosis, stunting, purple coloration of the main stem and petioles usually a lesion at the base of the plant (stem splitting) and premature necrosis. Tentative Zn toxicity critical values were set at 12 and 220 mg/kg for soil (Mehlich 1 extractable) and tissue, respectively. Tissue Zn concentration increased 15 mg/kg for each 1 mg/kg increase in soil Zn. Liming reduced Zn uptake and stunting but did not change the level of Mehlich 1 extractable Zn in soil. Davis-Carter *et al*. (1990) showed that leaf chlorosis and stem purpling were not well correlated with leaf Zn levels, and described new symptoms of horizontal leaf growth and leaf closure.

Rhoads *et al*. (1989) stated that peanut response to Zn appeared to be more dependent on soil Ca level than on soil pH. Up to 23 kg Zn/ha did not affect plant growth at soil Ca >400 mg/kg with soil pH of 6.5 to 6.8, but 8.1 kg Zn/ha reduced plant growth when soil Ca ranged from 150 to 200 mg/kg and pH was ≤6.6. Parker *et al*. (1990) studied data from growers' fields which indicated that a leaf Ca:Zn ratio ≤50 was required for Zn toxicity to groundnuts. Leaf Zn was affected more by soil pH than by soil Zn. A regression equation, including both factors, showed that an increase in soil Zn from 1 to 10 mg/kg increased leaf Zn by 202 mg/kg at soil pH 4.6 but only 9 mg/kg at pH 6.6. Cox (1990) used data from North Carolina and Georgia to predict plant Zn concentration from soil pH and soil Zn. Davis-Carter *et al*. (1991) stated that, since Mehlich 1 extraction of Zn from soil is not pH sensitive, it is necessary to include soil pH in any regressions predicting leaf Zn. They used such equations to calculate the probabilities for the development of Zn toxicity symptoms as a function of soil pH and soil Zn. Rhoads *et al*. (1991a) showed that increasing Zn rates in greenhouse studies decreased Ca concentration in groundnut tissues. In two greenhouse tests, the Ca:Zn ratio in tissue proved to be a good diagnostic tool for predicting dry matter yield; in one test, the critical ratio was 140:1, and in another it was 78:1.

Rhoads *et al*. (1989, 1991b) also noted cultivar differences in tolerance to Zn toxicity. Southern Runner had greater dry matter yield and lower plant Zn concentration than Sunrunner at the same soil Zn level. Davis-Carter *et al*. (1990) illustrated the influence of soil texture on critical levels. Groundnuts grown on clayey soils required lower soil pH and higher soil Zn levels to develop toxicity symptoms than those grown on sandy soils. On clayey soils, leaf Zn >470 mg/kg was related to toxicity. However, on sandy soils, plants with leaf Zn >350 mg/kg exhibited zinc toxicity symptoms.

7.9 CURRENT EMPHASIS AND RESEARCH NEEDS

Only a few scientists are now actively working to improve the mineral nutrition of groundnut. Many studies appear to be localized and indicate responses on a particular soil or in a local situation. Often studies provide only yield results from applications of various fertilizers applied at several rates, without providing adequate supporting soil and plant analytical results. Such studies can certainly provide local help but are difficult to extend beyond limited confines. Recent emphasis is placed on efficiency of production. Supplying inputs, such as fertilizers, at rates that do not exceed needs is consistent with both top economic returns and environmental concerns. For example, emphasis in the south-eastern USA peanut belt has recently been placed on reviewing fertilizer recommendations for groundnuts. The review indicated that responses to P and K are rare when groundnut is in a rotation with other crops to which fertilizer is applied. Reviews such as this can lead to increased efficiency; they also emphasize the fallacy of making general applications of all or most nutrients without consideration of plant needs via research-based recommendations.

Another major consideration for groundnut is quality. Seed quality and aflatoxins in groundnuts are affected by mineral nutrition. Some studies have attempted to relate quality to mineral nutrition, but more research may be quite fruitful.

Increasing interest in and need for better waste utilization throughout the world makes it imperative to study the effects of application of some wastes to agricultural lands. The effects of several by-products and waste products on groundnut are unknown but researchable.

In areas subject to erosion, conservation tillage methods are being adopted rapidly. Studies are needed to define and solve problems unique to tillage methods that provide maximum residue coverage and minimum soil disturbance.

Finally, increased linkage between scientists will lead to major improvements. Several possibilities can be cited, but one of them may be the improvement of N fixation in groundnut by linking breeders with plant nutritionists. Links between scientists are also needed to study groundnut nutrition as part of a total cropping system rather than as a single crop.

REFERENCES

Adams, Fred and Hartzog, D.L. (1980) The nature of yield of Florunner peanuts to lime. *Peanut Science*, **7**, 120–123.

Adams, James F., Hartzog, D.R. and Nelson, D.P. (1993) Supplemental calcium application on yield, grade and seed quality of runner peanut. *Agronomy Journal*, **85**, 86–93.

Allison, A.H. (1980) Agronomic recommendations and procedures, in *1980 Peanut Production Guide*, Virginia Polytechnic Institute and State University, Blacksburg, VA, pp. 3–6.

Alva, A.K., Gascho, G.J. and Hodges, S.C. (1989a) Peanut yield and grade vs. soil calcium indices in coastal plain soils. *Agronomy Abstracts*, p. 232.

Alva, A.K., Gascho, G.J. and Guang, Yang (1989b) Gypsum material effects on peanut and soil calcium. *Communications in Soil Science and Plant Analysis*, **20**, 1727–1744.

Alva, A.K., Gascho, G.J. and Guang, Yang (1991) Soil solution and extractable calcium in gypsum-amended coastal plain soils used for peanut culture. *Communications in Soil Science and Plant Analysis*, **22**, 99–116.

Anderson, O.E. (1964) Manganese, in *Micronutrients and Crop Production in Georgia*, (ed. R.L. Carter), University of Georgia College of Agriculture Bulletin N.S. 126, pp. 33–41.

Anon. (1965) *Agricultural Progress*, Virginia Agricultural Experiment Station Research Report 102, 70 pp.

Anon. (1972) *The Annual Report of the Agricultural Research Council of Malawi.*

Arrendell, S., Wynne, J.C., Elkan, G.H. and Isleib, T.G. (1985) Variation from nitrogen fixation among progenies of a virginia x spanish peanut cross. *Crop Science*, **25**, 865–869.

Arrendell, S., Wynne, J.C., Elkan, G.H. and Schneeweis, T.J. (1988) Selection among early generation peanut progeny for enhanced nitrogen fixation. *Peanut Science*, **15**, 90–93.

Asokan, S. and Raj, D. (1974) Effect of forms and levels of boron application on groundnut. *Madras Agricultural Journal*, **61**(8), 467–471.

Bahl, G.S., Baddesha, H.S., Pasricha, N.S. and Aulakh, M.S. (1986) *Indian Journal of Agricultural Science*, **56**, 429–433.

Ball, S.T., Wynne, J.C., Elkan, G.H. and Schneeweis, T.J. (1983) Effects of inoculation and applied nitrogen on yield, growth and nitrogen fixation of two peanut cultivars. *Field Crops Research*, **6**, 85–91.

Barber, S.A. (1984) *Soil Nutrient Bioavailability: A Mechanistic Approach*, John Wiley & Sons, New York, NY.

Bell, M.J. (1985a) Calcium nutrition of peanuts (*Arachis hypogaea* L.) on Cockatoo sands of the Ord River irrigation area. *Australian Journal of Experimental Agriculture*, **25**, 642–648.

Bell, M.J. (1985b) Phosphorus nutrition of peanut (*Arachis hypogaea* L.) on Cockatoo sands of the Ord River Irrigation Area. *Australian Journal of Experimental Agriculture*, **25**, 649–653.

Bell, R.W., Edwards, D.G and Asher, C.J. (1989) External calcium requirements for growth and nodulation of six tropical legumes grown in flowing solution culture. *Australian Journal of Agriculture Research*, **40**, 85–96.

Bell, R.W., Kirk, G, Plaskett, D. and Loneragan, J.F. (1990) Diagnosis of zinc deficiency in peanut (*Arachis hypogaea* L.) by plant analysis. *Communications in Soil Science and Plant Analysis*, **21**, 273–285.

Benac, R. (1976) Action de la concentration en manganese de la solution nutritive sur le comportement de l'arachide (*Arachis hypogaea* L.). *Oléagineux*, **1**(12), 539–543.

Beringer, H. and Taha, M.A. (1976) [45]Calcium absorption by two cultivars of groundnut (*Arachis hypogaea*). *Experimental Agriculture*, **12**, 1–7.

Bhan, S. (1977) Nutrient uptake by groundnut (*Arachis hypogaea* L.) as influenced by variety, spacing and soil fertility on desert soil. *Indian Journal of Agricultural Research*, **11**, 65–74.

Black, M.C., Pataky, J.K., Beute, M.K. and Wynne, J.C. (1984) Management tactics that complement host resistance for control of cylindrocladium black rot of peanuts. *Peanut Science*, **11**, 70–73.

Blamey, F.P.C. and Chapman, J. (1979) Boron toxicity in spanish groundnuts. *Agrochemophysica*, **11**, 57–59.

Blamey, F.P.C., Chapman, J. and Smith, M.F. (1981) Boron fertilization and soil amelioration effects on the boron nutrition of spanish groundnuts. *Crop Production*, **10**, 143–146.

Bledsoe, R.W., Comar, C.L. and Harris, H.C. (1949) Absorption of radioactive calcium by the peanut fruit. *Science*, **109**, 329–330.

Bledsoe, R.W. and Harris, H.C. (1947) Nutrition and physiology of the peanut. *University of Florida Agricultural Experiment Station Annual Report.*

Bledsoe, R.W. and Harris, H.C. (1948) Nutrition and physiology of the peanut. *University of Florida Agricultural Experiment Station Annual Report.*

Bledsoe, R.W. and Harris, H.C. (1949) Nutrition and physiology of the peanut. *University of Florida Agricultural Experiment Station Annual Report.*

Bockelee-Morvan, A. and Martin, G. (1966) Les besoins en soufe de L'Arachide effets sur les rendements. *Oléagineux*, **11**, 679–682.

Bolhuis, G.G. and Stubbs, R.W. (1955) The influence of calcium and other elements on the fruitification of the peanut in connection with absorption capacity of its gynophores. *Methods Journal of Agricultural Science*, **3**, 220–237.

Boote, K.J., Stansell, J.R., Schubert, A.M. and Stone, J T. (1982) Irrigation, water use, and water relations, in *Peanut Science and Technology* (eds H.E. Pattee and C.T. Young), American Peanut Research and Education Society Inc., Yoakum, TX.

Boswell, F.C. (1964) Copper, in *Micronutrients and Crop Production in Georgia*, (ed. R.L. Carter) University of Georgia Agricultural Experiment Stations Bulletin N.S. 126, pp. 22–28.

Boswell, F.C., Anderson, 0.E. and Welsh, L.F. (1967) *Molybdenum studies with peanuts in Georgia*, University of Georgia Agricultural Experiment Stations Research Bulletin 9.

Bouyer, S. and Collot, L. (1952) Oligoéléments et arachide. *Bulletin Agronomique de la Ministère de France d'outre mer*, **7**, 77–88 (*Field Crops Abstracts*, **6(3)**, 184).

Boyd, H.W. (1971) Manganese toxicity to peanuts in autoclaved soil. *Plant and Soil*, **34**, 133–144.

Branch, W.D. and Gascho, G.J. (1985) Screening for low fertility tolerance among peanut cultivars. *Agronomy Journal*, **77**, 963–965.

Carter, R.L. (1964) Zinc, in *Micronutrients and Crop Production in Georgia*, (ed. R.L. Carter), University of Georgia Agricultural Experiment Stations Bulletin N.S. 126, pp. 53–63.

Chahal, R.S. and Ahluwalia, S.P.S. (1977) Neutroperiodism in different varieties of ground-nut with respect to zinc and its uptake as affected by phosphorus application. *Plant and Soil*, **47**, 541–546.

Chauhan, Y.S., Jain, V.K., Kandekar, M.P., and Jain, P.C. (1988) Response of groundnut (*Arachis hypogaea*) varieties to phosphorus fertilization. *Indian Journal of Agricultural Science*, **58**, 359–361.

Chrudimsky, W.W. (1970) *Boron assimilation and its effect on the quality of Spanish peanuts*. Ph.D. thesis, Oklahoma State University, 92 pp.

Cope, J.T., Evans, C.E. and Williams, H.C. (1981) *Soil test fertilizer recommendations for Alabama crops*. Alabama Agricultural Experiment Station Circular 251, Auburn University, Alabama.

Cope, J.T., Starling, J.G., Ivey, H.W. and Mitchell, C.C. Jr (1984) Response of peanuts and other crops to fertilizers and lime in two long term experiments. *Peanut Science*, **11**, 91–94.

Cox, F.R. (1990) A note of the effect of soil reaction and zinc concentration of peanut tissue zinc. *Peanut Science* **17**, 15–17.

Cox, F.R., Adams, F. and Tucker, B.B. (1982) Liming, fertilization and mineral nutrition, Chapter 6 in *Peanut Science and Technology*, (eds H.E. Pattee and C.T. Young), American Peanut Research and Education Soc., Inc., Yoakum, TX.

Cox, F.R., and Reid, P.H. (1964) Calcium–boron nutrition as related to concealed damage in peanuts. *Agronomy Journal*, **56**, 173–176.

Cox, F.R., Sullivan, G.A. and Martin, C.K. 1976. Effect of calcium and irrigation treatments on peanut yield, grade, and seed quality. *Peanut Science*, **3**, 81–85.

Csinos, A.S. and Gaines, T.P. (1986) Peanut pod rot complex: A geocarposphere nutrient imbalance. *Plant Disease*, **68**, 61–65.

Davidson, J.I., Blankenship, P.D., Sanders, T.H. *et al.* (1983) Effect of row spacing, row orientation and gypsum on the production and quality of nonirrigated peanuts. *Proceedings of the American Peanut Research and Education Society*, **15**, 46–51.

Davis-Carter, J.G., Parker, M.B. and Gaines, T.P. (1990) Zinc toxicity symptoms in peanut. *Proceedings of The American Peanut Research and Education Society, Inc. July 10–13, 1990, Stone Mountain, GA*, p. 64.

Davis-Carter, J.G., Parker, M.B. and Gaines, T.P. (1991) Interaction of soil zinc, calcium and pH with zinc toxicity in peanuts, in *Utilization of Acidic Soils for Crop Production* (eds R.J. Wright, V.C. Baligar, and R.P. Murrmann), Kluwer Academic Publishers, pp. 339–347.

Dungarwal, H.S., Mathur, P.N. and Singh, H.G. (1974) Effect of foliar sprays of sulphuric acid with and without elemental sulphur in the prevention of chlorosis in peanut (*Arachis hypogaea* L.). *Communications in Soil Science Plant Analysis*, 5(4), 331–339.

Dwivedi, R.S., Joshi, Y.C., Shara, S.N. *et al.* (1987) Modeling of peanut (*Arachis hypogaea* L.) for higher yield on phosphorus deficient soil. *Oléagineux*, 42, 165–168.

Filonow, A.B., Melouk, H.A., Martin, M. and Sherwood, J. (1988) Effect of calcium sulphate on pod rot of peanut. *Plant Disease*, 72, 589–593.

Foster, H.L. (1980) The influence of soil fertility on crop performance in Uganda. II. Groundnuts. *Tropical Agriculture (Trinidad)*, 57, 29–42.

Gaines, T.P., Parker, M.B. and Walker, M.E. (1989) Runner and virginia type peanut response to gypsum in relation to soil calcium level. *Peanut Science*, 16, 116–118.

Ganesan, S. and Sundararajan, S.R. (1972) Studies on the effect of boron on the bunch groundnut in Parambikulam Alivar Project Region in Tamil Nadu. *Madras Agricultural Journal*, 59(5), 308.

Garren K.H.(1964) Landplaster and soil rot of peanut pods in Virginia. *Plant Disease Reporter*, 48, 349–352.

Gascho, G.J. (1992) Groundnut (Peanut), Chapter 5.2 in (*IFA World Fertilizer Use Manual*, eds D.J. Halliday, M.E. Trenkel and W. Wichmann), International Fertilizer Industry Association, Paris.

Gascho, G.J. and Alva, A.K. (1990) Beneficial effects of gypsum for peanuts. *Proceedings Third International Symposium on Phosphogypsum*, Vol. 1, pp. 376–393. Florida Institute of Phosphate Research, Miami, FL.

Gascho, G.J., Gaines, T.P. and Plank, C.O. (1990) Comparison of extractants for testing coastal plain soils. *Communications in Soil Science and Plant Analysis*, 21, 1051–1077.

Gascho, G.J., Guerke, W.R., Parker, M.B. and Gaines, T.P. (1992) Peanut germination related to potassium, calcium, and magnesium in seed, hulls and soils. *Proceedings of the American Peanut Research and Education Society*, 24, 33.

Gascho, G.J., and Hodges, S.C. (1991) Limestone and gypsum as sources of calcium for peanuts. *1990 Peanut Research-Extension Report*, pp. 61–64, University of Georgia, Coastal Plain Experiment Station, Tifton, GA.

Gascho, G.J., Hodges, S.C., Alva, A.K. *et al.* (1993) Calcium source and time of application for runner and virginia peanut. *Peanut Science*, 11 (in press).

Gillier, P. and Silvestre, P. (1969) Fertilization, in *L'Arachide*. G.P. Masonneuve et Larose, Paris.

Golakiya, B.A. (1989) In search of compromisation between calcium boron antagonism in the groundnut crop. *Journal of the Maharashtra Agricultural University*, 14(1), 123.

Golakiya, B.A. and Patel, M.S. (1986) Effect of calcium carbonate and boron on yield of groundnut. *Indian Journal of Agricultural Science*, 56(1), 41–44.

Goldsworthy, P.R. and Heathcote, K. (1963) Fertilizer trials with groundnuts in northern Nigeria. *Empire Journal of Experimental Agriculture*, 31, 351–365.

Gopal, N.H. (1968) Boron deficiency in groundnut (*Arachis hypogaea* L.). *Indian Journal of Agricultural Science*, 38(5), 832–834.

Gopal, N.H. (1969) Effect of boron toxicity on iron, heme enzymes and boron–protein complexes in groundnut. *Indian Journal of Experimental Biology*, 7(3), 187–189.

Gopal, N.H. (1970a) Antagonistic action of boron on copper in groundnut plant. *Current Science*, 39(2), 44–45.

Gopal, N.H. (1970b) Studies on recovery of groundnut plants from boron injury. *Turrialba*, 20, 198–203.

Gopal, N.H. (1971a) Influence of boron on the growth and yield in groundnut. *Turrialba*, 21, 435–441.

Gopal, N.H. (1971b) Effect of excess boron supply on accumulation of boron and nitrogen metabolism in groundnut plants. *Proceedings of the Indian Academy of Sciences*, B73(4), 192–201.

Gopal, N.H. (1975) Physiological studies on groundnut plants with boron toxicity. III. Effect on chlorophyll, iron and copper metabolism chlorosis. *Turrialba*, 25(3), 30–315.

Gopal, N.H. and Rao, I.M. (1972) Some agro-physiological aspects of boron nutrition in an Indian variety of groundnut. *Current Science*, **41**(19), 695–698.

Graham, R. (1979) The groundnut – (*Arachis hypogaea* L.). *Extension Newsletter – Department of Agricultural Extension, University of the West Indies (Trinidad & Tobago),* **10**(2), 4–6.

Haggin, J. and Koyumjisky, H. (1966) Effects of potassium fertilizers on peanuts in Israel. *Experimental Agriculture,* **2**, 295–298.

Hago, T.M. and Salama, M.A. (1987) The effects of elemental sulphur on shoot dry weight, nodulation and pod yield of groundnut under irrigation. *Experimental Agriculture,* **23**, 93–97.

Hallock, D.L. (1964) Effect of some chelated nutrients on peanut yield and seed size. *42nd Annual Meeting, Virginia Academy of Sciences*, Charlottesville, VA, May 6–9, 1964.

Hallock, D.L. (1966) Boron deficiency (hollow heart) in large seeded Virginia type peanuts. Virginia Academy of Science 17 N.S. (**4**), 243, *Proceedings for the Year 1965–1966, Minutes of the 14th Annual Meeting. May 4–7, 1966, Madison College.*

Hallock, D.L. (1979) Relative effectiveness of several Mn sources on Virginia-type peanuts. *Agronomy Journal,* **71**, 685–688.

Hallock, D.L. and Allison, A.H. (1980) Effect of three Ca sources applied on peanuts. I. Productivity and seed quality. *Peanut Science,* **7**, 19–25.

Hallock, D.L. and Garren, K.H. (1968) Pod breakdown, yield, and grade of Virginia type peanuts as affected by Ca, Mg, and K sulphates. *Agronomy Journal,* **60**, 253–257.

Hanlon, E.A. and Johnson, G.V. (1984) Bray/Kurtz, Mehlich 3, AB/D and ammonium acetate extractions of P, K, and Mg in four Oklahoma soils. *Communications in Soil Science and Plant Analysis,* **15**, 277–294.

Harigopal, N. and Rao, I.M. (1964) Physiological studies on boron toxicity in groundnut (*Arachis hypogaea*). *The Andhra Agricultural Journal,* **11**(4), 144–152.

Harigopal, N. and Rao, I.M. (1967) Agro-physiological studies on groundnut (*Arachis hypogaea* Linn.) with boron toxicity. *The Andhra Agricultural Journal,* **14**(1), 12–20.

Harris, H.C. (1952) Effect of minor elements, particularly copper, on peanuts. *University of Florida Agricultural Experiment Station Bulletin 494.*

Harris, H.C. (1959) Research on peanuts during the last twenty years. *Soil and Crop Science Society of Florida Proceedings,* **19**, 208–226.

Harris, H.C. (1963) Symptoms of nutritional deficiencies in plants. *Proceedings of Soil and Crop Science Society of Florida,* **23**, 139–152.

Harris, H.C. (1965) Nutrition and physiology of the peanut. *Florida Agricultural Experiment Station Annual Report,* p. 53.

Harris, H.C. (1968) Calcium and boron effects on Florida peanuts. *University of Florida Agricultural Experiment Station Bulletin 723,* 18 pp.

Harris, H.C. and Brolmann, J.B. (1966a). Effect of imbalance of boron nutrition on the peanut. *Agronomy Journal,* **58**(1), 97–99.

Harris, H.C. and Brolmann, J.B. (1966b) Comparison of calcium and boron deficiencies of the peanut. I. Physiological and yield differences. *Agronomy Journal,* **58**, 575–578.

Harris, H.C. and Brolmann, J.B. (1966c) Comparison of calcium and boron deficiencies in peanuts. II. Seed quality in relation to histology and viability. *Agronomy Journal,* **58**, 578–582.

Harris, H.C. and Gilman, R.L. (1957) Effect of boron on peanuts. *Soil Science,* **84**, 233–242.

Hartzog, D. and Adams, F. (1968) Soil fertility experiments with peanuts in 1967. *Auburn University Agricultural Experiment Station* Progress Report Series No. 89.

Hartzog, D. and Adams, F. (1971) Soil fertility experiments with peanuts in 1970. *Auburn University Agricultural Experiment Station* Progress Report Series No. 94.

Hartzog, D.L. and Adams, F. (1973a) Fertilizer, gypsum, and lime experiments with peanuts in Alabama. *Alabama Agricultural Experiment Station Bulletin 448.*

Hartzog, D. and Adams, F. (1973b) Soil fertility experiments with peanuts in 1972. *Auburn University Agricultural Experiment Station* Progress Report Series No. 101.

Hartzog, D.L. and Adams, F. (1988a) Relation between soil test P and K and yield response of runner peanuts to fertilizer. *Communications in Soil Science and Plant Analysis*, **19**, 1645–1653.

Hartzog, D.L. and Adams, J.F. (1988b) Soil fertility experiments with peanuts in Alabama, 1973–1986. *Alabama Agricultural Experiment Station Bulletin 594*, Auburn University, AL.

Hartzook, A., Eichman M. and Karstadt, D. (1971) The treatment of iron deficiency in peanuts cultivated in basic and calcareous soils. *Oléagineux*, **26**(6), 391–395.

Hartzook, A., Karstadt, D., Naveh, M. and Feldman, S. (1974a) Differential iron absorption efficiency of peanut (*Arachis hypogaea* L.) cultivars grown on calcareous soils. *Agronomy Journal*, **66**, 114–115.

Hartzook, A., Karstadt, D., Naveh, M. and Sander, N. (1974b) Groundnut (*Arachis hypogaea* L). cultivars for cultivation on calcareous soils. *Plant and Soil*, **41**, 685–688.

Heathcote, R.G. (1972) Potassium fertilization in the Savannah zone of Nigeria. *Potash Review*, **16**, 57.

Heinis, J.L. (1972) Methionine content of 25 peanut selections, and effect of molybdenum on methionine and nitrogen in peanut plants. *Oléagineux*, **27**(3), 147–152.

Hickey, J.M., Robertson, W.K., Hubbell, D.H. and Whitty, E.B. (1974) Inoculation, liming, and fertilization of peanuts on Lakeland fine sand. *Soil and Crop Science Society of Florida Proceedings*, **33**, 218–222.

Hill, W.E. and Morrill, L.G. (1974) Assessing boron needs for improving peanut yield and quality. *Soil Science Society America Proceedings*, **38**, 791–794.

Hill, W.E. and Morrill, L.G. (1975) Boron, calcium, and potassium interactions in Spanish peanuts. *Soil Science Society America Proceedings*, **39**, 80–83.

Hiltbold, A.E., Hartzog, D.L., Harrison, R.B. and Adams, F. (1983) Inoculation of peanuts in farmer's fields in Alabama. *Peanut Science*, **10**, 79–82.

Hodges, S.C., Gascho, G.J. and Kidder, G. (1994) Calcium, in *Research-based soil testing interpretation and fertilizer recommendations for peanuts on coastal plain soils*, (ed. C.C. Mitchell), Southern Cooperative Series Bulletin (in press).

Jadhao, P.N., Fulzele, G.R., Bhalerao, P.D. and Thorne, P.V. (1989) Response of peanut to boron application under shallow soils. *Annals of Plant Physiology*, **3**(1), 44–48.

Kafkafi, V. and Neumann, R.G. (1985) Correction of iron chlorosis in peanut (*Arachis hypogaea* Shulamit) by ammonium sulphate and nitrification inhibitor. *Journal of Plant Nutrition*, **8**(4), 303–309.

Katawatin, R., Ruaysoongnern, S., Keerati-Kasikorn, P. *et al.* (1989) Effect of phosphorus and copper application on copper uptake by peanut. *Khon Kaen Agriculture Journal*, **17**(6), 373–380.

Kayode, G.O. (1987) Potassium requirement of groundnut (*Arachis hypogaea*) in the lowland tropics. *Journal of Agricultural Science*, **108**, 643–647.

Keisling, T.C., Lauer, D.A., Walker, M.E. and Henning, R.J. (1977) Visual tissue, and soil factors associated with Zn toxicity of peanuts. *Agronomy Journal*, **69**, 765–769.

Kene, D.R., Pathey, M.K. and Thakare, K.K. (1988) Effect of graded levels of nitrogen and molybdenum on root nodulation and nitrogen fixation by *Rhizobium* in groundnut grown in Vertisol. *PKV Research Journal*, **12**(2), 155–157.

Kernick, M.D. (1961) in *Agricultural and Horticultural Seeds*, FAO, United Nations, Rome, pp. 345–348.

Kiat, T.B. (1979). *The influence of soil reaction (pH) and molybdenum on yield and nutrient uptake by peanuts (Arachis hypogaea) and red clover (Trifolium pratense)*. Thesis, Louisiana State University.

Kvien, C.S., Branch, W.D., Sumner, M.E. and Csinos, A.S. (1988) Genotypic factors influencing calcium concentrations in the seed and hull of peanut (*Arachis hypogaea* L.). *Crop Science*, **28**, 666–671.

Kvien, C.S., Weaver, R.W. and Pallas, J.E. (1986) Mobilization of nitrogen-15 from vegetative to reproductive tissue of peanut. *Agronomy Journal*, **78**, 954–958.

Lachover, D. and Ebercon, A. (1971) Iron deficiency problems in peanuts under irrigation. *World Crops*, July/August, **1971**, 202–204.

Lachover, D. and Ebercon, A. (1972a) Iron chlorosis in peanuts on a calcareous Jordan Valley soil. *Experimental Agriculture*, **8**, 241–250.

Lachover, D. and Ebercon, A. (1972b) The suitability of different physical forms of the chelate Sequestrene 138 for correcting iron-induced chlorosis in peanuts. *Oléagineux*, **27**(4), 205–209.

Lachover, D., Fichman, M. and Hartzook, A. (1970) The use of iron chelate to correct chlorosis in peanuts under field conditions. *Oléagineux*, **25**(2), 85–88.

Lakshminarasimhan, C.R., Andi, K and Surendran, R. (1977) Effect of zinc fertilisation for groundnut. *Oils and Oilseeds Journal*, **29**(3), 13.

Lal, R., and Saran, G. (1988) Influence of nitrogen and phosphorus on yield and quality of groundnut under irrigated conditions. *Indian Journal of Agronomy*, **33**, 460.

Lauter, D.J., Meiri, A. and Yermiyahu, V. (1989) Tolerance of peanut to excess boron. *Plant and Soil*, **114**, 35–38.

Longanathan, S. and Krishnamoorthy, K.K. (1977) Total uptake of nutrients at different stages of the growth of groundnut and the ratios in which various nutrient elements exist in groundnut plant. *Plant and Soil*, **46**, 565–570.

Luke, J.F. (1969) *Residual effects of high rates of fertilizer boron on a Norfolk sandy loam*. MS Thesis, North Carolina State University, Raleigh, NC.

Lund, Z.F. and Murdock, L.W. (1978) Effects of sulphur on early growth of plants. *Sulphur Agriculture*, **2**, 6–8.

Lynd, J.Q. and Ansman, T.R. (1989) Effects of phosphorus and calcium with four levels of potassium on nodule histology, nitrogenase activity, and improved 'Spanco' peanut yields. *Journal of Plant Nutrition*, **12**, 65–84.

Mali, A.L., Verma, R.R., Rathope, P.S. and Sharme, H. (1988) Nutrient uptake in groundnut as influenced by dates of planting and phosphorus and nitrogen application. *Madras Agricultural Journal*, **75**, 356–358.

Maliwal, G.L. and Tank, N.K. (1988) Effect of phosphorus in the presence and absence of sulphur and magnesium on yield and uptake of P, S, and Mg by Gaug-10 (*Arachis hypogaea*). *Indian Journal of Agricultural Science*, **58**, 557–560.

McGill, J.F. and Bergeaux, P.J. (1966) Boron for peanuts. *University of Georgia, College of Agriculture, Cooperative Extension Service, Peanut Release No. 49*.

Mengel, K. and Kirkby, E.A. (1982) *Principles of plant nutrition*. Potash Institute, Berne, Switzerland.

Mitchell, C.C. and Adams, J.F. (1994) Phosphorus and potassium, in *Research-based soil test interpretation and fertilizer recommendations for peanuts on coastal plain soils*, (ed. C.C. Mitchell), Southern Cooperative Series Bulletin (in press).

Mizuno, S. (1959) Physiological studies on the fruitification of peanut. I. Distribution of radioactive calcium administered to the fruiting zone on the fruiting organ. *Proceedings of the Crop Science Society of Japan*, **28**, 83–85 (Japanese–English Summary).

Morrill, L.G., Hill, W.E., Chrudimsky, W.W. *et al.* (1977) Boron requirements of Spanish peanuts in Oklahoma: Effects on yield and quality and interaction with other nutrients. *Oklahoma State University Report No. MNP-99*, 20 pp.

Morris, H.D., and Pierre, W.H. (1949) Minimum concentrations of manganese necessary for injury to various legumes in culture solutions. *Agronomy Journal*, **41**, 107–112.

Mupawose, R.M. (1978) Yield improvement in maize, rice and groundnuts grown on Chisumbanje basalt soils using zinc foliar sprays. *Rhodesia Agricultural Journal*, **75**(2), 37–40.

Muthuswamy, T.D. and Sundararajan, S.R. (1973) Effect of boron on bunch groundnut. *Madras Agricultural Journal*, **60**(6), 403.

Nambiar, P.T.C. and Anjaiah, V. (1989) Effect of manganese toxicity on growth and N_2 fixation in groundnut, *Arachis hypogaea*. *Annals of Applied Biology*, **115**, 361–366.

Nicholaides, J.J., and Cox, F.R. (1970) Effect of mineral nutrition on chemical composition and early reproductive development of virginia type peanuts (*Arachis hypogaea* L.). *Agronomy Journal*, **62**, 262–265.

Papastylianou, I. (1989) Effect of selected soil factors on chlorosis of peanuts grown in calcareous soils in Cyprus. *Plant and Soil*, **117**, 291–294.

Parker, M.B. (1964) Molybdenum in *Micronutrients and Crop Production in Georgia*, (ed. R.L. Carter) *University of Georgia Agricultural Experiment Stations Bulletin N.S.* **126**, pp. 42–52.

Parker, M.B., Gaines, T.P., Walker, M.E. *et al.* (1990) Soil zinc and pH effects on leaf zinc and the interaction of leaf calcium and zinc on zinc toxicity of peanuts. *Communications in Soil Science and Plant Analysis*, **21**, 2319–2332.

Parker, M.B., Gascho, G.J. and Gaines, T.P. (1983) Chloride toxicity of soybeans grown on Atlantic Coast Flatwoods Soils. *Agronomy Journal*, **75**, 439–443.

Parker, M.B. and Walker, M.E. (1986) Soil pH and manganese effects on manganese nutrition of peanut. *Agronomy Journal*, **78**, 614–620.

Parker, M.B., Walker, M.E. and Gaines, T.P. (1984), personal communication.

Patel, J.C., Vyas, M.N. and Malavia, D.D. (1988) Response of summer groundnut to irrigation under varying levels of nitrogen and phosphorus. *Indian Journal of Agronomy*, **33**, 56–59.

Patil, V.C., Radder, G.D. and Kudasomannavar, B.T. (1979) Effect of zinc, iron and calcium under varying levels of phosphorus on groundnut. *Mysore Journal of Agricultural Sciences*, **13**, 395–399.

Perkins, H.F. (1964) Iron, in *Micronutrients and Crop Production in Georgia*, (ed. R.L. Carter) University of Georgia College of Agriculture Bulletin N.S. **126**, pp. 29–32.

Perry, Astor. (1971) Boron – Peanuts' 'Big' Minor Element. *The Progressive Farmer*, May 1971, p. 6.

Phillips, T.D., Wynne, J.C., Elkan, G.H. and Schneeweis, T.J. (1989) Inheritance of symbiotic nitrogen fixation in two peanut crosses. *Peanut Science*, **16**, 66–70.

Piggot, C.J. 1960. The effect of fertilizers on the yield and quality of groundnuts in Sierra Leone. *Empire Journal of Experimental Agriculture*, **28**, 59–64.

Plank, C.O. (1989a) *Soil test handbook for Georgia*. Georgia Cooperative Extension Service, University of Georgia, Athens, GA.

Plank, C.O.(1989b) *Plant analysis handbook for Georgia*. Georgia Cooperative Extension Service, University of Georgia, Athens, GA.

Quintana, R.U. (1972) Zinc studies in peanuts (*Arachis hypogaea* L.). PhD Thesis, Texas A & M University, *Dissertation Abstracts International*, **32**(8), 4357B.

Rao, V.L., Narasimha, Krishna Murty, K.M. and Rao, M.P. Narasimha (1960) Groundnut – its response to applications of nitrogen, phosphorus, potassium, boron and molybdenum. *Second Conference of Oil Seeds Research Workers in Madras, India.*

Rao, J.K., Sahrawat, K.L. and Burford, J.R. (1987) Diagnosis of iron deficiency in groundnut, *Arachis hypogaea* L. *Plant and Soil*, **97**, 353–359.

Reddy, S.C.S. and Patil, S.V. (1980) Effect of calcium and sulphur and certain minor nutrient elements on the growth, yield and quality of groundnut (*Arachis hypogaea* L.). *Oléagineux*, **35**(11), 507–510.

Reid, P.H. and Cox, F.R. (1973) Soil properties, mineral nutrition and fertilization practices, in *Peanuts, Culture and Uses*. American Peanut Research and Education Association, Oklahoma State University, Stillwater.

Reid, P.H. and York, E.T., Jr (1958) Effect of nutrient deficiencies on growth and fruiting characteristics of peanuts in sand cultures. *Agronomy Journal*, **50**, 63–67.

Rhoads, F.M., Shokes, F.M. and Gorbet, G.W. (1989) *Response of two peanut cultivars to soil zinc levels*, University of Florida, Institute of Food and Agricultural Sciences, Research Report NF-89-2.

Rhoads, F.M., Shokes, F.M. and Gorbet, D.W. (1991a) *Dolomite and zinc interactions in two peanut cultivars*. University of Florida, North Florida Research and Education Center Research Report 91–6.

Rhoads, F.M., Shokes, F.M. and Gorbet, G.W. (1991b) *Response of Southern Runner peanuts to lime, gypsum, and zinc.* University of Florida, North Florida Research and Education Center Research Report 91-7.

Rich, C.I. (1956) Manganese content of peanut leaves as related to soil factors. *Soil Science,* **82,** 353–363.

Sahu, M.P. and Singh, H.G. (1987) Effect of sulphur on prevention of iron chlorosis and plant composition of groundnut on alkaline calcareous soils. *Journal of Agricultural Science,* **109,** 73–77.

Sankaran, N., Sennaian, P. and Morachan, Y.B. (1977) Effects of forms and levels of calcium and levels of boron on the uptake of nutrients and quality of groundnut. *Madras Agricultural Journal,* **64**(6), 384–388.

Saxena, H.K. and Mehrotra, O.N. (1985) Effect of boron and molybdenum in presence of phosphorus and calcium on groundnut. *Indian Journal of Agricultural Research,* **19,** 11–14.

Scarsbrook, C.E. and Cope, J.T. (1956) Fertility requirements of runner peanuts in southeast Alabama. *Alabama Experiment Station Bulletin,* **302.**

Schilling, R. and Hirsch, P.J. (1974) Chlorine nutrition of peanuts in Senegal. *Oléagineux,* **29,** 85–90.

Schmidt, J.P. and Cox, F.R. (1992) Evaluation of the magnesium soil test interpretation for peanuts. *Peanut Science,* **19,** 126–131.

Schneider, R.P. and Anderson, W.B. (1972) *Micronutrient nutrition of Spanish peanuts.* Association of Southern Agricultural Workers, Inc., 69th Annual Convention. Richmond, Virginia, Feb. 13–16, 1972.

Sellschop, J.P.F. (1967) Groundnuts – all aspects of cultivation. *Farming in South Africa,* February 1967, pp. 3–19.

Shiralipour, Aziz, Harris, H.C. and West, S.H. (1969) Boron deficiency and amino acid and protein contents of peanut leaves. *Crop Science,* **9,** 455–456.

Smal H., Sumner, M.E., Csinos, A.S. and Kvien, C.S. (1988) On the calcium nutrition of peanut (*Arachis hypogaea* L.). *Journal of Fertilizer Issues,* **5,** 103–108.

Smal H., Kvien, C.S., Sumner, M.E. and Csinos, A.S. (1989) Solution calcium concentration and application date effects on pod calcium uptake and distribution in Florunner and Tifton-8 peanut. *Journal of Plant Nutrition,* **12,** 37–52.

Snyman, J.W. (1972) *Nutritional studies with a spanish-type groundnut on an Avalon medium sandy loam soil.* PhD dissertation, University of Natal, Pietermaritzburg, South Africa (from Blamey *et al.*, 1981).

Sridhar, V., Soundararajan, M.S., Sudakara Rao, R. and Sreeramulu, C. (1985) Response of JL-24 groundnut to rates, times and methods of gypsum application. *Madras Agricultural Journal,* **72,** 47–53.

Stoller, E.W. (1966) The effect of boron nutrition on growth and protein and nucleic acid metabolism in peanut plants. PhD Thesis, North Carolina State University, *Dissertation Abstracts,* **27**(6), 1697B.

Sumner, M.E., Kvien, C.S., Smal, H. and Csinos, A.S. (1988) On the calcium nutrition of peanut (*Arachis hypogaea* L.). I. Operational model. *Journal of Fertilizer Issues,* **5,** 97–102.

Survanvesh, T. and Morrill, L.G. (1986) Foliar application of phosphorus to Spanish peanuts. *Agronomy Journal,* **78,** 54–58.

Swamy, N.R. and Reddy, P.R. (1983) Influence of growth regulator and nutrients on the quality of groundnut (*Arachis hypogaea* L.). *Madras Agricultural Journal,* **70**(11), 740–745.

Terry, R.E., Hartzook, A., Jolley, V.D. and Brown, J.C. (1988) Interactions of iron nutrition and symbiotic nitrogen fixation in peanuts. *Journal of Plant Nutrition,* **11,** 811–820.

Walker, M.E. (1967) Optimum rates of plant nutrients for peanut fertilization are shown, in

Serving Georgia through Research, *University of Georgia, Agricultural Experiment Stations Annual Report*, pp. 18–19.

Walker, M.E., Branch, W.D., Gaines, T.P. and Mullinix, B.G., Jr (1984) Response of nodulating and nonnodulating peanuts to foliarly applied nitrogen. *Peanut Science*, **11**, 60–63.

Walker, M.E. and Csinos, A.S. (1980) Effect of gypsum on the yield, grade and incidence of pod rot in five peanut cultivars. *Peanut Science*, **7**, 109–113.

Walker, M.E., Flowers, R.A., Henning, R.J. *et al.* (1979) Response of early bunch peanuts to calcium and potassium fertilization. *Peanut Science*, **6**, 119–123.

Walker, M.E., Gaines, T.P. and Henning, R.J. (1982) Foliar fertilization effects on yield, quality, nutrient uptake, and vegetative characteristics of Florunner peanuts. *Peanut Science*, **9**, 53–57.

Walker M.E., Gaines, T.P. and Parker, M.B. (1989) Potassium, magnesium, and irrigation effects on peanuts grown on two soils. *Communications in Soil Science and Plant Analysis*, **20**, 1011–1032.

Walker, M.E. and Keisling, T.C. (1978) Response of five cultivars to gypsum fertilization on soils varying in calcium content. *Peanut Science*, **5**, 57–60.

Walker, M.E., Keisling, T.C. and Drexler, J.S. (1976) Response of three peanut cultivars to gypsum. *Agronomy Journal*, **68**, 527–528.

Welch, L.F. and Anderson, O.E. (1962). Molybdenum content of peanut leaves and kernels as affected by soil pH and added molybdenum. *Agronomy Journal*, **54**, 215–217.

Wiersum, L.K. (1951) Water transport in the xylem as related to calcium uptake by groundnuts. (*Arachis hypogaea* L.). *Plant and Soil*, **3**, 160–169.

Wilson, D.M., Walker, M.E. and Gascho, G.J. (1989) Some effects of mineral nutrition on aflatoxin contamination of corn and peanuts, in *Management of Diseases with Macro- and Microelements*, (ed. A.W. Engelhard) APS Press, St. Paul, MN.

Wolt, J.D., and Adams, F. (1979) Critical levels of soil- and nutrient solution-calcium for vegetative growth and fruit development of Florunner peanuts. *Soil Science Society America Journal*, **43**, 1159–1164.

York, E.T., Jr and Colwell, W.E.(1951) Soil properties, fertilization, and maintenance of soil fertility, Chapter V in *The Peanut – The Unpredictable Legume*, National Fertilizer Association, Washington, DC, pp. 122–172.

Zaharieva, T. (1986) Comparative studies of iron inefficient plant with plant analysis. *Journal of Plant Nutrition*, **9**, 939–946.

Zaharieva, T., Kasabov, D. and Römheld, V. (1988) Responses of peanuts to iron–manganese interaction in calcareous soil. *Journal of Plant Nutrition*, **11**:1015–1024.

Zhang, J., Zhang, D., Jaing, Z and Liu, C. (1986) Boron nutrition and application of boron fertilizer to peanut crop. *Turang Tongbao*, **17**, 173–176.

CHAPTER 8

Nitrogen fixation

J. Sprent

8.1 INTRODUCTION

Groundnuts belong to the subfamily Papilionoideae of the family Leguminosae, using the classification system of Polhill and Raven (1981). Most but not all of the Papilionoideae form nitrogen fixing nodules with soil organisms known generally as rhizobia (Faria *et al.* 1989). General characteristics of advanced papilionoid nodules are listed in Table 8.1. Variations on this theme are discussed elsewhere (Sprent, 1989; Sprent and Sprent, 1990) except for characters specific to *Arachis* which are considered in the next section.

Within an active nodule, rhizobia carry out the reduction of nitrogen gas to ammonia using the enzyme complex nitrogenase:

$$N_2 + 8H^+ + 8e^- \rightarrow 2NH_3 + H_2$$

The reduction of protons to hydrogen gas appears to be an inevitable concomitant to the reduction of N_2 to ammonia. Nitrogenase can also use a number of other substrates, the best known of which is acetylene, C_2H_2: this reaction forms the basis of the acetylene reduction (AR) assay for nitrogenase. Reactions are coupled to the hydrolysis of ATP to ADP + Pi, with probable stoichiometry of 16 ATP for reduction of one molecule to N_2. This, coupled with the need for reducing power at a redox potential of around −430mV, means that the nitrogenase reaction is energy intensive. Further energy is required to assimilate the products of fixation into amino acids. However, this is true also for assimilation of soil ammonium and nitrate (Figure 8.1). Many people have attempted to compare the costs of nitrogen fixation and nitrate assimilation in theory and by experiment. Results vary with species and assumptions made, but it is generally accepted that nitrogen fixation is more expensive to the plant than nitrate assimilation and that both are more expensive than ammonium assimilation (Raven, 1988; Vance and Heichel, 1991).

The Groundnut Crop: A scientific basis for improvement. Edited by J. Smartt. Published in 1994 by Chapman & Hall, London. ISBN 0 412 408201.

TABLE 8.1   *Some general features of nodules of members of well-studied tribes of the Papilionoideae*

Character	Group		
	1	2	3
Bacterial partner	*Rhizobium*	*Bradyrhizobium* (except *Phaseolus*)	*Bradyrhizobium*
Infection process	*Via* root hairs and involving infection threads		*Via* lateral root junctions. Infection threads not formed
Nodule growth	Indeterminate	Determinate	Determinate
Infected region	Contains uninfected (interstitial) cells		Uniformly infected
Bacteroid shape	Pleomorphic	Rod-shaped	Spherical or rods
Predominant export product	Amides	Ureides	Amides

Group 1 is typical of the Vicieae and Trifolieae and is largely based on *Vicia, Pisum, Trifolium* and *Medicago*; Group 2 is typical of many genera of the Phaseoleae and Desmodieae and is largely based on *Glycine, Phaseolus* and *Vigna*; Group 3 is generally typical of the Aeschynomeneae and some Dalbergieae but details given are based solely on *Arachis*: variants are described in Sprent *et al.* (1989).

In addition to the direct biochemical costs of reducing N_2, legumes have to make an initial investment in nodule material (Sprent and Thomas, 1984) before they can reap any benefit. The construction costs also include production of mechanisms for controlling the diffusion of oxygen into nodules. Because nitrogenase is inactivated by free O_2 and yet the bacteria need O_2 to produce ATP to drive the reaction, nodules must maintain a high flux of O_2 at a low concentration from the soil atmosphere to the site of nitrogen fixation. The exact flux required will vary according to environmental conditions (e.g. temperature, water supply), which will affect not only nodule metabolism but also the rate at which the plant shoot system can supply nodules with photosynthate. In the few nodules which have been studied (e.g. soybean and clover) there appear to be both a fixed and a variable resistance to oxygen transport located in the nodule cortex (Witty *et al.*, 1986; Figure 8.6). Once past this barrier, oxygen transport is facilitated by an extensive intercellular space network and by the presence of the oxygen carrying pigment, haemoglobin, in the host cells surrounding the peribacteroid units. The appearance of haemoglobin, which gives active nodules a distinct pink colour, is closely coupled with the onset of nitrogenase activity. The latter is coupled to the repression, in the bacteria (now known as bacteroids), of the ability to assimilate ammonia. Instead

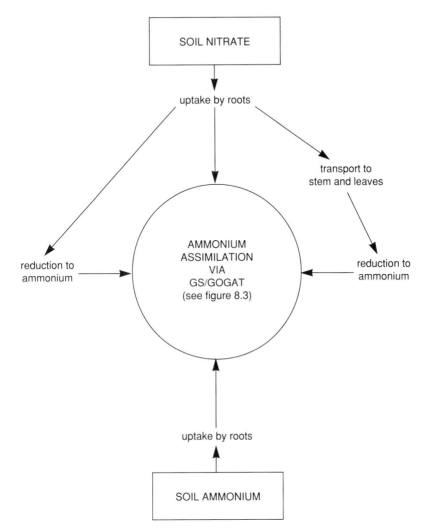

Figure 8.1 Major pathways of nitrogen assimilation in plants (see Figure 8.3 for terms GS, GOGAT).

the ammonia (or ammonium) passes to the host cell for assimilation. Very few rhizobia – and none known to infect groundnut – can be induced both to fix nitrogen *ex planta* and to assimilate the products of fixation.

8.2 *ARACHIS* AS A NODULATING GENUS

Within the Papilionoideae, *Arachis* is classified in the tribe Aeschynomeneae. In terms of its nodulation this tribe has a group of unusual and interesting characters, shared only (as far as we know) with a few members of the tribe Dalbergiae (Sprent *et al.*, 1989).

Arachis does not normally produce 'typical' root hairs, i.e. outgrowths from epidermal cells, even though it has the genetic material to do so – apparently normal root hairs can be induced by salt (Sprent and McInroy, 1984). Multicellular hairs are found near lateral root junctions and, although these are not involved in the infection process in the way that normal root hairs on legumes such as clover are, mutants lacking these hairs do not nodulate (Nambiar *et al.*, 1983). The original observations on nodulation in *Arachis* (Allen and Allen, 1940) were followed by various other studies, of which that of Chandler (1978) is the most comprehensive. The following description is based on Chandler's work. Infection was only found where cells at the base of the (usually) multicelluar root hairs at the root junction were swollen. It was suggested that this swelling may contribute to separation of cells at the root junction, permitting intercellular passage of rhizobia. Signs of intracellular penetration of both root hairs and basal cells were seen but these did not develop further. Intercellular penetration proceeded into the axillary/subtending root interface and some adjacent cortical cells (detectable by enlarged nuclei) were penetrated by rhizobia. No organized infection thread, characteristic of the root hair type of infection of soybean, clover etc., was seen at any stage. Instead, cell wall material appeared to be broken down and bacteria spread into the host cell, although surrounded at all times by an electron-dense matrix and a host-derived membrane. This rather amorphous structure was termed a zoogloea. When inside the host cell, the bacteria continued to divide frequently. The net result was a uniform population of infected cells, without the frequent uninfected interstitial cells characteristic of most other nodules studied (Figure 8.2).

When first described, the above process was thought to be unusual. However, on the basis of absence of interstitial cells and infection threads and the occurrence of nodules in axils of lateral roots, we believe it to be common in related legumes. The only other cases where the infection process has been studied in any detail are *Stylosanthes* (Chandler *et al.*, 1982) and *Aeschynomene* (Arora, 1954). These genera show broadly similar intercellular passage of rhizobia to that of *Arachis* but with a few minor variants such as lack of swollen basal cells in *Stylosanthes* (Sprent *et al.*, 1989).

All species of *Arachis* which have been examined form nodules and so it may be safe to assume that this nodulation is a generic attribute. Within genera this is usually the case, although there are exceptions, such as in the mimosoid genera *Acacia* and *Parkia*. As with a number of other genera, non-nodulating mutant forms of *A. hypogaea* have been generated and these are useful experimental tools (Giller *et al.*, 1987).

Distribution of nodules on plants of *Arachis* may vary with subspecies and variety. *A. hypogaea* var. *hypogaea* has a natural tendency to form

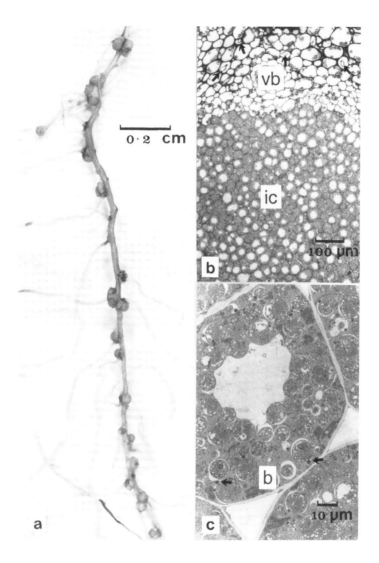

Figure 8.2 Photomicrographs of groundnut root nodules (b, bacteroids; ic, infected cells; vb, vascular bundle): (a) part of nodulated root (note that nodules are located in the axils of lateral roots); (b) light micrograph of cortex and part of infected region of a groundnut nodule; (c) electron micrograph of an infected cell. Cells in the outer layers of the cortex in (b) have larger cells with more intensely stained walls than those in the inner layers – compare also with Figures 8.5 and 8.7. Crystals (arrows) in outer layers are similar to the calcium oxalate crystals observed by Sutherland and Sprent (1984) in certain ureide-exporting legumes. Most of the infected cells in (b) are vacuolate. The bacteroids in (c) are spherical; the small black structures (arrows) are probably oleosomes (Jayaram and Bal, 1991). ((b) and (c) courtesy of K. Wilson and K.E. Giller.)

lateral roots on the hypocotyl and 13–18% of its nodules may be located in this region (Nambiar *et al.*, 1982). Nodules have also been reported by these workers to be associated with lateral roots arising at the base of the stem, a property which needs further investigation but is consistent with the position in the related genus *Aeschynomene* (e.g. Arora, 1954). Nodules on roots are arranged in rows according to protoxylem points. There are major genotypic differences in the vascular anatomy of the root/hypocotyl/stem regions, where the central root stele divides into discrete bundles (O'Brien and Sprent, unpublished). These differences may relate to the occurrence of nodules on the root/stem interface.

8.3 THE RHIZOBIAL COMPONENT

The taxonomy of rhizobia is currently undergoing major changes. All early classifications were based on host plant affinities but, increasingly, molecular characters such as 16s-rRNA are being taken into account (Young, 1992). These are showing up interesting affinities between rhizobia and other soil bacteria. Rhizobia from tropical soils have been less intensively studied than those from temperate soils, making it unwise to generalize too widely. Growth rate in culture is still a convenient preliminary. The old categories of 'slow-growing' and 'fast-growing' forms are still useful, but we now have additional 'extra slow-growing' forms (e.g. Barnet, 1988). Of those that have been properly categorized, slow-growing forms usually conform to the genus *Bradyrhizobium* and fast-growing forms to *Rhizobium*. *Arachis* is usually nodulated by slow-growing forms of the *Bradyrhizobium* sp. (cowpea) type, which generally have a wide host range and occur abundantly (more than 10 000 per gram) in most tropical soils. Some strains of fast-growing rhizobia can nodulate *Arachis* with various degrees of effectiveness (Rupela *et al.*, 1991), although most are ineffective (Wong *et al.*, 1988). Other legumes may also be nodulated by either fast- or slow-growing rhizobia and nodule development is different in some, such as *Lotus* (Wood *et al.*, 1985). Ineffective nodules formed by *Rhizobium* NGR 234 were found by Wong *et al.* (1988) to develop in a manner similar to effective nodules formed by *Bradyrhizobium* apart from bacteroid shape, which was polymorphic in the former and spherical (the norm for groundnut nodules) in the latter.

The widespread occurrence and promiscuity of rhizobia nodulating *Arachis* makes difficulties for field studies and inoculation programmes. Fortunately strains can sometimes be distinguished serologically and morphologically, enabling studies on competition and host/rhizobial interactions. One such study was carried out by Hadad and Loynachen (1986). Colony morphology was used as a basic criterion for separating strains, backed up by the more tedious serological characters. Clear-cut results were obtained in which a commercial inoculant strain outcompeted two strains isolated from Sudanese soils in occupying nodules on Sudanese cultivars. However, the colony morphology characters used included polysaccharide secretion. The inoculant strain produced abundant polysaccharide when grown on yeast-extract–mannitol agar, a standard medium for rhizobia (Bergersen, 1980; Peoples et al., 1989) whereas the two Sudanese soil isolates produced little.

It has recently become clear that extracellular polysaccharide (EPS) production by rhizobia can be important in successful infection of some legume roots by rhizobia (reviewed by Brewin, 1991). Of the species studied, all have infection through root hairs but subsequent nodule development falls into two types. Those nodules having an apical meristem (e.g. lucerne, alfalfa) and consequently of indeterminate growth do not become invaded by rhizobial mutants lacking EPS. Those with determinate growth, such as soybean, can be invaded by mutants lacking EPS. These two types of nodulated legumes fall into different taxonomic groups (tribes) and their nodules vary in metabolic as well as structural properties (Sprent, 1980). They are also nodulated by the genera *Rhizobium* and *Bradyrhizobium* respectively. *Arachis* is not closely related to either of these other types of legume: it has a different type of infection pathway (section 8.2) and a different nodule structure (section 8.4). Whether or not EPS molecules are involved in establishing the symbiosis is not known.

Exciting new developments are occurring in unravelling the signal exchange between plants and rhizobia which leads to successful invasion through root hairs. One of the first of these important developments was the discovery that the rhizobial *nod*D genes can be switched on by the flavonoid signals from the plant root (Peters et al., 1986). The *nod*D genes have an important regulatory role in the operation of several of other *nod* genes (Table 8.2). They are also a component of the very complex system which controls host/rhizobial specificity. Even though the infection process in *Arachis* appears rather simple, it might be expected that the rhizobial *nod* genes would still have to be triggered before nodulation could proceed. Evidence that this is indeed the case has been obtained by Smit and co-workers (1990). They used the now routine procedure of constructing genetically engineered rhizobia in which the *lacZ* gene is inserted in a position such that it is switched on when *nod*D is switched on. The *lacZ* gene codes for an enzyme, β galactosidase, which produces coloured products from suitable substrates. The induction of *nod*D can thus be studied on rhizobial cultures, without the host plant. In this way *Bradyrhizobium* sp. NC92, which nodulates *Arachis*, was compared with *B. japonicum* USDA110, which nodulates soybeans. In both, genistein was the most active inducer of those tested, but NC92 was 50 times more sensitive to it than USDA110. Five inducing compounds were isolated from cotyledons and seed coats of *Arachis*: all were active in NC92, none in USDA110.

Later on in the development of nodules the precise structure of rhizobial

TABLE 8.2   *Nodulation genes of* Bradyrhizobium *(extended from Barbour, Wang and Stacey, 1992)*

nod	
A, B, C	Involved in infection and early events in nodulation; interchangeable with relevant genes in *Rhizobium*, hence called 'common *nod* genes'.
D	May have several copies; involved in regulation and host specificity.
I	Function unknown.
J	Function unknown; membrane based.
K	Function unknown.
L, M, N	Determine host range in *Rhizobium* but not confirmed in *Bradyrhizobium*.
S, U	Confer broad host range in some strains.
V, W	Host range.
Y	Function unknown.
Z	Host range.

In addition to the *nod* genes listed, host-genotype specific genes also occur. Most information is from *B. japonicum* which nodulates soybean and *B.* spp. (*Parasponia*) which nodulates the non-legume *Parasponia*. Available information for *B.* sp. (cowpea), which nodulates both *Vigna* and *Arachis*, shows that similar genes occur. Symbiotic genes in *Bradyrhizobium* unlike those of *Rhizobium*, are chromosomal.

lipopolysaccharides (LPS) appears important, especially for endocytosis and bacteroid differentiation. Brewin (1991) suggested that, because determinate nodules undergo endocytosis at a relatively earlier stage and infected cells then divide, LPS mutants would have an earlier and more profound effect than in indeterminate nodules where endocytosis occurs later and infected cells do not subsequently divide. If this suggestion is correct, it would be expected that the effect of LPS mutants on *Arachis* nodule development would be at least as profound as on determinate nodules such as soybean.

Although most soybean rhizobia do not normally nodulate *Arachis*, some can cause production of 'nodule-like structures' (Devine *et al.*, 1983). Devine and colleagues, in studying host genetics and nodulation, have isolated a series of rhizobial strains which can nodulate soybeans carrying the allele rj_1 which normally restricts nodulation. These strains generally also produce the chlorosis-inducing substance rhizobitoxine. When tested on two cultivars of *Arachis*, they produced nodule-like swellings in the axils of lateral roots but did not produce chlorosis in the shoot. In one case the swellings were relatively large and rhizobia were isolated from them. *Arachis* did not form swellings with any of the non-chlorosis-inducing soybean strains. In this case the rhizobia are defective in their symbioses with these two different hosts and it was suggested that they may be fully compatible with another (unidentified) legume or that they may represent an active stage in evolution in which specificity is not yet fixed.

The situation is quite different with many cowpea strains of *Brady-rhizobium* which can form highly effective nodules on both *Arachis* and *Vigna unguiculata*. Sen and Weaver have studied nodules on these two hosts in some depth (e.g. 1984) and shown that the nodule structure is controlled by the host. This overriding effect of the host appears common; the same rhizobial strain may infect different plants in different ways and produce nodules with different structures and metabolic processes (Sprent, 1989).

8.4 FUNCTIONAL NODULES

All legume nodules which have been examined assimilate the product of nitrogen fixation (ammonia) in plant cells using the GS:GOGAT system (Figure 8.3). The first organic product is thus glutamine. This may be exported from nodules or further reactions may occur, so that other compounds are exported, such as asparagine, citrulline, or the ureides allantoin and allantoic acid (Sprent and Sprent, 1990). The compounds exported by groundnut have been a matter of controversy for some years. In addition to asparagine and glutamine, methylene glutamine and ureides have been reported (review by Schubert, 1986). Several groups have failed to confirm ureides as a product of N fixation in *Arachis*. One such study (Peoples *et al.*, 1986) also resolves the source of 4-methylene glutamine in the xylem sap of *Arachis* plants. These workers used nodulated plants, non-nodulated plants given nitrate and nodulated plants with the root system purged with ArO_2 instead of air. The latter treatment did not affect nitrogenase (AR) activity but no fixation of nitrogen occurred in the absence of N_2 and thus N compounds in the xylem sap must have arisen from other metabolic pathways. The overall conclusion was that 4-methylene glutamine is a product of cotyledonary metabolism and its concentration in xylem sap is unrelated to nitrogen fixation (see also Goto *et al.*, 1987). A further complication is that unidentified compound(s) in xylem sap of *Arachis* species may interfere with the standard ureide assay (Peoples *et al.*, 1991a). Nitrogen fixing and nitrate assimilating plants export both glutamine and asparagine, but the amounts vary in a way which may serve as an indicator of nitrogen fixation (section 8.5).

Legume seeds are well known for their production of lectins – proteins with pronounced haemagglutination properties. Lectins also figure prominently as a possible constituent of the host-rhizobial recognition system (Sprent and Sprent, 1990). Strong evidence that they play a part in root hair infection systems has recently been obtained for white clover (Diaz *et al.*,1989), but this does not appear to be the case for *Arachis* (Pueppka *et al.*, 1980), which does not have a root hair infection process and which is very promiscuous in its relations with different rhizobia. However, lectins may be involved in nodule development in *Arachis*. Lectins extracted from nodules vary from those extracted from seeds: although each organ produces separate molecules capable of binding to galactose and to mannose, nodule lectins differ from seed lectins in being glycosylated (i.e. in being glycoproteins) and, in the case of the mannose binding (ML) one, in amino acid composition and antigenic proteins (Law *et al.*, 1988). Within nodules, the ML and galactic/binding lectins (GL) had different distributions (Figure 8.4), which were interpreted as suggesting a storage function (Law and van Tonder, 1992). This logical suggestion raises the question as to why nodules should store proteins. Since *Arachis* nodules are of determinate growth, storage would not appear to be an insurance against a need for further growth and yet in earlier work (Kishinevsky *et al.*, 1988) total lectin concentration in *Arachis* nodules reached a peak and then declined, indicating usage or at least breakdown.

Not only are *Arachis* nodules unusual in their protein storage properties but also in their lipid metabolism. Seed oleosomes (lipid bodies) have been known and studied for many years but recent work, mainly from Bal and co-workers in Newfoundland, has shown that nodules have oleosomes which differ in size and other properties from those of seeds (Table 8.3; Jayaram and Bal, 1991). It was suggested that nodule oleosomes are transient storage organelles which can be metabolized, whereas seed oleosomes are for long-term storage. This suggestion is backed up by studies with de-topped and darkened plants. Both treatments led to a gradual decrease in number of oleosomes, coupled with evidence of the β-oxidation pathway and glyoxylate cycle in nodule homogenates (Figure

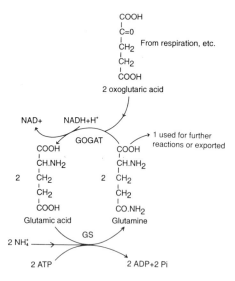

Figure 8.3 Assimilation of ammonium into glutamine using the enzymes glutamine synthetase (GS) and glutamate synthase, commonly known as GOGAT (glutamine-2-oxoglutarate-amino-transferase).

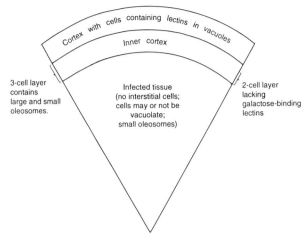

Figure 8.4 Specialized layers in groundnut nodules.

TABLE 8.3 *Some differences between nodule and seed derived oleosomes (lipid bodies) (after Jayaram and Bal, 1991)*

		Nodule	Seed
Approximate % with diameter			
	< 1 μm	79	52
	1–2 μm	17	32
	> 2 μm	4	16
% total lipid in certain fractions	C 16:0*	19	12
	C 18:0	28	3
	C 18:1	14	51
	C 18:2	20	29
	C 18:3	14	2
Electron dense rim with lipolytic activity		Present	Absent

* These values indicate the number of C atoms, followed by the number of unsaturated linkages (C=C bonds)

8.5). Compared with cowpea nodules formed with the same strain (*Brady-rhizobium* sp. 32H1), groundnut nodules were able to maintain nitrogenase (AR) activity for 48 hours after de-topping and darkening, whereas activity in cowpea fell within hours (Siddique and Bal, 1991). Ineffective nodules accumulated excess lipid (Bal and Siddique, 1991). All these data are consistent with a functional role of oleosomes in nodules.

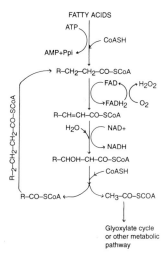

Figure 8.5 Simplified pathway for β-oxidation of fatty acids (AMP, adenosine monophosphate; ATP, adenosine triphosphate; CoASH, coenzyme A; FAD, flavin adenine dinucleotide; Ppi, pyrophosphate; R, a long chain $CH_3(CH_2)_n$). The final product, acetyl CoA, usually enters the glyoxylate cycle in fat-storing seeds, such as groundnut, and may do the same in nodules. Note the production of hydrogen peroxide (H_2O_2) which may be broken down by catalase (Figure 8.7).

A further point of considerable interest is the location of nodule oleosomes. The smaller 'typical' nodule form was largely located in the infected cells, whereas the larger form, which constitutes only a small fraction of nodule oleosomes but which is more frequent in seeds, was located mainly in the three layers of cortical cells immediately exterior to the infected zone (Jayaram and Bal, 1991). This is the same region (Figure 8.4) in which Law and van Tonder (1992) found a lack (or only low concentrations) of cells containing ML in vacuoles. Thus the inner cortical layers appear to be metabolically distinct from layers exterior to them (as well as from the infected region – see also Figure 8.2). Distinct layers can also be observed in other nodules, such as soybean (e.g. Parsons and Day, 1990; James *et al.*, 1991; Sutherland and Sprent, 1984) and may be related to location of oxygen diffusion resistance control mechanisms (Figure 8.6). The diffusion barrier has both fixed and variable components, and enables the supply of oxygen to be tailored to demand without inactivating nitrogenase; it also enables the plant to control nodule activity (for example, when under stress). Barriers have not yet been shown to exist for groundnut but comparable physiological studies on the latter are under way in our laboratory.

It has long been known that bacteroids in groundnut nodules are spherical (e.g. Sen and Weaver, 1984) whereas the same strain of *Bradyrhizobium* in cowpea produces rod-shaped bacteroids, which are

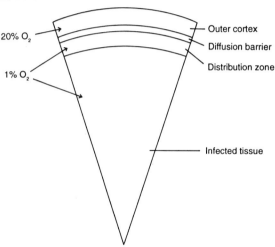

Figure 8.6 Location of oxygen diffusion barrier in nodules of the soybean type.

very common in legumes generally, though some tribes (Vicieae, Trifolieae) have nodules with enlarged pleomorphic bacteroids (Sprent, 1980). The spherical bacteroids of *Arachis* (Figure 8.2) appear to be yet another apparently unique feature of its nodules. Originally the shape was thought to be the result of the bacteroid having little or no cell wall material (to confer shape), so that it adopted a protoplast-like spherical structure (Staphorst and Strijdom, 1972). More recent studies have shown that a distinct wall layer is present throughout (Bal *et al.*, 1985). In some ineffective nodules the bacterial spheres are greatly enlarged (Wilson and Giller, personal communication). Bacteroids in groundnut nodules occur singly, each surrounded by a peribacteroid membrane (PBM) and forming a peribacteroid unit (PBU), sometimes now termed a symbiosome (e.g. Brewin, 1991). PBUs containing single bacteroids are found in other species, such as peas (in which the bacteroids are greatly enlarged and polymorphic, e.g. Newcomb, 1981), whereas soybeans (for example) may have eight or more bacteroids per PBU (e.g. Sprent, 1972). The outer membrane of the PBU is the functional interface between the symbiotic partners and has many characteristic properties, such as a membrane-bound ATPase (Brewin, 1991). In the groundnut, the PBM is often very close to the bacteroid surface and may even touch it. Further, in many cases where the PBM and bacteroid surface are close or touching, oleo-somes are found to be close or touching (Bal *et al.*, 1989). In the narrow peribacteroid space (i.e. between the bacteroid surface and PBM) several amorphous, electron-dense bodies can be found. These 'dense bodies' (Bal *et al.*, 1989) remain attached to the bacterial surface after removal of the PBM and are also seen 'inside' bacteroids, following invagination of the surface membrane. Often they are seen in intact PBU, near oleosomes. In these nodules there is a close correlation between nitrogenase and catalase activities (Figure 8.7). Since dense bodies show a pH-sensitive positive reaction with diaminobenzidine (DAB), characteristic of catalase, it is suggested that they are a site of catalase activity. Also showing the DAB reaction are microbodies in host cells, both near the PBU and towards the periphery. Since catalase is often observed (e.g. in lipid-storing seeds, Tolbert, 1981) to be associated with the β oxidation pathway (Figure 8.5), these microbodies and dense bodies may be an integral part of lipid metabolism in the nitrogen-fixing groundnut nodule.

8.5 MEASUREMENT OF NITROGEN FIXATION IN THE FIELD

From a practical point of view, farmers and agronomists need to know the value of nitrogen fixation to the production of an economic yield of grain. In order to assess this, they require a reliable way of measuring nitrogen fixation in the field. Numerous techniques have been tried, with varying degrees of success, and this section will discuss some possibilities.

8.5.1 Acetylene reduction

Although greeted with much enthusiasm and used with success in many experiments, the acetylene reduction assay has recently been shown to

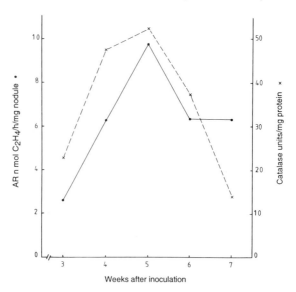

Figure 8.7 Pattern of nitrogenase (acetylene-reducing) activity in groundnut nodules in relation to that of catalase (Figure 8.5). The data give catalase activity in bacteroids; similar patterns, but at lower levels, were found for host cells. (After Bal *et al.*, 1989.)

have many problems as an accurate estimator of N_2 fixed (e.g. Witty *et al.*, 1986). In many species it has been shown to be reliable only for pot-grown plants subjected to continuous-flow analytical techniques. Clearly this precludes assay of field-grown material. The major problems are that many factors, from plant disturbance to acetylene itself, may result in a decline in nitrogenase activity. However, species studied in sufficient detail have so far been few and have not included groundnut. In view of the unique structure of its nodules, and the fact that nitrogenase activity is retained far longer after drastic treatments such as de-topping than in other species such as cowpea (Siddique and Bal, 1991), groundnut may be a species that is less sensitive to manipulation than other grain legumes. Until more studies have been carried out it would be unwise to use acetylene reduction techniques for anything other than a general guide to nitrogenase activity. Even if it were quantitative, activity is only measured for a short period (usually minutes) and so there remains the problem of integration over a season.

8.5.2 Xylem sap analysis

This last objection can also be levelled at assays based on xylem sap analysis. On the other hand 'instant' measurements are useful for examining seasonal and diurnal patterns of nodule activity. It is important to know, for example, which host/rhizobial combinations commence activity early in plant growth and which persist into the grain-filling period. Are xylem sap methods therefore any more reliable than acetylene reduction? As discussed earlier, there are no unique products of nitrogen fixation in groundnuts, unlike soybeans and related plants where ureides are exported from nodules. In most agricultural systems, nitrate is the most abundant form of mineral nitrogen in soil. Nitrate is absorbed by roots and may be assimilated there or passed to the shoot for assimilation (Figure 8.1). When the latter occurs, nitrate can be extracted from xylem sap. In a detailed study of xylem sap composition of cv. Virginia Bunch, in which N_2-fixation was assayed independently by isotope dilution (a method described next), Peoples *et al.* (1986) found that, as plants derived less of their N_2:

(1) the concentration of an asparagine in sap fell;
(2) the ratio of asparagine:nitrate fell; and
(3) the ratio of nitrate:total amino acid rose.

These data suggest that xylem sap analysis could be a valid assay for nitrogen fixation in *Arachis*. The method has been tested on field-grown groundnuts in Malaysia. Results from percentage asparagine in sap and percentage nitrate in sap are similar (Norhayati *et al.*, 1988), mean values being given in Table 8.4. In this work, inoculation improved N_2 fixation but N fertilizer depressed it.

TABLE 8.4   Possible use of xylem sap analysis to estimate nitrogen fixation in the field

Site	% total xylem sap as		Estimated % plant N derived from N₂ fixation using	
	Asparagine (1)	Nitrate (2)	(1)	(2)
A	71	2	96	90
B	66	1	88	95
C	57	9	67	73
D	53	14	57	65
E	41	30	33	40

Data are from 5 field sites in Malaysia, as reported by Norhayati *et al.* (1988), interpreted using solute data of Peoples *et al.* (1986). Site A was inoculated and site E given N fertilizer: sites B, C and D were neither inoculated nor fertilized.

8.5.3 Isotopes

It is more difficult to integrate N_2 fixation over longer periods – up to a whole season. The simplest case is where soils contain only trace amounts of combined nitrogen, in which case classic N analysis (usually by Kjeldahl digestion) may suffice. However, such conditions are relatively rare. The only reasonable alternatives involve use of isotopes of nitrogen. Of these ^{13}N is radioactive, but has a half-life of about 11 minutes, making it completely unsuitable for field measurements (and indeed many laboratory studies). The stable isotope ^{15}N has been the subject of much investigation and forms the basis of various techniques, two of which have been used to good effect with groundnut. More details can be found in Bergersen (1980) and Peoples *et al.* (1989).

8.5.4 The ^{15}N-dilution technique

Basically this technique involves the incorporation of ^{15}N-labelled substrate into soil and then the growing of fixing and non-fixing plants in this soil. N_2-fixing plants will take up some soil ^{15}N but the non-fixing plants will not use atmospheric N_2 (largely ^{14}N). Thus the dilution of ^{15}N in the fixing plants compared with the non-fixing ones can be used as a measure of N_2 fixation. There are many experimental problems with this technique, especially in the field – for example, proper incorporation of ^{15}N into all layers of soil to which roots have access, slow release of ^{15}N from insoluble to soluble forms during the growing season, and choice of a suitable 'reference' non-fixing plant (Witty, 1983).

Where non-nodulating genotypes are available, these form good reference plants: they are more likely to have similar rooting behaviour as their

TABLE 8.5   *Use of the ^{15}N dilution technique, using a non-nodulating mutant as a reference plant, to estimate N_2 fixation by groundnuts in the field in India (after Giller et al. 1987)*

Type	Genotype	Total N in plants kg/ha	Atom % excess ^{15}N	Estimated % N fixed
Virginia	Robut 33–1	162	0.024	92
	NC–17	150		
	ICGS–22	127	0.023	89
Spanish	Gargapuri	131	0.032	89
	J–11	111	0.029	90
Valencia	PI–259747	126	0.039	86
Non-nodulating mutant		41	0.039	0
Standard error		4.5	0.0028	1.0

Nodulated plants were grown in soil with 10 kg/ha $(NH_4)_2SO_4$ containing 5.123 atom % excess ^{15}N; non-nodulated plants received 100 kg N/ha at 0.512 atom % excess ^{15}N.

fixing relatives than the different genera often (inevitably) used. Giller *et al.* (1987) used a non-nodulating genotype as a reference plant to assess nitrogen fixation by seven groundnut genotypes, grown in an Indian alfisol which had previously been used for *Zea mays* to reduce soil N to a low level. Some of the data are shown in Table 8.5. It can be seen that the amount of N fixed varied from 100 to 152 kg/ha over the 89-day growing season, and that the virginia types generally fixed more than the valencia or spanish types. The percentage of N fixed by the different genotypes did not vary greatly. Recovery of soil N by the non-nodulating genotype was poor. These data not only show the utility of the ^{15}N dilution technique but also suggest that total plant N is a good way of comparing N-fixing potential of different groundnut genotypes.

The superior performance of virginia types has been shown elsewhere. For example, Arunachalam *et al.* (1984) looked at 19 characters in 21 genotypes in India. Of these genotypes, six were virginia bunch and five virginia runner types – and they all performed better for most characters than the five spanish and five valencia types. Nodule mass, estimated at the S2 stage of growth was found to be a good indicator of nitrogen fixing potential in this case, as was acetylene reduction activity measured in the 'old-fashioned' closed system assay.

8.5.5 'Natural abundance' ^{15}N method

This method has become more common during the last few years, based on the fact that (usually) soil N contains more ^{15}N than atmospheric N_2. Thus the difference ($\delta^{15}N$) in ^{15}N between plants reliant on soil N (or on the proportion of ^{15}N in soil itself) and those fixing N_2 can be used as an estimate of N_2 fixation. The method has been well discussed by Shearer and Kohl (1986). It is most useful where the ^{15}N content of soil is fairly uniform and where the difference between ^{15}N content of soil and atmosphere is greatest. It avoids the problem of incorporation of ^{15}N into soil but needs much more careful analysis and a more sensitive mass spectrometer than the isotope dilution technique. It may be the only feasible method for deep-rooted plants, especially trees, but has also been applied successfully to grain legumes.

For groundnut, values from 25% to 64% of plant N derived from fixation were quoted by Peoples *et al.* (1991b). The data given were from a variety of experiments involving different cultivars, crop rotations and levels of water supply. N_2 fixation generally supplied a lower proportion of plant N than in other studies quoted, but the higher proportions were correlated with higher amounts of N_2 fixed. Thus the low values may reflect the fact that treatments inhibited N_2 fixation more than assimilation of soil N, a common phenomenon (Sprent *et al.*, 1983).

8.6 EFFECTS OF ENVIRONMENTAL STRESS IN NODULATION AND NITROGEN FIXATION

Compared with those for other grain legumes, studies on the effects of environmental variables on nodulation and N_2 fixation by groundnuts are comparatively rare and comparison of these effects with utilization of soil N even more so. However, there are several reports which deal with the direct interactions between soil N and N_2 fixation. Although groundnuts can take up and use mineral nitrogen (as already explained), there is evidence that they do not do so as readily as some other grain legumes, such as soybean. For example, Nambiar *et al.*(1986) found that groundnut was a poor user of fertilizer N when compared with sorghum. They examined nitrate reductase activity and found that it was low in leaves of both nodulating and non-nodulating genotypes, even when leaf nitrate content was high. This leaf nitrate may have been in a non-metabolic pool (vacuole) as has been found in various plant species. These workers found that nitrogenase activity declined during pod-fill, but that addition of mineral N at this stage was without effect. Their results are consistent with those of Sung and Sun (1990), who found that nitrate reductase activity fell during pod fill but who also found that nitrogenase activity (AR) *increased* during pod-fill. Both genotypic and environmental effects may explain this discrepancy in nitrogen fixation during pod-fill but the results do suggest that selection for better N_2 fixation may be worthwhile.

It has been known for many years and in several agricultural legumes that nitrate affects root hair infection and nodule function (Sprent and

Sprent, 1990) but it is not known how the infection pathway found in groundnut responds, or how nitrate affects nodule functioning. In several species it has been shown that nitrate increases the oxygen diffusion resistance in the cortex of nodules and in this way inhibits (reversibly, within limits) N_2 fixation. However, as seen above, groundnut nodules have cortices and infected regions with characteristic structural features: these may respond differently to nitrate and other forms of stress. Detailed studies on this are urgently required.

8.6.1 Water

Water stress is often a problem in areas where groundnut is grown (Chapter 9). There is evidence that groundnut nodules are less sensitive than those of some other legumes to soil water deficit. Venkateswarlu et al. (1989) compared the responses of groundnut and cowpea. In cowpea, nitrogenase activity ceased when leaf water potential fell below -0.9 MPa whereas the corresponding value in groundnut was -1.7 MPa. Nodules on cowpea, but not on groundnut, were shed under stress. In the latter, stress resulted in a sharp increase in total soluble sugars and leghaemoglobin. Devries et al. (1989), comparing groundnut with soybean and pigeon pea, found that groundnut alone responded to drought by increasing its proportion of roots below 30 cm. These workers also found groundnut nitrogenase activity to be continued longer into the drought period than that of the other legumes and coupled this with a higher leaf water potential and stomatal conductance. The comparative drought tolerance of groundnut thus seems to result from both nodule and plant factors.

8.6.2 Iron and bicarbonate

Groundnuts are very susceptible to lime-induced iron deficiency (Chen and Barak, 1982). Legume nodules have a particularly high requirement for iron, because of its incorporation into both nitrogenase and haemoglobin. Iron-deficient soils in which groundnuts are grown are often alkaline, with a high bicarbonate content. Recent work has shown that both iron and high or low pH can separately affect nodulation and nitrogen fixation (O'Hara et al., 1988a,b; Tang et al., 1991). Effects are complex and involve both host and rhizobia. In common with many soil micro-organisms, bradyrhizobia can secrete siderophores to aid Fe uptake. However, strains vary in their ability to do this (O'Hara et al., 1988b). One strain which produces siderophores, NC92, was found to nodulate groundnuts better in low Fe conditions than strain TAL 1000, which does not (Tang et al., 1991). Nodulation, nodule development and nodule functioning were all reduced at low iron concentrations. As with most legumes, bicarbonate adversely affected nodulation and nodule functioning and effects were more pronounced on N_2-fixing plants than on those given mineral N. Table 8.6 gives some examples of the effects of Fe and bicarbonate on nodule parameters, taken from Tang *et al.*, (1991).

8.6.3 Salinity

Another soil factor often encountered by groundnut is salinity. This may affect nodule initiation, although interestingly NaCl causes normally bare roots to produce hairs (Sprent and McInroy, 1984). Arutyunova and Shevyakova (1984) looked at the effects of both NaCl and Na_2SO_4 on growth of nodulated groundnuts fed with NH_4^+ and NO_3^-. However, the work concentrated on mineral N and it was found that ammonium could partially offset the adverse effects of salinity. On plants fed with ammonium, nodules were formed in the presence of 34 mol/m^3 of NaCl, but 14 mol/m^3 for Na_2SO_4. The reverse was true for nitrate-grown plants. These data show clearly that the effects of salinity on nodulation are complex but little light is shed on the mechanism of its action.

8.6.4 Phosphate

In many soils, nodulation is limited by available phosphate, although not all legumes are equally sensitive. Possession of endomycorrhizas generally

TABLE 8.6 *Effect of iron on nodulation and nitrogen fixation by groundnut cv. Tainar, inoculated with strain NC92 (after Tang et al., 1991)*

Age at harvest (days)	Concentrations of Fe (mmol/m^3)	Number of nodules/ plant	Plant dry weight (g)	N in shoot (mg)
20	0.05	14	0.42	
	0.4	26	0.40	
	7.5	40	0.49	
	7.5 + HCO$_3^-$	2	0.41	
30	0.05	133	0.83	12
	0.4	208	0.97	16
	7.5	202	1.07	15
	7.5 + HCO$_3^-$	26	1.01	11
50	0.05	172	0.97	16
	0.4	374	2.03	40
	7.5	389	2.57	49
	7.5 + HCO$_3^-$	101	1.33	19

Plants were grown in solution culture pH 5.5 in a greenhouse. In one treatment bicarbonate was given by including CaCO$_3$ and NaHCO$_3$, each at 5 mol/m^3: the pH of this solution was 8.6–8.7.

TABLE 8.7 *Effect of inoculation with the mycorrhizal fungus* Glomus clarum *on nodulation and shoot dry weight of 5 genotypes of groundnut grown in sand containing rock phosphate for 90 days*

Type	Genotype	Nodule rating ($^{+}$M/$^{-}$M) × 100	Shoot D.W. g	
			−M	+M
Virginia	Robut 33–1	138	1.54	2.18
	ICGS 15	131	1.74	2.24
Spanish	Gangapuri	156	1.13	1.60
	J–11	145	1.25	1.86
Valencia	PI 1259747	147	1.78	2.41

+M: mycorrhizal plants
−M: non-mycorrhizal plants
Note that (as in Table 8.4) the spanish types performed less well.

aids uptake of P from relatively insoluble forms such as rock phosphate. This is certainly true for groundnut. In a detailed study of several geno-types, Daft (1991) found that mycorrhizal infection increased nodulation and shoot dry weight of plants grown on rock phosphate from a variety of sources. However, genotypes differed markedly in response: increases of shoot dry weight varied from 40% to 360%. Detailed inspection of the nodulation data showed far more variation in the absence than in the presence of mycorrhizal infection (Table 8.7). These interesting data suggest that assessment of nodulation potential might best be carried out on mycorrhizal plants. In low-P soils, plants are likely to be naturally mycorrhizal.

8.7 PROSPECTS FOR IMPROVEMENT

Many possibilities have been touched on in earlier sections. Selection of both host and rhizobial genotypes can give improved N_2-fixation in labora-tory experiments. However, one problem with nodulation in the field is competition from indigenous rhizobia, and this has been studied at ICRISAT over many years. Nambiar *et al.* (1984) give data for various inoculant strains, host cultivars, seasons (rainy, dry) and methods of inoculation. Not surprisingly, a range of responses was found, but there were consistent increases in economic yield with cultivar Robut 33–1 given inoculant strain NC92.

There are various practical problems associated with inoculation. In the study quoted above, liquid inoculants were sometimes found to be better than granular forms, but this depended greatly on soil moisture. Further, these workers point out that groundnut seed may be too fragile for seed inoculation (as generally used for soybeans, for example) and fungicide treatments may be toxic to rhizobia. Other management practices which may affect nodulation include weed control. In two seasons of field trials in India, Malavia and Patel (1989) found that pre-emergence application of three weed-control chemicals adversely affected groundnut nodulation.

Taking improvements in a more general sense to include soil fertility, can groundnuts in a crop rotation significantly reduce N-fertilizer requirements for non-fixing crops? This is a complex question for all legumes, especially grain forms, where much of the N is removed with crop. If more N is removed than fixed, then soil N will be depleted and this sometimes occurs. Further, the C:N ratio of legume residues (as with any plant residues) will affect how quickly N is mineralized and made available to subsequent crops (Sprent, 1987). With these cautions in mind, there is evidence that groundnuts can have a positive effect on soil N in subsequent crops. Dakora *et al.* (1987) compared groundnut and cowpea in the field in northern Ghana. Both legumes nodulated well with indigenous rhizobia. Groundnut fixed 101 kg N/ha (estimated by difference in N content of groundnut and maize grown in the same soil), of which 46% was removed with grain. The corresponding values for cowpea were 201 kg/ha and 34% removed. Thus cowpea had far more N in its residue than groundnut (150 compared with 68 kg/ha). In spite of this the two legumes had approximately equal effects on a subsequent maize crop, equivalent to 60 kg N/ha. In other words a larger proportion (60%) of N in the groundnut residue was made available to the subsequent maize crop than cowpea residue (27%).

ACKNOWLEDGEMENTS

I should like to thank Dr M.B. Peoples and Dr I. Law for permission to quote data in advance of publication. Production of this chapter relied on the technical skills of Ms M Gruber (photographs) and Mrs I Pimbley (diagrams).

REFERENCES

Allen, O.N. and Allen, E.K (1940) Response of the peanut plant to inoculation with rhizobia, with special reference to morphological development of the nodules. *Botanical Gazette*, **102**, 121–142.

Arora, N. (1954) Morphological development of the root and stem nodules of *Aeschynomene indica* L. *Phytomorphology*, **4**, 211–216.

Arunachalam, V., Pungle, G.D., Dutta, M. *et al.* (1984) Efficiency of nitrogenase activity and nodule mass in predicting the relative performance of genotypes assessed by a number of characters in groundnut (*Arachis hypogaea*). *Experimental Agriculture*, **20**, 303–309.

Arutyunova, N.V. and Shevyakova, N.I. (1984) Role of nitrogen source in improving salt resistance of the peanut *Arachis hypogaea* L. *Biological Bulletin of the Academy of Sciences of the USSR*, **10**, 495.

Bal, A.K., Sen, D. and Weaver, R.W. (1985) Cell wall (outer membrane) of bacteroids in nitrogen-fixing peanut nodules. *Current Microbiology*, **12**, 353–356.

Bal, A.K., Hameed, S. and Jayaram, S. (1989) Ultrastructural characteristics of the host–symbiont interface in nitrogen-fixing peanut nodules. *Protoplasma*, **15**, 19–26.

Bal, A.K. and Siddique, A.B.M. (1991) Fine structure of peanut root nodules induced by Nod$^+$Fix$^-$ strains of *Bradyrhizobium* with special reference to lipid bodies. *Annals of Botany*, **67**, 309–315.

Barbour, W.M., Wang, S.P. and Stacey, G. (1992) Molecular genetics of *Bradyrhizobium* symbioses, in *Biological Nitrogen Fixation* (eds G. Stacey, R.H. Burris and H.-J. Evans), Chapman & Hall, N.Y., 648–684.

Barnet, Y.M. (1988) Nitrogen-fixing symbioses with Australian native legumes, in *Microbiology in Action* (eds W.G. Murrell and I.R. Kennedy), Research Studies Press/John Wiley and Sons Inc., Chichester, 81–92.

Bergersen, F.J. (ed.) (1980) *Methods for Evaluating Biological Nitrogen Fixation*, John Wiley and Sons, Chichester.

Brewin, N.J. (1991) Development of the legume root nodule. *Annual Review of Cell Biology*, **7**, 191–226.

Chandler, M.R. (1978) Some observations on infection of *Arachis hypogaea* L. by *Rhizobium*. *Journal of Experimental Botany*, **29**, 749–755.

Chandler, M.R., Date, R.A. and Roughley, R.J. (1982) Infection and root nodule development in *Stylosanthes* species by *Rhizobium*. *Journal of Experimental Botany*, **33**, 47–57.

Chen, Y. and Barak, P. (1982) Iron nutrition of plants in calcareous soils. *Advances in Agronomy*, **35**, 217–240.

Daft, M.J. (1991) Influence of genotypes, rock phosphate and plant densities on mycorrhizal development and the growth responses of five different crops. *Agricultural Ecosystems and Environment*, **35**, 151–169.

Dakora, F.D., Aboyinga, R.A., Mahama, Y. and Apaseku, J. (1987) Assessment of N_2 fixation in groundnut (*Arachis hypogaea* L.) and cowpea (*Vigna unguiculata* L. Walp) and their relative N contribution to a succeeding maize crop in Northern Ghana. *Mircen Journal*, **3**, 389–399.

Devine, T.E., Kuykendall, L.D. and Breithaupt, B.H. (1983) Nodule-like structures induced on peanut by chlorosis producing strains of *Rhizobium* classified as *R. japonicum*. *Crop Science*, **23**, 394–397.

Devries, J.D., Bennett, J.M., Albrecht, S.L. and Boote, K.J. (1989) Water relations, nitrogenase activity and root development of three grain legumes in response to soil water deficits. *Field Crops Research*, **21**, 215–226.

Diaz, C.L., Melcher, L.S., Hooykaas *et al.* (1989) Root lectin as a determinant of host-plant specificity in the *Rhizobium*–legume symbiosis. *Nature*, **338**, 579–81.

Faria, S.M. de, Lewis, G.P., Sprent, J.I. and Sutherland, J.M. (1989) Occurrence of nodulation in the Leguminosae. *New Phytologist*, **111**, 607–619.

Giller, K.E., Nambiar, P.T.C., Rao, B.S. *et al.* (1987) A comparison of nitrogen fixation in genotypes of groundnut (*Arachis hypogaea* L.) using ^{15}N-isotope dilution. *Biology and Fertility of Soils*, **5**, 23–25.

Goto, S., Inanaga, S. and Kumazawa, K (1987) Xylem sap composition of nodulated and non-nodulated groundnut plants. *Soil Science and Plant Nutrition*, **33**, 619–627.

Hadad, M.A. and Loynachan, T.E. (1986) Groundnut nitrogen fixation by three serologically and morphologically distinct rhizobia. *Soil Biology and Biochemistry*, **18**, 161–166.

James, E.K., Sprent, J.I., Minchin, F.R. and Brewin, N.J. (1991) Intercellular location of glycoprotein in soybean nodules: effects of altered rhizosphere oxygen concentration. *Plant Cell and Environment*, **14**, 467–476.

Jayaram, S. and Bal, A.K (1991) Oleosomes (lipid bodies) in nitrogen-fixing peanut nodules. *Plant. Cell and Environment*, **14**, 195–203.

Kishinevsky, B.D., Law, I.J. and Strijdom, B.W. (1988) Detection of lectins in nodulated peanut and soybean plants. *Planta*, **176**, 10–18.

Law, I.J., Haylett, T. and Strijdom, B.W. (1988) Differences in properties of peanut seed lectin and purified galactose- and mannose-binding lectins from nodules of peanut. *Planta*, **176**, 19–27.

Law, I.J. and Tonder, H.J. van (1992) Localization of mannose- and galactose-binding lectins in an effective peanut nodule. *Protoplasma*, **167**, 10–18.

Long, S.R. (1992) Genetic analysis of *Rhizobium* nodulation, in *Biological Nitrogen Fixation* (eds G. Stacey, R.H. Burris and H.J. Evans), Chapman & Hall, New York, pp. 560–597.

Malavia, D.D. and Patel, J.C. (1989) Effect of cultural and chemical weed control on weed parameters, yield and nodulation of groundnut. *Indian Journal of Agronomy*, **34**, 205–208.

Nambiar, P.T.C., Dart, P.J., Rao, B.S. and Rao, V.R. (1982) Nodulation in the hypocotyl region of groundnut (*Arachis hypogaea*). *Experimental Agriculture*, **18**, 203–207.

Nambiar, P.T.C., Nigam, S.N., Dart, P.J. and Gibbons, R.W. (1983) Absence of root hairs in non-nodulating groundnut, *Arachis hypogaea* L. *Journal of Experimental Botany*, **34**, 484–488.

Nambiar, P.T.C., Dart, P.J., Rao, B.S. and Ravishankar, H.N. (1984) Response of groundnut (*Arachis hypogaea* L.) to *Rhizobium* inoculation. *Oléagineux*, **39**, 149–154.

Nambiar, P.T.C., Rego, T.J. and Rao, S.B. (1986) Comparison of the requirements and utilization of nitrogen by genotypes of sorghum (*Sorghum bicolor* (L.) Moench), and nodulating and non-nodulating groundnut (*Arachis hypogaea* L.). *Field Crops Research*, **15**, 165–179.

Newcomb, W. (1981) Nodule morphogenesis and differentiation. *International Review of Cytology Supplement*, **13**, 247–298.

Norhayati, M., Noor, S.M., Chong *et al.* (1988) Adaptation of methods for evaluating N_2 fixation in food legume cover crops. *Plant and Soil*, **108**, 143–150.

O'Hara, G.W., Dilworth, M.J., Boonkeid, N. and Parkpian, P. (1988a) Iron deficiency specifically limits nodule development in peanut inoculated with *Bradyrhizobium* sp. *New Phytologist*, **108**, 51–57.

O'Hara, G.W., Hartzook, A., Bell, R.W. and Loneragan, J.F. (1988b) Responses to *Bradyrhizobium* strains of peanut cultivars grown under iron stress. *Journal of Plant Nutrition*, **11**, 843–852.

Parsons, R. and Day, D.A. (1990) Mechanisms of soybean nodule adaptation to different oxygen pressures. *Plant, Cell and Environment*, **13**, 501–512.

Peoples, M.B., Pate, J.S., Atkins, C.A. and Bergersen, F.J. (1986) Nitrogen nutrition and xylem sap composition of peanut (*Arachis hypogaea* L. cv Virginia Bunch). *Plant Physiology*, **82**, 946–951.

Peoples, M.B., Faizah, A.W., Rerkasem, B. and Herridge, D.F. (1989) *Methods for evaluating nitrogen fixation by nodulated legumes in the field*, Australian Centre for International Agricultural Research, Canberra.

Peoples, M.B., Atkins, C.A., Pate, J S. *et al.* (1991a) Re-evaluation of the role of ureides in the xylem transport of nitrogen in *Arachis* species. *Physiologia Plantarum*, **83**, 560–567.

Peoples, M.B., Bergersen, F.J., Turner, G.L. *et al.* (1991b) Use of the natural enrichment of ^{15}N in soil mineral N for the measurement of symbiotic N_2 fixation, in *Proceedings of International Symposium on the use of stable isotopes in plant nutrition. soil fertility and environmental studies*, FAO/IAEA, Vienna, pp. 117–129.

Peters, N.K., Frost, J.W. and Long, S.R. (1986) A plant flavone, luteolin, induces expression of *Rhizobium meliloti* nodulation genes. *Science*, **223**, 977–980.

Polhill, R.M. and Raven, P.H. (eds) (1981) *Advances in legume systematics*, part 1. Royal Botanic Gardens, Kew.

Pueppke, S.G., Freund, T.G., Schulz, B.C. and Friedman, H.P. (1980) Interactions of lectins for soybean and peanut with rhizobia that nodulate soybean, peanut or both plants. *Canadian Journal of Microbiology*, **26**, 1489–1497.

Raven, J.A. (1988) Acquisition of nitrogen by the shoots of land plants: its occurrence and implications for acid–base regulation. *New Phytologist*, **109**, 1–20.

Rupela, O.P., Kumar Rao, J.V.D.K., Sudanshana, M.R. *et al.* (1991) *Rhizobium* germplasm resources at ICRISAT Center. *Research Bulletin No. 15*, ICRISAT.

Schubert, K.R. (1986) Products of biological nitrogen fixation in higher plants: synthesis, transport and metabolism. *Annual Review of Plant Physiology*, **37**, 539–574.

Sen, D. and Weaver, R.W. (1984) A basis of different rates of N_2-fixation by some strains of *Rhizobium* in peanut and cowpea root nodules. *Plant Science Letters*, **34**, 239–246.

Shearer, G. and Kohl, D.H. (1986) N_2-fixation in field settings: estimations based on natural ^{15}N abundance. *Australian Journal of Plant Physiology*, **13**, 699–756.

Siddique, A.M. and Bal, A.K. (1991) Nitrogen fixation in peanut nodules during dark periods and detopped conditions with special reference to lipid bodies. *Plant Physiology*, **95**, 896–899.

Smit, G., Buckner, D., Puvanesarajah, V. *et al.* (1990) Purification of *Bradyrhizobium* nod-gene inducers from peanut, in *Nitrogen fixation: Achievements and Objectives*, (eds P.M. Gresshoff, L.E. Roth, G. Stacey and W.E. Newton), Chapman & Hall, New York, p. 273.

Sprent, J.I. (1972) The effects of water stress on nitrogen fixing root nodules II. Effects on the fine structure of detached soybean nodules. *New Phytologist*, **71**, 443–450.

Sprent, J.I. (1980) Root nodule anatomy, type of export product and evolutionary origin of some Leguminosae. *Plant, Cell and Environment*, **3**, 35–43.

Sprent, J.I. (1987) *The Ecology of the Nitrogen Cycle*, Cambridge University Press.

Sprent, J.I. (1989) Which steps are essential to the formation of functional legume nodules? Tansley Review No 15. *New Phytologist*, **111**, 129–153.

Sprent, J.I., Minchin, F.R. and Thomas, R.J. (1983) Environmental effects on the physiology of nodulation and nitrogen fixation, in *Temperate Legumes: Physiology, Genetics and Nodulation*, (eds D.G. Jones and D.R. Davies) Pitman, London, 269–318.

Sprent, J.I. and McInroy, S.G. (1984) Effects of salinity on growth and nodulation of *Arachis hypogaea*, in *Advances in Nitrogen Fixation Research*, (eds C. Veeger and W.E. Newton), Nijhoff/Junk Pudoc, p. 546.

Sprent, J.I. and Thomas, R.J. (1984) Nitrogen nutrition of seedling grain legumes: some taxonomic, morphological and physiological constraints. *Plant, Cell, and Environment*, **7**, 637–645.

Sprent, J.I., Sutherland, J.M. and Faria, S.M. de (1989) Structure and function of nodules from woody legumes, in *Advances in Legume Biology*, (eds C.H. Stirton and J.L Zarucchi), *Monographs of Systematic Botany*, Missouri Botanic Gardens, **29**, 559–578.

Sprent, J.I. and Sprent, P. (1990) *Nitrogen fixing organisms: pure and applied aspects*, Chapman & Hall, London.

Staphorst, J.L. and Strijdom, B.W. (1972) Some observations on the bacteroids in nodules of *Arachis* spp. and the isolation of rhizobia from these nodules. *Phytophylactica*, **4**, 87–92.

Sung, F.J.M. and Sun, Y.W. (1990) Seasonal patterns of nitrate reductase and nitrogenase activities in *Arachis hypogaea*. *Field Crops Research*, **25**, 215–222.

Sutherland, J.M. and Sprent, J.I. (1984) Calcium oxalate crystals and crystal cells in determinate root nodules of legumes. *Planta*, **161**, 193–200.

Tang, C., Robson, A.D. and Dilworth, M.J. (1991) Inadequate iron supply and high bicarbonate impair the symbiosis of peanuts (*Arachis hypogaea* L.) with different *Bradyrhizobium* strains. *Plant and Soil*, **138**, 159–168.

Tolbert, N.E. (1981) Metabolic pathways in peroxisomes and glyoxysomes. *Annual Review of Biochemistry*, **50**, 133–157.

Vance, C.P. and Heichel, G.H. (1991) Carbon in N_2 fixation: limitation or exquisite adaptation? *Annual Review of Plant Physiology and Plant Molecular Biology*, **42**, 373–392.

Venkateswarlu, B., Maheswari, M. and Saharan, N. (1989) Effects of water deficit on N_2 (C_2H_2) fixation in cowpea and groundnut. *Plant and Soil*, **114**, 69–74.

Witty, J.F. (1983) Estimating N_2-fixation in the field using ^{15}N-labelled fertilizer: some problems and solutions. *Soil Biology and Biochemistry*, **15**, 631.

Witty, J.F., Minchin, F.R., Skøt, L. and Sheehy, J.E. (1986) Nitrogen fixation and oxygen in legume root nodules. *Oxford Survey of Plant Molecular and Cell Biology*, **3**, 275–314.

Wong, C.H., Patchamuthu, R., Meyer, H.A. *et al.* (1988) Rhizobia in tropical legumes: ineffective nodulation of *Arachis hypogaea* L. by fast-growing strains. *Soil Biology and Biochemistry*, **20**, 677–681.

Wood, S.M., Layzell, D.B. and Pankhurst, C.E. (1985) A morphometric study of effective nodules induced by *Rhizobium loti* and *Bradyrhizobium* sp. (*Lotus*) on *Lotus penduncula-tus*, *Canadian Journal of Botany*, **63**, 43–53.

Young, J.P.W. (1992) Phylogenetic classification of nitrogen-fixing organisms, in *Biological Nitrogen Fixation* (eds G. Stacey, R.H. Burris and H.J. Evans), Chapman & Hall, New York, pp. 43–86.

CHAPTER 9

Groundnut water relations

G.C. Wright and R.C. Nageswara Rao

9.1 INTRODUCTION

About 80% of the world groundnut production comes from seasonally rainfed areas in the semi-tropics, where climate is characterized by the low and erratic rainfall. Drought is recognized as one of the major constraints limiting groundnut productivity in these regions (Gibbons, 1980). This chapter reports various aspects of groundnut water relations, including the effect of water deficits on a range of physiological and morphological processes which affect shoot and root growth, and ultimately pod yield. Extensive reviews on this subject have been published over the past decade (for example, Boote et al., 1982; Boote and Ketring, 1990). The topics discussed in this chapter are intended to supplement and update, where necessary, the literature reviewed in these excellent papers. These reviews concentrate particularly on groundnut irrigation management and so readers are referred to them for detailed information in that area. This chapter therefore concentrates more on potential genetic and management solutions to the problem of drought in groundnut.

9.2 GROUNDNUT WATER STATUS

Leaf water status affects numerous physiological processes which contribute to plant growth and yield. The status of water in plants represents an integration of atmospheric demand, soil water potential, rooting density and distribution, and is therefore a true measure of drought stress in plants (Kramer, 1969). The water status of a crop plant is usually defined in terms of its water content, water potential, or its components, viz osmotic and turgor potential (Turner, 1986a).

The Groundnut Crop: A scientific basis for improvement. Edited by J. Smartt. Published in 1994 by Chapman & Hall, London. ISBN 0 412 408201.

9.2.1 Water content

Leaf relative water content (RWC) has been successfully used to monitor water content and status in groundnuts (e.g. Bennett *et al.*, 1981, 1984; Slatyer, 1955). Sinclair and Ludlow (1985) argue that RWC is a more useful integrator of plant water balance than leaf water potential and should provide universal relationships between physiological traits and level of drought stress. RWC values in well-watered groundnuts are typically in the range of 85–98% (Bhagsari *et al.*, 1976; Joshi *et al.*, 1988; Bennett *et al.*, 1981, 1984; Prabowo *et al.*, 1990) Under drought conditions, RWC as low as 29% has been measured (Bhagsari *et al.*, 1976), indicating that groundnut has a very low lethal water status. This attribute should contribute to high levels of dehydration tolerance and leaf survival in groundnut during intermittent drought stress (Ludlow and Muchow, 1988), in a similar fashion to that reported for pigeon pea (Flower and Ludlow, 1986).

Care needs to be taken when measuring RWC in severely stressed groundnut leaves. Wright (unpublished data) found that the standard 4 h floating time, recommended by Barrs and Weatherley (1962) and Bennett *et al.* (1984), was insufficient for the full rehydration of severely stressed groundnut leaves. For these leaves a floating time of between 8 and 12 h was required to achieve full turgidity. Thus, errors in RWC of up to 25% were recorded when 4 rather than 12 h floating time was provided.

9.2.2 Plant water potential

The establishment of plant water status on a sound thermodynamic basis by the introduction of the concept of water potential has led to the widespread adoption of total water potential as a measure of plant water status (Slatyer and Taylor, 1960). The total water potential, usually in leaves because of ease of measurement, can be partitioned into the major components of osmotic and turgor potential (Turner, 1986a). In groundnut, problems with phloem sap exudation can make determination of the end-point difficult when the pressure chamber technique is used (Pallas *et al.*, 1979; Bennett *et al.*, 1981, 1984). This problem has limited leaf water potential research in groundnut (Bennett *et al.*, 1981) and led to more widespread use of thermocouple psychrometers for leaf water potential measurement. Wright *et al.* (1988b) compared thermocouple psychrometer and pressure chamber techniques and concluded that the pressure chamber over-estimated leaf water potential in groundnut by about 0.4 MPa. This error can lead to large errors in leaf turgor potential, where the pressure chamber is used for measurement of total potential component. It was suggested, however, that the pressure chamber technique may be appropriate in comparative studies of groundnut water status. A procedure was described to aid end-point identification when using groundnut leaves in the pressure chamber. Rajendrudu *et al.* (1983) also reported that the hydraulic press was suitable for leaf water potential measurement in groundnut.

Leaf water potential (ψl) measured about midday in field-grown groundnuts under well-watered conditions has been reported to range widely from about -0.6 to -2.85 MPa (Joshi *et al.*, 1988; Pallas *et al.*, 1979; Patel *et al.*, 1983; Bennett *et al.*, 1984; Allen *et al.*, 1976; Erickson and Ketring, 1985; Stone *et al.*, 1985). The lower (more negative) values of ψl measured under well-watered conditions (Stone *et al.*, 1985; Erickson and Ketring, 1985) occurred in semi-arid regions where plants were presumably under some environmental stress. Ong *et al.* (1985, 1987) have shown that ψl decreases linearly with increasing vapour pressure deficit.

In droughted crops, ψl in groundnut decreases markedly in response to declining soil water availability and atmospheric demand. Values in the range of -3.0 to -4.5 MPa have been reported in drought stressed crops in the field (Bennett *et al.*, 1984; Joshi *et al.*, 1988; Pallas *et al.*, 1979; Stone *et al.*, 1985; Erickson and Ketring, 1985), while some reports suggest that groundnut can reduce its ψl to even lower values such as -6.3 MPa (Sarma, 1984), -8.3 MPa (Allen *et al.*, 1976) and -10.0 MPa (Ludlow and Muchow, 1988).

While transpiration rate and related processes are closely coupled to ψl, it has become evident in recent times that physiological and morphological growth processes are more closely coupled with the osmotic ($\psi\pi$) and turgor (P) pressures of tissues (Begg and Turner, 1976; Turner, 1986a). In particular it is believed that P is the transducer of water deficits in the cell (Hsiao, 1973) and physiological and morphological processes such as stomatal closure and leaf rolling occur at zero P (Turner, 1974; Wright *et al.*, 1983b; Ludlow *et al.*, 1985). There is also recent evidence that the root acts as a sensor of plant water deficit, possibly through turgor pressure regulation in the growing root (Turner, 1986a).

Turgor pressure is usually calculated as the difference between ψl and $\psi\pi$. In groundnut, relationships among water potential components have been measured and indicate that zero P occurs at ψl of between -1.3 and -2.0 MPa in field (Bennett *et al.*, 1981, 1984; Erickson and Ketring, 1985) and controlled environment (Black *et al.*, 1985) studies. Variation in the ψl at which zero P occurs is indicative of osmotic adjustment, a common adaptive response which allows turgor maintenance through active solute accumulation and lowering in $\psi\pi$ (Begg and Turner, 1976). There are, however, conflicting reports concerning the extent of osmotic adjustment in groundnut. Bennett *et al.* (1981, 1984) reported it was small in cv. Florunner, although it was suggested that some adjustment in the order of 0.3 to 0.6 MPa occurred in response to midday tissue water deficits and naturally occurring soil water deficits. Similarly, Black *et al.* (1985) found a small degree of adjustment (0.3 to 0.4 MPa) in cv. Robut 33-1, while Ong *et al.* (1985), in a similar controlled environment facility, found no significant long-term adjustment. On the other hand, Erickson and Ketring (1985) reported substantial osmotic adjustment in field-grown groundnut, ranging from 0.6 to 0.9 MPa. There was also evidence that there were significant cultivar differences in the extent of adjustment. Stirling *et al.* (1989b) showed that, while substantial osmotic adjustment between 0.84 and 1.58 MPa occurred in expanding leaves, it was virtually non-existent in mature leaves. This response allowed expanding leaves to maintain higher turgor levels during periods of stress. Interestingly, turgor remained high in pegs despite severe plant water deficits experienced during early and late reproductive development. The situation regarding the importance, extent and possible cultivar variation in osmotic adjustment in groundnut is therefore unclear. Further research is needed to assess the influence of other factors, such as severity of stress and radiation levels, on the expression of osmotic adjustment in groundnut.

9.3 EFFECT OF WATER DEFICIT ON SOME PHYSIOLOGICAL AND MORPHOLOGICAL PROCESSES

9.3.1 Photosynthesis and stomatal conductance

Bhagsari *et al.* (1976) found that net photosynthesis (Pn) in potted groundnut plants was not affected by increasing stress until shortly before wilting commenced. Pn had declined to between 8% and 13% of well-watered controls (maximum Pn rate 37.7 mg CO_2/dm/h) by the time ψl reached -3.0 MPa. Their data indicated the major effect of drought on Pn was exerted through stomatal closure and suggested that stressed plants stomatal conductance (g_l) is often higher than is necessary to satisfy photosynthetic CO_2 requirements. Similarly, Pallas and Samish (1974) and Pallas *et al.* (1974) reported parallel diurnal trends in Pn and transpiration which were apparently due to stomatal effects.

Unfortunately there is a lack of published information on the influence of water deficits on photosynthesis in groundnut grown under field conditions. The situation with regard to g_l is somewhat better, no doubt due to the lower cost and availability of porometers compared with portable photosynthesis chambers. Sivakumar and Sarma (1986) made diurnal measurements of g_l in groundnut on a range of water stress treatments in Central India. They showed that both the time of day and the degree of drought stress influenced g_l. In moderately stressed plants g_l increased until 1300 h, followed by a dramatic decline in the afternoon. In severely droughted plants, following relatively high g_l in the early morning, stomatal closure occurred by 1100 h in response to reduced soil water availability. Such a stomatal response, which enables stomata to open briefly and photosynthesise during the morning when evaporative demand is low and to close stomata when demand increases later in the day, is an important adaptive feature. Similar stomatal adaptations in response to water deficits have been reported for other crops (Turner, 1974; Turner et al., 1978).

Maximum stomatal conductance in groundnut of about 3–4 cm/s under well watered conditions, and of greater than 2 cm/s in wilted leaves are exceptionally high compared with many temperate species (Black and Squire, 1979; Black et al, 1985; Devries et al., 1989a). The rapid and almost complete recovery to high stomatal conductance following severe stress has been widely reported in groundnut (Allen et al., 1976; Pallas et al., 1979; Black et al., 1985; Joshi et al ., 1988). Presumably plant water status and leaf photosynthesis also return to near normal following severe stress. This capacity to recover fully from stress represents an important adaptive response in groundnut.

Bennett et al. (1984) showed that stomatal conductance in groundnut was curvilinearly related to plant water status, with g_l declining rapidly below a ψl of -1.4 MPa and a P of 0.1 MPa. Other workers report that g_l declines linearly with ψl (Stone et al., 1985; Black et al., 1985) and that g_l is poorly correlated with bulk leaf P. The ability to maintain high g_l at low ψl, known as stomatal adjustment, is a common adaptive response to drought (Ludlow, 1980). Such adjustment has been observed in groundnut (Black et al., 1985) and within groundnut cultivars (Joshi et al., 1988); however, the physiological mechanisms responsible for this adaptation are unknown. Black et al. (1985) argue that osmotic adjustment, which can maintain high turgor pressure around the stomatal cavities in other species (Brown et al., 1976; Wright et al., 1983b; Turner et al., 1978), is not responsible for stomatal adjustment in groundnut.

Vapour pressure deficit (VPD) has also been shown to influence g_l and photosynthesis in groundnut (Black and Squire, 1979; Tsuno, 1975; Stone et al., 1985; Erickson et al., 1986; Azam-Ali, 1984). Radiation-use efficiency (RUE, dry matter production per unit of light intercepted) in groundnut declines in response to increasing VPD, largely due to stomatal closure arising from lowered ψl at high VPD (Ong et al., 1987).

9.3.2 Leaf area development and display

A reduction in water use in groundnut can be achieved by limiting leaf area development and via paraheliotropic leaf movements, in addition to that achieved through stomatal control (mentioned above). Regulation of water use through reductions in leaf area development, or changes in leaf orientation, rather than through reductions in g_l is possible when drought is imposed slowly. In this situation high g_l permits high assimilation rates to be sustained in the remaining leaves for extended periods of water deficits (Black et al., 1985).

The reduction in leaf area development in groundnut in response to water deficits can arise from reduced rates of leaf initiation (Ochs and

Wormer, 1959; Billaz and Ochs, 1961; Boote and Hammond, 1981; Ong *et al.*, 1985) and reduced rates of leaf expansion (Pandey *et al.*, 1984b; Ong *et al.*, 1985; Ong 1986). Leong and Ong (1983) found that initiation of leaves was more sensitive to mild water deficits than were flower, peg or pod initiation. Leaf number has been shown to be greatly reduced by stress (Ong *et al.*, 1985). A linear decline in the relationship between the thermal time to produce one leaf and ψl below -0.6 MPa was shown by Harris *et al.* (1988). This relationship was similar for four cultivars used in this study.

Soil water deficit inhibits leaf expansion through a reduction of RWC (Slatyer, 1955; Allen *et al.*, 1976) or leaf turgor potential (Ong *et al.*, 1985). In the absence of temperature limitations, the rate of leaf extension (and hence expansion) was linearly related to P in the range of 1.0 to 0 MPa, or as ψl declined from -0.5 to -1.5 MPa (Ong, 1984; Ong *et al.*, 1985). Variation in the water content of leaves was the main factor controlling P, since diurnal changes in $\psi \pi$ were small, and no significant long-term osmotic adjustment occurred in expanding leaves. Black *et al.* (1985) suggests that persistent small differences in water status may therefore induce large changes in final leaf size, since the duration of expansion is almost unaffected by water stress. Assimilate shortages during the night can further reduce leaf-expansion in water-stressed groundnut leaves (Ong *et al.*, 1985).

Leaf adaptations that change leaf angle and orientation decrease the radiation load on groundnut leaves. Such movements reduce leaf temperature and the vapour pressure gradient at a time when stomata are nearly closed and leaves are least able to lose heat by transpirational cooling (Gates, 1968). Mathews *et al.* (1988b) have shown that the degree of leaf movement was influenced by ψl, with significant movement (as measured by the ratio of fractional radiation interception to the square root of leaf area index) occurring between -0.8 and -1.2 MPa. The observations that stomatal conductance and leaf expansion cease at similar ψl levels in the field (Bennett *et al.*, 1984; Ong *et al.*, 1985) suggests that all of these processes provide an adaptive capability for optimizing water loss and restricting leaf temperature over a wide range of soil water deficits. Boote *et al.* (1982) give a more comprehensive review of the literature on leaf display, and description of likely mechanisms responsible for paraheliotropic movements in groundnut.

Leaf senescence, or abscission, is often considered to be a mechanism for conserving water during periods of soil water deficits. Leaf senescence can occur in groundnut, particularly when exposed to late season drought (Pandey *et al.*, 1984b), but the crop is considerably more able to retain leaf area despite severe soil water deficits than many other grain legumes (Pandey *et al.*, 1984b; Devries *et al.*, 1989a). Few attempts have been made to quantify leaf senescence in groundnut, although Boote *et al.* (1986) included a leaf senescence function in the PNUTGRO model. Leaf senescence is an irreversible process; and, although it is often considered a mechanism for conserving water during dry periods, it constitutes a loss of potential production if the stress is subsequently relieved.

9.3.3 Nitrogen fixation

In groundnut, nitrogen (N) is vitally important to the process of leaf photosynthesis (Hubick, 1990), radiation use efficiency and biomass accumulation (Wright *et al.*, 1993b). Information on the influence of water deficits on nodulation and N fixation in groundnut is limited (Boote and Ketring, 1990). Recent data has demonstrated that N fixation is extremely sensitive to periods of soil water deficits (Venkateswarlu *et al.*, 1989) and is reduced well before visible stress symptoms appear (Devries *et al.*, 1986; 1989a). Nitrogenase activity has been shown to be severely reduced as leaf and nodule ψ decrease below -1.4 MPa (Devries, 1989a). Little is known about the relative sensitivity of N fixation and photosynthesis in groundnut. In soybean, the effects of soil dehydration on N fixation occur at much higher fractions of transpirable soil water than do the effects on the carbon accumulation processes (Sinclair, 1986; Sinclair and Ludlow, 1986), indicating that this response may be widespread in legumes. Clearly, greater sensitivity to soil water deficits by N accumulation compared with carbon accumulation could result in greater yield reductions than those expected based on carbon limitation to drought alone. More research on groundnut productivity and N accumulation in response to drought is clearly warranted.

9.4 EFFECT OF WATER DEFICITS ON ROOT AND SHOOT GROWTH

9.4.1 Root growth and soil water extraction

The extent and function of roots are influenced by seasonal, genetic and environmental factors (Klepper, 1987). They are discussed below in relation to observations made in groundnut.

(a) Seasonal factors

The apparent root depth progression (ARDP), estimated by examining the time taken by a crop to extract soil water from particular soil layers, provides a method to evaluate seasonal progression of groundnut root growth. Boote *et al.* (1982) reported that the ARDP for cv. Florunner progressed steadily at 2.2–2.8 cm/day until 130 days after planting. In contrast, ARDP rates in four groundnut cultivars ranged between 1.0 and 1.2 cm/day (Mathews *et al.*, 1988a). Plant population can markedly affect ARDP, with values ranging from 1.2 to 2.0 cm/day in a subtropical environment (Wright and Bell, 1992b), and 0.85 to 2.3 cm/day in a semi-arid environment (Simmonds and Williams, 1989). It is evident that the rate of root penetration into the soil appears to depend on the soil type, temperature, cultural practices and the aerial environment (Boote and Ketring, 1990). For instance, Simmonds and Ong (1987) report that ARDP was 2.0 compared with 1.5 cm/day for plants exposed to a VPD of 2.5 and 1.0 kPa, respectively.

There are conflicting reports about the seasonal pattern of root growth in groundnut. In some studies, the proportion of biomass allocated to roots declines markedly as the season progresses. For instance, while 37% of the total biomass was contained in roots at 21 days after planting, only 1.5% was recorded at harvest (McCloud, 1974). Similarly, in the rainy season crop studied by Gregory and Reddy (1982) total root length did not increase after 55 days after planting. In contrast, total root length nearly doubled between *c.* 60 and 90 days after planting in groundnut crops grown under droughted conditions (Robertson *et al.*, 1980; Nageswara Rao *et al.*, 1989b). It appears that groundnut is capable of a large investment in root growth throughout the season, although Nageswara Rao *et al.* (1989b) suggest significant root growth during pod filling may only be maintained when water is severely limited.

(b) Genetic factors

The superior ability of groundnut to maintain favourable water status during periods of soil water deficits, in comparison to soybean and pigeon-pea was related to greater proliferation of roots in deeper regions of the rooting zone (Devries *et al.*, 1989a). Similarly, Pandey *et al.* (1984b) concluded that the higher root densities in groundnut at lower soil depths conferred superior drought tolerance compared to soybean and mungbean.

Groundnut cultivar differences in a number of rooting characteristics have been shown to exist in glasshouse and field studies. The extent of the root depth and root length density becomes very important for soil water extraction during periods of prolonged water deficits. The extension rate of the tap root up to 32 days of age, and the rate of descent of the soil water extraction front for the entire season, varied between four cultivars under a prolonged drought in the field (Mathews *et al.*, 1988a).

However, only minor differences in patterns of water extraction and total water use were observed among cultivars. Based on their limited cultivar range, they concluded that the extent to which root characteristics and water extraction patterns contribute to yield variation in groundnut was small. In another field study, Wright *et al.* (1991) have shown substantial cultivar variation in water extraction at depth, and total water use, exists in groundnut. The cultivar Virginia Bunch was able to extract *c.* 40 mm more soil water at depths below 70 cm compared to three other cultivars (Figure 9.1). Similarly, Chapman (1989) found that cv. Virginia Bunch extracted significantly greater amounts of soil water (c.12–30 mm) than a range of other cultivars subjected to a drought in the pegging stage on a deep krasnozem soil. These studies have clearly demonstrated that considerable variation in water extraction from deep soil layers exists among groundnut cultivars. Based on these results, it is clear the effort required to develop large-scale screening methods for breeding programmes may well be justi-fied. Deep-rootedness and faster extraction may be very appropriate in tropical environments where groundnut is grown solely on stored moisture in the dry season (often following rice) on deep and high water-holding capacity soils (Prabowo et al., 1990).

Substantial cultivar variation in a number of root characteristics of groundnut has been shown under controlled conditions. Ketring et al. (1982) and Ketring (1984) found that cultivars differed in root volume, root dry weight, root length and number. Similar cultivar variation was measured in 22 cultivars grown in hydroponics, where large differences in root length, volume and dry weight were apparent (Pandey and Pendleton, 1986). A strong correlation between root dry weight and shoot dry weight seems to exist in groundnut as evidenced from data reported from glas-shouse (Ketring, 1984; Pandey and Pendleton, 1986) and field studies (Wright et al., 1994). Figure 9.2 presents this data, which suggests that indirect selection for large root systems may be possible under field conditions.

There have been few quantitative studies showing that the differences in rooting characteristics mentioned above are important in allowing one cultivar to perform better than another under a specific drought regime. Higher yield and improved drought tolerance have been associated with larger and deeper root systems in wheat (Hurd, 1974), sorghum (Wright and Smith, 1983) and rice (Steponkus et al., 1980). Presumably a similar association between root size and yield exists in groundnut, although no definitive data are available which categorically support this hypothesis. Considerable research effort is required to show that such an association exists in groundnut. Simple, rapid and inexpensive methods to screen for rooting characters which correlate with observed field patterns need to be developed. For instance, Ketring et al. (1985) and Ketring (1986) report that 'root-effectiveness' can be quantified by measuring the apparent sap velocity (A_v). Under well-watered conditions in the glasshouse, A_v ranged from 0.8 to 1.2 cm/min, while under stress it declined to less than 0.5 cm/min. Cultivar variation in A_v was apparent, and plant growth measurement indicated that the greater A_v was probably due to differences in root function rather than root mass.

(c) Environmental factors

The depth and distribution of root length in groundnut is dependent on soil type (Boote and Ketring, 1990). In sandy soils, groundnut roots have been

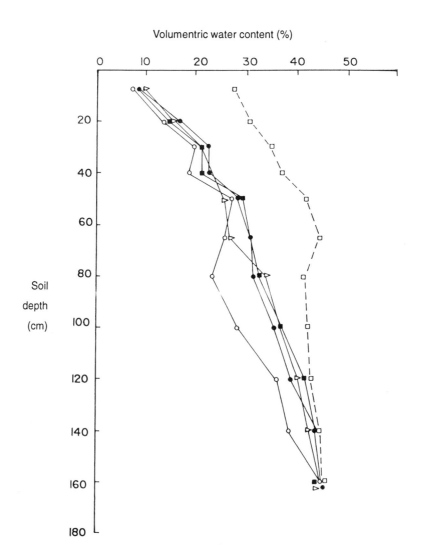

Figure 9.1 End-of-season volumetric soil water content profile for four cultivars grown under a terminal drought stress: (○) Virginia Bunch; (●) Q18801; (△) McCubbin; (■) Red Spanish. The initial volumetric soil water content measured 2 weeks after planting is also shown (□ - - - □). (Reproduced with permission, Wright *et al.*, 1991.)

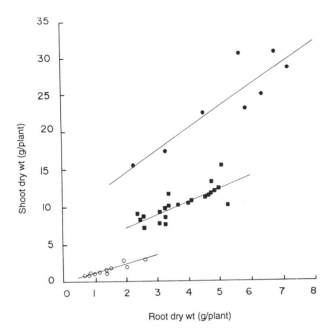

Figure 9.2 Relationship between root dry matter and shoot dry matter in a number of glasshouse and field experiments. Data from: (○) Ketring, 1984; (■) Pandey *et al.*, 1984b; (●) Wright *et al.*, 1992.

measured to a depth of 150 cm (Robertson *et al.*, 1980) and greater (Hammond *et al.*, 1978). On loam and clay soils, roots and soil water extraction have been recorded at depths in excess of 100 cm (Nageswara Rao *et al.*, 1989b; Mathews *et al.*, 1988a; Prabowo *et al.*, 1990; Wright *et al.*, 1991). Ludlow and Muchow (1988) warn that measurements of rooting depth and root length density in crops do not necessarily give an indication of a genotype's ability to extract soil water. Measurements of soil water extraction can, however, provide useful indirect information on root function (Ketring, 1986).

Soil water content can affect root depth and distribution in groundnut. As surface layers dry out, the pattern of water extraction shifts and a proportionately smaller part of the root system in the subsoil is responsible for water uptake (Mantell and Goldin, 1964; Allen *et al.*, 1976). Root length density profiles reported by Devries *et al.* (1989a) and Pandey *et al.* (1984b) are shown in Figure 9.3, and illustrate how drought increased root length density in the lower profile compared with well-watered crops. By contrast, Robertson *et al.* (1980) concluded that root length density in groundnut was not affected by differential water management. Total root length in a 150 cm profile was not significantly different (range of 68–

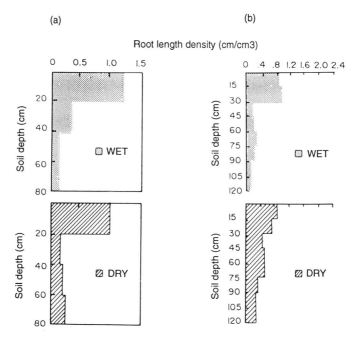

Figure 9.3   Root length density distribution for groundnut under well-watered and water-stressed regimes at: (a) day 68; (b) day 55. (After Pandey *et al.*, 1984b; Devries *et al.*, 1989a.)

108 cm/150 cm) between irrigation treatments. It is possible that the relatively high rainfall (330 mm) which occurred during this experiment did not cause equivalent crop water deficits, and hence root responses, to that observed by Pandey *et al.* (1984b) and Devries *et al.* (1989a).

The growth rate of roots is markedly affected by soil water content through its effect on soil strength and soil water potential (Klepper, 1987). Groundnut tap root elongation decreased from 60 mm/day to less than 10 mm/day as soil strength, as measured by penetrometer resistance, increased from 0 to 5 MPa (Taylor and Ratliff, 1969). Controlled environment studies indicate that root elongation in groundnut appears to be less severely affected by compact soil than is cotton (Taylor and Ratliff, 1969), pea (Gerard *et al.*, 1972) and pigeon pea (Kirkegaard *et al.*, 1992). Field evidence also indicates that groundnut roots had superior root penetration and development through a compacted soil zone when compared with soybean and pigeon pea (Devries *et al.*, 1989a).

Plant population can influence the rate of soil water extraction and final rooting depth in groundnut. Nageswara Rao *et al.* (1989b) found that total length and weight of the root system decreased with wider rows and reduced plant density, although significant amounts of root biomass were still invested in the inter-row spaces of the sparse plantings (12 plants/m^2). Soil water extraction from depth commenced sooner; it was more rapid in the denser stands and was associated with greater root production at deeper layers in the soil (Simmonds and Williams, 1989). Similar ground-nut plant population effects on the pattern of water extraction and use in response to crop water deficits imposed during early crop growth have been reported elsewhere (Bhan and Misra, 1970; Ishag et al., 1985; Wright and Bell, 1992b).

9.4.2 Shoot growth

(a) Vegetative growth

Soil water deficits reduce leaf and stem growth through effects on plant water status, photosynthesis and leaf expansion (sections 9.2 and 9.3). Boote et al. (1982) and Boote and Ketring (1990) have extensively reviewed the literature on this aspect, and give more detailed descriptions.

In groundnut, reproductive growth consists of three distinct stages: production of flowers, development of pegs that carry the ovary below ground, and the subsequent formation and filling of pods.

(b) Flowering

Water deficits during flowering can result in a decrease in flower number and a delay in time to flower. The timing and severity of stress largely influence these effects (Boote et al., 1982). For instance, Chapman (1989) showed that the initiation of flowers ceased within five days of applying an early stress, at which time about 65% of extractable soil water had been used. Since only 15–20% of flowers result in pods that contribute to yield (Smith, 1954), reductions in flower numbers arising from water deficits do not directly influence pod yield (Nageswara Rao et al., 1988). Also, groundnut can compensate for reduced flower numbers resulting from water deficits by producing a flush of flowers once the stress has been relieved (Pallas et al., 1979; Nageswara Rao et al., 1988; Harris et al., 1988).

(c) Pegging and podset

Soil water deficits during pegging and podset decrease yield primarily by reducing pod number rather than kernel weight per pod (Skelton and Shear, 1971; Ono et al., 1974; Boote et al., 1976; Pallas et al., 1979). Harris et al. (1988) wisely point out that this is true only when sufficient water is available later in the season to allow adequate production of assimilates. Numerous experiments reporting the influence of mid-season drought yield have been conducted, with general pod yield reductions in the range of 15–30% of well-watered controls (Pallas *et al.*, 1979; Stansell and Pallas, 1985; Nageswara Rao *et al.*, 1985; Chapman, 1989; Wright *et al.*, 1991).

A reduction in soil water content can have a dual effect on peg and pod development owing to the subterranean fruiting habit of groundnut. On the one hand, root zone water content can directly affect plant water status, photosynthesis and hence assimilate supply to developing pegs and pods. On the other hand, water content in the pegging and podding zones (soil surface to a depth of 5 cm) can affect reproductive growth independently of root zone moisture content. For instance, peg penetration and conversion into pods, as well as pod calcium and water uptake, are influenced by pod zone water content (Boote and Ketring, 1990). Because the effects of soil water content on reproductive development are not always separable, much of the research on this topic refers to soil water status without reference to root or podding zones (Boote *et al.*, 1982).

Considering the direct effect of soil water content first, Chapman (1989) has shown that peg initiation and peg elongation in four groundnut cultivars ceased when about 80% of the extractable soil water had been exhausted. Figure 9.4 shows the sensitivity of peg initiation and elongation to soil water deficits applied during early reproductive development (46–67 days after planting, cv. Robut 33-1). Interestingly, pegs initiated prior to or during the drought period had the capacity to renew elongation upon rewatering. This attribute occurred in all four cultivars studied and appears to be an important adaptation in intermittent drought situations. Stirling *et al.* (1989b) studied the physiological mechanisms controlling the partitioning of assimilate to pegs and pods using labelled ^{14}C. It was found that the rate at which new assimilate was allocated to pegs declined before there was a detectable fall in the turgor potential of pegs. This finding implies that the primary mechanism restricting elongation of pegs during the drought was indirect, and was probably mediated by plant hormones.

Cultivars differ substantially in their pegging and pod set response to soil water deficits. Harris *et al.* (1988) showed that the cultivar TMV-2 was able to maintain a peg production efficiency (ratio of pod number to peg number) of about 0.8 irrespective of a drought applied during the early reproductive phase (17–72 days after planting). In contrast, three other cultivars (Kadiri-3, NCAC 17090 and EC 76446 (292)) had an efficiency of only 0.15 during the drought. Upon relief of the stress (73 days after planting) all cultivars increased their efficiency to 0.8; however, Kadiri-3 took longer to reach this level owing to a renewed flush of flowers, pegs and pods. These differing reproductive patterns in response to drought provide useful information on the physiological basis of adaptation to drought among cultivars. For instance, the yield strategy of TMV-2 was to produce and fill pods at a moderate but constant rate, irrespective of changes in drought intensity. In contrast, the drought-sensitive character of Kadiri-3 resulted in a late flush of pods following rewatering which could

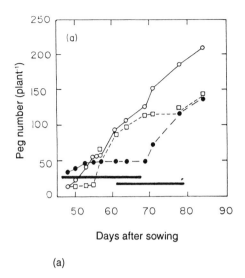

(a)

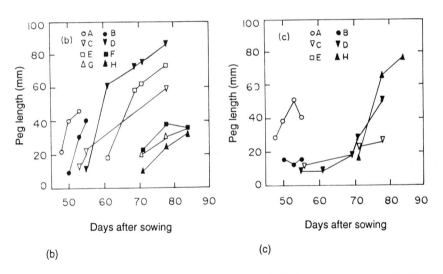

(b)   (c)

Figure 9.4 (a) Production of pegs (cv. Robut 33–1) in response to non-limiting water (○—○), water stress between 46 to 67 days after planting (●—●) and 61 to 78 days after planting (□—□). Elongation of pegs under (b) well-watered conditions and (c) when water was withheld from 46 to 67 days after planting. Each symbol represents a different age class, i.e. pegs that appeared on the same day. (After Chapman, 1989.)

not mature prior to harvest. Similar cultivar responses have been reported elsewhere (Wright *et al.*, 1991).

We now consider the indirect effect of podding zone water content on reproductive development. The situation where a dry surface soil overlays a moist subsoil is particularly common in groundnut producing areas where both intermittent and protracted droughts occur. Even where irrigation is employed on sandy soils, adequate water may be applied to supply the root zone for a period of a week or so but the surface podding zone may dry out rapidly in only a few days. Very little research has been conducted to understand the effects of dry podding zone soil on peg penetration and subsequent development into pods. It seems that peg and pod development under these conditions may be influenced by podding zone water content directly, by lack of calcium (Ca) uptake in developing pods in dry soil, and by greater soil strength associated with dry soil.

A few investigators have used techniques to separate the rooting and podding zones in order to study the effects of pod and root zone water content on reproductive development (Bennett *et al.* (1990) give a review of these techniques). Underwood *et al.* (1971) showed that, although pod development was unaffected in podding zone soil maintained at higher than -1.5 MPa, less than 20% of pods developed in air-dry soil compared with well-watered control plants. Skelton and Shear (1971) found that a dry podding zone significantly reduced the number of pegs which developed into pods, as well as increasing the number of dead pegs. Wright (1989) demonstrated the existence of cultivar differences in peg and pod development in response to podding zone water content. Where the root zone was kept moist and the podding zone was air-dry, seed yield of cv. Robut 33-1 was the same as a treatment whose podding zone was kept at field capacity, while seed yield in cvv. McCubbin and Gajah was reduced by 43% and 35% respectively. A decrease in the number of pegs that developed into fully grown pods along with a reduction in the number of seeds per pod were responsible for the yield reductions. These results suggest that selection for cultivars capable of maintaining normal reproductive development in dry podding zone soil may be possible. Bennett *et al.* (1990) found that a dry podding zone reduced the percentage of cv. Florunner pegs which developed into full pods from 61% to 48%, and also reduced seed yield by 27% relative to the well-watered control plants.

It is well documented that low podding zone water content can decrease Ca uptake and thus induce Ca deficiency in groundnuts (Hallock and Allison, 1980; Rajendrudu and Williams, 1987). The interaction between soil water deficit and Ca requirement for developing pods occurs as a result of the subterranean nature of groundnut fruits. In aerial fruiting plants, Ca is absorbed by roots and transported via the xylem vessels to the fruit. After being conducted upwards and absorbed in the shoot, it then remains highly immobile (Hanson, 1984). In groundnut, early studies showed that Ca cannot move from one side of the plant to the other, or be translocated from the root to the developing pod (Bledsoe *et al.*, 1949; Brady, 1947). The phloem is unable to conduct sufficient Ca to the developing carpophore, although it does conduct all other substances necessary for pod growth (Wiersum, 1951). After carpophore enters the soil, the developing pod has no functional evaporative surface, and therefore no direct access to root-absorbed, xylem-transported Ca (Skelton and Shear, 1971; Wiersum, 1951; Beringer and Taha, 1976). The Ca requirement for developing pods is therefore mainly derived from absorption (through the pods) direct from the soil solution.

Considering the requirement for direct soil uptake of Ca to developing pods, it is difficult to explain how groundnut pods develop at all under very dry pod zone conditions. Bennett *et al.* (1990) suggest that most of the available evidence supports the hypothesis that carpophores enter a dry podding zone and lose water by transpiration from peg or developing pod. Under these conditions Ca can move in the transpiration stream to the developing fruit. Underwood *et al.* (1971) suggests that this pathway may provide sufficient water (and presumably Ca) for normal pod development down to a soil water potential of -1.5 MPa, but not in air-dry soil. The physiological mechanisms underlying Ca uptake and pod development in dry soils is not well understood. The fact that cultivar differences in pod development and seed calcium concentration in air dry pod zone soil exist in groundnut (Wright, 1989) means that comparative physiological studies to elucidate such mechanisms should be possible.

Calcium, usually applied as gypsum, has been widely shown to increase yield of groundnut under drought conditions occurring during early pod set (Hallock and Allison, 1980; Cox *et al.*, 1976; Rajendrudu and Williams, 1987; Radder and Biradar, 1973; Balasubramanian and Yayock, 1981). Rajendrudu and Williams (1987) suggested that gypsum application, which resulted in more rapid establishment of pod numbers, provided an effective drought escape mechanism by getting more pods past the critical pod set sage. Under well-watered conditions, this effect can provide greater pod synchrony and even more maturity.

The inability to relate groundnut yield response to soil test Ca levels across a wide range of soil water availabilities (Daughtry and Cox, 1974) is no doubt related to reduced Ca availability and uptake in dry podding zones. Large-seeded virginia cultivars have frequently been shown to be more sensitive to drought-induced Ca deficiency than smaller seeded types (Stansell *et al.* 1976; Slack and Morrill, 1972; Beringer and Taha, 1976). The sensitivity of the large-seeded cultivars to Ca supply was suggested to be related to the smaller surface-to-volume ratios of the large pod when compared with a small pod (Boote *et al.*, 1982). Kvien *et al.* (1988) later confirmed that this was the case by studying the influence of several pod characteristics on Ca accumulation and concentration in eight groundnut genotypes. The ability of a cultivar to translocate Ca from shells to developing seeds also appears to be important in determining cultivar sensitivity to soil Ca levels. Beringer and Taha (1976) found that, although the large-seeded cultivar, Makulu Red, absorbed more Ca than the smaller seeded Natal Common, it was less effective in translocating Ca from shells to seeds. Similarly, Kvien *et al.* (1988) concluded that thin, light shells trap less Ca than thick, dense shells, and hence promote higher Ca concentrations in the seed. Recent investigations by Webb and Hansen (1989) have also shown that carpophores and pods produce unicellular structures resembling root hairs up to 0.75 mm long, which reach very high densities. These structures, which were evident at the R5-6 growth stage (Boote, 1982) and degenerated by maturity, are considered to be an adaptation for calcium and water uptake. Wright (1989) also found that cv. Robut 33-1 had more pod hairs and higher seed Ca concentrations compared with two cultivars whose pods were grown in air-dry soil. It was hypothesized that these hairs may have increased the effective pod absorptive surface area and facilitated Ca uptake under drought conditions.

Surface soil physical condition is of considerable importance to peg and pod development in groundnut, considering that the peg must penetrate the soil surface to a depth of 2–7 cm to allow normal pod development (Smith, 1954). It is surprising that there is only one report in the literature pertaining to this topic. Underwood *et al.* (1971) showed that, with soil water at field capacity, the ability of pegs to penetrate depends on the bulk density of the soil. Figure 9.5 shows that the number and total weight of pods per plant decreased asymptotically as penetration resistance (and bulk density) increased. They also showed that most pegs which penetrated 1.0–1.5 cm developed pods, although development was slower when pods were near the soil surface. This solitary set of data clearly establishes that soil physical factors have a marked influence on peg penetration and subsequent pod development.

Wright *et al.* (1992) studied the influence of surface soil strength on reproductive development in groundnut. Pegs were allowed to enter small cores which had been packed to different bulk densities. The depth of peg entry and pod growth (dry weight) were measured in two cultivars (Robut 33-1 and McCubbin) 21 days after pegs were introduced into the cores. Figure 9.6 shows the relationship between depth of peg penetration and soil strength, as measured by penetrometer resistance. It was clear that depth of peg penetration (and subsequent pod growth) declined linearly as penetrometer resistance increased from 0.1 to about 2.0 MPa, with pegs ceasing to penetrate and develop further at resistances greater than 2.0 MPa. There was little evidence of cultivar differences in peg response to soil strength; however, further experiments assessing cultivar variation in the 0–2.0 MPa region need to be conducted with greater precision. The peg response to soil strength agrees well with the data of Underwood *et al.* (1971), which showed the number of developed pods, and pod yield per plant, ceased at penetrometer resistance > 2.0 MPa.

Figure 9.6 also shows how pegs appear to be considerably more sensitive

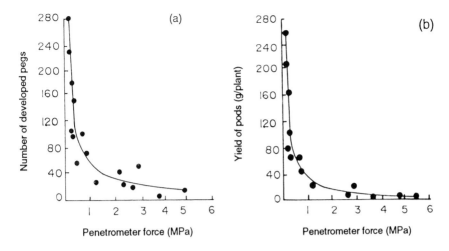

Figure 9.5 (a) Pod yield per plant and (b) number of pegs that developed to diameters greater than 3 mm, as a function of penetrometer resistance in the surface 1.5 cm of soil. (Reproduced with permission, Underwood *et al.*, 1971.)

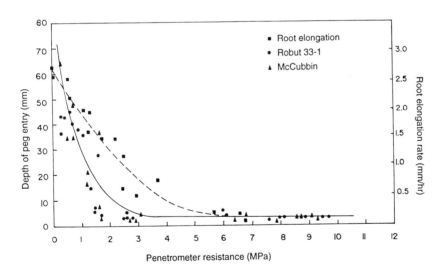

Figure 9.6 Depth of peg entry as a function of penetrometer resistance in the surface 1.5 cm of soil for two cultivars measured during a 21-day pegging period (Wright and So, unpublished data). The broken line shows the effect of penetrometer resistance on root elongation (after Taylor and Ratliff, 1969).

to soil compaction than do roots, as Taylor and Ratliff (1969) showed groundnut roots ceased growth at a penetrometer resistance of >5.0 MPa. The results of this experiment have demonstrated that peg entry, pod growth and development are extremely sensitive to soil hardness, which commonly increases with declining surface soil water content (Sivakumar and Sarma, 1986; So and Woodhead, 1987). It therefore follows that high soil strengths during the pegging and pod set phase must severely inhibit reproductive efficiency and reduce harvest index and pod yield in groundnut.

(d) Pod filling

Water deficits during pod fill generally reduce pod and kernel weight (Pallas *et al.*, 1977, 1979; Pandey *et al.*, 1984a; Lenka and Misra, 1973; Nageswara Rao *et al.*, 1985; Wright *et al.*, 1991). The timing of water deficit in the pod filling stage largely influences pod growth, pod size and maturation. There are numerous reports available on the effect of drought during podfilling but the findings vary enormously (as reviewed by Boote *et al.*, Boote and Ketring, 1990). This is not surprising, considering that some work has investigated the effect of water deficits within each growth phase separately, with water being non-limiting during earlier phases (e.g. Nageswara Rao *et al.*, 1985), while other studies have considered protracted water deficits throughout the season (Mathews *et al.*, 1988a). There are therefore many possible interactions between plant water status and stage of growth, which influence the establishment of pod numbers and the availability of assimilate to fill those pods.

Late-season drought in longer-season maturity types has been shown to reduce pod yields more severely than in shorter-season types, largely through reductions in pod numbers and seed size relative to well-watered controls (Boote *et al.*, 1976; Pallas *et al.*, 1979; Wright *et al.*, 1991). Shorter-season types are able to initiate and develop pods earlier and thereby escape severe soil water deficits late in the season.

Water deficits occurring during the pod formation stage can delay the establishment of a full pod load, as discussed in the previous section. Upon relief of this stress, groundnut has often been observed to initiate a flush of flowers, peg and pods which can delay harvest maturity by 10 or more days (Stansell and Pallas, 1985; Boote *et al.*, 1976; Wright *et al.*, 1991; Shorter and Simpson, 1987). In such a situation, two or more flushes of pod formation can occur, and farmers using mechanical harvesters may need to make decisions on which flush of pods is the most suitable to harvest. In many environments full yield compensation through delayed harvesting may be difficult, as overmature pods may remain in the soil, and in subtropical environments low temperatures late in the season may reduce pod growth rates and hence quality (Bell *et al.*, 1992). Cultivars which have more synchronous pegging and podding patterns, despite the occurrence of soil water deficits, can overcome this problem (Harris *et al.*, 1988; Chapman, 1989; Wright *et al.*, 1991).

9.4.3 Effects of water deficits on partitioning and harvest index

Harvest index (H), defined as the proportion of pod to total biomass, can vary enormously depending on the timing and severity of water deficit relative to pod set (Ong, 1986). Duncan *et al.* (1978) suggested that H can be considered as a function of the crop growth rate (CGR), partitioning of assimilate to pods (p), and the effective duration of podfilling phase (t). Mathews *et al.* (1988a) showed that H can be expressed as the mean value over the season of p weighted by CGR, described by the expression:

$$H = \frac{(p\ CGR\ t)}{W} \tag{1}$$

where W is total biomass, adjusted for energy content in the pods (Duncan *et al.*, 1978), at maturity. This simple analysis allows the investigation of specific variables influencing treatment and cultivar differences in H in response to water supply. The above framework has been used by a number of workers to analyse cultivar and treatment differences in ground-nut pod yield performance under a range of soil water regimes. For instance, Duncan *et al.* (1978) concluded that cultivar differences in p, rather than in CGR and t, were largely responsible for pod yield improve-ment of recently developed cultivars in the USA under well-watered conditions. Mathews *et al.* (1988a) and Harris *et al.* (1988) found that cultivar differences in pod yield during a terminal drought were due to variation in H, which was associated with differences in p and t. Similarly, Wright *et al.* (1991) showed that CGR was similar among four cultivars; however, differences in p and t were responsible for H and pod yield variation in a terminal drought situation. The positive pod yield response to drought during the preflowering phase, compared with a fully irrigated treatment, observed by Nageswara Rao *et al.* (1985, 1988) could largely be explained by greater synchrony of pod set which allowed a longer period of podfilling (i.e. greater t). There were only minor differences in p and CGR among irrigation treatments.

During podfilling the size, number and strength of reproductive sinks, relative to other sinks in the plant, determine the final degree of partition-ing to pods. It has been shown in many determinate crops that pre-anthesis biomass can be remobilized to reproductive sinks from other parts of the plant, particularly in response to water stress (e.g. Constable and Hearn 1978; Bidinger *et al.*, 1977; Blum *et al.*, 1983; Wright *et al.*, 1983a). Ludlow and Muchow (1988) suggest this trait can improve yield stability by acting as a buffer against the effects of water deficits on current assimilation. Duncan *et al.* (1978) and Ong (1986) have suggested that no translocation of vegetative assimilate to pods occurs in groundnut, as partitioning ratios

(*PR*, ratio of pod growth rate to crop growth rate) were always less than 100% (Table 9.1). In contrast, Bell (1986) recorded *PR*s well in excess of 100% in irrigated groundnuts (cv. Early Bunch) in a tropical environment, which suggested that substantial remobilization of assimilate can occur. Other recent data (Chapman, 1989; Wright *et al.*, 1991) has shown *PR*s in excess of 100% can occur, and also that cultivar differences in *PR* were apparent in groundnut subjected to water deficits during the podfilling phase (Table 9.1). Based on this data, cultivar variation in the degree of remobilization during podfilling may exist, and more research into its role in improving *H* and pod yield in groundnut under drought conditions seems warranted.

9.5 FACTORS AFFECTING EVAPOTRANSPIRATION

Evapotranspiration (*ET*), or crop water use, refers to the combined processes of evaporation and transpiration which account for the consumptive water loss during crop growth. While transpiration (*T*) is responsible for water used by the plant, soil evaporation (*Es*) represents water loss from the soil surface that does not pass directly through the plant. While *T* may be readily measured in container studies in which *Es* can be minimized, it is much more difficult to measure, calculate or estimate soil evaporation in field studies (Turner, 1986a). Most field data that have been reported for groundnut simply combine *T* and *Es*. This has made generalizations about crop production as related to *ET*, rather than *T*, a very difficult task (Ritchie, 1983). Boote *et al.* (1982) have discussed experimental techniques available for estimating groundnut *ET* in the field.

ET is influenced by many factors, which can be broadly categorized into the effects of canopy development and soil water deficits. These are briefly discussed below and have been exhaustively reviewed by Boote *et al.* (1982) and Boote and Ketring (1990).

9.5.1 Effect of canopy development on *ET*

Stansell *et al.* (1976) showed that daily *ET* in groundnut increased from 2 mm/day at 20 days after planting (DAP) to a maximum of 6 mm/day at between 70 and 95 DAP, depending on cultivar. *ET* then declined gradually to about 3 mm/day at maturity, in response to plant senescence and decreasing evaporative demand. Similar seasonal *ET* responses have been observed elsewhere (Dancette and Forest, 1986; Ishag *et al*, 1985). Figure 9.7 shows a typical seasonal pattern of daily *ET* for a virginia type (cv. Ashford) grown under well-watered and water-limited conditions in the Sudan, Africa (Ishag *et al.*, 1985). Higher maximum daily *ET* rates of 7–10 mm/day have been recorded under conditions of high evaporative demand (Mantell and Goldin, 1964; Grosz, 1986).

The seasonal pattern of *ET* in groundnut is closely related to the pattern of canopy, and hence LAI, development (Ishag *et al.*, 1985: Dancette and Forest, 1986; Wright *et al.*, 1991). When crop canopy is establishing, a large proportion of *ET* can be lost through soil evaporation (*Es*) (Cooper *et al.*, 1983). Ritchie and Burnett (1971) found that in the early stages of crop

TABLE 9.1 *Partitioning ratios (PR) in groundnut cultivars, as influenced by water supply*

Cultivar	Remarks		PR (%)	Reference
Dixie Runner	Irrigated		40.5	Duncan *et al.* (1978)
Early Runner	"		75 7	"
Florunner	"		84.7	"
Early Bunch	"		97.8	"
Spancross	"		66.2	"
Early Bunch	Irrigated, planted	7/12	91.3	Bell (1986)
	" "	29/12	85.7	"
	" "	18/1	128.7	"
	" "	8/2	125.0	"
	" "	1/3	143.5	"
Virginia Bunch	Terminal drought		82.0	Wright *et al.* (1991)
UF78114–1	"		110.0	"
McCubbin	"		88.0	"
Red Spanish	"		88.0	"
Virginia Bunch	Irrigated		99.0	Chapman (1989)
McCubbin	"		107.0	"
UF78114–1	"		107.0	"
Robut 33–1	"		98.0	"
Virginia Bunch	Early stress		33 0	"
McCubbin	"		25.0	"
UF78114–1	"		50.0	"
Virginia Bunch	Late stress		102.0	"
McCubbin	"		122.0	"
Robut 33–1	"		118.0	"

*PR*s were calculated as the ratio of pod growth rate (adjusted for oil content) to crop growth rate × 100.

growth, until an LAI of 1.0 is reached, *Es* can account for well over 50% of the total *ET*. Simmonds and Williams (1989), for instance, found that when groundnut canopy coverage was small, about 16 mm was lost by soil evaporation during one week following an irrigation, with almost half of this loss occurring in the first two days. The rate of *Es* is dependent on surface soil wetness, and the degree of shade provided by the canopy. When the surface soil is very wet, the rate of evaporation is governed by the supply of energy, and upon further drying *Es* is limited by soil hydraulic conductivity (Philip, 1957).

To obtain greater understanding of the effect of management or cultivar on crop water use, it is necessary to split *ET* into its components – namely *T* and *Es*. The direct measurement of *Es* using small trays filled with fully drained soil and sunk in between rows has been attempted in groundnut (Simmonds and Williams, 1989). Trays were removed daily and weighed, with the difference in weight being assumed to be water lost by evaporation. The technique is unreliable for estimating water loss after the first few days after wetting (ODA, 1987). Ritchie (1972) and Tanner and Jury

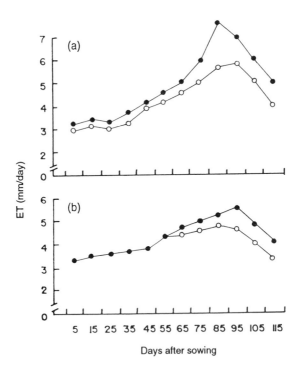

Figure 9.7 Mean daily evapotranspiration as a function of time under (a) well-watered and (b) water-limited conditions, in (●) dense and (○) sparse stands of groundnut. (After Ishag *et al.*, 1986.)

(1976) developed empirical models to describe the temporal change in water loss from an uncropped soil and the effect of crop canopy development on radiant energy interception. These models are predictive in nature, and have been successfully used in larger crop models (e.g. Boote *et al.*, 1986). Cooper *et al.* (1983) extended this approach, and described a technique whereby standard field measurements of *ET*, evaporation from uncropped soil and canopy cover (via canopy light interception, or LAI and canopy extinction coefficient measurements) allow reliable estimates of *Es* and *T*. Wright *et al.* (1994) used this technique to estimate *T* and *E* in mini-lysimeters installed in groundnut canopies in the field. In a terminal drought stress treatment, *Es* was 45 mm out of a total *ET* of 185 mm, when averaged over four cultivars.

9.5.2 Effect of soil water deficits on *ET*

ET will continue at an energy limited rate until water content of a large part of the root zone drops below a critical level, when stomatal closure restricts water loss. There is little published information on the critical level of available water content below which *ET* declines in groundnut. Wormer and Ochs (1959) found that *T* of groundnut remained constant until two thirds of the available soil water had been depleted, after which it decreased to the wilting point. In a range of crops and locations, Ritchie *et al.* (1976) found that the ratio of *ET* to potential evaporation from freely evaporating surface remained relatively constant if extractable soil water was greater than 25%, then decreased linearly between 25% and 0%. The generality of this response across a wide range of environments has been questioned. Following the theoretical analyses by Jordan and Miller (1980) it was shown that the decline in *ET* is reached at higher soil water contents under high evaporative demand conditions because steeper gradients of soil water potential are required to sustain flow. Denmead and Shaw (1962) demonstrated this effect with corn plants grown in containers in the field under differing meteorological conditions. The implications of these findings are that the magnitude of the total extractable soil water varies, depending on the evaporative conditions experienced by the crop. It also depends on rooting depth, root length and density, and root resistance (Sheriff and Muchow, 1984).

9.6 YIELD RESPONSES TO EVAPOTRANSPIRATION AND TRANSPIRATION

Based on a review of the available literature, Boote *et al.* (1982) and Boote and Ketring (1990) concluded that about 600 mm of water is required for optimal pod yield performance. The presented relationships between pod yield and *ET* showed significant variability, which they attributed to factors other than water use. An example of the type of variability observed is illustrated in Figure 9.8(a), which compares the relationship between pod yield and ET for data from Georgia, USA and Hyderabad, India. While pod yield increased linearly from approximately 1.5 t/ha to 6 t/ha in both regions, the amount of ET used differed substantially, ranging from 200 to 600 mm for the USA data, compared with 450–800 mm for the Indian data. Hanks (1983) and Kanemasu (1983) discussed some of the factors influencing the pod yield/ET relationship in a range of crops and concluded that (a) improvements in the correlation are observed when T rather than ET is used; (b) total dry matter production (TDM) and ET are more closely correlated than pod yield and ET; and (c) pod yield/ET relationships can vary from location to location, year to year and crop to crop. These factors as they relate to groundnut yield/water use relationships are briefly discussed below.

9.6.1 Transpiration versus evapotranspiration

De Wit (1958) concluded there was a close relation of TDM to T across a wide range of species and conditions. This response is not surprising: TDM and photosynthesis are closely related, and photosynthesis and transpiration are also related via the diffusional resistances (Tanner and Sinclair, 1983). In field studies, the problems of evaluating the amount of T from the generally measured ET (soil evaporation plus T) can introduce considerable error into T estimates, and hence the ET/TDM relationship. Similarly, drainage below the root zone and/or runoff can cause considerable error in ET measurements. (Methods of estimating T from ET measurements are discussed in section 9.5.)

9.6.2 TDM versus pod yield

Figures 9.8(a) and (b) illustrate how there was a considerably stronger correlation between TDM and ET ($r^2 = 0.61$) compared with pod yield and ET ($r^2 = 0.38$) for groundnut crops exposed to different timings of soil water deficits (Nagweswara Rao et al., 1985). Indeed in a narrow range of nearly the same total ET (500–600 mm), pod yield varied from 0.8 t/ha to over 5 t/ha, depending on the timing of water deficit. Stansell and Pallas (1985) also showed that different pod yields occurred at similar total ET values, as a result of timing of water deficits. Attempts have been made in other crops to relate harvest index to the timing and severity of water stress in order to improve the prediction of ET/pod yield relationships (Slabbers et al., 1979; Stewart et al., 1977). Kanemasu (1983) warns that the ET/pod yield relationships are not unique because of the complex interactions between development, assimilate partitioning and environment, and considers it is doubtful that an ET/pod yield relationship can be extended to climatically diverse regions.

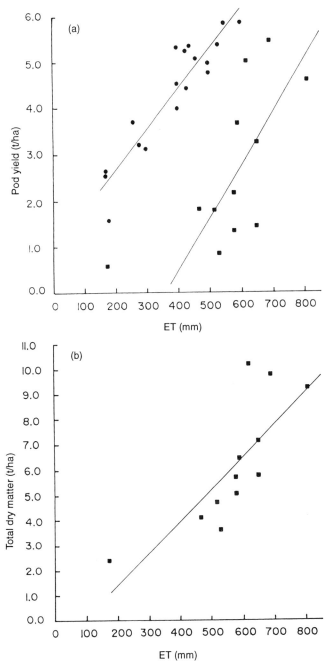

Figure 9.8 Relationship between (a) pod dry matter and evapotranspiration (ET) for data derived from (■) Stansell *et al.* (1976) and (●) Nageswara Rao *et al.* (1985); and (b) total dry matter and evapotranspiration for data derived from Nageswara Rao *et al.* (1985).

9.6.3 Environmental effects

Much of the variation in the ET/pod yield relationship between the USA and Indian crops shown in Figure 9.8(a) can be explained by site differences in evaporative demand. Ong *et al.* (1987) showed that in groundnut T increased, and TDM increased linearly as vapour pressure deficit increased from 1 to 2.0 kPa. Differences in evaporative demand from year to year have been shown to change the ET/pod yield relationship in other crops (Hanks, 1983). Other possible reasons for variations in the relationship between ET and pod yield include diseases, insects, nematodes, nutritional stress, soil water capacity and rooting depth. Boote and Ketring (1990) suggest that a modelling approach capable of integrating these interacting effects may enable better prediction of pod yield response to water application.

9.6.4 Water use efficiency and transpiration efficiency

Water use efficiency can be looked at from many perspectives, and care is needed in defining its use. Water use efficiency (WUE) is generally defined as dry matter production (total or pod) per unit of ET, while transpiration efficiency (TE) is dry matter production per unit of crop transpiration. The distinction between these terms is important in the context of genetic and agronomic solutions to the problem of drought in groundnut. For instance, treatments that suppress soil evaporation can improve WUE but need not improve TE (Tanner and Sinclair, 1983). Several models have been successfully employed to predict dry matter production from ET and T in a range of crops (de Wit, 1958; Arkley, 1963; Bierhuizen and Slatyer, 1965; Tanner and Sinclair, 1983). Such models allow opportunities for improving the efficient use of water to become more clearly defined. All of the models are largely derived from the analysis of de Wit (1958) which proposed that, in closed canopies:

$$\frac{TDM}{T} = \frac{m}{E_0} \tag{2}$$

where E_o is the potential evaporation from pan data or the Penman equation, and m is a constant governed mainly by species. The later models used vapour pressure deficit rather than E_o to improve prediction across contrasting environments. Tanner and Sinclair (1983) concluded that WUE can mainly be improved by modifying the vapour pressure environment in which plants grow by, for instance, applying mist irrigation or planting crops earlier or later to coincide with lower vapour pressure deficits. Similarly, management options such as plant population which modify evaporation loss relative to transpiration can influence WUE, with little effect on TE.

It has been recognized for many years that the internal method of utilizing carbon can markedly influence TE of plants with C_4 plants having about twice the TE of C_3 plants (Tanner and Sinclair, 1983). The constant m in the above equation is used to account for such variations. Reviews of the literature often concluded that the exploitable variations in TE among cultivars within a species is small and the potential for improvement by breeding is limited (Fischer and Turner, 1978; Fischer, 1981; Tanner and Sinclair, 1983). Recent evidence in groundnut (and many other species) has overturned this conclusion, with Hubick *et al.* (1986) reporting that the variation in TE of up to 60% exists in a range of groundnut cultivars grown in large pots in the glasshouse. Large cultivar differences in TE have subsequently been demonstrated in the field under well-watered (Wright *et al.*, 1988a) and water-limited conditions (Wright *et al.*, 1994), using mini-lysimeters installed in small canopies. Table 9.2 shows the extent of this variation, with Tifton-8 having the highest TE range at 3.1–3.7 g/kg, compared with the lowest values of 1.8–2.0 g/kg for Chico. Interestingly, variation in TE among cultivars was largely due to differences in biomass rather than to differences in water use. This result suggests that photosynthesic capacity, rather than leaf stomatal conductance, dominates the TE response in groundnut cultivars. Similar groundnut cultivar differences in TE have been reported in the field by Mathews *et al.* (1988a). In their study, cv. Kadiri-3 had the highest (2.17 g/kg) and cv. EC 76446(292) the lowest TE (1.71 g/kg). It is clear that considerable scope exists to improve TE and ultimately pod yield under water-limited conditions by selection for this trait in breeding programmes.

9.7 AMELIORATING THE EFFECT OF WATER DEFICITS ON YIELD

About 80% of groundnut production occurs in seasonally rainfed areas of the semi-arid tropics, where climate is characterized by low and erratic rainfall. Drought is recognized as one of the major constraints for groundnut productivity in these regions (Gibbons, 1980). Although effects of supplementary irrigation on yields are significant, this option is only feasible on a small proportion of the total arable area. Therefore other methods of improving or stabilizing groundnut yields in drought-prone environments are necessary. Genetic and management options to alleviate effects of water deficits on groundnuts are considered in this section.

9.7.1 Genotypic improvement

Despite the importance of drought as one of the major constraints to groundnut production, the research effort into genetic improvement of drought tolerance in groundnut has been limited. This situation has largely arisen because of a lack of simple screening technologies to assess the drought tolerance of genotypes, and a lack of access to a large groundnut

TABLE 9.2 *Biomass (incl. roots), water use and* TE *in groundnut cultivars under well watered conditions (Wright* et al., *1988a) and two levels of water limited conditions (Wright* et al., *1992b)*

Study	Cultivar	Biomass (g)	Water use (kg)	TE (g/kg)
Well-watered	Tifton-8	63.1	17.0	3.71
	VB-81	46.9	16.2	2.90
	Robut 33-1	55.3	19.0	2.91
	Shulamit	51.6	16.8	3.07
	McCubbin	48.6	16.9	2.88
	Cianjur	43.4	16.3	2.66
	Rangkasibitung	41.6	16.9	2.46
	Pidie	47.3	16.6	2.85
lsd P = 0.05		7.0	1.5	0.3
Water-limited (intermittent stress)	Tifton-8	37.5	12.2	3.07
	Shulamit	35.7	12.8	2.79
	McCubbin	36.3	13.4	2.71
	Chico	20.5	11.4	1.80
(Terminal stress)	Tifton-8	31.3	10.0	3.13
	Shulamit	29.0	9.9	2.93
	McCubbin	26.8	10.0	2.68
	Chico	17.8	8.8	2.02
lsd P = 0.05		5.6	1.9	0.4

Measurements of the change in biomass and water use were between 40 and 90 DAP.

germplasm pool. Bidinger *et al.* (1982) and Garrity *et al.* (1982) argued that genetic improvement in yields of crops can be brought about if attributes that confer yield advantage under drought conditions can be identified and used as tools in breeding programmes to enable identification of drought-tolerant genotypes. Many traits (physiological and biochemical) have been proposed for improving the yield performance of crops under drought conditions (Seetharama *et al.* 1983; Turner, 1986a). There are, however, few examples where the trait-based approach has been successfully employed in large-scale drought resistance breeding programmes. Ludlow and Muchow (1988) critically evaluated traits for improving crop yields in water-limited environments.

(a) Screening for drought tolerance

Drought is a complex syndrome with three main components: its timing, duration and intensity, which can all vary widely during a cropping cycle.

The extreme variability in these three components has made it difficult to define plant attributes required for improved performance under all drought situations. The success of a crop genotype in drought situations is a syndrome of effects manifested by the genotypic attributes contributing to the productivity. In addition, droughts interact with other problems, enhancing or reducing their significance. For example, the involvement and interaction of drought in calcium deficiency, induced seed abortion, pod rots and aflatoxin contamination in groundnut are well recognized. This complexity means that the plant breeder relies on an approach which seeks to select genotypes solely on the basis of increased yield and stability in drought-prone environments.

Systematic attempts to screen groundnut genotypes for drought tolerance have only been made at IRHO, Senegal (Gautreau, 1967, 1969, 1970; Bocklee-Morvan, 1983), and at the International Crops Research Institute for the Semi-Arid Tropics (ICRISAT) in India (ICRISAT, 1983). A common approach has been used in both screening programmes, with yield performance of genotypes as the major selection criterion. This approach is not surprising since no simple screening traits for the identification of drought-tolerant genotypes have been developed to date. The breeding methodologies adopted by the researchers have varied substantially. For example, Gautreau (1967) evaluated genotypes in Senegal, under rainfed conditions in drought-prone environments. At ICRISAT, genotypes are evaluated for drought tolerance based on their total dry matter and pod yield performance under a range of water deficits created by a line-source sprinkler (Hanks et al., 1976) during critical growth stages.

Each approach has its limitations. In the Senegal programme, only limited numbers of entries could be included in multi-locational experiments, due to the high costs of maintaining these trials. Further, variability in environmental conditions between sites, arising from variation in amount and distribution of rainfall, confounds the intrinsic genotypic traits contributing to superior drought tolerance. The ICRISAT approach allows large numbers of genotypes to be screened but the line-source technique has some limitations. For instance, screening has to be conducted in seasons where interference from rainfall is minimized (i.e. during the dry season). Also strong wind during irrigation can influence the systematic nature of water deficits created, requiring complex statistical techniques for data analysis (Murarisingh et al., 1991).

However, the line-source system offers certain advantages, in that it allows large numbers of genotypes to be evaluated at varying intensities of drought in a given environment. The sprinkler irrigation technique simulates rainfall of varying intensities which subsequently wets the soil to different depths, a factor that is particularly important for groundnut with its subterranean podding habit. The genotypes identified as being superior yielding types under this technique do, however, have to be re-evaluated under rainfed conditions in drought-prone environments.

TABLE 9.3 *Groundnut genotypes identified as drought tolerant by various researchers*

Research programme	Varieties	Botanical group	Dura-tion	Reference
IRHO, Senegal	55-437	Spanish	90	Gautreau (1967, 1969, 1970, 1978)
	73-30	Spanish	95	
	Te-3	Spanish	90	Bocklee Morvan (1987)
	Ts 32-1	Spanish	90	
	KH-241-D	Spanish	90	
	73-33	Virginia	110	
USDA, ARS, USA	Tifton-8	Virginia	140	USDA, 1984
ICRISAT	ICG-1697	Spanish	110	ICRISAT, 1986
	ICGV-86707	Spanish	110	ICRISAT, 1990
	ICGV-86635	Spanish	110	

Table 9.3 summarizes the list of genotypes identified as being drought tolerant by the various research programmes in different countries. The genotypes were selected as being drought tolerant based on their yield performance under imposed drought or low rainfall situations.

9.7.2 Selection of genotypes for drought-prone environments

Because yield is the most important trait of a genotype, farmers prefer to plant those that have high yield potential in normal to good rainfall years, but which are also capable of producing reasonable yields in the event of drought (i.e. types that have low sensitivity to drought). Nageswara Rao *et al.* (1989a) examined the relationship between genotypic sensitivity to drought and yield potential (yields achieved under adequately irrigated conditions) in 22 groundnut genotypes (Table 9.4) under a range of single and multiple drought patterns which were simulated using a line-source sprinkler technique (Hanks *et al.*, 1976). The sensitivity of the crop to a given drought pattern was examined by regressing the mean pod yield of the 22 genotypes against the water deficit created across the line source gradient within each drought pattern (Table 9.5). The slope (Sd) of this regression represents the sensitivity of the crop to water deficits within each drought pattern. The results suggested that the crop is sensitive to drought under almost all drought patterns, as indicated by the negative Sd term. The magnitude of sensitivity to drought also varied among the drought patterns. The crop's sensitivity to drought was relatively less when drought occurred early in the season (P1 in Table 9.5). It seems that, for some drought patterns, there is little variation among genotypes in the sensitivity to drought (*Sd*) other than that associated with yield potential (*Yp*), while for other drought patterns the association between *Sd* and Yp is weak (column r^2 in Table 9.5). For the early and mid-season droughts, correlation between *Sd* and *Yp* was not as strong as it was for the droughts which occurred during the pod-filling phase (Figure 9.9). Thus, breeding for drought resistance without sacrificing yield potential may only be feasible for some drought patterns (for example, early and mid-season droughts); in other cases, drought resistance may require the sacrifice of yield potential in good environments.

In areas where the probability of end-of-season drought is high, the best strategy may be to select for earliness to enable escape from this type of drought. The ability of a genotype to recover from mid-season drought when water again becomes available also plays a dominant role in genotype adaptation to a drought pattern where deficits are relieved by intermittent rains (Harris *et al.*, 1988; Nageswara Rao *et al.*, 1988). In this type of drought, selection in well-watered environments is unlikely to identify genotypes with greater recovery responses. Although early water stress reduces initial shoot growth and development, synchronous renewal of vegetative and reproductive development is often observed when the drought is relieved (Stirling *et al.*, 1989a,b). An increase in ^{14}C translocation into stem apices and pegs has been observed when a crop was

TABLE 9.4 *Timing and duration of droughts applied to create the different drought patterns (Pn) (from Nageswara Rao* (et al.,*1989a)*

DAS+	P1	P2	P3	P4	P5	P6	P7	P8	P9	P10	P11	P12
1	U	U	U	U	U	U	U	U	U	U	U	U
15	U	U	U	U	U	U	U	U	U	U	U	U
29	LS	U	U	U	LS	U	LS	LS	U	LS	U	LS
39	LS	U	U	U	LS	U	LS	LS	U	LS	U	LS
51	LS	U	U	U	LS	U	LS	LS	U	LS	U	LS
57	LS	LS	U	U	LS	U	LS	U	LS	U	U	U
66	U	LS	U	U	LS	U	LS	U	LS	LS	LS	LS
72	U	LS	U	U	LS	U	LS	U	LS	LS	LS	LS
82	U	U	U	U	U	LS	LS	LS	U	LS	LS	LS
93	U	U	LS	LS	U	LS	LS	LS	U	U	U	U
100	U	U	LS	LS	U	LS	LS	LS	U	U	LS	LS
111	U	U	LS	LS	U	LS	LS	U	LS	U	LS	LS
118	U	U	U	LS	U	LS	LS	U	LS	U	LS	LS
129	U	U	U	U	U	U	U	U	U	U	U	U

(header: Drought patterns)

das+ = days after sowing
U = uniform irrigation using sprinklers
LS = line-source irrigation to create eight intensities of water deficit

re-watered following a drought during the early reproductive phase. However, it was interesting to note that there was no interrelationship between ^{14}C partitioning and plant water status, which meant that the changes in ^{14}C assimilate partitioning to stem apices and reproductive structures preceded any detectable changes in bulk leaf turgor levels. These results suggest that factors other than cell turgor such as cell wall extensibility (Barlow, 1986) and plant growth substances (Davies et al., 1986; Turner, 1986b; ICRISAT, 1990) may mediate recovery responses of developmental processes. Significant variation among groundnut genotypes in ability to recover from mid-season drought has been observed by Harris et al. (1988) (Figure 9.10). The physiological factors responsible for genotypic variation in recovery patterns is unclear at this stage.

Another approach in the selection and identification of drought-adapted genotypes is to model the yield of genotypes in drought-prone environments using historical weather data and information on genotypic sensitivity to a given drought pattern (Bailey, 1990). It is argued that future research in genotype selection lies more in improved modelling of cultivar performance in varying environments. This approach, however, requires basic and detailed understanding of the effect of environmental factors and their interactions on crop growth and yield in order to assess the potential value of new cultivars in drought-prone environments.

Current research is assessing variability among groundnut germplasm for putative traits related to drought resistance which may ultimately be used

TABLE 9.5 *Regression coefficients of the linear relationship between yield potential (Yp) and genotypic sensitivity to water deficit (Sd) in 22 genotypes (from Nageswara Rao et al., 1989a)*

Drought pattern	Regression coefficients	r^2
P1	Yp = 1.56–0.007 Sd	0.15 NS
P2	Yp = 1.50–0.014 Sd	0.51*
P3	Yp = 0.94–0.013 Sd	0.64**
P4	Yp = 0.95–0.013 Sd	0.87**
P5	Yp = 1.67–0.011 Sd	0.52*
P6	Yp = 0.003–0.013 Sd	0.91**
P7	Yp = 0.14–0.017Sd	0.91**
P8	Yp = −0.30–0.008 Sd	0.38 NS
P9	Yp = 1.61–0.016Sd	0.90**
P10	Yp = 1.57–0.015 Sd	0.86**
P11	Yp = 1.30–0.017 Sd	0.92**
P12	Yp = 0.15–0.015 Sd	0.82**

*, significant at 5% level
**, significant at 1% level

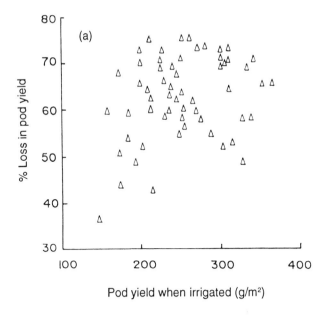

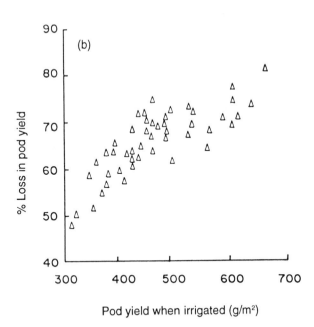

Figure 9.9 Relationship between % pod yield loss due to drought and pod yield potential under irrigated conditions in 60 groundnut genotypes. Yield loss due to: (a) midseason drought; (b) end-of-season drought. (After Nageswara Rao *et al.*, 1989a.)

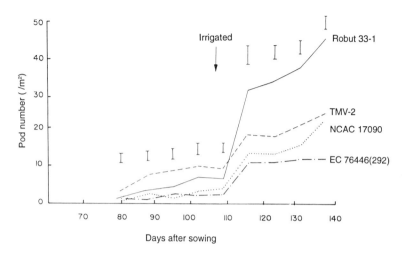

Figure 9.10 Number of pods developed over time by four groundnut genotypes during, and following release from, an early season drought, ICRISAT Center, post-rainy season, 1982–83. Standard errors are presented as vertical bars. (After Harris *et al.*, 1988.)

as selection criteria in breeding programmes. For instance, substantial genetic variation exists for partitioning of dry matter to pods. (Mathews *et al.*, 1988a; Nageswara Rao *et al.*, 1992), and efficient root systems with capacity to penetrate deeper soil layers (Ketring *et al.*, 1985; ICRISAT, 1990; Wright *et al.*, 1991; Wright and Nageswara Rao 1993). As discussed in section 9.6.4, recent evidence has also shown that large cultivar differences in transpiration efficiency (*TE*) exist in groundnut grown under glasshouse and field conditions (Hubick *et al.*, 1986, 1988; Wright *et al.*, 1988a, 1994). The difficulty in accurately measuring *TE* in glasshouse and field situations means that it is virtually impossible to include such a trait in large-scale breeding programmes. This difficulty may have been overcome following the elegant work of Farquhar *et al.* (1982), who proposed on the basis of theory that *TE* and isotopic discrimination against ^{13}C during CO_2 assimilation Δ should be correlated in plants with the C_3 photosynthetic pathway. Subsequent glasshouse and field experiments have show that *TE* and Δ are indeed well correlated in groundnut (Hubick *et al.*, 1986, 1988; Wright *et al.*, 1988a, 1993a, 1994) and have raised the possibility of using Δ as a rapid, non-destructive and relatively inexpensive technique for selection of *TE* in groundnut breeding programmes. Other related research has shown that specific leaf area (*SLA*, cm^2/g) is well correlated with *TE* (and Δ) over a wide range of cultivars and environments (Nageswara Rao and Wright, 1994). This observation has now raised the possibility of using *SLA* as an even more rapid and inexpensive technique for selection of *TE*. Continuing research is assessing the value of this trait in identifying drought-adapted groundnut germplasm in large-scale breeding programme (Wright, 1993).

Lawn (1989) argued that simple procedures should be developed to identify sources for the putative traits, followed by divergent selection for and against the trait in breeding programmes. Crop models will have to be further developed to assess the value of individual traits in contributing to a genotype's performance in a given drought environment (Muchow *et al.*, 1991; Shorter *et al.*, 1991).

9.7.3 Drought management options

(a) Supplementary irrigation

Irrigation is a popular production practice wherever groundnut is culti-vated under high input conditions. Pod yields as high as 10 t/ha have been achieved under large-scale commercial conditions where irrigation has been employed (Hildebrand, 1980), reflecting the high yield potential of the crop. Various aspects of irrigation management, including irrigation scheduling, have been described in detail by Boote *et al.* (1982) and Boote and Ketring (1990). In regions where the crop is grown under rainfed conditions, supplementary irrigation is available only on a fraction of the arable land. Where supplementary irrigation facilities are available, irriga-tion water must be used efficiently to ensure the best returns to the farmer. Increases in frequency or intensity of droughts, declining water tables and increased costs of irrigation have led to considerable research effort on irrigation management in the groundnut crop.

Variable results on the response of groundnuts to irrigation have been obtained, depending upon several factors including the timing and amount of irrigation, method of irrigation, the intensity of water deficit experi-enced by the crop, and climatic conditions. Nageswara Rao *et al.* (1985) observed that drought during the pre-flowering phase, followed by ade-quate water availability, resulted in pod yields of between 13% and 19% greater than fully irrigated crops in a two-year study. This study showed a significant interaction between *ET* and pod yield with different timing of drought. The yield advantage in the pre-flowering drought treatment occurred due to the better synchrony in pod set and pod filling following release from the drought, which in turn resulted in an increase in the number of mature pods at harvest (Nageswara Rao *et al.*, 1988). Several reports have indicated that the pod-filling phase is most sensitive to drought (Boote *et al.*, 1976; Pallas *et al.*, 1979; Nageswara Rao *et al.*, 1985).

Effects of irrigation timing and intensity on yield of groundnut have been examined in 22 groundnut genotypes by subjecting them to 12 different types of single and multiple droughts (Table 9.4). As mentioned earlier, the sensitivity of the crop to drought was less in early and mid-season

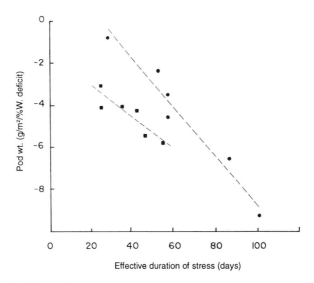

Figure 9.11 The effect of (■) drought or (●) non-limiting water during the preflowering stage on the sensitivity of groundnut (mean of 22 genotypes) to subsequent droughts of varying duration, ICRISAT Center, post-rainy season, 1982–83. (After Nageswara Rao and Williams, 1984.)

drought patterns compared with late-season droughts. However, it is interesting to note that Nageswara Rao and Williams (1984) found that exposing the crop to a short drought during the early vegetative phase reduced the impact of a second drought at the seedling phase, indicating an adaptive response of groundnuts to drought (Figure 9.11). It is possible that early drought may enhance root development and reduce transpirational losses by limiting leaf area development, which subsequently allows the use of soil moisture from deeper in the soil profile.

The beneficial effect of subjecting the crop to an early drought has considerable practical implications for irrigation management. Since the crop can endure long droughts during the early growth phase without major yield losses, supplementary irrigation, when available, should be used largely to support crop growth during the reproductive stage. For instance, in a five-year study in India, sowing of groundnut with one supplementary irrigation prior to the arrival of the monsoon (in July) resulted in moderate crop water deficits during the early growth stage, and an increase in yield of up to 22% (Pasricha et al., 1987).

(b) Adaptation of genotypes to drought-prone environment

Manipulation of crop phenology to match the duration of soil water availability is one of the most important ways of influencing crop performance in a given environment. One can match crop phenology to the environment by either manipulating sowing date or selecting genotypes with appropriate phenology to suit the environment.

Matching the environment to crop phenology
Selection of appropriate sowing dates to match crop growth with soil water availability is an important management practice where groundnut depends on seasonal rainfall (e.g. in India and Africa). Yield and phenological traits vary significantly with changes in sowing dates, which can influence soil and environmental components (Murthy and Nageswara Rao, 1986; Kvien and Bergmark, 1987; Tsai *et al.*, 1987; Bell, 1986; Bell *et al.*, 1991a). Several studies in India have shown that yields of groundnut grown during the rainfed season (July–October) decline as the sowing date is delayed beyond the end of July (Murthy and Nageswara Rao, 1986; Hosmani *et al.*, 1989; Dhoble *et al.*, 1990). Here, the rainy season in most areas ends by October, so that sowing beyond July exposes the crop to terminal drought conditions.

Matching the genotype to the environment
Significant diversity exists in the groundnut germplasm for various phenological traits (Wynne and Coffelt, 1982) and offers the possibility of selecting genotypes with desired phenology for use in breeding programmes. In regions where the growing season is longer, cultivars belong to the virginia botanical group (subspecies *hypogaea*) are generally selected; in regions where the season is shorter, spanish and valencia types (subspecies *fastigiata*) are selected. With perceivable changes in global climate (principally temperature and rainfall patterns), it may become necessary to match genotype more carefully to the length of growing season. For example, groundnut production in Nigeria has declined remarkably over the past few years due to severe droughts. This has prompted the need for revision of recommendations on longer season genotypes (Harkness and Dadirep, 1978, cited by Gibbons, 1978). Agroclimatological analysis of major rainfed groundnut-growing environments in the semi-arid tropics (SAT) clearly indicates that growing areas in the SAT are characterized by short growing seasons, i.e. 75–110 days (Virmani and Piara Singh, 1986). This explains why short duration genotypes are generally successful in West African regions (Gautreau, 1967; Bocklee-Morvan, 1983) as well as in some parts of India (S.N. Nigam, Principal Groundnut Breeder, ICRISAT, personal communication).

(c) Manipulation of plant density

Adjustment of plant population to match crop water demand to soil water availability is another management option for growing groundnuts under rainfed conditions. It is well known that dry matter accumulation under

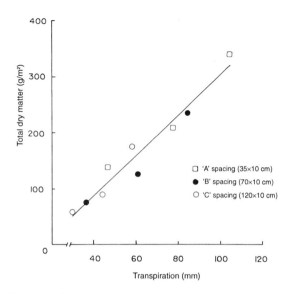

Figure 9.12 Relationship between transpiration and total dry matter of groundnut grown at three plant densities under drought conditions, ICRISAT Center, post-rainy season, 1981–82. (After Azam-Ali *et al.*, 1989.)

non-limiting conditions is proportional to the amount of radiation intercep-tion by the foliage (Biscoe and Gallagher, 1978; Azam-Ali *et al.*, 1989). Similarly, Azam-Ali *et al.* (1989) have shown that the dry matter produced is also tightly linked with crop transpiration (Figure 9.12). The tight linkage between transpiration and dry matter production indicates that yield cannot be enhanced by limiting transpiration, and hence that any trait or management practice which improves the transpirational component of total *ET* should improve yields. Obviously, manipulation of plant density is an important way of optimizing leaf area to water supply and environ-mental demand.

The relationship between groundnut productivity and plant density is complex, particularly under water-limited conditions (Jones, 1986). The optimum plant density under variable water availability will involve a delicate compromise between maximizing water extraction and reducing soil evaporation on the one hand, and conserving soil water for use during pod filling on the other. Both the timing and severity of crop water deficits occurring in response to rainfall availability and distribution, and soil water holding capacity, will interact to determine the plant population/yield response. For instance, groundnut crops grown at high plant population under severe end-of-season drought stress depleted the soil water rapidly and led to severe crop water deficits during reproductive growth, and reduced pod yield, compared with low density crops (Wright and Bell,

1992b). Indeed, the optimum plant density for groundnut has been shown to decline as the severity of drought increases (Wright and Bell, 1992a): while pod yield was maximized at over 100 000 plants/ha under full irrigation, only 40 000 plants/ha were needed under a terminal drought stress. These results indicate that significant savings in seed costs, and hence increased profitability, could be achieved by reductions in plant population under specific drought regimes.

Much of the information on plant density response in groundnut is empirically based, and few attempts have been made to understand the physiological factors responsible for the variation often observed in different genotypes, drought patterns, soil types and localities. Nageswara Rao *et al.* (1989b) studied root and shoot growth in response to varying plant density for groundnut (cv. TMV-2) grown on stored soil water. They found that total weight and length of roots per unit area increased with plant density, but proportional increases in shoot weight were greater, such that the root to shoot ratio increased from 0.3 at high plant population (35 × 10 cm spacing) to 0.5 under low plant population (120 × 10 cm spacing) (Table 9.6). The influence of plant population on the relative size of roots and shoots has important implications for the water relations of the crop in a given environment.

There have so far been limited efforts at predicting the plant population requirements from knowledge of weather and soil parameters. Climatic, soil and management factors, and the interaction between supply and demand for water, determine the optimum plant population at which maximum productivity can be obtained. Azam-Ali *et al.* (1993) reported that the optimum plant population requirements for various climates can be determined using environmental features such as seasonal soil water supply, saturation vapour pressure deficit, and crop parameters such as water-use efficiency and harvest index. Bell *et al.* (1991b) developed a simple model, originally proposed by Gardner and Gardner (1983), for predicting the relationship between groundnut pod yield and plant density under non-limiting water conditions. The model provided a useful method for summarizing key components of the pod yield/plant density relationship, and served to highlight important characteristics associated with different density responses among cultivars. Wright and Bell (1992a) extended this approach to characterize the relationship under water-limited conditions. Some of the assumptions made in the model were shown to be invalid under drought conditions; however, it was suggested that these crop constants may be able to be related to either crop or soil water status, in order to improve the predictive capability of the model under water-limited conditions. The model provides a useful analytical tool for intercepting groundnut plant density responses, by defining key physiologically meaningful parameters. It has also been used recently to interpret the differing groundnut population response observed between subtropical and tropical regions (Wright, 1992).

TABLE 9.6 *The mean weights of shoots (including pods) and the total weight and length of the root systems per unit land area (from Nageswara Rao et al., 1989b)*

Spacing ++	Days after sowing	Shoot plus pod weight (g/m^2)	Root weight W_a (g/m^2)	Root length l_a (km/m^2)	Root: total weight ratio
A	47	49	–	–	–
A	60	134	58	3.8	0.30
A	76	206	68	4.7	0.24
A	90	267	143	6.6	0.36
A	97	230	–	–	–
B	47	30	–	–	–
B	60	68	41	2.7	0.38
B	76	113	58	3.7	0.33
B	90	155	120	5.6	0.43
B	97	147	–	–	–
C	47	16	–	–	–
C	60	44	44	2.8	0.49
C	76	71	49	3.2	0.41
C	90	105	101	4.6	0.49
C	97	100	–	–	–
Analysis of variance					
Main effects					
Spacing		***	**	**	***
Date		***	***	***	**
Interactions					
Spacing × date		***	NS	NS	NS

*** and ** denote significance at the 1 and 2% levels, respectively;
'NS' denotes lack of significance at the 5% level.
Spacing: A = 35 × 10 cm; B = 70 × 10 cm; C = 120 × 10 cm.

(d) Intercropping

Intercropping of groundnut with various legume and cereal crops is a popular practice, particularly under rainfed conditions where farmers have limited access to irrigation. Many studies have reported advantages due to intercropping in terms of land resource use, cost benefit ratio, and sustainable yields (Willey, 1981). Yield advantages in an intercropping system can occur if the component crops have variable phenology and hence differing resource requirements (primarily light and water) in space and time. The extent of yield advantage, however, depends on the environmental resource capture, resource use and the resource conversion efficiency. There are several reports on yield advantages of intercropping groundnut with cereals (Reddy and Willey, 1981; Harris *et al.*, 1987), although the factors causing yield advantage varied depending on the situation. In the intercropping of sorghum with groundnut, shading of groundnut by sorghum ameliorated the effects of high temperature and water deficit to a certain extent, especially under drought (Harris and Natarajan, 1987). It was suggested that alleviation of high temperature effects on flower production resulted in a greater proportion of pegs forming pods in the intercropped groundnut compared with the sole crop, leading to the observed increase in harvest index of groundnuts. Despite a serious drought in 1984 in the savannah zone of the Sudan, Arya *et al.* (1988) reported that intercropping of millet with groundnut reduced soil loss due to wind erosion, and increased yields compared with the sole-cropped groundnuts.

The significant variability in phenology and duration among different botanical groups of groundnut genotypes makes it possible to intercrop short and long duration genotypes (Nageswara Rao *et al.*, 1990). Experiments on intercropping of these in a ratio of 1:1 under a range of water-deficit conditions imposed during the pod-filling phase resulted in an increase in land equivalent ratio (*LER*) of up to 1.25. However, it was interesting to note that specific combinations of cultivars were necessary to maximize *LER*. Although the intercropping of early and late maturing genotypes resulted in absolute benefits for total dry matter, there was no apparent advantage in pod yields because of low partitioning in the case of the long duration component genotype used in the study. The total yields of this system could probably be improved by using a long season genotype with greater partitioning of dry matter to pods. These results suggest that there may be scope for achieving greater productivity in environments with variable season length by intercropping late and early maturing groundnut cultivars. Further research into the factors limiting exploitation of water by the long season component of the intercrop following harvest of the early genotype is needed to optimize the temporal and spatial utilization of light and water resources. Such an option may be advantageous in regions where hand harvesting is practised but may present management problems in sowing and harvesting operations where mechanical cultivation is practised.

(e) Mulching

As mentioned earlier, the linear relation between dry matter production and amount of water transpired by the crop (Azam-Ali *et al.*, 1989) means that any practice or trait which increases transpiration can thereby increase dry matter production. Under water-limited conditions, another management option to enhance a crop's productivity is to increase the amount of transpirable water, by using soil water conservation practices to reduce the soil evaporation component of total evapotranspiration. One of the most popular methods for reducing soil evaporation is by the use of soil mulches. Araya *et al.* (1988) reported that, despite serious drought in the 1984 cropping season, mulching with surface residues and no till planting increased soil moisture, biomass and seed yields on the cracking vertisols and stabilized sands in the savanna zones of the Sudan.

Plastic mulches have been widely adopted in China for production of groundnut under dryland conditions (Wang and Han, 1990). Besides causing increased surface soil temperatures, which allows earlier sowing times, the plastic mulch significantly reduces soil evaporative losses. For instance, soil water content in the surface 10 cm was up to 5% higher during early seedling growth under plastic mulch compared with uncovered soil. Pod yield responses of greater than 50% have been recorded using plastic mulch under rainfed conditions in Chinese groundnut production systems.

9.8 CONCLUSIONS

Drought is recognized as one of the major constraints limiting groundnut productivity in rainfed production areas. Plant water deficits strongly influence physiological processes such as photosynthesis, stomatal conductance, leaf expansion and nitrogen fixation, which all ultimately have an impact on pod yield performance. Groundnut is capable of withstanding extremely low relative leaf water content (as low as 29%) and low leaf water potential (as low as -10.0 MPa), indicating substantial dehydration tolerance. The maintenance of turgor pressure through osmotic adjustment has been demonstrated in groundnut leaves and pegs. However, the situation regarding the importance, extent and possible cultivar variation in osmotic adjustment in groundnut is unclear.

Water deficits strongly influence root and shoot growth in groundnut. Cultivar differences in root growth and soil water extraction capacity under drought stress have been shown to exist. Further research effort is needed to show that higher yield and improved drought tolerance are associated with larger and deeper root systems. Simple, rapid and inexpensive methods to screen for rooting characters which correlate with observed field performance are needed. Pod yields can be severely reduced by water deficits occurring during the pegging and podset phase. Soil water content can influence peg and pod development through direct effects on plant water status and assimilate supply, or through effects on peg penetration and calcium uptake in dry surface soil. Peg entry and subsequent pod development have also been shown to be extremely sensitive to soil hardness, which commonly increases with declining soil water content. The partitioning of biomass to pods can vary widely, depending on the timing and severity of water deficits, and strongly influences final pod yield. Both genetic and environmental factors influence partitioning patterns in groundnut.

There are a number of management and genetic options available to ameliorate the effects of water deficits in groundnuts. Management options, such as supplementary irrigation and manipulating crop phenology (by cultivar selection and/or sowing date) to match the duration of soil water availability, are widely employed in many groundnut producing regions. Current research has also shown that plant density manipulations and intercropping of groundnut with other crops (or with different maturity cultivars) can ameliorate the detrimental effects of drought by delaying the onset of severe crop water deficits.

The genetic solution to drought is considered the best long-term strategy to improve pod yield performance in water-limited environments. Systematic attempts to screen groundnut genotypes for drought tolerance have only been made at IRHO (Senegal) and ICRISAT (India). In these breeding programmes, adapted genotypes have been selected solely on the basis of increased pod yield and stability in drought-prone environments. Future genotypic selection based on desirable drought tolerance traits contributing directly or indirectly to superior performance under drought conditions might be possible. For instance, current research is assessing the potential of traits such as transpiration efficiency, partitioning of dry matter to pods, and efficient root systems to improve the efficiency of selection of drought-adapted germplasm in groundnut breeding programmes.

ACKNOWLEDGEMENTS

We thank the Queensland Department of Primary Industries (QDPI) the International Crops Research Institute for the Semi-Arid Tropics (ICRISTAT) and the Australian Centre for International Agricultural Research (ACIAR) for their support during the production of this chapter, which was submitted as ICRISAT journal article No. JA 1457.

REFERENCES

Allen, L.H. Jr., Boote, K.J. and Hammond, L.C. (1976) Peanut stomatal diffusion resistance affected by soil water and solar radiation. *Soil Crop Science, Florida*, **36**, 42–6.

Arkley, R.J. (1963) Relationship between plant growth and transpiration. *Hilgardia*, **34**, 559–584.

Araya, L.M., Woldetatios, T. and Riley, J.J. (1988) Potentials for increasing crop productivity through soil and water conservation in the Sahelian and Savannah zones of Sudan, in *Arid Lands: Today and Tomorrow*, (eds E.E Whitehead, C.F. Hutchinson, B.N. Timmermann and R.G. Varady), Westview Press, Boulder, Colorado, USA, pp. 459–476.

Azam-Ali, S.N. (1984) Environmental and physiological control of transpiration by groundnut crop. *Agricultural and Forestry Meteorology*, **33**, 129–40.

Azam-Ali, S.N., Simmonds, L.P., Nageswara Rao, R.C. and Williams, J.H. (1989) Population growth and water use of groundnut maintained on stored water. III. Dry matter, water use and light interception. *Experimental Agriculture*, **25**, 77–86.

Azam-Ali, S.N., Nageswara Rao, R.C., Craigon, J. *et al.* (1993) A method for calculating the population/yield relations of groundnut (*Arachis hypogaea L.*) in semi-arid climates. *Journal of Agricultural Science*, **121**, 213–232.

Balasubramanian, V. and Yayock, J.V. (1981) Effect of gypsum and moisture stress on growth and podfill of groundnuts (*Arachis hypogaea L.*) *Plant and Soil*, **62**, 209–19.

Bailey, E. (1990) The use of risk analysis in the evaluation of genotype performance in drought prone areas. *Dissertation Abstracts International, A. Humanities and Soil Sciences*, **49(10)**, 3806, Cornell University, 289pp.

Barlow, E.W.R. (1986) Water relations of expanding leaves. *Australian Journal of Plant Physiology*, **13**, 45–58.

Barrs, H.D. and Weatherley, P.E. (1962) A re-examination of the relative turgidity technique for estimating water deficits in leaves. *Australian Journal of Biological Sciences*, **15**, 413–28.

Begg, J.E. and Turner, N.C. (1976) Crop water deficits. *Advances in Agronomy*, **28**, 161–217.

Bell, M.J. (1986) Effect of sowing date on growth and development of irrigated peanuts, *Arachis hypogaea L.* cv. Early Bunch, in a monsoonal tropical environment. *Australian Journal of Agricultural Research*, **37**, 361–73.

Bell, M.J., Shorter, R. and Mayer, R. (1991a) Cultivar and environmental effects on growth and development of peanuts (*Arachis hypogaea L.*) 1. Flowering. *Field Crops Research*, **27**, 35–49.

Bell, M.J., Harch, G.R. and Wright, G.C. (1991b) Plant population studies on peanut (*Arachis hypogaea L.*) in subtropical Australia. 1. Growth under fully irrigated conditions. *Australian Journal of Experimental Agriculture*, **31**, 535–543.

Bell, M.J., Wright, G.C. and Hammer, G.L. (1992) Night temperature affects radiation use efficiency in peanut. *Crop Science*, **32**, 1329–1335.

Bennett, J.M., Boote, K.J. and Hammond, L.C. (1981) Alterations in the components of peanut leaf water potential during desiccation. *Journal of Experimental Botany*, **32**, 1035–43.

Bennett, J.M., Boote, K.J. and Hammond, L.C. (1984) Relationships among water potential components, relative water content, and stomatal resistance of field grown peanut leaves. *Peanut Science*, **11**, 31–35.

Bennett, J.M., Sexton, P.J. and Boote, K.J. (1990) A root tube-pegging pan apparatus: Preliminary observations and effects of soil water in the pegging zone. *Peanut Science*, **17**, 68–72.

Beringer, H. and Taha, M.A. (1976) Calcium absorption by two cultivars of groundnut (*Arachis hypogaea*). *Experimental Agriculture*, **12**, 1–17.

Bhagsari, A.S., Brown, R.H. and Scheper, J.S. (1976) Effect of moisture stress on photosynthesis and some related physiological characteristics in peanut. *Crop Science*, **16**, 712–15.

Bhan, S. and Misra, O.K. (1970) Water utilisation by groundnut (*Arachis hypogaea L.*) as influenced due to variety, plant population and soil fertility under arid zone conditions. *Indian Journal of Agronomy*, **15**, 158–63.

Bidinger, F.R., Musgrave, R.B. and Fischer, R.A. (1977) Contribution of stored pre-anthesis assimilates to grain yield in wheat and barley. *Nature*, **270**, 431–3.

Bidinger, F.R., Mahalakshmi, V., Talukdar, B.S. and Alagarswamy, G. (1982) Improvement of drought resistance in pearl millet, in *Drought resistance in crops with emphasis on rice*, Los Banos, Laguna, Philippines: International Rice Research Institute, pp. 357–375.

Bieurhuizen, J.F. and Slatyer, R.O. (1965) Effect of atmospheric concentration of water vapour and CO_2 in determining transpiration–photosynthesis relationships of cotton leaves. *Agricultural Meteorology*, **2**, 259–70.

Billaz, R. and Ochs, R. (1961) Stades de sensibilités de l'arachide à la secheresse. *Oléagineux*, **16**, 605–11.

Biscoe, P.V. and Gallagher, J.N. (1978) A physiological analysis of cereal yield. I. Production of dry matter. *Agricultural Progress*, **53**, 34–50.

Black, C.R. and Squire, G.R. (1979) Effects of atmospheric saturation deficit on the stomatal conductance of pearl millet (*Pennisetum typhoides* S. *and* H.) and groundnut (*Arachis hypogaea L.*) *Journal of Experimental Botany*, **30**, 935–45.

Black, C.R., Tang, D.Y., Ong, C.K. *et al.* (1985) Effects of soil moisture stress on the water relations and water use of groundnut stands. *New Phytologist*, **100**, 313–28.

Bledsoe, R.W., Comar, C.L. and Harris, H.C. (1949) Absorption of radioactive calcium by the peanut fruit. *Science*, **109**, 329–30.

Blum, A., Poiarkova, H., Gozlan, G. and Mayer, J. (1983) Chemical desiccation of wheat plants as a simulator of post-anthesis stress. I. Effects of translocation and kernel growth. *Field Crops Research*, **6**, 51–8.

Bocklee-Morvan, A. (1983) The different varieties of groundnut. Geographical and climatic distribution, availability. *Oléagineux*, **38**, 73–116.

Boote, K.J. (1982) Growth stages of peanut (*Arachis hypogaea L.*). *Peanut Science*, **9**, 35–40.

Boote, K.J. and Hammond, L.C. (1981) Effect of drought on vegetative and reproductive development of peanut. *Proceedings of American Peanut Research and Education Society*. **13**, 86 (Abstract).

Boote, K.J. and Ketring, D.L. (1990) Peanut, in *Irrigation of Agricultural Crops*, (eds B.A. Stewart and O.R. Nielsen), *Agron. Monograph 30*, pp. 625–717.

Boote, K.J., Jones, J.W., Mishoe, J.W. and Wilkerson, G.G. (1986) Modelling growth and yield of groundnut, in *Agrometeorology of Groundnut*, Proceedings of International Symposium, ICRISAT Sahelian Centre, Niamey, Niger, ICRISAT, Patancheru, Andhra Pradesh, India, pp. 243–54.

Boote, K.J., Stansell, J.R., Schubert, A.M. and Stone, J.F. (1982) Irrigation, water use and water relations, in *Peanut Science and Technology*, (eds H.E. Pattee and C.T. Young.), American Peanut Research and Education Association, Yoakum, Texas, pp. 164–205.

Boote, K.J., Varnell, R.J. and Duncan, W.G. (1976) Relationships of size, osmotic concentration, and sugar concentration of peanut pods to soil water. *Proceedings Crop and Soil Science Society, Florida*, **35**, 47–50.

Brady, N.C. (1947) The effect of calcium supply and mobility of calcium in the plant on peanut fruit filling. *Soil Science Society of America Proceedings*, **12**, 336–41.

Brown, K.W., Jordan, W.R. and Thomas, J.C. (1976) Water stress induced alteration in the stomatal response to leaf water potential. *Physiologia Plantarium*, **37**, 1–5.

Chapman, S.C. (1989) The effect of drought during reproductive development on the yield of cultivars of groundnut (*Arachis hypogaea L.*). PhD Thesis, University of Queensland, Brisbane.

Constable, G.A. and Hearn, A.B. (1978) Agronomic and physiological responses of soybean and sorghum crops to water deficits. I. Growth, development and yield. *Australian Journal of Plant Physiology*, **5**, 159–67.

Cooper, P.J., Keatinge, J.D. and Hughes, G. (1983) Crop evapotranspiration – a technique for calculating its components by field measurements. *Field Crops Research*, **7**, 299–312.

Cox, F.R., Sullivan, G.A. and Martin, C.K. (1976) Effect of calcium and irrigation treatments on peanut yield, grade and seed quality. *Peanut Science* **3**, 81–5.

Dancette, C. and Forest, F. (1986) Water requirements of groundnuts in the semi-arid tropics, in *Agrometeorology of Groundnut*, Proceedings of International Symposium, ICRISAT Sahelian Centre, Niamey, Niger, ICRISAT, Patancheru, Andhra Pradesh, India, pp. 69–82.

Daughtry, J.A. and Cox, F.R. (1974) Effect of calcium source, rate, and time of application on soil calcium level and yield of peanuts (*Arachis hypogaea L.*). *Plant and Soil*, **62**, 209–19.

Davies, W.J., Metcalfe, J., Lodge, T.A. and Costa, A.R. (1986) Plant growth substances and the regulation of growth under drought. *Australian Journal of Plant Physiology*, **13**, 105–123.

de Wit, C.T. (1958) *Transpiration and crop yields*. Versl. Landbouwk. Orderz. 64.6 Institute of Biological and Chemical Research on Field Crops and Herbage, Wageningen, The Netherlands.

Denmead, O.T. and Shaw, R.H. (1962) Availability of soil water to plants as affected by soil moisture content and meteorological conditions. *Agronomy Journal*, **54**, 385–9.

Devries, J.D., Bennett, J.M., Boote, K.J. *et al.* (1986) Effect of soil water on water relations, nitrogen fixation and nitrogen accumulation of peanut and soybean. *Proceedings of Annual Society for Peanut Research and Education Society*, **18**, 39.

Devries, J.D., Bennett, J.M., Albrecht, S.L. and Boote, K.J. (1989a) Water relations, nitrogenase activity and root development of three grain legumes in response to soil water deficits. *Field Crops Research*, **21**, 215–26.

Devries, J.D., Bennett, J.M., Boote, K.J. *et al.* (1989b) Nitrogen accumulation and partitioning by three grain legumes in response to soil water deficits. *Field Crops Research*, **22**, 33–44.

Dhoble, M.V., Thete, M.R. and Khating, E.A. (1990) Productivity of some kharif crops as influenced by varying dates of planting under rainfed conditions. *Indian Journal of Agronomy*, **35**, 190–98.

Duncan, W.G., McCloud, D.E., McGraw, R.L. and Boote, K.J. (1978) Physiological aspects of peanut yield improvement. *Crop Science*, **18**, 1015–20.

Erickson, P.L. and Ketring, D.L. (1985) Evaluation of peanut genotypes for resistance to water stress *in situ*. *Crop Science*, **25**, 870–76.

Erickson, P.I., Stone, J.F. and Garton, J.E. (1986) Critical evaporative demands for differential stomatal action in peanut grown in runner and wide row spacings. *Agronomy Journal*, **78**, 254–58.

Farquhar, G.D., O'Leary, M.H. and Berry, J.A. (1982) On the relationship between isotope discrimination and the intercellular carbon dioxide concentration in leaves. *Australian Journal of Plant Physiology*, **9**, 121–37.

Fischer, R.A. (1981) Optimising the use of water and nitrogen through breeding of crops. *Plant and Soil*, **58**, 249–78.

Fischer, R.A. and Turner, N.C. (1978) Plant productivity in the arid and semi-arid zones. *Annual Review of Plant Physiology*, **29**, 277–317.

Flower, D.J. and Ludlow, M.M. (1986) Contribution of osmotic adjustment to the dehydration tolerance of water-stressed pigeonpea (*Cajanus cajan L. millsp.*) leaves. *Plant, Cell and Environment*, **9**, 33–40.

Gardner, W.R. and Gardner, H.R. (1983) Principles of water management under drought conditions. *Agricultural Water Management*, **7**, 143–55.

Garrity, D.P., Sullivan, C.Y. and Ross, W.M. (1982) Alternative approaches to improving grain productivity under drought stress, in *Drought resistance in crops with emphasis on rice*, Los Banos, Laguna, Philippines: International Rice Research Institute, pp. 339–356.

Gates, D.M. (1968) Transpiration and leaf temperatures. *Annual Review of Plant Physiology*, **12**, 21–238.

Gautreau, J. (1967) Varietal research on drought resistance of groundnut. *Oléagineux*, **22**, 25–29.

Gautreau, J. (1969) Measuring suction pressure in groundnut plants. *Oléagineux*, **24**, 338–42.

Gautreau, J. (1970) Comparative study of the relative transpiration of two groundnut varieties. *Oléagineux*, **25**, 23–28.

Gerard, C.J., Mehta, H.C. and Hinojosa, E. (1972) Root growth in a clay soil. *Soil Science*, **114**, 37–49.

Gibbons, R.W. (1978) Adaptation and utilisation of groundnuts in different environments and farming systems, in *Advances in Legume Science*, (eds R.J. Summerfield and A.H. Bunting), Royal Botanic Gardens, Richmond, Surrey, England.

Gibbons, R.W. (1980) The ICRISAT Groundnut Program, in *Proceedings of International Workshop on Groundnuts*. International Crops Research Institute for Semi-Arid Tropics (ICRISAT), 13–17 Oct. 1980, Patancheru, Andhra Pradesh, India, pp. 12–16.

Gregory, P.J. and Reddy, M.S. (1982) Root growth in an intercrop of pearl millet/groundnut. *Field Crops Research*, **5**, 241–52.

Grosz, G.D. (1986) *Simulation of peanut growth in Oklahoma*. M.S. Thesis. Oklahoma State University, Stillwater, USA.

Hallock, D.L. and Allison, A.H. (1980) Effect of three calcium sources applied on peanuts in Virginia. I. Productivity and seed quality. *Peanut Science*, **7**, 19–25.

Hammond, L.C., Boote, K.J., Varnell, R.J. and Robertson, W.K. (1978) Water use and yield of peanuts on a well drained sandy soil. *Proceedings of American Peanut Research and Education Association*, **10**, 73 (Abstract)

Hanks, R.J. (1983) Yield and water-use relationships: an overview, in *Limitations to Efficient Water Use in Crop Production*, (eds H.M. Taylor, W.R. Jordan and T.R. Sinclair), American Society of Agronomy, Crop Science Society of America, and Soil Science Society of America, Madison, Wisconsin, USA, pp. 393–411.

Hanks, R.J., Keller, J., Rasmussen, V.P. and Wilson, G.D. (1976) Line-source sprinkler for continuous variable irrigation crops production studies. *Soil Science Society of America Journal*, **40**, 426–429.

Hanson, J.B. (1984) The functions of calcium in plant nutrition, in *Advances in Plant Nutrition*, Volume 1, (eds P.B. Tinker and A. Lauchi.), Praeger Publishers, New York, pp. 149–208.

Harris, D. and Natarajan, M. (1987) Physiological basis for yield advantage in sorghum/groundnut intercrop exposed to drought. 2. Plant temperature, water status and components of yield. *Field Crops Research*, **17**, 273–288.

Harris, D., Natarajan, M. and Willey, R.W. (1987) Physiological basis for yield advantages in sorghum and groundnut intercrop exposed to drought. 1. Dry-matter production, yield, and light interception. *Field Crops Research*, **17**, 259–272.

Harris, D., Mathews, R.B., Nageswara Rao, R.C. and Williams, J.H. (1988) The physiological basis for yield between four genotypes of groundnut (*Arachis hypogaea*) in response to drought. III Developmental processes. *Experimental Agriculture*, **24**, 215–26.

Hildebrand, G.L. (1980) Groundnut production, utilization, research problems and further research needs in Zimbabwe, in *Proceedings of International Workshop on Groundnuts*, International Crops Research Institute for the Semi-Arid Tropics (ICRISAT), 13–17 Oct. 1980, Patancheru, Andhra Pradesh, India, pp. 290–296.

Hosmani, M.N., Jayakumar, B.V. and Sharma, K.M.S. (1989) Performance of different crops in relation to varieties and sowing dates under rainfed 213 conditions in Shimoga tract. *Mysore Journal Agricultural Science*, **23**, 156–58.

Hsiao, T.C. (1973) Plant responses to water stress. *Annual Review of Plant Physiology*, **24**, 519–70.

Hubick, K.T. (1990) Effects of nitrogen source and water limitation on growth, transpiration efficiency and carbon-isotope discrimination in peanut cultivars. *Australian Journal of Plant Physiology*, **17**, 413–30.

Hubick, K.T., Farquhar, G.D. and Shorter, R. (1986) Correlation between water-use efficiency and carbon isotope discrimination in diverse peanut (*Arachis*) germplasm. *Australian Journal of Plant Physiology*, **13**, 803–816.

Hubick, K.T., Farquhar, G.D. and Shorter, R. (1988) Heritability and genotype x environment interactions of carbon isotope discrimination and transpiration efficiency in peanut (*Arachis hypogaea* L.). *Australian Journal of Plant Physiology*, **15**, 779–813.

Hurd, E.A. (1974) Phenotype and drought tolerance in wheat. *Agricultural Meteorology*, **14**, 39–55.

ICRISAT (1983) *Annual Report 1982*, Patancheru, India, pp. 202–207.

ICRISAT (1990) *Annual Report 1989*, Patancheru, India, pp. 143–144.

Ishag, H.M., Osman, A.Fadl., Adam, H.S. and Osman, A.K. (1985) Growth and water relations of groundnuts (*Arachis hypogaea* L.) in two contrasting years in the irrigated Gezira. *Experimental Agriculture*, **21**, 403–8.

Jones, M.J. (1986) Maize population densities and spacings in Botswana. *Tropical Agriculture*, *(Trinidad)*, **63**, 25–29.

Jordan, W.R. and Miller, F.R. (1980) Genetic variability in sorghum root systems.

Implications for drought tolerance, in *Adaptation of Plants to Water and High Temperature Stress*, (eds N.C. Turner and P.J. Kramer), Wiley, New York, pp. 383–399.

Joshi, Y.C., Nautiyal, P.C., Ravindra, V. and Snetic Dwivedi, R. (1988) Water relations of two cultivars of groundnut (*Arachis hypogaea L.*) under soil water deficit. *Tropical Agriculture, (Trinidad)*, **65**, 182–4.

Kanemasu, E.T. (1983) Yield and water-use relationships: Some problems of relating grain yield to transpiration, in *Limitations to Efficient Water Use in Crop Production*, (eds H.M. Taylor, W.R. Jordan and T.R. Sinclair), American Society of Agronomy, Crop Science Society of America, and Soil Science Society of America, Madison, Wisconsin, USA, pp. 413–417.

Ketring, D.L. (1984) Root diversity among peanut genotypes. *Crop Science*, **24**, 229–32.

Ketring, D.L. (1986) Physiological response of groundnut to temperature and water deficits-breeding implications, in *Agrometeorology of Groundnut*, Proceedings of International Symposium, ICRISAT Sahelian Centre, Niamey, Niger, ICRISAT, Patancheru, India, pp. 135–143.

Ketring, D.L., Erickson, P.I. and Stone, J.F. (1985) Apparent sap velocity in peanut. *Proceedings of American Peanut Research and Education Society Incorporated*, **17**, 68 (abstract).

Ketring, D.L., Jordan, W.R., Smith, O.D. and Simpson, C.E. (1982) Genetic variability in root and shoot growth characteristics of peanut. *Peanut Science*, **9**, 68–72.

Kirkegaard, J., So, H.B. and Troedgson, R. (1992) Effect of soil strength on growth of pigeon pea radicles and seedlings. *Plant and Soil*, **140**, 65–74.

Klepper, B. (1987) Origin, branching and distribution of root systems, in *Root Development and Function*, (eds P.J. Gregory, J.V. Lake and D.A. Rose). Society of Experimental Biology, Seminar Series 30, Cambridge University Press, pp. 103–124.

Kramer, P.J. (1969) *Plant and soil water relationships: a modern synthesis*, McGraw Hill, New York.

Kvien, C.S. and Bergmark, C.L. (1987) Growth and development of the Florunner peanut cultivar as influenced by population, planting date and water availability. *Peanut Science*, **14**, 11–16.

Kvien, C.S., Branch, W.D., Sumner, M.E. and Csinos, A.S. (1988) Pod characteristics influencing calcium concentrations in seed and hull of peanut. *Crop Science*, **28**, 666–71.

Lawn, R.J. (1989) Agronomic and physiological constraints to productivity of tropical grain legumes and prospects for improvement. *Experimental Agriculture*, **25**, 509–528.

Lenka, D. and Misra, P.K. (1973) Response of groundnut (*Arachis hypogaea L.*) to irrigation. *Indian Journal of Agronomy*, **18** 492–7.

Leong, S.K. and Ong, C.K. (1983) The influence of temperature and soil water deficit on the development and morphology of groundnut (*Arachis hypogaea L.*). *Journal of Experimental Botany*, **34**, 1551–61.

Ludlow, M.M. (1980) Adaptive significance of stomatal responses to water stress, in *Adaptation of Plants to Water and High Temperature Stress*, (eds N.C. Turner and D.J. Kramer), Wiley and Sons, New York, pp. 123–138.

Ludlow, M.M. and Muchow, R.C. (1988) Critical evaluation of the possibilities of modifying crops for high production per unit of precipitation, in *Drought Research Priorities for the Dryland Tropics*, (eds F.R. Bidinger and C. Johansen). pp. 179–211.

Ludlow, M.M., Fisher, M.J. and Wilson, J.R. (1985) Stomatal adjustment to water deficits in three tropical grasses and a tropical legume grown in controlled conditions in the field. *Australian Journal of Plant Physiology*, **12**, 131–49.

Mantell, A. and Goldin, E. (1964) The influence of irrigation frequency and intensity on the yield and quality of peanuts (*Arachis hypogaea*). *Israel Journal of Agricultural Research*, **14**, 203–10.

Mathews, R.B., Harris, D., Nageswara Rao, R.C. *et al.* (1988a) The physiological basis for yield differences between four genotypes of groundnut (*Arachis hypogaea*) in response to drought. 1. Dry matter production and water use. *Experimental Agriculture*, **24**, 191–202.

Mathews, R.B., Harris, D., Williams, J.H. and Nageswara Rao, R.C. (1988b) The physiological basis for yield differences between four genotypes of groundnut (*Arachis hypogaea*) in response to drought. II. Solar radiation interception and leaf movement. *Experimental Agriculture*, **24**, 203–13.

McCloud, D.E. (1974) Growth analysis of high yielding peanuts. *Proceedings of Soil and Crop Science Society, Florida*, **33**, 24–26.

Muchow, R.C., Hammer, G.L. and Carberry , P.S. (1991) Optimising crop and cultivar selection in response to climatic risk, in *Climatic Risk in Crop Production: Models and management for the semi-arid tropics and sub-tropics*, (eds R.C. Muchow and J.A. Bellamy), CAB International, Wallingford, UK, pp. 235–262.

Murarisingh, Nageswara Rao, R.C. and Williams, J.H. (1991) A statistical assessment of genotypic sensitivity of groundnut (*Arachis hypogaea L*) to drought in line-source sprinkler experiments. *Euphytica*, **57**, 19–25.

Murthy, P.S.S. and Nageswara Rao, R.C. (1986) Physiological basis of variation in rainfed groundnut (*Arachis hypogaea L.*) under different dates of sowings. *Indian Journal of Agronomy*, **81**, 106–108.

Nageswara Rao, R.C. and Williams, J.H. (1984) Effects of duration, timing and intensity of single and multiple droughts on peanuts, in *Proceedings of American Peanut Research and Education Society*, pp. 31 (abstract).

Nageswara Rao, R.C., Singh, S., Sivakumar, M.V.K. *et al.* (1985) Effect of water deficit at different growth phases of peanut. I. Yield responses. *Agronomy Journal*, **77**, 782–6.

Nageswara Rao, R.C., Williams, J.H., Sivakumar, M.V.K. and Wadia, K.D.R. (1988) Effect of water deficit at different growth phases of peanut. II. Responses to drought during preflowering phase. *Agronomy Journal*, **80**, 431–8.

Nageswara Rao, R.C., Williams, J.H. and Murari Singh (1989a) Genotypic sensitivity to drought and yield potential of peanut. *Agronomy Journal*, **81**, 887–893.

Nageswara Rao, R.C., Simmonds, L.P., Azam-Ali, S.N. and Williams, J.H. (1989b) Population, growth and water use of groundnut maintained on stored water. I. Root and shoot growth. *Experimental Agriculture*, **25**, 51–61.

Nageswara Rao, R.C., Wadia, K.B.R. and Williams, J.H. (1990) Intercropping of short and long duration groundnut genotypes to increase productivity in environments prone to end of season droughts. *Experimental Agriculture*, **26**, 63–72.

Nageswara Rao, R.C., Wadia, K.D.R., Hubick, K.T. and Farquhar, G.D. (1992) Crop growth, water-use efficiency and carbon isotope discrimination in groundnut (*Arachis hypogaea L.*) genotypes under end-of-season drought conditions. *Annals of Biology*, **122**, 357–67.

Nageswara Rao, R.C. and Wright, G.C. (1994) Stability of the relationship between specific leaf area and carbon isotope discrimination across environments in peanut. *Crop Science*, **34**, 98–103.

Ochs, R. and Wormer, T.M. (1959) Influence de l'alimentation en eau sur la croissance de l'arachide. *Oléagineux* **14**, 281–91.

ODA (1987) *Microclimatology in Tropical Agriculture Vol 1. Introduction, Methods and Principles*, Overseas Development Administration Final Report, London, p. 136.

Ong, C.K. (1984) The influence of temperature and water deficit on the partitioning of dry matter in groundnut (*Arachis hypogaea L.*). *Journal of Experimental Botany*, **35**, 746–55.

Ong, C.K. (1986) Agroclimatological factors affecting phenology of groundnut, in *Agrometeorology of Groundnut*, Proceedings of International Symposium, ICRISAT Sahelian Centre, Niamey, Niger, ICRISAT, Patancheru, Andhra Pradesh, India, pp. 115–126.

Ong, C.K., Black, C.R., Simmonds, L.P. and Saffell, R.A. (1985) Influence of saturation deficit on leaf production and expansion in stands of groundnut (*Arachis hypogaea L.*) grown without irrigation. *Annals of Botany*, **56**, 523–36.

Ong, C.K., Simmonds, L.P. and Mathews, R.B. (1987) Responses to saturation deficit in a stand of groundnut (*Arachis hypogaea* L.) 2. Growth and development. *Annals of Botany*, **59**, 121–8.

Ono, Y., Nakayama, K. and Kubota, M. (1974) The effects of soil temperature and soil moisture in podding zone on pod development of peanuts. *Crop Science Society of Japan Proceedings*, **43**, 247–51.

Pallas, J.E., Jr and Samish, Y.B. (1974) Photosynthetic response of peanut. *Crop Science*, **14**, 478–82.

Pallas, J.E., Jr, Samish, Y.B. and Willmer, C.M. (1974) Endogenous rhythmic activity of photosynthesis, transpiration, dark respiration and carbon dioxide compensation point of peanut leaves. *Plant Physiology*, **53**, 903–11.

Pallas, J.E., Jr, Stansell, J.R. and Bruce, R.R. (1977) Peanut seed germination as related to soil water regime during pod development. *Agronomy Journal*, **69**, 381–3.

Pallas, J.E., Jr, Stansell, J.R. and Koske, T.J. (1979) Effects of drought on Florunner peanuts. *Agronomy Journal*, **24**, 355–9.

Pandey, R.K. and Pendleton, J.W. (1986) Genetic variation in root and shoot growth of peanut in hydroponics. *Philippines Journal of Crop Science*, **11**, 189–193.

Pandey, R.K., Herrera, W.A.T. and Pendleton, J.W. (1984a) Drought responses of grain legumes under irrigation gradient. 1. Yield and yield components. *Agronomy Journal*, **76**, 549–53.

Pandey, R.K., Herrera, W.A.T., Villegas, A.N. and Pendleton, J.W. (1984b) Drought response of grain legumes under irrigation gradient. III. Plant growth. *Agronomy Journal*, **76**, 557–560.

Pasricha, N.S., Aulakh, M.S., Baddesha, H.S. and Bahl, G.S. (1987) Early sowing of groundnut give higher yield. *Indian Farming*, **37**, 23–24.

Patel, C.L., Padalia, M.R. and Babaria, N.B. (1983) Growth and plant water relations in groundnut grown under different soil moisture stress. *Indian Journal of Agricultural Science*, **56**, 340–5.

Philip, J.R. (1957) Evaporation and moisture and heat fields in the soil. *Journal of Meteorology*, **14**, 354–66.

Prabowo, A., Prastowo, B. and Wright, G.C. (1990) Growth, yield and soil water extraction of irrigated and dryland peanuts in South Sulawesi, Indonesia. *Irrigation Science*, **11**, 63–8.

Radder, G.D. and Biradar, B.M. (1973) Effect of gypsum application and topping of main shoot on pod development and yield of groundnuts. *Oilseeds Journal*, **3**, 11–13.

Rajendrudu, G. and Williams, J.H. (1987) Effect of gypsum and drought on pod initiation and crop yield in early maturing groundnut (*Arachis hypogaea*) genotypes. *Experimental Agriculture*, **23**, 259–71.

Rajendrudu, G., Singh, M. and Williams, J.H. (1983) Hydraulic press measurement of leaf water potential in groundnuts. *Experimental Agriculture*, **19**, 287–93.

Reddy, M.S. and Willey, R.W. (1981) A study of pearlmillet/groundnut intercropping with particular emphasis on the efficiencies of leaf canopy and rooting pattern, in *Proceedings of International Workshop on Intercropping* (ed. R.W. Willey), International Crops Research Institute for the Semi-Arid Tropics (ICRISAT), Patancheru, Andhra Pradesh, India, pp. 502 324. India, pp. 202–209.

Ritchie, J.T. (1972) Model for predicting evaporation from a row crop with incomplete cover, *Water Resources Research*, **8**, 1204–1213.

Ritchie, J.T. (1983) Efficient water use in crop production: Discussion on the generality of relations between biomass production and evapotranspiration, in *Limitations to Efficient Water Use in Crop Production*, (eds H.M. Taylor, W.R. Jordan and T.R. Sinclair), American Society of Agronomy, Crop Science Society of America, and Soil Science Society of America, Madison, Wisconsin, USA, pp. 29–44.

Ritchie, J.T. and Burnett, E. (1971) Dryland evaporative flux in a subhumid climate. II. Plant influences. *Agronomy Journal*, **63**, 56–62.

Ritchie, J.T., Rhodes, E.D. and Richardson, C.W. (1976) Calculating evapotranspiration from native grassland watersheds. *Transactions of American Society of Agricultural*

Engineers, **19**, 1098–1103.

Robertson, W.K., Hammond, L.C., Johnson, J.T. and Boote, K.J. (1980) Effects of plant-water stress on root distribution of corn, soybeans and peanuts in sandy soil. *Agronomy Journal*, **72**, 548–50.

Sarma, P.S. (1984) Soil plant water relations, growth and yield of groundnut under moisture stress, Ph.D. thesis, Andhra Pradesh Agricultural University, Hyderabad, India.

Seetharama, N., Sivakumar, M.V.K., Bidinger, F.R. *et al.* (1983) Physiological basis for increasing and stabilising yield under drought in sorghum. *Proceedings of Indian National Science Academy, Part B*, **49**, 498–523.

Sheriff, D.W. and Muchow, R.C. (1984) The water relations of crops, in *The Physiology of Tropical Field Crops*, (eds P.R. Goldsworthy and N.M. Fisher), Wiley, New York, pp. 39–83.

Shorter, R. and Simpson, B.W. (1987) Peanut yield and quality variation over harvest dates, and evaluation of some maturity indicies in south-eastern Queensland. *Australian Journal of Experimental Agriculture*, **27**, 445–53.

Shorter, R., Lawn, R.J. and Hammer, G.L. (1991) Improving genotypic adaptation in crops. A role for breeders, physiologists and modellers. *Experimental Agriculture*, **27**, 155–175.

Simmonds, L.P. and Ong, C.K. (1987) Responses to saturation deficit in a stand of groundnut (*Arachis hypogaea* L.). 1. Water use. *Annals of Botany*, **59**, 113–123.

Simmonds, L.P. and Williams, J.H. (1989) Population, water use and growth of groundnut maintained on stored water. II. Transpiration and evaporation from soil. *Experimental Agriculture*, **25**, 63–75.

Sinclair, T.R. (1986) Water and nitrogen limitations in soybean grain production. I. Model development. *Field Crops Research*, **15**, 125–41.

Sinclair, T.R. and Ludlow, M.M. (1985) Who taught plants thermodynamics? The unfulfilled potential of plant water potential. *Australian Journal of Plant Physiology*, **12**, 213–17.

Sinclair, T.R. and Ludlow, M.M. (1986) Influence of soil water supply on the plant water balance of four tropical grain legumes. *Australian Journal of Plant Physiology*, **13**, 329–41.

Sivakumar, M.V.K. and Sarma, P.S. (1986) Studies on water relations of groundnut, in *Agrometeorology of Groundnut*, Proceedings of International Symposium, ICRISAT Sahelian Centre, Niamey, Niger, ICRISAT, Patancheru, Andhra Pradesh, India, pp. 83–98.

Skelton, B.J. and Shear, G.M. (1971) Calcium translocation in peanut (*Arachis hypogaea* L.). *Agronomy Journal*, **63**, 409–412.

Slabbers, P.J., Herrendorf, V.S. and Stapper, M. (1970) Evaluation of simplified water-crop yield models. *Agricultural Water Management*, **2**, 95–129.

Slack, T.E. and Morrill, L.G. (1972) A comparison of a large-seeded (NC2) and a small-seeded (Starr) peanut (*Arachis hypogaea* L.) cultivar as affected by levels of calcium added to the fruit zone. *Proceedings of Soil Science Society of America*, **36**, 87–90.

Slatyer, R.O. (1955) Studies on the water relations of crops plants grown under natural rainfall in northern Australia. *Australian Journal of Agricultural Research*, **6**, 365–77.

Slatyer, R.O. and Taylor, S.A. (1960) Terminology in plant–soil–water relations. *Nature*, **187**, 922–4.

Smith, B.W. (1954) *Arachis hypogaea*. Reproductive efficiency. *American Journal of Botany*, **41**, 607–16.

So, H.B. and Woodhead, T. (1987) Alleviation of soil physical limits to productivity of legumes in Asia, in *Food Legume Improvement in Asian Farming Systems*, (eds E.S. Wallis and D.E. Byth), Proceedings of International Workshop at Khon Kaen, Thailand. *ACIAR Proceedings*, **18**, 112–120.

Stansell, J.R. and Pallas, J.E., Jr (1985) Yield and quality response of Florunner peanut to drought at several growth stages. *Peanut Science*, **12**, 64–70.

Stansell, J.R., Shepherd, J.L., Pallas, J.E., Jr *et al.* (1976) Peanut responses to soil water variables in the Southeast. *Peanut Science*, **3**, 44–8.

Steponkus, P.J., Cutler, J.M. and O'Toole, J.C. (1980) Adaptation to water stress in rice, in *Adaptation of Plants to Water and High Temperature Stress*, (eds N.C. Turner and D.J. Kramer), Wiley and Sons, New York, pp. 401–418.

Stewart, J.I., Hagan, R.M., Pruitt, W.O. *et al.* (1977) *Optimising crop production through control of water and salinity levels in the soil*, Utah Water Research Laboratory, Utah State University, Logan. Publication No. PRWG 151-1.

Stirling, C.M., Ong, C.K. and Black, C.R. (1989a) The response of groundnut (*Arachis hypogaea L.*) to timing of irrigation. I. Development and growth. *Journal of Experimental Botany*, **40**, 1145–1153.

Stirling, C.M., Black, C.R. and Ong, C.K. (1989b) The response of groundnut (*Arachis hypogaea L.*) to timing of irrigation. II. ^{14}C-Partitioning and plant water status. *Journal of Experimental Botany*, **40**, 1363–73.

Stone, J.F., Erickson, P.I. and Abdul-Jabbar, A.S. (1985) Stomatal closure behaviour induced by row spacing and evaporative demand in irrigated peanuts. *Agronomy Journal*, **77**, 197–202.

Tanner, C.B. and Sinclair, T.R. (1983) Efficient water use in crop production: Research or re-search, in *Limitations to Efficient Water Use in Crop Production*, (eds H.M. Taylor, W.R. Jordan and T.R. Sinclair), American Society of Agronomy, Crop Science Society of America, and Soil Science Society of America, Madison, Wisconsin, USA, pp. 1–27.

Tanner, C.B. and Jury, W.A. (1976) Estimating evaporation and transpiration from a row crop during incomplete cover. *Agronomy Journal*, **68**, 239–43.

Taylor, H.M. and Ratliff, L.F. (1969) Root elongation rates of cotton and peanuts as a function of soil strength and soil water content. *Soil Science*, **108**, 113–19.

Tsai, S.L., Chu, T.M. and Lin, J.L. (1987) Production and distribution of dry matter in relation to yield in Virginia type groundnut. *Journal of Agricultural Association of China*, **139**, 30–43.

Tsuno, T. (1975) The influence of transpiration upon the photosynthesis of several crop plants. *Proceedings of Crop Science Society of Japan*, **44**, 129–40.

Turner, N.C. (1974) Stomatal behaviour and water status of maize, sorghum and tobacco under field conditions. *Plant Physiology*, **53**, 360–5.

Turner, N.C. (1986a) Crop water deficits: A decade of progress. *Advances in Agronomy*, **39**, 1–51.

Turner, N.C. (1986b) Adaptations to water deficit, a changing perspective. *Australian Journal of Plant Physiology*, **13**, 175–190.

Turner, N.C., Begg, J.E., Rawson, H.M. *et al.* (1978) Agronomic and physiological responses of soybean and sorghum crops to water deficits. III. Components of leaf water potential, leaf conductance, CO_2, photosynthesis, and adaptation to water deficits. *Australian Journal of Plant Physiology*, **5**, 179–194.

Underwood, C.V., Taylor, H.M. and Hoveland, C.S. (1971) Soil physical factors affecting peanut pod development. *Agronomy Journal*, **63**, 953–4.

Venkateswarlu, B., Maheswari, M. and Saharan, N. (1989) Effects of water deficit in N_2 (C_2H_2) fixation in cowpea and groundnut. *Plant and Soil*, **114**, 69–74.

Virmani, S.M. and Piara Singh (1986) Agroclimatology characteristics of the groundnut growing regions in the semi-arid tropics, in *Agrometeorology of Groundnut*, Proceedings of International Symposium, ICRISAT Sahelian Centre, Niamey, Niger, ICRISAT, Patancheru, Andhra Pradesh, India, pp. 35–46.

Wang, G.S. and Han, L.M. (1990) Economic evaluation of dryland peanut growing with perforated plastic mulch, in *Dryland Management: Economic Case Studies*, (eds J.A. Dixon, D.E. James and P.B. Sherman), Earthscan Publications, London, pp. 72–85.

Webb, A.J. and Hansen, A.P. (1989) Histological changes of the peanut (*Arachis hypogaea*) gynophore and fruit surface during development, and their potential significance for nutrient uptake. *Annals of Botany*, **64**, 351–7.

Wiersum, L.K. (1951) Water transport in the xylem as related to calcium uptake by groundnuts (*Arachis hypogaea L.*). *Plant and Soil*, **3**, 81–5.

Willey, R.W. (1981) A scientific approach to intercropping research, in *Proceedings of International Workshop on Intercropping*, (ed. R.W. Willey), International Crops

Research Institute for the Semi-Arid Tropics, ICRISAT, Patancheru, Andhra Pradesh, India. pp. 4–14.

Wormer, T.M. and Ochs, R. (1959) Humidité du sol, ouverture des stomates et transpiration du palmier à huile et l'arachide. *Oléagineux*, **14**, 571–80.

Wright, G.C. (1989) Effect of pod zone moisture content on reproductive growth in three cultivars of peanut (*Arachis hypogaea*). *Plant and Soil*, **116**, 111–14.

Wright, G.C. (1992) The influence of plant population on peanut growth and yield under variable water supply. *ACIAR Food Legumes Newsletter*, **17**, 3–6.

Wright, G.C. (1993) Selection for water-use efficiency in food legumes – a new ACIAR project. *ACIAR Food Legumes Newsletter*, **19**, 4–7.

Wright, G.C. and Smith, R.C.G. (1983) Differences between two grain sorghum genotypes in adaptation to drought stress. II. Root water uptake and water use. *Australian Journal for Agricultural Research*, **34**, 627–36.

Wright, G.C., Smith, R.C.G. and McWilliam, J.R. (1983a) Differences between two grain sorghum genotypes in adaptation to drought stress. I. Crop growth and yield responses. *Australian Journal for Agricultural Research*, **34**, 615–26.

Wright, G.C., Smith, R.C.G. and Morgan, J.M. (1983b) Differences between two grain sorghum cultivars in adaptation to drought stress. III. Physiological responses. *Australian Journal for Agricultural Research*, **34**, 637–51.

Wright, G.C., Hubick, K.T. and Farquhar, G.D. (1988a) Discrimination in carbon isotopes of leaves correlates with water-use efficiency of field grown peanut cultivars. *Australian Journal of Plant Physiology*, **15**, 815–825.

Wright, G.L., Rahmianna, A. and Hatfield, P.M. (1988b) A comparison of thermocouple psychrometer and pressure chamber measurements of leaf water potential in peanuts. *Experimental Agriculture*, **24**, 355–9.

Wright, G.C., Hubick, K.T. and Farquhar, G.D. (1991) Physiological analysis of peanut cultivar response to timing and duration of drought stress. *Australian Journal for Agricultural Research*, **42**, 453–70.

Wright, G.C. and Bell, M.J. (1992a) Plant population studies on peanut (*Arachis hypogaea* L.) in subtropical Australia. 2. Water limited conditions. *Australian Journal of Experimental Agriculture*, **32**, 189–196.

Wright, G.C. and Bell, M.J. (1992b) Plant population studies on peanut (*Arachis hypogaea* L.) in subtropical Australia. 3. Growth and water use during a terminal drought stress. *Australian Journal of Experimental Agriculture*, **32**, 197–203.

Wright, G.C., Adisarwanto, T., Rahmianna, A. and Syarifuddin, D. (1992) Drought tolerance traits conferring adaptation to drought stress in peanut, in *Peanut Improvement: A Case Study in Indonesia*, (eds G.C. Wright and K.J. Middleton), *ACIAR Proceedings*, **40**, 74–84.

Wright, G.C. and Nageswara Rao, R.C. (1993) Variation in root growth and water uptake in peanut cultivars, in *Proceedings 7th Australian Agronomy Conference, Adelaide, S.A.* pp. 92–95.

Wright, G.C., Hubick, K.T., Farquhar, G.D. and Nageswara Rao, R.C. (1993) Genetic and environmental variation in transpiration efficiency and its correlation with carbon isotope discrimination and specific leaf area in peanut, in *Isotopes and Plant Carbon–Water Relations*, (eds J.R. Ehleringer, A.E. Hall, and G.D. Farquhar), Academic Press, San Diego. pp. 246–67.

Wright, G.C., Bell, M.J. and Hammer, G.L. (1993b) Leaf nitrogen content and minimum temperature interactions affects radiation use efficiency in peanut. *Crop Science*, **34**, 476–481.

Wright, J.C., Nageswara Rao, R.C. and Farquhar, G.D. (1994) Peanut cultivar variation in water-use efficiency and carbon isotope discrimination under drought conditions in the field. *Crop Science*, **34**, pp. 92–97.

Wynne, J.C. and Coffelt, T.A. (1982) Genetics of *Arachis hypogaea* L., in *Peanut Science and Technology*, (eds H.E. Pattee and C.T. Young), American Peanut Research and Education Association, Yoakum, Texas, pp. 50–94.

CHAPTER 10

Diseases

K.J. Middleton, S. Pande, S.B. Sharma and D.H. Smith

A well-balanced groundnut research team will always include a plant pathologist, such is the importance of diseases in the crop. However, the importance of a particular disease in a crop depends on a number of environmental factors. Under cool wet conditions, sclerotinia blight and web blotch may be important; bacterial wilt will be serious only under hot wet conditions. The leaf spot diseases and nematodes are important when the crop is grown regularly on the same soil; rust can be important under drier conditions than those favouring leaf spots. Aflatoxin contamination is worst under drought conditions, while those virus diseases spread by aphids will be severe when groundnuts are growing well but other crops in the area are suffering from water shortage. Other viruses which frequently infect weed hosts are more serious when those weeds are prevalent in the crop or its surroundings. The damping off of seedlings is most serious when poor quality seed is planted without fungicidal protection. Under most conditions, some incidence of disease can be expected, and in most production areas disease control requires research input.

In this chapter, it is assumed that the reader knows the identity of the disease causing concern and is consulting this volume to obtain more information about it. Sometimes, the disease has been well covered in another treatise, and in these cases the reader will be referred to the other sources of information, but all diseases covered here are treated in adequate detail to suit most purposes. To simplify discussion on the interrelationships between similar diseases, the chapter is organized so that similar diseases are in proximity.

10.1 DISEASES CAUSED BY BACTERIA

10.1.1 Bacterial wilt

Bacterial wilt is caused by *Pseudomonas solanacearum* (E.F. Smith), an aerobic Gram-negative rod shaped bacterium (Hayward, 1964). Virulent

The Groundnut Crop: A scientific basis for improvement. Edited by J. Smartt. Published in 1994 by Chapman & Hall, London. ISBN 0 412 408201.

isolates are fluidal, while avirulent isolates are butyrous in nature (Porter *et al.*, 1984; Cook *et al.*, 1989). This is the only important bacterial pathogen of groundnut. When this disease is severe, infection causes such rapid wilting of stems and foliage that the affected leaves remain green. When damage is less, due to lower disease pressure, the decline of the affected plants is slower, resulting in general chlorosis. Entire plants or individual branches may be affected and the plants often die, although recovery from the wilted state is possible. The root system of infected plants may show external and internal discoloration. The internal discoloration usually extends into the lower stems. This symptom coupled with visible oozing or streaming of bacteria from stems placed in water are diagnostic characteristics.

Yield losses are closely related to plant death; quality is rarely affected as the diseased plants generally have zero yield. In southern China, up to 10% of the planted area can experience yield losses of 30%, with extremes of 80% loss (Sun *et al.*, 1981). The groundnut crop grown on land not used for paddy rice production is most seriously affected. In Indonesia, up to 25% of the yield of susceptible cultivars can be lost, particularly in South Sumatra, West Java and South Sulawesi (Machmud, 1986). In Uganda, yield losses of 24% have been recorded (Opio and Busolo-Bulafu, 1990). This disease has also been recorded causing some crop loss in Mauritius (Felix, 1971), Zimbabwe, Malaysia (Hayward, 1986), USA (Jenkins *et al.*, 1966) and recently in the Philippines (Natural *et al.*, 1988), Fiji (Iqbal and Kumar, 1986), Sri Lanka (Jayasena *et al.*, 1988), Papua New Guinea (Tomlinson and Mogistein, 1989) and India (Devi and Menon, 1979).

P. solanacearum has a wide host range of more than 30 families including Solanaceae, Musaceae, Asteraceae and Fabaceae (Kelman, 1953). Both weed and crop species are recorded as hosts, although some may not always express symptoms. There are considerable differences in the host range of the three recognized races of the pathogen: race 1 has a wide host range, race 2 is restricted to *Musa* and a few related perennial hosts, and race 3 is restricted to potato and a small number of other alternative hosts (Buddenhagen and Kelman, 1964). A fourth race attacking mulberry has been reported (He *et al.*, 1983). In addition, four biovars have been delineated by biochemical tests (Hayward, 1990). Groundnuts are normally attacked by biovar 3, occasionally by biovar 4, and by biovar 1 only in the USA.

The pathogen can be introduced into clean fields through the use of infected planting material of alternative crops, particularly vegetatively propagated crops such as cassava. Contaminated soil and water could introduce the organism, but this method of dissemination is not considered to be important (Persley, 1986). Seed transmission was implied by Palm (1922) but has not been reliably demonstrated, although Machmud and Middleton (1990) produced sufficient evidence to urge caution and further research in this important area.

Pathogen survival is usually enhanced in wet, heavy clay soils, especially where the crop is grown continuously (Sun *et al.*, 1981) but may also be high in sandy soil (He, 1990). Young plants are more rapidly diseased than older ones (Miller and Harvey, 1932) and high soil temperatures favour disease development (He, 1990). Low soil temperature favours survival in soil (Akiew, 1986). High levels of soil organic matter may lead to a decline in population of some races. Persistence between susceptible crops is enhanced by the presence of weed hosts, which are common under the environmental conditions and farming systems in those areas of the world where bacterial wilt is an important groundnut disease (Yeh, 1990). It is considered that some form of root injury may increase infection. When the infected plants suffer water stress, the extent of infection rapidly becomes apparent.

Crop rotations provide some measure of control, especially when the rotations include a paddy rice phase (Kelman and Cook, 1977). As the pathogen has a wide host range, the effect of rotations on disease incidence is likely to be due to the management conditions imposed when different crops are grown, rather than the presence of hosts more or less favourable to pathogen persistence. Obviously, continuous cropping to groundnuts will increase the disease problem.

It was with groundnuts that host resistance against *P. solanacearum* was first successful (Porter *et al.*, 1984) The resistance in Indonesian cultivar Schwarz 21 has continued to be useful for almost 60 years, with no apparent change in virulence of the pathogen. Cultivars developed in China have greater resistance than Schwarz 21 (Yeh Wei-Lin, personal communication) and Chinese breeding programmes aimed at bacterial wilt resistance are better developed than any other (Boshou *et al.*, 1990; Tan and Boshou, 1990). Host plant resistance has also been used as a control strategy in Uganda (Simbwa-Bunnya, 1972). However, resistance can be eroded in a breeding programme unless efforts to test progeny against field populations of the pathogen are made regularly. Most groundnut cultivars developed in production areas not affected by bacterial wilt are susceptible to the disease.

Biological control using rapidly growing bacteria has reduced the incidence of bacterial wilt in other crops, e.g. potatoes, but control in groundnuts using these methods has been attempted only recently (He, 1990).

10.1.2 Bacterial leaf spot

It was stated in the previous section that bacterial wilt is the only important bacterial disease of groundnut. While that statement is true, a symptom commonly seen on groundnut leaves has recently been attributed to a bacterial pathogen. Many workers have seen a characteristic leaf spot, usually when the crop is growing well, often when the crop is young.

The symptoms have been described as 'eye spot', due to the presence of a small (up to 5 mm) tan spot on the upper surface of the leaflet, partly surrounded by one or two semi-circular, lens-shaped tan lines. These may eventually surround the central spot. Symptoms on the lower surface are less distinct than for either of the two fungal leaf spots. In the past, this symptom has been attributed to the effect of atmospheric pollutants, but Joshi (pers. comm.) has confirmed that this disease is caused by *Xanthomonas* sp.

The disease does not cause serious losses in the groundnut crop and it is included in this chapter solely to assist in removing past confusion about its cause.

10.2 DISEASES CAUSED BY VIRUSES AND MYCOPLASMAS

10.2.1 Peanut mottle virus (PMV)

First reported in USA (Kuhn, 1965), this potyvirus is commonly found in groundnut production areas of eastern USA, South America, East Africa, Asia and Australia. It is sometimes referred to under the synonyms groundnut mottle virus, peanut mild mosaic virus or peanut severe mosaic virus (Brunt *et al.*, 1990). These names are evidence of the diversity of symptoms produced by PMV in groundnuts. Most of these symptoms are systemic mottling and mosaic of the foliage, with occasional necrosis. In other hosts, however, the virus can cause more serious symptoms and can be lethal (Behncken, 1970; Bock and Kuhn, 1975). Occasionally, the symptoms can be so inconsequential (for example, wavy margins of leaf-lets) that even experienced groundnut virologists have difficulty in detecting them (Kuhn and Demski, 1984).

The natural host range of PMV includes a number of other legume crops such as peas (*Pisum sativum*), beans (*Phaseolus vulgaris*), lupins (*Lupinus* spp.) and soybeans (*Glycine max*). These alternative crop hosts have implications for control of the virus in groundnuts and for the effects of groundnuts on other legumes in diverse cropping mixtures. *Cassia* spp. and *Vigna* spp. are often infected, aiding in persistence of the virus within a locality (Brunt *et al.*, 1990).

The virus is naturally transmitted by aphids in a non-persistent manner, although some vectors can remain infective for several hours (Paguio and Kuhn, 1976). Seed transmission is also important (Kuhn, 1965) and is often the means of introduction of the virus into a new production area, or a new crop cycle in situations where groundnut production is seasonal. Seed transmission rates vary from 0.02% to 2% in *Vigna unguiculata* (Demski *et al.*, 1983) and from 0% to 8.5% in groundnut (Reddy, 1991). PMV is also seed transmitted in mung bean and cowpea but not in soybean and pea.

The effects of PMV on yield have been examined and reported on several occasions. Abdelsalam *et al.* (1987) reported that the major effect of PMV infection was reduced seed size, not reduced seed number. Crop losses are difficult to estimate, as yield responses depend upon a dynamic interplay between the age of the plants when they become infected and the ratio of infected plants to healthy ones. On an individual plant basis, yield reduction of a plant infected within 5 weeks of sowing can be severe (*c.* 50%) but the effect declines with delayed infection (Paguio and Kuhn, 1974). In general, it is estimated that crop losses due to PMV infection of commercial crops are under 5%.

Control of seedborne viruses non-persistently transmitted by aphids is difficult. Insecticidal control of the vector can be counterproductive, as the presence of insecticide residues on the plants increases the feeding activity of the vectors and hence the number of plants infected. Disease-free seed is difficult to produce in commercial quantities when alternative sources of the virus exist. Host-plant resistance remains the only economically and environmentally sound approach to control.

Bijaisoradat *et al.* (1988) reported that none of 70 cultivars tested were resistant to the four isolates of PMV used. However, low seed transmission has been located among a few germplasm entries (Reddy, 1991) and could be used to produce varieties which would no longer carry a reservoir of virus into each field at planting. This avenue appears to be the only possible option of control available. However, the apparent low yield losses and the possibility that this form of resistance could be overcome by aphid transmission from an external source has led to PMV resistance being allotted a low priority in most breeding programmes.

For detailed information on the virus and its properties, see Brunt *et al.* (1990).

10.2.2 Peanut stripe virus (PStV)

During the late 1980s, this potyvirus assumed the status of a major threat to groundnut production in Asia and a potential threat to all production areas. This is due to increased awareness of its distribution, the potentially high crop losses which it may cause and its relatively high level of seed transmission. Before the identity of the pathogen was clarified, it became known under various synonyms including peanut mild mottle virus, peanut chlorotic ring virus and sesame yellow mosaic virus. However, an international ad hoc committee published justification for adopting 'peanut stripe' as the correct nomenclature (Demski *et al.*, 1984a; Demski *et al.*, 1988).

The symptoms produced on groundnuts infected by PStV vary considerably, depending on the strain of the virus (Demski *et al.*, 1988). These include mild mottle (produced by isolates from Thailand, China, Indonesia, Myanmar, India, Philippines and USA), blotch (Thailand, Philippines, USA and Japan), stripe (Thailand and Philippines), blotch-stripe (Thailand and Indonesia), chlorotic ring mottle (Thailand), chlorotic line pattern (USA) and necrotic (Thailand). This list does not suggest that a particular strain occurs only in countries shown above for that symptom. Limited investigations into the effect of symptom variants have been undertaken (Rechcig *et al.*, 1989).

The virus is widely distributed throughout Asia, from China to Indonesia and Myanmar, and has been accidentally introduced to the USA and India in recent years (Demski *et al.*, 1984a, 1984b; Rao *et al.*, 1988a, 1988b, 1989). It has not been reported from Africa or Australasia. The situation in Papua New Guinea is unclear.

PStV is transmitted by seed, and consequently is of great concern in those countries where it is currently not known. Reported transmission rates vary from 0.3% (Saleh *et al.*, 1989) and 8% (Xu *et al.*, 1989) to 29% (Rao *et al.*, 1988b), depending on botanical type and cultivar, and age of the plants at the time of infection. The virus is efficiently transmitted by several species of aphids in a non-persistent manner.

As well as groundnuts, lupins (*Lupinus albus*), soybeans, *Dolichos lablab*, *Vigna unguiculata* and sesame are natural hosts of PStV and there may be more as yet undetected (Wongkaew and Dollet, 1990). Differentiation of PStV from PMV can be achieved through mechanical inoculation of *Chenopodium amaranticolor* (PStV gives local lesions, PMV does not infect), groundnut and *Phaseolus vulgaris* cv. Top Crop (PStV does not infect, PMV gives local lesions). PStV infects *P. vulgaris* cv. Bataaf and cv. Kintoki (Brunt *et al.*, 1990; Demski *et al.*, 1984a).

PStV is capable of causing serious losses. Although one set of experiments (Lynch *et al.*, 1988) showed no effect of the virus on growth, yield and grade of Florunner, subsequent results showed that decreased nutritive value and changed flavour characteristics may follow PStV infection (Ross *et al.*, 1989). Careful observation in South Sulawesi, Indonesia, during the 1986 dry season (Middleton and Saleh, 1988) suggested yield losses of 70% in individual plants, and almost all plants were infected. Xu (1988) reported greenhouse tests showing 23% loss and estimated annual production loss of 200 000 tonnes in northern China. Adalla and Natural (1988) reported up to 24% yield loss in the Philippines, depending on cultivar. Designed plant loss experiments (Saleh *et al.*, 1989) have shown that losses are dependent upon the age of plants when infected, with 50% loss when plants are infected by 2 weeks after sowing (i.e. seed transmission), 25% loss if plants are infected from 3 to 8 weeks of age, and 14% loss if plants are infected between 9 and 10 weeks of age. In the same report, crop losses of 10% to 40% were measured, again dependent upon age at the time of infection. Wakman *et al.* (1989) reported yield losses of 15% to 28%. Wongkaew (1989) reported variation in measured crop losses from year to year, also dependent on cultivars used. He showed, however, that the 'chlorotic ring' isolate could reduce pod yield up to 42% in all cultivars. There can be no doubt that PStV can cause serious losses of groundnut yield in Asian dry season production.

However, control of non-persistently aphid-transmitted viruses which have a wide host range is very difficult once they become endemic (Brunt, 1989). Quarantine restrictions are obviously important to limit the spread of PStV into those countries where it has not been found. Imported germplasm can be non-destructively tested for the presence of PStV, using serological techniques. Imported trade product from locations where PStV is endemic should not be grown, either deliberately or by accident, as the probabilities for vector transmission from seedlings are high.

Production of sufficient quantities of PStV-free seed is one possible way of minimizing the effects of this virus, as seed transmission appears to be the major source of infection – dispersal by aphids occurs over only limited distances (Xu, 1988). However, there are great difficulties in producing and distributing virus-free seed in countries where groundnut production is normally conducted by many farmers each producing on a small scale.

Plastic film mulch, as used for groundnut production in China, appears to reduce infections of isolated plots, possibly because of its insect repellent action (Xu, personal communication). However, when used widely, it may contribute towards a uniform high incidence of PStV in fields.

The most economic, reliable and environmentally acceptable method of control is by the use of cultivars which are either immune or have high levels of resistance to the virus and/or its vectors. The only immunity located to date, despite testing of over 11 000 germplasm accessions from ICRISAT's world collection, is in the wild species *Arachis duranensis* (Saleh, personal communication) and in *A. cardenasii* (Rao *et al.*, 1991) after mechanical, aphid and graft inoculation. Several laboratories are currently attempting to produce plants with high levels of resistance, using genetic techniques.

Another avenue of genetic control is that of non-seed transmission, i.e. to find cultivars in which the virus is not transmitted by seed. Genotypes showing very little seed transmission have been identified (Zeyong *et al.*, 1989) and are being used in breeding programmes to develop cultivars with good agronomic traits. The success of the non-seed transmission approach depends upon the extent of virus infection in other hosts in the region and the activity of insect vectors.

For detailed information on the virus and its properties, see Brunt *et al.* (1990).

10.2.3 Peanut stunt virus (PSV)

First reported in USA (Miller and Troutman, 1966), this virus is widely distributed in the south-eastern USA, east and south-east Asia, Europe and Africa (Reddy, 1991). This cucumovirus causes severe stunting and dwarfing of plant parts in the USA but does not cause severe stunting in China (Zeyong *et al.*, 1989). Leaves are malformed; infected plants are chlorotic and small; malformed pods with a split pericarp wall are produced. PSV occurs mainly in beans and forage legumes. Host range includes a number of legume crops and plant species in the Chenopodiaceae, Compositae, Cucurbitaceae and Solanaceae (Fischer and Lockhart, 1978).

The virus is transmitted in a non-persistent manner by aphids (Hebert, 1967). It is readily transmitted by sap inoculation and is also transmissible by dodder (Miller and Troutman, 1966). Seed transmission in groundnut is less than 0.1% (Troutman *et al.*, 1967).

Control of PSV is difficult. There have been no reports of groundnut genotypes with resistance to PSV (Tolin, 1984) but genotypes with less severe symptoms are available (Culp and Troutman, 1968).

10.2.4 Peanut clump virus (PCV)

The appearance of stunted plants, often with mottling, mosaic or chlorotic ring spot symptoms on foliage, in the same part of a field each time groundnuts are grown suggests the presence of peanut clump virus. This rod-shaped furovirus was first recorded from Burkina Faso, West Africa (Thouvenel *et al.*, 1974), and has since been recorded in Ivory Coast, Niger, Senegal, South Africa and India (Brunt *et al.*, 1990). The Indian strain of PCV is sometimes referred to as Indian peanut clump virus (IPCV).

The virus may be introduced into uninfested fields by sowing infected seeds (Thouvenel and Fauquet, 1981). The seed transmission rate in groundnuts may be 5–11% and the virus is also seed transmitted (10%) in *Setaria italica*, a symptomless host. After introduction, the virus is transmitted to further groundnut plants by the soil-inhabiting fungus, *Polymyxa graminis*. This accounts for the slow enlargement of affected areas and the regular recurrence of the symptoms with each groundnut crop. Soil pH affects transmission by *P. graminis*.

Other hosts of PCV include *Sorghum arundinaceum*, *S. bicolor*, *Pennisetum glaucum*, *Eleusine coracana* (symptomless natural hosts); *Chenopodium amaranticolor* (chlorotic local lesions, ringspots, and line patterns extending along veins); wheat (systemic mosaic); *Phaseolus vulgaris*, *Vigna unguiculata*, *V. mungo* (systemic ringspots and/or mosaic), and *Nicotiana benthamiana* (systemic ringspots and leaf distortion).

The Indian isolates collected from various parts of that country produce differing symptoms on experimental hosts, although they are indistinguishable on the basis of particle size and produce identical symptoms on groundnut, mung bean (*Vigna radiata*) and *Vicia faba*. The hosts *Canavalia ensiformis* and *Nicotiana clevelandii* × *glutinosa* hybrid were the most useful for separating these isolates on the basis of symptom expression (Nolt *et al.*, 1988). Serological differences between IPCV isolates were demonstrated.

The effects of PCV on yield depend on the age of the plant when infection occurs. Early infected plants produce a few small pods, and plants infected later can suffer yield losses of 60% (Nolt and Reddy, 1984).

Control of this virus relies largely upon planting seeds from areas not infested with PCV, although selection of planting date can allow crops to escape infection (Nolt and Reddy, 1984). Some cultivar resistance has been observed but screening for resistance should be done at a number of different locations to assess reaction to more than one isolate (Nolt et al., 1988).

Development of an improved PCV detection system is in progress (D.V.R. Reddy, personal communication). For detailed information on the virus and its properties, see Brunt et al. (1990).

10.2.5 Groundnut rosette virus (GRV)

GRV is one of the most important viruses on groundnut in Africa, where it probably originated and to which it is apparently restricted. The virus infection causes chlorotic rosette, mosaic rosette and green rosette symptoms (Reddy, 1991). Chlorotic rosette, characterized by severe stunting of plants with isolated flecks or dark green coloured leaves, is more prevalent in East and Central Africa. Mosaic rosette is recorded in East and Central Africa: younger leaflets show conspicuous mosaic symptoms and stunting is rather less pronounced than for chlorotic rosette. Green rosette, characterized by slight mottling of young leaflets and presence of yellow leaves with green veins, occurs only in West Africa and Uganda. Plants infected with chlorotic rosette bear smaller, curled and distorted leaflets; the stunting of plants depends on time of the infection.

Rosette is caused by a complex of two viruses and a satellite RNA (Hull and Adams, 1968; Dubern, 1980; Murant et al., 1988). GRV is mechanically transmissible but requires the presence of groundnut rosette assistor virus for aphid transmission (Hull and Adams, 1968). It is transmitted in a persistent manner and over 92% transmission occurs within 10 minutes of the beginning of the inoculation access period (Dubern, 1980).

High-yielding rosette resistant groundnut cultivars RMP 12, RMP 91 and KH-149A are available. Resistance is controlled by two independent recessive genes and resistant genotypes are reported in Malawi, Nigeria and West Africa (Berchoux, 1960; Gibbons, 1977; Harkness, 1977; Gillier, 1978). All available rosette resistant cultivars are late-maturing but early-maturing ones are being developed. The disease is partially controlled by destruction of volunteer groundnut plants, early planting, the use of high plant density and the application of insecticides to control aphids.

For detailed information on GRV, see Reddy (1991).

10.2.6 Spotted wilt (tomato spotted wilt virus, TSWV)

The tomato spotted wilt virus has an extremely wide host range, including many crops and weed species. It is occasionally known as pineapple yellow spot virus. On groundnut, it can cause a wide range of symptoms (one strain produces 'peanut yellow spot') and consequently is a virus that requires careful diagnosis (Reddy *et al.*, 1990).

For full coverage of possible symptoms, including colour plates, refer to Reddy *et al.* (1990). The range of symptoms extends from general chlorosis, through chlorotic mottles, lines or spots, to necrotic rings, spots or streaks, and can lead to necrosis of the terminal apices or buds – a symptom which has inspired adoption of the secondary name 'bud necrosis' for the disease caused by TSWV. However, this symptom does not occur in all situations and it is most common in India and mid-western USA. In addition to the foliage symptoms, plants may be stunted and axillary shoots may proliferate. Seed produced on plants infected early in life can be shrivelled, and the testa is often discoloured red or purple.

TSWV has spread worldwide (Brunt *et al.*, 1990). Spotted wilt disease of groundnuts was first reported from Brazil in 1941 (Costa, 1950) and subsequently from South Africa (Dyer, 1949), Australia (Helms *et al.*, 1961), India, Nepal, Nigeria, People's Republic of China, Thailand and USA (Reddy, 1984; Reddy *et al.*, 1990).

This virus is not transmitted by seed, by plant-to-plant contact, or by pollen. It can be mechanically inoculated into groundnuts but the main method of transmission is by thrips. Transmission is persistent, i.e. an insect, once it has fed upon an infected plant, retains the capacity to transmit the virus throughout the remainder of its life. No helper virus is required for successful transmission (Brunt *et al.*, 1990).

Crops and weeds that are hosts of both the virus and its vectors and are often found in association with groundnuts include *Amaranthus* spp., *Datura stramonium*, *Glycine max*, *Helianthus annuus*, *Lactuca sativa*, *Lycopersicon* spp., *Nicotiana tabacum*, *Phaseolus vulgaris*, *Solanum* spp., *Vicia faba* and *Vigna* spp. (Brunt *et al.*, 1990). *Lycopersicon esculentum* produces systemic symptoms (Reddy, 1984). Diagnostic hosts are cowpea (*Vigna unguiculata*) and petunia (*Petunia hybrida*). On cowpea cvs. C-152 or California Black Eye, TSWV produces concentric chlorotic and necrotic lesions on leaves 4–5 days after inoculation, while necrotic lesions are produced on petunia 3 or 4 days after inoculation (Reddy *et al.*, 1990). Precise confirmation of the presence of TSWV can be made with ELISA procedures.

Serious yield losses from TSWV in groundnut have been reported from India since the mid-1960s, from Australia since mid 1970s and from the USA since the mid 1980s (Black, 1988). Of these three countries, the bud necrosis phase is rare in Australia, and appears to be associated with high temperature (Reddy *et al.*, 1990).

Yield losses due to TSWV depend upon the incidence of infected plants, their age when infected and environmental conditions. In India and the USA, plants infected when under one month of age frequently die. TSWV is currently considered to be the most important virus in Indian groundnuts (Amin, 1985).

Reducing losses due to TSWV in groundnuts depends on integration of vector management, the virus and cropping practices. Control of vectors with insecticides can be difficult as they habitually exist in parts of the plant where insecticidal contact is difficult, and the species have high potential for mass migration. However, use of systemic insecticides may be successful in reducing TSWV effects, although unwanted side effects (such as other insect problems) may follow. The vectors are parasitized at times and use of insecticides could interfere with this process. Planting groundnuts at a time to avoid migratory thrips can be most helpful, if practicable under local conditions (Reddy et al., 1990).

Management of the virus inoculum is most readily achieved by elimination of weed hosts from the planting and its surroundings (Purss, 1962) but intercropping with cereals can also be beneficial. Planting at a density to give rapid canopy closure reduces the proportion of affected plants (Reddy et al., 1990).

Host-plant resistance to TSWV has not been found, despite exhaustive testing. However, 'field resistance' has been reported (Amin, 1985) and is thought to be due to non-preference for particular cultivars by the vectors. This aspect of control is being strenuously pursued.

For greater detail on the virus, its purification and antiserum production, and on the vectors of TSWV, see Reddy et al. (1990).

10.2.7 Other viruses

Groundnut crinkle, groundnut eye spot, peanut green mosaic, passion fruit woodiness, peanut veinal chlorosis, peanut yellow spot and big bud viruses have been reported on groundnut. However, these do not seem to be of major importance at present. For details, see Reddy (1991).

10.3 DISEASES OF FOLIAGE, CAUSED BY FUNGI

10.3.1 Early leaf spot

The presence of clearly defined brown lesions on the leaflets of groundnuts suggests that one or both of the two common leaf spot diseases are attacking the crop. These are among the most serious of groundnut diseases when considered on a worldwide scale (Smith, 1984; McDonald et al., 1985). It is common for both leaf spot diseases to occur simultaneously on the same crop but this does not always occur.

Early leaf spot is caused by the fungus Cercospora arachidicola Hori., which has as its perfect state Mycosphaerella arachidis Deighton. The symptoms of early leaf spot include circular or irregularly shaped lesions up to 10 mm in diameter, although they may coalesce to form larger areas of necrotic tissue, particularly when lesions include a major leaflet vein. Necrosis extends to the lower surface of the leaflet. The lesions are medium brown in colour on the upper surface and light brown on the lower leaf surface. A chlorotic halo is often present around lesions but may be absent under some environmental conditions or on some cultivars. Fungal sporulation is sparse and difficult to see with the unaided eye; most occurs randomly distributed on the upper surface (Smith, 1984; McDonald *et al.*, 1985). Early leaf spot is distinguished from late leaf spot by the lighter colour of the lesion's lower surface and by the random, sparse distribution of conidiophores on the upper lesion surface. Microscopic examination of conidia clearly separates the two diseases, when *C. arachidicola* will be found to produce subhyaline olivaceous conidia measuring 35–110 µm × 3–6 µm; they are much more slender than the conidia of the late leaf spot pathogen, which are 20–70 µm × 4–9 µm and have a finely roughened wall (Mulder and Holliday, 1974). The pathogen may also attack petioles, stems and pegs (carpophores). In these instances, lesion shape is determined by that of the infected structure.

Early leaf spot is usually found wherever groundnuts are grown (Commonwealth Mycological Institute, 1966). It is common for both leaf spot diseases to occur on the same crop with early leaf spot usually being present before late leaf spot. Many factors determine whether early leaf spot or late leaf spot predominates in a particular crop, but there appears to be a tendency for late leaf spot to dominate over early leaf spot when there has been a prolonged series of heavily diseased crops (McDonald *et al.*, 1985).

The damage caused depends upon cropping practices, rainfall and other environmental variables and may vary from crop to crop. Loss of production (or its equivalent in cost of control) from the combined effects of the two leaf spot diseases can range from 10% to 50% (Smith, 1984; McDonald *et al.*, 1985).

C. arachidicola survives in crop debris from previously infected crops. Seed transmission is unimportant, at worst being mere surface contamination (McDonald *et al.*, 1985). Conidia of *C. arachidicola* are dispersed aerially from debris and infected crops, with rain increasing spore dispersal (Smith and Crosby, 1973). No other hosts of *C. arachidicola* are known (Smith, 1984).

The perfect state of *C. arachidicola* is not uncommon but plays no vital role in the development of disease epidemics. Infection is favoured by temperatures of 25–30 °C and high relative humidity (Smith, 1984; McDonald *et al.*, 1985). Early leaf spot develops under slightly lower temperatures than does late leaf spot. Earliest infections are usually on older leaves close to the soil surface (inoculum source) and, given continued environmental conditions suitable for infection, the epidemic will continue until most of the foliage and stems are infected and defoliation occurs. As a result, pod yield and quality can suffer (Smith, 1984; McDonald et al., 1985).

The literature on control of the leaf spot diseases of groundnuts is vast, and both diseases are controlled simultaneously by most control measures. Biological control through the use of mycoparasites has been suggested but not seriously pursued (McDonald et al., 1985). Cultural practices which may make a significant impact on the damage due to leaf spot include crop rotation, destruction or removal of crop residues from previous groundnut crops and controlled planting schedules – all designed to minimize the exposure of the crop to inoculum at least until it has reached more advanced stages of growth (Smith, 1984; McDonald et al., 1985).

Fungicides have been developed for the control of leaf spot; they are widely used in developed countries and with increasing frequency in developing countries. Most are applied as an aqueous foliar spray, although dusting is still used in some situations. The earliest fungicides for this purpose have a broad spectrum of biological activity but only provide protection against infection. This means that it is necessary to apply them before the crop is exposed to inoculum, under weather conditions suitable for infection of the plant. Under most conditions suitable for groundnut production, the period between fungal invasion and visible symptoms of leaf spot is 8–10 days, and some modern fungicides have the capacity to control infections after invasion, until the first symptoms are appearing. Some of the earliest systemic fungicides effective for leaf spot control affected the growth of pathogen in a very simple way and, after widespread use, 'fungicide resistance' developed: the pathogen changed its physiology sufficiently to be able to grow in the presence of the fungicide (Smith, 1984).

Some wild *Arachis* spp. are either immune or highly resistant to early leaf spot (McDonald et al., 1985) and a long-term goal of many groundnut breeding programmes is to include resistance to early leaf spot as a trait in released lines. Progress is being made towards multiple resistances of a level high enough to eliminate or significantly reduce the need for fungicides in lines which are also of acceptable marketable type and which produce high yields, but to date the use of fungicides has not been superseded.

10.3.2 Late leaf spot

Late leaf spot, caused by the imperfect fungus *Phaeoisariopsis personata* (Berk. & Curt.) V. Arx. [*Cercosporidium personatum* (Berk. & Curt.) Deighton], is widely distributed throughout the world (Commonwealth Mycological Institute, 1967). Late leaf spot along with early leaf spot (*Cercospora arachidicola* Hori.) can cause pod yield losses of up to 50%, worldwide (McDonald and Fowler, 1977; Smith, 1984; McDonald et al., 1985). In India, leaf spots and rust (*Puccinia arachidis* Speg.) normally occur together and yield losses as high as 70% have been attributed to their combined attack (Subrahmanyam *et al.*, 1980).

Symptoms of the disease have been described by several workers (Woodroof, 1933; Jenkins, 1938; Jackson and Bell, 1969; Feakin, 1973; Garren and Jackson, 1973; Porter *et al.*, 1982, McDonald *et al.*, 1985). Recently Smith *et al.* (1992) reviewed this disease in greater detail. The symptoms of the disease are influenced by host genotype and environmental conditions. Late leaf spots are first recognizable as small necrotic flecks that enlarge and become light to dark brown measuring from 1 to 10 mm in diameter. Lesions formed by *P. personata* tend to be smaller, more nearly circular, and darker than those formed by the early leaf spot pathogen. On the abaxial surface the lesions are black and slightly rough in appearance. Late leaf spot fungus usually sporulates on the abaxial surface of the leaflet and conidial tufts of *P. personata* are macroscopically visible as raised circles. The yellow halo which is often present around early leaf spots is often less conspicuous in or absent from late leaf spot lesions. Although symptoms are useful diagnostic features, positive identification of the late leaf spot requires microscopic examination of the conidiophores and conidia. The late leaf spot pathogen also produces lesions on petioles, stems, stipules and pegs in the later stage of an epidemic. These are oval to elongate and have more distinct margins than the leaflet lesions.

The perfect state of the late leaf spot pathogen is *Mycosphaerella berkeleyii* W.A. Jenkins. However, the nomenclature of the anamorph of this fungus has gone through several changes in the literature. Until recently the combination *Cercosporidium personatum* (Berk. & Curt.) Deighton was commonly used. In 1983, Von Arx proposed the new combination *Phaeoisariopsis personata* (Berk. & Curt.) v. Arx. and described the pertinent morphological characters of the anamorph as follows: stroma dense, pseudoparenchymatous, up to 13 μm in diameter, conidiophores numerous, pale to olivaceous brown, smooth, 1–3 geniculate, 10–100 × 3.0–6.5 μm in size, conidial scars conspicuous, prominently 2–3 μm wide; conidia medium olivaceous, cylindrical, obclavate, usually straight or slightly curved, wall usually finely roughened, rounded at the apex, base shortly tapered with a conspicuous hilum, 1–9 septa not constricted, mostly 3–4 septate, 20–70 × 4–9 μm in size. Jenkins (1938) described the teleomorph of the fungus.

Although the teleomorph of the late leaf spot pathogen is known, the ascigerous and spermagonial states are not generally regarded as important sources of primary inoculum. The principal source of initial inoculum is probably conidia that are produced on groundnut crop residues in the soil. Following early rains, conidia are blown or rain-splashed on the leaves of young groundnut plants, where they initiate the disease cycle. Temperatures in the range of 25–30°C and high relative humidity favour infection and disease development (Lyle, 1964; Jensen and Boyle, 1965). Lesions develop within 10–14 days of initial infection. Duration of leaf wetness is another important factor in late leaf spot infection. In favourable conditions, disease progress continues throughout the season and may lead to total defoliation of the infected plants. Spatial and temporal spread of late leaf spot have been studied by several workers (Venkataraman and Kazi, 1979; Knudsen *et al.*, 1987; Alderman *et al.*, 1989) and used in developing a computer simulation model to produce schedules for fungicide spraying (Phipps and Powell, 1984; Smith, 1986; Knudsen *et al.*, 1988; Alderman *et al.*, 1989). The pathogen may also survive from season to season on volunteer groundnut plants and on 'ground-keepers' (Hemingway, 1954). No authentic host species outside the genus *Arachis* are reported. Long-distance dispersal of *P. personata* is not known. Heavy application of nitrogen and phosphorus fertilizers increases severity of late leaf spot (Shanta, 1960).

Late leaf spot can be partially controlled with crop management practices that reduce initial inoculum (Smith and Littrell, 1980; Smith, 1984). Additionally, fungicidal control of the disease has proved effective and economical. Large numbers of fungicides have been developed and extensively evaluated both in developed and developing countries (McDonald *et al.*, 1985; Subrahmanyam and Ravindranath, 1988; Jackson and Damicone, 1991). Possible effects of fungicides on non-target organisms should be considered. Exclusive use of chlorothalonil on a calendar schedule has been shown to increase sclerotinia blight (Porter, 1980a). Similarly, higher levels of southern stem rot caused by *Sclerotium rolfsii* Sacc. have occurred in fields of Florunner groundnuts sprayed with benomyl (Backman *et al.*, 1975, 1977).

Breeding resistant cultivars is one of the best means of reducing crop yield losses from late leaf spot. Resistance levels are not great enough to discontinue the use of foliar fungicides but several fungicide applications may be eliminated when moderate levels of resistance are integrated with chemical control (Shokes and Gorbet, 1991). The decision as to whether or not disease control should be recommended has to be made at the local level, as it takes several factors into consideration such as extent of loss, the cost of control measures, and economic and other returns expected. Several mycoparasites have been reported to parasitize the late leaf pathogen but no serious attempts have been made to use them in the integrated management of leaf spots (Subrahmanyam and Ravindranath, 1988).

10.3.3 Rust

Rust caused by *Puccinia arachidis* Speg. is one of the most destructive fungal foliar diseases of groundnuts in many groundnut-producing countries of the world (Bromfield, 1971; Hammons, 1977; Subrahmanyam and McDonald, 1983; Smith, 1984; McDonald and Subrahmanyam, 1992). Yield losses from rust are over 50% (Subrahmanyam *et al.*, 1980a; Ghuge *et al.*, 1981). The effect of rust on groundnut yields is likely to increase with improvement in the production situation with more favourable environmental conditions. However, damage increases with multiple pathosystems particularly where rust and late leaf spot are involved (Savary and Zadoks, 1992a, 1992b).

Rust disease can be readily recognized by the appearance of orange-coloured pustules or uredinia (uredia) on the abaxial surface of the groundnut leaves. Uredinia then rupture to expose masses of reddish-brown urediniospores (uredospores), which land on the leaf surface. If temperatures are in the 15–30 °C range and leaves are wet with dew or rain, the urediniospores germinate and produce appressoria and infection hyphae, which penetrate the leaf through stomata (Subrahmanyam and McDonald, 1983; Butler and Jadhav, 1991). The uredinia appear within 8–20 days but chlorotic flecks are usually visible macroscopically 2 or 3 days before the appearance of erumpent sori (pustules). Incubation period is greatly influenced by environmental factors and host genotype (Mallaiah, 1976; Sokhi and Jhooty, 1982; Subrahmanyam et al., 1983). Pustules appear first on the abaxial surface but in extended periods of high humidity are sometimes formed on the adaxial surface. In highly susceptible cultivars, the primary pustule is encircled by colonies of secondary pustules. The pustules, which develop on all aerial plant parts except flowers and pegs, are usually circular and range from 0.3 to 1.4 mm in diameter. In contrast to rapid defoliation associated with leaf spots, leaves infected with rust become necrotic and dried, and remain attached to the plants for several days (Subrahmanyam and McDonald, 1983).

In general, rust fungi are known for their complex life-cycles. Some species at times produce five different kinds of spores and require two unrelated hosts to complete the life-cycle. However, groundnut rust is known almost exclusively by its uredinial stage. Uredinia are predominantly located on the underside of the leaf, scattered or irregularly grouped; they are round, ellipsoid or oblong, and dark cinnamon brown when mature. The ruptured epidermis is conspicuous: urediniospores are broadly ellipsoid or obovoid, $16–22 \times 23–29$ μm in size, brown-walled, $1–2.2$ μm thick, and finely echinulate. They have usually two but occasionally three or four germ pores, nearly equatorial, often in flattened areas.

A few reports of the occurrence of teliospores, mainly from South America (Arthur, 1934; Jackson and Bell, 1969; Hennen et al., 1976; Cummins, 1978), suggest that telia chiefly occur on the underside of leaves; they are 0.2–0.3 mm in diameter, scattered, prominent, naked and chestnut or cinnamon brown, becoming grey or almost black. Ruptured epidermis is prominent; teliospores are oblong, obovate, ellipsoidal or ovate with a rounded to acute and thickened apex. They are constricted in the middle, gradually attenuate at both ends, smooth walled, light yellow or golden yellow to chestnut brown, predominantly two-celled, sometimes three or four cells, $38–42 \times 14–16$ μm with walls 0.7–0.8 μm thick at the sides and 2.5–4.0 μm thick at the top. The apical thick part is almost hyaline. The pedicel is thin-walled, usually collapsed laterally, hyaline, up to 35–65 μm long and normally detached at the spore base. It germinates at maturity without dormancy. *Puccinia arachidis* has not been observed to produce metabasidia and basidiospores.

Rust disease is known to perpetuate, spread and cause severe epidemics by means of urediniospores. There are a few records of the occurrence of the telial stage on cultivated groundnuts in South America (Spegazzini 1884; Hennen *et al.*, 1976) and on wild *Arachis* species (Guarch, 1941; Bromfield, 1971). It is not known if the fungus produces spermatia and aecia or if any alternate host is involved in the life-cycle. Subrahmanyam and McDonald (1982) have examined various common crop and weed plants as possible hosts for the groundnut rust pathogen, but no case of infection has been recorded.

Urediniospores are short-lived in infected crop debris (Mallaiah and Rao, 1979; Subrahmanyam and McDonald, 1982). It is therefore unlikely that the fungus is perpetuated from season to season in crop debris under hot climatic conditions when there is a break of over 4 weeks between crop seasons (Subrahmanyam *et al.*, 1980b). The pathogen may survive from season to season on volunteer groundnut plants. Long-distance dissemination of the disease may occur via airborne urediniospores and/or by the movement of infected crop debris (van Arsdel and Harrison, 1972). There is no authenticated report of rust being internally seedborne or spread by germplasm exchange.

Cultural practices, host-plant resistance and fungicide usage are integrated into effective management strategies to minimize initial levels of disease and slow down the progress of rust epidemics. Wherever possible, successive groundnut crops should not be planted where viable urediniospores are present. Volunteer groundnut plants and 'ground-keepers' should be eradicated. If cropping systems permit, times of sowing should be adjusted to avoid infection of the crop from outside and to avoid environmental conditions conducive to rust build up. Strict plant quarantine restrictions should be enforced to avoid the spread of rust conidia on pods or seeds to areas where the disease is not present.

A number of fungicides developed for the control of leaf spots have been evaluated for effectiveness against rust (Subrahmanyam and McDonald, 1983; Subrahmanyam *et al.*, 1984). However, the old established Bordeaux mixture has given useful control of both rust and leaf spots. Among the new generation systemic fungicides, chlorothalonil gives good control of both the diseases but calixin is effective only against rust, while benomyl is effective against leaf spots but not against rust.

The rust resistance available in the cultivated groundnut is of the 'slow rusting' type, i.e. resistant genotypes have increased incubation periods, decreased infection frequency, reduced pustule size and spore production and reduced spore viability (Subrahmanyam *et al.*, 1980b; Subrahmanyam and McDonald, 1983). When rust resistance is combined with resistance to leaf spots and judicious use of fungicides, these diseases can be controlled effectively and economically.

10.3.4 Web blotch

Web blotch was first recognized on dead groundnut leaves in Russia (Woronichin, 1924). Similar disease symptoms were observed on groundnut leaves in Brazil (Cruz *et al.*, 1962), Rhodesia (Rothwell, 1962), Argentina (Frezzi, 1969), and the United States (Pettit *et al.*, 1973). Since then it has been reported from Canada, China and Australia (Taber, 1984). The disease is also known as Phoma leaf spot, Ascochyta leaf spot, spatselviek, muddy spot and net blotch. It appears to be of increasing importance in the United States (Phipps, 1985), and in South Africa it is considered as one of the most important foliar diseases of groundnut. It has been observed in groundnut-producing countries of Africa since 1946 (Taber, 1984).

Though the disease has been largely recognized to be caused by the imperfect fungus *Phoma arachidicola* Marasas, Pauer, and Boerema, generic distinctions between *Phoma* and *Ascochyta* and among *Mycosphaerella*, *Didymosphaeria* and *Didymella* need to be clarified before assigning proper anamorphic and teleomorphic names (Taber *et al.*, 1981).

Web blotch is characterized by a webbing or net-like pattern of tan to bronze colour on the adaxial surface of the leaves. Another pertinent symptom is the production of larger and nearly circular tan to dark brown blotches on the adaxial leaf surface. Lesions coalesce to encompass the whole leaf area. Rarely, a leaf flecking is observed. Older lesions become dry and darker in colour than the younger ones. Lesions on the abaxial surfaces are less prominent and pale brown in colour. The disease is capable of causing severe defoliation, especially when associated with late leaf spot.

Morphological features of *P. arachidicola* have been described in detail by several researchers (Marasas *et al.*, 1974; Taber *et al.*, 1981). Pycnidia which form on the infected host tissue, are 80–120 μm in diameter. Pycnidial cell walls are translucent. The mycelium is 2–8.0 μm in diameter, olivaceous to brown and septate. It also forms dark chlamydospores (7–19 × 8 × 20 μm), both aerial and submerged. Pycnidiospores are formed in basipetal succession on short conidiogenous cells and are single-celled. Pseudothecia have also been observed. They are dark brown, globose, short-beaked to unbeaked, single, and 60–147 μm in diameter. Asci (36–60 × 10–17 μm) are bitunicate and cylindric to cylindric–clavate. Ascospores (12.5–16 × 5.7 μm) are divided into two sections. Ascospores are two-celled, smooth hyaline to pale yellow and become darker with age.

In 1975, Philley studied web blotch in detail and found that pycnidiospores, ascospores and chlamydospores are capable of initiating disease. These inoculum propagules infect leaflets, petioles, stipules and stems.

Germ tubes penetrate directly through the cuticle. Appressoria with small infection pegs are formed near the germinated conidia. Epidermal penetration is intercellular. The disease appears to be most severe during cool seasons with high relative humidity. Groundnut is the only natural host. Sweet clover and hairy vetch were found to be the most susceptible when six leguminous genera were artificially inoculated (Philley, 1975). There is evidence that airborne inoculum carried from South America during Hurricane David introduced the disease to North Carolina and southeastern Virginia (Phipps, 1985).

Web blotch can be effectively controlled by commonly recommended fungicides such as chlorothalonil and mancozeb plus benomyl. Crop rotations and planting of less susceptible cultivars can decrease the incidence of the disease (Phipps, 1985).

10.3.5 Pepper spot and leaf scorch

The pepper spot and leaf scorch disease of groundnuts caused by the perfect fungus *Leptosphaerulina crassiasca* (Sechet) Jackson & Bell, has been reported from Madagascar (Sechet, 1955), Taiwan (Yen *et al.*, 1956), India (Nayudu, 1963), Argentina (Frezzi, 1965) and USA (Luttrel and Boyle, 1960; Pettit *et al.*, 1968). Since then it has been reported from Mauritius (Anon., 1960), Malawi (Mercer, 1977), Nigeria (McDonald, 1978) and Nepal (Pande *et al.*, 1993). In comparison with early leaf spot, late leaf spot and rust, this disease is considered of minor economic importance throughout the groundnut-growing countries of the world (Porter *et al.*, 1982).

The disease is principally recognized by two types of symptoms. Pepper spots are dark brown or black and restricted to the adaxial leaflet surface. These lesions later enlarge and measure up to 1.0 mm in diameter. Lesions are irregular to circular in shape and sometimes depressed. The leaf scorch phase of the disease is commonly identified as wedge-shaped lesions with a yellow zone adjacent to tips or margins of leaflets.

The causal agent of this disease is observed only in the sexual state (Jackson and Bell, 1968). Pseudothecia (perithecia) are initially submerged, later erumpent, amphigenous, yellowish-brown, spherical, thin-walled, and 60–120 μm in diameter. The ostiole is short and papillate. Pseudothecia contain 8–20 asci. Asci (50–80 × 25–55 μm) are hyaline, ovoid to broadly clavate, and have short stripes. Paraphyses are absent. Each ascus contains eight ascospores, hyaline initially but becoming yellowish-brown with age. Ascospores are oblong to ellipsoidal and measure 23–40 × 11–17 μm, with transverse septa, and commonly constricted at the septa.

The pathogen survives in necrotic leaflets and produces abundant ascospores. Ascospores are ejected up to 7.5 μm in experimental conditions (Mallaiah and Rao, 1976). Peak dispersal periods of ascospores occur at the end of the dew period and at the onset of rainfall (Smith and Crosby, 1973). Yen *et al.* (1956) observed that cardinal temperatures for *in vitro* growth are 8, 28 and 35 °C.

The fungus is known to infect only members of the genus *Arachis*. The disease is apparently controlled by the fungicides commonly used to control early and late leaf spots (McGill and Samples, 1965). Host-plant resistance is not available and most of the groundnut lines evaluated were found to be susceptible (Porter *et al.*, 1971).

10.3.6 Anthracnose

Several species of *Colletotrichum* have been reported to cause anthracnose disease on groundnuts. However, *C. mangenoti* Chevaugeon, *C. arachidis* Sawada and *C. dematium* Pers. ex. Fr. Grove are the commonly associated pathogens of *Arachis* spp. The disease has been reported on groundnuts in the United States, Uganda, Taiwan, Senegal, India, Tanzania (Tanganyika) and Argentina (Porter *et al.*, 1982; Smith, 1984). At present the disease is considered of minor importance throughout groundnut-growing countries.

On plants infected with *Colletrotichum mangenoti* the leaflets, petioles and stems display grey to brownish-grey lesions which are generally marginal to elongate (Chevaugeon, 1952). Plants infected with *C. dematium* have small water-soaked yellow spots, 1–31 mm in diameter (Saksena *et al.*, 1967). Symptoms induced by *C. arachidis* are scattered lesions (3–5 mm), circular to irregular, with greyish-white centres surrounded by brown borders (Sawada, 1959).

Morphological details and relationships of these three species of *Colletotrichum* have been discussed in greater detail by several workers (Frezzi, 1965; Jackson and Bell, 1969; Porter *et al.*, 1982) who have also described the *Colletotrichum* species that infect wild groundnuts.

Very little is documented on disease management. Copper fungicides, Bordeaux mixture (Chohan, 1974) and mancozeb (Singh *et al.*, 1975) have been effective in controlling the disease.

10.3.7 Scab

Scab caused by *Sphaceloma arachidis* Bit. & Jenk. was first observed in diseased groundnut plant materials collected in 1937 in São Paulo, Brazil (Bitancourt and Jenkins, 1940). It has also been reported from Argentina (Ojeda, 1966).

Primary symptoms appear as small spots ($\leqslant 1$ mm diameter), round to irregular in shape, on the adaxial leaflet surfaces. Spots are tan with narrow, brown marginal lines. As the disease progresses, spots appear on both sides of the leaf and become light tan with raised margins and sunken centres on the upper leaf surface (Bitancourt and Jenkins, 1940; Giorda, 1984). On the petioles and stems, spots are larger and more irregular in appearance than on the leaves and become cankerous with age. In late stages of disease development, lesions coalesce and the infected plants grow in a sinuous fashion and become stunted. Some lesions acquire a cork-like appearance and cover nearly all the plant parts, including pegs.

The disease is endemic under both dry and humid conditions. However, the fungus produces fruiting structures under conditions of high humidity and these cover the lesions with an olivaceous velvety growth.

Host-plant resistance has been reported from Argentina (Ojeda, 1966) and Brazil (Soave *et al.*, 1973). The fungus persists in crop debris. Cultural practices such as rotation may be useful in reducing the initial inoculum and disease. Benomyl is found effective in controlling scab. Some evidence suggests the seedborne and seed-transmitted nature of the disease (Giorda, 1984).

10.3.8 Pestalotiopsis leaf spot

A leaf spot of groundnut caused by *Pestalotiopsis arachidis* Satya was first observed in India (Satya, 1964). Similarly, *P. adjusta* and *P. vericolor* have been reported to cause leaf spot diseases in Nigeria (McDonald, 1978). Lesions caused by *P. arachidis* are dark brown and circular, and are surrounded by a yellow halo. They are commonly restricted to either side of the midrib. The centres of diseased leaf tissues are marked by the presence of black, spherical acervuli. The fungus produces pink mycelium. Conidia are ellipsoid–fusoid, four-celled and $17 \times 6.9\,\mu\text{m}$ in size. At present, this is not considered an important disease of the groundnut.

10.3.9 Phoma blight

Phoma blight, caused by *Phoma microspora* Balasubramanian and Narayanasamy, was first observed in Tamil Nadu in India (Balasubramanian and Narayanasamy, 1980). Symptoms of the disease appear as scattered light-brown lesions on the lower leaves of a crop more than 50 days old. However, in artificial inoculations the fungus is capable of producing typical lesions even on apical leaves, petioles and stem and can cause dieback. Lesions on the stem are dark brown or greyish with irregular margins. Numerous pycnidia have been observed on dead leaf tissues. Currently the disease is restricted to Tamil Nadu, India, where it causes considerable damage to the groundnut crop. Dhanju and Chohan (1974) found *P. glomerata* associated with bud blight affected groundnut plants. Little is known about the biology of these pathogens or the epidemiology of the diseases caused by them.

10.3.10 Powdery mildew

Powdery mildew of groundnut, caused by *Oidium arachidis* Chorin, has been reported from Mauritius, Portugal, Tanganyika and Israel (Porter *et al.*, 1982). Chorin (1961) named the powdery mildew fungus. Symptoms of the disease are mainly restricted to the upper leaf surface. Oidia vary in size and measure $31-44 \times 13-15 \mu m$. Conidiophores bear one or two oidia in dry weather, but chains of three or four oidia are common in undisturbed humid conditions. Subspherical pyriform haustoria develop in epidermal cells (Chorin and Frank, 1966). In Israel, development of the disease on groundnut foliage is rapid at 25 °C.

10.3.11 Phyllosticta leaf spot

Leaf spots caused by *Phyllosticta* spp., though geographically widely distributed, are among the minor diseases of groundnut (Jackson and Bell, 1969). Disease has also been reported from India (Rao, 1963), Barbados (Norse, 1974) and Nigeria (McDonald, 1978).

Frezzi (1960) described the symptoms of the disease as circular to oval lesions of 1.5–5.0 mm in diameter with definite borders, halos absent, reddish-brown on the perimeters and becoming lighter or tawny in the centre. Shot-hole symptoms develop in some spots.

In India, Rao (1963) described *Phyllosticta arachidis-hypogaea* Vasant Rao as the cause of a leaf spot. *P. sajoecola* Massal was reported to cause a similar leaf spot in Africa (Chevaugeon, 1952). The morphological description of these fungi is given in greater detail in the monograph by Jackson and Bell (1969).

The fungus can remain viable in mycelial form for $\leqslant 1$ year on infected crop debris and is the primary source of initial infection (Patil, 1981). Optimum temperature and pH for spore germination and growth of *P. arachidis-hypogaea* are 25–30 °C and 5.5–6.5, respectively (Patil, 1981).

Authenticated disease management practices are not known. However, a few fungicides can inhibit the growth of the fungus *in vitro* (Patil, 1981).

10.3.12 Alternaria leaf spot

Leaf spot diseases of groundnut caused by both known and unknown species of *Alternaria* have been reviewed recently by Subrahmanyam and Ravindranath (1988), but currently most of them are regarded as minor diseases worldwide. However, leaf spots caused by *Alternaria arachidis* Kulk. and *A. alternata* (Fr.) Keissler are of growing importance on the post-rainy season irrigated groundnut crop in southern India (Subrahmanyam *et al.*, 1981). Generally orange-brown necrotic spots in the interveinal areas of leaves and extending into veins and veinlets are characteristic of infection caused by *A. alternata*. The lower surface of the lesions is sometimes covered by a light to heavy olive brown-green sporulation. Porter *et al.* (1982) indicated that *Alternaria* spp. are frequent colonizers of necrotic groundnut leaf tissues. Several fungicides are being evaluated in various research centres in India (Subrahmanyam and Ravindranath, 1988) and it appears that some modern fungicides have the capacity to control alternaria leaf spot.

10.3.13 Zonate leaf spot

A minor foliar disease caused by *Cristulariella moricola* (Hino) Redhead (Syn. *C. pyramidalis* Waterman and Marshall) was reported by Smith (1972). The perfect state of the fungus is *Sclerotium cinnamomi* Sawada. The disease is observed late in the growing season. Symptoms appear as necrotic spots on the abaxial leaf surface. Lesions range from 2 to 12 mm in diameter. Small lesions have a light brown centre surrounded by a brownish ring of necrotic tissue and characteristic zonations are present on both leaf surfaces. Pyramidal heads (conidia) form on both sides of the leaf. They develop only on necrotic tissues, sometimes singly but rarely in clusters of up to 45 per lesion. Sclerotia are produced on potato dextrose agar and V8-juice agar.

10.3.14 Groundnut leaf blight

Subrahmanyam (1979) reported *Myrothecium roridum* Tode ex Fr. as the causal agent of a leaf blight disease of groundnut in India. Leaf lesions are round to irregular, 5–10 mm in diameter, grey in colour and surrounded by a chlorotic halo. Diseased leaves have a blighted appearance after lesions coalesce. Black fruiting bodies, often arranged in concentric rings, form on the adaxial and abaxial leaf surface. Currently the disease is not economically important in the groundnut-producing countries of the world.

10.3.15 Melanosis

Melanosis of groundnut leaves caused by *Stemphylium botryosum* Wallr. has been reported in Argentina (Frezzi, 1960). The symptoms of the disease are dark brown, small (0.5–1.0 mm), irregularly circular and solitary lesions on the abaxial leaflet surface. Initially these lesions are submerged but later become raised and crust-like. Conidia of the fungus are $12–23 \times 15.5–23.0$ μm in size, borne terminally, solitary, muriform, black, and sometimes slightly rugose. Conidiophores are dark in colour and have a globose apex and may or may not be branched.

10.3.16 Choanephora leaf spot

Mukiibi (1975) reported *Choanephora* spp. causing leaf spots on groundnut in Uganda. The brown lesions originate at the leaflet margins and spread to cover almost the entire leaflet. Abundant sporulation occurs on the adaxial and abaxial leaflet surfaces. Faint concentric circles are present in the lesions.

10.4 DISEASES OF STEMS, ROOTS AND PODS, CAUSED BY FUNGI

10.4.1 Stem rot

Stem rot caused by the fungus *Sclerotium rolfsii* Sacc. is one of the most widely recognized soilborne diseases of groundnut (Backman, 1984). Other names for this disease include white mould, southern blight, sclerotium rot, sclerotium blight, sclerotium wilt, root rot and foot rot (Jackson and Bell, 1969). Yield losses vary between 25% and 80% (Rodriguez-Kabana *et al.*, 1975). The fungus attacks and kills the crown of the plant, destroying entire plants and the pods they are producing.

First symptoms of the disease are usually the yellowing and wilting of a branch or, if the main stem becomes infected, of the whole plant. The leaves turn brown and sometimes are shed prematurely. Sheaths of white mycelium can be seen at or near the soil line around affected plant parts. Mycelial growth is rapid under favourable environmental conditions and quickly spreads to other branches and plants. Spherical sclerotia (0.5–2.0 mm in diameter) are produced abundantly both on plant parts and on the soil surface. They are initially white but later turn dark brown. Lesions produced on branches and pegs are initially light brown, becoming dark brown as disease develops. Infected pods usually rot and may occur on plants without visible symptoms on the above-ground plant parts. *S. rolfsii* produces oxalic acid, a phytotoxin that gives a purple discoloration of seeds and is also responsible for chlorosis and necrosis of foliage in the early stages of disease development.

The fungus does not produce asexual spores. The basidial stage, *Aethalia rolfsii* (Cruzi) Tu & Kimber, is rarely seen. *S. rolfsii* grows well on a wide range of cultural media and is characterized by the presence of white mycelium and hard, round, brown sclerotia.

S. rolfisi has a broad host range of more than 200 plant species (Aycock, 1966). Sclerotia are produced on many infected crops and weeds. Even in the case of non-hosts, the debris from these plants may serve as an organic food base for attack of subsequent crops. Warm, moist conditions favour disease development. The fungus has a high demand for oxygen. Therefore, the overwintering sclerotia are only activated when they occur in the upper regions of the soil. Continuous cropping year after year may cause inoculum build-up. The fungus is seedborne, in the form of mycelium present both internally and externally, and may be carried through seed to areas where the disease is not present. However, soilborne inoculum is the most important source of initial infection. Factors that tend to increase or maintain soil moisture, such as a dense foliage canopy, increase the disease severity (Backman *et al.*, 1975).

Unfortunately, control of leaf spot diseases can directly increase the severity of stem rot. Maintaining a complete canopy allows a moist subcanopy that is favourable to stem rot development. However, the results of not controlling leaf spot diseases would be far worse (Backman *et al.*, 1975). Thus the application of chemicals to the soil is the most appropriate approach to control this disease. It has been shown that application of PCNB and carboxin could control the disease satisfactorily (Cooper, 1956; Harrison, 1966), though at some locations responses with PCNB have been erratic (Harrison, 1967; Indulkar and Grewal, 1970; Thompson, 1978). Rodriguez-Kabana *et al.* (1979) obtained significant control by soil fumigation using dibromochloropropane.

Several cultural practices are effective for stem rot control. Deep ploughing of soil after harvest to bury crop residues is commonly followed in most of the groundnut-growing states in USA. One-year crop rotation with maize (*Zea mays*) and grain sorghum (*Sorghum bicolor*) is effective in preventing severe infestation. Rotations of 2–4 years are necessary if a severe infestation has developed (Backman, 1984).

Porter *et al.* (1982) gave a detailed account of the biological control of *S. rolfsii* in groundnut. *Trichoderma harzianum* Rifai grown on granules containing molasses has been reported to be effective in controlling the disease under field conditions (Backman and Rodriguez-Kabana, 1975).

10.4.2 *Aspergillus* species and aflatoxin

This subject is also considered in Chapter 13. In the early 1960s, mycotoxins of *Aspergillus flavus* Link were found in groundnut meal. Feed prepared from this meal caused the death of 100 000 turkeys in Great Britain, and the disease was named Turkey X disease. Later a similar disease caused deaths in ducklings in Kenya and Uganda. An extensive piece of research indicated the presence of toxic Brazilian groundnuts used to feed the birds at these places (Blount, 1961).

Groundnut pods and seeds are vulnerable to several soil fungi but species of *Aspergillus*, *Fusarium* and *Penicillium* are the predominant colonizers (McDonald, 1969; Mehan and McDonald, 1983) and produce mycotoxins (Ciegler, 1978). Four mycotoxins – aflatoxin, citrinin, ochratoxin A and zearalenone – occur naturally in groundnuts and, of these, aflatoxins produced by *A. flavus* and *A. parasiticus* Speare are the most important and ubiquitous throughout the groundnut-growing areas of the world (FAO/WHO/UNEP, 1977). Aflatoxins are toxic and even a small amount of them in feed (10–20 ppb) can produce fatal liver cancer in young animals and humans (Wogan, 1965; Diener *et al.*, 1982).

A. *flavus* thrives in the soil as a saprophyte and invades groundnut seeds, seedlings and pods during pre- and post-maturity stages in the field. Yellow mould, a disease of ungerminated seeds and seedlings, and related afla-toxin contaminations of groundnut seed can develop during harvest and in storage (Dickens, 1977; Mehan and McDonald, 1983).

Seedling infection with A. *flavus* is characterized by necrotic lesions on the emerging plumule and cotyledons. Lesions become covered with masses of yellow-green spore heads of A. *flavus* group. Infection may spread to the emerging radicle. Infected seedlings lack a secondary root system, a condition known as 'aflaroot' (Chohan and Gupta, 1968; Aujla *et al.*, 1978). Infected seedlings sometimes die. The use of high-quality seed treated with protectants greatly minimizes the severity of this disease.

A. *flavus* also attacks groundnut pods and seeds as they mature in the soil before digging, especially when plants are drought-stressed. Yellow-green mouldy colonies of *Aspergillus* develop on pods and seeds which are overmature. Additional pods and seeds may become contaminated during the curing process. The fungi invade maturing pods, seeds and cotyledons. Hyphal colonies are apparent between the cotyledons. Aflatoxin contami-nation during harvest, after harvest and in storage has had a tremendous impact on the entire groundnut industry and on consumers. These aspects of aflatoxin contamination have been adequately reviewed by several workers (Mehan and McDonald, 1982; Mehan, 1988; Mehan *et al.*, 1991), who have also suggested possible control measures to minimize aflatoxin contamination.

The taxonomy and morphology of the genus *Aspergillus* were described by Raper and Fennell (1965). A. *flavus* and A. *parasiticus* produce hyphae that are colourless, septate and branched. Conidiophores are 300–500 μm long. A vesicle is borne at the end of each conidiophore. On the vesicle, rows of strigmata bear chains of yellow-green to blue-green conidia (7–9 × 3.5–5.0 μm). Strigmata of A. *flavus* are borne in single or double series on an elongate to subglobose vesicle (15–25 μm in diameter). The radiated heads split with maturity. The strigmata of A. *parasiticus* are closely packed in a single series on a flask- or pestle-shaped vesicle 20–35 μm in diameter, and the heads are loosely radiate. Cultures of A. *flavus* produce only aflatoxins B_1 and B_2, whereas cultures of A. *parasiticus* produce all four aflatoxins (Diener *et al.*, 1982).

The disease cycle and epidemiology of yellow mould and aflatoxin contamination have been discussed in greater detail by Jackson and Bell (1969) and Pettit (1984). A. *flavus* and A. *parasiticus* are saprophytic inhabitants of many soils and survive in crop residues. Their ability to cause plant disease is related to their competition with other soil micro-organisms, the availability of susceptible plant tissues and occurrence of favourable environmental conditions for infection and development of the disease. In general, the A. *flavus* group of fungi is more competitive in a declining soil moisture regime (Pettit *et al.*, 1971) and when the atmospheric relative humidity remains ⩾90%. These fungi can grow over a temperature range of 17–42 °C but the optimum temperature for aflatoxin production is 25–35 °C.

An aflatoxin detection system has been developed in the United States to prevent groundnut seed contaminated by aflatoxin from entering the food chain of humans and animals (Dickens and Welty, 1969; Aibara and Nobumitsu, 1977; Dickens, 1977). The FDA has set levels of aflatoxin for the edible products trade.

Control of yellow mould and management of aflatoxin contamination of groundnuts is achieved by preventing the *A. flavus* group from entering groundnut tissues, by either destroying or diverting the contaminated seeds and adopting improved crop husbandry (Dickens, 1977; Mehan, 1988). Recommendations include:

- avoiding mechanical damage to the crop during cultivation, harvesting and subsequent processing;
- harvesting at full maturity;
- drying the produce in the field as rapidly as possible;
- preventing re-wetting during or after drying;
- removal of damaged or mouldy pods;
- drying to a safe moisture level (8%) before placing in storage;
- storage at low temperature and humidity.

Such measures have been applied with considerable success by large-scale groundnut farmers in the USA and other developed countries but are neglected in developing countries (Mehan and McDonald, 1983).

Genetic resistance to seed invasion has been identified by several workers (Mixon and Rogers, 1973; Bartz *et al.*, 1978; Mehan *et al.*, 1987) but it depends upon the presence of an undamaged seed testa: any damage to the testa greatly reduces the levels of resistance (Mehan, 1988).

10.4.3 Aspergillus crown rot

Aspergillus crown rot, a fungal disease caused by *Aspergillus niger* van Tieghem, is found on groundnuts in all the major growing areas of the world (Porter *et al.*, 1982). The appearance of this disease on groundnut is generally sporadic and unpredictable.

Although *A. niger* attacks groundnuts at all stages, seedlings and young plants are particularly susceptible. Sudden loss of leaf turgidity and wilting of plants are characteristics of crown rot. The crown of an infected plant becomes swollen and the tissues become brittle and corky. Infected parts are profusely covered with black masses of mycelium and conidia. Additional symptoms of aspergillus crown rot are given by Garren and Jackson (1973).

The fungus *A. niger* is widely distributed in soils throughout the world. It colonizes and survives in organic matter. Growth and sporulation of the fungus are usually favoured by warm and moist conditions, and under such conditions it produces an abundance of large, black conidial heads that reach 700–800 μm in diameter. Soil type has not been correlated consistently with prevalence of crown rot (Joffe and Lisker, 1970; Abdalla, 1974) but the disease is more prevalent in soils low in organic matter (Morwood, 1953). *A. niger* is also capable of initiating disease through infected seeds (Agnihotri and Goyal, 1971; Vaziri and Vaughan, 1976).

Most of the commonly grown cultivars are susceptible to the crown rot fungus. However, bunch cultivars are usually less susceptible. Fungicidal seed protectants such as Thiram and Captan provide some control when used under conditions and practices that favour rapid germination and seedling emergence.

10.4.4 Blackhull

This disease is caused by the fungus *Thielaviopsis basicola* (Berk. and Br.) Ferraris. In the United States the disease was first observed in 1963 (Mason, 1964) but there had been earlier reports from other countries (Caccarone, 1949; Frezzi, 1960). More recently it has been reported from South Africa (Prinsloo, 1980), where extensive hull discoloration, severe pod and peg rots and up to 80% loss in pod yields were attributed to blackhull disease.

Blackhull is characterized by scattered black dots on pod surfaces. As individual lesions increase in size, they coalesce and form large blackened areas. Large masses of chlamydospores embedded in the sclerenchymatous hull tissues give rise to black discoloration of the hull.

Seed inside infected pods sometimes become discoloured and the causal agent can often be isolated from these tissues. Tabachnik *et al.* (1979) observed necrosis of the tap root and pegs.

Lucas (1965) described the morphological and physiological characteristics of the causal organism. The fungus produces both endoconidia and chlamydospores. Endoconidia are produced by specially modified conidiophores (4–6 × 9–14 μm) composed of bulbous bases having a restricted cylindric barrel.

T. basicola persists indefinitely in the soil as a saprophyte. It can overwinter as chlamydospores in the endocarp of the fruit (Mason, 1964). Resting chlamydospores germinate and produce mycelium which grows along the young pod surface and penetrates the endocarp. Later, chlamydospores are produced abundantly from the mycelium in a characteristic black granular mass which may fill the intercellular spaces of the mesocarp. Hsi (1965) found no constant relationship between seed from infected pods and subsequent disease incidence.

Crop rotation with grain sorghum controls blackhull disease (Hsi, 1978), and it can be suppressed if benomyl and thiophanate-methyl are applied into the soil (Hsi and Ortiz, 1980). However, no commercial groundnut cultivars are resistant to this disease.

10.4.5 Botrytis blight

Botrytis blight of groundnuts is caused by *Botrytis cinerea* Pers. ex. Fries. It was first reported in Japan (Suematu, 1924) and has been observed in most of the groundnut-producing countries worldwide (Porter, 1984c). Occasional outbreaks do substantially reduce yields. The disease is also called 'grey mould' and in South Africa it is known as 'botrytis shoot disease' (Dyer, 1951).

All mechanically wounded groundnut plant parts are invaded by this soilborne fungus. Tissues weakened by other fungi (Rothwell, 1962) or injured by frost are vulnerable to colonization by *B. cinerea*. Under favourable environmental conditions, the fungus colonizes plant tissues rapidly and moves from above-ground plant parts into underground parts. It produces abundant conidia and sclerotia on infected tissue. Conidia are hyaline, ellipsoid to ovoid, single-celled, measuring 9–12 × 6.5–10 μm. Conidiophores are usually unbranched septate and 11–23 μm thick. The ascomycetous stage of the fungus is *Botryotinia fuckeliana* (de Bary) Whetzel.

Conidia are frequently disseminated by air currents. Low temperatures (≤20 °C) accompanied by heavy dews or excessive rainfall favour infection and disease development. The fungus overwinters in the sclerotial form in the soil. Although apothecia have been reported, the primary source of inoculum appears to be mycelium originating from germinating sclerotia or omnipresent conidia.

Fungicides such as benomyl, chlorothalonil and dicloran offer some control when sprayed on groundnut foliage. Early-maturing groundnut cultivars may escape frost damage and thus reduce disease severity in some countries.

10.4.6 Charcoal rot

Charcoal rot, caused by a common soilborne fungus, *Macrophomina phaseolina* (Tassi) Goid., is widely distributed throughout the world's groundnut-producing countries. The fungus has a wide host range and in groundnuts it causes seed and seedling rots, wilt, root and stem rot, leaf spots, and rotting of developing pods and seeds. The term 'charcoal rot' is used to describe the damage done to roots and stems of seedlings and older plants. It is a disease of minor importance but can greatly reduce seedling stands (Hoffmaster *et al.*, 1943; Mathur *et al.*, 1967; Shanmugam and Govindaswamy, 1973a).

Groundnut seedlings are usually attacked by the fungus at the soil line,

and water-soaked lesions appear on the hypocotyl. Later, infected hypocotyls are girdled and the seedlings die. Similar symptoms are observed on
the older plants at the soil line, although all plant parts at all stages of
growth are vulnerable to attack. Stem and root lesions appear water-
soaked at first, but infected tissues later become a dull light brown. The
infection extends downwards into the taproot and up into the stem and
branches. Where lesions girdle the stem, the plant wilts and the fungus
rapidly colonizes the branches. Jackson and Bell (1969) published an
excellent summary of symptoms caused by *M. phaseolina*.

Pycnidia of *M. phaseolina* are membranous to sub-carbonaceous, first
immersed and then at least partially erumpent, globose or flattened globose
with inconspicuous truncate ostioles, 100–200 μm in diameter. Walls are
composed of several layers of dark thin-walled angular cells, 9 μm in
diameter, and are lined with a hyaline layer two or three cells thick, bearing
simple rod-shaped conidiophores, 10–15 μm long. Conidia (14–33 × 6–12
μm) are single-celled, hyaline and elliptic or oval. The sterile mycelial phase
of the fungus is commonly known as *Rhizoctonia bataticola* (Taub.) Butl.

Charcoal rot is both seedborne (Subrahmanyam and Rao, 1977; Mridha
and Fakir, 1978) and soilborne (Garrett, 1956). Mycelium in seeds and
mycelium and sclerotia in plant debris are the primary sources of inoculum.
The sclerotia can remain viable in dry soil for many years (Smith, 1966),
but soon lose viability in very wet soils (Shokes *et al.*, 1977). The disease
can readily be disseminated by infected groundnut seeds and pods. High
soil temperatures, low soil osmotic potential and reduced plant vigour
favour the development of charcoal rot (Odvody and Dunkle, 1979).

Crop rotation is generally ineffective in reducing the soil inoculum
because the fungus can grow saprophytically and has a wide host range.
Adequate fertilization and soil water to ensure good crop growth reduce
the chances of infection and charcoal rot development. Soil drenching with
PCNB has given some control (Shanmugam and Govindaswamy, 1973b).
No immune or highly resistant genotypes have been identified.

10.4.7 Diplodia collar rot

Diplodia collar rot, caused by *Diplodia gossypina* (Cke.) McGuire and
Cooper, occurs sporadically throughout groundnut-growing countries of
the world but rarely causes economically important losses. The disease is
also known as 'diplodia blight' and 'collar rot'.

Groundnut seedlings and older plants are attacked at or near soil level
and the fungus quickly invades the stem. The first obvious symptom of the
disease is wilting of a lateral branch or the entire plant. Wilt develops
rapidly and plants generally die within a few days. Lesions that develop on
the above-ground tissues are characterized by elongated necrotic areas
having light brown centres and dark brown margins. Infected roots become
slate grey to black and shred easily. Pycnidia on the infected host tissue

appear as small black dots. The fungus is capable of causing concealed damage to the seeds (Porter, 1984b).

Morphology and physiology of the causal organism are described by Jackson and Bell (1969). Although several species of Diplodia have been isolated from infected groundnut parts, *D. gossypina* is probably the most common and the causal agent of collar rot disease. Simple or compound pycnidia occur either singly or in groups over the necrotic tissues. Pycnidia range up to 400 μm in diameter. Conidia are borne on short conidiophores; mature conidia (17–34 × 10–18 μm) are elliptic with one septation. They are brown with longitudinal striations.

D. gossypina survives as mycelium and mature conidia in the soil and in crop residues for a long time. Upon germination, mature conidia may initiate primary infection. Groundnut plants predisposed by heat injury are more vulnerable but predisposition is not a prerequisite for infection. Once the infection site is established, the fungus moves quickly into the adjacent healthy tissue. Under field conditions hot dry weather favours disease development.

Crop rotations with non-host crops such as cotton or soybean can reduce disease incidence. Resistance to *D. gossypina* is not found in commercial cultivars but some breeding lines that possess high levels of tolerance to the disease are available in the USA (Porter and Hammons, 1975; Porter, 1984b). Predisposition of groundnuts to *D. gossypina* by heat injury can be minimized by controlling foliage diseases so that a dense foliage canopy shade is maintained throughout the growing season.

10.4.8 Pythium diseases

Several species of *Pythium* cause pod breaking (pod rot), pre-emergence and post-emergence damping-off, vascular wilt and root rot diseases of groundnuts throughout the world. All *Pythium* species are cosmopolitan in soils and attack a wide range of crop species.

Pod breakdown or pod rot (Garren, 1966) is widespread in groundnut-growing areas of the world and often causes economic losses that are difficult to define, because infection by *Pythium* spp. and other related fungi usually does not produce well-defined above-ground symptoms. *P. myriotylum* Drechs. is the major pod-rotting pathogen in certain groundnut-growing areas in the USA. However, several other species of *Pythium*, *Rhizoctonia solani* and *Fusarium solani* can cause pod rot either singly or in combination. The synergistic effects of *P. myriotylum* and *F. solani* in a pod breakdown complex are known in Israel (Frank, 1972). In Israel, *Pythium* spp. are thought to precede *Fusarium* spp. (Frank, 1968), while in the United States *Fusarium* spp. usually precede *Pythium* spp. (Garren, 1966). *P. myriotylum* attacks groundnuts at all stages of growth (Frezzi, 1956).

Pod breakdown symptoms caused by the above-mentioned fungi, either singly or in combination, are often indistinguishable. Both mature and immature pods may be attacked. Infected pods exhibit degrees of discoloration from apparent rusting to browning. In wet soil, pegs and pods infected by *P. myriotylum* become water-soaked. They later become dark brown to black and rot quickly (Jackson and Bell, 1969). *F. solani* and *R. solani* produce distinct lesions under dry soil conditions. Pod rot caused by *R solani* progresses much more slowly than that caused by *P. myriotylum* (Garren, 1970; Garcia and Mitchell, 1975a).

Occasionally groundnut plants without apparent pythium root rot symptoms wilt suddenly (Porter, 1970). Shortly after the appearance of wilt symptoms, leaflets become chlorotic or light green. This condition of the diseased plants is commonly referred to as 'vascular wilt'. The plants have some evidence of root deterioration; vascular tissues of the tap root are dark brown. Fibrous roots are also vulnerable to *P. myriotylum* and infected roots become dark brown to black (Jackson and Bell, 1969).

P. myriotylum was described by Drechsler (1930), and a study of the genus *Pythium* as plant pathogens was made by Hendrix and Campbell (1973). *Pythium* spp. are easy to isolate from infected tissues and they are characterized by the presence of coenocytic mycelium, from which develop asexual reproductive structures (sporangia) that differ in size and shape between species. The sporangia of *P. myriotylum* may be terminal or intercalary and germinate by germ tube or by producing zoospores. Zoospores and sporangia are short-lived. The primary survival structures of *P. myriotylum* in soil are oospores (sexual spores), which are 12–37 μm in diameter and do not fill oogonia. Oospores have a single reserve globule, walls up to 2 μm thick and pale golden content.

Pythium spp. are natural inhabitants of the soil and can survive indefinitely as saprophytes. *P. myriotylum* has a wide host range that includes cereal crops used in rotation with groundnuts (McCarter and Littrell, 1968). Mycelia of *P. myriotylum*, produced by zoospores or germination of oospores, form appressoria and penetrate epidermal cells of groundnut pods directly (Jones, 1975). Penetration occurred in 2 hours at 30–34 °C but there was no penetration below 22 °C. The optimum temperature for mycelial growth of *P. myriotylum* is 35 °C.

A significant positive correlation between increasing soil moisture and frequency of pythium pod rot has been demonstrated in Israel (Frank, 1974). Frequent irrigation of sandy soils increased the severity of pod rot caused by a combination of *P. myriotylum* and *F. solani*, whereas less frequent and heavier irrigations reduced its severity. The soil fauna, including mites and springtails (Shew and Beute, 1979), nematodes (Garcia and Mitchell, 1975a) and southern corn root-worm (Porter and Smith, 1974), enhances pod breakdown disease caused by *P. myriotylum*.

Moderate levels of resistance to pod rot caused by *P. myriotylum* and high yield potentials have been reported (Porter *et al*, 1975).

Since more than one organism is involved in pod breakdown, wide-spectrum fungicides or combinations of fungicides are needed to control pod rot. Use of effective nematicides can be important, since some types of nematodes intensify pod rot in certain locations (Garcia and Mitchell, 1975b). Pod breakdown caused by *P. myriotylum* can be significantly suppressed by the application of high rates of gypsum (Garren, 1964; Hallock and Garren, 1968; Boswell and Thames, 1976). Control of pythium diseases of groundnuts through crop rotation and field management have been difficult.

10.4.9 Cylindrocladium black rot

Cylindrocladium root, peg and pod rot caused by the fungus *Cylindrocladium crotalariae* (Loos) Bell and Sobers (perfect stage, *Calonectria crotalariae* (Loos) Bell and Sobers) is reported to occur in the United States (Bell and Sobers, 1966), Japan (Misonou, 1973), Australia (Jackson and Bell, 1969) and India (Sharma *et al.*, 1978). The disease has the capacity to inflict heavy yield losses (Garren *et al.*, 1972; Lewis *et al.* 1977).

The disease was reviewed in 1980 (Beute, 1980). The first symptoms of cylindrocladium black rot (CBR) are wilting and chlorosis (yellowing) of infected leaves and stems. Usually the central stem dies. As the disease progresses, the lateral branches also die. Occasionally the whole plant becomes chlorotic and stunted. All underground plant parts may develop CBR symptoms. Hypocotyl, roots and pods become black and necrotic. A diagnostic sign of CBR is the occurrence of small, reddish-orange perithecia of the pathogen in dense clusters on stems, pegs and pods. These structures form only on tissues within a few millimetres above or below the soil surface during periods of wetness and high humidity. If these symptoms are lacking, a positive diagnosis of CBR requires a laboratory assay. When CBR is seen for the first time in a field, the infested area develops in one or more localized spots, which sometimes measure 10–30 m in diameter.

The fungus grows well on potato-dextrose agar, producing light yellow to white web-like aerial mycelium and a burnt orange to dark brown submerged growth. Morphology and taxonomy of both sexual and asexual stages have been described by Bell and Sobers (1966).

Microsclerotia of *C. crotalariae* are produced abundantly in roots and *Rhizobium* nodules. As the infected tissue decomposes, microsclerotia are released into soil and disseminated by tillage equipment. They overwinter in soil, and soil temperature during winter months is a primary determinant of longevity. Their population declines sharply in frozen soil (Phipps and Beute, 1977) but they can survive for several years in the soil without a host crop. Development and ejection of ascospores from perithecia of *C. crotalariae* was optimum at 25 °C (Rowe and Beute, 1975). Ascospores are extremely susceptible to desiccation and thus play only a minor role in disease spread. Wind and water can contribute to long-range spread of the pathogen.

The development and release of CBR-resistant cultivars of groundnut appears to be one of the most promising strategies for CBR control. In general, spanish cultivars are most resistant, valencia are the least resistant and virginia cultivars are intermediate.

Crop rotations that include non-legumes such as corn, sorghum or forage grasses minimize the development of new destructive races of the pathogen. Sanitation can be used to minimize infection by *C. crotalariae*. Movement of the pathogen from field to field can be reduced by cleaning tillage equipment (Krigsvold *et al.*, 1977) and by leaving groundnut debris on the soil surface for maximum winter kill of microsclerotia.

10.4.10 Rhizoctonia diseases

Groundnut diseases caused by *Rhizoctonia solani* Kuhn (teleomorphic stage, *Thanatephorus cucumeris* [Frank] Donk) are commonly known as damping off, root and stem rot, pod breakdown (pod rot) and foliage blight. They occur throughout the world (Porter *et al.*, 1982). The economic importance of these diseases is difficult to assess because they usually occur simultaneously with other soilborne diseases and foliar pathogens. However, diseases caused by *R. solani* have become increasingly important with the greater use of irrigation and cultural practices that produce dense canopy growth.

Groundnut seeds are frequently infected by *R. solani* before or shortly after germination. It may cause damping off of the seedling or dry sunken cankers on the hypocotyl (sore skin) and taproot. Later in the growing season the fungus forms infection cushions on the hypocotyl, penetrates the epidermal and cortical cells and causes the tissues to collapse.

Roots infected by *R. solani* develop small sunken lesions of a light to mid brown colour. The root cortex of young seedlings is decayed and necrosis frequently envelops the entire root system, extending up the hypocotyl and killing the plant.

Branches of the mature plant that are in contact with the soil are freely colonized by *R. solani*. Lesions, which are sunken and circular at the site of infection, elongate and girdle the stem. The infected branch wilts and dies. With abundant moisture, the fungus often grows to the top of the foliar canopy, destroying leaves and stems. Leaf lesions are light to dark brown and have distinct zonate patterns.

R. solani frequently infects the tips of pegs just below the soil surface and destroys the pod-forming region. If soil moisture and temperature are favourable, the fungus may grow up the pegs and produce sunken and elongated lesions. Pegs are girdled and tissue becomes shredded with exposed vascular bundles.

Pods may be infected by *R. solani* at any stage of development. Infected pods have dull coloured (light to dark brown) sunken lesions in contrast to those infected with *P. myriotylum*, which have lesions that appear black and greasy. This phase of the disease is frequently referred as pod breakdown (pod rot).

R. solani is usually found without a distinct spore form and thus is put in the group of fungi known as *Mycelia sterilia* (Ainsworth, 1963). It produces many sclerotia in host tissue. Sclerotia, which are aggregates of thickwalled hyphae, can persist in the soil in the absence of the host crop (Blair, 1943; Papavizas and Davey, 1960). Sclerotia germinate when stimulated by exudates from a susceptible host or by the addition of organic matter to the soil. If the soil contains adequate organic matter, the fungus can grow saprophytically. Since many crops and weeds are susceptible to *R. solani* and organic debris is periodically added to the soil, long-term survival of the fungus may occur in many soils. *R. solani* is a common contaminant of groundnut seed and may initiate seedling diseases. Movement of soil or infested plant residue also facilitate the spread of the fungus (Garren and Jackson, 1973).

Resistance to seedling and pod diseases caused by *R. solani* is available in wild groundnuts but not in commercial cultivars. Chemical seed protectant may provide some control of seedborne infection (Jackson, 1963). The systemic fungicides benomyl and carboxin (Abu-Arkoub, 1973), TCMTB (Apdou and Khadr, 1974) and a combination product of thiram, picloram and carboxin provide moderate to good control of seed and seedling diseases caused by *R. solani*. Cultural practices also aid in controlling groundnut diseases (Garren and Jackson, 1973).

10.4.11 Sclerotinia blight

Sclerotinia blight was first observed on groundnut plants in Argentina in 1922. It is now present in most groundnut-producing countries of the world (Porter, 1980b). Yield losses of 10% are common, and in fields showing severe symptoms of disease pod losses often exceed 50% of the expected yield (Porter, 1980b).

Sclerotinia minor Jagger is the causal agent but *S. sclerotiorum* (Lib) de Bary has also been found associated with this disease. The first symptoms are lighter green to yellowing of foliage of plant tops and loss of turgidity (flagging or wilting). Examination of the lower canopy of the infected plant early in the morning reveals the presence of fluffy cottony mycelia of the fungus on the main stem or lateral branches and other affected parts. Initial loci are characterized by small, light green, water-soaked lesions, which later appear sunken and become elongated and light tan. Older lesions are dark brown with a distinct zonation between infected and healthy tissues.

Foliage of the infected branches becomes chlorotic, turns dark brown, and withers. Once the stem is girdled, the branch dies. These symptoms result in the blighted appearance for which the disease is named. Shredding of infected stems, branches and pegs is a characteristic sign of this disease; another prominent characteristic is numerous sclerotia on and in infected plant tissues.

In nature, *S. sclerotiorum* is always found in conjunction with *S. minor*. In artificial inoculations, either species can produce blight lesions. Apothecia of *S. minor* are rarely observed in soils during growing season. However, apothecia are generally observed on crop debris; they are pale orange to white and have a funnel-shaped or flat top. Kohn (1979) used several characteristics to resolve the taxonomic position of the genus *Sclerotinia*. Additional taxonomical and morphological characteristics to support the separation of *Sclerotinia* species are provided by Willetts and Wong (1980).

Sclerotia of the fungus can persist in field soils for 4–5 years in the absence of groundnuts. One sclerotium in 100 g of soil is reported to be sufficient to initiate the disease (Porter, 1980b). This may result in severe disease in groundnut when environmental conditions are conducive to its development. Infection with *S. minor* is favoured by cool temperatures (18–20 °C), moist soil, high rainfall and high relative humidity (95–100%) in the lower crop canopy (Porter, 1980b).

Tillage practices such as ploughing distribute sclerotia throughout the field. Mechanically injured groundnut foliage is very susceptible to colonization by *S. minor* (Porter and Powell, 1978). Plants injured by tractor tyres during pesticide application were attacked at twice the frequency of noninjured plants.

Effective control measures are not yet available (Porter, 1984a). The use of some systemic fungicides, soil fumigants and biocides can aid in reducing losses due to disease (Beute *et al.*, 1975; Porter, 1977; Porter, 1980a) but systemic fungicides such as chlorothalonil or captafol, when applied to control leaf spot fungi, increase the severity of sclerotinia blight (Porter 1980a). Cultural practices such as crop rotation with a non-leguminous crop (corn or sorghum), and avoiding excessive irrigation during the growing season when temperatures are cool, help to reduce the soilborne inoculum of the pathogen.

10.4.12 Verticillium wilt

A vascular wilt of groundnuts caused by a soilborne fungus *Verticillium dahliae* Kleb. was reported from Asia (Golovin, 1937; Morwood, 1953), USA (Smith, 1960) and Australia (Purss, 1961; Purss, 1962). The disease occurs in other groundnut-growing countries but is not considered of economic importance worldwide. However, it can cause yield losses as high as 50% (Smith, 1960; Purss, 1961).

The first symptoms of verticillium wilt appear at flowering time but older plants are also infected. Symptoms consist of marginal chlorosis of leaflets, loss of leaf turgidity and curling of leaves. Leaflets of the infected plants usually turn dull green. The root systems of infected plants appear normal except for a brown-to-black vascular discoloration. Yellowing, leaf necrosis, wilting, defoliation, stunting and eventual dehydration of infected plants become more pertinent as the disease progresses.

Though *V. dahliae* is considered as the primary causal agent of verticillium wilt, *V. albo-atrum* Reinke and Berth has also been found to cause wilt in groundnuts (Smith, 1960). Both the species have a broad host range and are widely distributed. The taxonomic basis for separation of the species has been questioned (Jackson and Bell, 1969).

V. dahliae can survive in the soil for long periods by forming black microsclerotia which can resist adverse environmental conditions. These microsclerotia remain dormant until groundnut root exudates stimulate them to germinate. The fungus initially invades plants through their root system, then develops rapidly and systemically, spreading throughout the vascular system and all plant parts. The pathogen spreads by movement of farm machinery from infected to non-infected fields and also by wind and water-borne movement of infected soil and plant tissue. Another possible mode of spread is through infected seeds.

Populations of wilt pathogens can be suppressed by rotation with non-host crops such as grain sorghum and alfalfa (Hsi, 1967). Removing and burning of crop residues helps in reducing the inoculum density of the pathogen in field soils. Use of clean verticillium-free seed is also recommended. Chemical control of the disease has not been satisfactory in the USA but fungus was effectively controlled in sandy soils in Israel with metma-sodium applied through sprinkler irrigation (Krikun and Frank, 1981). Adequate levels of resistance to this disease are not available (Khan *et al.*, 1972).

10.5 DISEASES CAUSED BY NEMATODES

Diseases caused by plant-parasitic nematodes are among the least understood problems currently confronting agriculture. Yield losses from nematode diseases are generally due to their feeding on the roots, which weakens the plants and reduces their vigour. Groundnut is one of the few crops in which nematodes reduce crop yield directly by damaging pegs, pods and seeds, and indirectly by feeding on roots and weakening the plants. On a worldwide basis, annual groundnut yield losses caused by nematodes are estimated at approximately 12%, while monetary losses are estimated at just over one billion US dollars (Sasser and Freckman, 1987). Although more than 90 species of plant-parasitic nematodes have been reported in association with groundnut (Sharma and McDonald, 1990a), only a few species are known to cause economically important diseases (Subrahmanyam *et al.*, 1990). Species of *Meloidogyne, Pratylenchus,*

Aphasmatylenchus, *Scutellonema*, *Belonolaimus*, *Criconemella*, *Tylencho-rhynchus*, *Aphelenchoides* and *Ditylenchus* can cause appreciable damage to groundnut and are considered to be economically important in several groundnut-producing regions.

10.5.1 Pod and root-knot disease

This disease, caused by four species of *Meloidogyne* (*M. arenaria* (Neal) Chitwood, *M. hapla* Chitwood, *M. incognita* Kofoid and White, and *M. javanica* (Treub) Chitwood), is the most important nematode disease of groundnut. It is widespread over the groundnut-producing regions of the world. *M. arenaria* and *M. javanica* are more common in the warmer regions, and *M. hapla* in the cooler regions. *M. incognita* occurs in the Mediterranean region (McDonald and Raheja, 1980).

M. arenaria is the most widespread and serious of the groundnut root-knot nematodes. In Senegal, it does not successfully parasitize groundnut; juveniles penetrate the roots and subsequently die (Netscher, 1975). However, groundnuts sown in soils heavily infested with this nematode have crop establishment problems as numerous juveniles penetrate the roots, causing browning and necrosis of tissues and in some cases leading to death of the young plants.

Root-knot damage to groundnut is often not suspected until roots are examined and the typical galls are observed on roots and pods. Galling can occur on all underground parts including the pods, which appear warty. Nematode infection disrupts the vascular system, resulting in devitalized roots, retarded growth, yellowing of foliage, stunting and wilting of plants. Gall size varies depending on the nematode species involved. *M. hapla* produces small galls and extensive root proliferation – roots often form a dense mat when infection is severe. Production of roots just above the gall is generally associated with *M. hapla* infection. Galls produced by *M. arenaria and M. javanica* infection are very similar to one another. Early infection of pegs prevents full development of the fruits so that diseased plants have many necrotic pegs and few mature pods.

Damage to groundnut roots by *M. arenaria* at an early stage in plant development impairs growth of the main taproot. As the root system is small at this time, this affects the plant growth more severely than damage which occurs later in the season when roots are well developed and are capable of sustaining the nematode population without serious effects on plant growth. Loss in yield can occur even in soils harbouring as few as 0.5 juveniles/cm^3 of soil.

Nematode infection reduces the number and efficiency of *Rhizobium* nodules. *M. arenaria* infection increases the incidence of damping-off disease of groundnut caused by *Pythium myriotylum* Dreschler (Garcia and Mitchell, 1975b). Inoculation of groundnuts with *M. arenaria* and *Fusarium solani* (Mart.) Sacc. together causes earlier appearance of wilting of groundnut than when *F. solani* is inoculated alone (Patel *et al.*, 1985). The nematode also enhances development of cylindrocladium black rot [*C. crotalariae* (Loos) Bell and Sobers] on the resistant variety NC 3033. *M. hapla* similarly enhances the development of cylindrocladium black rot (Diomonde and Beute, 1981). Soil population densities of *M. arenaria* at more than 0.01 individual/cm^3 in the USA and 1.0 individuals/cm^3 in India at sowing can cause appreciable damage (Sharma and McDonald, 1990b).

Production losses due to *M. arenaria* in major groundnut-producing states of the USA range from 0.5% to 5.4%, and for *M. hapla* range from 0.3% to 4.7% (Anon., 1987). Estimates of economic losses caused by pod and root-knot are not available from many countries but the disease is considered important in Australia, the People's Republic of China, Egypt and India (Sharma and McDonald, 1990b).

Farmers in Egypt and in the USA use nematicides to control pod and root-knot disease. Aldicarb, carbofuran, oxamyl and phenamiphos at the rate of 1.0–3.4 kg a.i./ha are generally used (Rodriguez-Kabana and King, 1979; Rodriguez-Kabana and King, 1985; Rodriguez-Kabana *et al.*, 1981). Seed treatments with 3% aldicarb sulfone and 6% carbofuran reduce the damage by *M. arenaria* (Patel *et al.*, 1986). Formulations of fumigant nematicides are either not available in developing countries or their use is restricted.

Rotations with non-hosts or poor hosts can be effective in reducing the damage to groundnut. The root-knot species have a wide host range but monocotyledons are, in general, non-hosts or poor hosts of the species that attack groundnut, so that maize, sorghum, pearl millet, and rice are good crops to rotate with groundnut. *Sesamum indicum* L., *Ricinus communis* L., *Aeschynomene indica* L., *Cassia fasiculata* Michx., *Indigofera hirsuta* L., and *Paspalum notatum* Flugge are also good crops to include in rotations for reducing *M. arenaria* populations (Rodriguez-Kabana and Morgan-Jones, 1987; Sasser and Kirby, 1979). Appropriate rotations should reduce population densities below the damage threshold. Breaks of three years or more between susceptible crops in the rotation should give good control of root-knot disease. Care should be taken to remove weed hosts of the root-knot species during this break period.

The ideal method for controlling root-knot nematodes on groundnut would be to use resistant cultivars; however, only limited information is available on this subject. In most countries where the disease is important, no efforts are being made to identify resistance. Resistances in wild *Arachis* spp. have been identified to *M. arenaria* and *M. hapla* (Baltensperger *et al.*, 1986; Castillo *et al.*, 1973; Holbrook and Noe 1990; Nelson *et al.*, 1988). Eleven groundnut lines (NC 8C, NC 6, NC 343, NC 3033, NC 17921 × NC 18016, Tainan 9, CES 103, CS 20, NC Ac 17090, NC 18230 × NC 2 and NC 18231 × NC 2) have been reported to be resistant to *M. arenaria* and *M. hapla* (Anon., 1985). Grewal *et al.* (1986) reported JSP 1, ICGS 2, NC 4X and CGC 4 to be resistant to *M. arenaria*, and Sakhuja and

Sethi (1985) found four genotypes resistant to *M. javanica*. Resistance of these varieties needs to be tested under field conditions.

10.5.2 Pod and root-lesion disease

This is an important nematode disease of groundnut in Australia, Brazil, Thailand and Zimbabwe. Species of *Pratylenchus*, *Tylenchorhynchus*, *Criconemella*, and *Belonolaimus* cause lesions on pods and/or roots, but *Pratylenchus* is the most important. Steiner (1949) first reported *Pratylenchus brachyurus* (Godfrey) Filipjev and Sch. Stekh. on groundnut in the USA, and this species is an economic deterrent to groundnut production in Texas (Smith *et al.*, 1978). The nematode was recently reported in India from one of the major groundnut-producing areas of Andhra Pradesh (Varaprasad and Sharma, 1990) but its distribution in other Indian states is not known.

Belonolaimus longicaudatus Rau occurs in sandy soils along the Atlantic coastal plain of the USA from New Jersey to Florida and westward to Texas and Arkansas (Minton and Baujard, 1990). This nematode has not been found outside the USA.

Criconemella ornata (Raski) de Grisse and Loof causes lesions on pods and roots of groundnuts in the USA and has been associated with groundnuts in the African countries of Burkina Faso, Egypt and the Gambia, and in India (Germani and Dhery, 1973; Ibrahim and El-Saedy, 1976; Sharma, 1988). *Tylenchoryhnchus brevilineatus* Williams damages groundnuts in the Nellore and Chittoor districts of Andhra Pradesh, India (Reddy *et al.*, 1984). Distribution of this nematode-caused disease is apparently limited to Andhra Pradesh.

P. brachyurus feeds on cortical cells. The enzymes produced by the nematode hydrolyse storage compounds to toxic products which cause lesions. These start as small brown flecks where the nematode enters, and increase in size with time. Once a lesion appears, fungi and bacteria attack the dead tissue and cause rotting. *Pratylenchus* spp. are migratory endo-parasites and are carried in roots and pegs, and in the shells of mature pods. Several hundred nematodes may be present in a single lesion. It is strongly suspected that *P. brachyurus* increases the severity of diseases caused by *Sclerotium rolfsii* Sacc., *Pythium* sp. and *Aspergillus flavus* Link: Fr. (Jackson and Minton, 1968; Sharma and McDonald, 1990b).

B. longicaudatus is an ectoparasite and feeds on root tips as well as on young pegs and pods. Small necrotic lesions may be observed on the roots, pegs and pods (Owens, 1951). The nematode-infected plants are severely stunted and chlorotic (Minton and Baujard, 1990). Heavy infestations cause stubby lateral roots. Roots, pegs and pods attacked by *C. ornata* are severely discoloured with brown necrotic lesions and many lateral root primordia are killed (Minton and Bell, 1969).

T. brevilineatus infection causes brownish black discoloration of the pod surface. Small brownish-yellow lesions appear on pegs and young developing pods. Discoloration on roots is less severe than on pods (Reddy *et al.*, 1984). Above-ground symptoms of *T. brevilineatus* are often deceptive; foliage is dark green and this distinguishes the disease from that caused by *C. ornata* which leads to yellowing of foliage on diseased plants.

Damage thresholds for *P. brachyurus* are not well established. Application of nematicides in infested soils in the USA results in significant increase in yield (Boswell, 1968; Minton and Morgan, 1974). Although most infestations cause only moderate to low damage, the nematode is considered important in the USA because of its widespread distribution in groundnut soils (Ingram and Rodriguez-Kabana, 1980) and production losses due to *Pratylenchus* spp. range from 0.1% to 2.0% (Minton and Baujard, 1990). No information has been obtained from other countries. Sasser *et al.* (1960) obtained yield increases of more than 100% with nematicide application in fields infested with *B. longicaudatus*. Economic losses in the USA due to this nematode range from 0.25% to 0.5% (Anon., 1987). Losses due to *C. ornata* are not well established and damage in the field often goes undetected. A population density of less than one *C. ornata* individual/cm^3 soil can cause significant loss in yield (Barker *et al.*, 1982) and one *T. brevilineatus* individual/cm^3 of soil causes significant loss in yield.

Nematicides that control the pod and root-knot diseases also control the pod and root-lesion diseases but chemical control is not commonly practised. For *B. longicaudatus*, however, nematicide application is a major means of control (Minton and Baujard, 1990). Crop rotations are not entirely effective because of the wide host ranges of the nematodes. Fallowing for more than 6 weeks greatly reduces *P. brachyurus* populations (Brodie and Murphy, 1975). There are no commercial cultivars with resistance to *P. brachyurus*, *C. ornata* or *T. brevilineatus*. Three plant introductions to the USA – P1 1290606, P1 1295233 and P1 1365553 – are resistant to *P. brachyurus* (Smith *et al.*, 1978; Starr, 1984) and preliminary field screening work indicates the availability of 23 sources of resistance to *T. brevilineatus* in groundnut genotypes (Siva Rao *et al.*, 1986; Mehan and Reddy, 1987).

10.5.3 Groundnut chlorosis

This disease, caused by *Aphasmatylenchus straturatus* Germani, is confined to south-west Burkina Faso where it has infested approximately 2600 ha of groundnut crops (Minton and Baujard, 1990). The nematode causes interveinal chlorosis and stunting of plants. Chlorotic plants have reduced root systems and poor nodulation (Germani and Dhery, 1973).

Disease symptoms occur in fields with less than one *A. straturatus* individual/cm^3 soil. Yield reductions due to groundnut chlorosis range from 30% to 70%.

Application of dibromochloropropane (DBCP) at sowing gives satisfactory control of the disease (Dhery *et al.*, 1975). The nematode attacks many economically important crops in Burkina Faso (Germani and Dhery, 1973). There is no information on crop rotations likely to reduce disease severity, nor are there reports of varietal resistance.

10.5.4 Yellow patch disease/crop growth variability

This is a problem in the Sahelian Zone of West Africa and two species of *Scutellonema* have been implicated as major biotic factors (Germani *et al.*, 1985; Sharma *et al.*, 1990). *Scutellonema cavenessi* Sher in Senegal and *S. clathricaudatum* Whitehead in Niger cause serious damage to groundnut crops and the symptoms in each country are similar. These species are also reported from other countries in West Africa where they are suspected of causing damage to groundnuts (Sharma, 1989). Foliage of groundnut plants grown in soil infested with the *Scutellonema* spp. is chlorotic with reduced leaf size and canopy development. Lateral roots are also reduced in size. The poorly growing plants appear in randomly distributed patches which persist and increase in size during the subsequent growing season.

The extent of crop loss in Senegal and Niger has not been fully evaluated. Application of nematicides to the nematode-infested fields in these two countries have increased pod and haulm yields by more than 100% (Germani *et al.*, 1985). An initial population density of more than one *S. clathricaudatum* individual/cm^3 soil causes stunting of foliage. The nematode-caused damage appears to be aggravated by low soil pH and aluminium toxicity in the soil (Sharma, 1989). Applications of DBCP at 15 l/ha and carbofuran at 6 kg a.i./ha reduce the nematode populations in soil and increase crop growth and yield (Germani and Gautreau, 1976; Sharma, 1989). The are no known groundnut cultivars resistant to *S. cavenessi*, but screening of cultivars for resistance or tolerance to *S. clathricaudatum* has shown that some genotypes (e.g. ICGS 41) are less affected by the disease than is the commonly grown groundnut cultivar 55–437.

10.5.5 Seedborne nematode diseases

Aphelenchoides arachidis Bos and *Ditylenchus destructor* Thorne infect the seeds of groundnut. *Aphelenchoides arachidis* distribution is limited to a few isolated areas in northern Nigeria (Bos, 1977). *Ditylenchus destructor* is reported from major groundnut-producing regions of the Republic of South Africa and can destroy 40–60% seeds in heavily infested fields (De Waele *et al.*, 1988). These diseases have not been reported on groundnut in other parts of the world.

A. arachidis is a facultative endoparasite of groundnut and it parasitizes tissues of pods, testas, roots and hypocotyls, but not the cotyledons, embryos or other parts of the plant (Bos, 1977; Bridge *et at.*, 1977). Testas infested with *A. arachidis* are thicker and more uneven than normal. Heavily infested seeds are light brown; examined at lifting, their testas are translucent and have dark vascular strands within them. Testas of the infected dried seeds are dull in colour and often are wrinkled. Nematode damage has little effect on seed germination.

D. destructor is found in roots, pegs, shells and seeds (de Waele *et al.*, 1988). The initial symptom of the disease is development of brown necrotic tissue at the pod base, at the juncture of peg and pod. Infected pods lack the lustre of healthy pods. Infected seeds are usually shrunken and the micropyles are dark brown to black. Infected embryos are usually olive green to brown.

Very limited information is available on the extent of losses caused by these nematodes, their interrelationships with other micro-organisms and their control (Bridge *et al.*, 1977; Minton and Baujard, 1990).

No groundnut cultivars resistant to either *A. arachidis* or *D. destructor* have been reported. Immersing seeds infested with *A. arachidis* in four times their volume of water and heating at 60 °C for five minutes gives complete control of the nematode without adversely affecting germination. Sun-drying of pods after harvest in very dry conditions reduces the number of viable nematodes in the pods (Bridge *et al.*, 1977). One method of managing *D. destructor* is rotation of groundnut with non-hosts and less susceptible crops. Maize and grain sorghum crops should not precede groundnut as a rotational crop in fields infested with the nematode. Wheat is apparently a poor host (Basson *et al.*,1990). Unless appropriate precautions are taken, the seedborne nematodes could be disseminated with seeds and become serious pests worldwide.

REFERENCES

Abdalla, M.H. (1974) Mycoflora of groundnut kernels from Sudan. *British Mycological Society Transactions*, **63**, 353–359.

Abdelsalam, A.M., Khalil, E.M., Fahim, M.M. and Ghanem, G.A. (1987) The effect of peanut mottle virus infection on growth and yield of peanuts. *Egyptian Journal of Phytopathology*, **19**, 127–132.

Abu-Arkoub, M.M. (1973) Seedling damping-off and pod rot disease of cultivated peanut in El-Tahrin Province (A.R.E.) and its control. M.S. Thesis, Plant Pathology Department, College of Agriculture, Cairo University.

Adalla, C.B. and Natural, M.P. (1988) Peanut stripe virus disease in the Philippines, in *Summary proceedings of the First Meeting to Coordinate Research on Peanut Stripe Virus Disease of Groundnut*, 9–12 June 1987, Malang, Indonesia. ICRISAT, Patancheru, India, p. 9.

Agnihotri, J.P., and Goyal, J.P (1971) Groundnut vs collar rot. *Intensive Agriculture*, **9**, 10.

Aibara, K. and Nobumitsu, Y. (1977) New approach to aflatoxin removal, in *Mycotoxins in Human and Animal Health*, (eds J.V. Rodricks, C.W. Hesseltine and M.A. Mehlamn), Pathotox Publishers Inc., Park Forest South, IL., pp. 151–161.

Ainsworth, G.C. (1963) *Ainsworth and Bisby's Dictionary of the Fungi*, Commonwealth Mycological Institute, Kew.

Akiew, E.B. (1986) Influence of soil moisture and temperature on the persistance of *Pseudomonas solanacearum*, in *Bacterial Wilt disease in Asia and the South Pacific*, (ed. G.J. Persley). Proceedings of an international workshop held at PCARRD, Los Baños, Philippines, 8–10 October, 1985. ACIAR Proceedings No. 13.

Alderman, S.C., Nutter, F.W., Jr and Labrinos, J.L. (1989) Spatial and temporal analysis of spread of late leaf spot of peanut. *Phytopathology*, **79(8)**, 837–844.

Amin, P.W. (1985) Apparent resistance of groundnut cultivar Robut 33–1 to bud necrosis disease. *Phytopathology*, **69**, 718–719.

Anon. (1985) *Report of co-operative research between the International Meloidogyne project (IMP) and the International Crops Research Institute for the Semi-Arid Tropics (ICRI-SAT)*, North Carolina State University, Raleigh.

Anon. (1987) Bibliography of estimated crop losses in the United States due to plant-parasitic nematodes. *Journal of Nematology*, **15**, 6–12.

Anon. (1960) Plant pathology report, Department of Agriculture, Mauritius, 1959. pp. 41–49 (vide *Review of Applied Mycology*, **40**, 651–651, 1961).

Apdou, Y.A. and Khadr, A.S. (1974) Systemic control of seedling and pod rot disease of peanuts (*Arachis hypogaea*). *Plant Disease Reporter*, **58**, 176–179.

Arthur, J.C. (1934) *Manual of the Rusts in the United States and Canada*, Prudue Research Foundations, Lafayette, Indiana, 438 pp..

Aujla, S.S., Chohan, J.S. and Mehan, V.K. (1978) The screening of peanut varieties for the accumulation of aflatoxin and their relative reaction to the toxigenic isolate of *Aspergillus flavus* Link ex. Fries. *Journal of Research, Punjab Agricultural University*, **15**, 400–403.

Aycock, R. (1966) Stem rot and other diseases caused by *Sclerotium rolfsii*. *North Carolina Agricultural Experiment Station Technical Bulletin 174*, 202 pp.

Backman, P.A. (1984) Stem rot, Pages 15–16 in *Compendium of Peanut Diseases*, (eds D.M. Porter, D.H. Smith and R. Rodriguez-Kabana), American Phytopathology Society, St. Paul, Minnesota, pp. 15–16.

Backman, P.A. and Rodriguez-Kabana, R. (1975) A system for the growth and delivery of biological control agents to the soil. *Phytopathology*, **65**, 819–821.

Backman, P.A., Rodriguez-Kabana, R. and Williams, J.C. (1975) The effect of peanut leaf spot fungicides on the non-target pathogen, *Sclerotium rolfsii*. *Phytopathology*, **65**, 773–776.

Backman, P.A., Rodriguez-Kabana, R., Hammond, J.M. *et al.* (1977) Peanut leaf spot research in Alabama. *Auburn University Bulletin No. 489*.

Balasubramanian, R. and Narayanasamy, P. (1980) A note on a new blight disease of groundnut caused by *Phoma microspora* sp nov. in Tamil Nadu. *Indian Phytopathology*, **33**, 133–136.

Baltensperger, D.D., Prine, G.M. and Dunn, R.A. (1986) Root-knot nematode resistance in *Arachis glabrata*. *Peanut Science*, **13**, 78–80.

Barker, K.R., Schmitt, D.P. and Campos, V.P. (1982) Response of peanut, corn, tobacco and soybean to *Criconemella ornata*. *Journal of Nematology*, **14**, 576–581.

Bartz, J.A., Norden, A.J., LaPrade, J.C. and DeMuynk, T.J. (1978) Seed tolerance in peanuts (*Arachis hypogaea* L.) to members of the *Aspergillus flavus* group of fungi. *Peanut Science*, **5(1)**, 53–56.

Basson, S., Waele, D. de and Meyer, A.J. (1990) An evaluation of crop plants as hosts for *Ditylenchus destructor* isolated from peanut. *Nematropica*, **20**, 23–29.

Behncken, G.M. (1970) The occurrence of peanut mottle virus in Queensland. *Australian Journal of Agricultural Research*, **21**, 465–472.

Bell, D.K. and Sobers, E.K. (1966) A peg, pod, and root necrosis of peanuts caused by a species of *Calonectria*. *Phytopathology*, **56**, 1361–1364.

Berchoux, Chr. de. (1960) La rosette de l'arachide en Haute-Volta: Comportement de lignes resistantes. *Oléagineux*, **15**, 229–233.

Beute, M.K. (1980) Cylindrocladium black rot (CBR) disease of peanut (*Arachis hypogaea*), in *Proceedings of the International Workshop on Groundnuts*, (eds R.W. Gibbons and J.W. Mertin) at ICRISAT, 13–17 October. ICRISAT, Patancheru, pp. 171–176.

Beute, M.K., Porter, D.M. and Hadley, B.A. (1975) Sclerotinia blight of peanut in North Carolina and Virginia and its chemical control. *Plant Disease Reporter*, **59**, 697–701.

Bijaisoradat, M., Kuhn, C.W. and Benner, C.P. (1988) Disease reaction, resistance, and viral content in six legume species infected with eight isolates of peanut mottle virus. *Plant Disease*, **72**, 1042–1046.

Bitancourt, A.A. and Jenkins, A.E. (1940) Novas especies de *Elsinoe* e *Sphaceloma* sobre hospedes de importancia economica. *Arquivos Instituto Biologico de São Paulo*, **11**, 45–58.

Black, M.C. (1988) Pathological aspects of TSWV in south Texas. *Proceedings of the American Peanut Research and Education Society, Inc.*, **19**, 66.

Blair, I.D. (1943) Behaviour of the fungus *Rhizoctonia solani* Kuhn. in the soil. *Annals of Applied Biology*, **30**, 118–127.

Block, K.R. (1973) Peanut mottle virus in East Africa. *Annals of Applied Biology*, **74**, 171–179.

Blount, W.P. (1961) Turkey 'X' disease. *Journal of the British Turkey Federation*, **9**(2), 52, 55–58, 61–77.

Bock, K.R. and Kuhn, C.W. (1975) *Peanut mottle virus*. CMI Descriptions of Plant Viruses No. 141. Commonwealth Mycological Institute, Kew.

Bos, W.S. (1977) *Aphelenchoides arachidis* n. sp. (Nematoda Aphelenchoidea) an endoparasite of the testa of groundnut in Nigeria. *Zeitschrift für Pflanzenkrankheit*, **84**, 95–99.

Boshou, L., Yuying, W., Xingming, X. *et al.* (1990) Genetic and breeding aspects of resistance to bacterial wilt in groundnut, in *Bacterial wilt of groundnut* (eds K.J. Middleton and A.C. Hayward). Proceedings of an ACIAR/ ICRISAT collaborative research planning meeting held at Genting Highlands, Malaysia, 18–19 March, 1990. ACIAR Proceedings No. 31. Australian Centre for International Agricultural Research, Canberra, pp. 39–43.

Boswell, T.E. (1968) Pathogenicity of *Pratylenchus brachyurus* to spanish peanut. PhD Dissertation, Texas A & M. University.

Boswell, T.E. and Thames, W.H., Jr (1976) Pythium pod rot control in South Texas. *Proceedings of the American Peanut Research and Education Association*, **8**, 89 (Abstract).

Bridge, J., Bos, W.S., Page, L.J. and McDonald, D. (1977) The biology and possible importance of *Aphelenchoides arachidis*, a seedborne endoparasitic nematode of groundnut from northern Nigeria. *Nematologica*, **23**, 253–259.

Brodie, B.B. and Murphy, W.S. (1975) Population dynamics of plant nematodes as affected by combinations of fallow and cropping sequences. *Journal of Nematology*, **7**, 91–92.

Bromfield, K.R. (1971). Peanut rust – a review of literature. *Journal of American Peanut Research and Education Association, Inc.*, **3**, 111–121.

Brunt, A.A. (1989) Keynote Address: Tropical legume viruses and their control, in *Summary proceedings of the Second Coordinators' Meeting on Peanut Stripe Virus*, (eds S.R. Beckerman, D.V.R. Reddy and D. McDonald), 1–4 August 1989, ICRISAT Center. ICRISAT, Patancheru. pp. 15–16.

Brunt, A., Crabtree, K. and Gibbs, K. (1990). *Viruses in tropical plants*, CAB International, Wallingford.

Buddenhagen, I. and Kelman, A. (1964) Biological and physiological aspects of bacterial wilt caused by *Pseudomonas solanacearum*. *Annual Review of Phytopathology*, **2**, 203–230.

Butler, D.R. and Jadhav, D.R. (1991) Requirements of leaf wetness and temperature for infection of groundnut by rust. *Plant Pathology*, **40**, 395–400.

Caccarone, A. (1949) Rotting of groundnut pods due to *Thielaviopsis basicola* L. *Italian Agriculture*, **86**, 741–743.

Castillo, M.B., Morrison, L.S., Russel, C.C. and Banks, D.J. (1973) Resistance to *Meloidogyne hapla* in peanut. *Journal of Nematology*, **5**, 281–285.

Chevaugeon, J. (1952) Maladies des plantes cultivées en Moyenne–Casamance et dans le delta Central Nigérien. *Revue de pathologie végétale et d'entomologie agricole de France*, **31**, 3–51.

Chohan, J.S. (1974) Recent advances in diseases of groundnut in India, in *Current Trends in Plant Pathology*, Lucknow University Press, Lucknow, Uttar Pradesh, pp. 171–184.

Chohan, J.S. and Gupta, V.K. (1968) Alfaroot, a new disease of groundnut caused by *Aspergillus flavus* Link ex. Fries. *Indian Journal of Agricultural Sciences*, **38**, 568–570.

Chorin, M. (1961) Powdery mildew on leaves of groundnuts. *Bulletin Research Council, Israel*, **10D**, 148–149.

Chorin, M. and Frank, Z.R. (1966) *Oidium arachidis* Chorin. Powdery mildew of groundnut foliage. *Israel Journal of Botany*, **15**, 133–137.

Ciegler, A. (1978) Fungi that produce mycotoxins: conditions and occurrence. *Mycopathologia*, **65**, 5–11.

Commonwealth Mycological Institute (1966) Distribution of *Mycosphaerella arachidicola*. *CMI Map No. 166*, Kew, Surrey.

Commonwealth Mycological Institute (1967) Distribution of *Mycosphaerella berkeleyii*. *CMI Map No. 152*, Kew, Surrey.

Cook, D., Barlow, E. and Sequeira, L. (1989) Genetic diversity of *Pseudomonas solanacearum*: detection of restriction fragment length polymorphisms with DNA probes that specify virulence and the hypersensitive response. *Molecular Plant–Microbe Interactions*, **2**, 113–121.

Cooper, W.E. (1956) Chemical control of *Sclerotium rolfsii* in peanuts. *Phytopathology*, **4**, 9–10 (Abstract).

Costa, A.S. (1950) Mancha anular do amendoim causado pelo virus de viracabeca. *Bragantia*, **10**, 67–78.

Cruz, B.P.B., Figueiredo, M.B. and Almelda, E. (1962) Principals doencas e pragas do amedoim no Estado de São Paulo. *O. Biologico*, **28**, 189–195.

Culp, T.W. and Troutman, J.L. (1968) Varietal reaction of peanuts, *Arachis hypogaea*, to stunt virus disease. *Plant Disease Reporter*, **52**, 914–918.

Cummins, G.R. (1978) *Rust fungi on legumes and composites in North America*, University of Arizona Press, Tucson, pp. 181–182.

Demski, J.W., Reddy, D.V.R. and Sowell, G. (1984a) Stripe diseases of groundnuts. *FAO Plant Protection Bulletin*, **32**, 114–115.

Demski, J.W., Reddy, D.V.R., Sowell, G. and Bays, D. (1984b) Peanut stripe virus – a new seedborne potyvirus from China infecting groundnut (*Arachis hypogaea*). *Annals of Applied Biology*, **105**, 495–501.

Demski, J.W., Reddy, D.V.R., Wongkaew, S. *et al.* (1988) Naming of peanut stripe virus. *Phytopathology*, **78**, 631–632.

Demski, J.W., Wells, H.D., Millar, J.D. and Khan, M.A. (1983) Peanut mottle virus epidemics in lupines. *Plant Disease*, **67**, 166–168.

Devi, L.R. and Menon, M.R. (1979) Additional hosts of *Pseudomonas solanacearum*. Indian Phytopathology, **32**, 452–453.

de Waele, D., Jones, B.L., Bolton, C. and van den Berg, E. (1988) *Ditylenchus destructor* in hulls and seeds of peanut. *Journal of Nematology*, **21**, 10–15.

Dhanju, A.S. and Chohan, J.S. (1974) *Phoma glomerata* – in relation to bud-blight-affected plants of groundnut. *Indian Phytopathology*, **27**, 665–666.

Dhery, M., Germani, G. and Giard, A. (1975) Resultats de traitments nematicides contre la chlorose et le rabougrissement de l'arachide en Haute Volta. *Cah ORSTOM Ser. Biol.*, **10**, 161–167.

Dickens, J.W. (1977) Aflatoxin *Aspergillus flavus*, occurrence and control during growth, harvest and storage of peanuts in *Mycotoxins in Human and Animal Health* (eds J.V. Rodericks, C.W. Hessltine and M.A. Mehlman), Pathotox Publishers Inc., Park Forest South, Illinois, pp. 99–105.

Dickens, J.W. and Welty, R.W. (1969) Detecting farmers' stock peanuts containing aflatoxln by examination for visible growth of *Aspergillus flavus*. *Mycopathology-Mycology Applied*, **37**, 65–69.

Diener, U.L., Pettit, R.E. and Cole, R.J. (1982) Aflatoxins and other mycotoxins in peanuts in *Peanut Science and Technology*, (eds H.E. Pattee and C.T. Young), American Peanut Research and Education Society Inc., Yoakum, Texas, pp. 486–519.

Diomonde, M. and Beute, M.K. (1981) Effects of *Meloidogyne hapla* and *Macroposthonia ornata* population on cylindrocladium black rot in peanuts. *Plant Disease*, **65**, 339–342.

Douine, L. and Devergne, J.C. (1978) Isolement en France du virus de rabougrissement de l'arachide (peanut stunt virus). *Annual Review of Phytopathology*, **10**, 79–92.

Drechsler, C. (1930) Some new species of Pythium. *Journal of the Washington Academy of Sciences*, **20**, 398–418.

Dubern, J. (1980) Mechanical and aphid transmission of an Ivory Coast strain of groundnut rosette virus. *Phytopathologische Zeitschriff*, **99**, 318–326.

Dyer, R.A. (1949) Botanical surveys and control of plant diseases. *Farming in South Africa*, **24**, 1.

Dyer, R.A. (1951) Plant classification and control of crop diseases. *Farming in South Africa*, **26**, 488–490.

FAO/WHO/UNEP (1977) *Report of the Joint FAO/WHO/UNEP Conference on Mycotoxins*, 19–27 September 1977, Nairobi, Kenya.

Feakin, S.D. (1973) *Pest control in groundnut*, PANS manual No. 2, Centre for Overseas Pest Research, Foreign and Commonwealth Office, Overseas Development Administration, London, p. 197.

Felix, S. (1971) Maladies de la pomme de terre et de l'arachide à Mauriée: resistance et épidémiologie. *Revue agricole et sucrii de l'ile Maurice*, **50**, 241–247.

Fischer, H.U. and Lockhart, B.E.L. (1978) Host range and properties of peanut stunt virus from Morocco. *Phytopathology*, **68**, 289–293.

Frank, Z.R. (1968) Pythium pod rot of peanut. *Phytopathology*, **58**, 542–543.

Frank, Z.R. (1972) *Pythium myriotylum* and *Fusarium solani* as cofactors in a pod-rot complex of peanut. *Phytopathology*, **62**, 1331–1334.

Frank, Z.R. (1974) Effect of constant moisture levels on Pythium rot of peanut pods. *Phytopathology*, **64**, 317–319.

Frezzi, M.J. (1956) Especies de Pythium fitopathogens identificades en la Republica Argentina. *Revista de Investigaciones Agricolas*, **10**, 113–241.

Frezzi, M.J. (1960) Enfermedades del mani en la provincia de Cordoba (Argentina). *Revista de Investigaciones Agricolas*, **14**, 113–155.

Frezzi, M.J. (1965) "Quemadura" de las hojas causada por *Leptosphaerulina arachidicola* Y. Otros hongos, en manies silvesres (grupo rhizomatosa) de distinta procedencia. *Revista de Investigaciones Agricolas*, Series 5, Patologia Vegetale, **2**, 13–24.

Frezzi, M.J. (1969) *Ascochyta arachidis* Woron., hongo parasito del mani (*Arachis hypogaea* L.) y su forma sexual *Mycosphaerella argentinensis* n. sp. en Argentina. *Revista de Investigaciones Agropecuaria*, **6**, 147–153.

Garcia, R. and Mitchell, D.J. (1975a) Interactions of *Pythium myriotylum* with *Fusarium solani*, *Rhizoctonia solani* and *Meloidogyne arenaria* in pre-emergence damping-off of peanut. *Plant Disease Reporter*, **59**, 665–669.

Garcia, R. and Mitchell, D.J. (1975b) Synergistic interactions of *Pythium myriotylum* with *Fusarium solani* and *Meloidogyne arenaria* in pod rot of peanut. *Phytopathology*, **65**, 832–833.

Garren, K.H. (1964) Recent developments in research on peanut pod rot, in *Proceedings of the third National Peanut Research Conference*, July, Auburn, Alabama, pp. 20–27.

Garren, K.H. (1966) Peanut (groundnut) microfloras and pathogenesis in groundnut pod rot. *Phytopathologische Zeitschrift*, **55**, 359–367.

Garren, K.H. (1970) Antagonisms between indigenous *Pythium myriotylum* and introduced *Rhizoctonia solani* and peanut pod breakdown. *Phytopathology*, **60**, 1291. (Abstract).

Garren, K.H. and Jackson, C.R. (1973) Peanut diseases, in *Peanuts: Culture and Uses*,

American Peanut Research and Education Association, Stone Printing Co., Roanoke, Virginia, pp. 429–494.

Garren, K.H., Beute, M.K.and Porter, D.M. (1972) The cylindrocladium black rot of peanuts in Virginia and North Carolina. *Journal of American Peanut Research and Education Association*, **4**, 67–71.

Garrett, S.D. (1956) *Biology of Root Infecting Fungi*, Cambridge University Press, London, 293 pp.

Germani, G., Baujard, P. and Luc, M. (1985). *La lutte contre les nematodes dans le bassin arachidier sénégalais*, ORSTOM, Dakar.

Germani, G. and Dhery, M. (1973) Observations et expérimentations concernant le rôle des nematodes dans deux affections de l'arachide en haute Volta: la 'chlorose' et le 'clump'. *Oléagineux*, **28**, 235–242.

Germani, G. and Gautreau, J. (1976) Resultats agronomiques obtenus par des traitements nematicide sur arachide au Sénégal. *Cahiers* ORSTOM *Serie Biologique*, **11**, 193–202.

Ghuge, S.S., Mayee, C.D. and Godbole, G.M. (1981) Development of rust and leaf spots of groundnuts as influenced by foliar application of carbendazim and tridemorph. *Pesticides*, **25(6)**, 16–19.

Gibbons, R.W. (1977) Groundnut rosette virus, in *Diseases, Pests and Weeds in Tropical Crops* (eds. J. Kranz, H. Schmutterer and W. Koch), Verlag Paul Parey, Berlin and Hamburg, pp. 19–21.

Gillier, P. (1978) New horizons for growing groundnuts resistant to drought and rosette. *Oléagineux*, **33**, 25–28.

Giorda, L.M. (1984) Scab, in *Compendium of Peanut Disease* (eds. D.M. Porter, D.H. Smith and R. Rodriguez-Kabana), American Phytopathological Society, St. Paul, Minnesota.

Golovin, P. (1937) Diseases of south oil plant cultures. *Acta Univ. Asiae. Med. Series 8b*, **35**, 1–77.

Grewal, P.S., Chhabra, H.K. and Kaul, V.K. (1986) Screening of groundnut germplasm against root-knot nematode, *Meloidogyne arenaria*. *Indian Journal of Nematology*, **17**, 151–152.

Guarch, A.M. (1941) Communicaciones fitopatologicas. Revista de la Faculted de agronomia de la Universidad de la Republica, Uruguay 23, 14–16. (vide *Review of Applied Mycology*, **21**, 129–30, 1942).

Hallock, D.L. and Garren, K.H. (1968) Pod breakdown, yield, and grade of virginia type peanuts as affected by Ca, Mg, and K sulfates. *Agronomy Journal*, **60**, 253–257.

Hammons, R.O. (1977) Groundnut rust in United States and the Caribbean. *PANS*, **23**, 300–304.

Harkness, C. (1977) *The breeding and selection of groundnut varieties for resistance to rosette virus disease in Nigeria*. Report submitted to African Groundnut Council, 45 pp.

Harrison, A.L. (1966) Pentachloronitrobenzene for the control of southern blight of peanuts. *Plant Disease Reporter*, **50**, 855–858.

Harrison, A.L. (1967) Increasing peanut yields by cultural and chemical means. *Plant Disease Reporter*, **51**, 441–443.

Hayward, A.C. (1964) Characteristics of *Pseudomonas solanacearum*. *Journal of Applied Bacteriology*, **27**, 265–277.

Hayward, A.C. (1986) Bacterial wilt caused by *Pseudomonas solanacearum* in Asia and Australia: an overview, in *Bacterial Wilt disease in Asia and the South Pacific*, (ed. G.J. Persley), Proceedings of an international workshop held at PCARRD, Los Baños, Philippines, 8–10 October, 1985. ACIAR Proceedings No. 13, Burnett Printing Co. Pty. Ltd. Kingaroy, Australia.

Hayward, A.C. (1990) Diagnosis, distribution and status of groundnut bacterial wilt, Pages 12–17 in *Bacterial wilt of groundnut* (eds K.J. Middleton and A.C. Hayward), Proceedings of an ACIAR/ICRISAT collaborative research planning meeting held at Genting Highlands, Malaysia, on 18–19 March, 1990. ACIAR Proceedings No. 31, Burnett Printing Co. Pty. Ltd., Kingaroy, Australia.

He, L.Y., Sequeira, L. and Kelman, A. (1983) Characteristics of strains of *Pseudomonas solanacearum* from China. *Plant Disease*, **67**, 1357–1361.

He, L.Y. (1990) Control of Bacterial Wilt of groundnut in China with emphasis on cultural and biological methods, in *Bacterial wilt of groundnut* (eds K.J. Middleton and A.C. Hayward), Proceedings of an ACIAR/ICRISAT collaborative research planning meeting held at Genting Highlands, Malaysia on 18–19 March, 1990. ACIAR Proceedings No. 31. Burnett Printing Co. Pty. Ltd., Kingaroy, Australia, pp. 20–25.

Hebert, T.T. (1967) Epidemiology of the peanut stunt virus in North Carolina. *Phytopathology*, **57**, 461 (Abstract).

Helms, K., Grylls, N.E. and Purss, G.S. (1961) Peanut plants in Queensland infected with tomato spotted wilt virus. *Australian Journal of Agricultural Research*, **12**, 239–246.

Hemingway, J.S. (1954) Cercospora leaf spots of groundnut in Tanganyika. *East African Agricultural Journal*, **19**, 263–271.

Hendrix, F.F., Jr and Campbell, W.A. (1973) Pythium as plant pathogens. *Annual Review Phytopathology*, **11**, 77–98.

Hennen, J.F., Figueiredo, M.B., Ribeiro, I.J.A. and Soave, J. (1976) The occurrence of teliospores of *Puccinia arachidis* (uredinales) on *Arachis hypogaea* in São Paulo State, Brazil. *Summa Phytopathologica*, **2**, 44–46.

Hoffmaster, D.E., McLaughlin, J.H., Winfred, R.W. and Chester, K.S. (1943) The problem of dry root rot caused by *Macrophomina phaseoli* (*Sclerotium bataticola*). *Phytopathology*, **33**, 1113–1114 (Abstract).

Holbrook, C.C., and Noe, J.P. (1990) Resistance to *Meloidogyne arenaria* in *Arachis* spp. and the implications on development of resistant peanut cultivars. *Peanut Science*, **17**, 35–38.

Hsi, D.C.H. (1965) Blackhull disease of Valencia peanuts. *New Mexico Agricultural Experimental Station Research Report, No. 110*, 10 pp.

Hsi, D.C.H. (1967) Relationship between crop sequence and several diseases of Valencia peanuts. *Phytopathology*, **57**, 416 (Abstract).

Hsi, D.C.H. (1978) Effect of crop sequence, previous peanut blackhull severity, and time of sampling on soil population of *Thielaviopsis basicola*. *Phytopathology*, **68**, 1442–1445.

Hsi, D.C.H. and Ortiz, M., Jr (1980) Suppression of *Thielaviopsis basicola* by two fungicides applied to sandy loam soils in New Mexico. *Plant Disease*, **64**, 1011–1012.

Hull, R. and Adams, A.N. (1968) Groundnut rosette and its assistor virus. *Annals of Applied Biology*, **62**, 139–145.

Ibrahim, I.K.A. and El-Saedy, M.A. (1976) Plant-parasitic nematodes associated with peanut in Egypt. *Egyptian Journal of Phytopathology*, **8**, 31–35.

ICRISAT (1992) *Groundnut monitoring tour report of the ICRISAT Legumes Unit.* ICRISAT, Patancheru.

Indulkar, A.S. and Grewal, J.S. (1970) Studies on chemical control of *Sclerotium rolfsii*. *Indian Phytopathology*, **23**, 455–458.

Ingram, E.G. and Rodriguez-Kabana, R. (1980) Nematode parasitic on peanuts in Alabama and evaluation of methods for detection and study of population dynamics. *Nematropica*, **10**, 21–30.

Iqbal, M. and Kumar, J. (1986) Bacterial wilt in Fiji, in *Bacterial wilt disease in Asia and the South Pacific*. Proceedings of an international workshop held at PCARRD, Los Baños, Philippines, 8–10 October, 1985, (ed. by G.J. Persley), ACIAR Proceedings No. 13.

Jackson, C.R. (1963) Seed-treatment fungicides for control of seed-borne fungi in peanut. *Plant Disease Reporter*, **47**, 32–35.

Jackson, C.R. and Bell, D.K. (1968) *Leptosphaerulina crassiasca* (Sechet) comb. nov., the cause of leaf scorch and pepper spot of Peanut. *Oléagineux*, **23**, 387–388.

Jackson, C.R. and Bell, D.K. (1969) *Diseases of Peanut (groundnut) caused by fungi.* University of Georgia Agricultural Experiment Station Research Bulletin 56, 137 pp.

Jackson, K. and Damicone, J. (1991) Evaluation of foliar fungicides on peanut for control of cercospora leaf spot, in *Results of 1990 plant disease control field studies*, (eds K.E.

Jackson, J.P. Damicone, E. Williams *et al.*) State University Press, Stillwater, Oklahoma (Research Report, Agricultural Experiment Station, Oklahoma State University, No. P-920), pp. 18–24.

Jackson, C.R. and Minton, N.A. (1968) Pod invasion by fungi in the presence of lesion nematodes in Georgia. *Oléagineux*, **23**, 531–534.

Jayasena, K.W., Lekha, D.P.D. and Rajapakasa, R.H.S. (1988) Wilt disease of groundnut (*Arachis hypogaea*) caused by *Pseudomonas solanacearum* in Sri Lanka. *Abst. 5th International Congress of Plant Pathology*, Kyoto, Japan.

Jenkins, W.A. (1938) Two fungi causing leaf spots of peanut. *Journal of Agricultural Research*, **56**, 317–332.

Jenkins, S.F., Jr, Hammons, R.O. and Dukes, P.D. (1966) Disease reaction and symptom expression of seventeen peanut cultivars to bacterial wilt. *Plant Disease Reporter*, **50**, 520–523.

Jensen, R.E. and Boyle, L.W. (1965). The effect of temperature, relative humidity and precipitation on peanut leaf spot. *Plant Disease Reporter*, **49**, 975–978.

Joffe, A.Z. and Lisker, N. (1970) Effect of crop sequences and soil types on the mycoflora of groundnut kernels. *Plant and Soil*, **32**, 531–533.

Jones, B.L. (1975) The mode of *Pythium myriotylum* Drechsler penetration and infection in peanut pods. *Proceedings of American Peanut Research and Education Association*, 7:79 (Abstract).

Kelman, A. (1953) *The bacterial wilt caused by Pseudomonas solanacearum*. North Carolina Agricultural Experiment Station Technical Bulletin 99.

Kelman, A. and Cook, R.J. (1977) Plant Pathology in the People's Republic of China. *Annual Review of Phytopathology*, **17**, 409–429.

Khan, B.M., Wadsworth, D.F. and Kirby, J.S. (1972) Screening peanut germplasm for resistance to Verticillium wilt. *Journal of American Peanut Research and Education Association*, **4**, 145–147.

Knudsen, G.R., Spurr, H.W., Jr and Johnson, C.S. (1987) A computer simulation model for cercospora leaf spot of peanut. *Phytopathology*, **77**(8), 1118–1121.

Knudsen, G.R., Johnson, C.S. and Spurr, H.W., Jr (1988) Use of a simulation model to explore fungicide strategies for control of cercospora leaf spot of peanut. *Peanut Science*, **15**(1), 39–43.

Kohn, L.M. (1979) A monographic revision of the genus *Sclerotinia. Mycotaxon*, **9**, 365–444.

Krigsvold, D.T., Garren, K.H. and Griffin, G.J. (1977) Importance of field cultivation and soybean cropping in the spread of *Cylindrocladium crotalariae* within and among peanut fields. *Plant Disease Reporter*, **61**, 495–499.

Krikun, J. and Frank, Z.R. (1981) Metham sodium applied by sprinkler irrigation to control pod rot of peanuts. *Plant Disease*, **66**, 128–130.

Kuhn, C.W. (1965) Symptomatology, host range, and effect on yield of a seed-transmitted peanut virus. *Phytopathology*, **55**, 880–884.

Kuhn, C.W. and Demski, J.W. (1984) Peanut mottle, in *Compendium of Peanut Diseases* (eds D.M. Porter, D.H. Smith and R. Rodriguez-Kabana), American Phytopathological Society, St. Paul, Minn. USA.

Lewis, J.S., Powell, N.L., Garren, K.H. *et al.* (1977) Detection by remote sensing of cylindrocladium black rot in peanut fields during 1974 and 1976. *Proceedings of the American Peanut Research and Education Association*, **9**, 28 (Abstract).

Lucas, G.B. (1965) *Diseases of Tobacco*, The Scarecrow Press, Inc., New York, 778 pp.

Luttrel, E.S. and Boyle, L.W. (1960) Leaf spot of peanut in Georgia caused by *Leptosphaerulina arachidicola. Plant Disease Reporter*, **44**, 609–611.

Lyle, J.A. (1964) Development of cercospora leaf spot of peanut. *Journal of the Alabama Academy of Sciences*, **30**, 9.

Lynch, R.E., Demski, J.W., Branch, W.D. *et al.* (1988) Influence of peanut stripe virus on growth, yield, and quality of Florunner peanut. *Peanut Science*, **15**, 47–52.

Machmud, M. (1986) Bacterial wilt in Indonesia, in *Bacterial wilt disease in Asia and the*

South Pacific (ed. G.J. Persley), Proceedings of an international workshop held at PCARRD, Los Baños, Philippines, 8–10 October, 1985. ACIAR Proceedings No. 13. Australian Center for International Agricultural Research, Canberra, pp. 30–34.

Machmud, M. and Middleton, K.J. (1990) Seed infection and transmission of *Pseudomonas solanacearum* on groundnut, in *Bacterial wilt of groundnut* (eds K.J. Middleton and A.C. Hayward). Proceedings of an ACIAR/ICRISAT collaborative research planning meeting held at Genting Highlands, Malaysia, 18–19 March 1990. ACIAR Proceedings No. 31. Australian Center for International Agricultural Research, Canberra, pp. 5–7.

Mallaiah, K.V. (1976) A note on the seasonal changes in the incubation time of groundnut rust. *Current Science*, **45**, 26.

Mallaiah, K.V. and Rao, A.S. (1976); Some observations of pepper spot disease of groundnut. *Current Science*, **45**, 737–738.

Mallaiah, K.V. and Rao, A.S. (1979) Survival of groundnut rust in India. *Indian Journal of Microbiology*, **19**, 209–213.

Marasas, W.F.O., Pauer, G.D. and Boerema, G.H. (1974) A serious leaf blotch disease of groundnuts (*Arachis hypogaea* L.) in southern Africa caused by *Phoma arachidicola* sp. nov. *Phytophylactica*, **6**, 195–202.

Mason, J.L. (1964) *Thielaviopsis basicola* (Berk & Br.) Ferraris, a pathogenic fungus on the peanut *Arachis hypogaea* L. MS Thesis, New Mexico State University, Las Cruces, 61 pp.

Mathur, S.B., Singh, A. and Joshi, L.M. (1967) Varietal response to *Sclerotium bataticola*. *Plant Disease Reporter*, **51**, 649–651.

McCarter, S.M. and Littrell, R.H. (1968) Pathogenicity of *Pythium myriotylum* to several grass and vegetable crops. *Plant Disease Reporter*, **52**, 179–183.

McDonald, D. (1969) The influence of the developing groundnut fruit on soil mycoflora in Nigeria. *Transactions of the British Mycological Society*, **53**, 394–406.

McDonald, D. (1978) *The groundnut disease situation in Northern Nigeria*. Miscellaneous Paper, Institute of Agricultural Research, Ahmadu Bello University, Samaru, Zaria, Nigeria.

McDonald, D. and Fowler, A.M. (1977) Control of cercospora leaf spot of groundnuts in Nigeria. *Nigerian Journal of Plant Protection*, **2**, 43–59.

McDonald, D. and Raheja, A.K. (1980) Pests, disease and crop protection in groundnuts, in *Advances in Legume Science*, (eds J. Summerfield and A.H. Bunting), Royal Botanic Gardens, Kew.

McDonald, D., Subrahmanyam, P., Gibbons, R.W. and Smith, D.H. (1985) *Early and late leaf spots of groundnut*. Information Bulletin No. 21, ICRISAT, Patancheru.

McDonald, D. and Subrahmanyam, P. (1992) Rust of groundnut, in *Plant Diseases of International Importance*, (eds U.S. Singh, A.N. Mukhopadhyay, J. Kumar and H.E. Chaube), Prentice Hall Inc., Englewood Cliffs, New Jersey, pp. V.2:272–284.

McGill, J.F. and Samples, L.E. (1965) *Peanuts in Georgia*. Agricultural Extension Service Bulletin 640, University of Georgia, 30 pp.

Mehan, V.K. (1988) The Mycotoxin Problem in *Groundnut*, (ed. P.S. Reddy), Publication and Information Division, Indian Council of Agricultural Research (ICAR), New Delhi, India, pp. 526–541.

Mehan, V.K. and McDonald, D. (1982) Mycotoxin-producing fungi in groundnuts – Potential for mycotoxin contamination, in *Proceedings of the International Symposium on Mycotoxins and Phycotoxins*, 1–3 September 1982, Vienna, Austria, pp. 98–101.

Mehan, V.K. and McDonald, D. (1983) Mycotoxin contamination in groundnut – Prevention and Control, in *Proceedings of the symposium on Mycotoxins in Food and Feed* (eds K.S. Bilgrami, T. Prasad and K.K. Sinha), Bhagalpur, 11–12 February 1983. ICRISAT, Patancheru, pp. 237–250.

Mehan, V.K., McDonald, D., Haravu, L.J. and Jayanthi, S. (1991) The groundnut aflatoxin problem: review and literature database, Patancheru. ICRISAT.

Mehan, V.K., McDonald, D. and Rajgopalan, K. (1987) Resistance of peanut genotypes to seed infection by *Aspergillus flavus* in field trials in India. *Peanut Science*, **14**(2), 17–21.

Mehan, V.K. and Reddy, D.D.R. (1987) Investigations on a nematode disease of groundnut caused by *Tylenchorhynchus brevilineatus*. *Groundnut Pathology Progress Report* No. 19. ICRISAT, Patancheru, 64 pp.

Mercer, P.C. (1977) Pests and diseases of groundnuts in Malawi. I.Virus and foliar diseases. *Oléagineux*, **32**, 483–488.

Middleton, K.J. and Saleh, N. (1988) Peanut stripe virus disease in Indonesia and the ACIAR project, *Summary proceedings of the First Meeting to Coordinate Research on Peanut Stripe Virus Disease of Groundnut*, 9–12 June 1987, Malang, Indonesia. ICRISAT, Patancheru, pp. 4–6.

Miller, J.M. and Harvey, H.W. (1932) Peanut wilt in Georgia. *Phytopathology*, **22**, 371–383.

Miller, L.I. and Troutman, J.L. (1966) Stunt disease of peanuts in Virginia. *Plant Disease Reporter*, **50**, 139–143.

Minton, N.A. and Baujard, P. (1990) Nematode parasites of peanut, in *Plant-parasitic Nematodes in Subtropical and Tropical Agriculture*, (eds M. Luc, R.A. Sikora and J. Bridge) CAB International, Wallingford, pp. 285–320.

Minton, N.A. and Bell, D.K. (1969) *Criconemoides ornatus* parasitic on peanuts. *Journal of Nematology*, **1**, 349–351.

Minton, N.A. and Morgan, L.W. (1974) Evaluation of systemic and non-systemic pesticides for insect and nematode control in peanut. *Plant Science*, **1**, 91–98.

Mridha, A.U. and Fakir, G.A. (1978) Seed transmission of *Macrophomina phaseolina* and *Sclerotium rolfsii* in groundnut (*Arachis hypogaea* L.) in Bangladesh. *Bangladesh Journal of Botany*, **7**, 31–34.

Misonou, T. (1973) New black root rot disease in soybeans and peanuts caused by *Cylindrocladium crotalariae*. *Shokubutsu Boeki*, **27**, 77–82 (translated from the Japanese).

Mixon, A.C. and Rogers, K.M. (1973) Peanut accessions resistant to seed infection by *Aspergillus flavus*. *Agronomy Journal*, **65**, 560–562.

Morwood, R.B. (1953) Peanut pre-emergence and crown rot investigtions. *Queensland Journal of Agricultural Science*, **10**, 222–236.

Mulder, J.L. and Holliday, P. (1974) Mycosphaerella arachidis. *CMI descriptions of pathogenic fungi and bacteria*, No. 411. Commonwealth Mycological Institute, Kew.

Mukiibi, J. (1975) Minor diseases of groundnuts in Uganda. *East African Forestry Journal*, **62**, 164–165.

Murant, A.F., Rajeshwari, R., Robinson, D.J. and Raschke, J.H. (1988) A satellite RNA of groundnut rosette virus that is largely responsible for symptoms of groundnut rosette disease. *Journal of General Virology*, **69**, 1479–1486.

Natural, M.P., Valencia, L.D. and Pua, A.R. (1988) Peanut – a natural host of *Pseudomonas solanacearum* in the Philippines. *ACIAR Bacterial Wilt Newsletter*, **3**, 3.

Nayudu, M V (1963) *Leptosphaerulina arachidicola* on groundnut. *Indian Phytopathology*, **16(4)**, 384–386.

Nelson, S.C., Starr, J.L. and Simpson, C.E. (1988) Resistance to *Meloidogyne arenaria* in exotic germplasm of the genus *Arachis*. *Journal of Nematology*, **20**, 651.

Netscher, C. (1975) Studies on the resistance of groundnut to *Meloidogyne* spp. in Senegal. *Cahiers ORSTOM Ser. Biol.*, **10**, 227–232.

Nolt, B.L. and Reddy, D.V.R. (1984) Peanut clump, in *Compendium of Peanut Diseases* (eds D.M. Porter, D.H. Smith and R. Rodriguez-Kabana), American Phytopathological Society, St. Paul, Minn., USA, pp. 50–51.

Nolt, B.L., Rajeshwari, R., Reddy, D.V.R. *et al.* (1988) Indian peanut clump virus isolates: Host range, symptomatology, serological relationships, and some physical properties. *Phytopathology*, **78**, 310–313.

Norse, D. (1974) *Plant diseases in Barbados*. Phytopathology Paper No. 8, Commonwealth Mycological Institute Kew, p. 4.

Odvody, G.N. and Dunkle, L.D. (1979) Charcoal stalk rot of sorghum : Effect of environment on host–parasite relations. *Phytopathology*, **69**, 250–254.

Ojeda, H.R. (1966) La 'Sarna' o 'Verrugosis' del mani enfermedad observada por primera vez en Argentina. *Boletin Catedra Genetica Fitotecnia*, **2**, 1–6.

Opio, A.F. and Busolo-Bulafu, C.M. (1990) Status of bacterial wilt on groundnut in Uganda, in *Bacterial wilt of groundnut*, (eds K.J. Middleton and A.C. Hayward), Proceedings of an ACIAR/ICRISAT collaborative research planning meeting held at Genting Highlands, Malaysia, 18–19 March 1990, ACIAR Proceedings No. 31, Australian Centre for International Agricultural Research, Canberra, pp. 54–55.

Owens, J.V. (1951) The pathological effects of *Belonolaimus gracilis* on peanuts in Virginia. *Phytopathology*, **41**, 29.

Paguio, O.R. and Kuhn, C.W. (1974) Incidence and source of inoculum of peanut mottle virus and its effect on peanut. *Phytopathology*, **64**, 60–64.

Paguio, O.R. and Kuhn, C.W. (1976) Aphid transmission of peanut mottle virus. *Phytopathology*, **66**, 473–476.

Palm, B.T. (1922) Aanteekeningen over slizmziekte in *Arachis hypogaea* (Katjang tanah). *Inst. V. Plantenziekten, Meded.*, **52**, p. 41 (English summary).

Pande, S., Sharma, B.P., Ranga Rao, G.V., Kumar Rao, J.V.D.K. *et al.* (1993) First report of pepper spot and leaf scorch on groundnut in Nepal. *International Arachis Newsletter*, **13**, 8–9.

Papavizas, G.S. and Davey, C.B. (1960) Rhizoctonia disease of bean as affected by decomposing green plant materials and associated mycofloras. *Phytopathology*, **50**, 516–522.

Patel, H.R., Vaishnav, M.U. and Dhruj, I.U. (1985) Interaction of *Meloidogyne arenaria* and *Fusarium solani* on groundnut. *Indian Journal of Nematology*, **15**, 98–99.

Patel, H.R., Vaishnav, M.U. and Dhruj, I.U. (1986) Efficacy of aldicarb sulfone and carbofuran flowable seed treatment on plant growth and against *Meloidogyne arenaria* on groundnut. *Pesticides*, **20**, 29–31.

Patil, A.S. (1981) Taxonomy, physiology and control of phyllosticta leaf spot of groundnut (*Arachis hypogaea* L.) *Seminar on Rust and other Foliar Diseases of Groundnut*, Mahatma Phule Agricultural University and Coordinated Research Project on oilseed crops. Agricultural Research, Jalgaon, Maharashtra, India, p. 5.

Persley, G.J. (1986) Ecology of *Pseudomonas solanacearum*, the causal agent of bacterial wilt, in *Bacterial Wilt disease in Asia and the South Pacific*, (ed. G.J. Persley) Proceedings of an international workshop held at PCARRD, Los Baños, Philippines, 8–10 October, 1985, ACIAR Proceedings No. 13.

Pettit, R.E. (1984) Yellow-mold and aflatoxin, in *Compendium of Peanut Diseases* (eds D.M. Porter, D.H. Smith and R. Rodriguez-Kabana), American Phytopathological Society, St. Paul, Minn. USA, pp. 35–36.

Pettit, R.E., Taber, R.A. and Harrison, A.L. (1968) *Leptosphaerulina-cercospora on peanuts in Texas (Abstract)*. *Phytopathology*, **58**, 1063.

Pettit, R.E., Taber, R.A. and Harrison, A.L. (1973) Ascochyta web blotch of peanuts. *Phytopathology*, **63**, 447 (Abstract).

Pettit, R.E., Taber, R.A., Schroeder, H.W. and Harrison, A.L. (1971) Influence of fungicides and irrigation practices on aflatoxin in peanuts before digging. *Applied Microbiology*, **22**, 629–634.

Philley, G.L. (1975) Peanut web-blotch: Growth, pathogenesis, and hosts of the causal agent, *Mycosphaerella argentinensis*. PhD Thesis, Texas A & M University, College Station, 114 pp.

Phipps, P.M. (1985) Web blotch of peanut in Virginia. *Plant Disease*, **69**, 1097–1099.

Phipps, P.M. and Beute, M.K. (1977) Influence of soil temperature and moisture on the severity of cylindrocladium black rot in peanut. *Phytopathology*, **67**, 1104–1107.

Phipps, P.M. and Powell, N.L. (1984) Evaluation of criteria for the utilization of peanut leaf spot advisories in Virginia. *Phytopathology*, **74**, 1189–1193.

Porter, D.M. (1970) Peanut wilt caused by *Pythium myriotylum*. *Phytopathology*, **60**, 393–394.

Porter, D.M. (1977) The effect of chlorothalonil and benomyl on the severity of *Sclerotinia sclerotiorum* blight of peanuts. *Plant Disease Reporter*, **61**, 394–395.

Porter, D.M. (1980a) Increased severity of sclerotinia blight in peanuts treated with captafol and chlorothalonil. *Plant Disease*, **64**, 394–395.

Porter, D.M. (1980b) Sclerotinia blight of peanut : A disease of major importance in the USA, in *Proceedings of the International Workshop on Groundnuts*, (eds R.W. Gibbons and J.V. Mertin) at ICRISAT, 13–17 October ICRISAT, Patancheru, pp. 177–185.

Porter, D.M. (1984a) Sclerotinia blight, in *Compendium of Peanut Diseases*, (eds D.M. Porter, D.H. Smith and R. Rodriguez-Kabana), The American Phytopathological Society, St. Paul, Minnesota, pp. 17–18.

Porter, D.M (1984b) Diplodia collar rot, in *Compendium of Peanut Diseases*, (eds D.M. Porter, D.H. Smith and R. Rodriguez-Kabana), American Phytopathological Society, St. Paul, Minnesota, pp. 28–29.

Porter, D.M. (1984c) Botrytis blight, in *Compendium of Peanut Diseases* (eds D.M. Porter, D.H. Smith and R. Rodriguez-Kabana), American Phytopathological Society, St. Paul, Minnesota, pp. 32–33.

Porter, D.M., Garren, K.H. and van Schaik, P.H. (1975) Pod breakdown resistance in peanuts. *Peanut Science*, **2**, 15–18.

Porter, D.M., Garren, K.H., Mozingo, R.W. and van Schaik, P.H. (1971) Susceptibility of peanuts to *Leptosphaerulina crassiasca* under field conditions. *Plant Disease Reporter*, **55**, 530–532.

Porter, D.M. and Hammons, R.O. (1975) Differences in plant and pod reaction of peanut lines to infection by *Diplodia gossypina*. *Peanut Science*, **2**, 23–25.

Porter, D.M. and Powell, N.L. (1978) Sclerotinia blight development in vines injured by tractor tires. *Peanut Science*, **5**, 87–90.

Porter, D.M. and Smith, J.C. (1974) Fungal colonization of peanut fruit as related to southern corn root-worm injury. *Phytopathology*, **64**, 249–252.

Porter, D.M., Smith, D.H. and Rodriguez-Kabana, R. (1982) Peanut Diseases, in *Peanut Science and Technology*, (eds H.E. Pattee and C.T. Young), American Peanut Research and Education Society, Yoakum, Texas, pp. 326–410.

Porter, D.M., Smith, D.H. and Rodriguez-Kabana, R. (1984) *Compendium of Peanut Diseases*, American Phytopathological Society, St. Paul, Minnesota, USA.

Prinsloo, G.C. (1980) *Thielaviopsis basicola* associated with a pod disease of groundnuts in South Africa. *Phytophylactica*, **12**, 25–26.

Purss, G.S. (1961) Wilt of peanuts (*Arachis hypogaea* L.) in Queensland, with particular reference to verticilliam wilt. *Queensland Journal of Agriculture Science*, **18**, 453–462.

Purss, G.S. (1962) Peanut diseases in Queensland. *Queensland Agriculture Journal*, **88**, 540–553.

Rao, V.G. (1963) The genus *Phyllosticta* in Bombay – Maharashtra. *Sydowia*, **16**, 275–283.

Rao, R.D.V.J.P., Chakrabarty, S.K. and Reddy, A.S. (1988a) First report of peanut stripe virus from India. *Plant Disease*, **72**, 912.

Rao, R.D.V.J.P., Reddy, A.S. and Chakravarty, S.K. (1988b) Survey for peanut stripe virus in India. *Indian Journal of Plant Protection*, **16**, 99–102.

Rao, R.D.V.J.P., Reddy, A.S., Chakrabarty, S.K. *et al.* (1989) Peanut stripe virus research in India, *Summary proceedings of the Second Meeting to Coordinate Research on Peanut Stripe Virus Disease of Groundnut*, (eds S.A. Beckerman, D.V.R. Reddy and D. McDonald). 1–4 August 1989, ICRISAT Center. ICRISAT, Patancheru, pp. 8–9.

Rao, R.D.V.J.P., Reddy, A.S., Chakrabarty, S.K. *et al.* (1991) Identification of peanut stripe virus resistance in wild *Arachis* germplasm. *Peanut Science*, **18**, 1–2.

Raper, K.B. and Fennell, D.I. (1965) *The genus Aspergillus*, Williams and Wilkins Co., Baltimore, Maryland, 608 pp.

Rechcig, N.A., Tolin, S.A., Grayson, R.L. and Hooper, G.R. (1989) Ultrastructural comparison of peanut infected with stripe and blotch variants of peanut stripe virus. *Phytopathology*, **79**, 156–161.

Reddy, D.V.R. (1984) Tomato spotted wilt virus, in *Compendium of Peanut Diseases*, (eds D.M. Porter, D.H. Smith and R. Rodriguez-Kabana), American Phytopathological Society, St. Paul, Minnesota.

Reddy, D.V.R. (1991) Groundnut viruses and virus diseases: distribution, identification and control. *Review of Plant Pathology*, **70**, 665–677.

Reddy, D.V.R., Subrahmanyam, P., Sankar Reddy, G.H. *et al.* (1984) A nematode disease of peanut caused by *Tylenchorhynchus brevilineatus*. *Plant Disease*, **68**, 528–529.

Reddy, D.V.R., Wightman, J.A., Beshear, R.J. (1990) *Bud necrosis: a disease of groundnut caused by tomato spotted wilt virus*, Information bulletin No. 31, ICRISAT, Patancheru.

Rodriguez-Kabana, R., Backman, P.A. and Williams, J.C. (1975) Determination of yield losses to *Sclerotium rolfsii* in peanut fields. *Plant Disease Reporter*, **59**, 855–858.

Rodriguez-Kabana, R. and King, P.S. (1979) Relation between the method of incorporation of systemic nematicides into soil and their effectiveness against root-knot nematodes on peanut. *Nematropica*, **9**, 167–172.

Rodriguez-Kabana, R. and King, P.S. (1985) Evaluation of selected nematicides for control of *Meloidogyne arenaria* in peanut: a multi-year study. *Nematropica*, **15**, 155–164.

Rodriguez-Kabana, R. and Morgan-Jones, G. (1987) Novel rotations and organic materials show promise for management of nematodes. Alabama Agricultural Experiment Station, *Highlights of Agricultural Research*, **34**, 13.

Rodriguez-Kabana, R., Backman, P.A. and Williams, J.C. (1975) Determination of yield losses to *Sclerotium rolfsii* in peanut fields. *Plant Disease Reporter*, **59**, 855–858.

Rodriguez-Kabana, R., Beute, M.K. and Backman, P.A. (1979) Effect of dibromochloropropane fumigation on the growth of *Sclerotium rolfsii* and the incidence of southern blight in field-grown peanuts. *Phytopathology*, **69**, 1219–1222.

Rodriguez-Kabana, R., King, P.S. and Pope, M.H. (1981) Comparison of in-furrow applications and banded treatments for control of *Meloidogyne arenaria* in peanuts. *Nematropica*, **11**, 53–67.

Ross, L.F., Lynch, R.E., Conkerton, E.J. *et al.* (1989) The effect of peanut stripe virus infection on peanut composition. *Peanut Science*, **16**, 43–45.

Rothwell, A. (1962) Diseases of groundnuts in Southern Rhodesia. *Rhodesia Agriculture Journal*, **59**, 199–201.

Rowe, R.C. and Beute, M.K. (1975) Ascospore formation and discharge by *Calonectria cerotalariae*. *Phytopathology*, **65**, 393–398.

Sakhuja, P.K. and Sethi, C.L. (1985) Screening of groundnut germplasm for resistance to root-knot nematode, *Meloidogyne javanica*. *Indian Journal of Nematology*, **15**, 128–130.

Saksena, H.K., Singh, G.P. and Nath, S. (1967) Blight disease of groundnut in India. *Indian Phytopathology*, **20**, 67–69.

Saleh, N., Middleton, K.J., Baliadi, Y. *et al.* (1989) Research on peanut stripe virus in Indonesia, *Summary proceedings of the Second Coordinator's Meeting on Peanut Stripe Virus*, (eds S.R. Beckerman, D.V.R. Reddy and D. McDonald), 1–4 August 1989, ICRISAT Center, ICRISAT, Patancheru, pp. 9–10.

Sasser, J.N., Cooper, W.E. and Bowery, T.G. (1960) Recent development in the control of sting nematode *Belonolaimus* on peanut with 1,2-dibromo-3-chloropropane and EN 18133. *Plant Disease Reporter*, **44**, 733–737.

Sasser, J.N. and Freckman, D.W. (1987) A world perspective on nematology: The role of society, in *Vistas on nematology: commemoration of the twenty-fifth anniversary of the Society of Nematologists*, (eds J.A. Veech and D.W. Dickson), Society of Nematologists, Hyattsville, Maryland, pp. 7–14.

Sasser, J.N. and Kirby, M.F. (1979) *Crop cultivars resistant to root-knot nematodes, Meloidogyne spp., with information on seed sources*, Department of Plant Pathology, North Carolina State University and US Agency of International Development, Raleigh, NC. 24 pp.

Satya, H.N. (1964) A new species of *Pestalotiopsis* on *Arachis hypogaea* L. from Bhopal. *Current Science*, **33**, 57.

Savary, S. and Zadoks, J.C. (1992a) Analysis of crop loss in the multiple pathosystem groundnut-rust-late leaf spot. I. Six experiments. *Crop Protection*, **11**, 99–109.

Savary, S. and Zadoks, J.C. (1992b) Analysis of crop loss in the multiple pathosystem groundnut-rust-late leaf spot. II. Study of the interactions between diseases and intensification in factorial experiments. *Crop Protection*, **11**, 110–120.

Sawada, K. (1959) *Descriptive catalogue to Taiwan (Formosan) fungi*, Part XI, National Taiwan University College of Agriculture Special Publication No. 8, p. 268.

Sechet, M. (1955) A *Pleospora* parasitic on groundnut leaves. *Oléagineux*, **10**, 414.

Shanmugham, N. and Govindaswamy, C.V. (1973a) Varietal susceptibility of groundnut (*Arachis hypogaea* L.) to *Macrophomina* rot. *Madras Agriculture Journal*, **60**, 591.

Shanmugham, N. and Govindaswamy, C.V. (1973b) Control of macrophomina root rot of groundnut. *Madras Agricultural Journal*, **60**, 500–503.

Shanta, P. (1960) Studies on cercospora leaf spot of groundnuts (*Arachis hypogaea*). *Journal of Madras University*, B **30**, 167–177.

Sharma, N.D., Vyas, S.C. and Jain, A.C. (1978) A groundnut disease new to India. *Science and Culture*, **44**, 119–120.

Sharma, S.B. (1988) Survey for nematode diseases of groundnut in some coastal and Rayalaseema districts of Andhra Pradesh, India. *International Arachis Newsletter*, **3**, 11–13.

Sharma, S.B. (1989) Further investigations on the role of plant-parasitic nematodes in crop growth variability of groundnut in Niger. *Legumes Pathology Progress Report* 8, ICRISAT, Patancheru, 61 pp.

Sharma, S.B. and McDonald, D. (1990a) A world list of plant-parasitic nematodes associated with groundnut. *International Arachis Newsletter*, **7**, 13–18.

Sharma, S.B. and McDonald, D. (1990b) Global status of nematode problem of groundnut, pigeonpea, chickpea, sorghum and pearl millet and suggestions for future work. *Crop Protection*, **9**, 453–458.

Sharma, S.B., Subrahmanyam, P. and Starr, E. (1990) Plant-parasitic nematodes associated with groundnut in Niger. *Tropical Pest Management*, **36**, 71–72.

Shew, H.G. and Beute, M.K. (1979) Evidence for the involvement of soilborne mites in Pythium pod rot of peanut. *Phytopathology*, **69**, 204–207.

Shokes, F.M. and Gorbet, D.W (1991) Commercial fungicide use for leaf spot on a partially-resistant peanut (*Arachis hypogaea*) cultivar, *Proceedings, Soil and Crop Science Society of Florida*, 50. North Florida Research and Education Center, Quincy pp. 37–40.

Shokes, F.M., Lyda, S.D. and Jordan, W.R. (1977) Effect of water potential on the growth and survival of *Macrophomina phaseoli*. *Phytopathology*, **67**, 239–241.

Simbwa-Bunnya, M. (1972) Resistance of groundnut varieties to bacterial wilt (*Pseudomonas solanacearum*) in Uganda. *East African Agricultural and Forestry Journal*, **37**, 341–343.

Singh, B.P., Shukla, B.N. and Dubery, K.M. (1975) Studies on varietal susceptibility, assessment of loss, and control of anthracnose of groundnuts (*Arachis hypogaea* L.). *Pesticides*, **5**, 36–37.

Siva Rao, D.V., Srinivasan, S. and Raja Reddy, C. (1986) Reaction of selected groundnut cultivars to nematode infection (*Tylenchorhynchus brevilineatus*) under field conditions. *Tropical Pest Management*, **32**, 168–169.

Smith, D.H. (1972) *Arachis hypogaea*. A new host of *Cristulariella pyramidalis*. *Plant Disease Reporter*, **56**, 796–797

Smith, D.H (1984) Early and late leaf spots, in *Compendium of Peanut Diseases* (eds D.M. Porter, D.H. Smith and R. Rodriguez-Kabana), American Phytopathological Society, St. Paul, Minnesota, pp. 5–7.

Smith, D.H. (1986) Disease-forecasting method for groundnut leaf spot diseases, in *Proceedings of an International Symposium – Agrometerology of groundnut*, 21–26 August 1985, ICRISAT, Sahelian Center, Niamey, Niger. ICRISAT, Patancheru, pp. 239–242.

Smith, D.H. and Crosby, F.L. (1973) Aerobiology of two peanut leaf spot fungi. *Phytopathology*, **63**, 703–707.

Smith, D.H. and Littrell, R.H. (1980) Management of peanut foliar diseases with fungicides. *Plant Disease*, **64**, 356–361.

Smith, D.H., Pauer, G.D.C. and Shokes, F.M. (1992) Cercosporidium and cercospora leaf spots of peanut (Groundnut), in *Plant Diseases of International Importance*, (eds U.S. Singh, A.N. Mukhopadhyay, J. Kumar and H.S. Chaube), Prentice Hall Inc., Englewood Cliffs, New Jersey, pp. V.2:285–304.

Smith, O.D., Boswell, T.E. and Thames, W.H. (1978) Lesion nematode resistance in peanuts. *Crop Science*, **18**, 1008–1011.

Smith, R.S., Jr (1966) Effect of diurnal temperature fluctuations on the charcoal rot disease of *Pinus lambertiana*. *Phytopathology*, **56**, 61–64.

Smith, T.E. (1960) Occurrence of Verticillium wilt on peanuts. *Plant Disease Reporter*, **44**, 435.

Soave, J., Paradela Filho, O., Riebeiro, I.J.A. *et al.* (1973) Avaliacao da resistancea de variedade de amendoim (*Arachis hypogaea* L.) a verrugose (*Sphaceloma arachidis* Bit. & Jenk.) em condicoes de Campo. *Revista Agricolo*, **48**, 129–132.

Sokhi, S.S. and Jhooty, J.S. (1982) Factors associated with resistance to *Puccinia arachidis*. *Peanut Science*, **9**, 96–97.

Spegazzini, C.L. (1884) Fungi Guaranitici. *Anales Sociedad Cientifica Argentina*, **17**, 69–96 and 119–134. *Puccina arachidis* Speg. (n.sp.), p. 90.

Starr, J.L. (1984) Expression of resistance in peanuts, *Arachis hypogaea* to *Pratylenchus brachyurus*: Impact on screening for resistance. *Journal of Nematology*, **16**, 404–406.

Steiner, G. (1949) Plant nematodes the grower should know. *Proceedings of Soil Crop Science Society of Florida*, 1942, **4**, 72–117.

Subrahmanyam, P. (1979) Leaf blight of groundnut caused by *Myrothecium roridum*. *Food and Agriculture Organisation Plant Protection Bulletin* **27**, 95–96.

Subrahmanyam, P., Gibbons, R.W., Nigam, S.N. and Rao, V.R. (1980a) Screening methods and further sources of resistance to peanut rust. *Peanut Science*, **7**, 10–12.

Subrahmanyam, P. and McDonald, D. (1982) Groundnut rust – its survival and carry-over in India. *Proceedings of the Indian Academy of Sciences, (Plant Science)*, **91**, 93–100.

Subrahmanyam, P. and McDonald, D. (1983) *Rust disease of groundnut*, Information Bulletin No. 13 (ICRISAT), Patancheru.

Subrahmanyam, P., McDonald, D., Gibbons, R.W. and Subba Rao, P.V. (1983) Components of resistance to *Puccinia arachidis* in peanuts. *Phytopathology*, **73**, 253–255.

Subrahmanyam, P., McDonald, D. and Hammons, R.O. (1984) Rust, in *Compendium of Peanut Diseases*, (eds D.M. Porter, D.H. Smith and R. Rodriguez-Kabana), American Phytopathological Society, St. Paul, Minnesota, pp. 7–9.

Subrahmanyam, P., McDonald, D., Siddaramaiah, A.L. and Hedge, R.K. (1981) Leaf spot and veinal necrosis disease of groundnut in India caused by *Alternaria alternata*. *Food and Agriculture Organization, Plant Protection Bulletin*, **29**, 74–76.

Subrahmanyam, P., Mehan, V.K., Nevill, D.J. and McDonald, D. (1980b) Research on fungal diseases of groundnut at ICRISAT, in *Proceedings of the International Workshop on Groundnuts*, ICRISAT Center, 13–18 October 1980, ICRISAT, Patancheru, pp. 193–198.

Subrahmanyam, P. and Rao, A.S. (1977) Fungal infection of groundnut pods and aflatoxin accumulation before harvest. *Proceedings of the Indian Academy of Sciences*, **85**, 432–443.

Subrahmanyam, P. and Ravindranath, V. (1988) Fungal and Nematode Diseases, in *Groundnut* (ed. P.S. Reddy), Publication and Information Division, Indian Council of Agricultural Research (ICAR), New Delhi, pp. 453–554.

Subrahmanyam, P., Reddy, D.V.R., Sharma, S.B. *et al.* (1990) *A world list of groundnut diseases*, Legumes Pathology Progress Report 12, ICRISAT, Patancheru.

Suematu, N. (1924) Ueber eine Botrytiskrankheit der Erdnuss (*Arachis hypogaea* L.) (A botrytis disease of peanuts). *Japan Journal of Botany*, **2**, 35–38.

Sun, D., Chen, C. and Wang, Y. (1981) Resistance evaluation of bacterial wilt (*Pseudomonas solanacearum* E.F. Sm.) of peanut (*Arachis hypogaea* L.) in the People's Republic of China. *Proceedings of American Peanut Research and Education Society, Inc.*, **13**, 21–28.

Tabachnik, M., Devay, J.E. and Wakeman, R.J. (1979) Influence of soil inoculum concentrations on host range and disease reactions caused by isolates of *Thielaviopsis basicola* and comparison of soil assay methods. *Phytopathology*, **69**, 974–977.

Taber, R.A., Pettit, R.E. and Philley, G.L. (1981) Identity of the peanut web blotch fungus in the United States (Abstract). *Journal of American Peanut Research and Education Society*, **13**, 99.

Taber, R.A. (1984) Web blotch, in *Compendium of Peanut Diseases* (eds D.M. Porter, D.H. Smith and R. Rodriguez-Kaban), American Phytopathological Society, St. Paul Minnesota pp. 9–10.

Tan, Y. and Boshou, L. (1990) General aspects of groundnut bacterial wilt in China, in *Bacterial wilt of groundnut*, (eds K.J. Middleton and A.C. Hayward), Proceedings of an ACIAR/ICRISAT collaborative research planning meeting held at Genting Highlands, Malaysia, 18–19 March 1990. ACIAR Proceedings No.31. Australian Center for International Agricultural Research, Canberra, pp. 44–47.

Thompson, S.S. (1978) Control of southern stem rot of peanuts with PCNB plus fensulfothion. *Peanut Science*, **5**, 49–52.

Thouvenel, J.C. and Fauquet, C. (1981) Further properties of peanut clump virus and studies and its natural transmission. *Annals of Applied Biology*, **97**, 99–107.

Thouvenel, J.C., Germain, G. and Pfeiffer, P. (1974) Preuve de l'origine virale du robougrissement au 'clump' de l'Arachide en Haute-Volta et en Sénégal. *C.R. Labdoniadaire Seanus. Academy of Sciences*, **278**, 2807–2849.

Tolin, S.A. (1984) Peanut stunt virus, in *Compendium of peanut diseases* (eds D.M. Porter, D.H. Smith and R. Rodriguez-Kabana), pp. 45–48. American Phytopathological Society, St. Paul, Minnesota, USA.

Tomlinson, D.L. and Mogistein, M. (1989) Occurrence of bacterial wilt of peanut (*Arachis hypogaea*) caused by *Pseudomonas solanacearum* and opportunistic infection of aibika (*Abelmoschus manihot*) in Papua New Guinea. *Plant Pathology*, **38**, 287–289.

Troutman, J.L., Bailey, W.K. and Thomas, C.A. (1967) Seed transmission of peanut stunt virus. *Phytopathology*, **57**, 1280–1281.

van Arsdel, E.P. and Harrison, A.L. (1972) Possible origin of peanut rust epidemics in Texas. *Phytopathology*, **62**, 794 (Abstract).

Varaprasad, K.S. and Sharma, S.B. (1990) First report of the lesion nematode, *Pratylenchus brachyurus* on groundnut in India. *Indian Journal of Plant Protection*, **18**, 140.

Vaziri, A. and Vaughan, E.K. (1976) *Aspergillus niger* crown rot of peanut in Iran. *Plant Disease Reporter*, **60**, 602.

Venkatarama, S. and Kazi, S.K. (1979) A climatic disease calendar for 'tikka' of groundnut. *Journal of Maharashtra Agricultural University*, **4**, 91–94.

von Arx, J.A. (1983) *Mycosphaerella* and its anamorphs. *Proceedings of the Konenklijke Nederlandse Akademie van Wetenschappen, Series (86)*, **1**, 32–43.

Wakman, W., Pakki S. and Hasanuddin, A. (1989) Yield loss of groundnut due to peanut stripe virus in *Summary proceedings of Second Coordinator's Meeting on Peanut Stripe Virus* (eds S.R. Beckerman, D.V.R. Reddy and D. McDonald), 1–4 August 1989, ICRISAT, Patancheru.

Willetts, H.J. and Wong, J.A. (1980) The biology of *Sclerotinia sclerotiorum, S. trifolium* and *S. minor* with emphasis on specific nomenclature. *Botanical Review*, **46**, 101–165.

Wogan, G.N. (1965) *Mycotoxins in Foodstuffs*, MIT Press, Cambridge, Massachusetts, 291 pp.

Wongkaew, S. (1989) Groundnut virus research in Thailand, *Summary proceedings of the Second Coordinators' Meeting on Peanut Stripe Virus*, (eds S.R. Beckerman, D.V.R. Reddy and D. McDonald), 1–4 August 1989, ICRISAT, Patancheru, pp. 12–13.

Wongkaew, S. and Dollet, M. (1990) Comparison of peanut stripe virus isolates using symptomatology on particular hosts and serology. *Oléagineux*, **45**, 267–278.

Woodroof, N.C. (1933) Two leaf spots of peanut (*Arachis hypogaea* L.). *Phytopathology*, **23**, 627–640.

Woronichin, N.H. (1924) New species of fungi from the Caucasus. *III Not. Syst. ex Inst. Crypt. Hort. Bot. Reipubl. Rossicae*, **3**, 31–32 (In Russian).

Xu, Z. (1988) Research on peanut stripe virus disease in the Peoples' Republic of China in *Summary Proceedings of the First Meeting to Coordinate Research on Peanut Stripe Virus Disease of Groundnut*, 9–12 June 1987, Malang, Indonesia. ICRISAT, Patancheru, pp. 6–7.

Xu, Z., Zhang Zong Xi, Chen Kunrong and Chen Jinxen (1989) Groundnut virus research in China, in *Summary proceedings of the Second Meeting to Coordinate Research on Peanut Stripe Virus Disease of Groundnut*, (eds S.R. Beckerman, D.V.R. Reddy and D. McDonald), 1–4 August 1989, ICRISAT Center, India. ICRISAT, Patancheru, pp. 7–8.

Yen, J., Chen, M.J. and Huang, K.T. (1956) Leaf scorch of peanut (a new disease). *Journal of Agricultural Forestry* (Taiwan) **10**, 1–24 (in Chinese, English summary).

Yeh, W.L. (1990) A review of bacterial wilt on groundnut in Guandong Province, People's Republic of China, in *Bacterial wilt of groundnut* (eds K.J. Middleton and A.C. Hayward). Proceedings of an ACIAR/ICRISAT collaborative research planning meeting held at Genting Highlands, Malaysia, 18–19 March 1990. ACIAR Proceedings No. 31, Australian Center for International Agricultural Research, Canberra, pp. 48–51.

Zeyong, X., Zhong, Z.Y., Chan, K.R. and Chen, J.X. (1989) Groundnut virus research in China, in *Summary proceedings of second meeting to coordinate research on Peanut Stripe Virus Disease of Groundnut*, (eds S.R. Beckerman, D.V.R. Reddy and D. McDonald) 1–4 August, 1989. ICRISAT, Patancheru, pp. 7–8.

Published as ICRISATS Publication J.A. No 1468

CHAPTER 11

Groundnut pests

J.A. Wightman and G.V. Ranga Rao

11.1 INTRODUCTION

There have been four major reviews of the literature discussing the insects living on groundnut plants since 1973 (Feakin, 1973; Smith and Barfield, 1982; Wightman *et al.*, 1990; Gahukar, 1992). There are also several more concise accounts dealing with pest problems in general (e.g. Wightman and Amin, 1988; Wightman *et al.*, 1989; Lynch and Douce, 1992), specific topics, such as host plant resistance (Lynch, 1990), and discrete geographical zones such as India (Amin, 1988), developed countries (Biddle *et al.*, 1992) and southern Africa (Wightman, 1988a; 1989; Sohati and Sithanantham, 1990; Sithanantham *et al.*, 1990). Feakin (1973), Redlinger and Davis (1982), Dick (1987a,b) and Wightman *et al.* (1990) provide details of the insect pest problems associated with stored groundnut and their management in developed and developing countries. There is little more to add to what has already been recorded about the post-harvest pests of groundnut – the limited coverage given to them in this chapter should not be taken as an indication that they lack importance. Thus, although the general literature up to 20 years ago was somewhat sparse (despite the publication of a large body of information in primary sources), there have since been attempts to redress the situation.

Smith and Barfield (1982) extended Feakin's (1973) pioneer work by providing an invaluable list of the pest species associated with the crop. There is not yet enough verified or verifiable data to distinguish between those insects that merely live on, under or around groundnut stands without causing appreciable damage, and those that are capable of causing significant or economic reductions in crop yields when their populations reach a particular intensity. We call the latter **pests**, restricting the term to mean 'insects' as opposed to all biotic constraints. For the sake of simplification, we include all yield-reducing arthropods (in particular, myriapods and arachnids) with the insects. The term **intensity** is adopted to indicate that yield loss can be influenced by pest density, the duration of the

The Groundnut Crop: A scientific basis for improvement. Edited by J. Smartt. Published in 1994 by Chapman & Hall, London. ISBN 0 412 408201.

exposure of a crop to a cohort of insects and the combination of the two. **Insect-days** is a convenient measure of insect intensity.

11.1.1 The taxonomic status of groundnut pests

Smith and Barfield's list includes more than 360 species of insect that were known to be associated with the crop in the early 1980s. It could be assumed that this list could be extended considerably after 20 years and that it would be appropriate to include an updated version in this chapter. However, this has not happened because progress is uneven. Lists that supplement Smith and Barfield (1982) are included in this chapter where specialized input has made them available.

The slow progress in extending our knowledge of the groundnut fauna is partly attributable to the fact that many groundnut-growing areas have not been systematically surveyed for insects, especially in Asia and Africa. Secondly, there remains the problem of locating taxonomists and paying for the identification of the insects located during surveys (Wightman, 1988b). This is likely to be a major stumbling block because there is reason to believe, as is explained below, that significant sectors of the all-important soil fauna of the agroecological zones involved have not been described. Thus even if surveys are carried out, many of the insects collected may be new to science. The same problems – lack of knowledge and lack of experts – are associated with the natural enemies of the potential pests. Unfortunately, the trend for the withdrawal of the public money needed to support prestigious centres of insect systematics does not give hope that the situation will improve in the future.

It is admitted that it is not necessary to attach a latin binomial to an insect to know that it is destroying part of a farmer's crop. Indeed the taxa (above the level of genus) most likely to be damaging groundnut crops are well known and are described briefly in section 11.2; Wightman *et al.* (1990) give a more extensive account. However, a certain amount of information is needed about distribution and the biology of insects before management strategies can be developed. Pest species often coexist with closely related, but harmless, members of the same taxon. Specialist input from taxonomists is needed to ensure that non-taxonomists are able to distinguish pests and non-pests at all stages of their life-cycles, preferably in field conditions. These comments reflect a theme that will reappear elsewhere in this chapter: what do we (still) need to know?

11.1.2 Where are the groundnut entomologists?

Three sectors have generated and collated the information about management of groundnut pests: university and government scientists of the national agricultural research systems (NARS) of the countries in which groundnut is important, bilateral research agencies, and the Consultative Group for International Agricultural Research (CGIAR).

The lines of demarcation are blurred. For instance, entomologists from the US Department of Agriculture and two state universities have been part of a bilateral programme of the Peanut Cooperative Research Support Program (P-CRSP) in south-east Asia and West Africa. This has involved American scientists working in the target countries, as well as scientists from less developed countries visiting the USA to tackle some of its endemic peanut research problems. The International Crops Research Institute for the Semi-Arid Tropics (ICRISAT), which is based in peninsular India, is the CGIAR Institute with a mandate including groundnut. The current focus of its two groundnut entomologists is almost entirely on Asia but the ICRISAT Center in India accommodates many scientists from elsewhere, including entomologists from other centres of excellence, such as the United Kingdom's Natural Resources Institute (NRI), to work on mandate-related problems of mutual interest. Much of the germplasm tested for pest resistance by the American scientists in Africa and Asia was supplied through the Genetic Resources Unit at ICRISAT. However, many of the test lines originated in US laboratories and came originally from the farms of many countries. We freely and gratefully acknowledge the international effort that lies behind the production of much of the information contained in this chapter.

A considerable amount of new information has become available in the last 10 years or so. Some of it is had not been published formally at the time of writing (1993) and is therefore presented in more depth than would normally be the case, with due reference and deference being made to its originators.

11.1.3 Emphasis

There is a bias in this chapter towards emphasizing information that is available or that is needed to develop pest management systems for the groundnut farmers of less developed countries. Stress is given to information that is needed to understand the ecology of the insects (section 11.3) and the application of the pest management tools that farmers have at their disposal (the subsequent sections). The next step is to evaluate how effective these tools are, their relationship with other aspects of farm management and how they may be handled to support the farming communities. This is discussed in the final section.

Groundnut insects are not unique just because they live on this rather special crop. We have therefore drawn on examples related to other crops and insects to illustrate specific points.

11.1.4 Integrated pest management: scientists and farmers

This chapter leans towards discussing groundnut pests from the point of view of integrated pest management (IPM). This concept is seen as a general move away from complete reliance on synthetic insecticides for managing (i.e. killing) insects that may have a detrimental effect on crop yields. There may also be a case for suggesting changes to the management of the many farms that grow groundnut without insecticide application. The alternatives are combinations of:

- the provision of crop varieties that are in some way resistant to the most important pest(s);
- sowing crops in combinations or patterns that result in pest outbreaks being diminished in intensity and frequency: such modifications may involve the physical environment, and/or
- changes in the farm environment that encourage the natural enemies of potential pests to aggregate within the cropping system;
- enhancing natural control processes by releasing biological agents such as specific insect pathogens, parasites or predators in the farm environment;
- applying natural or synthetic insecticides only when they are needed and in such a way that the impact of predators and parasites is not diminished.

A common feature of these intertwined alternatives is the emphasis on the prevention of pest outbreaks. It is also clear that they are not necessarily crop specific, and not always even farm specific. There can be good reasons for IPM being the business of the community.

These concepts are the basis of the IPM tool box and involve making management decisions beyond the traditional or conventional gamut of what has been known as pest management. The people who are going to make these decisions are the farmers. The role of the scientists (of several disciplines) and extension workers is to evaluate procedures that may not have occurred to farmers and to demonstrate alternative approaches to procedures that the advisors see as being detrimental. Farmers in developing countries can rarely employ specialists (soil analysts, IPM scouts, disease forecasters, etc.) to help them make decisions. Thus, although a team of scientists can and should ensure that IPM procedures 'fit' particular sets of environments, it is the farmer who has to make the day-to-day and season-by-season decisions according to his or her own perceptions of the resource base of the farm or family, its requirements and aspirations.

Once they have their initial data sets, scientists genuinely interested in IPM should probably consider the need for interaction between the laboratory and the land, the farmer, the advisor and the researcher. General principles can be worked out on research stations but do not gain relevance until they are established as being viable on a number of farmers' fields. However, IPM is knowledge intensive, and some of the available knowledge is presented below.

11.2 THE INSECTS

This section describes the insect taxa most likely to be associated with reduced groundnut production, together with an indication of their distribution and the kind of damage they cause. Further details of their biology and ecology are available in Wightman *et al.* (1990) and in the other publications indicated below.

11.2.1 Soil insects

The insects that live in the soil of groundnut fields are responsible for higher levels of yield loss than foliage feeders. They attack pods and roots, and the foliage via the roots. Anitha (1992) has constructed a key to the pod borers based on damage symptoms.

Soil insects are difficult to manage because farmers usually do not know that they are present until plants die or until the crop is harvested. The prophylactic application of insecticides at sowing is not generally a feasible proposition because of the non-availability of suitable products in many countries, their high cost, and the residue problems many create in the seed. The latter point refers specifically to the lipophilic but highly effective cyclodienes such as dieldrin, aldrin, endrin and heptachlor. The unacceptability of organochlorine residues in the environment in general, and in the oil of groundnut seeds in particular, has meant that farmers have lost access to the only persistent insecticides that give good control of soil insects at a low price. However, they did present a risk to the health of the applicators and of other non-target organisms. Progress is being made in other areas of soil pest management following international recognition that a widespread problem exists.

(a) Isoptera – termites

Termites are pests of groundnut throughout Africa, and in western and southern Asia. Several of the most troublesome species are distributed throughout this rather large region. Their attacks are usually associated with periods of drought and therefore tend to be most serious at the end of the growing season. They are less serious in Asia as a whole because they prefer sandy or at least light, well-drained soil. Their life style is thus not compatible with the lowland paddy systems that dominate much of the agricultural landscape of this continent. However, as Wood and Cowie (1988) list three of the termite genera mentioned below as being pests of upland rice in West Africa, it is apparent that the cultivation of this crop is not a complete barrier to termites. They are also rated as being a major groundnut pest in Nepal (personal observation) and are recognized as pests in southern Asia and Thailand (Wightman *et al.*,1990; Logan *et al.*, 1992).

In addition to being post-harvest pests (section 11.2.3), termites deplete stand density and damage pods and seeds. *Microtermes* spp., especially *obesi* and *lepidus*, penetrate the root system and quickly cause plant death, certainly by their own feeding activity but also by exposing the plant tissue to soil pathogens. Root penetration is usually as a result of their own burrowing but it can occur via the lesions left by white grubs, wire worms, etc. *Nasutitermes* apparently causes similar damage to *Microtermes*. *Ancistrotermes latinotus* also penetrates the roots, but is more likely to attack closer to the crown than *Microtermes* and then tunnel upwards through the parenchyma tissue of the stems. In India, we have also found *Odontotermes obesi*, *wallonensis* and *brunneus* penetrating the roots of groundnut plants, causing damage similar to (but distinguishable from) that of *M. obesus*.

Odontotermes spp, (especially *obesus* and *brunneus*) carry soil up the stems and leaves and remove living tissue from within the shelter of this soil sheeting, which presumably protects the workers from solar radiation and natural enemies.

In Africa, *Macrotermes* spp.and *Hodotermes mossambicus* feed on stems at the level of the crown. They can 'fell' whole plants but more often cut out up to about half of the stems before moving on to another plant, which is not necessarily next to the one they previously attacked. The economic importance of the so-called harvester termites is often underestimated because their activities are likely to go unnoticed. The fallen stems quickly disappear as a result of the activities of other insects, especially soil sheeting termites such as *Odontotermes* spp.

Microtermes spp. also attack pods. Sometimes they penetrate the pods and remove the seeds; more often, they 'scarify' the shell by removing the corky tissue from between the veins of the pod. This weakens the shell considerably, resulting in exposure of the seed and its damage during harvest and subsequent handling. More importantly, scarification also increases the probability that the seed will be contaminated by aflatoxins – carcinogens produced by common soil fungi of the genus *Aspergillus*, especially *A. flavus*.

(b) Dermaptera – earwigs

A little more is now known about earwigs than was indicated by Wightman *et al.* (1990). On the vertsiols (black or clay cotton soils) of the ICRISAT farm in peninsular India, *Euborella annulipes*, *E. plebeja* and *Forcipula quadrispinosa* (Anitha, 1992) can reach population densities that result in damage to more than half of the harvestable pods in a stand. No satisfactory insecticidal control method is documented.

Pitfall trapping in Alabama revealed that predacious earwigs, *Labidura riparia*, were the dominant species living on the soil surface in the peanut fields studied over a two-year period (Kharboutli and Mack, 1991). Thus not all earwigs are potential pests.

(c) Hemiptera – Homoptera

Pseudococcidae – mealybugs Mealybugs, such as the cosmopolitan *Pseudococcus solani*, are frequently found on the roots of groundnut plants. The comparatively small size of colonized plants is an indication that they are capable of reducing per-plant yield and that they may be a pest but further surveys are needed to establish their importance. The pineapple mealybug, *Dysmicoccus brevipes* (Cockerell), is occasionally found in Australian groundnut crops (Brier and Rogers, pers. comm.).

Tettigometridae – hilda(bug) *Hilda patruelis* is one of the most serious constraints to groundnut production in sub-Saharan Africa. This is not only because of the across-farm plant mortality it causes but also because the uncontrollable outbreaks can turn farmers away from this crop for many years.

It lives on the roots of groundnut plants but is equally at home on the stems of some of its other hosts, which include common herbaceous weeds and woody ornamentals. The eggs are parasitized by *Psyllechthrus oophagus* (Hymenoptera: Encyrtidae) but the rate of attack falls off markedly in the rainy season, which is when the groundnut crops are in the ground. There is thus a case for obtaining a better understanding of the ecology of the parasite. Outbreaks of hilda are most likely to occur when crops suffer from drought stress, especially when farmers sow after pre-season showers that are followed by a dry spell. This species is remarkably mobile for an insect that is adapted to life underground.

Hypochthonellidae *Hypochthonella caeca* was recognized as a sporadic pest of groundnut in Zimbabwe by Rose (1962). It has since been found on groundnut roots in Zambia by Wightman. In a letter published in the *International Arachis Newsletter* (No. 9, May 1991, p. 4) Dr M.R. Wilson, International Institute of Entomology, London, indicated that this species is no longer considered to be a member of the Tettigometridae but is the sole member of the Hypochthonellidae.

(d) Hymenoptera

Formicidae – ants Most of the ants encountered in groundnut fields are predators, especially of termites and caterpillars, although many of them tend aphids and other homopterans in exchange for honeydew. On balance, the inevitable presence of ants in groundnut fields (at least in Africa) is, by virtue of their role as predators, a good reason for not promoting the use of insecticides in these farming systems (Wightman and Wightman, 1988).

The only species of ant known to be groundnut pests are *Dorylus orientalis* and *D. labitus* (doryline, blind or red ants). The ants hollow out the pods as they approach maturity, entering by regular holes 2–3 mm in diameter.

Both species have been known as pests of groundnut in India and Malaysia (Dammerman, 1929) for many years. Only in Thailand is the problem sufficiently serious for action to be taken to manage *Dorylus* spp. (poisoned coconut meat baits). It is recently believed to have been found in Philippine groundnut fields (ICRISAT, 1991a). There is thus a good case for a concerted search of groundnut fields in the countries between the Philippines and India to determine its range and perhaps the site characteristics that govern its presence or absence in Asia. Wightman and Wightman have found *Dorylus* sp. under groundnut in Malawi, Zambia and Zimbabwe and have detected pod damage in Malawi.

(e) Coleoptera

Buprestidae – jewel beetles   The jewel beetle *Sphenoptera indica*, which is a root borer, has been known as a widespread resident of groundnut fields in India for many years (Rai, 1979). Information about its potential pest status has recently become available (Logan *et al.*, in press). In a rain-fed field on the ICRISAT Research Farm 20 days before the 1986 rainy-season harvest, it was found that 23% of the groundnut plants had *S. indica* in their roots as larvae, pupae or adults. As the three larval instars tunnel through the parenchyma of the root, a high rate of mortality or severe wilting among the attacked plants was anticipated. On one side of the field, where 76% of the plants were attacked, 82% of the dead plants and 64% of the living plants were host to this species. A survey of groundnut plants growing in three irrigated fields on the ICRISAT farm in the 1986 rainy season revealed infestation rates of 0.4%, 9% and 14%. This indicates that irrigation does not eliminate the risk of attack by this species. Attacks in subsequent years have been sporadic, and spread across the 1300 ha farm; they have been sufficiently heavy to permit us to screen for resistance to this species among *Arachis* spp. A survey in Andhra Pradesh and Karnataka (southern India) in the rainy season of 1992 revealed that up to 10% of the plants growing on red soils (alfisols) had been killed by this species.

Scarabaeidae – white grubs (as larvae); cockchafers, May bugs or June beetles (as adults)   The general importance of white grubs as pests of groundnut in India and in parts of southern Africa has been recognized only recently (Wightman *et al.*, 1990). Since then, further research and discussion with experts from Asian countries has revealed that white grubs are associated with yield loss from India to China and the Philippines. They are also a problem in moister areas of West Africa, such as Senegal.

Appert (1956) noted that 'underground larvae' (i.e. white grubs, Table 11.1) can cause as much damage as millipedes, which are usually recognized as a major constraint to groundnut production in West Africa.

It has therefore been possible to extend the list of white grub species associated with groundnut (Table 11.1; cf. Table 5.10 in Wightman *et al.*, 1990) to include further information from Asia and southern Queensland, Australia, where the identity of the main peanut pest species has been resolved and a key provided (Rogers *et al.*, 1992).

It appears that there are some 70 named species (including subspecies) of white grub known to be associated with the groundnut crop. However, the current maximum exceeds 100 because of the number of undescribed species. This compares with *c.* 52 species of termites (Wightman *et al.*, 1990) and more than 60 lepidopteran species (Smith and Barfield, 1982).

White grubs feed mainly on the taproots and/or the peripheral rootlets. Either way, the net effect is to restrict the growth of the plant. This is particularly marked in sandy soils in drought prone areas such as occur in groundnut fields of the Middle Veldt of Zimbabwe. Seedlings can be killed outright if the phenologies of the crop and pest result in large larvae and small plants occupying the field at the same time. In the Sudan, white grubs have been associated with aflatoxin contamination of groundnut left in the ground for 6 weeks after it should have been harvested (Ahmed *et al.*, 1989).

Heteronyx piceus is the species most likely to cause damage in Queensland. It forms 90% of the scarab population under groundnut in the main growing area. It is unusual in that it attacks the pods but not the roots. Population densities of up to 30 larvae per metre of row nave been reported (Brier and Rogers, unpublished).

Kalshoven (1981) indicated that there are many species of *Anomala* in the Indonesian archipelago. Supriyatin (1991) reported that white grubs attack the roots of groundnut plants in upland crops in that country. As members of this genus are known to eat groundnut roots in other countries, it is likely that there may be an undefined white grub problem in East and/or West Java, the centres of groundnut production in Indonesia. Cadapan and Escano (1991) indicated that *Leucopholis irrorata* has recently been associated with the groundnut crop in the Philippines. This ties in with the major pest status awarded to this species in maize–rice systems (Litsinger *et al.*, 1983).

A survey by Nath and Singh (1987) of cropped fields (mainly groundnut and sugar cane) in a relatively small area of eastern Uttar Pradesh, northern India, added 16 species to the list. They indicated that these species were common to all crops but were most numerous on groundnut and sugarcane. Several of these species were also found during a detailed study of white grubs many miles away in semi-arid Rajasthan, particularly around Jaipur (Yadav, 1981). However, Yadav did not associate them specifically with the groundnut crop in his report, which concentrates on

TABLE 11.1 *Scarabaeidae associated with the groundnut crop in the larval (white grub) stage*

Species	Location	References
Adoretus cribrosus	Zimbabwe	Smith and Barfield, 1982
A. decanus	India	Nath and Singh, 1987
A. laisopygos	India	Nath and Singh, 1987
A. limbatus	India	Nath and Singh, 1987
A. umbrosus	'Africa'	Smith and Barfield, 1982
A. versutus	India	Nath and Singh, 1987
Adoretus spp.	Malawi, Zambia	Wightman
(up to 4 spp.)	Zimbabwe	
Anomala antiqua	Burma	Smith and Barfield, 1982
A. atrovirens	Indonesia	Smith and Barfield, 1982
A. bengalensis	India	Nath and Singh, 1987
A. corpulenta	China (PR)	Xu, B.C., 1982
A. dorsalis	India	Yadav, 1981
A. dorsalis	India	Nath and Singh, 1987
var. *fraterna*		
A. plebeja	Senegal	Appert, 1956
	Burkina Faso	IRAT, 1976
A. punjabensis	India	Yadav, 1981
A. ruficapilla	India	Nath and Singh, 1987
A. rufocuprea	Korea (Rep)	Cho *et al.*, 1989
Anomala spp.	Botswana, India,	Wightman; Yadav, 1981;
(up to 11 spp. in	Indonesia, Malawi,	Kalshoven, 1981
southern Africa)	Zambia, Zimbabwe	
Anomala sp.	Burma	R. Milner, personal communication, not *A. antiqua*
Apogonia cribricollis	India	Nath and Singh, 1987
A. ferruginea	India	Nath and Singh, 1987
A. roucca	India	Nath and Singh, 1987
A. uniformis	India	Nath and Singh, 1987
Autoserica atratula	India	Nath and Singh, 1987
A. insanabilis	India	Nath and Singh, 1987
A. nathani	India	Nath and Singh, 1987
Dyscinetus trachipygus	India	Nath and Singh, 1987
Crator cuniculatus	Burkina Faso	IRAT, 1971
Eulepida mashona	'Africa'	Smith and Barfield, 1982
Heteroligus claudius	Nigeria	Smith and Barfield, 1982
Heteronyx brevicollis	Australia	Smith and Barfield, 1982
H. piceus	Australia	Rogers *et al.*, 1992
H. rugosipennis	Australia	Rogers *et al.*, 1992
H. sp. nr.	Australia	Rogers *et al.*, 1992
rugosipennis	Australia	Rogers *et al.*, 1992
H. diomphalia	China (PR)	Shang *et al.*, 1981
	Korea (Rep)	Cho *et al.*, 1989
Holotrichia consanguinea	India	Smith and Barfield, 1982
H. formosana	China (PR)	Lu *et al.*, 1987

TABLE 11.1 *Cont.*

Species	Location	References
H. morosa	Korea (Rep)	Cho *et al.*, 1989
H. oblita	China (PR)	Wang *et al.*, 1986
H. parallela	China (PR)	Wang *et al.*, 1986
H. sauteri	China (PR)	Huang and Lin, 1987
H. serrata	India	Smith and Barfield, 1982
Lachnosterna caudata[1]	Australia	Smith and Barfield, 1982
L. fissa	India	Smith and Barfield, 1982
Lepidiota sp.	Australia	Gough and Brown, 1988
L. crenita	Australia	D.J. Rogers and H.B. Briers, personal communication
Leucopholis irrorata	Philippines	Cadapan and Escano, 1991
Maladera orientalis	China (PR)	Wang *et al.*, 1986
	Korea (Rep)	Cho *et al.*, 1989
Maladera sp.	Thailand	Sathorn Sirisingh, personal communication
Neodon pecuarius (= *Trissodon puncticollis*)	Australia	Smith, 1946
Oxycetonia versicolor	India	Smith and Barfield, 1982
Pentodon idiota	USSR	Smith and Barfield, 1982
Phyllophaga ephilida	'Americas'	Smith and Barfield, 1982
P. armicans	'Americas'	Smith and Barfield, 1982
Podalgus (Crator) cuniculus	'Africa'	Smith and Barfield, 1982
Popillia japonica	China (PR), N. America	Smith and Barfield, 1982
Pseudoheteronyx basicollis	Australia	P.G. Allsopp, in Rogers *et al.*, 1992
Rhopaea magicornis	Australia	Smith and Barfield, 1982
Schizonycha africana	NE Africa	Smith and Barfield, 1982
	Senegal	Appert, 1956
S. fusca	Malawi	Wightman
S. ruficollis	India	Nath and Singh, 1987
S. straminea	Malawi	Wightman
Schizonycha spp. (up to 8)	Malawi	Wightman
Sericesthis ino	Australia	Rogers *et al.*, 1992
S. suturalis	Australia	Rogers *et al.*, 1992
Schizonycha spp. (up to 3)	Malawi	Wightman
Strigoderma arboricola	USA	Smith and Barfield, 1982
Trochalus pilula	Senegal	Appert, 1956
Trochalus sp.	Malawi	Wightman
Xylotrupes gideon	Burma	Smith and Barfield, 1982
Tribe: Sericini 8 indet. spp.	Malawi, Zambia, Zimbabwe	Wightman

1. not recognized in Australia, possibly *Lepidiota caudata*, see Gough and Brown (1988) (P.G. Allsopp, personal communication)

the predominant *Holotrichia consanguinea*. The predominant species in southern India is *H. serrata* (Veeresh, 1977).

Elateridae – Click beetle larvae or wireworms; Tenebrionidae – false wireworms Wireworms and false wireworms can be treated together because they have a similar morphology and the symptoms of their pod boring activities are indistinguishable: both make symmetrical holes, 2–3 mm in diameter, in the maturing and mature pods. Elaterid larvae have also been discovered eating the tissues of germinating seeds on the ICRISAT research farm in peninsular India. This activity could lead to diseased and malformed plants.

Appert (1956) indicated that wireworms ('taupins'), mentioning *Cardiophorus subspinosus* and *C. cognatus*, can be predators but, together with false wireworms, they can reduce the density of seedling stands by as much as 10%.

The survey carried out by Wightman in southern Africa in 1987 revealed a major problem with both of these taxa: the difficulty of identifying them to species from the larval stages and sometimes the adults. Table 11.2 shows species in the Elateridae and Tenebrionidae added by Wightman (with determinations by the British Museum of Natural History) to those listed by Smith and Barfield (1982).

Chrysomelidae Members of this large family will almost certainly be found tasting the foliage of groundnut wherever it is grown, but they probably cause little damage. The same cannot be said for members of the genus *Diabrotica*, especially *D. undecimpunctata howardi*, the spotted cucumber beetle, in the southern states of the USA. The adults cause relatively minor damage to the terminal leaves but the subterranean larvae can cause major injury to the pods and pegs. As such, it has been recognized as a major pest of this crop throughout this century (Smith and Barfield, 1982).

Curculionidae – Weevils The only species of weevil that has had a high profile as a pest of groundnut is the white fringed weevil, *Graphognathus leucoloma* (Feakin, 1973), especially in the Americas and, more recently, in Australia. The larvae eat the roots and cause stunting of the stems and plant death. The adults (parthenogenetic females) eat the foliage. Each can lay 1000–2000 eggs, hence the high damage potential of this species.

Adult weevils of other species are often numerous in groundnut crops and can often be seen eating the edges of the leaflets. In southern Asia, the ash grey or grey cotton weevil *Myllocerus undecimpustulatus maculosus* is often found at densities of up to 10 per plant. *Systates* spp., *Mesoleurus dentipes* and *Diaecoderus* sp. can reach even higher densities in southern Africa (Jepson, 1948; Rose, 1962; Broad, 1966). Jepson (1948) associated *S. articollis* with the 'yellowing and failure' of a young plantation. This is a

TABLE 11.2 *Additions to species lists (Smith and Barfield, 1982) in Elataridae and Tenebrionidae*

Species	Locations
Elataridae (two previously listed)	
Agrypninae, Agrypnini (larvae) indet.	Mtopwa, Tanzania
Agrypninae, Monocrepidiini (larvae) indet.	Mawengo, Zimbabwe
Elaterina (larvae) possibly 6 spp. indet.	Malawi and Zambia
Tenebrionidae (five previously listed)	
Pimeliinae (larvae) possibly 13 spp.	From 20 locations in Malawi, Zambia and Zimbabwe
Tenebrioninae (larvae) possibly 3 spp.	Malawi and Zimbabwe
Zophosis sp. (adults)	Malawi
Gonocephalum nr *simplex* (adults)	Malawi
Anchophthalamus plicipennis (adults)	Malawi
Drosochrus sp. (adults)	Malawi

significant observation because it is likely that the larval stages of weevils caused the yellowing (by attacking the nodules, cf. *Sitona* spp.) and crop failure by eating the lateral roots. The importance of weevil larvae as subterranean groundnut pests is worthy of further consideration.

(f) Lepidoptera

The caterpillars of several lepidopteran species that live at the soil surface damage groundnut plants – *Agrotis* spp. and *Feltia* spp. feed at the crown, *Spodoptera litura* (in India) and *S. littoralis* (in southern Africa) are pod borers. In Australia, the larvae of *Etiella behrii*, the lucerne seed web moth, penetrate the pods and feed on the seed (Brier, personal communication).

The only true soil-dwelling lepidopteran to cause major damage to groundnut is *Elasmopalpus lignosellus*, the lesser corn stalk borer, which is restricted to the New World (Smith and Barfield, 1982). This species is regarded as a major pest of groundnut and other crops. Larvae feed at or close to the soil surface in the first two instars, paying particular attention to the flower and vegetative buds. The older stages feed on the underground parts of the plant, and often scarify the pods. This results in a high risk of seed contamination with aflatoxin (Lynch and Wilson, 1991).

(g) Myriapoda – millipedes

Millipedes, often more than 30 cm long, are the most important pests of groundnut in the drier areas of West Africa, where they attack the pods, mainly before they have hardened (ICRISAT, 1988). They are also recognized as pod borers of lesser importance in southern Africa, where they damage 5–10% of the immature pods, some of which would not reach maturity by harvest time (Wightman, 1989).

Their importance in West Africa is related to the amount of damage they do, the unpredictable nature of the attack and because there is no control method available, irrespective of the socio-economic status of the farmers who have to contend with them. The literature reviewed by Wightman *et al.* (1991) indicates that the impetus of research carried out on the ecology and control of millipedes in the 1970s has not been maintained.

11.2.2 Insects that live on the leaves and flowers

(a) Orthopteroid orders – grasshoppers, locusts, crickets, mantids

Orthopteroid insects are frequently found in groundnut crops but, with the exception of locust plagues, no record of them achieving pest status has been located. Observation indicates that their presence in groundnut foliage is associated more with sunbathing than feeding. Mantids appear to be an exception. They can achieve relatively high densities in southern Africa (Wightman).

(b) Thysanoptera – thrips

Smith and Barfield (1982) list 18 species that have been associated with groundnut crops. Several species can be added to this list:

Megalurothrips usitatus is frequently encountered in the flowers of groundnut in Asia and southern Africa (Palmer *et al.*, 1990; Wightman *et al.*, 1990).

Scirtothrips aurantii was detected in terminal (folded) leaflets in Malawi (Wightman). *S. oligochaetus* has been found on groundnut in India (Palmer *et al.*, 1990).

Thrips palmi has been found in the terminal leaflets of groundnut in India and other Asian countries (Palmer *et al.*, 1990) and also lives in Australia and the 'Pacific' but has not been recorded from New Zealand and New Guinea (Houston *et al.*, 1991). *T. setosus* is known only from Japan and Korea (Reddy *et al.*, 1991).

Each species has a preferred niche – presumably the feeding site – within a plant. However, flower dwellers can be found in other parts of the plant before flowering occurs (Table 11.3).

Thrips can have pest status in groundnut crops as virus vectors and as leaf-eaters. We believe that thrips are also of considerable pest status because the (largely cosmetic) damage they cause induces farmers to apply insecticides unnecessarily (Lynch *et al.*, 1984; Ranga Rao and Shanower,

TABLE 11.3 *The distribution of thrips between flowers and leaf buds in the rainy and post-rainy seasons at ICRISAT Center, 1990–1991 (pooled data)*

| | Total number of thrips observed (% in parentheses) | | | |
| | Rainy season | | Post-rainy season | |
	Leaf	Flower	Leaf	Flower
Scirtothrips dorsalis	9366 (97)	455 (26)	5274 (85)	153 (6)
Thrips palmi	148 (1.5)	102 (6)	792 (13)	361 (13)
Frankliniella schultzei	145 (1.5)	1202 (68)	118 (2)	2183 (81)

1988; Reddy *et al.*, 1991). This results in outbreaks of other pests because of interference in the natural insect density control processes.

Thrips as virus vectors
Thrips can have pest status in groundnut crops as virus vectors and as leaf-eaters. They are vectors of the tomato spotted wilt virus (TSWV) and its Indian variant, the bud necrosis virus (BNV). The various isolates of this disease can cause widespread damage to groundnut in many countries (Reddy *et al.*, 1991).

Not all groundnut thrips are virus vectors. In India the most important vector is *T. palmi* (not *Frankliniella schultzei*, as was previously reported, although this species can be forced to transmit at a very low rate in laboratory conditions). *Scirtothrips* spp. are not vectors in India (Palmer *et al.*, 1990; Vijayalaksmi pers. comm.).

The most likely vectors in the southern USA are *F. occidentalis* and *F. fusca* (Reddy *et al.*, 1991; Culbreath *et al.*, 1992). Although *T. tabaci* is a vector of TSWV in other crops, it has not been linked with outbreaks of this disease in groundnut crops in the USA or elsewhere (Reddy and Wightman, 1988; Reddy *et al.*, 1991).

Thrips as leaf-eaters
Thrips can cause damage as a result of their feeding activity. The leaf-eating species are usually found between the folded leaflets at the stem tips. They cause little visible damage at the time of feeding. As the leaflets grow, the small lesions and patches of dead cells left by the thrips do not expand at the same rate (if at all) as the undamaged cells. This means that the most conspicuous leaflets at the top of the plant are contorted and have small holes in them. This damage is most apparent in young plants when

the temperature is not high enough to promote rapid growth, e.g. during spring in North America and the post-rainy season in India. The most acute case is that of *Enneothrips flavens* in Brazil. Increases in yield following insecticide application and attributed to the control of this species range from 35% to 50% (Smith and Barfield, 1982).

A debate about the economic status of thrips (mainly *T. fusca*) in southern USA seems to have reverberated around the peanut industry for many years (Lynch *et al.*, 1984). Turnjit (1988) added further light to the situation by indicating that in North Carolina no yield loss will occur until 40–50% of the leaflets are damaged, according to the variety, but then only in plants less than 4 weeks old.

The situation in India is not clear, mainly because the populations of the main leaf damagers (*F. schultzei* and *S. dorsalis*) cohabit with jassids (*Empoasca kerri*) and there is no method of separating out the relationships between density and yield losses attributable to each taxon. The impression is that thrips cause little yield loss, at least on the ICRISAT farm.

(c) Homoptera

Aphididae – aphids The aphid species most frequently linked with groundnut is *Aphis craccivora*, the groundnut or cowpea aphid. It occurs throughout the tropics and subtropics and has many hosts. It normally appears on groundnut crops in the early rainy season, when it can cause considerable damage to young plants. In recent years a second outbreak has occurred in the post-rainy season on the ICRISAT farm and this may be linked with a perceived air pollution problem (Dohman *et al.*, 1984).

This species is of particular significance because it is the vector of the (persistent) groundnut rosette virus complex in Africa. This disease can be crippling but has become less common in recent years, especially in southern Africa, where the widespread adoption of regulatory cultural practices (especially earlier and denser sowing than was once practised) may have limited its potential effects.

A. craccivora is capable of reducing the yield of groundnut crops by means of its feeding activity alone (Mayeux, 1984; Bakhetia and Sidhu, 1976). Our experience in southern India indicates that populations are regulated by coccinellids and other predators and rarely survive a spell of persistent rain.

As representatives of the aerial plankton that overflies agricultural areas, many 'other species' – such as *A. robiniae*, *A. gylycines*, *A gossypii*, *A. solanella*, *Myzus persicae*, *Macrosiphum euphorbiae*, *M. avenae*, *Rhophalosiphum padi* and *Lipaphis erysimi* – are implicated in the non-persistent transmission of most other groundnut virus diseases, including peanut stripe, cucumber mosaic, mottle and peanut stunt viruses. *A. gossypii* transfers groundnut streak necrosis disease (= sunflower yellow blotch virus) to groundnut in the Rift Valley area of southern Africa from *Tridax procumbens*, a common tropical weed (Saleh, 1991; Wightman *et al.*, 1990; Xu Zeyong *et al.*, 1991).

Cicadellidae or Jassidae – jassids or leaf-hoppers   Jassids are a common feature of groundnut crops in most parts of the world. Perhaps they are so familiar that entomologists have not bothered to collect them, because only 20 species have been associated with the groundnut crop (Smith and Barfield, 1982; Wightman *et al.*, 1990). The economically significant genus *Empoasca* predominates – *E. fabae* in the Americas, *E. kerri* in India – but also *Jacobiasca formosana* in south-east Asia (but the latter two inadvertently omitted or not clearly referred to by the aforementioned authors) and *Austroasca alfalfae* in Queensland. The symptoms commonly include pronounced chlorosis followed by peripheral necrosis, a condition known as hopper burn.

There are no definitive accounts of the relationships between their density and crop yield. The main reason for this is the almost inevitable cohabitation with thrips and other insects in the early stages of crop development. The consensus in the USA, where experimentation has been in progress for about 50 years, is that the damage caused by these insects is cosmetic. Smith *et al.* (1985) indicated that the range of reduction in photosynthetic area among 14 groundnut genotypes was 3.8–28%. This is unlikely to have a direct effect on yield in view of groundnut's high leaf area index (> 5 after 40 days). However, the shading of the lower leaflets by damaged leaflets, which are usually on top of the canopy, may have a greater effect than anticipated.

Aleyrodidae – whiteflies   Feakin (1973) does not mention whiteflies and Smith and Barfield (1982) indicate, correctly, that *Bemisia tabaci* is cosmopolitan but add no more. This confirms our suspicion that the observed outbreaks of whiteflies on groundnut in India and southern and south-eastern USA are a recent phenomenon (Shanower and Ranga Rao, 1988, Lynch and Simmons, 1993).

Bemisia tabaci epidemics have created problems associated with the overuse of insecticides in cotton crops in coastal Andhra Pradesh. The appearance of this insect on groundnut is considered to be associated with this general problem but may have been created by the local overuse of insecticides in groundnut. The situation is being monitored in view of the ability of this species to debilitate crops and spread virus diseases.

The situation appears to be rather different in the USA where the cotton strain (biotype A) has been replaced by the poinsettia strain (biotype B), which is capable of defoliating groundnut plants. This was first recognized in 1987 in Florida. Since then, the pest status of this species on groundnut has worsened as populations of the 'new strain' have increased in density and spread through Georgia and Texas (Lynch and Simmons, 1993).

TABLE 11.4 *Miridae associated with the groundnut crop in addition to those listed by Smith and Barfield, 1982*

Species	Location	Source
Creontiades pallidus	Lilongwe, Malawi	J.A. Wightman (collected by sweep net, 1987; det. M.R. Wilson, CIE)
Creontiades sp.	Queensland, Australia	Rogers and Brier (personal communication)
Creontiades pallidifer	New Delhi, India	Singh *et al.*, 1990
Helopeltis sp.	Lilongwe, Malawi	J.A. Wightman
?*Taylorilygus* sp.	Lilongwe, Malawi	J.A. Wightman

(d) Heteroptera

Miridae Mirids can often be found in groundnut crops at low densities – perhaps less than one per plant. Smith and Barfield (1982) list 11 species belonging to eight genera from Africa, India and the USA. Table 11.4 shows additions to that list.

It is not often realized that this taxon can cause considerable damage at low densities. For instance, in an Australian glasshouse experiment set up in Queensland to demonstrate this point, Rogers and Brier (personal communication) found that two adult *Creontiades* sp. per plant reduced flower production by 86% over a 3-week period. Peg initiation showed a corresponding decrease of 87%. The plants resumed flower production 7 days after the mirids were removed, and produced more flowers and pegs than the unexposed control plants during the subsequent 4 weeks. However, it is envisaged that compensation is unlikely to be possible under the conditions of sustained attack that are likely to be encountered in field conditions, so that there are serious implications in finding mirids in a groundnut field. This is especially so in dryland agriculture where the cohort of flowers that produces the main crop results from a particular rainfall event that may not be repeated.

Damage in the Queensland experiment was typical of mirid attack to legume crops in that the flower buds were attacked at their earliest appearance and quickly became necrotic (Sorenson, 1936; Wightman and Whitford, 1982; Clifford *et al.*, 1983). The vegetative buds and other tissue were apparently not attacked in this way so that there was no distortion of the stems and leaves.

The potential of this species to cause damage is demonstrated by Singh *et al.* (1990). The mirid populations that they detected on groundnut in New Delhi peaked at just over one per plant in the rainy seasons of 1987 and 1988 and more than two per plant in the intervening summer crop. This species was present from the second week after sowing until about the pod maturation stage. It is unwise to transfer the conclusions derived from the data of Rogers and Brier working in a glasshouse in Australia to field conditions in India but we suggest that the mirids detected in New Delhi had a marked effect on the rate of flowering and the subsequent yield. Clearly the mirids living in groundnut crops need to be looked at rather closely.

Pentatomidae and Lygaeidae Members of these families are often conspicuous in groundnut crops but we have not attributed yield loss to the 'big bugs'. Their feeding activity appears to be concentrated on the vegetative tissue, especially the growing points, which take on a limp appearance – hence the common name 'tip wilters'. Groundnut entomologists can ignore them, unless they are found to be attacking the reproductive tissues.

(e) Lepidoptera

There are many species of leaf-eating caterpillars found on groundnut plants – Smith and Barfield (1982) list more than 60 – and no doubt a concerted search would find many more. Of these, relatively few are of economic importance or limited to the groundnut crop. We wish to play down the importance of defoliators to groundnut crop production because natural control processes usually keep them at densities well below the economic threshold. However, apparently spontaneous flare-ups can occur and these are usually associated with the injudicious use of insecticides or other examples of suboptimal management.

Spodoptera spp. (the armyworms) are prominent in the list of potential pests, with *S. frugiperda, ornithogalli, latifascia, sunia* and *eridania* predominating in the New World. *S. littoralis* and *exempta* are associated with the groundnut crop in Africa and northern Asia, while *S. litura* extends across the remainder of Asia. *S. exigua* is cosmopolitan.

The heliothine genera can also be pests of groundnut over most of the crop's range, e.g. *Helicoverpa zea* in North America and *Helicoverpa armigera* in Asia, Africa and Australia. *H. punctigera* also feeds on groundnut in Australia but, unlike *H. armigera* (a flower and peg feeder), it is primarily a defoliator (H. Brier, personal communication). The arctiid hairy caterpillars *Amsacta* spp. and *Diacrisia obliqua* are sporadic defoliators in southern Asia. They can appear in devastatingly high numbers in newly emerged rainy season crops. They are polyphagous and, fortuitously, usually have one generation per year.

Gelechiids predominate among the leaf miners, rollers, webbers and tiers. *Aproaerema modicella* (= *A. nertaria, Stomopteryx subsecivella, S. nertaria, Anacampsis nertaria* and *Biloba subsecivella*), the groundnut leaf miner (Mohammad, 1981; Shanower *et al.*, 1993; Wightman *et al.*, 1990), causes widespread damage and is fairly cosmopolitan within Asia. It is known as a sporadic but potentially devastating pest with up to four generations in one crop cycle. It is oligophagous, with a clear preference for leguminous species, especially soybean and groundnut. Other hosts of commercial importance include lucerne, pigeon-pea, mung bean and lab-lab. On groundnut, the first three larval instars are leaf miners: when it gets too large for this habit it becomes a leaf tier (or leaf folder). It pupates between the folded leaflets.

Anarsia ephippias (groundnut leaf webber) can be a pest in northern India. It webs the growing points and its feeding activity on the younger leaflets results in 'shot-holing' as the leaves mature (Bakhetia, 1977).

Acarina – mites

Smith and Barfield (1982) have 17 entries under Acarina, 10 of which are for *Tetranychus* spp. They are potentially a world-wide problem. The high reproductive rate and short generation time of these mites mean that they have a high potential for rapid population increase if the natural control processes are disrupted or if specific environmental conditions are severe. This happens when fungicides reduce the effectiveness of an entomopha-gous fungus, when inappropriate irrigation methods are applied and/or when insecticides kill other natural enemies (Campbell, 1978; G.V. Ranga Rao *et al.*, 1990). In Australia, the peanut mite *Paraplanobia* sp. appears in groundnut fields during periods of prolonged drought. Population densities crash after heavy rainfall (H. Brier, personal communication).

11.2.3 Post-harvest pests

The ecology and management of the post-harvest pests of groundnut have been discussed in full by Dick (1987b) and in Wightman *et al.* (1990). In general, the storage pests *per se* and the approaches to their management are common to many other products and so this aspect is not dwelt upon here. The emphasis is on several pests of the post-harvest situation in the groundnut crop that are particularly relevant to Africa and Asia.

Attention is drawn to the period after the crop is harvested and before it is 'plucked' and stored or bagged pending transfer to the market. During this time it is usually stacked or windrowed to allow it to dry in the sun and wind. This stage is critical in the production of a crop with high yield and good seed quality and one that will not become the origin of contamination with aflatoxin.

Unfortunately, the groundnut crop is not exempt from insect damage during this time. Termites can come from below and remove significant proportions (perhaps 30–40%) of the seeds and as much hay. *Odontotermes* spp. are the most conspicuous in this regard, especially in Africa (Logan *et al.*, 1990; Logan *et al.*, 1992). Pod damage at this stage can add to the risk that stored material will become infected with *Aspergillus flavus*, the fungal source of aflatoxin.

The other pest specific to the drying stage is the 'wang', *Elasmolomus sordidus* (= *Aphanus sordidus*), a lygaeid, which can be found in surprisingly high numbers within stacks of drying groundnut plants. All stages feed on the drying seeds by penetrating the pods with their mouthparts. The net effect is a marked decline in quality caused by the build-up of moulds and a rancid taste associated with the oxidation of oils to fatty acids. This insect can be transferred into storage structures, where it continues to cause seed deterioration.

The third species of note is the groundnut bruchid *Caryedon serratus*, which is restricted to groundnut as a post-harvest pest. The only other known host of commercial importance is the tamarind tree (*Tamarindus indica*), the pods of which are traded mainly within southern Asian communities. Groundnut stored in unbroken pods is usually safe from attack by most insects and diseases, except with this species. The eggs are laid on the pod, through which the neonate larvae dig to reach the seeds. Warehouse contaminations probably originate from field infestations. Although this pest was dealt with routinely in the heyday of West African groundnut exports, its significance in India has only come to light in recent years (Dick, 1987a).

11.3 APPLIED ECOLOGY AND ECONOMICS

Section 11.2 indicates that there are many different kinds of insect living on or under groundnut crops and that some are undoubtedly influencing the yield of the crop. The taxa and sometimes the species most likely to be reducing yield have been mentioned with more or less detail. The sectors of the many communities who have a stake in these matters (rich farmers, poor farmers, female farmers and mothers, teachers, agroindustrialists, extension specialists, researchers) will have a range of attitudes to these insects according to their vested interests.

Some would take every opportunity to get rid of them with pesticides, irrespective of their potential pest status; others would do so if they had the means to purchase pesticides. Perhaps a few would ponder upon methods of managing them without this drastic approach or would just like to have the time to sit in a field and watch them. We are mainly concerned with the section of this spectrum of interest that excludes the first category.

When trying to determine the status of the insects that live on a crop and deciding whether they are pests, neutrals or beneficials, for instance, it is necessary to have knowledge of their ecology and their influence on the yield of the crop. This section reviews what is known about those aspects of the applied ecology of the key pests that are fundamental to developing what we regard as rational management strategies.

11.3.1 Flight activity

Most of the insects colonizing a crop originate from parental stock that reproduced after flying into the field or its bordering vegetation. The same is often true for virus vectors. Thus, when considering the risk of pest attack, it is necessary to know about the times of year when economically important insects are most likely to be flying, their pattern of movement within and between localities, the relationship between flight intensity and insect population density and yield loss, and the most effective methods of monitoring their flight. This has a direct bearing on pest avoidance (section 11.7).

(a) Monitoring flight activity

The most effective method of catching flying insects is to lure a fraction of the aerial population into a trap. Southwood (1978) discussed many ways of doing this. Non-specific methods such as light traps (groundnut leaf miner, Shanower, 1993), suction traps (for thrips, C.S. Gold, ICRISAT, personal communication) and yellow water (pan-) traps for aphids (Farrell, 1976b) have yielded reproducible results for insects living in groundnut fields.

Even better are traps that are attractive to single species. With regard to groundnut, this category is currently restricted to Lepidoptera, in particular *Spodoptera litura* (Ranga Rao *et al.*, 1991a,b), *S. frugiperda*, *S. exigua*, *Helicoverpa zea*, *Heliothis virescens* (Lynch and Douce, 1992) and the groundnut leaf miner (ICRISAT 1991b) for which synthetic pheromones are available. Traps enclosing virgin female moths have also been used to lure male groundnut leaf miner moths but are considerably less effective than the traps baited with the synthetic pheromone developed by NRI (Chatham, UK) (Table 11.5; and Nandagopal and Reddy, 1990). Such traps are satisfactory if it can be proved, or if it is accepted, that the catch of males represents the activity of the population as a whole.

(b) Flight activity and its implications

The main flight period of *Aphis craccivora*, the vector of the groundnut rosette virus, can be expected some 6 weeks after the first 'planting' rains in southern Africa. This was the basis of the recommendation that groundnut crops should be sown 'early'. This advice is tempered by the observation that a crop, once it has been established for 40 days or more, is much less likely to be adversely affected by this virus than are younger crops. Adherence to the appropriate management practices is probably the reason for the virtual non-appearance of this disease in southern Africa in the last 20 years.

Ranga Rao *et al.* (1991a) compared the data from light and pheromone traps set on the ICRISAT farm in their study of the flight activity of *S. litura*. The former was the standard monitoring procedure until the mid

TABLE 11.5 *Mean number of groundnut leaf miner caught in delta traps baited with experimental pheromone lures at ICRISAT Center during rainy season, 1991*

Lure type	Mean number of moths trapped in different weeks of exposure								
	1	2	3	4	5	6	7	8	9
'Small septa'									
Fresh	696	545	650	245	50	69	74	34	23
Continuously exposed	654	703	630	360	58	70	78	22	14
'Big septa'									
Fresh	922	800	856	468	72	108	137	38	22
Continuously exposed	990	804	573	326	32	30	22	34	19
Virgin female	195	185	40	84	11	52	14	9	6
Empty trap	7	20	7	5	2	16	3	1	3
SE	± 32.1	± 37.8	± 64.6	± 41	± 8.1	± 13.9	± 15.7	± 7.3	± 3.4
CV (%)	12	17	31	37	48	62	64	71	51

TABLE 11.6 *Relationship between log of number of moths caught (x) and log of number of egg masses (y) per 100 plants (Kumari, 1989)*

Site	y	r^2
1	0.019 + 0.445x	0.798
2	−0.220 + 0.535x	0.826
3	−0.070 + 0.474x	0.838

x = mean of three traps per site and two nights' data combined
Data cover 38 observation periods between 1 February and 16 April 1989.

1980s. The pheromone traps consistently demonstrate a distinct peak of flight activity in March (just before harvest of the post-rainy season crop) that was not present in the light trap data. However, the implications are not clear because there was no corresponding increase in oviposition in groundnut crops at that time. This suggests that the females accompanying the males were 'flying through' the farm. Perhaps they did not recognize groundnut as a suitable host, or they may have been undergoing a long-distance dispersal flight and were not attuned to respond to a potential food plant.

In addition, the pheromone trap catches indicated that there was no clear cessation in flight activity during the hottest time of the year. This suggests that a period of aestivation cannot be assumed to be a normal event, as suggested by the light trap catches. A periodicity of flight activity corresponding to the generation length was also detected. This could be of relevance to the timing of sowing in *S. litura* endemic areas, as it is best not to sow at a time that will result in the colonization of the seedlings, which tend to be the only stage susceptible to this pest.

One such endemic area is the coastal strip of Andhra Pradesh, India, where groundnut is grown under irrigation from November to March. *S. litura* is the only insect pest during this period. It can cause extensive defoliation but this appears to be largely as a result of the disruption of natural control processes associated with excessive insecticide application. This area has been targeted as a test area for implementing IPM procedures based on damage forecasting via a pheromone trap network (Table 11.6).

Kumari (1989) also demonstrated similar relationships between moth catch and the density of small larvae (instars 1–3), large larvae and the number of damaged leaflets per 100 plants. There was a delay of 4 days between the appearance of the first moths in the traps and the detection of the first egg masses. The delay was 8 days for the small larvae and 20 days for the large larvae.

It is hoped that similar relationships will be established for the groundnut leaf miner now that a synthetic pheromone is available and trap technology has been made effective (Hall, Cook, Ranga Rao and Wightman, unpublished). Attention will be directed towards the establishment of determining relationships between the number of male moths caught per trap and the density of larvae in the next generation. The feasibility of mating disruption exercises can then be investigated.

The groundnut leaf miner has not been observed to undergo long migratory flights. Our observations on the ICRISAT farm indicate that it will move only a matter of 50 m from a high concentration area (for instance, in a soybean crop) to colonize groundnut. This means that distributions can be extremely uneven, even'within a field.

Cockchafers (May beetles) are the adult stage of white grubs and have a species characteristic flight pattern that is related to mating, feeding and dispersal. The typical pattern is for adults to emerge at dusk over a period of about 3 weeks or more at a precise time related to the time that the sun goes over the horizon. First emergences occur over a period of 5–10 days. Individuals do not appear every night; the females presumably spend a day or so laying eggs in chambers, 10 cm or more below the soil surface (Farrell and Wightman, 1972). The median first emergence date may be related to temperature *per se* (day degree accumulation), a period of chilling followed by warming in temperate climes (Wightman, 1974) or, in the tropics, the onset of the rainy season (Yadav, 1981, 1991). In varying combinations and sequences, according to the species, weather conditions or the locality, the beetles:

- emerge from the ground (usually males before females);
- mate on the soil surface or on low-lying vegetation;
- undertake significant dispersal ('beeline') flights in a straight line for >50 m away from the original emergence hole;
- feed on surface vegetation;
- fly towards and aggregate around markers, such as tall trees or telegraph poles, and seek mates;
- fly to trees of a small range of species, and feed and mate on their foliage;
- either drop off or fly away from the tree when satiated or cold and, in the case of females, lay eggs wherever they land.

This stage in the life-cycle of the Scarabaeidae is of economic importance because it is during this period that the population is at its lowest density and greatest accessibility to humans, i.e. it is the best time to attempt a control strategy, if one can be devised.

Data sets on the host preferences and behaviour patterns of the adults are needed. This information would, for instance, guide farmers as to which tree species should be avoided or selected (as attractants or repellents) in farm-forestry projects in areas where white grubs pose a high risk.

TABLE 11.7 *Degree-days (°D) required for completion of development from eggs through to pre-oviposition for* S. litura *(Ranga Rao et al., 1989) and groundnut leaf miner (Shanower et al., 1992a)*

Stage	S. litura[a]	Groundnut leaf miner	
	°D	°D	Lower threshold temperature (°C)
Eggs	64	60.1	(12.3)
Larvae	303	327	(8.9)
Pupae	155	72.3	(14.7)
Pre-oviposition period	29		
Total			
Range	508–631		
Mean	(543)		

[a] Bred from egg masses found on the ICRISAT farm.
Temperature summation estimated from laboratory experiments carried out under constant temperatures.

11.3.2 Environmental factors

Environmental factors can work directly on the insect:

- Temperature regulates the development rate and can be a cause of mortality.
- Rain washes aphids off plants and promotes the high humidity that stimulates entomophagous fungi.
- Wind influences the stimulus, distance and direction of migratory flights.

The physical environment can have many indirect and often complex interactions via the soil and, through the soil and its water or nutrient content, to the plant and then the insect. Similarly, interactions between temperature, relative humidity and cloud cover or levels of ultraviolet radiation have observable effects (but without parameters) on the virulence of insect pathogens.

(a) Direct effects

One of the basic sets of information that should be available for key pests is the relationship between its development rate and temperature (Table 11.7). This is needed for matching insect damage or population models with plant or crop development models. Biotype differences can also be detected by comparing the relationship between temperature and the development rate of isolated or transient populations.

Light, well-drained soils, as opposed to heavy, waterlogged soils, favour the activities of *Elasmopalpus lignosellus*, especially when the weather is hot and dry. Similarly, white grubs in general prefer sandy soil (Smith and Barfield, 1982). An exception is *Holotrichia serrata*, which is most likely to be found in the heavier soils of southern India, in contrast to its northern counterpart, *H. consanguinea*, a denizen of the light soils of the Gangetic Plain.

The opposite pertains for the southern corn rootworm (*Diabrotica undecimpunctata howardi*). It proved particularly susceptible to dry soil, especially in the egg and first instar stage and when living in a sand medium (Brust and House, 1990).

(b) Indirect effects

Wheatley *et al.* (1989) investigated the response of groundnut to drought stressed hosts during the post-rainy season at ICRISAT Center. Four varieties of groundnut were grown across a drought stress gradient. The groundnut leafminer was most abundant on the most stressed plants. Jassids showed the reverse trend and this preference for a non-stressed host was also observed for jassids living on lucerne (Hoffman and Hogg, 1991). Thrips were at first most abundant on the least stressed plants, reversed this trend in mid-season and then reversed it back again before harvest, by which time the physical condition of the hosts had deteriorated. A further study might well investigate more closely the relative abundance and feeding site of the thrips species involved. Of the four varieties tested, NC Ac 343 (ICG 2271) proved to be the best to grow in times of potential drought stress *vis-à-vis* insect attack.

Several other aspects of this study were also of note:

- It was found that the groundnut leaf miner was able to withstand a midday canopy temperature of 47 °C.
- The source of the irrigation water – overhead or furrow – did not influence the distribution of the insects.
- An inspection of the bamboo plot pegs indicated that termites (*Microtermes* sp.) had a preference for a soil moisture of about the wilting point (12%).

Observations in another year on a similar drought gradient (Ranga Rao *et al.*, 1991c) showed that *Aphis craccivora* was most abundant on plants that were not drought stressed. This was in spite of being pounded with water from an overhead irrigation system.

Wheatley *et al.* (1989) found that the the groundnut leaf miner was at an advantage when its host was severely drought stressed. Extremes in precipitation also appear not to influence the survivorship of the larvae of this species. Shanower *et al.* (1992a) found no evidence that the extreme fluctuation in population density that is characteristic of this species was in any way related to rainfall events. Supplementary experiments showed that egg and larva survival were not influenced by artificial rain for periods and intensities that exceeded the normal range of field conditions.

11.3.3 Influence of insects on groundnut yield and economics

The development of an understanding of the relationships between the feeding activity of insect populations and the yield of the host, with respect to the characteristics of the farm system within which it is growing, is a fundamental task facing all applied insect ecologists. In many cases, the importance of natural enemies in the life system of a potential pest means that the dynamics of three trophic levels have to be taken into consideration (Shanower *et al.*, 1992b). This implies the need for the construction of predictive models. Despite the importance of such sets of information, there are few reliable and published accounts relevant to pests of groundnut crops. This is unfortunate but is not surprising in view of the difficulty of assembling and analysing the necessary data sets.

The relevance of this topic to insect pest management as a whole is that these data are needed to establish parameters for the **economic threshold** – the density (or intensity, in insect days) of an insect population at which its activity results in the decline in the potential yield of the harvestable component(s) (pods and/or haulm) of a crop.

The economic threshold can be distinguished from the **action threshold**, which is the insect population density at which activity is needed to prevent the population density exceeding the economic threshold at some time in the future. Depending on the pest complex and the cropping system, the 'activity' could be:

- the commencement of hand picking or trapping the offending pest;
- the application of an insecticide;
- crop rotation; or
- deep ploughing, in the case of soil insect life systems.

A number of approaches have been adopted to secure relevant information depending on the circumstances. The 'circumstances' are often in fact governed by the inability of the entomologist to locate appropriate pest populations.

(a) Energetics and laboratory feeding experiments

Most animals can be considered as machines that convert food (consumption = C) into new body tissue (production = P), a process that exploits part of the potential energy of the food to fuel the animal's metabolism (respiration = R) of the food. Measurable by-products of this process also include unassimilated ingesta (faeces = F) and excreta (U). The relationship is:

$$C = P + R + F + U$$

In the last 20 years there have been many studies of this relationship, especially with respect to the establishment of the trophic relationships of phytophagous insects. They usually include estimates of the gross ecological efficiency (P/C). This and other ecological efficiencies can be discussed in terms of dry matter, energy and, more rarely, essential nutrients, especially nitrogen, depending on the context of the study. For instance, because energy supply is rarely limited in terrestrial systems, trophic dynamics or system structure can be discussed in terms of the energy flow from one trophic level to the next (Grimm, 1973; Axelsson et al., 1975; Axelsson, 1977; Schroeder 1978; Wightman, 1979; Bellows et al., 1983). Energy (or carbon) units, rather than units of mass, are adopted as the common denominators when describing such systems because the 'concentration' of energy – i.e. the number of joules per unit mass – is species and possibly system specific. An extreme example is that of bruchid beetle larvae living on dried pulse seeds. The cotyledon of the host, the larval food, has an energy equivalent of 18.7 J/mg whereas bruchid larvae have up to 27.8 J/mg (Wightman, 1978).

The constraints to energy flow or the 'bottle necks' in biological systems are most likely to be revealed by studies of nutrient cycles. This is because (after water-related problems) nutrient excesses or shortages are the most common constraints on the components of biological systems.

Measurements of the mass of system components are applicable to studies where only changes over time are critical or where one component is of primary importance. For instance, leaf mass consumed by an insect can easily and accurately be considered in terms of the leaf area removed (Schroeder, 1984) because this is directly related to the amount of light energy intercepted by the host. Dry matter, or preferably ash-free dry matter, is usually determined because variations in the water content of most organisms (as influenced by environmental conditions) affect the precision of biomass estimates.

Over the years it has become apparent that when a phytophagous insect is feeding on a suitable host that is growing in stable conditions, the gross ecological efficiency is in the region of 0.14 (mass) and 0.18 (energy) (reviews by Edwards and Wightman, 1984; Wightman and Rogers, 1978). This ratio will vary according to host species (and the genotype, if there are variations in the level of allelochemicals within the species), the nutrient status of the soil in which the host is growing and the degree of drought (or water) stress (e.g. Mansour, 1981; Scriber, 1979a,b; Crawley, 1989). Thus if the mass of the insect is known, it is possible to calculate the amount of leaf material it has removed up to the time a measurement of mass (or length, which is a function of mass) is made. It is then a matter of

TABLE 11.8 *Leaflet area, fresh weight, dry weight and dry:fresh (d/f) ratio of 20*
mature leaflets from five groundnut genotypes (Wightman, unpublished data)

	Mean cm^2	fresh wt g	dry wt g	d/f	mg (dry) cm^{-2}
Shulamit	199	4.88	1.49	0.305	7.48
VB	248	6.17	1.85	0.300	7.56
NC 7	196	4.98	1.42	0.285	7.23
A 46 L 10	215	5.43	1.62	0.297	7.52
A 81 L 18	184	4.62	1.45	0.313	7.88
Mean	208.4	5.22	1.56	0.300	7.53

arithmetic to determine the effect of an insect population on the biomass or
leaf area index of a plant population or crop. This is the area in which plant
or crop growth models and insect models interface.

Estimation of larval damage

As an example, in the case of *Spodoptera litura* larvae feeding on groundnut foliage, their maximum length is *c.* 40 mm, which means that they
weigh 0.3 g dry (Rogers *et al.*, 1976) which is equivalent to about 1.4 g live
weight). If P/C (mass) = 0.14, then C = 2.14 g.

Adopting the energy route gives a similar answer. Phytophagous insects
have an energy content of about 23 kJ/g (e.g. Edwards and Wightman,
1984; Schroeder 1977, 1978, 1984), thus P = 0.3 × 23 kJ = 6.9 kJ. If P/C
(energy) = 0.18, then C = 38.3 kJ. As leaves have *c.* 18.4 kJ/g energy
content (Petrusewicz and Macfadyen, 1970), this indicates that one larva
consumes 2.08 g of (dry) leaf during its development. As there are 7.53 mg
(dry) per cm^2 of groundnut leaf (Table 11.8) the mean (2.11 g) of these two
estimates is equivalent to 280 cm^2. Thus one larva consumes 20–30 leaflets
during its development, depending on the size of the leaf and proportional
mass of non-consumed vascular tissue.

Garner and Lynch (1981) measured the area of groundnut leaflet consumed by *S. frugiperda* larvae (fall armyworm) in Georgia, USA. Their
main experiment showed that the mean leaf area consumed during the
larval period was 94.6 cm^2 (which, following data in Table 11.8, is equivalent to 712.3 mg dry leaf). This is close to the cumulative consumption data
of ±100 cm^2 for the same species indicated by Smith and Barfield (1982).

The mean pupal mass was 177.5 mg. This means that the maximum
larval weight was about 213 mg, because pupae weigh *c.* 20% less than fully
grown larvae (Hagvar, 1975; Mackay, 1978; Wightman 1978). The dry
weight to live weight ratio of phytophagous insect larvae is normally about
0.2 (personal observation) so that the dry weight of the fully grown larva

was *c*. 43 mg. A P/C ratio of 0.14 indicates consumption of 304 mg, which is considerably less than the observed.

Data from ancillary experiments carried out by Garner and Lynch (1981) indicate that the disparity may be due to the age of the foliage with which the larvae were fed. Adopting the data from follow-up experiments certainly brings the model and experimental data closer to agreement. Larvae fed on 2-day-old leaves ate up to twice the area of leaflet, probably bringing the P/C ratio closer to 14%. Furthermore, the larvae developed more quickly, had a much lower mortality rate and finished somewhat larger than counterparts fed on leaves up to 40 days old. This indicates that antibiosis may have developed in the older leaves and was the cause of the low estimated P/C ratio.

However, the possibility of a disparity between a model and the experimental data calls for a re-examination of both. In this case we need to look at several factors:

- Is the P/C really 6% for *S. frugiperda* or is this an artifact related to the insects being fed old (and excised) leaflets?
- Are the Australian host data transferable to the US genotype?
- What are the equivalent experimental data for *S. litura* and other *Spodoptera* spp.?

It is certainly an indication that model data should be applied with circumspection, and preferably with experimental verification. Huffman and Smith (1979) present data indicating that *Helicoverpa zea* consumes 176 cm^2 of foliage of the cv. Starr. However, we do not have access to biometric data for this species.

(b) Simulation of defoliator damage in cage and field experiments

Continuing on the theme of conventional defoliators (which have attracted most attention from experimentalists), several groups have evaluated the effects of leaf removal by human or insect agencies on groundnut yield (e.g. Greene and Gorbet, 1973; Enyi, 1975; Smith and Barfield, 1982). The latter authors present a defoliation level (0–100%), by time (35–110 days after emergence) and by yield reduction (0–50%) response surface for a spanish variety. They conclude from this and other data: 'peanut is most susceptible to defoliation from 70–80 days post emergence and practically immune to yield reductions from defoliation prior to bloom initiation and near harvest.' Certainly, the figure they present and the other data they review support this conclusion.

At ICRISAT we released specially reared fourth instar *Spodoptera litura* larvae onto plants that were surrounded by a 20 cm high metal barrier that stopped their escape. This procedure was adopted to avoid the possibility of delivering a systemic shock to the plants' system by abrupt hand or mechanical ablation. In these experiments we have consistently found that plants defoliated after the seedling stage or early flowering stage were tolerant to attack. This was most marked in the rainy season. Plants even produced pods (amounting to 50% of the control yield) when they were completely defoliated from 10 days after emergence by the addition of two larvae per plant at 20-day intervals (Wightman *et al.*, 1990).

Similar experiments carried out by Sathorn Sirisingh and Manochai Keerati-Kasikorn (1986) in Thailand indicated similar insensitivity to defoliation by groundnut plants.

The clue to the disparity in the results from North America and Asia may lie in the genotype, their partitioning coefficients and sowing pattern. In the experiments in India we expect canopy closure after 20–30 days. The genotypes we experimented with (mainly spanish bunch) have a leaf area index of >4 by the time the vegetative stage is complete. This means that, in practical terms, after the crop has been in the ground for more than 30 days it can suffer at least 50% loss of leaf area before its photosynthetic capability is reduced. This statement ignores the photosynthetic capacity of the stems and petioles, which may be more than anticipated.

The principle behind cage experiments is to isolate plants from as many insects and natural enemies as possible and then infest them with members of the required species at whatever range of stage, time and density is required. Primary assumptions are that the introduced insects, whether reared or collected from another site, are healthy and are otherwise 'normal', as are the plants grown in cages. This approach has proved satisfactory for establishing the relationships between the densities of *Nezara viridula* and *Riptortus serripes* on soybean yield (Brier and Rogers, 1991) and *Helicoverpa armigera* on chickpea yield (ICRISAT Legumes Program, 1991b). However, the cage experiments with groundnut leaf miner at ICRISAT are believed to have given non-representational results because the fine cage netting needed to secure this species intercepted too much light, thereby changing the physiology of the plants.

(c) Field experiments

Experimental approach
Experiments in Thailand carried out by Sathorn Sirisingh and Manochai Keerati-Kasikorn (1986) indicate that the groundnut leaf miner is capable of more drastic yield reductions than *S. litura*. Their data indicate that, in the season under discussion, a heavy infestation (*c.* 40 per plant) in the first or second generation (flowering stage) reduced the yield from a potential 1.28 t/ha (site 1) or 0.95 t/ha (site 2) to 0.63 and 0.65 t/ha. The data also show that the control of fungal diseases increased the above potential yield by 25–50%.

The determination of the relationship between the density of groundnut leaf miner populations and groundnut yield has also been the objective of experiments carried out on the ICRISAT farm for 15 seasons (two per

TABLE 11.9 *Groundnut yield, groundnut leaf miner intensity and GLM parasitism as influenced by insecticide application (III° generation)*

Treatment	Stand yield (pods) t/ha	% parasitism		insect 'days'
		8 Apr 85	19 Apr 85	
Dimethoate (400 g) (8 applications)	1.78	3.1	0	76.9
Dimethoate (200 g) (3 applications)	1.70	13.6	33.9	752.4
Diflubenzuron (250 g) (3 applications)	1.43	20.5	42.5	1443.2
Dichlorvos (3 applications)	1.58	16.3	50.0	742.8
Control (no insecticides)	1.15	23.0	61.0	1617.0
SE	± 0.06	± 3.5	± 10.4	± 67.0

The number of insecticide applications refers to the whole season (110 days).
Insecticides were applied in 350 l water/ha.
Rates (g) are a.i./ha.

year). Although this species was present in most seasons, it only reached a density that could be described as damaging on three occasions. The essence of these experiments is to allow the population to build up to beyond our concept of an economic threshold and then to apply insecticides in such a way that we achieve a range of population intensities and control efficiencies. We then relate pest intensity to pod yield. These experiments cover about 1 ha and are conducted in such a way that there is a minimum of season-to-season variation.

With this species there is the possibility of several combinations of events because it has three or four generations per growing season. It can therefore appear (and disappear) at any density between emergence and harvest and display considerable variations in density change during a season. We are, however, accumulating data that will enable us to put together a rational management outline (Table 11.9). They indicate that an insecticide application (e.g. dimethoate at 200–350 g a.i./ha) should be applied when there are five or more new mines per plant during the seedling stage, 10 new mines at flowering and 15–20 mines per plant up to 2 weeks before harvest, after which insecticide application will have little impact on yield.

Jassids and thrips often occur concurrently. Even if no other pest is present, this makes it difficult to determine whether there is a relationship between their intensity and pod or haulm yield. However, in the 1992 rainy season at ICRISAT there was heavy infestation of jassids with minimal thrips densities. The yield data (Table 11.10) indicate that there was a high

TABLE 11.10 *Cumulative effect of jassid* (Empoasca kerri) *damage (eight days before harvest) on haulm and pod yield of susceptible groundnut variety ICGS 44 (post-rainy season 1991/92, ICRISAT Center)*

Insecticide applications (*n*)	Leaflets with jassid damage	Dry haulm mass	Dry pod mass
	%	t/ha	t/ha
0	60.0	3.26	1.73
1	31.3	3.20	1.91
4	17.0	3.55	2.14
SE	± 2.7	± 0.22	± 0.08

level of leaf damage by 8 days before harvest in unsprayed plots. However, the reduction in the number of damaged leaflets by means of four insecticide applications had no effect on haulm yield and little effect on pod yield.

Indications of the effects of soil insects on crop yields can be deduced from the results of experiments involving the application of insecticides to the soil. For instance, data provided by Kumawat and Yadava (1990) indicate a linear relationship between the density of larval *Holotrichia consanguinea* (white grubs) and plant mortality in experimental conditions in Jaipur, Rajasthan, northern India. There was a log–log relationship between density and pod yield.

Wightman *et al.* (1994) simulated white grub attack by cutting through the roots of groundnut plants (White Spanish) 30 or 51 days after emergence (in glasshouse conditions) (Table 11.11). The root systems regrew when the plants were cut after 30 days, although there was a considerable energy cost in terms of reduced pod yield. Plants cut at the later date did not regrow their roots. This was even more accentuated when the plants were drought stressed. The plants with roots cut 51 days after emergence were close to death at the end of this experiment.

This experiment demonstrated differential debilitation as a result of root damage from flowering to harvest. But seedlings can also be killed when attacked by white grubs (Bakhetia, 1982; Kumawat and Yadava, 1990). Apart from the loss in yield, the farmers' profits are further reduced because weeds are able to grow in the gaps. This reduces the opportunity for compensatory growth by the plants next to the spaces left by the killed seedlings. The potential role of compensatory growth following stand thinning may, in any case, be overestimated.

The extent and cost of seedling mortality and its management in northern India have been estimated by Bakhetia (1982). He demonstrated 1.3–2.6% plant mortality where seeds had been dressed with insecticides compared with 10.1% mortality where seeds were untreated. The overall yield was at least doubled by using insecticide seed dressings, indicating

TABLE 11.11 *Mean weight (n = 5) of pods produced when groundnut plants (variety White Spanish) were cut through the root at 10, 15 or 20 cm below the soil surface (0 cm = uncut control), 30 or 51 days after emergence when grown under drought stress or fully irrigated conditions*

Cut	Mean pod weight g/plant ± SE	
	Drought stress	No drought stress
Cut 30 days after emergence		
Depth of cut		
0 cm	12.92 ± 1.64	22.96 ± 0.86
10 cm	9.81 ± 0.14	14.19 ± 0.57
15 cm	11.17 ± 0.36	15.11 ± 1.10
20 cm	12.02 ± 0.09	15.91 ± 0.56
Cut 51 days after emergence		
Depth of cut		
0 cm	12.92 ± 1.64	22.96 ± 0.86
10 cm	7.73 ± 0.73	10.98 ± 1.08
15 cm	9.36 ± 1.52	13.93 ± 0.47
20 cm	8.99 ± 0.18	14.18 ± 1.50

that there was considerable additional sublethal yield loss that was avoided by introducing insecticides into the soil as a seed coating.

Gough and Brown (1988) indicated that groundnut crops in the Atherton Tablelands of north-east Australia were equally sensitive to attack by white grub (*Lepidiota* sp.) Their data indicate that one larva per metre row (6–8 plants) of cv Virginia Bunch reduced crop yield by 381 kg/ha.

Within-stand compensation following plant mortality
Wightman and Wightman (1987) found that, in conditions typical of farming systems in Malawi, there was no within-row compensation by plants in stands that suffered up to 50% mortality once the stand had been above ground for 26 days. ('Compensation' is defined as an increase in the pod and/or haulm yield of plants in depleted stands, relative to plants in control stands that are not depleted.) Up to this time there was within-stand compensation for plant death only if there was >30% mortality after 17 days and >50% mortality at 26 days. These data refer to a crop that was harvested after 5 months. They indicate, at least in the conditions of this experiment, that compensatory growth of the pods and haulms in response to the death of neighbouring plants occurs only in earliest stages of stand development.

The empirical or trial-and-error approach
Because the end point of the process under discussion is the development of pest-management methods that are appropriate for the given

system, it is feasible to derive action thresholds by trial and error. For instance, in São Paulo State, Brazil, where thrips are the major pest and a serious constraint, farmers are advised to apply a suitable insecticide only when there are more than three thrips per leaflet on 20% of a random 200 leaflets taken from 1 ha. This procedure reduced insecticide applications from seven to two per season with an increase in yield and profitability (Snhr Dalmo Lasca, Director, São Paulo State Extension Service, personal communication). There is no experimental data to support this procedure, but it works.

Similarly, at ICRISAT, we set up in 1984–85 a series of ad hoc action thresholds to assist the farm manager's plant protection team before we had supportive data. For instance, we recommended that an insecticide (normally dimethoate at 200–350 g a.i./ha) should be applied for groundnut leaf miner control if the density exceeds five mines per seedling, 10 mines per plant at the flowering stage and 15–20 mines per plant up to 3 weeks before harvest, after which insecticide application is likely to have little benefit. Our experimental data (above) indicate that this rule of thumb had some merit.

11.3.4 Dynamic programming as a tool to guide research orientation

The action thresholds for the groundnut leaf miner just mentioned were the basis of a series of modelling exercises carried out to evaluate management scenarios for this pest. The exercise was based on a population dynamics model of this insect and contemporary village-level fixed and variable costs for southern India (Dudley *et al.*, 1989).

The scenarios covered issues such as: 'If a farmer has available varieties with 0, 10, 20 . . . 90% host plant resistance, how much natural mortality is required at each level of host plant resistance to eliminate the need for insecticide application?' The role of the market value in determining the optimal number of sprays was also investigated. The final conclusions from this piece of work are realistic and pointed to a difference between pest management in developed and less developed countries. They point to our need to make assessments of the effectiveness of farmers' insecticide application activities and, implicitly, the role and effectiveness of natural enemies (assumed to be inversely related to insecticide application activity) before it is possible to work out what level of host plant resistance to a given insect pest is needed.

11.4 HOST PLANT RESISTANCE

Host plant resistance, where it exists, can be made available to farmers as an effective and environmentally friendly component of pest management that involves little or no extra cost or effort than the normal purchasing, sowing and keeping of seed – 'technology in the seed'. It is thus part of the applied entomologist's job to detect and exploit it where feasible and rational.

As inferred, the cost to the farmer is small; however, the institutional costs are considerable and include the development and support of a germplasm collection, research farm and research facilities, many years of screening and selecting germplasm and probably about 10 seasons of resistance screening and selection of breeders' material. This is followed by on-farm testing and evaluation by farmers.

Of course, there is no guarantee that the traits sought exist in the germplasm of the crop species, its close relatives or, in the context of contemporary biotechnological feasibilities, any other species. There is also no guarantee that the breeders and other gene-shifters will be able to ensure that the desirable genes to manifest themselves in a variety that is adapted to the target environment. Thus, breeding for host plant resistance is primarily an activity of organizations that are stable, mission-orientated, well endowed and non-profit-making.

Many groundnut genotypes have characteristics that protect them from herbivores. This observation is based on the relatively small number of pests (as opposed to insects) associated with the above-ground parts of this species. (We have yet to come to terms with resistance to root-eating arthropods.) Lynch (1990) has made a notable contribution to groundnut science by listing the genotypes that are known to have resistance to many of the most important pests.

If a particular insect is established as being a major constraint in one or several agroecological zones, it is rational to consider the inclusion of host plant resistance in a groundnut management programme that includes the provision for breeding or selecting adapted varieties. If no source of resistance genes is known, it is necessary first to develop guaranteed screening procedures for distinguishing between resistant, susceptible and 'escape' plants or genotypes. Rationally, this process should also enable other characteristics of the screened genotypes to be assessed. Screening methods are not covered here but the general principles are described by Smith (1989).

Once resistance to one or several potential pests has been found, it is usually necessary to work with breeders to combine the relevant genes with a 'background' that is agronomically suited to the target environment. This procedure is made more efficient if it is possible to supplement the field screening of progeny by monitoring the presence or absence of the physical or chemical markers associated with the mechanism of resistance (Lundgren et al., 1981, 1982).

11.4.1 What is host plant resistance?

Host plant resistance is a phenomenon that has evolved in most higher plants to permit them to coexist with or to avoid the many species of potential herbivores that could exploit them as food. Viewed broadly, it can take several forms that are usually connected with the feeding activity of the free-living forms of the herbivores or the provision of food and shelter for their progeny.

1. **Repellence** (antixenosis, 'non-preference' or the turning away) of herbivores before they come into contact with the plant. This can be associated with, for instance, the release by the plant of physiologically active chemicals (kairomones) into its air space or with a physical factor that influences the herbivores' visual response to a plant (or group of plants), perhaps to the extent that it is not recognized as a potential host. Physical characteristics such as the presence or absence of trichomes on leaf or stem can also influence the way that an insect reacts to a plant when first approaching it.
2. **Antibiosis**, where the plant contains chemicals that, when tasted or ingested by a herbivore, prove to be antimetabolites (e.g. insect growth hormone analogues), repellents, antifeedants, or toxins (including the toxic manifestations of the genes from other organisms, such as *Bacillus thuringiensis*).
    (a) **Latent antibiosis** awaits being switched on by a challenge from a herbivore or by a systemic (within plant) or pheromonal (between plant) message. Latent resistance has not (yet) been detected in groundnut although it may exist; it is a factor in the resistance of tomato plants to *Spodoptera littoralis* (Edwards *et al.*, 1985). Further information about this phenomenon can be found in Kogan (1986) and Edwards and Wratten (1987).
    (b) **Temporary antibiosis** is only present during a particular stage in the development of a given organ or the phenology of a plant.
    (c) **Permanent antibiosis** is a characteristic of a given plant species or organ.
    Antibiosis can also take the form of the absence or masking of a feeding stimulant.
3. **Tolerance**, where the plant can continue to develop and reproduce despite being attacked by herbivores. The misuse of this term to denote low levels of antibiosis or an undefined aspect of the resistance phenomenon, in general, often leads to unnecessary confusion.
4. **Physical** (structural), where the plant has structures (trichomes, thorns) or surface characteristics (thick or waxy cuticle, or even a layer of water – Nwanze *et al.*, 1990) that interfere with a herbivore's ability to exploit it.

This is a development of the conventional view of host plant resistance in plants to herbivores, based on Painter (1951). The following could be added because they can be complementary or confounding:

5. **Avoidance (seasonal)**, where the plant's phenology (or a crop's sowing

pattern) is such that its life-cycle (or a sensitive developmental stage such as flowering or seed swelling) does not coincide with the time of year when a key herbivore is active.

6. **Avoidance (spatial)**, where a plant has evolved into forms that grow (or are sown) outside the biogeographic or climatic range of a key herbivore.

However, in practical terms, avoidance is best considered as being an aspect of cultural control.

The genus *Arachis* displays several of these resistance factors.

11.4.2 Host plant resistance in *Arachis hypogaea*

(a) Resistance to *Aphis craccivora*

Although seedling pests in their own right, groundnut aphids have been highlighted as primary groundnut pests in Africa because of their ability to transmit the groundnut rosette virus complex (GRV). In one season (1975) they converted Nigeria from one of the world's leading groundnut exporting nations to relative obscurity in this regard.

Considerable energy and expertise has been devoted to breeding for GRV resistance *per se* but it is surprising (with the benefit of 40 years of hindsight) that the colonial authorities of the time did not follow up the discovery made by Evans (1954) in Tanzania. He found that several varieties, especially Asiriya Mwitunde (a name indicating that the variety belongs to the Mwitunde people of northern Tanzania), carried comparatively small aphid populations compared with the other varieties tested. This was associated with fewer and smaller GRV primary infestation sites. This clear indication of virus management by vector control was accompanied by significant yield advantages.

Unfortunately, screening of the East African genotypes in the ICRISAT germplasm collection has only revealed comparatively low levels of aphid resistance. Apart from the implication that important germplasm has not yet been collected or has been lost, this is now of less importance because two genotypes with high levels of aphid resistance, ICG 5240 (= EC 36892) and ICG 5725, have been detected (Wightman *et al.*, 1990). Field tests of ICG 5240 in Africa (Wightman *et al.*, 1990; Sithanantham *et al.*, 1991), India (Padgham *et al.*, 1990a,b) and China (Dr Xu Zeyong, personal communication) have shown that the level of aphid resistance remains high across continents. These experiments have shown sufficient intercontinental variation in aphid response to demonstrate the existence of biotypes of *A. craccivora*. Resistance is manifested by longer generation time and considerably diminished fecundity.

Field tests in the rigorous conditions of a GRV screening nursery in Malawi showed that after 40 days exposure (after which GRV has a minor effect on yield) 14% of the 248 ICG 5240 plants were infected (control = 70% infestation, $n = 749$). At harvest time, 4 months after emergence, the susceptibles had a >98% infestation, and ICG 5240 only 44%.

The next step in exploiting this resistance was to determine the mechanism. This has given us access to a method for screening the progeny of crosses between the aphid resistant genotypes and lines with agroecological adaptation that would complement or replace our bioassay technique.

These studies (carried out mainly at ICRISAT by scientists from the Natural Resources Institute, Chatham, UK) have shown that there is only evidence for antibiosis. It is temporary in that the factor is concentrated in the areas where the aphid is most likely to feed, i.e. the terminal leaf buds, leaflets and petioles. Electronic monitoring showed that the aphids fed for half the time on the phloem of ICG 5240 compared with TMV 2 (susceptible control). This is likely to reduce considerably the chances of absorption of the GRV virus by the vector (now known to be a complex of viruses, e.g. Murant, 1990; Murant and Kumar, 1990), which has an acquisition time of >4.5 h (Padgham et al., 1990a; Padgham et al., 1990b).

The analysis of phloem extracts from the petioles of ICG 5240 had 2–8 times the concentration of procyanidin than did the phloem of TMV 2. Further tests showed that there was a strong negative correlation ($r^2 = -0.86$) between the log procyanidin concentration in seven groundnut genotypes and the intrinsic rate of increase of aphids on those genotypes. Assays showed that a concentration of only 0.005% procyanidin in an artificial diet reduced honeydew production (equivalent to diet ingestion) by 50% (Grayer et al., 1992; Kimmins personal communication). This implicates procyanadin in the resistance process, though not necessarily as the active component: it is, however, a convenient quantitative marker that is being used as an index of resistance level.

Thus resistance to the vector of GRV has been detected and the possible mechanism of resistance described. Vector resistance has been shown to be a valid method of managing the disease and there is a relatively simple method of detecting minute quantities of the resistance factor (or a precursor or breakdown product) in plant sap.

(b) Resistance to thrips and jassids

Lynch (1990) and Wightman et al. (1990) list more than 100 genotypes with resistance to thrips and/or jassids. Subsequent screening at ICRISAT has revealed several more – not detailed here, except that it is worth mentioning that ICG 5240 has high resistance to jassids as well as to aphids (above) and foliar diseases (Sithanantham et al., 1990). However, at this stage in the development of pest management procedures for groundnut, it is more appropriate to develop varieties with high yield potential that also have resistance to key pests and groups of pests than to seek more resistant genotypes when plenty have already been identified. Table 11.12 indicates

TABLE 11.12   *The pedigree, yield and jassid score of 10 out of 18 groundnut varieties bred for jassid resistance and high yield, tested in the 1990 rainy season at ICRISAT Center (communicated by Dr S.L. Dwivedi)*

Variety (ICGV)	Pedigree	Pod yield t/ha	Jassid score*
87745 (SB)	ICG 799 × (ICG 799 × NC Ac 2214)	2.54	3.3
86455 (SB)	ICG 799 × (ICG 156 × NC Ac 2214)	2.40	3.0
86393 (VB)	ICG 1326 × (ICG 156 × NC Ac 2214)	2.48	3.0
86462 (SB)	ICGS 1 × NC Ac 2240	2.31	2.3
86522 (SB)	ICG 799 × (M 13 × NC Ac 2214)	2.20	3.0
86518 (SB)	ICG(PRS) 12 × NC Ac 2214	2.20	2.3
87430 (SB)	ICGS 6 × (ICG 799 × NC Ac 2214)	2.16	3.0
87252 (SB)	ICGS 7 × NC Ac 2214	2.15	2.0
87495 (SB)	F334A-B-14 × NC Ac 2214	2.12	3.0
87468 (SB)	ICGS 24 × NC Ac 2214	2.04	2.7
Controls			
High yielding susceptible:			
ICGV 87128 (SB)		2.36	7.0
Resistant parents			
NC Ac 2214 (ICG 5040) (VR)		1.51	1.0
NC Ac 2240 (ICG 5043) (VR)		0.39	1.0
Trial mean (total of 18 entries)		2.25	3.5
SE ±		0.17	0.2

Notes: SB = spanish bunch, VB = virginia bunch, VR = virginia runner;
*Jassid score 1 = no damage (chlorosis), 9 = >75% of leaves are chlorotic;
Data in parentheses are the ranking of yields within the experiment.

that this procedure has led to the combination of jassid resistance genes from ICG 5040 (low yield potential) and ICG 5043 (very low yield potential) with backgrounds conferring relatively high yield potential in research station conditions (Dr S.L. Dwivedi, Groundnut Breeder, ICRISAT). They now have to be tested in farmers' fields in appropriate areas.

Screening for resistance to thrips and jassids in Thailand over a number of years has indicated that (NC Ac 343 × NC 17367) had the highest yields of the lines tested and multiple insect resistance. NC Ac 343, (NC 1107 × (NC 2232 × NC 2214)) and (NC 6 × NC 3033) had slightly lower yields but high levels of pest resistance (investigators were W.V. Campbell, Manochai Keerati-Kasikorn and Turnjit Satayavirut; the latter communicated this information). This agrees with findings previously reported for resistance to *Frankliniella fusca* in North Carolina in which NC Ac 343 and its derivative NC 6 were linked with thrips resistance (Campbell and Wynne, 1980).

TABLE 11.13 *Incidence of* Thrips palmi *and bud necrosis disease (BND) in an experimental trial carried out at Rajendranagar, near Hyderabad, 1992 post-rainy season (means of four replicates, data of Vijaya Lakshmi (unpublished); Jl 24 = susceptible control)*

Days after sowing	Genotype	*Thrips palmi* $n/25$ terminal	BND incidence % of all plants
29	ICGV 86031	9.5	0
	ICGV 86338	9.8	0
	Jl 24	29.8	0
44	ICGV 86031	9.8	0
	ICGV 86338	13.8	0
	Jl 24	14.8	1.2
57	ICGV 86031	4.5	0
	ICGV 86338	4.3	0.6
	Jl 24	5.3	3.6
70	ICGV 86031	1.5	0
	ICGV 86338	1.8	3.6
	Jl 24	6.0	8.0
85	ICGV 86031	0.5	0
	ICGV 86338	0.5	3.6
	Jl 24	1.0	18.7
99	ICGV 86031	0	0.6
	ICGV 86338	0	3.6
	Jl 24	3.8	24.8
114	ICGV 86031	0	0.6
	ICGV 86338	1.3	3.6
	Jl 24	2.0	33.6
132	ICGV 86031	0.3	1.2
	ICGV 86338	0	4.2
	Jl 24	0	39.1

Resistance to the thrips vector (*Thrips palmi*) of the bud necrosis virus (BNV – a variant of the tomato spotted wilt virus (TSWV) that is found in the Indian sub-continent) is viewed as being an important key to the management of this disease in groundnut (Amin, 1985a; Reddy and Wightman, 1988; Reddy *et al.*, 1991). A variety of great potential in this respect is ICG 86031 which has resistance to the vector and to other insects (below) and has unconfirmed resistance to the virus. Field trials have indicated that ICGV 86388 also has resistance to the vector which results in low levels of BNV incidence (Table 11.13). In the USA, Southern Runner suffers less from TSWV than Florunner (Culbreath *et al.*, 1992).

High trichome density, distribution and length have been shown to be important resistance factors in genotypes such as ICG 5040 (NC Ac 2214) and ICG 5043 (NC Ac 2240) (Campbell *et al.*, 1976; Dwivedi *et al.*, 1986).

However, the important line ICG 2271 (NC Ac 343) and its derivatives are not particularly hairy, so that presumably there is a chemical basis to its antibiosis. As in the wild species (below), a flavone glucoside has been linked with antibiosis in the cultivated species (Holley *et al.*, 1984).

(c) Resistance to the lesser cornstalk borer (LCB)

In view of the importance of this insect to the peanut industry in the USA, it is not surprising that resistance to this species has been sought on several occasions. Smith *et al.* (1980a,b) screened 490 accessions in artificial conditions and indicated that varieties Early Runner, Florigiant, Florunner and Virginia Bunch were among the resistant lines. Field tests in North Carolina of an initial 120 lines with natural infestations (i.e. screening was carried out in realisitic conditions) were carried out from 1976 to 1981. PI 269116 ranked first or second for lowest peg and pod damage in the four seasons in which infestations were heavy enough to give good screening conditions (Stalker *et al.*, 1984). Tests carried out in North Carolina indicated that several lines are promising.

(d) Resistance to *Spodoptera litura*

The development of resistance to *S. litura* in suitable varieties has been regarded as being of high priority for Asian groundnut farmers for a number of years. The results of experiments carried out in 1986 and 1987 (Table 11.14) indicated the possibility that ICGV 86031 (breeder = S.L. Dwivedi, ICRISAT) had some resistance to *S. litura* combined with high yield in the post-rainy season. This hope was substantiated in further tests on the ICRISAT research farm and in farmers' fields in coastal Andhra Pradesh (southern India). In the limited trials so far carried out, farmers had sufficient confidence to grow this variety without protecting it with insecticides. They were rewarded with higher yields and lower variable costs than neighbours who grew locally acceptable varieties but applied insecticides to kill defoliators. PI 269116, PI 269118 and PI 262042 had resistance to this insect but none were outstanding (Campbell and Wynne, 1980).

Bioassays carried out with larvae as preliminaries to detecting the mechanism of resistance (independent tests by Ranga Rao and Dr D.E. Padgham, NRI) revealed no antibiosis effect on II–VI instar larvae when fed mature leaves of ICGV 86031. The main mechanism of resistance is currently thought to be tolerance, manifested as the enhanced ability of vegetative tissue to regrow following defoliation.

However, first instar larvae suffered 56% mortality when fed on ICGV 86031 compared with 12% mortality when fed on susceptible ICG 221. Padgham also found that newly hatched larvae had a marked propensity to vacate the leaves of this variety in the first two hours of free life. This

TABLE 11.14 *Effect of releasing two fourth instar* Spodoptera litura *larvae on the percentage defoliation (%d) (assessed after the larvae had pupated) and subsequent mean plant yield (y) in g per plant of five groundnut genotypes (the data are the means of five replicates; percentages were transformed to arc-sines for ANOVA)*

Genotype	Days after emergence										
	10		30		50		70		Control	
	%d	y	%d	y	%d	y	%d	y	%d	y
Post-rainy season										
ICGV 86031	86	10.0	58	11.7	37	15.0	22	18.8	–	18.3
ICG 5240	61	8.4	70	8.3	44	6.9	11	14.5	–	14.5
ICGV 86535	83	10.2	68	11.7	54	10.4	12	12.8	–	12.5
ICG 156	100	6.1	75	13.3	41	13.1	28	17.1	–	17.9
ICG 221	100	4.4	91	7.9	50	10.0	18	12.8	–	13.7

SE for: %d = ±1.6 (calculated via arc-sine transformation)
  g = ±1.3

	10		30		50		70		Control	
	%d	y	%d	y	%d	y	%d	y	%d	y
Rainy season										
ICGV 86031	100	4.5	45	9.2	64	10.8	35	10.8	–	11.4
ICG 5240	100	10.5	54	10.0	67	13.3	24	12.0	–	14.7
ICGV 86535	100	3.9	47	9.2	64	10.9	26	9.6	–	12.0
ICG 156	100	4.7	44	10.0	60	10.2	38	12.5	–	10.4
ICG 221	100	1.7	50	4.0	69	5.3	39	5.4	–	5.4

SE for: %d = ±0.5 (calculated via arc-sine transformation)
  g = ±1.1

10 days after emergence (DAE) = seedling stage. 30 DAE = flowering, 50 DAE = pegging, 70 DAE = pod filling. Control plants had no insects and were insect-free throughout the experiment.

suggests that the resistance factor that influences the neonates is associated with the leaf surface, because their feeding activity is restricted to scraping the leaf surface. The antixenosis demonstrated by ICGV 86031 is likely to increase the first instar mortality that is characteristic of r-strategist noctuids (Kyi *et al.*, 1991) and will therefore contribute to the determination of the level of damage caused by the older larvae, among which mortality is comparatively low.

(e) Resistance to the corn earworm (CEW)

Campbell and Wynne (1980) report resistance to the CEW in Early Bunch and NC 6. The resistance in NC 6 affects larval development and is most likely to be indicative of antibiosis.

(f) Resistance to the southern corn rootworm (SCR)

Experiments in North Carolina carried out over nearly 20 years resulted in the development of NC 6 (= NC Ac 343 x VA 61R) which, with other NC Ac 343 crosses, competed favourably with Florigiant in terms of quality and price. It also had 10–20% of the damage caused by SCR and did not need protection from this pest under high infestation pressure (Campbell and Wynne, 1980).

(g) Resistance to the groundnut leaf miner

Resistance to the groundnut leaf miner has been as difficult to locate. This is mainly because of the sporadic nature of the infestations and difficulties in performing realistic screens on a large number of genotypes with artificially reared insects. The difficulties include the need to exclude the parasites but to allow plants to grow in a natural light regime without shading. The inference is that fine net cages (needed to exclude parasites) cut down the amount of light reaching the plants, and that supplementary artificial light appears to cause variations in the leaf chemistry that modify resistance factors (P.J. Moss, unpublished MSc thesis). An indoor screening process that gives a satisfactory degree of consistency has been developed. Progress has also been made with contributions by P.W. Amin, Dr R.V. Satyanarayana Rao, (Indian Agricultural Research Institute) and Ms P.J. Moss (University of Bath, UK) in that resistance (tolerance, antibiosis and physical) to this species has been detected. Tests have reached the stage where trials in farmers' fields are called for to evaluate the significance of this resistance in relevant conditions. Noteworthy genotypes are ICG 2271 (NC Ac 343), ICG 1697 (NC Ac 17090) and ICG(FDRS) 4. Anderson *et al.* (1990) found that NC Ac 2821 (as well as NC Ac 17090 and PI 405132) had resistance to the groundnut leaf miner when tested in Khon Kaen, Thailand.

In contrast to jassids and thrips, it appears that groundnut leaf miner moths are attracted to hairy leaves, as opposed to shiny (glabrous) ones for oviposition (R.V. Satyanarayana Rao and G.V. Ranga Rao, unpublished). Females are attracted to NC Ac 2214 (ICG 5040) for oviposition but the larvae that hatch subsequently suffer higher levels of mortality because of the antibiotic properties of this genotype. The latter appear to be associated with the exudation of relatively large volumes of sap when the plant is injured. The sap is 'gummy', inferring the possibility that the activity of the larvae may be impaired, but it also contains comparatively high concentrations of polyphenols which have been associated with resistance to *Spodoptera litura* in groundnut.

(h) Resistance to soil insects

Amin *et al.* (1985) reported that several of the lines that have resistance to jassids and thrips also suffered less from termite scarification than the other genotypes tested. Subsequent testing in Burkina Faso confirmed this finding in field conditions (Lynch *et al.*, 1986). Important lines include ICG 2271, ICG 5043, ICG 5044 and ICG 5045 (= NC Ac 343, NC Ac 2240, NC Ac 2242 and NC Ac 2243).

Resistance to white grubs has not yet been located in *Arachis*. Trials carried out with the above termite resistant lines in Australia, testing for resistance to the pod feeding *Heteronyx*, revealed no resistance (H. Brier, personal communication). However, resistance to white grubs has been found in other crops. Crocker *et al.* (1990) recorded considerable levels of resistance to *Phyllophaga congrua* in one wheat and four oat cultivars. Lucerne, lupins and *Lotus* are highly resistant to *Costelytra zealandica* (Farrell and Sweeney, 1974; Kain and Atkinson, 1970). Lucerne has been sown in New Zealand as a cleansing crop in dry, lowland pastures that have been badly affected by this pest since the beginning of this century. Lucerne is also resistant to *Heteronyx arator*, another white grub pest of pasture in New Zealand and Australia (King *et al.*, 1975).

11.4.3 Host plant resistance in *Arachis* spp.

Many *Arachis* spp., the 'wild species', have levels of insect resistance that approach immunity, i.e. insects walk away from them or, if confined on the leaves, die if they eat them or die of starvation rather than eat them. This phenomenon has been known for many years (Smith and Barfield, 1982; Amin, 1985b) but, unfortunately, the exploitation of this knowledge has been insufficient in view of the economic importance of some of the insect species involved.

The extent of resistance in the wild species is shown in Table 11.15 which summarizes the results of tests carried out at ICRISAT Center between

1988 and 1992. This represents a formidable amount of potentially exploitable material, especially when it is realized that the levels of resistance far exceed that detected in *A. hypogaea* to the more intractable insects. The inclusion of a root feeding species (jewel beetle) is an indication that a search for resistance to the soil insect pests of groundnut among the wild species may be justified.

It is also noteworthy that 6 of 18 (*A. hypogaea* × *A. cardenasii*) interspecific derivatives were resistant to the groundnut leaf miner in research station (open field) conditions in Tamil Nadu, India (Kalaimani *et al.*, 1989).

However, progress has been made in other directions in that a start has been made in determining the mechanisms of resistance in 14 species (Stevenson *et al.*, 1993a, b; Kimmins *et al.*, 1993). Bioassays (Table 11.16) indicated that the survival and growth rates were, in all cases, significantly lower than in the susceptible *A. hypogaea* control (ICG 221 = TMV 2). Estimates of leaf toughness ('biteability') indicated that physical factors may be components of this resistance phenomenon.

Bioassays of solvent extracts of the leaves of the most resistant species, *A. paraguariensis* and *A. chacoensis* and an F_1 hybrid (*A. chacoensis* × *A. hypogaea*) indicated the presence of biologically active (antibiotic) fractions which were quercetin glycosides that resembled chlorogenic acid. Subsequent tests have indicated the flavonoid diglycosides that are present in the leaves and may be the main resistance factors (Table 11.17).

Thus resistance has been found within the genus to reduce the effects of the serious above-ground pests in groundnut. In several cases, there is no scientific reason why this phenomenon should not be helping farmers in a number of countries. The gaps in our knowledge point to the need to find resistance to soil insects and to exploit the wild species.

11.5 NATURAL ENEMIES OF GROUNDNUT INSECTS

Not much is known about the dynamics of the natural enemies of groundnut pests. There are exceptions, but they are from the southern states of the USA and emanate from institutions which provide strong linkages between high quality research organizations and extension services that are well tuned to the needs of local farming systems. Elsewhere in the world we are somewhere between a zero knowledge base and the stage where research reveals the need to carry out more research rather than the solutions to problems.

11.5.1 North America

Smith and Barfield (1982) and Lynch and Douce (1992) are positive about the potential for the natural control of *Heliothis zea* (corn earworm) and

TABLE 11.15 Arachis *spp. tested at ICRISAT Center for resistance to* Aphis craccivora *(screen house), groundnut leaf miner (GLM) (field), jewel beetle larvae (field), and* Spodoptera litura *(laboratory assay) with an indication of the level of resistance*

Collection number	*Arachis* species	Groundnut aphid	Groundnut leaf miner	Jewel beetle	Army worm
10002	*apressipila*	HR	S	–	HR
9990	*apressipila*	HR	R	R	–
9993	*apressipila*	HR	S	HR	–
30003	*apressipila*	HR	R	HR	–
30009	*apressipila*	HR	R	HR	–
30080	*batizocoi*	–	HR	S	–
30079	*batizocoi*	HR	–	–	–
30081	*batizocoi*	HR	HR	–	–
GKP 9667(316)	*batizogaea*	–	–	–	HR
36034Y0-1	*cardenasii*	HR	S	S	–
36034YF	*cardenasii*	HR	–	R	S
36019-1	*cardenasii*	–	–	R	–
36033Y	*cardenasii*	HR	–	R	–
GKP 10017	*cardenasii*	HR	–	–	–
10602	*chacoensis*	S	S	–	R
10602-5	*chacoensis*	–	R	–	–
10602(5)	*chacoensis*	–	–	–	HR
30109	*paraguariensis*	–	R	–	–
565-6	*paraguariensis*	–	S	–	–
HLKHe565-6	*paraguariensis*	–	–	HR	–
30134	*paraguariensis*	–	–	HR	HR
KCF11462	*paraguariensis*	–	–	R	–
30109	*paraguariensis*	–	–	–	HR
GKPSC30124	*paraguariensis*	–	–	–	HR
9634	*pseudovillosa*	–	–	–	R
12922	*pusilla*	HR	R	HR	–
114	*rigonii*	–	S	–	–
30007	*Arachis* sp.	R	HR	–	R
A77/113	*Arachis* sp.	–	–	–	HR
GKP9578(312)	*Arachis* sp.	–	–	–	S
GKP9797	*Arachis* sp.	–	–	–	HR
GKPSCS30135(21)	*Arachis* sp.	–	–	–	HR
GKP9893(321)	*Arachis* sp.	–	–	–	HR
GKP9629(315)	*Arachis* sp.	–	–	–	HR
GKP9572(311)	*Arachis* sp.	–	–	–	HR
9921(100)	*Arachis* sp.	–	–	–	HR
KG30012(339)	*Arachis* sp.	–	–	–	HR
GKPSC30116/19	*Arachis* sp.	–	–	–	HR
GKPSC30114	*Arachis* sp.	–	–	–	HR
GKPSCS30144(18)	*Arachis* sp.	–	–	–	HR

TABLE 11.15 *Cont.*

Collection number	*Arachis* species	Groundnut aphid	Groundnut leaf miner	Jewel beetle	Army worm
GKP9553(90)	*Arachis* sp.	–	–	–	HR
GKPSCS30135	*Arachis* sp.	–	–	–	HR
GKPSCS30138(3)	*Arachis* sp.	–	–	–	HR
GKP 10602	*chacoense*	–	–	–	IR
36025-1	*chiquitana*	–	S	HR	–
9530	*correntina*	–	S	R	–
K 7988	*duranensis*	HR	S	S	–
30074	*duranensis*	–	R	–	–
30065	*duranensis*	–	S	–	–
30067	*duranensis*	–	S	–	–
30070	*duranensis*	HR	–	–	–
HL 189	*glabrata*	–	–	–	HR
HLK He 571(91B)	*glabrata*	–	–	–	HR
A45/114	*hagenbeckii*	–	–	–	HR
A27/117	*hagenbeckii*	–	–	–	HR
2A5	*hagenbeckii*	–	–	–	HR
GK 30006	*hoehnei*	R	–	–	–
30085	*kemf-mercadoi*	—	R	HR	HR
35001	*kemf-mercadoi*	HR	–	–	–
30035	*khulaminii*	HR	S	HR	–
30063	*monticola*	–	S	–	–
7264	*monticola*	–	–	–	R
30008	*otavoi*	R	S	–	–
30017	*otavoi*	HR	S	–	–
GKPSBS30132(1)	*Arachis* sp.	–	–	–	HR
GKPSC30122	*Arachis* sp.	–	–	–	HR
GKPSC30120(13)	*Arachis* sp.	–	–	–	HR
Manfredi-5	*Arachis* sp.	–	–	–	HR
10038LL	*spegazzini*	–	S	–	–
30126	*stenophylla*	HR	HR	–	HR
HLK 410	*stenosperma*	HR	HR	S	–
HLK 408	*stenosperma*	HR	HR	S	S
HLK 409	*stenosperma*	–	HR	R	–
30011	*valida*	–	S	R	–
–	*villosulicarpa*	–	HR	S	–
–	*villosa*	–	–	–	HR
TMV 2	*hypogaea*	S	S	S	S
EC 36892	*hypogaea*	HR	S	S	S

HR = highly resistant; R = resistant; S = susceptible

TABLE 11.16 *Insects and mites to which resistance has been located among the* wild Arachis *spp. and interspecific derivatives (details are in Lynch (1990) or in Table 11.15*

Thrips	Campbell and Wynne (1980)
	Amin (1985b)
Empoasca spp.	Campbell and Wynne (1980)
	Amin (1985b)
Tetranychus spp.	Johnson *et al.* (1977)
	Leuck and Hammons (1968)
Lesser corn stalk borer	Kamal (1978)
	Stalker *et al.* (1984)[1]
Groundnut leaf miner	Table 11.15
	Kalaimani *et al.* (1989)
Spodoptera spp.	Lynch *et al.* (1981)
	Kimmins *et al.* (1993)
	Table 11.15
Heliothis zea	Stalker and Campbell, (1983)
Southern corn rootworm	Stalker and Campbell (1983)
Root feeders	Table 11.15
Aphids	Table 11.15
	Amin (1985b)

[1]The authors noted that cultivated genotypes were as resistant as the wild species tested and suggested that there was no need to attempt to exploit the wild species for resistance to this species.

TABLE 11.17 *Flavonoid diglycosides in leaves of most resistant* Arachis *species*

Species	Flavonoid diglycosides	Concentration in fresh leaf (mM/g)
A. chacoensis	Quercetin 3-arabinosylgalactoside	3.10
	Quercetin 3-digalactoside	1.50
A. chacoensis × *A. hypogaea* hybrid	Total	2.50
A. paraguariensis	Quercetin 3-rhamnosylgalactoside	1.80
	Kaempferol 3-rhamnosylgalactoside	1.56
A. hypogaea (TMV2)	Quercetin 3-digalactoside	0.46

H. virescens (tobacco budworm) in the south-eastern USA. Egg parasites (*Trichogramma* sp.) giving 3–83% mortality, larval parasites (*Microplitis croceipes* and *Eucelatoria armigera*) and nuclear polyhedrosis virus combine to maintain defoliator population densities below economic status.

Smith and Johnson (1989) undertook a 3-year study, covering six discernible generations, of the population dynamics of *Elasmopalpus lignosellus* (lesser cornstalk borer, LCB) in Comanche County, Texas, basing themselves in fields of cv. Starr that had no insecticide applied, no irrigation and a history of LCB attack. They found that within-generation mortality ranged from 87.1% to 96.5%. Although larval mortality could be ascribed to an entomopox virus, a fungal disease, 13 primary parasite species and five species of predator, the main (key) mortality factors were 'unidentified' and density independent, and they influenced the eggs (average 7.4%) and first instar (average 53.8%). This is not an unusual feature of the population dynamics of R-strategist Lepidoptera (Kyi *et al.* 1991).

The implication of density independent mortality factors as being more important than parasites, predators and pathogens is consistent with the observation by Lynch and Douce (1992) to the effect that parasites and predators maintain LCB population densities at sub-pest levels except in seasons that are abnormally dry and hot. The host larvae are able to tolerate these conditions whereas the natural enemies are not. This results in outbreaks of the species in conditions where the crops are potentially least tolerant to additional stress. We now need to know the nature of the unidentified mortality factors and how or if they are influenced by the temperature and moisture content of the soil within the larval zone of activity. The possibility that changes in nutrient status of the plants that are associated with drought stress also favour the proliferation of this species (Wheatley *et al.*, 1989) has apparently not been considered.

The potential importance of spiders in groundnut fields is highlighted by a very detailed study carried out in Texas on irrigated and dryland fields that had not been treated with insecticides during the 1981 and 1982 seasons (Agnew and Smith, 1989). More than 25 000 spiders were collected, belonging to 18 families and 79 genera. Hunting spiders outnumbered web spinners by about 10 to 1. The list of prey is interesting because it includes a number of potential pests, such as mirids, larval *Heliothis* sp., jassids, and thrips. However, of the 220 prey records, 72 are of predacious insects (excluding Hymenoptera), including 38 spiders, and 21 are Hymenoptera (14 ants and 6 parasitica). This indicates that spiders confounded ecologist's concepts by operating in two trophic levels.

Earwigs (*Labidura riparia*) predominated in a 2-year study of arthropod predators in peanut fields in Alabama (Kharboutli and Mack, 1991). They ate caterpillar larvae (LCB, CEW and FAW). The exponential increase and decrease in their density was interpreted to be associated with the rapid exploitation of their food source and its subsequent exhaustion. The voracious and polyphagous red imported fire ant *Solenopsis invicta* was the next most abundant predator.

The stable and efficient operation of such natural control systems within agroecosystems is dependent on the presence or provision of off-season refuges that support a reservoir of hosts and their food plants. Such systems are also easily disrupted by insecticides to which, in our experience, spiders have a particularly low tolerance. We wonder to what extent the status quo for *Heliothis* spp. of the early 1970s still exists in commercial areas, and whether it could be improved upon by the exploitation of host plant resistance.

The results of recent research carried out at ICRISAT make us approach this notion with caution for two reasons. Firstly, it does not include reference to the potential confounding effect of host plant genotype on the rate of parasitism. Secondly, the fourth trophic level, which includes the parasites of parasites (hyperparsites), can also influence the success of natural control processes.

R.E. Lynch and J.J. Hamm (unpublished) demonstrated that, in the southern USA, the nuclear polyhedrosis virus (NPV) of *H. zea* was compatible in a tank mix with chlorothalonil applied for leaf spot control. The viability of the virus was reduced by about 60% but this did not prevent the initiation of epizootics in field trials. Subsequent population densities of the target species were maintained at below the economic damage level because natural enemies were not affected by the virus.

11.5.2 Asia

There is little doubt that, in insecticide-free conditions, natural control processes can have a marked effect on the densities of the pan-Asian defoliators. A list of the natural enemies of *Spodoptera litura* has been compiled (Ranga Rao *et al.*, 1993). It includes:

69 parasite species in 7 hymenopterous and 2 dipteran families
36 species of predacious insect in 14 families
12 species of spider in 6 families
4 species of protozoan
4 species of fungal pathogen
7 bacteria
4 viruses
5 nematode species

This probably represents the tip of the iceberg, because it represents what happens when a suitably qualified person with the means to observe, rear, collect, identify and report is in the right place.

The levels of larval and egg parasitism in *S. litura* are monitored on the ICRISAT farm (100–400 individuals or egg rafts) during each season. The data indicate that the tachinid *Paribaea orbata* is the most common larval parasite and that *Ichneumon* spp. and *Exorista xanthopis* also contribute to larval mortality. The rate of larval parasitism is 9.0% over eight seasons, but this includes two seasons where it reached 26.2% and 15.0%. Egg parasitism is not known to have exceeded 27.0%. The associated factors are the relatively low density of *S. litura* on this experimental farm (indicating that only the parasites with a highly efficient searching ability will have any impact) and the effectiveness of the insecticide applications.

Birds are known to predate *S. litura* larvae on the ICRISAT research farm. Cattle egrets removed 62% of the larvae released into unnetted enclosures (compared with netted enclosures). This may be an exaggeration of what happens in farmers' fields because the birds have learned that they may get food where they see the enclosures on our research farm. Cattle egrets are certainly present in farmers' fields in Asia and Australia and are known to include noctuid caterpillars in their diet (Siegfried, 1971). There is a need to learn more about the role of these and other birds as predators in groundnut and other crops.

S. litura is susceptible to viral diseases. We have heard verbal accounts of farmers in India spraying diluted suspensions made from diseased larvae to their crops. However, we have no details of the larval equivalents per unit area or the effectiveness of such activities.

Shanower *et al.* (1992) list 38 species of parasites known to be associated with the groundnut leaf miner. Seventeen of these were reared from larvae collected on the ICRISAT farm, and approximately half were known to be primary parasites. There were changes in dominance among parasite species during a season. For instance, in the post-rainy season 1987–1988, *Sympiesis dolichogaster* emerged from 26% of the parasitized larvae of the first groundnut leaf miner generation and declined to 12% and 16% by the third and fourth generation. *Stenomesius japonicus* emergences increased from 6% to 22% during the same period. The other species involved, which also emerged at higher rates as the season progressed, were *Goniozus* sp., *Chelonus* sp. and a group of three braconids – *Apanteles*, *Avga choaspes* and *Bracon* sp. The 'other species' group, at 40%, predominated by generation 4 and included unidentifiable species, some of which could have been hyperparasites.

This study also showed that:

- parasitism levels were not influenced by the genotypes – ICG 1697 (NCAc 17090) and ICG 799 (Kadiri 3 = Robut 33–1);
- insecticide application (dimethoate at 240 g/ha in 350 l water, once early in each caterpillar generation) reduced parasitism levels by 7–14%; and
- diseases accounted for up to 30% of the larval mortality (a new finding).

The latter observation indicated that a degree of larval mortality remained unaccounted for. This was assumed to be predation by carabids, e.g. *Chlaenius* sp. (Shanower and Ranga Rao, 1990) and spiders. The remaining

TABLE 11.18 *Influence of groundnut genotype on the rate of groundnut leaf miner parasitism at ICRISAT Center, rainy season 1992 (d = host density, % = percent parasitism) (data collected by Research Fellow M.L.J. Sison)*

Genotype	Leaf miner larvae per m^2 (d)	Parasitization[1] (%)
ICG 221	42	16
ICG 156	28	26
ICG 5240	26	26
ICG 2271	21	36
ICG 5040	78	80
ICGV 86031	31	22
ICG 799	51	30
ICG 5044	154	26
SEm	± 13.1	–

[1] Sample size: 50 larvae in each treatment.

questions about the interactions between host density, primary and secondary parasites and the mortality caused to all three levels by a range of insecticides, possibly applied at different times during the life-cycle of the caterpillar, remains as an example of a research project revealing further sets of problems. The contribution of the hyperparasites to the population dynamics of the herbivore is an important issue because we believe that the maintenance of a high parasitism rate (which can exceed 90%) is a lead factor in the management of the groundnut leaf miner.

Although Shanower *et al.* (1992) found that host plant genotype had no effect on the rate of parasitism (all species), this is not necessarily always the case. Observations at ICRISAT Center, on crops growing in an area that has never been treated with insecticides, indicate that host genotype and species can influence the level of groundnut leaf miner parasitism. Table 11.18 shows that, in a comparison of four genotypes, the rate of groundnut leaf miner parasitism was particularly low on aphid and jassid resistant ICG 5240, which is not known to be resistant to this caterpillar. Only ICG 156 is recognized as having a degree of resistance to the groundnut leaf miner.

Turlings and Tumlinson (1991) provide information to indicate that it is possible for chemicals released from a plant as a result of herbivore damage to attract parasites to the scene of the activity and thus to their host(s). There is a good case for following up this matter for groundnut in view of the clear effect of genotype on groundnut leaf miner parasitism.

The rate of parasitism (all species) also proved to be higher by 8% or

TABLE 11.19 *Density of groundnut leaf miner larvae and larval parasitism on sole (s-) and intercropped (i-) groundnut (gnt) and soybean (soy)[1]: ICRISAT Center, data for 22 August (rainy season)*

	s-gnt	s-soy	i-gnt	i-soy	±SE
Larvae per 25 plants	3.1	4.7	3.3	6.3	1.3
Larvae/m² soil surface	46.0	183.0	52.0	243.0	22.0
Larvae/100 cm² leaf	9.1	19.4	9.7	26.7	1.4
% parasitism	36.0	47.7	31.4	38.7	2.6

[1]Plot size = 15 × 15 m, with seed sown in four rows on 1 m wide raised beds. Alternate beds of the two crop species were sown in the intercropped treatments. There were five replicates of the three treatments.

12% on soybean than on groundnut, depending upon whether the crops were sole crops or intercrops (Table 11.19). Host density was higher on the soybean, so that the higher rate of parasitism could be a function of the parasites' searching efficiency, i.e. under the relatively low host density in this experiment, the parasites were able to sting more hosts on soybean because they needed to search fewer leaves to find a caterpillar – a complex aspect of parasite ecology in its own right (Hassell, 1982).

11.5.3 Africa

Weaving (1980) found that up to 56% of the eggs of *Hilda patruelis* are parasitized by *Psyllechthrus oophagus* (Hymenoptera: Encyrtidae). However, the rate drops considerably during the rainy season, which is when groundnut is grown.

Colonies of *Aphis craccivora* living on newly established groundnut crops (1–4 weeks after emergence) in Malawi were virtually without exception accompanied by (unidentified) syrphids, coccinellids and lacewings. Aphidioid mummies were also present. The aphid colonies died out within a week of this observation. The impression, as no data were collected, was that the predators were mainly responsible for this decline in aphid population density (observation by Wightman).

Later in the season, Wightman also noted the presence of high densities (up to one per 10 plants) of mantids in groundnut crops throughout southern Africa. They were of sufficient size (<7 cm body length) and density to warrant an investigation of their contribution to the natural control of potential groundnut pests in Africa.

The predominant arthropod life form, however, appeared to be ants. Members of genera such as *Pachycondyla*, *Myrmicaria* and *Platythyrea* were often abundant in groundnut fields and were seen in Malawi and Tanzania carrying caterpillars and other prey to their nests. Reimer (1988) found that thrips were eaten by ants (and anthocorid bugs, which are also common in groundnut fields in southern Africa).

Even less is known about the natural control of soil insects. In Malawi, ants were seen dragging a dropped live white grub larva, weighing approximately 3 g, along the furrow between two rows of groundnut plants. It is not known what happens when such an encounter between ant and white grub takes place underground. Wireworms are also recognized as the natural enemies of white grubs, but they can damage pods as well.

Ants are the natural foe of termites (Logan *et al.*, 1990). Wightman and Wightman (1988) found that the admixture of insecticides with the soil in groundnut fields can disrupt ant activity, whilst leaving termites unaffected. Whilst sampling the soil of groundnut fields for insects, Wightman found larvae belonging to the dipteran families (unidentifiable to species from larvae by the British Museum of Natural History) Scenopenidae, Mydidae and Therevidae. They are all known to prey on either Coleoptera larvae, *per se*, or arthropods and earthworms (Therevidae), and may, therefore, include white grubs in their prey.

A fungal disease, *Metarhyzium anisopliae*, the green muscardine fungus, is currently under evaluation for the control of the white grubs that attack groundnut in Queensland (Milner, Rogers and Brier, personal communication; Milner, 1989, 1992). After many years of evaluation, the selection of strains that are highly pathogenic to scarab larvae has allowed the application of this technique to proceed to the commercialization phase for pasture and sugar pests. The propagation and dispersal of the fungus is relatively simple. The successful outcome of the trials in Australia and the extension of the technology to Asia and Africa is the kind of breakthrough needed to give leverage on the almost intractable white grub problem.

11.5.4 Comments on the natural control of insects living on groundnut crops

It is known that the potential for natural control to make a major contribution in maintaining the herbivores living on groundnut at levels lower than action thresholds is high. Unfortunately, considerably less is known about the dynamics of the various processes. This is especially true of predation, where there is ample scope for applying immunological techniques to determine the nature and quantity of the prey of suspected arthropod predators (e.g. Giller, 1984; Stuart and Greenstone, 1990). Presumptive vertebrate predators could be studied by more conventional techniques involving trapping and faecal analyses.

11.6 INSECTICIDES

We are, in general, somewhat diffident about emphasising the application of insecticides for reducing the density of foliage feeding insects, except at the seedling stage, and perhaps to the soil. The data available indicate that mirids, as destroyers of flower buds, are also an exception and, where feasible, should be eliminated if present in crops during the flowering stage, at least in the more determinate varieties.

Since 'peanut entomological literature is replete with the effectiveness of insecticides in reducing pest populations and to increasing peanut yield,' (to quote Lynch and Douce, 1992), we shall not go into the details of the many insect/insecticide options. We shall instead look at areas that are not covered so well.

Although no documentation referring specifically to groundnut has been located, the 'Green' environmental lobby in the USA is putting pressure on the agroindustry as a whole to reduce pesticide usage. Without entering the rights and wrongs of such matters, this will undoubtedly push the groundnut industry towards exploiting the resistance that already exists in NC 6 and the genotypes and species discussed in section 11.4. It is fortuitous that excellent co-operation between discipline scientists in the south-east of the USA has led to the identification of such a choice of material. This review also signals an increase in interest in the process of natural control among groundnut entomologists in the USA since the late 1980s.

Our impression is that *A. hypogaea* is naturally resistant to many insects and that many of the insect pest problems on groundnut are induced by insecticides (or suboptimal management procedures). Tait and Napompeth (1987) provide many examples of how insecticide use in the less developed countries has created more problems than it has solved. Circumstantial evidence for this is Wightman's observation, during a survey of groundnut fields in southern Africa, that he only found significant defoliation (by *Helicoverpa* sp. and *Spodoptera littoralis*) on two research stations where insecticides had been applied in response to minor defoliation.

Amin (1988) stressed that, up to the mid 1960s, there were many fewer insects recognized as groundnut pests in India than there are today. This must be attributed to changes in groundnut management, and one such change is the increase in insecticide application.

Ranga Rao and Shanower (1988) found that 70% of farmers visited in a post-rainy season survey of groundnut fields in Andhra Pradesh, India, had applied insecticides to their groundnut crops, mainly in response to the appearance of groundnut leaf miner. Heavy users had crops with bad defoliation by *S. litura* and *Helicoverpa armigera*, but their crops were in no better condition than neighbours who had not applied insecticides. The latter had little defoliator injury but may have had reduced yields (though not necessarily submaximal profits) as a result of groundnut leaf miner activity. This was attributed to the destruction of the natural control process by insecticides. Our experience at ICRISAT is that areas of groundnut fields that have had insecticides applied to them always suffer more from defoliators (Table 11.20). The failure by farmers to eliminate

TABLE 11.20 *Damage by defoliators to groundnut in plots treated and untreated with insecticides: consolidated data from ICRISAT Center (experimental plots) in post-rainy season only*

	Percentage of leaflets with defoliator damage	
	With insecticides	No insecticide
1984–85	7.0 (8 × dimethoate)	0.3
1985–86	9.0 (8 × dimethoate)	5.3
1986–87	20.7 (8 × dimethoate)	5.8
1987–88	5.7 (2 × monocrotophos)	2.9
1988–89	2.8 (1 × monocrotophos)	1.7
1989–90	1.7 (2 × monocrotophos)	1.5
1990–91	1.9 (2 × monocrotophos)	0.3

the target Lepidoptera implies that these defoliators were no longer susceptible to the insecticides (carbamates, organophosphates and pyrethroids applied singly or as ad hoc cocktails). Insecticide resistance in both of these species in this area has been documented (Ramakrishnan *et al.*, 1984; Armes *et al.*, 1992).

Unfortunately, it is not possible to rerun history and perform controlled experiments on such matters, but it is possible to attempt to reverse the process. Prior to 1984, insecticides were applied to groundnut crops on the ICRISAT farm in response to the perceived needs of the scientists (breeders, physiologists etc.) involved, reacting to what was often minor damage. This resulted in a treadmill effect, to the extent that eight or nine applications were being made per season (D.S. Bisht, personal communication). In the 1984 post-rainy season this process ceased and insecticides have only been applied at the behest of a groundnut entomologist, according to thresholds that have been slowly relaxed. The result is that there has been a steady decline in the average number of applications made per season and the number of pest outbreaks (Table 11.21). The unfortunate result is that it is now difficult for breeders, entomologists and virologists to screen for resistance in realistic conditions. Table 11.21 indicates a reversal in the downward trend in the amount of insecticide applied to groundnut in the 1989 rainy season. This corresponded to increased levels of insecticide application in response to heavy outbreaks of *H. armigera* on chickpea and pigeon-pea that probably diminished the level of the natural control process across the whole farm.

11.6.1 Flower pests

What is not shown in Table 11.21 is that, often, only one low rate application of dimethoate was applied at the start of a season (5–12

TABLE 11.21   *Insecticide (kg or l formulated product) applied to groundnut fields on the ICRISAT Center Research farm (source D.S. Bisht, Farm Manager)*

Season		Total area (ha)	Total applied (kg or l)	Mean ha^{-1}	Materials
Up to 1984		20–30	>200	*c.*9	
1984	Rainy	20.00	102.10	5.10	Di, En, Dv, Me
	Post-rainy	30.00	123.50	4.12	Di, En, Ca, Dv
1985	Rainy	13.90	50.00	3.57	Di, En, Dv, Me
	Post-rainy	29.73	65.75	2.19	Di, En, Dv, Me
1986	Rainy	22.00	44.63	2.03	Di, En, Dv, Fe, Lo, Me
	Post-rainy	31.41	31.20	1.00	Di, En, Me, Fe
1987	Rainy	29.90	32.00	1.07	Di, Dv
	Post-rainy	34.75	36.25	1.00	Di, Dv
1988	Rainy	17.14	20.57	1.20	Di
	Post-rainy	26.90	31.55	1.17	Di
1989	Rainy	20.00	66.80	3.30	Di, Qu, En, Ca, Dv
	Post-rainy	19.95	36.5	1.92	Di, Qu, Dv
1990	Rainy	No data			
	Post-rainy	24.00	62.4	2.60	Di, En

Ca = carbaryl, Di = dimethoate, Dv = dichlorvos, En = endosulphan, Fe = fenvalerate, Lo = Lorsban, Me = metasystox, Qu = quinalphos

days after emergence) to kill thrips and jassids. Whilst even this may not have been necessary in economic terms, it served certain cosmetic needs of a research station. It is suspected that this regime would eliminate the mirids, which attack early and can have a marked impact on the flowering, without jeopardizing the natural control process. This is because the major potential defoliators have not established themselves by this stage so that there is no attendant cohort of parasites and predators. (These comments arise from discussion with D.J. Rogers and H.B. Brier, Queensland Department of Primary Industries.)

11.6.2 Soil insects

Currently there is no proven alternative to insecticides for the control of soil insects, other than the promise presented by certain resistant genotypes and wild species (NC 6 against *Diabrotica* is the exception). *Elasmopalpus* remains a problem insect in the USA, especially in dry seasons, but can be controlled by granular formulations of a range of insecticides (Smith and Barfield, 1982). However, as they are best incorporated in the soil prior to sowing, they must be applied prophylactically. This is also the conclusion arrived in India for the control of white grubs where phorate is the preferred insecticide (Yadav, 1991 publication), although Bakhetia (1982) found that seed dressings of other insecticides were effective for white grub control. The best yield response was achieved when the insecticide was mixed as seed dressing with thiram, for the control of *Aspergillus niger*, the cause of collar rot (Bakhetia *et al.*, 1982).

Logan *et al.* (1992) sought alternatives to cyclodiene insecticides for soil insect control by comparing the effects of aldrin, chlorpyrifos and carbosulphan on termites, white grubs and other insects living on or under groundnut stands in India and the Sudan. Chlorpyrifos was, in general, the most effective in that it gave as good results as aldrin, especially in a slow release formulation. The only problem was that residues were found in the seeds after harvest. Isofenphos granules were not as effective as chlorpyrifos for controlling soil insects, but could have a place in pest management programmes because the systemic effect of this material considerably reduced the incidence of groundnut leaf miner caterpillars.

11.7 CULTURAL CONTROL

The concept of cultural control covers all management activities a farmer could execute that are not included under the headings of the preceding three sections. At the same time, cultural control may also include these activities because they are very much intertwined and invariably involve management decisions. It is when we come to this area that we should begin to be able to consider catalysis in the form of insect densities stabilized below economic levels resulting from the simultaneous application of two or more pest management tools. However, before integration is considered it is necessary to review some of the management options that could be considered by groundnut farmers.

11.7.1 Intercropping and habitat diversification

Risch *et al.* (1983) reviewed the relationships between agricultural diversity and pest incidence. They concluded that the benefits, in terms of species abundance and outbreak suppression, were more likely to be associated with interference with insect dispersal activity than enhanced natural control. This may point to the need to design systems specifically to assist the proliferation of natural enemies. This means providing food (nectar sources) for parasitoids and refuges for arthropod predators (sometimes called 'weeds'), and considering an agroforestry component in the farming system to encourage birds, apart from the other benefits. This is why we stress that there is a close link between cultural control and natural control.

Low (or no) technology farmers have diversified their farm systems probably since farming began, largely to ensure that they have something to harvest at the end of the season. Sowing groundnuts between rows of a cereal (maize, pearl millet or sorghum) is common in parts of Africa but is not apparently practised so often in Asia. The indications are that the density of thrips, jassids and groundnut leaf miners on groundnut is reduced by this practice (Wightman and Amin, 1988; Muthiah *et al.*, 1991). Farrell (1976c) found that, in Malawi, a groundnut–bean (*Phaseolus*) intercrop reduced the incidence of GRV because the aphids became hooked on leaf hairs on the bean leaves as they moved from one row to another. Farmers in southern India sow castor at a low density in groundnut fields to attract female *Spodoptera* moths to its leaves for oviposition. The egg masses are easily detected on castor leaves and can be destroyed by hand.

It is our experience that cowpea and groundnut are not good companions because of the risk of the spread of *Aphis craccivora* from cowpea to groundnut. Also, the practice of juxtaposing soybean and groundnut, which is common in Indonesia, could lead to a bad attack of groundnut leaf miner in the groundnut crop. This insect prefers soybean (Table 11.19) but can transfer if the time of soybean harvest corresponds to the pupal or adult stage. We exploit this process by growing 1 m wide beds of soybean around our experimental plots. They are pulled up and left to dry on the soil surface during the pupal stage so that the moths leave the dying plants and oviposit on the groundnut plants.

A stand can also be diversified by sowing more than one genotype of the same species in alternate rows or beds, as a mixture or in some other pattern. To our knowledge, this process has not yet been attempted with groundnut, but it appears to be viable in chickpea where a high yielding but pest susceptible variety was 'protected' by a lower yielding variety with high pest resistance (ICRISAT, 1992).

11.7.2 Mulching

(a) Organic mulches to protect the harvested product

Farmers in Africa commonly windrow the newly harvested plants to permit sun-drying. *Odontotermes* frequently penetrate the piles of drying plants and damage the haulm and seeds. This can reduce the pod yield by as much as 30% or 40% and, of equal importance, increase markedly the subsequent levels of aflatoxin contamination.

Ipomaea fistulosa (morning glory bush) is a common roadside plant in India. Unlike most plants in this environment, the leaves are not eaten by insects or goats. Observation of this fact led to a possible approach

TABLE 11.22 *Effect of organic mulches on the percentage and level of pod scarification by termites (*Odontotermes *sp. and* Microtermes obesi)

	Pods scarified (%)		Scarification mean score (0–4)
Neem cake	2	(0.02)	0.04
Ipomaea	7	(0.07)	0.14
Celosia	17	(0.17)	0.38
Sunn hemp	59	(0.66)	1.69
Bare ground control	36	(0.37)	0.88
SE ±		(0.02)	0.06
F-value		156.4**	118.7**

Plot size = 200 × 50 cm
20 replicates
Neem cake mulch = 2.5 cm thick; others = 5 cm thick
Scoring: 0 = no scarification
           1 = 1–25% of shell surface scarified
           2 = 26–50%   "
           3 = 51–75%   "
           4 = 76–100%   "
(Figures in parentheses are arcsine-transformed value of radians)

to protecting pods as they dry on the ground after harvest. The idea of using a mulch made of the chopped *Ipomaea* leaves and stems was extended to the possibility of using other plants that are not attacked by termites for the same purpose. The results (Gold *et al.*, 1989; Table 11.22) indicated that the *Ipomaea* mulch and neem cake gave drying pods a high level of protection from termites. A mulch made of *Celosia argentea*, a common weed that survives in the vicinity of areas of high termite activity on the ICRISAT farm, gave some protection, but sunn hemp increased the level of termite activity, compared with bare ground. A subsequent experiment (ICRISAT, 1991a) indicated that the long-term benefits of using *Ipomaea* to discourage the surface activity of termites were negligible compared with a neem cake mulch. A neem cake mulch applied to row crops was then shown to have no beneficial effect on yield or foliar insect management – and was very costly (Ranga Rao *et al.*, 1991).

The data available indicates that selected organic mulches could reduce the termite and aflatoxin problem associated with field drying that has concerned farmers and traders alike for many years. There is no evidence to link such mulches with benefits to the growing crop. However, our experiments were not exhaustive in this respect and there is ample scope to investigate the link between enhanced crop water economy (reduced soil evaporation) and weed management associated with mulching and the ability of the plant to tolerate insect attack.

Grainge *et al.* (1985) indicate that *Ipomaea* spp. have anti-fungal and anti-insect (bruchids, 'bugs', various Homoptera and flea beetles) properties. This is clearly a fertile area for further investigation.

(b) Synthetic mulch to reduce virus vector incidence

The ability of a shiny (reflective) or white mulch to protect crops from aphids and the viruses they transmit has been known for many years. The theory and practice are well established (Kennedy *et al.*, 1961; Kring, 1972; Smith and Webb, 1969). Winged aphids respond positively to a short-wave radiation source (such as the sky) when they are ready to initiate their dispersal flight and whilst they have sufficient stored energy to sustain this flight. When 'fatigued' or 'hungry' they react positively to long-wave radiation such as that reflected by vegetation. Thus, whilst they are in the early phases of their flight they respond to the major source of short-wave radiation – the sky – but are confused and repelled when they receive a (reflected) sky signal from the ground when in the end-of-flight host-seeking stage. Presumably, they then fly on to the field of a non-mulching neighbour.

When it comes to protecting a crop from a non-persistent virus (e.g. peanut stripe virus) there are only two options: growing a resistant variety, if one is available, or preventing the vector coming into contact with the crop. Clearly the information about aphids' flight physiology is of fundamental value for protecting a crop from a non-persistent virus, which can be transmitted during just a few seconds' probing by a viruliferous aphid. With persistent viruses (e.g groundnut rosette virus) there are more options but avoidance is still one of them (A'Brook, 1964).

Xu Zeyong *et al.* (personal communication) have shown that plastic mulches may be a viable method of protecting groundnut crops from the non-persistent peanut stripe virus, for which farmers currently have no satisfactory management procedure. The data so far available (Table 11.23) indicate the potential benefits of this approach in terms of reducing the number of aphids alighting in a crop, virus incidence and yield, even though the reporting years were quite different climatically. Dr Xu Zeyong has indicated that the plastic material is available at prices that are within the economic bounds of groundnut farmers in the People's Republic of China. The benefits in yield increase indicated in Table 11.23 are almost certainly related to the lower virus incidence but may also be associated with lower levels of fungal disease (some foliar diseases are initiated by spores splashing up from the soil surface onto the leaves), weed management and a reduction in the amount of water evaporating from the soil surface. These matters are open for further evaluation.

Reflective mulches or aluminium strips painted on a black plastic mulch have recently been shown to protect narrow-leafed lupin and summer squash from non-persistant viruses by repelling the aphid vectors (Jones,

TABLE 11.23 *Influence of sheets of silver and transparent plastic used as a mulches on the number of aphids caught in yellow pan traps placed in the plots, peanut stripe virus (PStV) incidence and yields in field trials in Wuchang, Peoples Republic of China, communicated by Dr Xu Zeyong, Oil Crops Research Institute*

	Silver plastic mulch	Transparent plastic mulch	Bare soil control
Yellow pan catches (aphids per season)			
1990	37	31	268
1991	5	59	404
Virus incidence (%) 1990			
8 weeks after sowing	14.7	18.1	37.7
12 weeks after sowing	88.6	89.0	99.5
1991			
8 weeks after sowing	1.5	3.9	21.6
12 weeks after sowing	5.5	10.0	42.9
14 weeks after sowing	17.8	27.4	93.6
Yield (t/ha)			
1990	3.5	3.6	2.7
1991	1.8	2.1	0.7

1991; Lamont 1990). Shiny plastic mulches are used in Thailand (for instance, in vegetable seed nurseries) for weed control and water conservation.

It is likely that cultural practices will need to be modified to accommodate a reflective mulch if they include recommendations to ensure rapid canopy closure. This is because the mulch can only work if a sufficient area is exposed to the sky. However, as we anticipate that the mulch will also act as a weed control agent, this should not matter. Early canopy closure is usually recommended to reduce the number of weedings needed – and to make the crop less attractive to immigrant aphids, a role we anticipate will be taken on by the shiny mulch (A'Brook, 1964; Farrell, 1976a).

This is a good example of why and how pest management should be approached in a multidisciplinary manner. The preimplementation (research) phase needed for the further development of the shiny mulch technique clearly needs input from agronomists, weed scientists, virologists and entomologists.

11.7.3 Irrigation management

The likes of Smith and Johnson (1989) make it clear that the lesser corn stalk borer is not likely to achieve pest status under irrigated conditions.

The same is partially true of the groundnut leaf miner (Wheatley *et al.*, 1989) but not for the groundnut aphid, which proliferated in the wetter end of the drought stress gradient despite being pounded by water from an overhead irrigation system (Ranga Rao *et al.*, 1991c).

It also appears that the source of irrigation is important. The only outbreak of red spider mite on the ICRISAT farm in recent years was associated with furrow irrigation combined with insecticide application. Neighbouring fields, which were either irrigated with sprinkler and sprayed with a similar insecticide regime or irrigated with a sprinkler and received no insecticide, had no mite outbreaks (densities ranged from 13–110 mites per 50 leaflets). Overhead irrigation reduced the mite density from 6323 to 1282 per 50 leaflets in the outbreak field in a matter of days (Ranga Rao *et al.*, 1990) without recourse to pesticide application. This reduction was sufficient to allow the labourers to harvest without suffering from skin irritation caused by the mite.

11.7.4 Avoidance

Groundnut, being an annual or short season crop, depending upon the genotype and where it is grown, relies mainly upon immigration to provide its quota of foliage feeders. (Possible exceptions are the noctuids and hairy caterpillars that aestivate or hibernate as pupae in the soil.) In line with our data indicating that the older a crop is, the less likely it is to be damaged by insects, it would appear to be a rule of thumb to sow groundnut crops early, before pests have a chance to multiply on their winter/hot/dry season hosts and start the migration process. This is by now conventional wisdom for groundnut in southern Africa, where crops sown with the break in the dry season avoid groundnut rosette virus (section 11.7.2(b)). It is likely that co-ordinated early sowing by government behest in Malawi is responsible for the virtual elimination of this disease in that country.

A similar phenomenon exists in the USA where the corn earworm passes from corn to groundnut, soybean and cotton in late July and August. Sowing the crops early (early to mid April) diminishes the damage caused by this insect (Lynch and Douce, 1992). Early sowing is also likely to reduce the risk of damage being caused by the lesser corn stalk borer (Mack and Backman, 1990).

Our experience with thrips at ICRISAT is mixed. It is not unusual for early sown crops to be badly affected by bud necrosis virus and later sown crops to be unscathed. At the other end of the season, early or timely harvest is called for to avoid termite attack and to reduce the time that the mature crop is exposed to other pod-eating pests, including mammals (Lynch *et al.*, 1986). A delay in harvest is therefore likely to increase aflatoxin incidence.

Some cultural practices are believed to promote pest outbreaks. For instance, the development of irrigation systems in many areas of the semi-arid tropics during the last 30 years has allowed farmers to extend their operations beyond the rainy season. This unquestionably has had a major positive impact on agricultural production. Unfortunately there are some negative effects that would not have been anticipated by the planners. For instance, Yadav (1981) notes the proliferation of white grub damage in northern India where supplementary irrigation prevents soil drying out. Desiccation and soil heating as a result of drought probably affect the grubs more than groundnut plants.

Until perhaps 30 years ago, in southern India *Spodoptera litura* was limited to being a pest of tobacco. It is presumed (Amin, 1988) that its adaptation to groundnut and other crops in southern India was a result of its being exposed to these erstwhile single season crops for more of the year than was the norm under rainfed conditions. It is not unusual to see crops at all stages of development from June to March. A short close season (April to May) gives some respite, except in parts of Tamil Nadu where groundnut is grown all through the year. Groundnut leaf miner and *Spodoptera litura* are particular problems in these areas, as is *Helicoverpa armigera*, which has only recently been recognized as another (induced) pest of groundnut in southern India.

11.7.5 Soil preparation and related cultural practices

Yadav (1981) indicated that deep ploughing can reduce white grub damage, especially during the pupal stage. The 'mode of action' clearly involves the physical damage to the insects as well as disruption of the grubs' environment and their exposure to the sun. Birds are often seen following the plough, presumably to pick up exposed insects (Syamsunder Rao, 1992). The practice of ploughing in the cool of the night, which is made possible with the introduction of the tractors that are a characteristic of the 'white grub belt' of northern India, does not contribute to white grub control. Firstly, birds do not hunt at night; secondly, the grubs are not exposed to the burning and desiccating effects of solar radiation.

Mack and Backman (1990) compared the effects of conventional tillage (ploughing and disking before sowing), reduced tillage (sowing into wheat stubble with a combined subsoiler and seed drill) and 'no-till' (sowing into burned wheat stubble) on insect densities. The tillage system had no effect on the population density of the pest and beneficial insects sampled.

Farmers in Nigeria said they were able to control termites by a number of processes: drumming; burying dead animals, cassava meal, fish guts or the contents of torch batteries in their fields; and introducing soldier ants into the termite nests (Malaka, 1972). There were several other methods involving growing plants or their extracts (*Sanseveria libericum*, *Ocimum basilicum*, *Parkia clappertoniana*, *Vetivera nigritana*, *Digitaria*

sp., *Cymbopogon* sp., and *Pennisetum purpureum*) that are worth evaluation.

11.7.6 Other methods

Farmers may use a wide variety of other methods of managing their pests
that may be beyond the ability of scientists to evaluate under controlled
conditions. Such practices include making smoky bonfires around fields to
disrupt the evening post-eclosion flights of *Amsacta* spp., burning hand
picked blister beetles on fires lit close to fields (the beetles may release
an alarm pheromone) and collecting and killing cockchafers during the
crepuscular feeding and mating forays to specific trees. Such practices,
when combined with other cultural activities, may help to sustain insect
densities at levels below which they become pests. The benefits of other,
technology based approaches, such as trapping noctuid moths in ultraviolet
light traps, releasing egg parasites or catching male moths in pheromone
traps, have also yet to be proved of value.

A new approach to aphid control has been described by Harrewijn *et al.*
(1991). They achieved perfect or almost perfect protection from insects
and the vectored virus diseases of potato crops by covering the growing
plants with an ultrafine polypropylene net. The management of non-
persistent viruses, such as the peanut stripe virus in Asia, remains an
unsolved problem, especially, as in this case, the virus is seed-transmitted.
Clearly, the technique of covering crops with such a fine web, which is still
under investigation in Holland, could become a viable protection method
for farmers' groundnut crops in Asia. However, details of cost and availability have not reached us and we suspect that both could be constraints to
its widespread adoption. The technique may have a special application in
seed or research farms because of the need to provide virus-free seed for
farmers and researchers.

11.8 RESEARCH RELATED TO INSECT PESTS OF STORED GROUNDNUT

A number of insects feed on stored groundnut but perhaps only one is
specifically associated with this product, at least in Africa and Asia –
Caryedon serratus, the groundnut bruchid, borer or 'weevil'. The biology
and management of this and the other species have been discussed in detail
by Dick (1987b) and Wightman *et al.*, 1990. This section simply updates
what has gone before.

Ranga Rao *et al.* (1987) reported that a black carpet beetle, *Attagenus
fasciatus* (Dermestidae), was found in groundnut stores in India. This
species is one of the few capable of penetrating the pod (as a neonate).

Most groundnut storage pests gain access to the seed of the unhulled product via broken shells and testa.

11.8.1 Resistance

One way of limiting the activity of such storage pests is to seek genotypes with resistance factors in the shell. Entomologists have located such genotypes and it is for others to decide whether to attempt to breed such traits into commercial varieties.

The pyralids *Plodia interpunctella* (the Indian meal moth) and *Cadra cautella* (the almond moth) are of particular concern in commercial and farm stores in the south-eastern states of the USA. Kashyap and Campbell (1990) tested 39 *A. hypogaea* hybrids (including established varieties) for resistance to these species. They found, with a little variation, that breeding line 10-P10-B1-B1-B1-B1-B2, variety NC 7 and entries with NC 343, NC 2214 and NC 2232 in their parentage were among the most resistant to both species. Assays were made of oviposition preference, larval development and damage.

Advanced breeding lines have been screened for resistance to *Caryedon serratus* (256 lines) and *Corcyra cephalonica* (306 lines) at ICRISAT Center to discover whether a given line is likely to be more or less resistant to storage pests (Mittal, Wightman and Dwivedi, unpublished). Most varieties tested were neither particularly resistant nor susceptible to either insect. Seventeen were more resistant than the most resistant control to *Caryedon serratus* and 26 had resistance to *Corcyra cephalonica*.

11.8.2 Protecting stored seed

Farmers in developing countries often have to pay a large premium when buying groundnut seed because they are not able to store their own seed from one season to the next without loss of quality as a result of mould and insect infestation. An attapulgite-based clay dust applied to groundnut pods (0.5% w/w) limited the ability of *Corcyra cephalonica* to reproduce to any significant extent (Mittal and Wightman, 1989). This was not a total surprise as the principle that an inert dust can protect stored food stuffs from insects has been demonstrated for a number of scenarios (Shawir *et al.*, 1988). As fungi also reduce the quality of stored seed, we substituted a dust formulation of a number of fungicides (at 3 g/kg) for the clay dust. *C. cephalonica* did not propagate with these treatments (ICRISAT, 1991, pp. 87–88). *Caryedon serratus* was not as responsive to the fungicide dusts, in general, as it was to the attapulgite clay dust. However, the formulation of thiram (75 WDP) that was tested could be substituted for the dust. This experimentation was not continued in on-farm conditions but the information was made available to the Indian National Agricultural Research Program and to the general public via a newspaper article stressing that seed from pods treated with a fungicide should not be eaten unless thoroughly cleaned.

11.9 INTEGRATED CONTROL OF GROUNDNUT PESTS

We define **integrated pest management** as one or more activities that are carried out by farmers that result in the density of potential pest populations being maintained below levels at which they become pests, without endangering the productivity of the farming system as a whole, the health of the farm family and its livestock, and the quality of the adjacent and downstream environments.

One such farmer activity could be sowing an alternative to a particular crop because the latter could not be harvested without using excessive amounts of pesticide or without a high risk of suffering excessive crop loss. Examples in the case of groundnut are few but include the situation in parts of Tanzania and Malawi where this crop is not grown because of the risk of hilda attack.

The keys to the management of insect pests of groundnut are minimizing insecticide application (or, in the case of many African farms, maintaining the current zero to minimal level) and exploiting host plant resistance, combined with cultural practices, especially those that enhance natural control processes. For instance, Campbell and Wynne (1985) demonstrated that NC 6, which has resistance to thrips, jassids, *Heliothis zea* and *Diabrotica undecimpunctata howardi*, can be grown without yield loss penalty (in North Carolina) with no or minimal insecticide application – compared with Florigiant.

Unfortunately, there may always be pests that will cause damage in certain years. Examples are *Elasmopalpus lignosellus* in the USA and the groundnut leaf miner in Asia. Perhaps it is only coincidence but the biological success of both of these insects is favoured by hot, dry conditions. Both can be controlled with granular insecticides incorporated into the soil at sowing, i.e. well before a pest attack can be forecast. The answer to the management of these insects may have to wait until farmers can rely on long-term (6–8 week) weather forecasts.

11.9.1 IPM for groundnut farmers – current and future prospects

As has been made clear, we do not exclude insecticides from IPM. If this is generally accepted, it follows that pest management specialists are obliged to provide farmers or their advisors with the means of deciding when to apply insecticides selectively and which are most appropriate to alleviate or avoid a given condition. Such a scheme has been in place in Georgia (USA) for over a decade (Douce, 1982). It involves information exchange at farmer meetings, publications, field scouting and pheromone traps for monitoring key insects. Lynch and Douce (1992) indicated that, in its early stages, participants benefitted by a yield bonus of nearly 1 t/ha (27%) for an increased outlay of $22/ha per season for pesticides. A more recent evaluation showed that the major participants were spending less on pesticides than non-users and were maintaining high yields.

This indicates two features that should be components of well run, extension lead IPM schemes: routine monitoring of impact of recommendations, and continuous attempts to improve the advice provided, especially if faced with a labile pest scene.

The development of IPM schemes for groundnut is not limited to developed countries. In the Philippines (San Mariano, Isabella) an IPM programme involving pesticides, an egg parasite and *Bacillus thuringiensis* and resistant variety BPI Pn-9 gave higher yields with less pesticide application than no treatment controls and farmers' practices (Campbell *et al.*, unpublished).

The process of developing a similar programme for southern India has been initiated. In many districts there is a clear seasonal sequence of events:

1. Farmers apply insecticides early in the season in response to jassid and thrips symptoms.
2. This reduces the density of parasites and predators (including birds that will only stay in or around a farm if there are at least a few insects to eat).
3. Because of 2, groundnut leaf miner densities increase.
4. Farmers apply insecticide to kill the leaf miners.
5. The level of natural control is reduced even further.
6. Outbreaks of *Spodoptera litura* appear.
7. More insecticides are applied.
8. At this stage insecticide resistance almost certainly occurs so that higher concentrations, cocktails or over-frequent applications have been reported.
9. More severe *S. litura* outbreaks occur, accompanied by the appearance of *Helicoverpa armigera* and white flies.
10. Stages 7–9 are repeated until the crop is harvested or abandoned.

A parallel situation has been reported from Vietnam (ICRISAT, 1991a) where farmers have been known to apply insecticide to groundnut every day and still have crops that are completely defoliated by *S. litura*.

We believe that this treadmill effect can be avoided at the very beginning. Farmers apply insecticides because they want to protect their large financial investment in groundnut seed from (cosmetic) jassid and thrip damage. There is available a variety (ICGV 86031) that has high yield potential along with resistance to jassids and thrips (and, therefore, bud necrosis virus) and defoliators, as well as foliar disease. We are testing this variety in key areas on the farms of lead farmers. Although our experience at ICRISAT indicates that pest outbreaks can be reduced by holding back on insecticide application (Table 11.21), we see the need to give farmers the confidence to get off the treadmill. We hope that the provision of a new, suitably adapted variety will be accepted as an alternative to a heavy insecticide application regime.

We are not as confident of being able to subdue the groundnut leaf miner with host plant resistance at the same time as the other potential pests in this environment. However, we have evidence that our action thresholds for insecticide application for leaf miner control are realistic and can be handled by farmers, and that the parasite cadre is sufficiently robust to survive limited and selective insecticide application (Table 11.5). Continuing research will cover the role of pheromone trapping in dealing with this problem and the possibility of exploiting wild species genes.

This reflects the importance of developing varieties with multiple pest constraint resistance. Even though there may be varieties with higher yield potential, they may never achieve it on farm because pests and the farmers' reaction to them would not allow it. Similarly, a single constraint approach to pest management, followed to the exclusion of consideration of the needs of the system as a whole, can also lead to sub-optimal solutions. For instance, there is little logic or acknowledgement of the principles of IPM to be found in developing a variety that has resistance to a single constraint, such as a virus disease, that hits seriously in a particular zone in only one season in 10 or 20. It is more sensible to concentrate on multiple insect resistant lines that reduce the risk of one-in-two season insect outbreaks and give perhaps 70% virus control via vector resistance. We have also indicated that a general need for groundnut is foliar disease management, with insect resistance being required in specific agroecological zones.

We agree with Lynch and Douce (1992) that the future of IPM for groundnut crops will involve computer modelling. This will emphasize, even more, the need for IPM researchers and practitioners to adopt multidisciplinary approaches. For instance, those interested in pest damage need to be able to discuss leaf eaters with physiologists in terms of depleted leaf area indices and disturbed source–sink relationships in the case of root feeders.

It will be noted that there is in this chapter data that form the basis of a forecasting system for a defoliator (in this case *S. litura*) management. Here are its components, with cross-references to appropriate sections.

1. Flight intensity of migrant moths is monitored by means of a district-wide network of pheromone traps (section 11.3.1(a)). Trap data are reported daily to a central facility.

    (Alternatively, scouts could visit fields to count eggs or egg masses. This is not a preferred procedure because it gives farmers less lead time to prepare to treat their fields during the first instar.)

2. Data for each trap or group of traps are entered into a software package that, on the basis of those data and on a given day for a daily cohort, will estimate:
   (a) the density and date of first day fourth instar larvae and of the moult from third instar to fourth instar based on current prediction models and daily mean air temperatures for the district (sections 11.3.1(b) and 11.3.2(a));
   (b) the total leaf area removed per day of that cohort during instars IV–VI (from the energetics model in section 11.3.3(a));
   (c) the effect of the insect-induced reduction in leaf area on pod production (via a crop model running off the same meteorological data as the insect population density model) (section 11.3.3(b)).
   (d) whether the data for the cohort of that day, combined with the information from previous cohorts, will result in a yield reduction caused by the cumulative activity of the IV–VI larvae (note that the level of resistance to the defoliator in the various genotypes growing in a particular district would have to be taken into account in this stage and if antibiosis retards larval development in 2(a);
3. Output would be in the form of advice to farmers in the village or watershed to apply insecticide, covering:
   (a) the selection of materials to be applied;
   (b) the preferred date;
   (c) the rate, as well as linked advice about harvest dates, etc.

Certain strategies should be built into this third stage. The first is **the application of low rates of insecticides to conserve natural enemies, early in the season**. Ideally, farmers should apply insecticide just as the eggs hatch. This is when the larvae are at their most susceptible to insecticides, so that much lower rates than would be used against larger larvae can be applied. This strategy is fundamental to the conservation of natural enemies. However, it is not easy to achieve, because sharp eyes and constant surveillance are required. Farmer participation would be needed to build confidence in the timing of this procedure,which would be based on a physiological constant.

We are aware that populations can explode later in the season if the correct action is not taken early enough. This means that fall-back strategies are required. We look to **the exploitation of insect pathogens and perhaps selected natural insecticides to supplement them**.

We are also aware that the misuse of insecticides has resulted in the development of chronic levels of insecticide resistance in developing countries. This can be offset, with the co-operation of the agrochemical industry, by a planned approach to the phasing of given insecticides and **the alternation or rotation of active ingredients to avoid or manage insecticide resistance**.

There is a final component in this management system:

4. The information should be returned to the farmers within hours by whatever means are available: radio, television, telephone to village leaders of 'pest management clubs' or, simply, a person on a bicycle.

This covers an approach to managing *S. litura* that could drastically reduce the amount of insecticide applied in several parts of Asia, without endangering the productivity of the farming systems concerned. The basic data is available but needs integrating and verifying in farmers' fields. Others crops that are susceptible to *S. litura*, such as tobacco and vegetables, could also be integrated. Virtually all of the relevant information is available for the groundnut leaf miner which could be 'added on', perhaps even including a function relating pheromone-trap catch data to larval densities once it has been established. Soil insect management requires a different approach involving soil sampling and will not be discussed here.

11.10 CONCLUSIONS

Throughout this chapter we have indicated where there is scope for further research activity and where significant progress has been made. We have distinguished the need to set two kinds of priority – research and economic – on the grounds that specific insects can remain of great (potential) economic importance, even though researchers have provided all the information (within the bounds of the current technology) needed to manage them. Clearly, if and as the fruits of research become accepted as on-farm practice, the economic status of an insect should diminish: *vide Aphis craccivora*.

This is why we have divided the priority ranking of pests in Table 11.24 between economic and research. This table, which is presented as a basis for thought stimulation, challenge and discussion, as much as for information, indicates the need to pay considerably more attention to the soil zone than it has been given in the past. A comparison of Table 11.24 (here) with Table 6 in Smith and Barfield (1982) indicates where progress has been made – for instance, in our understanding of thrips and their relationship with groundnut management – and where recent research has uncovered the need to put emphasis in new directions, for instance white grubs.

Since Feakin completed the third edition of her comprehensive handbook in 1973, there has been a considerable consolidation in our knowledge of groundnut insects. It is clear that, in the USA, groundnut IPM has passed from the research to the implementation and extension phase. Looking at the American scene from a long distance, there appears to be a need for integration across political boundaries and deciding where common goals have been achieved, where voids in knowledge appear and who can best fill them.

In less developed countries, it is possible to detect all levels of progress.

TABLE 11.24 *Major groundnut pests of the world with an indication of their ranking in terms of the need to carry out research on them and their economic importance in the relevant agroeconomic zones together with an indication of current and potential management solutions*

	Importance		Current solution	Potential solution
	Research	Economic		
Millipedes	1	1	None	?
White grubs	1	1	Insecticides (India)	Resistant genes in *Arachis* spp.
			None in Africa	Slow release insecticide formulations
				Metarrhyzium
				Cultural control
Termites	1	1	None in Africa	Resistance to scarification
			Insecticides in Asia	
Hilda	1	1	None	?
Pod borers (wireworms, false wireworms, doryline ants, earwigs)	1	2	None	?
Elasmopalpus lignosellus	1	2	Granular insecticides	
			Irrigation	
Diabrotica	2	2	Insecticides	Host plant resistance
White fringed weevil	2	2	?	?
Miridae	1	2	None	?
Thrips:				
Brazil	2	1	Insecticides	Host plant resistance
as TSWV/BNV vectors	3	2	Host plant resistance	Host plant resistance
Spodoptera litura	1	2	Insecticides (excess)	Host-plant resistance, especially via wild spp. genes
				Natural and cultural control via IPM
S. frugiperda and *S. exigua*	2	2	IPM via pheromone traps monitoring (USA)	
Heliothis zea	2	2		Potential for natural control plus cultural control to obviate need for insecticide
Groundnut leaf miner	1	1	Insecticides	Host plant resistance, especially via wild species genes
				IPM, including pheromone trap monitoring
Hairy caterpillars	2	2	Insecticides and cultural practices	?
Aphis craccivora	3	3	Cultural	Host plant resistance
Jassids	3	3	Insecticides (if anything)	Host plant resistance

Ranking (scale of 1–5): 1= of greatest importance

Unfortunately, the pest spectrum and the farming systems (and their economics) in the USA (and Latin America) are so different to those in Africa and Asia that it is not easy to see direct ways of transferring suitably modified technology. On the other hand, recognition that the high levels of insect resistance in NC Ac 343, 2240, 2214, etc. (which were developed at North Carolina State University from irradiated material) are effective outside the USA and are heritable has given entomologists and breeders a 'flying start'. These and other more recently identified sources of host plant resistance, and hopefully genes from the wild species, will surely form the basis of the successful management of groundnut pests in the future.

ACKNOWLEDGEMENTS

Most of this chapter was compiled while J.A. Wightman was on sabbatical leave and we thank Dr L.D. Swindale (Director General *Emeritus*) and Dr J.G. Ryan (Director General) of ICRISAT for permitting the sabbatical and also Dr J.G. Miller (Director General of the Queensland Department of Primary Industries) and Professor Marcos Kogan (Director of the Integrated Plant Protection Center, Oregon State University) for their hospitality. Professors M.E. Irwin and Gail Kampmeier (Illinois Natural History Survey) provided invaluable information, just as it was needed. Special thanks go to Dr Robert Lynch (USDA, Georgia) for drawing our attention to relevant sections of the literature and for providing unpublished information. We also thank Dr Ali Niazee (Oregon State University) and Dr Lynch for their detailed comments on the manuscript. We are aware that we have absorbed the thoughts and experience of colleagues in many countries and embodied them perhaps rather tersely in this manuscript: their anonymous contribution is duly acknowledged.

REFERENCES

A'Brook, J. (1964) The effect of planting date and spacing on the incidence of groundnut rosette disease and the vector, *Aphis craccivora*, at Mokwa, Northern Nigeria. *Annals of Applied Biology*, **54**, 199–208.

Agnew, C.W. and Smith, J.W. (1989) Ecology of spiders (Araneae) in a peanut agroecosystem. *Environmental Entomology*, **18**, 30–42.

Ahmed, N.E., Yunis, Y.M.E. and Malik, K.M. (1989) *Aspergillus flavus* colonization and aflatoxin contamination of groundnut in Sudan, in *Aflatoxin contamination of groundnut*. Proceedings of the international workshop, ICRISAT, Patancheru, pp. 255–261.

Amin, P.W. (1985a) Apparent resistance of groundnut cultivar Robut 33–1 to bud necrosis disease. *Plant Disease*, **69**, 718–719.

Amin, P.W. (1985b) Resistance of wild species of groundnut to insect and mite pests, in *Proceedings of an International Workshop on the Cytogenetics of Arachis*, 31 Oct–2 November 1983. ICRISAT, Patancheru, pp. 57–69.

Amin, P.W. (1988) Insect and mite pests and their control, in *Groundnut*, (ed. P.S. Reddy), Indian Council of Agricultural Research, New Delhi, pp. 393–452.

Amin, P.W., Singh, K.N., Dwivedi, S.L. and Rao, V.R. (1985) Sources of resistance to the jassid (*Empoasca kerri* Pruthi), thrips (*Frankliniella schultzei* Trybom) and termites (*Odontotermes* sp.) in groundnut (*Arachis hypogaea* L.). *Peanut Science*, **12**, 58–60.

Anderson, W.F., Patanothai, A., Wynne, J.C. and Gibbons, R.W. (1990) Assessment of diallel cross for multiple foliar pest resistance in peanut. *Oléagineux*, **45**, 373–378.

Anitha, V. (1992) Studies on the groundnut pod borers. MSc Thesis, Andhra Pradesh Agricultural University, India.

Appert, J. (1956) Les insectes de'arachide au Sénégal, in *Annual Report of the Centre de Recherche de Bambey* (unpublished) p. 67.

Armes, N.J., Jadhav, D.R., Bond, G.S. and King, A.B.S. (1992) Insecticide resistance in *Helicoverpa armigera* in South India. *Pesticide Science*, **34**, 355–364.

Axelsson, B. (1977) Applicability of laboratory measurements of bioenergetic efficiencies to field populations of *Operophthera fagata* Scharf. and *O. brumata* L. (Lep., Geometridae). *Zoon*, **5**, 147–156.

Axelsson, B., Lohm, U, Nilsson, T.P. and Tenow, O. (1975) Energetics of a larval population of *Opherophthera* spp. (Lep., Geometridae) in Central Sweden during a fluctuation low. *Zoon*, **3**, 71–84.

Bakhetia, D.R.C. (1977) *Anarsia ephippias* (Meyrick) (Lepidoptera: Gelechiidae) damaging the groundnut crop in the Punjab. *Journal of Research of the Punjab Agricultural University*, **14**, 232–233.

Bakhetia, D.R.C. (1982) Studies on the white grub, *Holotrichia consanguinea* (Blanchard) in the Punjab. IV. Control in groundnut through seed treatment with insecticides. *Indian Journal of Entomology*, **44**, 310–317.

Bakhetia, D.R.C. and Sidhu, A.S. (1976) Studies on the chemical control of the groundnut aphid *Aphis craccivora* Koch. *Pesticides*, **10**, 22–24.

Bakhetia, D.R.C., Sukhija, H.S., Brar, K.S. and Narang, D.D. (1982) Studies on the white grub, *Holotrichia consanguinea* (Blanchard) in the Punjab. II. chemical control of the grub damage to groundnut crop. *Indian Journal of Entomology*, **44**, 63–70.

Bellows, T.S., Owens, J.C. and Huddleston, E.W. (1983) Model simulating consumption and economic injury level for the range caterpillar (Lepidoptera: Saturniidae). *Journal of Economic Entomology*, **76**(6), 1231–38.

Biddle, A.J., Hutchins, S.H. and Wightman, J.A. (1992) Pests of Leguminous Crops, in *Vegetable Crop Pests*, (ed. R.G. McKinlay), Macmillan, London, pp. 162–212.

Brier, H.B. and Rogers, D.J. (1991) Susceptibility of soybeans to damage by *Nezara viridula* (L.) (Hemiptera: Pentatomidae) and *Riptortus serripes* (F.) (Hemiptera: Alydidae) during three stages of pod development. *Journal of the Australian Entomological Society*, **30**, 123–128.

Broad, G.H. (1966) Groundnut Pests. *Rhodesian Journal of Agriculture*, **63**, 114–117.

Brust, G.E. and House, G.J. (1990) Effects of soil moisture, texture, and rate of soil drying on egg and larval survival of the southern corn rootworm (Coleoptera: Chrysomelidae). *Environmental Entomology*, **19**, 697–703.

Cadapan, E.P. and Escano, C. (1991) Status and research needs of peanut and other legume insect pests in the Philippines, in *Summary Proceedings of the Workshop on Integrated Pest Management and Insecticide Resistance Management (IPM/IRM) in Legume Crops in Asia*, 19–22 March 1991, Chiang Mai, Thailand. Legumes Program, ICRISAT, Patancheru.

Campbell, W.V. (1978) Effect of pesticide interactions on the twospotted spider mite on peanuts. *Peanut Science*, **5**, 83–86.

Campbell, W.V., Emery, D.A. and Wynne, J.C. (1976) Resistance of peanuts to the potato leafhopper. *Peanut Science*, **3**, 40–43.

Campbell, W.V. and Wynne, J.C. (1980) Resistance of groundnuts to insects and mites, in

Proceedings of the International Workshop on Groundnuts, 13–17 October 1980. ICRISAT, Patancheru, pp. 149–157.

Campbell, W.V. and Wynne, J.C. (1985) Influence of the insect-resistant peanut cultivar NC 6 on performance of soil insecticides. *Journal of Economic Entomology*, **78**, 113–116.

Cho, S.S., Kim, S.H. and Yang, J.S. (1989) Studies on the seasonal occurrence of the white grubs and the chafers, and on the species of chafers in the peanut fields in Yeo-Ju area. *Research Reports of the Rural Development Administration, Crop Protection* **31(3)**, 19–26 (in Korean).

Clifford, P.T.P., Wightman, J.A. and Whitford, D.N.J. (1983) Mirids in 'Grasslands Maku' lotus seeds crops: friends or foes? *Proceedings of the New Zealand Grassland Association*, **44**, 42–46.

Conway, J.A. (1976) The significance of *Elasmolomus sordidus* (F.) (Hemiptera: Lygaeidae) attacking harvested groundnuts in The Gambia. *Tropical Science*, **18**, 187–190.

Crawley, M.J. (1989) Insect herbivores and plant population dynamics. *Annual Review of Entomology*, **34**, 531–64.

Culbreath, A.K., Todd, J.W. and Chamberlain, J.R. (1992) Disease progress of spotted wilt in peanut cultivars Florunner and Southern Runner. *Phytopathology*, **82**: 766–771.

Dammerman, K.W. (1929) *The agricultural zoology of the Malay Archipelago*, J.H. de Bussy, Amsterdam, 473 pp.

Dick, K.M. (1987a) Losses caused by insects to groundnuts stored in a warehouse in India. *Tropical Science*, **27**, 65–75.

Dick, K.M. (1987b) *Pest management in stored groundnuts*, Information Bulletin 22, ICRISAT, Patancheru.

Dohman, G.P., McNeill, S. and Bell, N.J.B. (1984) Air pollution increases *Aphis fabae* pest potential. *Nature (London)*, **307**, 52–53.

Douce, G.K. (1982) *1981 Georgia integrated pest management program facts*, Georgia Cooperative Extension Service, University of Georgia, 66 pp.

Dudley, N.J., Mueller, R.A.E. and Wightman, J.A. (1989) Application of dynamic programming for guiding IPM on groundnut leafminer in India. *Crop Protection*, **8**, 349–357.

Dwivedi, S.L., Amin, P.W., Rasheedunnisa, Nigam, S.N., *et al.* (1986) Genetic analysis of trichome characters associated with resistance to jassid (*Empoasca kerri* Pruthi). *Peanut Science*, **13**, 15–18.

Edwards, P.B. and Wightman, J.A. (1984) Energy and nitrogen budgets for larval and adult *Paropsis charybdis* Stal (Coleoptera: Chrysomelidae) feeding on *Eucalyptus viminalis*. *Oecologia (Berlin)*, **61**, 302–310.

Edwards, P.J. and Wratten, S.D. (1987) Ecological significance of wound-induced changes in plant chemistry, in *Insects – Plants*, (eds V. Labeyrie, G. Fabres and D. Lachaise), Junk, Dordrecht.

Edwards, P.J., Wratten, S.D. and Cox, H. (1985) Wound-induced changes in the acceptability of tomato to larvae of *Spodoptera littoralis*: a laboratory bioassay. *Ecological Entomology*, **10**, 155–158.

Enyi, B.A.C. (1975) Effects of defoliation on growth and yield of groundnut (*Arachis hypogaea*), cowpeas (*Vigna unguiculata*), soybean (*Glycine max*), and green gram (*Vigna aureus*). *Annals of Applied Biology*, **79**, 55–66.

Evans, A.C. (1954) Groundnut rosette disease in Tanganyika. I. Field studies. *Annals of Applied Biology*, **41**, 189–206.

Farrell, J.A.K. (1976a) Effects of groundnut sowing date and plant spacing on rosette virus disease in Malawi. *Bulletin of Entomological Research*, **66**, 159–171.

Farrell, J.A.K. (1976b) Effects of groundnut crop density on the population dynamics of *Aphis craccivora* Koch (Hemiptera, Aphidae) in Malawi. *Bulletin of Entomological Research*, **66**, 317–329.

Farrell, J.A.K. (1976c) Effects of intersowing with beans on the spread of groundnut rosette virus by *Aphis craccivora* in Malawi. *Bulletin of Entomological Research*, **66**, 331–333.

Farrell, J.A.K. and Wightman, J.A. (1972) Observations on the flight and feeding activity of adult *Costelytra zealandica* (White) (Col., Scarabaeidae) in Nelson Province. *New Zealand Journal of Agricultural Research*, **15(4)**, 893–903.

Feakin, S.D. (1973) *Pest control in groundnuts*, PANS Manual No 2 (3rd ed), UK Centre for Overseas Pest Research, London, 197 pp.

Gahukar, R.T. (1992) Groundnut Entomology: retrospect and prospect. *Agricultural Zoology Reviews*, **5**, 139–199.

Garner, J.W. and Lynch, R.E. (1981) Fall armyworm leaf consumption and development on Florunner peanuts. *Journal of Economic Entomology*, **74**, 191–193.

Giller, P.S. (1984) Predator gut state and prey detectability using electrophoretic analysis of gut contents. *Ecological Entomology*, **9**, 157–162.

Gold, C.S., Wightman, J.A. and Pimbert, M. (1989) Mulching effects on termite scarification of drying groundnut pods. *International Arachis Newsletter*, **6**, 22–23.

Gough, N. and Brown, J.D. (1988) Insecticidal control of white grubs (Coleoptera: Scarabaeidae) on the Atherton Tableland, with observations on crop losses. *Queensland Journal of Agricultural and Animal Sciences*, **45**, 9–17.

Grainge, M., Ahmed, S., Mitchell, W.C. and Hylin, J.W. (1985) *Plant species reportedly possessing pest-control properties – an EWC/UH database*. Resource Systems Institute, Honolulu.

Grayer, R.J., Kimmins, F.M., Padgham, D.F. *et al.* (1992) Condensed tannin levels and resistance of groundnut (*Arachis hypogaea*) against *Aphis craccivora*. *Phytochemistry*, **31**, 3795–3800.

Greene, G.L. and Gorbet, D.W. (1973) Peanut yields following defoliation to assimilate insect damage. *Journal of the American Peanut Research and Education Association*, **5**, 141–142.

Grimm, R. (1973) Zum Energieumsatz phytophager Insekten im Buchenwald. 1. Untersuchungen an Populationen der Ruesselkaefer (Curculionidae) *Rhynchaenus fagi* L., *Strophosomus* (Schoenherr) and *Otiorrhyncus singuaris* L. *Oecologia (Berlin)*, **11**, 187–262.

Hagvar, S. (1975) Energy budget and growth during the development of *Melasoma collaris* (Coleoptera). *Oikos*, **26**, 140–146.

Harrewijn, P., den Ouden, H. and Piron, P.G.M. (1991) Polymer webs to prevent virus transmission by virus in seed potatoes. *Entomologia Expermentia et Applicata*, **58**, 102–107.

Hassell, M.P. (1982) What is searching efficiency? *Annals of Applied Biology*, **101**, 170–175.

Hoffman, G.D. and Hogg, D.B. (1991) Potato leafhopper (Homoptera: Cicadellidae) in water stressed alfalfa: population consequences and field tests. *Environmental Entomology*, **20(4)**, 1067–1073.

Holley, R.N., Weeks, W.W., Wynne, J.C. and Campbell, W.V. (1984) Screening peanut germplasm for resistance to corn earworm. *Peanut Science*, **11**, 105–108.

Houston, K.J., Mound, L.A. and Palmer, J.M. (1991) Two pest thrips (Thysanoptera) new to Australia with notes on the distribution and structural variation of other species. *Journal of the Australian Entomological Society*, **30**, 231–232.

Huang, C.Y. and Lin, B.X. (1987) A preliminary study on *Holotrichia sauteri* Moser. *Insect Knowledge*, **24**, 33–34.

Huffman, F.R. and Smith, J.W. (1979) Bollworm: peanut foliage consumption and larval development. *Environmental Entomology*, **8**, 465–467.

ICRISAT (1988) *Annual Report 1987*, Patancheru.

ICRISAT (1991a) *Summary Proceedings of the Workshop on Integrated Pest Management and Insecticide Resistance Management*, Chiang Mai, Thailand, March 1991. Legumes Program, ICRISAT, Patancheru.

ICRISAT (1991b) *Legumes Program Annual Report 1990*. Legumes Program, ICRISAT, Patancheru.

ICRISAT (1992) *Legumes Program Annual Report 1991*. Legumes Program, ICRISAT, Patancheru.

IRAT (1976) Report of the Project de Reinforcement de la Protection en Haute-Volta (unpublished).

Jepson, W.F. (1948) An annotated list of insects associated with groundnuts in East Africa. *Bulletin of Entomological Research*, **39**, 231–236.

Johnson, D.R., Campbell, W.V. and Wynne, J.C. (1977) Resistance of wild species of peanuts to the twospotted spider mite (Acari: Tetranychidae). *Journal of Economic Entomology*, **75**, 1045–1047.

Jones, R.A.C. (1991) Reflective mulch decreases the spread of two non-persistent aphid transmitted viruses to narrow-leafed lupins. *Annals of Applied Biology*, **118**, 79–85.

Kalaimani, S., Mahadevan, N.R., Manoharan, V. and Sethupathi Ramalingam (1989) Screening of interspecific derivatives for leaf miner (*Aproaerema modicella* Deventer) resistance in groundnut. *International Arachis Newsletter*, **5**, 15–16.

Kalshoven, L.G.E. (1981) *Pests of crops in Indonesia*, P.T. Ichtiar Bani van Hoeve, Jakarta, Indonesia, 701 pp.

Kamal, S.S. (1978) Resistance of species of *Arachis* to lesser corn stalk borer. Thesis (unpublished), Oklahoma State University.

Kashyap, R.K. and Campbell, W.V. (1990) Ovipositional preference and larval establishment of the Indian meal moth and almond moth on selected peanut genotypes. *Peanut Science*, **17**, 12–15.

Kennedy, J.S., Booth, C.O. and Kershaw, W.J.S. (1961) Host finding by aphids in the field. III. Visual attraction. *Annals of Applied Biology*, **49**, 1–21.

Kharboutli, M.S. and Mack, T.P. (1991) Relative and seasonal abundance of predaceous arthropods in Alabama peanut fields as indexed by pitfall traps. *Journal of Economic Entomology*, **84**, 1015–1023.

Kimmins, F.M., Stevenson, P.C., Padgham, D.E. and Grayer, R.J. (1993) Mechanisms of resistance in groundnut to homopteran and lepidopteran pests. *Proceedings of the 12th International Congress of Entomology, Beijing*.

Kogan, M. (1986) Plant defense strategies and host plant resistance, in *Ecological theory and integrated pest management* (ed. M. Kogan), John Wiley and Son, Chichester.

Kring, J.B. (1972) Flight behaviour of aphids. *Annual Review of Entomology*, **17**, 461–492.

Kumari, V.L.L. (1989) Sex pheromone systems of selected lepidopterous pests of groundnut. PhD Thesis (unpublished), Andhra Pradesh Agricultural University, Hyderabad, India.

Kumawat, S.K. and Yadava, C.P.S. (1990) Effect of various insecticides applied as soil treatment, seed coating and seed dressing on groundnut against white grub. *Indian Journal of Entomology*, **52**, 187–190.

Kyi, A., Zaluki, M.P. and Titmarsh, I.J. (1991) An experimental study of early stage survival of *Helicoverpa armigera* (Lepidoptera: Noctuidae) on cotton. *Bulletin of Entomological Research*, **81**, 263–271.

Lamont, W.J. (1990) Painting aluminum strips on black plastic mulch reduces mosaic symptoms on summer squash. *Horticultural Science*, **25**, 1305.

Leuck, D.B. and Hammons, R.O. (1968) Resistance of wild peanut plants to the mite *Tetranychus tumidellus*. *Journal of Economic Entomology*, **61**, 687–688.

Litsinger, J.A., Apostol, R.F. and Obusan, M.B. (1983) White grub, *Leucopholis irrorata* (Coleoptera: Scarabaeidae): pest status, population dynamics, and chemical control in a rice-maize cropping pattern in the Philippines. *Journal of Economic Entomology*, **76**, 1133–1138.

Logan, J.W.M., Cowie, R.H. and Wood, T.G. (1990) Termite (Isoptera) control in agriculture and forestry by non-chemical methods: a review. *Bulletin of Entomological Research*, **80**, 309–330.

Logan, J.W.M., Rajagopal, D., Wightman, J.A. *et al.* (1992) Control of termites and other soil pests of groundnuts with special reference to controlled release formulations of non-persistent insecticides in India. *Bulletin of Entomological Research*, **82**, 57–66.

Logan, J.W.M., Wightman, J.A. and Ranga Rao, G.V. (in press) Note on the biology and insecticidal control of jewel beetle larvae (Coleoptera: Buprestidae) living in peanut plants in peninsular India. *Peanut Science.*

Lu, D.Y., Dong, Y.H. and Liang, X.J. (1987) Studies on the occurrence and control of *Holotrichia formosana* Moser. *Insect Knowledge* **24(1)**, 33–34 (in Chinese).

Lundgren, L., Norelius, G. and Stenhagen, G. (1981) Selection of biochemical characters in the breeding for pest and disease resistance: A method based on analogy analysis of chromatographic separation patterns for emitted plant substances. *Hereditas*, **95**, 173–179.

Lundgren, L., Norelius, G. and Stenhagen, G. (1982) Prospects of a holistic approach to the biochemistry of pest and disease resistance in crop plants. *Hereditas*, **97**, 115–122.

Lynch, R.E. (1990) Resistance in peanut to major arthropod pests. *Florida Entomologist*, **73**, 422–445.

Lynch, R.E. and Douce, G.K. (1992) Implementation of integrated pest management in peanut: current status and future direction, in *Proceedings of the Second International Groundnut Workshop, November, 1991*, ICRISAT, Patancheru.

Lynch, R.E., Branch, W.D. and Garner, J.W. (1981) Resistance of *Arachis* species to the fall armyworm. *Peanut Science*, **8**, 106–109.

Lynch, R.E., Garner, J.W. and Morgan, L.W. (1984) Influence of systemic insecticides on thrips damage and yield of Florunner peanuts in Georgia. *Journal of Agricultural Entomology*, **1**, 33–42.

Lynch, R.E., Ouedrago, A.P. and Dicko, I. (1986) Insect damage to groundnut in semi-arid Africa, in *Agrometeorology of groundnut*. Proceedings of an International Symposium, 21–28 Aug. 1986 ICRISAT Sahelian Center, Niamey Niger. ICRISAT, Patancheru, pp. 175–183.

Lynch, R.E. and Simmons, A.M. (1993) Distribution of immature and sampling for adult sweetpotato whitefly, *Bemisia tabaci* (Gennadius) (Homoptera: Aleyrodidae), on peanut, *Arachis hypogaea* L. *Peanut Science*, in press.

Lynch, R.E. and Wilson, D.M. (1991) Enhanced infection of peanut, *Arachis hypogaea* L. seeds with *Aspergillus flavus* group fungi due to external scarification of peanut pods by the lesser cornstalk borer, *Elasmopalpus lignosellus* (Zeller). *Peanut Science*, **18**: 110–116.

Mack, T.P. and Backman, C.B. (1990) Effects of two planting dates and three tillage systems on the abundance of lesser corn stalk borer (Lepidoptera: Pyralidae), other selected insects, and yield in peanut fields. *Journal of Economic Entomology*, **83**, 1034–1041.

Mackay, A.P. (1978) Growth and bioenergetics of the moth *Cyclophragma leucosticta* Gruenberg. *Oecologia (Berlin)*, **32**, 367–376.

Malaka, S.L.O. (1972) Some measures applied in the control of termites in part of Nigeria. *Nigerian Entomologists Magazine*, **2**, 137–141.

Mansour, M.H. (1981) Efficiency of two allelochemicals on the conversion of ingested and digested food into the body tissues of *Spodoptera littoralis* (Boisd.) (Lepid., Noctuidae). Zusammung für Angewandte Entomologie, **91**, 493–499.

Mayeux, A. (1984) Le puceron de l'arachide – biologie et control. *Oléagineux*, **39**, 425–429.

Milner, R.J. (1989) Recent progress with *Metarhyzium anisopliae* for pest control in Australia, in *Proceedings of the 1st Asia Pacific Conference of Entomology*, Chiang Mai, 8–13 November, 1989, pp. 55–64.

Milner, R.J. (1992) The selection of strains of *Metarhyzium anisopliae* for control of Australian sugar cane white grubs, in *The use of pathogens in scarab pest management*, (eds T.A. Jackson and T.R. Glare), Intercept, London, pp. 209–216.

Mittal, S. and Wightman, J.A. (1989) An inert dust protects stored groundnut from insect pests. *International Arachis Newsletter*, **6**, 21–22.

Mohammad, A.B. (1981) The groundnut leafminer *Aproaerema modicella* Deventer (= *Stomopteryx subsecivella* Zeller) (Lepidoptera: Gelechiidae): a review of world literature. *Occasional Paper 3*, Groundnut Improvement Program, ICRISAT, Patancheru.

Murant, A.F. (1990) Dependence of groundnut rosette virus on its satellite RNA as well as on groundnut rosette assistor luteovirus for transmission by *Aphis craccivora*. *Journal of General Virology*, **71**, 2163–2166.

Murant, A.F. and Kumar, I.K. (1990) Different variants of the satellite RNA of groundnut rosette virus are responsible for the chlorotic and green forms of groundnut rosette disease. *Annals of Applied Biology*, **117**, 85–92.

Muthiah, C., Senthivel, T., Venkatakrishnan, J. and Sivaram, M.R. (1991) Effect of inter-cropping on incidence of pest and disease in groundnut (*Arachis hypogaea*). *Indian Journal of Agricultural Sciences*, **62**, 152–153.

Nandagopal, V. and Reddy, P.S. (1990) Report of sex pheromone in groundnut leaf miner. *International Arachis Newsletter*, **7**, 25–26.

Nath, P. and Singh, J. (1987) White grubs – a new pest problem in some pockets of eastern Uttar Pradesh. *Indian Journal of Entomology*, **49**, 181–185.

Nwanze, K.F., Reddy, Y.V.R. and Soman, P. (1990) The role of leaf surface wetness in larval behaviour of the sorghum shoot fly, *Atherigona soccata*. *Entomologia Experimentia et Applicata*, **56**, 187–195.

Padgham, D.E., Kimmins, F.M., Barnett, E.A. *et al.* (1990a) Resistance in groundnut and its wild relatives to *Aphis craccivora*, and its relevance to groundnut rosette disease management. *Proceedings of the Brighton Crop Protection Conference* (1990), pp. 191–196.

Padgham, D.E., Kimmins, F.M. and Ranga Rao, G.V. (1990b) Resistance in groundnut (*Arachis hypogaea* L.) to *Aphis craccivora* Koch. *Annals of Applied Biology*, **117**, 285–294.

Painter, R.H. (1951) *Insect Resistance in Crop Plants*. Macmillan, New York.

Palmer, J.M., Reddy, D.V.R., Wightman, J.A. and Ranga Rao, G.V. (1990) New information on the thrips vectors of tomato spotted wilt virus in groundnut crops in India. *International Arachis Newsletter*, **7**, 24–25.

Petrusewicz, K. and Macfadyen, A. (1970) *Productivity of terrestrial animals: principles and methods*. Blackwell Scientific Publications, Oxford.

Rai, B.K. (1979) *Pests of oilseed crops in India and their control*, Indian Council of Agricultural Research, New Delhi, 121 pp.

Ramakrishnan, N., Saxena, V.S. and Dhingra, S. (1984) Insecticide resistance in the population of *Spodoptera litura* (F.) in Andhra Pradesh. *Pesticides*, **18**, 23–27.

Ranga Rao, D.V., Singh, K.N., Wightman, J.A. and Ranga Rao G.V. (1991) Economic status of neem cake mulch for termite control in groundnut. *International Arachis Newsletter*, **9**, 12–13.

Ranga Rao, G.V., Wightman, J.A. and Ranga Rao D.V. (1993) World review of the natural enemies and diseases of *Spodoptera litura* (F.) (Lepidoptera: Noctuidae). *Insect Science and Application*, **14**, 273–284.

Ranga Rao, G.V., Ranga Rao, D.V., Prabhakar Reddy, M. and Wightman, J.A. (1990) Overhead irrigation and the integrated management of spider mites in a groundnut crop. *International Arachis Newsletter*, **8**, 23–24.

Ranga Rao, G.V. and Shanower, T.G. (1988) A survey of groundnut insect pests in Andhra Pradesh, India, postrainy season 1987/1988. *International Arachis Newsletter*, **4**, 8–10.

Ranga Rao, G.V., Surender, A., Wightman, J.A. and Varma, B.K. (1987) *Attagenus fasciatus* (Thunberg) [Coleoptera: Dermestidae] a new pest of stored groundnut. *International Arachis Newsletter*, **2**, 12–13.

Ranga Rao, G.V., Wightman, J.A. and Ranga Rao, D.V. (1989) Threshold temperatures and thermal requirements for the development of *Spodoptera litura* (F.) (Lepidoptera: Noctuidae) on groundnut leaves in laboratory and field conditions. *Environmental Entomology*, **18**, 548–551.

Ranga Rao, G.V., Wightman, J.A. and Ranga Rao D.V. (1991a) Monitoring *Spodoptera litura* (F) (Lepidoptera: Noctuidae) using sex attractant traps: effect of trap height and time of the night on moth catch. *Insect Science and Application*, **12**, 443–447.

Ranga Rao, G.V., Wightman, J.A. and Ranga Rao, D.V. (1991b) The development of a standard pheromone trapping procedure for *Spodoptera litura* (F) (Lepidoptera: Noctuidae) population in groundnut (*Arachis hypogaea* L.) crops. *Tropical Pest Management*, **37**, 37–40.

Ranga Rao, G.V., Wightman, J.A., Wadia, K.D.R. *et al.* (1991c) Influence of water stress on groundnut aphids. *International Arachis Newsletter*, **9**, 14.

Reddy, D.V.R. and Wightman, J.A. (1988) Tomato spotted wilt virus: thrips transmission and control, in *Advances in Disease Vector Research* (ed. K.V. Harris), Springer-Verlag, New York, pp. 203–220.

Reddy, D.V.R., Wightman, J.A., Beshear, R.J. *et al.* (1991) Bud necrosis: a disease of groundnut caused by tomato spotted wilt virus. Information Bulletin 31, International Crops Research Institute for the Semi-Arid Tropics, Patancheru.

Redlinger, L.M. and Davis, R. (1982) Insect control in postharvest peanuts, in *Peanut Science and Technology*, (eds H.E. Pattee and C.T. Young), American Peanut Research and Education Society, Yoakum, Texas, pp. 520–571.

Reimer, N.J. (1988) Predation on *Liothrips urichi* Karny (Thysanoptera: Phlaeothripidae): a case of biotic interference. *Environmental Entomology*, **17**, 132–134.

Richter, A.R. and Fuxa, J.R. (1990) Effect of *Steinernema feltiae* on *Spodoptera frugiperda* and *Heliothis zea* (Lepidoptera: Noctuidae) in corn. *Journal of Economic Entomology*, **83**, 1286–1291.

Risch, S.J., Andow, D. and Altieri, M.A. (1983) Agroecosystem diversity and pest control: data, tentative conclusions, and new research directions. *Environmental Entomology*, **12**, 625–629.

Rogers, D.J., Brier, H.B. and Houston, K.J. (1992) Scarabaeidae (Coleoptera) associated with peanuts in southern Queensland. *Journal of the Australian Entomological Society*, **31**, 177–181.

Rose, D.J.W. (1962) Pests of groundnuts. *Rhodesian Agricultural Journal*, **59**, 197–198.

Saleh, N. (1991) Bioecology of peanut stripe virus and its control in Indonesia, in *Summary Proceedings of the Workshop on Integrated Pest Management and Insecticide Resistance Management*, Chiang Mai, Thailand, March 1991. Legumes Program, ICRISAT, Patancheru, 15 pp.

Sathorn Sirisingh and Manochai Keerati-Kasikorn (1986) Management of arthropods on peanuts in Thailand, in *Proceedings of USAID – PCRSP Meeting*, Khon Kaen, 19–21 August, 1986, pp. 35–37.

Schowalter, T.D., Whitford, W.G. and Turner, R.B. (1977) Bioenergetics of the range caterpillar, *Hemileuca oliviae* (Ckll.). *Oecologia (Berlin)*, **28**, 153–161.

Schroeder, L.A. (1977) Caloric equivalents of some plant and animal material, the importance of acid corrections and comparisons of precision between the Gentry-Wiegert Micro and the Parr semi-micro bomb calorimeter. *Oecologia (Berlin)*, **28**, 261–267.

Schroeder, L.A. (1978) Consumption of black cherry leaves by phytophagous insects. *The American Midland Naturalist*, **100(2)**, 294–306.

Schroeder, L.A. (1984) Comparison of gravimetry and planimetry in determining dry matter budgets for three species of phytophagous larvae. *Entomologia Experimentia et Applicata*, **35**, 255–261.

Scriber, J.M. (1979a) The effects of sequentially switching food plants upon biomass and nitrogen utilization by polyphagous and stenophagous *Papilio* larvae. *Entomologia Experimentia et Applicata*, **25**, 203–215.

Scriber, J.M (1979b) Post-ingestive utilization of plant biomass and nitrogen by Lepidoptera: legume feeding by the southern armyworm. *New York Entomological Society*, **87**, 141–153.

Shang, X.H., Ma, L.J. and Zhang, S.J. (1981) Studies on the bionomics and control of *Holotrichia diomphalia*. *Acta Phytophylacica Sinica*, **8**, 95–100.

Shanower, T.G. and Ranga Rao, G.V. (1988) *Chlaenius* sp (Col.: Carabaeidae): a predator of groundnut leaf miner larvae. *International Arachis Newsletter*, **8**, 19–20.

Shanower, T.G., Gutierrez, A.P. and Wightman, J.A. (1992a) The effect of rainfall and pest management strategies on the population dynamics of the groundnut leaf miner, *Aproaerema modicella* (Deventer) (Lepidoptera: Gelechiidae), in India. In ms.

Shanower, T.G., Wightman, J.A., Gutierrez, A.P. and Ranga Rao, G.V. (1992b) Larval parasitoids and pathogens of the groundnut leaf miner, *Aproaerema modicella* (Deventer) (Lepidoptera: Gelechiidae), in India. *Entomophaga*, **37**, 419–427.

Shanower, T.G., Gutierrez, A.P. and Wightman, J.A. (1993) Biology of the groundnut leaf miner, *Aproaerema modicella* (Deventer) (Lepidoptera: Gelechiidae) in India. *Crop Protection*, **12**, 3–10.

Shawir, M., Patourel, Le G.N.J. and Moustafa, F.I. (1988) Amorphous silica as an additive to dust formulations of insecticides for stored grain pest control. *Journal of Stored Product Research*, **24** 123–130.

Siegfried, W.R. (1971) The food of the cattle egret. *Journal of Applied Ecology*, **8**, 447–459.

Singh, T.V.K., Singh, K.M. and Singh, R.N. (1990) Groundnut complex: III, incidence of insect pests in relation to agroclimatic conditions as determined by graphical super imposition technique. *Indian Journal of Entomology*, **52**, 686–692.

Sithanantham, S., Sohati, P.H., Syamasonta, M.B. and Kannaiyan, J. (1991) Screening for resistance to sucking insects among groundnut genotypes in Zambia. *Proceedings of the Fourth Regional Groundnut Workshop for Southern Africa*, 19–23 March 1990, Arusha, Tanzania. ICRISAT, Patancheru.

Smith, C.M. (1989) *Plant resistance to insects, a fundamental approach*, John Wiley, 286 pp.

Smith, J.H. (1946) Pests of the peanut crop. *Queensland Agricultural Journal*, **62**, 345–353.

Smith, J.W. and Barfield, C.S. (1982) Management of preharvest insects, in *Peanut Science and Technology*, (eds H.E. Pattee and C.T. Young), American Peanut Research and Education Society Inc., Yoakum, Texas, pp. 250–325.

Smith, J.W. and Johnson, S.J. (1989) Natural mortality of the lesser cornstalk borer (Lepidoptera: Pyralidae) in a peanut agroecosystem. *Environmental Entomology*, **18**, 69–77.

Smith, J.W., Posada, L. and Smith, O.D. (1980a) Greenhouse evaluation of 490 peanut lines for resistance to the lesser cornstalk borer. *Texas Agricultural Experiment Station*, **MP1464**. 42 pp.

Smith, J.W., Posada, L. and Smith, O.D. (1980b) Greenhouse screening for resistance to the lesser corn stalk borer. *Peanut Science*, **7**, 68–71.

Smith, J.W., Sams, R.L., Agnew, C.W. and Simpson, C.E. (1985) Methods of estimating damage and evaluating the reaction of selected peanut cultivars to the potato leafhopper. *Journal of Economic Entomology*, **78**, 1059–1062.

Smith, R.F. and Webb, R.E. (1969) Repelling aphids by reflective surfaces, a new approach to the control of insect-transmitted viruses, in *Viruses, Vectors, and Vegetation*, (ed. K. Maramorosch), Interscience Publishers, New York.

Sohati, P.H. and Sithananthum, S. (1990) Damage and yield loss caused by insect pests on groundnut in Zambia. *Proceedings of the Fourth Regional Groundnut Workshop for Southern Africa*, 19–23 March 1990, Arusha, Tanzania. ICRISAT, Patancheru.

Sorenson, C.J. (1936) Lygus bugs in relation to occurrence of shriveled alfalfa seed. *Journal of Economic Entomology*, **29**, 454–457.

Southwood, T.R.E. (1978) *Ecological Methods*, Chapman and Hall, London.

Stalker, H.T. and Campbell, W.V. (1983) Resistance of wild species of peanut to an insect complex. *Peanut Science*, **10**, 30–33.

Stalker, H.T., Campbell, W.V. and Wynne, J.C. (1984) Evaluation of cultivated and wild peanut species for resistance to the lesser cornstalk borer (Lepidoptera: Pyralidae). *Journal of Economic Entomology*, **77**, 53–57.

Stevenson, P.C., Anderson, J., Simmonds, M.S.J. and Blaney, W.M. (1993a) Developmental inhibition of neonate and 3rd instar larvae of *Spodoptera litura* (Fab.) by caffeic acid esters isolated from the foliage of a wild species of groundnut, *Arachis paraguariensis*. *Journal of Chemical Ecology*.

Stevenson, P.C., Simmonds, M.S.J., Blaney, W.M. and Wightman, J.A. (1993b) The evaluations and characterisation of resistance in 14 species of *Arachis* to *Spodoptera litura* (Fabr.) (Lepidoptera: Noctuidae). *Bulletin of Entomological Research*.

Stuart, M.K. and Greenstone, M.H. (1990) Beyond ELISA: a rapid, sensitive, specific immunodot assay for identification of predator stomach contents. *Annals of the Entomological Society of America*, **83**, 1101–1107.

Supriyatin (1991) Groundnut pests and their management in Indonesia, in *Summary Proceedings of the Workshop on Integrated Pest Management and Insecticide Resistance Management*, Chiang Mai, Thailand, March 1991. Legumes Program, ICRISAT, Patancheru, p. 12.

Syamsunder Rao, P. (1992) *Research Highlights of AICRP on Agricultural Ornithology*, Indian Council for Agricultural Research, New Delhi, p. 16.

Tait, J. and Napompeth, B. (1987) *Management of pests and pesticides: farmers' perceptions and practices*, Westview Press, Boulder and London.

Turlings, T.C.J. and Tumlinson, J.H. (1991) Do parasites use herbivore-induced plant chemical defenses to locate hosts? *Florida Entomologist*, **74**, 42–51.

Turnjit Satayavirut (1988). Thrips (*Frankliniella fusca* Hinds) population, damage and yield relationship for peanut types and selected peanut cultivars in Northern Carolina. PhD Thesis, North Carolina State University.

Veeresh, G.K. (1977) Studies on the root grubs in Karnataka: with special reference to bionomics and control of *Holotrichia serratus* Fabricius (Coleoptera, Melolonthinae). Bangalore, University of Agricultural Sciences, pp. vii + 77.

Wang, Z.R., Chu, Z.O. and Zhang, D.S. (1986) Underground distribution pattern of white grubs and sampling method in peanut and soybean fields. *Acta Entomologica Sinica*, **29**, 395–400.

Weaving, A.J.S. (1980) Observations on *Hilda patruelis* Stal. (Homoptera: Tettigometridae) and its infestation of the groundnut crop in Rhodesia. *Journal of the Entomological Society of South Africa*, **43**, 151–167.

Wheatley, A.R.D., Wightman, J.A., Williams, J.H. and Wheatley, S.J. (1989) The influence of drought stress on the distribution of insects on four groundnut genotypes grown near Hyderabad, India. *Bulletin of Entomological Research*, **79**, 567–577.

Wightman, J.A. (1974) Influence of low temperature on pupation induction in *Costelytra zealandica* (Coleoptera: Scarabaeidae). *New Zealand Journal of Zoology*, **1(4)**, 503–507.

Wightman, J.A. (1978) The ecology of *Callosobruchus analis* (Coleoptera: Bruchidae): morphometrics and energetics of the immature stages. *Journal of Animal Ecology*, **47**, 111–129.

Wightman, J.A. (1979) Energetics as an approach to estimating the economic impact of pasture pests. *New Zealand Journal of Zoology*, **6**, 509–517.

Wightman, J.A.(1988a) Soil insect problems in African groundnut crops, in *Advances in Management and Conservation of Soil Fauna*, (eds G.K. Veeresh, D. Rajagopal and C.A. Viraktamath), Proceedings of the 1988 International Colloquium of Soil Zoology. Bangalore, India, Oxford University Press and IBH Publishing Co., pp. 171–176.

Wightman, J.A. (1988b) Some solutions to insect identification problems. *International Arachis Newsletter*, **3**, 18–20.

Wightman, J.A. (1989) Contribution of insects to low groundnut yields in Southern Africa, in *Proceedings of the third Regional Groundnut Workshop for southern Africa*, Lilongwe, Malawi, March 1988. ICRISAT, Patancheru, pp. 129–140.

Wightman, J.A. and Amin, P.W. (1988) Groundnut pests and their control in the semi-arid tropics. *Tropical Pest Management*, **34**, 218–226.

Wightman, J.A., Brier, H.B. and Wright, G.C. (1994) The effect of root damage and drought stress on the transpiration rate and the yield components of groundnut plants. *Plant and Soil* (in press).

Wightman, J.A., Dick, K.M., Ranga Rao, G.V. *et al.* (1989) *Pests of groundnut in the semi-arid Tropics*, in *Insect Pests of Food Legumes*, (ed. S.R. Singh), John Wiley, pp. 243–322.

Wightman, J.A. Ranga Rao, G.V. and Pimbert, M.P. (1989) Pests of groundnut: some difficult nuts to crack. *Proceedings of the International DLB Symposium on Integrated Pest Management in Tropical and Subtropical Cropping Systems*, Bad Durkheim, Germany, 18–15 February, 1989, pp. 463–486.

Wightman, J.A. and Rogers, V.M. (1978) Growth, energy and nitrogen budgets and efficiencies of the growing larvae of *Megachile pacifica* (Panzer) (Hymenoptera: Megachilidae). *Oecologia (Berlin)*, **39**, 245–257.

Wightman, J.A. and Whitford, D.N.J. (1982) Integrated control of pests of legume seed crops. 1. Insecticides for mirid and aphid control. *Proceedings of the 3rd Australasian Conference on Grassland Invertebrate Ecology*, Adelaide 30 Nov–4 Dec 1981, (ed. K.E. Lee), South Australian Government Printer, Adelaide, pp 377–38.

Wightman, J.A. and Wightman, A.S. (1988) An evaluation of five insecticides for the control of foliage and soil insects in a groundnut crop in Malawi and some effects of soil insects on yield parameters. *Legumes Program report*, 25 pp, ICRISAT, Patancheru (limited circulation).

Wightman, J.A. and Wightman, A.S. (1987) The effect of within stand plant mortality on the yield of groundnut plants in Malawi.

Wood, T.G. and Cowie, R.H. (1988) Assessment of on-farm losses in cereals in Africa due to soil insects. *Insect Science and its Application*, **9(6)**, 706–717.

Xu, B.C. (1982) Field distribution pattern and sampling technique for the larvae of *Anomala corpulenta* Motsch. *Jiangsu Agricultural Science*, **11**, 26–29.

Xu Zeyong (1991) White grub, an important insect pest of peanut and their control in China, in *Summary Proceedings of the Workshop on Integrated Pest Management and Insecticide Resistance Management*, Chiang Mai, Thailand, March 1991. Legumes Program, ICRISAT, Patancheru, p. 15.

Xu Zeyong, Zhang Zongyi, Chen Kunrong, Middleton, K.J. and Reddy, D.V.R. (1991) Aphids – the vector responsible for epidemics of peanut virus diseases in China, in *Summary Proceedings of the Workshop on Integrated Pest Management and Insecticide Resistance Management*, Chiang Mai, Thailand, March 1991. Legumes Program, ICRISAT, Patancheru, p. 15.

Yadav, C.P.S. (1981) *Integrated control of white grub*, Department of Entomology, University of Udaipur (at Jobner), Rajasthan, India, ix + 219 pp.

Yadav, C.P.S. (1991) *White grub management in groundnut*, Indian Council of Agricultural Research, New Delhi, 14 pp.

CHAPTER 12

Industrial utilization and processing

J.J.K.B. Asiedu

12.1 UTILIZATION

The groundnut or peanut (*Arachis hypogaea* L.) is grown for its kernels, the oil and meal derived from them, and the vegetative residue (haulms). The kernels may be eaten raw, roasted or boiled, sometimes salted or made into a paste popularly known as peanut (or groundnut) butter. The tender leaves of the plant are used in certain parts of West Africa as a vegetable in soups. In Senegal, one of the leading countries in the production of groundnuts, roasting and selling of groundnuts are done by women (Figure 12.1).

Groundnut oil is perhaps the most important product of the crop. At present about 40% of the world crop is processed into oil, which has a multitude of domestic and industrial applications. It may be used for cooking, for margarines and vegetable ghee, for shortening in pastries and bread, for pharmaceutical and cosmetic products, as a lubricant and emulsion for insecticides, and as a fuel for diesel engines (Duke, 1981). The press cake containing 40–50% protein is used mainly as a poultry feed. Groundnut flour, produced from the cake, can be used for enhancing or enriching the nutritive value of tuber flours that are low in protein, such as cassava flour. The so-called 'Mysore Flour', which is a combination of 25 parts of groundnut flour and 75 parts of cassava flour and contains a protein content of 12%, is utilized as a partial substitute for cereals in India. Also in India, 'Paushtik atta', a blend of wheat flour, groundnut flour and cassava flour, is used for the preparation of chapattis – an unleavened bread.

The dry pericarp of the mature pods (known variously as shells or husks) may be used as fuel, as a soil conditioner, as a filler in feeds and as a source of furfural, or processed as a substitute for cork or hardboard, or composted with the aid of lignin-decomposing bacteria (Adams and Hartzog, 1980). The haulms of the groundnut crop are reported to be nutritionally

The Groundnut Crop: A scientific basis for improvement. Edited by J. Smartt. Published in 1994 by Chapman & Hall, London. ISBN 0 412 408201.

Figure 12.1   Roasting and selling groundnuts in Dakar, Senegal.

comparable with grass hay and are used in many countries (Gibbons, 1980).

Groundnut or peanut butter is a comminuted food product prepared from dry-roasted, clean, sound, mature groundnuts from which the seed coat and germs (hearts) are removed, and to which salt, hydrogenated fat and sometimes sugars, antioxidants and flavours are added. Peanut butter is by far the most important product made from groundnuts in the United States of America, consumption level reaching about 1.4 kg per person (Zamula, 1985). While more than 90% of the groundnut butter is stabilized with hydrogenated oil and antioxidants and consumed in homes or school lunch rooms, the remainder is unstabilized and is used commercially in confections, bakery goods, ice cream, breakfast cereals and in other minor ways. Groundnut butter also enjoys considerable popularity outside the United States. The countries in this respect include Canada, where consumption level is 0.2 kg per person, The Netherlands, with a consumption level of 0.16 kg per person and, to a lesser extent, the UK, France and Germany (Wilson, 1975).

A substantial amount of groundnuts finds its way into candies and confections. The characteristic flavour, crunchy texture and high protein content render roasted groundnuts especially suitable for use in candies and confectionery. Peanut candies and confections are considered to be the major growth area in groundnut consumption: 44.5% of chocolate bars and 3.7% of nut bars contain groundnuts (DEBS, 1984). Peanut candies and confections encompass a wide variety of product types which include (but are not limited to) nut roll candies, dragees, peanut brittles, peanut roll bars and spun candy bars.

Groundnuts are consumed in large quantities in the salted form. Salted groundnuts are shelled, roasted and salted to taste; they may or may not be blanched.

Among the products that have not yet obtained wide popularity are groundnut protein, groundnut milk, groundnut cheese, frozen boiled groundnuts, groundnut bread and partially defatted groundnuts.

12.2 PROCESSING

Processing is a unit operation or series of unit operations performed on a raw material to change its form or composition. Processing can be as simple as cleaning, grading and packing fresh produce or as sophisticated as using, for example, pure strains of enzymes to convert maize starch to high-fructose sweeteners. Traditional processing evolved from society's need to enhance the usefulness of agricultural products. The basic methods employed for this purpose are preservation and separation. Raw materials can be converted to a form less susceptible to deterioration, usually through drying, curing, fermentation or refrigeration. Some forms of preservation, such as fermentation and cooking, impart other desirable characteristics to the finished product that may become ends in themselves, but they remain essentially preservation processes. Irradiation and freeze-drying are the principal innovations in preservation during the past century.

Desirable components such as seed, oil or fibre can be isolated from the parent material occurring in nature. After the natural material has been disaggregated, usually through the application of mechanical force, separation is achieved by exploiting differences in the weight, specific gravity, size, shape, air resistance or solubility of the components. The use of catalysts and solvents has been the principal development in this type of processing in the past century.

Depending upon the desired product, groundnuts may be subjected to a number of unit operations which may include picking or stripping, cleaning, storing, shelling, blanching, roasting, salting, grinding, extraction or expression.

12.3 HARVESTING, STRIPPING AND DRYING

12.3.1 Harvesting

The timing of the groundnut picking or harvest is critical since it can significantly affect the yields and nut quality. Groundnuts are best harvested or picked when the soil is dry enough to fall freely from the stems and pods. However, if the soil is very dry and crusted many of the pods may break off and remain in the soil. In subsistence agriculture, harvesting consists of either digging or pulling up the plant manually from the soil, allowing the plant to wilt and then picking the pods from the vines. On large-scale farms, groundnuts are harvested by running a special groundnut digger or wing-type plough below the surface and under the plants.

Commercial peanut combines are also available. These consist essentially of a thresher, a pick-up feeder chute and an auxiliary engine. The pick-up feeder chute is attached to the front, while a long bagging elevator and a high bagging platform are fixed at the rear. The machine is powered by an auxiliary engine. The speed of the pick-up cylinder in such combines approximately equals the forward travel of the combine. The following types of combine harvesters are generally in use on commercial farms.

- Self propelled
- Tractor-drawn, p.t.o. driven
- Tractor-drawn, auxiliary engine driven
- Mounted, auxiliary engine driven.

12.3.2 Stripping

The process of separating groundnut pods from the plant or haulm is known as stripping, which may be accomplished manually or using mechanized means. Manual stripping, that is picking pod by pod, is time and labour consuming although this operation results in good physical appearance of the groundnuts (fewer pods with vine attached, and clean). The stripping capacity of such a practice varies from 2 to 6 kg/man-hour. In some areas farmers strip groundnuts by hitting or beating a handful of the harvested plants against a stick placed on top of a container. This method results in much higher stripping capacities – from 8 to about 18/kg man-hour, depending upon the moisture content of the plant. The major drawbacks associated with this method are high scattering loss and a substantial amount of impurities.

Groundnuts may be stripped in a Drum Stripper, in which a handful of groundnut plants are beaten over the rubber-covered rods of a revolving drum. The output of this stripper is about 16 kg/man-hour.

Over the last decade a number of other designs of stripper have been developed and tested and found appropriate for small-scale and medium-sized farms. These include:

- Engine-operated spike-tooth stripper consisting mainly of a rotating cylinder provided with appropriately spaced spike teeth and a concave.
- Hold-on mechanical stripper in which groundnut plants are held against stripping bars of a rotating cylinder until pods are detached.
- Manually operated paddle stripper/winnower. This unit consists of a stripping cylinder and a drive mechanism (Figure 12.2). Stripping is done by operating a foot on the paddle so that the stripping cylinder, which has eight steel rods as stripping bars, rotates outwards from the operator.
- Modified axial flow rice thresher. This consists of an upper concave with fins, and a rotating drum. Material fed in at one end of the drum passes along the drum as it is rotated, and is ejected at the other end. Stripping occurs as the groundnut plant passes between a revolving cylinder and a metal grate (concave), which covers part of the circumference of the drum.

12.3.3 Drying

At the time of digging, groundnuts generally contain 35–60% moisture. Until the moisture is reduced to below 10%, the nuts are prone to mould attack, especially at warm temperatures and high humidity. Groundnuts can be dried to about 8% in field stacks in 4–8 weeks and to 9% in windrows in 1–2 weeks. Mechanical driers using heated air may be used.

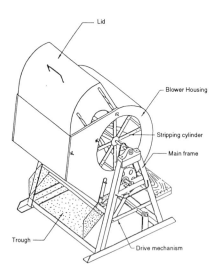

Figure 12.2   Paddle stripper/winnower (Khon Kaen University/IDRC, 1987).

The air should not be heated above 44 °C for groundnuts which are meant for table stock; temperatures up to about 52 °C can be used if the nuts are to be used for oil. On a continuous drier, with the groundnuts on the vines placed 0.6 m deep, the best drying and quality are obtained by using an air velocity of 0.30 m/s at 30–32 °C. The major problem in drying groundnuts is to prevent the splitting of the skin caused by rapid drying.

12.4 STORAGE

Groundnuts are known to be semi-perishable, which means that, on the one hand, they may be held for a number of years under optimum conditions. On the other hand, under unfavourable conditions they become inedible within a month due to mould, insects, absorption of foreign flavours or rancidity. For successful storage, the following requirements or conditions should be met as far as possible.

The storage temperature   should be low. In general, the lower the temperature, the longer the expected storage life of groundnuts, although no critical temperature has been established at which groundnuts must be held. Table 12.1 shows the relationship between temperature and the length of time the nuts retain their edible quality. At about 21 °C shelled groundnuts are susceptible to insect infestation, development of an 'amber' colour, staling and rancidity. At about 8 °C insect development is arrested and shelled nuts may be held for 6 months.

TABLE 12.1 *Relationship between temperature and storage time of groundnuts*

Temperature (°C)	Time of retaining edible quality	
	In-shell groundnut	Shelled groundnuts
21	6 months	4 months
8.3	9 months	6 months
0	3 years	2 years
−4	7–8 years	5 years
−12	15 years	10 years

The relative humidity   should be between 65% and 70%. High moisture in groundnuts is reported as being possibly the cause of more deterioration than any other single factor. Above 70% humidity, groundnuts are likely to develop mould. Below 65%, the nuts lose weight, become brittle and may split during handling.

The atmosphere in the store should be free of odours and flavours and well aerated. Loss of natural flavour of groundnuts increases with temperature and, likewise, absorption of foreign flavour increases with temperature.

12.5 SHELLING OR DECORTICATING

The terms 'shelling' or 'decorticating' are used in groundnut processing to describe the separation of the kernels or seeds from their shells or husks through the application of mechanical forces. In most cases compressive forces (sometimes, however, in combination with shear forces) are required to crush open the pod so that the kernels can be freed from the shells.

Shelling is usually carried out for two reasons:

1. Groundnuts in the shell are about 50% heavier than kernels alone and are therefore costlier to transport.
2. Groundnuts have to be shelled to facilitate further processing.

Shelling has the disadvantage that kernels removed from pods are more susceptible to attack by insects and moulds than in-shell kernels.

Smallholder farmers shell their groundnuts manually, using finger pressure. This is an arduous, boring, labour-intensive and low output operation which, in several cases, leads to the so-called 'thumb-sore syndrome'. The unique advantage of this method is its excellent outcome as regards cleanliness and minimum breakage and wastage of the shelled kernels. Waste is almost eliminated, and intactness and cleanliness of the kernels have great effect on the choice of groundnuts for seeds.

A number of groundnut shellers are available for various levels of production and powered by various means (manual, diesel engines, electric motors). Splitting or breakage of kernels can be as high as 22% during the operation of the more efficient shellers or decorticators. To reduce splitting and breakage to a minimum, groundnuts are shelled at 8% moisture (wet basis) or higher.

12.5.1 Hand-operated shellers

The basic design for hand-operated equipment is relatively simple and common to all makes. Essentially, the devices consist of a semi-cylindrical screen, a handle, a shelling grid which forms part of the outer casing, and spiked cast-iron rustlers attached to the handle. The action of the rustlers is effected by moving the handle back and forth. This causes the nuts to be rubbed between the rustlers and the shelling grid. The shelled kernels and crushed shells fall through the perforations of the shelling grid into a receptacle placed below it. This mixture of kernels and shells may be separated using air suction or a winnower.

The 'sliding' groundnut sheller is another hand-operated device consisting of a base, hopper box, adjustable throat slide, removable grid and decorticating bars. The box forming the hopper and accommodating the decorticating bars is pushed back and forth, using handles at both sides, above a removable grid. Shelling takes place in the clearance between the decorticating bars and the grid. The broken shells and the freed kernels fall through the gaps of the grid into a container placed below (Asiedu, 1987).

One of the most recently developed manually operated machines is the rubber tyre groundnut sheller (Khon Kaen University Report, 1987). This consists of a rubber tyre assembly, a concave, a hopper and a trough, with a crank or handle for the rotation of the tyre. The shelling takes place in the space between the tyre and the concave. Shelling capacity is reported to be 60–80 kg (pod) per hour, with a shelling efficiency of 95%.

12.5.2 Power decorticators/shellers

The basic elements of most of the power groundnut shellers are framework, drive shafts and pulley, hopper, electric motor or diesel engine and a shelling chamber, where the groundnuts are shelled by rotating flexible beaters. A fan, usually fixed below the wire-mesh chute, blows the shells upwards, and the kernels are recovered through a spout at the bottom of the machine. Depending upon the model, the hourly output of such decorticators varies from 150 kg to 400 kg.

A motorized version of the manual rubber tyre sheller enjoys considerable popularity among large-scale farmers in Thailand (Figure 12.3). The sheller is powered by a 1.49 kW electric motor and fitted with a blower and a three-layer sieve. It has a shelling capacity of 300 kg (pod) per hour and a shelling efficiency of 95%. However, its use is not recommended in shelling groundnuts for seed (Khon Kaen University Report, 1987).

12.6 THE PROCESSING OF GROUNDNUTS INTO PEANUT BUTTER

Peanut butter is a cohesive, comminuted and palatable food product prepared from dry-roasted, clean, sound (in terms of maturity and attack from insects and moulds) groundnuts from which the seed coats or skins and germs are removed and to which salt, hydrogenated fat and sugars, antioxidants and flavours are added. Although the manufacture of peanut butter appears to be relatively simple, a complex interaction occurs between the components. Essentially, the manufacture of peanut butter

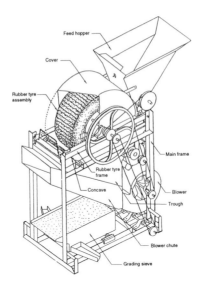

Figure 12.3 Motorized rubber tyre sheller (Khon Kaen University/IDRC, 1987).

consists of shelling, sorting, dry-roasting and blanching the nuts, followed by two stages of fine grinding and packing in hermetically sealed glass or tin containers. Ingredients such as salt, hydrogenated fat and dextrose, glycerin and lecithin or antioxidants are usually incorporated to improve flavour, prevent oil separation and control rancidity. A simplified flow chart for the processing of peanut butter is shown in Figure 12.4.

12.6.1 Processing and equipment

Shelled groundnuts are sorted to remove discoloured kernels and other undesirable contaminants before they are roasted. Roasting is done either on a conveyor belt passing through heated air, or in a rotary roaster. During roasting the kernels are heated to about 160–170 °C and are held at this temperature for 40–60 minutes to achieve the desired level of roast. After roasting the kernels are transferred to a perforated cylinder or cooler vat, where air is blown through the mass by fans.

Cooling is followed by dry blanching in a rotary cylinder or blancher. Blanching consists of removing the skins (seed coats) and the hearts of the kernels, which would otherwise give a bitter and rancid-like butter, by exposure to 126–145 °C heat for 5–20 minutes, followed by rubbing the kernels between soft surfaces and removing the skins by blowers and the hearts by screens. The blanched kernels are inspected and screened, and scorched or rotten ones are discarded.

The nuts are then ground, in one or two stages. In the first, they are

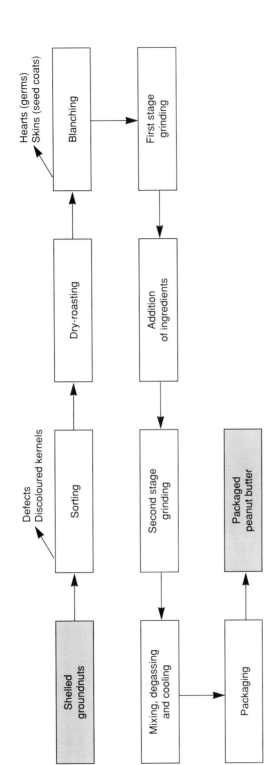

Figure 12.4 Manufacture of peanut butter.

ground into butter by attrition mills, homogenizers, hammer mills or colloidal mills. Ingredients like salt, hydrogenated oil, dextrose, corn syrup or honey and stabilizers are then added to improve smoothness, spreadability and flavour. A second, fine-grinding stage may follow and this fine-ground butter is mixed, degassed, and cooled in a rotating refrigerated cylinder before being packed, labelled and stored. Peanut butter contains about 1–2% moisture, 49–52% fat, 27–29% protein, 5–17% carbohydrates, 4% phosphorus and 2% fibre.

12.6.2 The role of ingredients and additives

A food additive is defined as:

> a substance or mixture of substances, other than a basic foodstuff, which is present in food as a result of any aspect of production, processing, storage and packaging. The term does not include chance contaminants. An additive may be either nutritive or non-nutritive; it may be physiologically active or inert; it may be present intentionally, to achieve some modification in the foods, or incidentally and serving no useful purpose in the final product
>
> (*National Research Council of the Food Protection Committee of the National Academy of Sciences, 1959*)

In general, there are seven major reasons for the use of additives: taste, appearance, texture and eating quality, processing, stability or preservation, nutritional benefit, and for special and dietary foods.

In particular, processing additives make possible the use of many ingredients by the food processor and the consumer not otherwise available at the desired quality level. For example, without the use of process additives such as monoglycerides, instant mashed potatoes would have the qualities of a sticky paste.

Before World War II, peanut butter is reported to have been prepared from groundnuts, salt and sugar 'only'. This resulted in oil separation, remixing difficulties, stickiness, graininess, poor spreadability, rancidity and poor shelf life. The addition of stabilizers prevented these defects and provided transportation and storage stabilities, and manufacturers have continued to rely on various stabilizers since 1945.

Glycerin and monoglycerides prepared from vegetable and groundnut oil are among stabilizers which prevent oil separation. Stabilizer selection generally depends upon the process and desired flavour release, grind, packing season, and fill and storage temperatures. Fine grinds produce more 'free' oil and require more stabilizer (0.1–0.15% or more). Single-stage grinds require a different stabilizer than two-stage grinds since the product leaves the mill at a higher temperature. Hard butters, which set rapidly after filling and cooling, can result when excessive stabilizer is used. At low fill temperatures of 35–44 °C, the stabilizer rapeseed oil hardfat

(1.6–1.8%) is preferred for summer packs and results in less 'pull away'. For higher fill temperatures (49–54 °C), followed by a chill tunnel, cottonseed hardfats (1.8–2.0%) are used. At even higher fill temperatures, monoglyceride stabilizers can be used along with glycerol monostearate (Weiss, 1983). Use of stabilizers can be avoided by refrigerator cooling (at less than 10 °C) or by adding absorbants like ground groundnuts, powdered sucrose, dried milk or partially defatted wheat germ (Woodroof *et al.*, 1945). Cooling and storing peanut butter at less than 10 °C, however, lead to increased storage and energy costs.

The choice of a sweetener depends upon the process. Maillard reaction between amino acids and reducing sugars can be accelerated by high processing temperatures to give dark peanut butters with altered flavour and texture. Sweetness and a smooth, non-sticky mouthfeel can be obtained with corn syrup solids (28–48% reducing sugars), while molasses imparts a unique flavour and consistency, although the high moisture content of honey (21%) favours lipid oxidation and rancidity. Also, honey or corn syrup, due to their high moisture, can lead to butters with high viscosity or pastiness (Weiss, 1983). The problems of lipid oxidation and the development of rancidity can be avoided using lecithin fatty acids or fatty acid polyglycerol esters at 0.15–0.85% of the weight of the honey. Lipid oxidation and rancidity can also be delayed by storing peanut butter in nitrogen atmospheres. Stickiness in peanut butter is reduced by adding partially hydrogenated vegetable oils.

12.7 THE PROCESSING OF ROASTED AND SALTED GROUNDNUTS

12.7.1 Roasted groundnuts

The unit operations involved in the making of roasted groundnuts (Figure 12.5) include roasting by heated air or oil cooking, blanching, sorting, salting/glazing and packaging. Shelled groundnuts are passed through cleaning screens to remove contaminants and other undesirables, separated by gravity tables or cyclones. After they have been white roasted (see below), the nuts are put into colour sorters or 'electric eyes' where discoloured and damaged nuts are rejected. The acceptable nuts are fed into blanchers, where the red skins are removed but (usually) not the germs. Dust, mould and other foreign material are also removed during blanching. The blanched nuts are again colour sorted, to discard unblanched and damaged ones, and then roasted.

The first effect of roasting by heated air is rapid drying during which the moisture content is reduced to about 0.5%. This is followed by the development of oily translucent spots on the surface of the cotyledons.

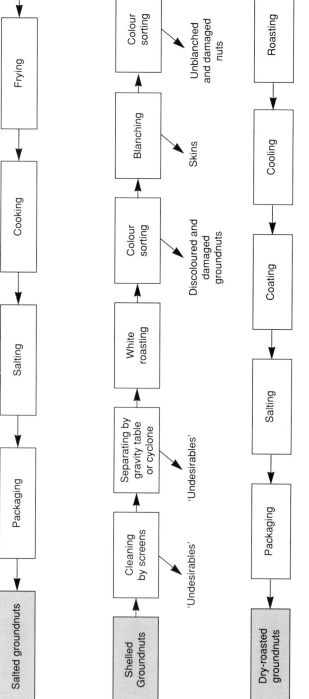

Figure 12.5 Processing of salted groundnuts and dry roasted groundnuts.

These 'steam blisters' are caused by the oozing of oil from the cytoplasm as free oil. Changes in nut colour, due to cell walls becoming wet with oil, are observed. This stage is usually referred to as **white roast**. The final stage of roasting is **brown roast**: the nuts develop a brown colour and roasting is completed.

Roasting is followed by cooling. In many industrial plants today the hot nuts emerging from the roasters are conveyed directly to a perforated cylinder or cooler box where fans blow air through the mass of the hot nuts.

After cooling, the nuts are coated. Coatings used include heated sorbitol–mannitol (7:3) blends or 5–6% acetic acid activated wheat glutins (Gutcho, 1973).

Coating is followed by salting. Electrostatically deposited flour salt of 70–200 mesh is reported to be preferable for air-roasted nuts (Matz, 1976). The type of salt applied in the processing of both roasted groundnuts and salted groundnuts is of considerable importance as regards rancidity and flavour. Groundnuts are rich in oils and since copper and iron (which are important impurities of salt) favour and accelerate rancidity of oils and fats, it is essential that a salt with minimal copper and iron impurities is used. Non-refined sea salt high in $CaCl_2$ is also reported to contribute to a harsh bitter flavour (Woodroof, 1983).

Finally, the roasted groundnuts are packed in airtight containers to retain 'freshness'. Methods include vacuum packing in cans and glass jars and the use of nitrogen-flushed laminated foil packs.

12.7.2 Salted groundnuts

Operations required for the processing of salted groundnuts are essentially the same as those for roasted groundnuts (Figure 12.5). The most conspicuous difference between the two products is the type of roasting they undergo. Salted groundnuts usually undergo oil-roasting or frying. This involves immersing the blanched and undamaged nuts for a period of 3–5 minutes in heated oil (coconut, groundnut or cottonseed oil) using a basket or submerged belt, or frying in solid fats (for instance, coconut oil). When nuts are fried in solid shortening, salt is applied before the nuts have cooled to room temperature (24–25 °C). Weiss (1933) reports that silicones (0.5–3 ppm) are often added to the heat transfer medium (coconut, groundnut or cottonseed oil) to raise the smoke point and reduce foaming. Cooking oil should be changed frequently to prevent rancidity. Rancidity can be detected within 30 days at 50% relative humidity (RH) at 24–28 °C for groundnuts cooked in used oil, compared with 60 days for those heated in fresh oil.

Salted groundnuts must be kept at less than 4% moisture, with 2–2.5% moisture giving the best texture, flavour and appearance retention. They

will remain fresh for up to 35 days at 50% RH but storage at 80% RH leads
to a loss in freshness within 15 days (Woodroof *et al.*, 1945).

12.8 THE PROCESSING OF PEANUT CANDIES AND CONFECTIONS

The use of roasted groundnuts in making different types of candies and
confections is due largely to their pronounced flavour, crunchy texture and
high protein content.

Unlike groundnut butter and roasted groundnuts, where the product has
a more or less uniform formula and processing technique, peanut confections encompass a wide spectrum of product types and processing techniques. However, the basic unit operations used in making peanut butter
and roasted peanuts still find their application in the processing of peanut
candies and confection. Figures 12.6 and 12.7 demonstrate flow charts for
making the different types commonly found on the market.

More than 50 types of candy are made from groundnuts – possibly the
most popular ingredient in American candies. For the preparation of
peanut brittles, a mixture of sugar, corn syrup and water (30:20:7) is boiled
and groundnuts are added until the mixture reaches 154–157 °C. Sodium
bicarbonate is added to aerate the brittle. The resulting batch is cooled to
about 3 °C prior to packaging. It is reported that the cooking of the nuts
within the batch gives improved flavour, through retention of volatile
flavours, but the high moisture content of the sugar solutions encourages
browning, development of rancidity and reduced shelf life (Janssen, 1978).

In dragees and hard-coated peanuts, the nut or a suitable peanut butter
formulation is coated with alternating supersaturated sugar/gum arabic
solutions (67%) and powdered granulated sugar. The final coating uses a
51% sucrose syrup, which is dried at a temperature of 37–46 °C before
packaging (Minifie, 1964). For the preparation of nut roll candies, groundnuts as the 'raw material' form a part of the nougat formula or are part of
the topping. Nougats are prepared by mixing sugar-corn syrup, egg white
and fat. This mixture is then whipped, extruded and finally coated with
chocolate.

Peanut candies and confections are relatively high in fats which tend to
shorten the shelf life. Stabilities differ, depending upon formula.
According to Labuza (1982), a typical shelf life is one year with distribution
times of 20–60 days. Uncoated peanuts – peanut brittles and peanut roll
bars – are more prone to rancidity. Brittles with 1.6% moisture were
reported to be stable for 1–4 weeks at 40–45% RH, while peanut rolls
having 5.89% moisture retained their original characteristics at 50–
55% RH. Stickiness is controlled by low humidity, while staleness is
checked by low temperature, i.e. below 0 °C. High temperatures (above
26 °C) and humidities over 50% lead to sticky, dark and runny candies.

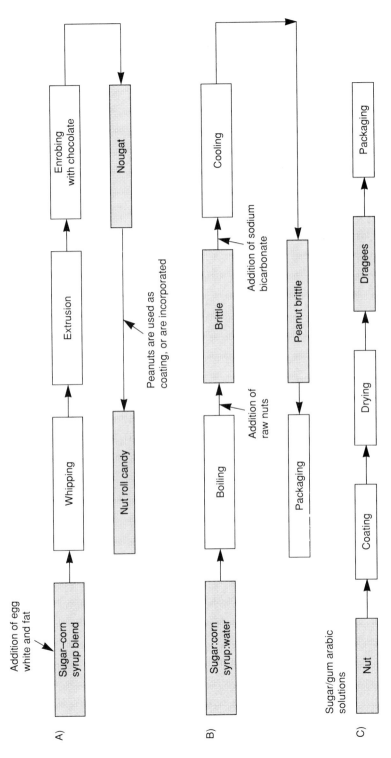

Figure 12.6 Making of different peanut candies and confections.

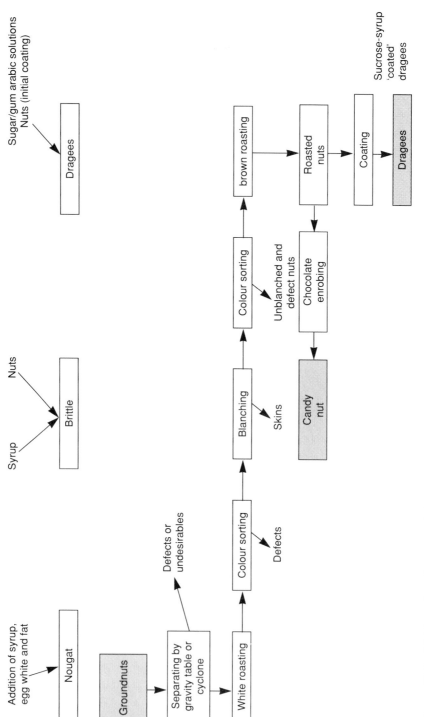

Figure 12.7 Making of candy nuts and dragees.

12.9 OIL EXPRESSION

Expression is the separation of liquids from solids by the application of mainly compressive forces. Three methods of expressing the liquid from the solid–liquid matrix are employed: hydraulic, roller pressing and screw pressing. Hydraulic and screw presses are widely used in processing fruit juice and oil seed; roller presses are universally used in expressing juices from sugar cane.

The efficiency of an expression process depends on several factors, including:

- the yield stress of the solid phase, i.e. its resistance to deformation;
- the porosity of the cake or mass formed during expression;
- the viscosity of the liquid expressed;
- the compressive force applied;
- the moisture content of the material to be expressed;
- material preparation or treatment prior to pressing.

In his study of factors affecting the yield of a number of vegetable oils, Koo (1942) developed the following general equation:

$$W = C * Wo * P^{1/2} * t^{1/6} * v^{-Z/2}$$

where W is the oil yield (in weight %), C is a constant for the type of oil seed (units consistent with unit analysis), Wo is the oil content of the seed (in weight %), P is the pressure (in MPa), t is the pressing time (in hours), v is the kinematic viscosity of the oil at press temperature (in m^2/s) and Z is the exponent of kinematic viscosity varying from $\frac{1}{6}$ to $\frac{1}{2}$. For groundnuts: C, Wo and Z values are respectively $19.4 * 10^3$; 51.9; $\frac{1}{3}$. For any one oilseed there is an optimum range of moisture content for maximum oil yield, from 5% to 13% (dry basis).

12.9.1 Processing groundnut for oil

The groundnut kernel is composed of approximately equal weights of fatty and non-fatty constituents, the relative amount of each depending upon variety and maturity. Most of the fatty constituents are found in the cotyledons, which make up the bulk of the kernel or seed. All cells of the groundnut embryo contain oil in an extremely fine emulsion. When kernels are bruised or broken, a sufficient number of cells are injured and, as a result, tiny drops of 'free' oil ooze out and collect on the surface of the kernels.

Oil may be obtained from groundnut kernels through mechanical pressing (expression) or by the use of a solvent. Mechanical pressing makes use of various types of equipment such as mortar and pestle, grinding stones, hydraulic presses, expellers or screw presses, crushing rolls and rotary ghanis. The rotary ghanis enjoy considerable popularity in the rural areas of India.

Solvent extraction using hexane or cyclohexane is capable of recovering nearly all the available oil from oilseed meal or flakes. In addition to the high degree of oil recovery, the solvent method produces oil with better qualities, and a higher protein meal. This method generally requires more capital investment than a mechanical process. Mechanical pressing or expression, though not as efficient as solvent extraction in terms of oil recovery, has the advantage of producing end-products (oil and meal or cake) free from dissolved chemicals and is a comparatively safer and less costly process, demanding less skill.

Among the parameters which influence oil yield and oil quality are pretreatment of groundnuts prior to pressing or extraction, operating temperature, pressing time and moisture content during expression. Post-extraction treatment of oil consists of refining and packaging.

(a) Hydraulic pressing

The use of hydraulic presses for oil extraction predominated until about 1945, since when the use of expellers has increased. For hydraulic pressing, shelled and crushed groundnuts with an optimum moisture content of 6% are heated by steam and then spread on press cloths, with the edges folded to prevent running. The containers holding the heated material are arranged one above the other (Figure 12.8). The racks are placed in tiers about 1 m high. When pressure is applied, oil is released and is collected beneath the tiers. As a press aid, a small amount of unshelled groundnut meal is mixed with the crushed mass prior to pressing. The press cake usually contains about 7% oil.

For **expeller pressing**, groundnuts are appropriately reduced in size and fed into an expeller, which is basically a screw rotating within a cylinder. The operation begins when the crushed groundnut mass is fed through a hopper into the larger end of the expeller chamber and pressure is exerted as the screw turns, forcing the mass towards the smaller or discharge end. Friction and pressure cause the mass to heat, which facilitates oil extraction. The groundnut oil passes through the perforated screen walls and is collected beneath the expeller chamber, while the press cake (which normally contains 5% oil) is extruded from the discharge end or cake outlet. The optimum moisture content of the groundnut mass for oil recovery by expeller is about 6%. The main parts of an expeller are shown in Figure 12.9. Expellers may be driven by an electric motor or via a pulley and V-belt from a separate diesel engine.

The types in use are single, duo, or duplex expellers. A **single expeller** is capable of pressing the assumed oil yield in a single pass. Such equipment may have a capacity of 45–55 kg per hour, a power requirement of 3 hp and

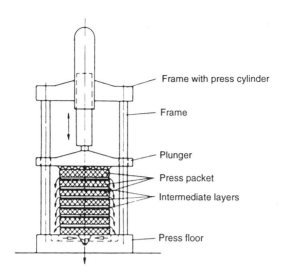

Figure 12.8 Multi-layer hydraulic press (Ulrich, 1967).

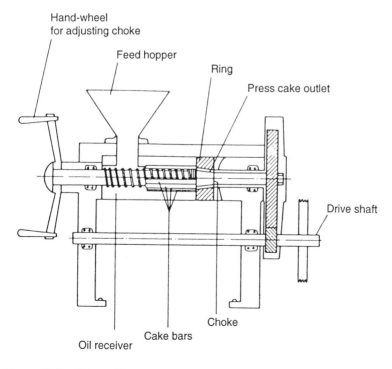

Figure 12.9 Oil expeller.

a rate of 300 revolutions per minute. In a **duo expeller** a 'first-pass' low pressure and a 'second-pass' high pressure expeller are combined to accomplish the extraction of the oil from the previously crushed groundnut. In a duo expeller the cake from the first pass is automatically conveyed to the second. A **duplex expeller** has screws of varying pitch and differently shaped chambers. This design enables pressure to be increased as the material is moved through the chamber. Duplex expellers normally achieve higher yields and faster throughput in a single operation.

In order to prevent the formation of groundnut 'butter' during extraction of oil in small expellers, groundnut shells may be added to the groundnut kernels.

(b) Solvent extraction

Oil extraction using expellers or hydraulic presses is always associated with (or stigmatized because of) losses which can be as high as 10% or more; a certain amount of oil always remains in the press cake, even with the most efficient expeller. With solvent extraction it is possible to reduce the residual oil in cake to less than 1%. The solvent extraction method for groundnut is similar to that for soya bean. Generally, groundnuts for extraction are shelled and winnowed to remove the fibre-rich shells, and whitened by removing the tannin-containing red skins. The nuts are cracked into pieces, conditioned to about 11% moisture at a temperature of 70 °C and then rolled to form flakes, which may be cooked before they are put into the extractor. Here the oil is extracted by means of a solvent. The solvent-laden flakes are passed through a desolventizer to recover the solvent. The extracted oil is clarified in a filter or frame press, after which it may be dehydrated.

Solvent extraction plants may be divided into two main groups: batch extraction plants and continuous extraction plants. The basic types of equipment are open percolation tanks, extraction batteries and continuous moving-bed extractors. Open mixing tanks, packed or plate columns and centrifuge contactors such as Podbielniak machines are used for continuous extraction. **Percolation tanks** are vessels with bottom outlets and valves: the solvent percolates through the solids by force of gravity. Percolation **extraction batteries** are tanks or vessels joined in series. They have the same design features (bottom outlet, valve, etc.) as the percolation tanks. The control valves are arranged so that a given vessel can be separated from the others in order to be emptied or refilled. In all **continuous extractors** the raw material moves through the vessels continuously against the flow of the solvent. Continuous extractors vary in design but the tower type is representative. This is equipped with travelling elements that contain the solids, sprays to deliver the solvent to the solids, and a feeder and discharger for the solids.

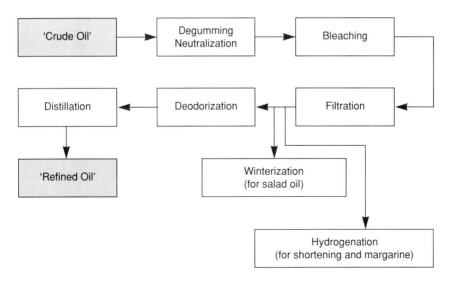

Figure 12.10   Sequence of operations in the refining process of groundnut oil.

12.9.2 Post-extraction/pressing treatment

Most vegetable oils can be consumed in their unrefined state, after simple mechanical pressing of the seed or fruit. But the colour and aroma of raw oils mask other food ingredients, and more sophisticated markets demand oils that do not dominate the sensory characteristics of food preparations. Simple refining consisting of filtration and water removal gives a product that has more acceptable cooking and shelf-life characteristics, but heat must be applied at a fairly specific temperature and a filtration process must be introduced. Today, however, most markets other than those in isolated rural areas demand oil that has been fully refined, through bleaching, deodorizing and the removal of free fatty acids to prolong shelf life. These processes cannot be accomplished without precise control of the temperature and material flow, and the use of caustic soda and bleaching and deodorizing compounds that must be thoroughly separated from the refined oil in additional stages of processing.

Groundnut oil obtained using a hydraulic press or expeller may be refined for various reasons as elucidated above. Refining (Figure 12.10) consists of alkali treatment, to neutralize the free fatty acids, using sodium hydroxide; bleaching, to improve flavour stability, using natural bleaching earth; and deodorizing, to remove odour, using steam distillation under vacuum. If groundnut oil is to be used as a salad oil, it undergoes the process of winterization prior to deodorization. **Winterization** is the removal of fats that crystallize out at about 0 °C.

For margarine-base oils and shortening, groundnut oil is subjected to

hydrogenation, the process by which elemental hydrogen combines catalytically with unsaturated organic compounds. It is conducted extensively in the edible fats and oils industry to modify the characteristics of natural fats and oils. Hydrogenation raises the melting point and increases the hardness, so that it is useful in changing vegetable oils from liquids to solids suitable for shortening and margarine.

Most hydrogenation is accomplished in batches, although equipment for continuous operation is also available. The typical **hydrogenator** is a carbon steel vertical pressure vessel. It is equipped with coils for heating with steam and for cooling with water, turbine type agitators, stationary baffles and a sparge pipe for admission of hydrogen. In operation, the vessel is filled to its operating level with the oil to be processed. Usually the feed has been refined and bleached and is free from water and dissolved gases. The oil is first heated to approximately 93 °C and a small amount removed to a catalyst mix tank, where catalyst is added. The oil and catalyst are then returned to the hydrogenator; where heating is continued to 121–135 °C, at which point hydrogen is admitted through the sparge pipe at a rate to maintain a predetermined pressure. The commonly used catalyst is finely divided metallic nickel.

The degree of **saturation of fats** is determined by their iodine value. The higher the iodine value, the greater the degree of unsaturation. The common edible vegetable oils have iodine values ranging from 100 to 150. Fats for shortening and margarine are hydrogenated to an iodine value of 70–90. These products are solid at ambient temperatures.

Groundnut oil is neutralized to dispose of free fatty acids which, when oxidized, would cause rancidity. **Neutralization** involves combining the free fatty acids with an alkali, usually caustic soda, to form a soap solution which can be separated from the oil. After the soap has been separated, the oil is washed and dried under vacuum. Then it is bleached with the aid of a specially treated form of fuller's earth, which is capable of absorbing colour pigments and residual oxidized material that would later affect the stability of the oil. After bleaching, the oil is passed through a filter press for removal of the fuller's earth and any other impurities, and is subjected to **deodorization**, a process of steam distillation under vacuum. High-pressure steam is blown through the oil so that the odoriferous materials, which are steam volatile, concentrate in the distillate.

12.9.3 Filtration

Filtration is a unit operation designed to separate, by means of a porous medium or screen, solids from liquids or liquids from liquids. Filtration is involved in many branches of the food industry, such as the making of fruit juices, vegetable oils, alcohol, sugar and beverages.

Filtration is accomplished by a filter medium through which a slurry is separated into solids and liquid by means of gravity, pressure, or vacuum. In the food industry the liquid is normally the valued or desired constituent and the solids are the undesirables.

For the filtration of groundnut oil, a filter press is generally used. It is made in two main forms: the plate and frame press, and the recessed plate or chamber press (Figure 12.11).

The **plate and frame press** consists of plates and frames arranged alternately and supported on a pair of rails. The plate has a ribbed surface and the edge stands slightly higher. The hollow frame is separated from the plate by the filter cloth, and the press is closed either by means of a hand screw or hydraulically; minimum pressure should be used in order to reduce wear on the cloths. A chamber is in this way formed between each pair of successive plates, (Figure 12.12). The oil to be filtered or clarified is introduced through a port in each frame and the filtrate passes through the cloth on each side so that two cakes are formed simultaneously in each chamber, and these join when the frame is full. The filtered oil drains into passages at the base of the plate and from there it is collected. In many filter presses, provision is made for steam heating so that the viscosity of the filtrate is reduced and a higher rate of filtration obtained.

The **chamber press** is similar to the plate and frame type, except that the use of frames is obviated by recessing the ribbed surface of the plates so that the individual filter chambers are formed between successive plates. In this type of press, therefore, the thickness of the cake cannot be varied. The feed channel usually differs from that used on the plate and frame press. All the chambers are connected by means of a hole in the centre of each plate and the cloths are secured in position by means of screwed unions. Slurries containing relatively large solid particles can readily be handled in this type of press without fear of blocking the feed channels.

For filtering vegetable oils, the filter press has several advantages:

- Owing to its simplicity, it is versatile and may be used for a wide range of materials under varying operating conditions of cake thickness and pressure.
- Maintenance is low.
- It provides a large filtering area on a small floor space.
- Most joints are external and leakage is easily detected.
- High pressures are easily obtained.
- It is equally suitable whether the cake or the liquid is the main or the desired product.

The disadvantages of the filter press include the following:

- It is intermittent in operation and continual dismantling is apt to cause high wear on the filter cloths.
- Labour requirements are high and the throughput is moderate.

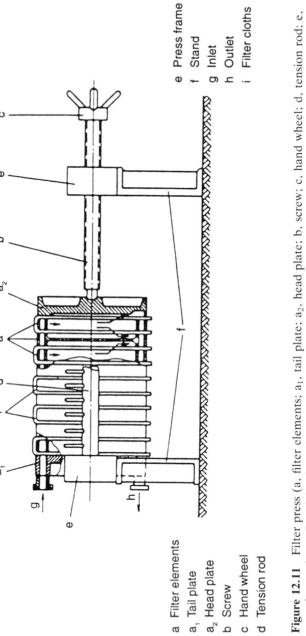

Figure 12.11 Filter press (a, filter elements; a₁, tail plate; a₂, head plate; b, screw; c, hand wheel; d, tension rod; e, press frame; f, stand; g, inlet; h, outlet; i, filter cloths). (Ulrich, 1967.)

a   Filter elements
a₁ Tail plate
a₂ Head plate
b   Screw
c   Hand wheel
d   Tension rod

e   Press frame
f   Stand
g   Inlet
h   Outlet
i   Filter cloths

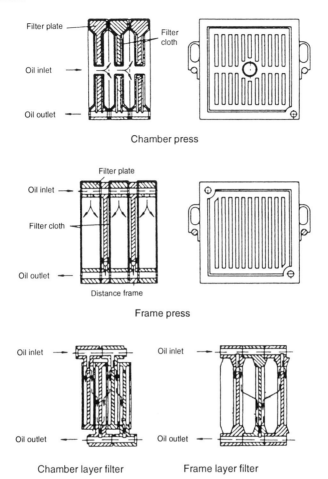

Filter plate

Filter cloth

Oil inlet

Oil outlet

Chamber press

Filter plate

Oil inlet

Filter cloth

Oil outlet

Distance frame

Frame press

Oil inlet   Oil inlet

Oil outlet   Oil outlet

Chamber layer filter   Frame layer filter

Figure 12.12   Filter elements of a chamber press and a frame press (Ulrich, 1967).

12.10 CONCLUSIONS

Groundnuts, for many years to come, will continue to occupy the status of both a nutritious food and a profitable cash crop. At present, groundnut crops are grown mainly for their kernels and the oil and the meal derived from them. The full potential of the crop has yet to be utilized. There exists great scope for processing the shells for economically useful purposes such as in the manufacture of activated charcoal, biogas, alcohol, extender resins, cork substitute and hardboards. The manufacture of adhesive glues, fire-extinguishing liquid and water-resistant powder from groundnut press cake has not yet been commercially exploited.

Against the background of the campaign for 'natural' and 'healthy' foods, partially defatted groundnut foods are becoming popular and will assume greater importance in the years ahead. Groundnut flours may contain 55–60% protein, minerals and vitamins and some carbohydrates. However, they are much lower in their content of the gas-forming oligosaccharides than are, for instance, soybean and cottonseed defatted flours. Despite their relatively low nutritional value because of their low levels of lysine, threonine and methionine, groundnut proteins have unique functional properties such as low solution viscosity at relatively high concentrations (5–10%), good compatibility with bread dough systems, white colour and bland flavour. In view of the above, opportunities exist for the food industry (using extrusion texturization) to manufacture defatted groundnut flours, groundnut protein isolates and concentrates as well as a wide range of food products which might include vitamin-fortified infant food, precooked dehydrated foods, groundnut bread, groundnut cheese and groundnut milk through the extraction process. In particular, texturized groundnut protein can provide an excellent yet economical substitute for expensive animal protein needed to meet the short-term and long-term requirements of nutritional improvement. The short-term nutritional improvement target for Africa, for example, requires an increase in the consumption of protein of high biological value from 11 g to 18 g per capita per day, while the long-term target calls for an increase to 25 g per capita per day.

The future pattern of industrial use of the groundnut crop will generally differ from the more industrialized to the less industrialized groundnut–producing countries. Most of the less industrialized countries have centralized large-scale processing facilities where groundnut oil and oil cake are produced from the kernels. These are invariably located in urban areas, far from the rural growing areas. In many instances the countries do not produce enough oil to satisfy their domestic markets. In such circumstances, it is usually the rural areas where the deficits occur, because of the difficulties and costs of transportation and distribution. The development and introduction of technologies for groundnut processing on a small or medium scale suitable for use in rural areas would help alleviate this problem.

This concept, which is beginning to take firm root in a number of countries, should be viewed as more than a technical solution to the need to enhance the nutritional status of people in rural areas. It should also be seen as a promoter of the overall development of rural areas – offering the prospect of creating rural employment, adding value to agricultural production, developing local engineering expertise and providing skills necessary for rural agro-industrial development. With income growth, urbanization and the need to satisfy the nutritional requirements of both human and animal populations, more effort is expected to be put in augmenting production as well as increasing the efficiency of post-production activities such as threshing, shelling, storage and expelling.

In the more industrialized countries, where relevant resources are available, the emphasis is likely to be placed more on the use of biotechnology for enhanced utilization of the groundnut crop.

The advent of biotechnology brings with it challenges and opportunities for the groundnut crop. Biotechnology will have an impact on the future pattern of industrial utilization of groundnut on at least three different stages:

- The development of more efficient techniques and the enzymatic modification of groundnut oil, such as fractioning, modifying and transferring fatty acids.
- The development of plant biotechnology.
- The microbial production of oils and fatty acids.

These three factors will lead to more flexibility in the supply and use of the groundnut and its oil. Also, as groundnut protein lacks essential amino acids like lysine and methionine, biotechnological research is likely to be carried out in order to improve the content of these amino acids. This again will expand the range of products which can be made from the crop.

In general biotechnological research on improvement of groundnut varieties will seek to improve yield and adaptation, enhance oil quantity, modify oil quality, and improve quantitatively and qualitatively changes in meal protein. However, the extent of utilization of future biotechnological research results will depend not only upon the biological characteristics of the crop but also on the socio-economic structure of production. The more established, large-scale groundnut growers in countries like the USA, Brazil, Argentina, India and China are expected to be more receptive to new high-technology inputs for production and processing. Small, less established growers, on the other hand, due to their very limited resources and inadequate facilities, are not likely to be able to exploit quickly and fully the benefits that biotechnology might bring.

With biotechnology research on other oilseed crops gaining momentum, it is uncertain what the demand for specific oil/protein crops and their products will be in future. But what is almost certain is that improvements in curing, storage, breeding and aflatoxin elimination will lead, in a measurable degree, to the use of more efficient known processing technologies, and also 'new' technologies such as enzymatic processing, extrusion texturization and biofermentation – and hence to additional and new groundnut products. Globally, it is suggested that consumption and industrial utilization of the groundnut crop will increase in future, partly due to the expected increase in consumption of ready-to-eat groundnut products in the former Eastern Bloc countries, including the former Soviet republics.

Groundnuts still have much to contribute to the food and industrial requirements of the world whether the need be for calories, proteins, fats, vitamins or cellulose. The science of biotechnology, which itself is experiencing a revolution, has the potential to precipitate a revolution in the production of the groundnut crop and in its industrial and domestic utilization.

REFERENCES

Adams, F. and Hartzog, D. (1980) *Peanut Science*, **7**, 120–123

Asiedu, J.J. (1987) Some selected engineering unit operations, in *Agricultural Processing Manual*, University of Malawi.

Coulson, J.M., Richardson, J.F., Backhurst, J.R. and Harker, J.H. (1978) *Chemical Engineering*. Vol. 2: *Unit operations*, 3rd ed, Pergamon Press, Oxford.

DEBS (1984) *Manufacturer Confectioner*, **63** (4), 53.

Duke, J.A. (1981) *Handbook of Legumes of World Economic Importance*, Plenum Press, New York.

Gibbons, R.W. (1980) The ICRISAT Groundnut Program, in *Proceedings of the Workshop on Groundnut*, ICRISAT, Patancheru, India, pp. 12–16.

Gutcho, M. (1973) *Prepared Snack Foods*, Noyes Data Co., Park Ridge, New Jersey.

Janssen, F. (1978) *Janssen's Tips*, p. 57.

Johnson, A.H. and Peterson, M.S. (1974) *Encyclopedia of Food Technology*, AVI Publishing Company, Inc., Westport, Connecticut, pp. 682–684.

Khon Kaen University (1987) *Final Report*: Groundnut Sheller/Stripper Project, submitted to the International Development Research Centre (IDRC), Canada.

Koo, E.C. (1942) Expression of vegetable oil: A general equation of oil expression. *Journal of Chemical Engineering*, China, **34**, 342–345.

Labuza, T.P. (1982) *Shelf-life Dating of Foods*, Food and Nutrition Press, Westport, Connecticut.

Matz, S.A. (1976) *Snack Food Technology*, AVI Publishing Co., Westport, Connecticut.

McGraw-Hill Encyclopedia of Science and Technology (1982) 5th edn, McGraw-Hill, Inc., USA.

Minifie, B.W. (1964) *Chocolate, Cocoa and Confectionery*, Leonard Hill, London.

Ulrich, H. (1967) *Mechanische Verfahrenstechnik*, Springer Verlag, Berlin.

Weiss, T.J. (1983) *Food Oils and their Uses*, AVI Publishing Co., Westport, Connecticut.

Wilson, R.J. (1975) The market for edible groundnuts, *Tropical Products Institute Bulletin G96*, TPI, London.

Woodroof, J.G. (1983) *Peanuts: Production, Processing, Products*, 3rd edn, AVI Publishing Co., Westport, Connecticut.

Woodroof, J.G., Cecil, S.R. and Thompson, H.H. (1945) The effects of moisture on peanuts and peanut products, *Bulletin 238*, Georgia Agricultural Experiment Station, Experiment, Georgia, USA.

Zamula, E. (1985) *Federal Drug and Administration Consumer*, February 1985, pp. 24–20.

CHAPTER 13

Mycotoxins in groundnuts, with special reference to aflatoxin

J.I. Keenan and G.P. Savage

13.1 INTRODUCTION

In 1961 reports started to appear of heavy losses in turkey poults and ducklings during the previous year. In the United Kingdom, 100 000 birds were reported to have died in approximately 500 outbreaks of what was named Turkey X disease. Infected birds showed characteristic histological lesions in the liver, which were initially thought to be related to seneciosis in fowls (Siller and Ostler, 1961). The common factor in the UK outbreaks was a diet containing 10% of certain consignments of Brazilian groundnut meal (Blount, 1961). During the same year a similar condition was reported in ducklings in Kenya and was traced to locally grown expeller-processed groundnut meal (Asplin and Carnaghan, 1961). Indian ground-nut meal, making up 6% of a diet, was linked to more deaths in the following year (Carnaghan and Sargeant, 1961). Also in 1961, a toxic factor was extracted with chloroform from the Brazilian meal (Allcroft *et al.*, 1961) and was shown to be extremely toxic for young ducklings, producing lesions histologically identical to those in Turkey X disease (Asplin and Carnaghan, 1961). It was also shown to be free from plant alkaloids (Sargeant *et al.*, 1961a). Toxicity testing of the Indian meal (Carnaghan and Sargeant, 1961) produced liver lesions identical to those produced by the Brazilian and East African toxic meals, though the toxicity of these samples was considerably less.

This prompted Sargeant *et al.* (1961b) to examine groundnuts imported from other areas. The toxin was found in samples from Nigeria, French West Africa, Uganda, Tanganyika, Ghana and The Gambia, the level of toxicity falling between the very toxic Brazilian and the moderately toxic Indian groundnuts. That the signs and symptoms were clinically and histologically indistinguishable from ragwort poisoning raised the question of incorrect diagnosis of poisoning incidents prior to 1961. This was answered

The Groundnut Crop: A scientific basis for improvement. Edited by J. Smartt. Published in 1994 by Chapman & Hall, London. ISBN 0 412 408201.

when it was shown that certain batches of Diet 18 (Paget, 1954), containing 15% groundnut meal, were associated with a non-infectious disease encountered in laboratory animals from 1951 onwards. Trials carried out by Paterson *et al.* (1962) with 15% toxic groundnut meal in an experimental diet (Rosetti Diet 18) produced results identical to those observed by Paget (1954).

Extensive studies were undertaken to identify the toxin and its source. A sample of highly toxic kernels, observed to be heavily contaminated with fungi, was cultured. Of the eight fungal isolates grown, only one produced an extractable toxin. This was subsequently shown to produce the typical liver lesions. The fungus was later identified as *Aspergillus flavus* Link ex Fries (Sargeant *et al.*, 1961c). It is now known that the groundnut seed (*Arachis hypogaea* L.), which develops in the soil, is particularly susceptible to invasion by fungi during its development. *A. flavus* and *A. parasiticus* Speare are two closely related species of ubiquitous soil fungi found in the tropics and subtropics. Certain environmental conditions predispose these two fungi to proliferate in the soil, with subsequent invasion of the groundnut (Pettit, 1986; Cole, 1989). Some strains have the ability to produce toxic metabolites, known as aflatoxins B1, B2, G1 and G2 (AFB1, AFB2, AFG1 and AFG2). They also have carcinogenic, mutagenic and teratogenic properties.

13.2 NATURAL OCCURRENCE OF AFLATOXINS

The four different aflatoxins are produced in widely varying amounts and proportions depending on the genetic capabilities of the fungus, the plant variety and the environmental conditions (Goldblatt, 1971; Jarvis, 1971; Schindler *et al.*, 1967; DiProssimo, 1976). The toxigenic strains generally produce only two or three aflatoxins under any given set of conditions. AFB1 is always produced; AFG1 is produced by the oxidation of AFB1. AFB2 and AFG2 are only ever produced in small amounts. They were originally thought to be produced from their respective precursors, AFB1 and AFG1. Yabe *et al.* (1988) showed AFB1 and AFG1 to be formed from different precursors to those required for AFB2 and AFG2 production.

Moisture, either as kernel moisture content (KMC) or relative humidity, has been shown to be the most important environmental parameter in the production of aflatoxin, followed closely by temperature. Different optimum temperatures have been shown for both the growth of the fungus and aflatoxin production, including changes in the ratio of AFB1 to AFG1 (Jarvis, 1971). Trace elements, particularly zinc, are also required for growth and sporulation of fungus, as well as for toxin production (Tupule, 1969). A toxigenic strain utilizes lipids in the biosynthesis of aflatoxin by methyltransferase and oxidoreductase enzymes (Yabe *et al.*, 1988). A zinc or copper deficiency thus causes an accumulation of lipid in toxigenic strains, rather than the synthesis of aflatoxin (Denning, 1987).

The competitive growth of fungi is reported to result in lower toxicities. Several studies (Joffe, 1969; Jarvis, 1971; Tsubouchi *et al.*, 1983) have shown non-aflatoxin-producing *A. flavus*, *A. niger*, *Rhizopus* spp. and strains of *Nocardia* to degrade aflatoxin. Nakazato *et al.* (1990) have now shown that non-aflatoxin-producing fungi also have the ability to reconvert aflatoxicol to AFB1.

13.3 PREHARVEST CONTAMINATION

Preharvest contamination is a major economic problem for the groundnut industry. This was supported by a study of various food crops by Schroeder and Boller (1973) which showed groundnuts to be the most susceptible and most heavily contaminated of those studied. There are two types of preharvest aflatoxin contamination in the crop. Groundnuts mechanically or biologically damaged in the soil are predisposed to invasion by fungi. The saprophytic fungus will live predominantly on dead or dying tissue, therefore it may infect the maturing kernel if the pod is damaged while still in the ground, especially when the growth rate of the plant is in decline during the later growing phase. Hot, dry soil conditions predispose to insect damage by termites which are known carriers of *A. flavus* spores. Alternate periods of rain and drought can lead to burst pods. High levels of aflatoxin are characteristically found in damaged kernels.

Another type of preharvest invasion can occur with no obvious kernel damage. An extensive series of studies has provided considerable insight into the cause of this (Blankenship *et al.*, 1984; Cole *et al.*, 1985; Dorner *et al.*, 1989; Hill *et al.*, 1983; Sanders *et al.*, 1985a).

The groundnut plant is unusual: it flowers above ground and, following fertilization, the peg is pushed into the soil and the fruit develops beneath the surface of the soil. Cole *et al.* (1986a) used colour mutants of both aflatoxin-producing strains, together with aflatoxin analysis of kernels, to prove that groundnuts are infected in the soil and not via flowers or aerial pegs. The study strongly indicated that the most significant route of invasion occurs after the groundnut peg penetrates the surface of the soil, some time during fruit development. The toxigenic fungi then remain inactive or dormant and under normal environmental conditions there is no significant aflatoxin contamination. This is because actively growing plants have a natural protective mechanism against disease. They are able to produce and accumulate phytoalexins, which are antimicrobial and antifungal compounds of low molecular weight (Strange, 1984). Three different phytoalexins have been found in the kernels and leaves of groundnuts. These have been named arachidins I, II and III. Water activity (A_W) of more than 0.97 in the kernel allows sufficient phytoalexin production to inhibit *A. flavus* growth. (A_W is defined as a measure of the available water in a food system, derived from the relative humidity of the atmosphere with which the food is in equilibrium.)

As the plant metabolism slows down during the maturation process, phytoalexin production falls. Below 0.95 A_W the plants are unable to produce phytoalexins, regardless of the temperature or maturity of the kernel, but fungal growth and aflatoxin production can occur down to 0.85 A_W. Cole (1989) describes this as a 'window of susceptibility' which leaves the groundnut kernels, especially the immature forms, without their primary resistance mechanism. The kernels seem more susceptible to aflatoxin contamination when the A_W is between 0.95 and 0.90. Below an A_W of 0.90, growth becomes restricted by reduced availability of water. Dorner *et al.* (1989) also showed that kernel moisture loss was accelerated at 29 °C, compared with 25 °C, and that this higher soil temperature promoted fungal growth and aflatoxin production once phytoalexin-producing capacity was lost. In support of this, Diener and Davis (1968) reported that old, overmature or moribund groundnut tissues have considerably less KMC and increased susceptibility to *A. flavus*.

Preharvest aflatoxin contamination occurs when there is extreme and prolonged drought stress during the last 4–6 weeks of the growing season. Bushnell's study in Rhodesia (1965) showed that in a drier year there was a high level of preharvest contamination with toxigenic strains of *A. flavus*, even with rapid drying of nuts. Groundnuts subjected to these conditions during the last 3–50 days were highly contaminated although visibly undamaged; a stress period of less than 20 days was insufficient to cause high levels of contamination (Cole *et al.*, 1986b). Drought stress causes the groundnut canopy to recede, allowing the soil temperature in the geocarposphere to increase. (The geocarposphere is defined as the fruiting zone of the groundnut plant, found 5 cm below the soil surface.)

Hot, dry soil conditions can encourage a rapid *A. flavus* build-up, possibly by eliminating microbial competitors. Moisture is lost from the kernels, the natural maturation process is accelerated and eventually the phytoalexin-producing capacity is lost (Dorner *et al.*, 1989). The soil temperature is also critical for aflatoxin production (Hill *et al.*, 1983; Blankenship *et al.*, 1984; Cole *et al.*, 1985; Cole *et al.*, 1986b). Under drought conditions, a minimum mean geocarposphere temperature of 26 °C is required for some contamination of sound kernels; lowering the temperature to 25.7 °C results in no aflatoxin production. A mean geocarposphere temperature of approximately 26.3–30.5 °C is necessary for preharvest contamination (optimum of 29.6–30.5 °C). A mean temperature of 31.3 °C is too high.

Adequate irrigation effectively prevents preharvest aflatoxin formation, even when soil temperatures are optimal for aflatoxin production and approximately 50% of the kernels are invaded by *A. flavus* (Hill *et al.*, 1983; Sanders *et al.*, 1985a). Also, *A. niger* grows vigorously in hot, moist conditions. This fungus is a known antagonist to *A. flavus*, with the ability to degrade aflatoxin.

Under stress conditions immature kernels become contaminated more easily and quickly than the larger, mature kernels that have passed beyond a certain developmental stage. It has been postulated that the capacity to produce phytoalexins at a lower A_W would be a good criterion for the selection of genotypes for aflatoxin resistance. However, under normal conditions the phytoalexin-producing capabilities of the mature kernels have been shown to be lower than those of immature kernels. This suggests that they possess some additional form of resistance, a theory which is supported by their resistance to preharvest contamination, even under conditions of severe and prolonged drought.

Groundnut cultivars vary in their resistance to invasion by *A. flavus* (Strange, 1984; Kannaiyan *et al.*, 1989; Siwela and Caley, 1989). Genotypes classed as resistant in the laboratory have become extensively colonized by *A. flavus* and contaminated by aflatoxins under stress conditions in the field (Sanders *et al.*, 1985b). Some genotypes have now been found that only support very low levels of aflatoxin production when invaded by *A. flavus* but this is dependent on the testa being entire and undamaged (Subrahmanyam *et al.*, 1980; Subrahmanyam, 1990; Manzo and Misari, 1987). All developing shells are susceptible to fungal invasion, but the degree of penetration through the pod cavity has been shown to vary between cultivars. Kernels possessing a thick waxy seed coat and small, covered hila are more resistant to invasion (Pettit, 1986; Pettit *et al.*, 1989). The presence of tannin-like compounds in the seed coat and inhibitory compounds in the cotyledon also have an inhibitory effect on *A. flavus* growth and aflatoxin production. It is thought that these features, together with the concentrations of phytoalexins and other compounds, are responsible for the variation in resistance found between cultivars (Azaiaeh *et al.*, 1990).

13.4 POST-HARVEST CONTAMINATION

While groundnut hulls may be invaded by *A. flavus* during growth in the soil, there is also potential for contamination to occur during the drying, transport and storage stages. Once again moisture is the single most important parameter, followed closely by temperature (Austwick and Ayerst, 1963; Llewellyn *et al.*, 1983; Llewellyn *et al.*, 1988).

At maturity the whole plant is uprooted and left in a windrow on the soil surface to dry naturally. (Pods left in the ground beyond maturity show a gradual increase in fungal invasion of the seeds.) Depending on conditions, it can take 2–3 days for the moisture content of the kernels to be reduced from 40–50% to 20–25%. The more quickly natural drying is achieved after lifting from the soil, the less aflatoxin contamination is observed. However, care must be taken that the nuts do not dry too rapidly. This can weaken the seed testa and testa damage decreases resistance to fungal penetration (Manzo and Misari, 1987).

Once the pods are field-dried, they are stripped from the plants. They should then be either sun-dried or artificially dried to <9% moisture prior to storage. The importance of this has been shown by the much higher incidence of kernel contamination in stored groundnuts harvested during the wet season and insufficiently dried (Tupule, 1969; Habish et al., 1971). Another factor that can lead to increased levels of contamination is that groundnuts are often sold at this stage – higher moisture levels mean more weight and increased profits for the grower.

Once dried to <9% KMC, groundnuts can be stored for long periods with little loss of quality if kept under controlled storage conditions, i.e. adequately cooled and well-ventilated (Baur, 1975; Aibara et al., 1985). These conditions will also prevent the spread of any preharvest contamination. Groundnuts stored in the shell are less likely to become contaminated during storage, due to the presence of the seed testa (Xiao, 1989). In a 2-year study of naturally contaminated groundnut materials, Bauer (1975) showed no statistically significant changes in aflatoxin levels with storage.

The use of modified atmospheres at ambient temperatures for bulk storage has been investigated with some success (Sanders et al., 1968). While it may offer a way to control aflatoxin accumulation by inhibiting the growth of toxigenic fungi, another study by Wilson et al. (1985) showed changes in the microflora of the stored nuts, depending upon the KMC and the modified atmosphere used. They also showed that something simple like diurnal temperature changes can cause moisture migration. This resulted in an increase from the 6–7% KMC in the original bulk sample to one of 11% in the top layer of groundnuts stored under 60% CO_2. This rise in moisture in the top layer resulted in a layer of visibly mouldy groundnuts.

13.5 PRESENCE IN FOODSTUFFS

In the USA, Canada, New Zealand and many European countries, regulations are applied to aflatoxin levels in imported groundnuts (Van Egmond, 1989). Most have declared a tolerance level of 5 µg/kg for aflatoxin B1 and 5–20 µg/kg for total aflatoxin in foods. A study in Czechoslovakia by Fukal et al. (1987) showed only 1.9% of the samples tested for aflatoxin gave a value above 5 µg/kg. No real difference between raw and roasted groundnuts was seen in this study. A survey in Norway by Yndestad and Underdal (1975) showed that 17.5% of groundnuts tested contained between 2 and 400 µg/kg aflatoxin but most were at the lower end of the range. A 6-year study in Canada carried out by Gelda and Luyt

(1977) showed that 90.1% of groundnuts and groundnut products had total levels of aflatoxin <5 μg/kg. In New Zealand, Stanton (1977) found aflatoxin B1 levels ranged from 5 to 6 μg/kg in two of 16 samples of imported groundnuts. However, a further 57 outwardly clean nuts from one sample developed *A. flavus* and *A. niger* when incubated in agar, which demonstrated the potential for toxin contamination during incorrect storage. The screening of groundnuts certified aflatoxin-free from the country of origin and imported in bulk into Trinidad showed aflatoxin levels to be <15 μg/kg (Chang-Yen and Felmine, 1987).

A study by Lim and Yeap (1966) showed one third of groundnut samples imported into Malaysia to be contaminated with aflatoxin G1, with no aflatoxin B1 being detected. This raises an interesting observation. If interpretation of the presence of aflatoxin is based solely on AFB1 levels, these samples would have been assessed as being aflatoxin free. In terms of oral toxicity this could be significant; the oral toxicity of AFB1 (0.35 mg/kg) is only 2.9 times higher than AFG1 (1.0 mg/kg). In contrast, an interesting study by Strzelecki *et al.* (1990) assessed the occurrence of aflatoxins in groundnut meal imported into Poland from a number of groundnut-producing countries (mainly India and Brazil) over an 8-year period. They found the mean contamination by AFB2, G1 and G2 to be 9.3% of the four main aflatoxins and their toxicity to be equal to 2% of AFB1. These experiments suggest that measuring the AFB1 levels in samples would be just as reliable, as well as being more rapid, easier and cheaper than measuring the full range of aflatoxins.

In those countries where groundnuts and groundnut products are locally grown and produced, and import regulations do not apply, the biological risk of exposure to aflatoxins is much higher owing to potentially much higher levels of contamination. Varsavsky and Sommer (1977) reported AFB1 contamination in 13.1% of samples grown and assayed over a two year period in Argentina, with highest levels reported in the 100–1000 μg/kg range. Keen and Martin (1971) compared samples of stored groundnuts in Swaziland and found 26% of those from rural areas to contain aflatoxins, whereas those stored in agricultural stations were aflatoxin free despite the presence of *A. flavus* in 7.6% of samples. They attributed the large difference in contamination to differences in storage containers and methods of storage. Different storage practices between Fiji and Tonga were reflected when 50% of Fijian samples were shown to be contaminated, compared with 9% of those from Tonga (Lovelace and Aalbersberg, 1989).

Groundnut products such as cake, meal and pellets, produced by expelling oil from groundnut kernels, can be consumed locally and used in the preparation of animal and poultry feeds. They can also contain high levels of aflatoxin (Abalaka and Elegbede, 1982; Akano and Atanda, 1990; Lim and Yeap, 1966; Natarajan *et al.*, 1975a; Strzelecki and Cader-Strzelecka, 1988). Once these toxins enter the food chain, there is the potential for secondary transmission. Eggs do not appear to contain detectable levels of aflatoxin, even when the pullets are fed contaminated meal (Allcroft and Carnaghan, 1963). Coker (1979) showed laying hens to excrete 70% of a dose of AFB1, which means that <30% is passed into the eggs. Platonow and Beauregard (1965) failed to show any significant changes in the tissues of ferrets fed meat from chickens raised on diet containing aflatoxin. Liver from an infected cow showed no toxicity (Allcroft and Carnaghan, 1962) but AFM1 has been found in the liver and kidney of sheep (Reed and Kasali, 1989). AFM1, a hydroxylated metabolite of AFB1, has also been found consistently in the milk of lactating animals exposed to dietary AFB1 (de Iongh et al., 1964; Fehr et al., 1968; Ferrando et al., 1984). This metabolite appears 12–14 hours after ingestion of AFB1 and remains constant throughout the feeding period, disappearing 3–4 days after the removal of the contaminated food (Allcroft and Roberts, 1968). AFM1 has almost identical toxicity to AFB1 but is 10 times less carcinogenic (Cullen et al., 1987; Groopman et al., 1988; Strzelecki et al., 1990). It is recognized as a potential health hazard and permitted levels are subject to legislation in many countries. In the USA, the Food and Drug Administration has set a practical action guideline of 0.5 μg/kg aflatoxin M1 for fluid milk. The major control is achieved through the regulation of AFB1 content of animal feed.

Another commodity being utilized increasingly in animal feed is ground-nut hulls. Sanders et al. (1984) showed that inoculated hulls can support the growth of the fungus but not the production of aflatoxin. They also noted that, when contamination was found in hulls from groundnuts, they were more likely to have been machine-hulled. Hulls from the same samples obtained by hand-shelling contained no detectable aflatoxin. This suggests that damaged kernels or small groundnuts were being mixed with the hulls in the machine-shelling process. Even so, they concluded that the risk associated with the use of groundnut hulls in animal feed would be relatively low.

A number of studies have been carried out on the occurrence and effect of aflatoxins in cooking oils. In many countries unrefined groundnut oil is consumed almost on a daily basis. For example, Fong et al. (1980) reported that 90% of Chinese households in Hong Kong repeatedly reuse groundnut oil for daily cooking. Studies have also shown that crude oil obtained by hydraulic pressing contains a much higher aflatoxin content than that obtained by solvent extraction (Parker and Melnick, 1966). Correct re-fining of groundnut oil should give a product free of aflatoxin, regardless of the quality of groundnuts used. The finding of aflatoxin B1 levels as high as 9 μg/kg in refined groundnut oil in Nigeria (Abalaka and Elegbede, 1982) therefore gives cause for concern.

Fong's study (1980) showed aflatoxin concentrations ranging from 95 to 1055 μg/kg in 10 samples of oil-grade groundnuts tested (obtained from local markets in Hong Kong and presumably the small and immature seeds); three oil samples had levels ranging from 98 to 150 µg/kg. This correlates with a study by Dwarakanath *et al.* (1969) in India. They assayed the aflatoxin content in unrefined oils prepared from freshly harvested nuts and from those stored for 6 months and found similar levels (100 µg/kg and 140 µg/kg respectively). Another Indian study by Sarnaik *et al.* (1988) showed the presence of toxigenic *A. flavus* strains in five of 26 oil samples tested. Included in this were seven samples of refined oil which showed no contamination by *A. flavus*, toxigenic or otherwise. Unfortunately this group did not report aflatoxin levels in the contaminated samples; however, they did demonstrate that the toxin must be present in the oil seed, by the inability of *A. flavus* group fungi to produce aflatoxin in samples of sterile oil. This indicates that the level of aflatoxin in the oil samples is directly related to the level of contamination in the groundnuts used to produce the oil.

The cumulative effects of the daily ingestion of small amounts of aflatoxin on the organs of the body is not known, but the incidence of hepatocarcinoma in humans shows a trend in parallel with aflatoxin B1 consumption in certain areas of the world. Fong *et al.* (1980) demonstrated mutagenicity in unrefined oil, related mainly to AFB1 content. Dwarakanath *et al.* (1969) found that the weight of oil absorbed by food fried in contaminated oil increased with the weight of the food; correspondingly the quantity of toxin in the food increased. However, studies also showed that repeated cooking with the oil decreased the aflatoxin content; e.g. heating contaminated oil at 120 °C for 10 minutes resulted in a 50% destruction of the aflatoxin content. Maximum destruction of toxin occurred in 20 minutes at 150 °C. A higher heating temperature was needed, rather than heating longer at this temperature, to further reduce levels.

13.6 THE CHEMISTRY AND METABOLISM OF AFLATOXIN

The aflatoxins are a family of difuranocoumarins (Figure 13.1) produced as secondary metabolites by strains of *A. flavus* and *A. parasiticus*. AFB1 and B2 were named because of their strong, blue fluorescence under ultraviolet light; whereas G1 and G2 fluoresce greenish-yellow. The B group are characterized by the fusion of a cyclopentenone ring to the lactone ring of the coumarin structure. The G group contain an additional lactone ring (Groopman *et al.*, 1988).

AFB1 (the most toxic and carcinogenic) and AFG1 both possess an unsaturated bond at the 8,9 position on the terminal furan ring. AFB2 and AFG2 lack this double bond. AFG1 is about 18% as active as AFB1 in a DNA-binding assay (Gurtoo *et al.* 1978), whereas AFB2 and AFG2 are essentially biologically inactive unless first metabolically oxidized to AFB1 and AFG1 *in vivo* (Groopman *et al.*, 1988).

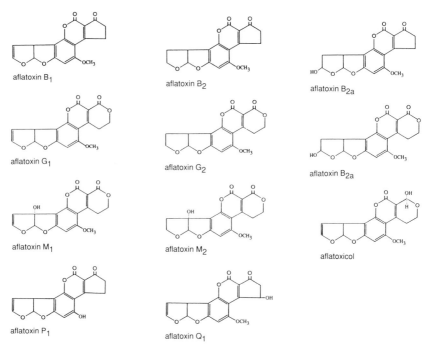

Figure 13.1   Structure of aflatoxins.

Aflatoxins are oxidatively metabolized by the microsomal mixed-function oxygenase system mainly localized on the endoplasmic reticulum of liver cells. This results in the formation of various hydroxylated derivatives (Figure 13.2), as well as an unstable, highly reactive epoxide metabolite, AFB1-8,9 epoxide. It is the epoxide metabolite (which can react covalently with the nucleophilic centres in cellular macromolecules such as DNA, RNA and protein) that is directly related to the susceptibility of a species to AFB1-induced hepatocarcinogenesis. A number of comparative studies have been made on the *in vitro* metabolism of AFB1 (Portman *et al.*, 1968; O'Brien *et al.*, 1983; Ramsdell and Eaton, 1990; Lipsky *et al.*, 1990). A microsomal enzyme, cytochrome P-450, has been shown to be the catalyst (Shimada and Guengerich, 1989). Roebuck and Wogan (1977) postulated that different cellular concentrations of the same microsomal enzyme could exist between species and individuals of a given species, to account for the differing effects of the toxin. Since then multiple species of P-450 have been shown to occur in hepatic mitochondria, as well as considerable (10-fold) inter-individual variation (Shayiq and Avadhani, 1989).

Mouse microsomes have a higher activation ratio (defined as the rate of epoxide formation divided by the sum of the rates of the oxidative AFB1 metabolites formed) than rat microsomes. However, they are resistant to

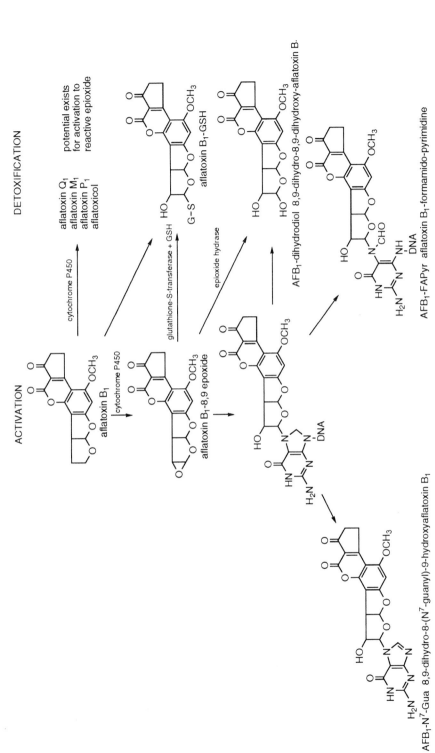

Figure 13.2 Metabolic transformations of aflatoxin B_1.

the carcinogenic effects of AFB1 due to the efficient conjugation of epoxide with GSH (reduced glutathione) (O'Brien *et al.*, 1983; Ramsdell and Eaton, 1990). The levels of glutathione S-transferase in mice are high enough to convert 12 times more epoxide to AFB1-GSH compared with male rats. O'Brien *et al.* (1983) also found female rats to produce less epoxide and to have higher levels of glutathione S-transferase than male rats, which would help account for the decreased susceptibility of females rats to primary liver cancer (PLC).

Several drugs and conditions have been shown to influence the levels and activity of cytochrome P-450. It is highly responsive to changes in diet and to inducers, e.g. phenobarbital. Rats fed a low protein diet became more sensitive to the acute lethal effects of aflatoxin (Tupule, 1969). Correspondingly, those fed aflatoxin, plus phenobarbital as an inducer, responded in such a way that suppressed the carcinogenic activity of a large portion of the ingested aflatoxin (McLean and Marshall, 1971). While phenobarbital treatment has been shown to increase AFB1 inactivation by hydroxylation, this route is now thought to play a minor role. Lotlikar *et al.* (1989) showed that rats fed aflatoxin plus phenobarbital as an inducer have much higher levels of glutathione S-transferase. Therefore, in species that convert more aflatoxin to the epoxide (e.g. rat and mouse), the amount of hepatic glutathione S-transferase activity is inversely related to the susceptibility of the species to AFB1-induced hepatocarcinogenesis.

The AFB1-8,9 epoxide interacts with the 7 position of guanine in DNA but spontaneous hydrolysis of the glycosyl bond means it is rapidly lost from nucleic acids *in vivo* (Bennet *et al.*, 1981). Essigmann *et al.* (1982) showed acute doses of the toxin to cause reversible inhibition of DNA and RNA synthesis. The adduct formed, AFB1-N^7-GUA [8,9-dihydro-8-(N^7-guanyl)-9-hydroxyaflatoxin B1], accounts for more than 80% of adducts found in the liver of rats 2 hours after administration, i.e. it is the major adduct formed *in vivo* and *in vitro*. In rats it is excreted in urine in a dose-dependent manner. The amount excreted over 48 hours represents a relatively fixed proportion (30–40%) of the total amount of AFB1-N^7-GUA initially formed in hepatic DNA (Bennett *et al.*, 1981; Autrup *et al.*, 1987). Another adduct, AFB1-FAPyr [aflatoxin B1-formamidiopyrimidine], has also been found (Lee *et al.*, 1989). This imidazole ring-opened adduct, the persistent form of AFB1-N^7-GUA, has been found in liver tissue and may play an important role in hepatocarcinogenesis (Hsieh *et al.*, 1988). AFB1-8,9 epoxide can also bind to protein. It has been shown that approximately 2% of ingested AFB1 binds covalently to lysine residues in albumin, i.e. it binds quantitatively in relation to dose (Wild *et al.*, 1990). Some epoxide is hydroxylated to AFB1-8,9-dihydrodiol and excreted.

In humans, AFQ1 is the major aflatoxin metabolite formed from AFB1, comprising 70–90% of the soluble metabolites (Roebuck and Wogan,

1977; Moss and Neal, 1985; O'Brien *et al.*, 1983; Lipsky *et al.*, 1990; Ramsdell and Eaton, 1990). AF-8,9 dihydrodiol (10–30%), AFM1 and aflatoxicol (AFL) are also detected; AFP1 is not present in measurable amounts (Wild *et al.*, 1990). The formation of these hydroxylated metabolites, rather than the reactive epoxide metabolite, means the production of compounds with substantially lower carcinogenic potency than the parent AFB1. However, the hydroxylated metabolites retain the unsaturated bond at the 8,9 position of the molecule, which means that the potential exists for further oxidation to an epoxide (Essigmann *et al.*, 1982; Groopman *et al.*, 1985). Some are partially detoxified by enzymatic conjugation with sulphate or glucuronic acid to form water-soluble sulphate or glucuronic esters, before being excreted in bile or urine (Groopman *et al.*, 1988). AFL, formed by an AFB1 reductase, can be reconverted to AFB1 by a dehydrogenase enzyme. Species with a high sensitivity to AFB1 appear to have a high ratio of AFB1 reductase activity to AFL dehydrogenase activity (Salhab and Edwards, 1977).

Of these hydroxylated metabolites, AFM1 is of particular interest. First discovered in cow's milk (Allcroft and Carnaghan, 1962), it has been shown to have a similar acute toxicity to AFB1 (0.32 mg/kg). Up to 3% of ingested AFB1 is excreted in the milk of lactating animals as AFM1 (de Iongh *et al.*, 1964; Fehr *et al.*, 1968; Ferrando *et al.*, 1984; Wild *et al.*, 1987). While proven to be a carcinogen in male rats and rainbow trout, it is a weak hepatic carcinogen compared with AFB1 (Cullen *et al.*, 1987).

Humans are capable of producing AFB-DNA adducts but at lower levels than rat cells. The fraction of AFB1 activated by human liver microsomes increases with decreasing AFB1 concentration. This is due to the high affinity for AFB1 of the cytochrome P-450 isoenzyme that forms the -8,9 epoxide, compared with the low AFB1 affinity of the AFQ1-forming species of enzyme. Compared with some species studied, human liver glutathione S-transferase appears to lack the substantial capacity for epoxide conjugation found in rats and mice (Moss and Neal, 1985) therefore the liver tissue may be relatively efficient at generating AFB-DNA adducts when exposed to dietary levels of AFB1 (Ramsdell and Eaton, 1990). Both rat and human hepatocytes display nucleolar segregation proportional to the level of AFB-DNA binding. This is thought to be a morphological manifestation of carcinogen-DNA binding (Lipsky *et al.*, 1990).

As with rats, human males have been shown to metabolize the aflatoxin 2 to 3 times faster than females. A 16-fold inter-individual variation in the level of AFB-DNA adduct formation has been shown (Lipsky *et al.*, 1990). There is also the possibility that the pattern of aflatoxin metabolism in compromised livers (damaged by alcohol or hepatitis B virus) may differ from that in normal livers.

TABLE 13.1   *Acute toxicity of AFB1 with a single dose (LD$_{50}$ mg/kg)*

	Wogan (1968)	Newberne and Butler (1969)	Roebuck and Wogan (1977)	Coker (1979)
Duckling	0.4	0.335		0.3
Rainbow trout	0.5			0.5–1.0
Pig		0.62		0.6
Baboon				2.0
Chicken				6 0–16.0
Rat (male)	1.0	0.56–17.9	1.0–5.0	7.0
Rat (female)				18.0
Mouse	63.0	9.0	15.0	9.0
Rabbit	0.5			
Dog	0.5	1.0		
Guinea pig	1.0	1.4		

13.7 AFLATOXICOSIS IN ANIMALS AND POULTRY

Aflatoxin B1 is the most toxic of the naturally occurring aflatoxins. It is reported to be almost twice as toxic as strychnine and potassium cyanide to rats (Strzelecki *et al.*, 1990). In all species studied, the liver appears to be the target organ, resulting in toxic hepatitis, cirrhosis or primary liver cancer (Reed and Kasali, 1989). The time taken for animals to succumb to aflatoxin appears to be related to the total dose of the toxin, rather than the rate of consumption.

The LD$_{50}$ of aflatoxin for most species of animal (Table 13.1) is within the range 0.3 mg/kg to 10 mg/kg body weight (Wogan, 1968; Newberne and Butler, 1969; Roebuck and Wogan, 1977; Coker, 1979; Akano and Atanda, 1990). The observed range of LD$_{50}$ values emphasizes the fact that the metabolism of aflatoxin varies quantitatively and qualitatively between animal species, thus causing marked differences in their susceptibility to the toxin. The tolerance to an acute (or subacute) dose can also vary widely within a given species. Factors governing this can include the age, sex, strain and condition of the animal, the level and type of aflatoxin in their diet and the time elapsed before the LD$_{50}$ is measured (Coker, 1979). Also, not all animals of a similar age on identical contaminated feeds are affected, which suggests genetic restriction (Loosmore and Markson, 1961; Sisk *et al.*, 1968; Hintz *et al.*, 1967; Murthy *et al.*, 1975). For all species of animal in which age has been recorded, the young have been shown to be more susceptible (Loosmore and Harding, 1961; Allcroft and Carnaghan, 1963; Horrocks *et al.*, 1965). Females are less susceptible than males (Wogan, 1968).

Subacute or chronic exposure affects growth feed efficiency and general well-being, especially during the early growing period (Duthie *et al.*, 1966).

In the early stages it may not be associated with overt clinical symptoms. It eventually leads to the development of a toxicity syndrome and liver damage. Clinical recovery can occur when the contaminated diet is removed; however, the livers of the affected animals will still show aflatoxin-induced changes (Duthie *et al.*, 1968).

Calves and pigs are the most susceptible domestic animals. Dairy cows are reported to have significantly reduced milk yields which coincide with the excretion of AFM1 in their milk (Allcroft and Carnaghan, 1963; Coker, 1979). Ray *et al.* (1986) reported abortion and symptoms suggestive of acute aflatoxicosis after pregnant cows ingested aflatoxin-contaminated groundnuts. Pigs fed aflatoxin during gestation did not abort but did show histological changes in their livers. In this study no biochemical effects were detected in their offspring (Gumbmann and Williams, 1969; Ray *et al.*, 1986). Sheep have been reported to tolerate a heavily contaminated diet for up to 3 years (Wogan, 1968). Some species, e.g. monkeys, calves and rabbits, can develop cirrhosis after chronic exposure. Chickens are less susceptible than turkey poults; ducklings are the most susceptible species known, which makes them ideal for biological screening purposes (Asplin and Carnaghan, 1961).

Both lipid and protein metabolism are affected by aflatoxin ingestion, manifested by an early decrease in feed conversion. The biochemical changes in the affected animals appear to be closely related to the degree of hepatic injury, which is in turn related to dose administered. The vacuolization and fatty changes which occur in the hepatocytes lead to necrosis, which is usually localised in one part of the hepatic lobule, depending on the species (Newberne and Butler, 1969; Syamasundara Rao, 1970). This accumulation of fat, seen in the early stages, is thought to be due to inhibition of fat removal from the liver (Reed and Kasali, 1989). Bile duct proliferation is a typical pathological change; hepatic veno-occlusive lesions are also common.

Because of its ability to modify DNA, AFB1 can cause a reduced rate of protein synthesis, which in turn increases the protein requirements of susceptible animals. Serum albumin decreases and gamma-globulin increases (Annau *et al.*, 1964). Therefore animals with an already reduced protein intake would have increased susceptibility to aflatoxin, due in part to the inability of the liver to synthesise the enzymes necessary for the metabolism and detoxification of the aflatoxin, leading to longer exposure and increased toxic effects (Tupule, 1969; Coffey *et al.*, 1989). Increasing the level of dietary protein or lipids above the normal requirement for optimum performance has been shown to prevent depressed performance due to aflatoxin exposure (Smith *et al.*, 1971). Coffey *et al.* (1989) showed increased dietary lysine or methionine can protect against aflatoxicosis. This is of special interest because it is now known that aflatoxin binds to the lysine residues of plasma albumin in a dose-related manner (Wild *et al.*, 1990).

Aflatoxin is also known for its immunosuppressive effect. Coker (1979) reported that poultry fed diets containing 250–500µg/kg were predisposed to attack by viruses and bacteria. This secondary effect could be particularly significant in livestock already compromised by poor nutritional status (Reed and Kasali, 1989).

The potential carcinogenicity of AFB1 is related to the proportion that ultimately binds to DNA. This is determined by both the amount activated *in vivo* and how much is subsequently detoxified (Ramsdell and Eaton, 1990; Lipsky *et al.*, 1990). Rats, which do not show signs of acute toxicity, are highly susceptible to aflatoxin-induced hepatomas (Lancaster *et al.*, 1961; Butler and Barnes, 1968). Rainbow trout and ducks are also highly susceptible (Wogan, 1968). Liver cancer has now been described in many species of vertebrate, although the dose and period of exposure varies widely. It is thought that chronic exposure to small doses of aflatoxin is more likely to produce hepatocellular carcinoma than exposure to the same amount in larger doses over a brief period of time (Wogan and Newberne, 1967). Even so, there are reports of rats surviving an LD_{50} dose and subsequently developing hepatoma (Carnaghan, 1967; Angsubhakorn *et al.*, 1990). Mice appear to be resistant to both subacute toxicity and carcinogenicity (Wogan, 1968).

A greater susceptibility of neonates to the carcinogenicity of B1 has been suggested (Cullen *et al.*, 1987). Males tend to develop tumours at a faster rate (Wogan, 1968). A difference in susceptibility between two strains of rat has also been demonstrated (Fong and Chan, 1981; Angsubhakorn *et al.*, 1990).

13.8 AFLATOXIN AND HUMAN DISEASE

Within a population, the character and intensity of human exposure to aflatoxin may vary depending on factors such as age, nutritional status, concurrent exposure to other agents (e.g. hepatitis B virus) as well as the level and duration of exposure (Dichter, 1984). Differences between individuals in a population in AFB1-FAPyr adduct formation (the persistent adduct) in liver tissue also suggests there is genetic restriction (Lee *et al.*, 1989).

The liver is the organ primarily affected by aflatoxin ingestion. The known metabolic effects of aflatoxin ingestion include inhibition of DNA, RNA and protein synthesis, reduction of certain enzyme activities, depression of glucose metabolism, inhibition of lipid synthesis and depression of clotting factor synthesis. Aflatoxins also tend to block steroid binding sites and are immunosuppressive. The pathological changes include fatty infiltration, biliary proliferation and toxic necrosis of the liver in acute severe poisoning. There are only four outcomes to toxic damage to the liver: recovery, cancer, cirrhosis and ongoing hepatitis (Coady, 1976). In man, studies have implicated aflatoxin in the aetiology of acute hepatitis, cirrhosis in malnourished children, kwashiorkor, Reye's syndrome and primary liver cancer (Denning, 1987, has written an excellent review).

To date there have been only a few documented cases of acute aflatoxicosis in man. These have ranged from an individual case to one outbreak in India in which 106 people were reported to have died (Serck-Hanssen, 1970; Ngindu *et al.*, 1982; Krishnamachari *et al.*, 1975). The livers of those affected showed signs of toxic hepatitis. AFB1 was found in the sera of two out of seven patients tested for the toxin (Krishnamachari *et al.*, 1975). Ngindu *et al.* (1982) reported AFB1 (38 and 89 µg/kg respectively) in two liver samples obtained. Contaminated foodstuffs other than groundnuts were implicated in each outbreak. Not all people exposed to the toxin were adversely affected.

The finding of aflatoxins in the breast milk, neonatal cord blood and the sera of pregnant women indicates that a high proportion of at-risk infants have not only prenatal but also continuing post-natal exposure to aflatoxin (Robinson, 1967; Lamplugh *et al.*, 1988; Denning *et al.*, 1990). Studies also indicate not only transplacental transfer but also concentration of aflatoxin by the foeto-placental unit (Lamplugh *et al.*, 1988; Denning *et al.*, 1990). Tupule (1969) showed rats and monkeys fed a protein-deficient diet were more susceptible to aflatoxin; therefore the combined exposure of a human foetus or infant to aflatoxin and malnutrition must severely diminish its chances of survival.

13.8.1 Kwashiorkor

Protein energy malnutrition (PEM) is implicated in increased susceptibility to aflatoxin by reducing the levels of cytochrome P-450 and therefore microsomal hydroxylation of aflatoxin in the liver (McLean and Marshall, 1971). In humans PEM is divided into two main groups: marasmus and kwashiorkor. Marasmus is due to chronic near-starvation. Kwashiorkor has an obscure pathogenesis; it is characterized by hypoalbuminaemia, fatty liver and immunosuppression, and is thought to be due to the deprivation of protein in the presence of adequate calorie intake (Anon., 1984). These metabolic and pathological consequences of kwashiorkor closely resemble those of aflatoxin ingestion. The occurrence of kwashiorkor also closely mirrors the geographical and seasonal prevalence of aflatoxin contamination in foodstuffs.

Aflatoxin is not thought to be responsible for initiating kwashiorkor. It is possible that a zinc deficiency, which becomes limiting before a protein deficiency, precipitates the condition by interfering with protein and nucleic acid synthesis. This is supported by the observation that in many cases the liver seems to suffer little pathological damage and responds well to zinc therapy (Golden and Golden, 1981). The ingestion of aflatoxin could exacerbate the problem; the inability of the malnourished liver to metabolize it could lead to the more severe liver lesions occasionally seen in children with kwashiorkor.

Breast-milk analyses have shown aflatoxins in >30% of samples, which can account for wholly breast-fed babies showing signs of kwashiorkor (Hendrickse, 1988). In these babies the reduced ability of the liver to synthesize albumin could predispose them towards the kwashiorkor end of protein energy malnutrition. Aflatoxin ingestion can be the final insult that precipitates the child into kwashiorkor. It is also postulated that aflatoxin causes a dose-related suppression of immune function, which would make these children even more susceptible to infectious disease mortality (Denning, 1987).

Serum measurements in children showed that aflatoxin was present more often and in higher concentrations in children with kwashiorkor than in those with marasmus or controls (Hendrickse *et al.*, 1982). They also found AFB1 or aflatoxicol in all the autopsy liver samples they obtained from children dying from kwashiorkor and never from those dying of other causes. It could be argued that this is the result of greater exposure to aflatoxin in these children but two other results have to be taken into account. While urinary metabolites were detected most often in samples from kwashiorkor patients, the concentrations were lower than in other groups. Aflatoxicol, a metabolic product of AFB1 that can be converted back to AFB1, was detected only recently in the sera of children with kwashiorkor. This suggests a difference in metabolism in these children, possibly linked to an inability to transport and excrete the aflatoxin (Hendrickse, 1988). A recent study by de Vries *et al.* (1990) supports this theory, and also shows that large amounts of aflatoxins can accumulate in the body tissue of these children. Their study also suggests that the metabolic fate of aflatoxin in children with kwashiorkor may differ from those with marasmic kwashiorkor.

13.8.2 Cirrhosis

There is a high prevalence of childhood hepatic cirrhosis in certain parts of India which peaks at 3 years of age (Tupule, 1969; Amla *et al.*, 1971). It has been directly linked (via breast-milk studies) with Indian mothers of cirrhotic children eating aflatoxin-contaminated foodstuffs (Robinson, 1967). Liver biopsies taken from Indian children with cirrhosis at different stages of severity have revealed varying degrees of cell destruction, fibrosis, inflammatory cell infiltration and regenerating nodules (Tupule, 1969). The presence of aflatoxin metabolites was also shown in the urine of these children. Amla *et al.* (1971) implicated aflatoxin as a cause when they showed that Indian children suffering from kwashiorkor, inadvertently treated with an aflatoxin-containing diet (<30 µg/kg daily), went on to develop cirrhosis. The hepatic lesions noted had a direct correlation with duration of toxin ingestion, which ranged from 5 days to 4 weeks. Despite this direct link with aflatoxin ingestion, studies of urine from cirrhotic children have been inconclusive; similar levels of AFB1 have been found in both subjects and controls (Amla *et al.*, 1970; Yadgiri *et al.*, 1970). This would support the theory that childhood hepatic cirrhosis has a multifactorial aetiology (Coady, 1976).

13.8.3 Reye's syndrome

Reye's syndrome is a childhood disorder of previously unknown aetiology. It is indistinguishable clinically from a condition found in Thai children, commonest at the end of the rainy season when aflatoxin contamination of food is highest. Both exhibit encephalopathy with fatty degeneration of the viscera. Because of the similarity of the lesions to aflatoxin poisoning, several studies were made on children dying from Reye's syndrome. Aflatoxin was identified in liver samples from cases of Reye's syndrome in New Zealand (Becroft and Webster, 1972), Czechoslovakia and the USA (Ryan *et al.*, 1979). In a subsequent study Rogan *et al.* (1985) were unable to isolate any aflatoxin metabolites or adducts in liver biopsy samples using more sensitive methods. Reye's syndrome is now known to occur after a viral illness. It is possible that in some cases aflatoxin could alter the host's response to the virus (Denning, 1987). More commonly a very strong link has been shown between Reye's syndrome and the intake of aspirin in children under 12 years of age (Glenn-Bott, 1987).

13.8.4 Primary liver cancer

Primary liver cancer (PLC) is a disease of striking geographic distribution and is one of the leading causes of cancer mortality in Asia and Africa. A number of epidemiological studies have shown what appears to be a good correlation between the level of ingested aflatoxin and incidence of PLC (Korobkin and Williams, 1968; Brudzynski *et al.*, 1978; Groopman *et al.*, 1988), which has raised the hypothesis that exposure to dietary aflatoxin is associated with an elevated risk of PLC. The indirect demonstration of AFB-DNA adducts, by the identification of the AFB-Gua 1 adduct in urine samples collected from people in different regions of the world where there is a high incidence of PLC, adds weight to this theory (Lipsky *et al.*, 1990). In contrast, the incidence of PLC is low in Central and South America, although environmental conditions match those of the high-risk areas of Africa and Asia (Wogan, 1968).

Recent work has found aflatoxin ingestion by itself to be insufficient to induce PLC (Stoloff, 1989; Campbell *et al.*, 1990). This would suggest a multifactorial aetiology, with factors such as individual susceptibility, age, nutritional status and concurrent exposure to other agents such as alcohol and hepatitis B virus (HBV), being important determinants in the development of PLC. Denning (1987) postulated that selenium or some other antioxidant deficiency could also contribute to aflatoxin-induced carcino-genesis.

Chronic hepatitis B virus is now thought by some authors to be the major aetiologic agent in PLC (Stoloff, 1989; Campbell et al., 1990). It is endemic in many areas of the world in which the incidence of liver cancer is high and PLC mortality has been shown to be positively correlated with HBsAg+ prevalence. Stoloff (1989) was even able to rule out the need for aflatoxin, or any similar cofactor, as a necessary condition for HBV-induced PLC. There is no disputing the fact that aflatoxin is activated in vivo by humans. This is confirmed by the detection of the AFB1-FAPyr in liver tissue (Lee et al., 1989) and AFB1-N7-Gua in urine (Lipsky et al., 1990). But while aflatoxin has the potential to act as a carcinogenic initiator, consumption must be high enough and prolonged enough for it to contribute a signifi-cant risk (Campbell et al., 1990).

The role of aflatoxin in the aetiology of PLC is not clear. Campbell et al. (1990) state that it is an unnecessary and insufficient cause; Stoloff (1989) postulates that it possibly has a late-stage effect in the development of PLC, acting in an acute (dose-dependent) manner on a liver already compromised by HBV infection. Denning (1987) postulates that HBV may be essential in lowering the threshold at which aflatoxin carcinogenesis occurs. Alternatively aflatoxin may act as an immune suppressive agent, rather than as a primary carcinogen (Lutwick, 1979). This would allow HBV to maintain itself more easily in the liver, producing more chronic infection. This may be critical for the initiation and development of either PLC, cirrhosis or ongoing hepatitis. It could be that these three pathologi-cal states may represent different responses to aflatoxin. It has also been postulated that exposure to AFB1/AFM1 via breast milk could be relevant in determining the integration and/or rearrangement of the HBV genome within the cellular genome (Chandra, 1970; Wild et al., 1987).

13.9 ANALYTICAL METHODS FOR THE DETECTION OF AFLATOXINS

Because of the potential for contamination, all groundnuts and groundnut products should be monitored for the presence of aflatoxin throughout all phases of their production. Originally biological methods were used to detect the presence of aflatoxin. Sargeant et al. (1963) found that aqueous suspensions of material extracted from toxic meals with solvents could be administered to day-old ducklings, to measure toxicity of the sample. Good agreement was found to exist between toxicity established by bio-logical assay and the amount of fluorescent material determined by paper chromatography. AFB1 was shown as a single spot, with a bright blue fluorescence under UV light, while AFG1 has a green fluorescence and is less mobile.

The standard chemical method, based on thin layer chromatography (TLC), was developed by the Association of Official Analytical Chemists (AOAC) in 1970. High performance liquid chromatography (HPLC) is also used for quantitative analysis. Several screening methods are now available. Immunochemical methods, such as the Enzyme-Linked ImmunoSorbent Assay (ELISA) and affinity column methods, are now being developed and offered as simpler, quicker alternatives.

13.9.1 Sampling

The difficulties in obtaining a representative sample of groundnuts for aflatoxin analysis have been established. Aflatoxin is not a natural constituent of groundnut tissue, and is present only in a proportion of kernels. In a sample of groundnuts with a high average content of aflatoxin, only a small fraction may be contaminated. Even these few groundnuts may have varying levels of aflatoxin, which makes representative sampling more difficult (Allcroft and Raymond, 1966; Cucullu *et al.*, 1966; Gardner *et al.*, 1968). For a given aflatoxin concentration, the proportion of lots accepted by any testing plan is dependent on the sample size, subsample size, number of analyses, method of sample preparation and acceptable level. The acceptable level is defined as the highest aflatoxin test result allowed for a lot to be accepted (Whitaker and Dickens, 1989). Some countries have a zero tolerance, which is in practice the limit of detection by the analytical method employed (Van Egmond, 1989).

The negative binomial distribution has been used for estimating the probabilities associated with sampling lots of shelled groundnuts for aflatoxin analysis (Whitaker *et al.*, 1972). It is characterized by a continuous decrease in the probability of finding a groundnut with increasing amount of aflatoxin (Whitaker *et al.*, 1985). This model has a property which enables distribution to be inferred for different sample sizes. However, this property applies only if all the variability may be attributed to sampling, i.e. variability due to subsampling and analysis may undermine the value of the model (Brown, 1984). With the large sample sizes (three 22 kg samples) and the high acceptance levels permitted in the USA (25 µg/kg total), the choice of statistical distribution model is not very important. The tolerance accepted in Europe is much lower (1–5 µg/kg). This, combined with a small sample, can be unfavourable to the consumer, as the chance of detecting a lot with a mean concentration greatly above the tolerance limit will be very small (Knutti and Schlatter, 1982; Whitaker and Dickens, 1989). The amount of sample necessary is inversely related to the parts per billion (µg/kg) aflatoxins (Cucullu *et al.*, 1966).

The particle size of the product being sampled is also a factor in detecting the true level of the contamination present (Waltking, 1980). A single groundnut can be severed into fewer than 50 particles to make chunky groundnut butter. The same groundnut can be divided into more than 300 million particles to make the creamy-style product. In the latter, there is a high probability of there being a uniform aflatoxin concentration (if present) which does not apply to the former.

13.9.2 Presumptive and screening methods

In the USA all of the kernels from each official grade sample (465 g) are examined visually. If one or more kernels are found with visible *A. flavus* growth, the lot is segregated for restricted oil processing, i.e. diverted from food use (Whitaker and Dickens, 1986).

Minicolumns have also been developed as a rapid, inexpensive screening method. They are all based on running a solvent-extract of a ground groundnut sample through a silica gel column. Holaday (1968) developed the original one; since then several modifications have been published (Holaday and Lansden, 1975; Davis *et al.*, 1980; Shotwell and Holaday, 1981). The fluorescence of the extracts under long-wave UV light (365 nm) is then compared with that of controls. However, the test is not specific for aflatoxins – some fluorescence from other fungal metabolites can occur.

The Romer All-Purpose Minicolumn method has been granted official first action for screening for aflatoxins in groundnuts (AOAC, 1984). It is applicable for detecting $\leq 10\ \mu g/kg$ in groundnuts, groundnut meals and peanut butter and $\leq 15 \mu g/kg$ total aflatoxins in mixed feeds. After extraction, the samples are treated with $FeCl_3$ and $CuCO_3$ to remove interfering substances. Lovelace and Aalbersberg (1989) showed that this method gave a good correlation with TLC. An inexpensive modification of this step substitutes $CuSO_4$ as the clarifying agent and still gives 100% agreement with Romer's method (Sylos and Rodrigues-Amaya, 1989).

The Holaday–Velasco method, which has been approved official first action for corn (AOAC, 1984), has also been recommended as a screening method for groundnuts. It uses methanol–water extraction and benzene for solvent partition. A collaborative trial showed that the limiting level for reliable aflatoxin detection using this method was between 5 and 13 $\mu g/kg$ (Shotwell and Holaday, 1981). A modification of this method is the SAM-aflatoxin tube. Using toluene instead of benzene for solvent partition, this phase is then passed through the SAM (Selectively Absorbed Mycotoxins) tube which removes other interfering compounds in a pre-absorption layer and selectively absorbs any aflatoxin at specific bands in the tip of the tube. The sensitivity of the assay is designed to give a yes/no answer at designated levels of aflatoxin. It is a rapid, sensitive, economical and chemically stable assay (Pettit *et al.*, 1990).

13.9.3 Quantitative methods

A number of methods exist for preparing a groundnut sample for quantitative analysis. The official AOAC Method I (CB method) (1984) is the standard by which other methods are judged. Based on a chloroform–water extraction, it gives better clean-up and higher recovery than other methods (Whitaker and Dickens, 1983). This makes it ideal for use with TLC but it is expensive and time-consuming. The AOAC Method II (BF method) uses a methanol–water (55:45) extraction and a solvent/groundnut ratio of 5:1 (ml/g) for the extraction of aflatoxin from 50 g samples of raw groundnuts (AOAC, 1984). This method is the faster and more economical of the two.

Modifications of Method II include the water slurry (WS) method. Whitaker et al. (1984) found the amount of aflatoxin extracted from raw groundnuts to be a function of both the methanol concentration and solvent/groundnut ratio. Increasing the methanol concentration to 60% and the solvent/groundnut ratio to 10.8:1 (ml/g) gave 12.1% more aflatoxin extracted. However, this increase required 2.36 times more methanol than the original AOAC Method II (Whitaker et al., 1986). Another modification is the Food Safety and Quality Service (FSQS) method. It has been developed for larger sized (21.8 kg) samples and is recommended for naturally contaminated lots where aflatoxin is unevenly produced.

A number of comparative studies have been made of the different extraction methods (Coon et al., 1972; Mechan et al., 1985; Pluyer et al., 1987; Van Egmonde and Wagstaffe, 1989). Minicolumn methods are also being used as rapid clean-up methods to replace the official quantitative methods (CB/BF), which are comparatively time-consuming and expensive. The purified extracts are suitable for TLC as well as HPLC and allow the detection of low levels of aflatoxin (Dorner and Cole, 1988). However, Mechan et al. (1985) compared several extraction methods with the Romer minicolumn method and found the latter to give considerably lower results. If used, the minicolumn has the advantage in that preliminary estimation of aflatoxin can be made which allows for dilution if necessary before quantitative analysis.

(a) Thin layer chromatography

Samples for TLC are spotted onto precoated silica gel plates. After development, they are detected under long-wave UV (365 nm) light. A visual comparison can be made with standards but fluorodensometric measurements give a much more sensitive result. Interfering fluorescent material in the samples, including oils, fats, sugars and protein not completely removed from the extracts by the clean-up procedures, can cause a loss of accuracy (Beljaars et al., 1972). This includes errors in measuring the proportion of the solvent solution analysed for aflatoxin and in measuring the proportion of the extracted aflatoxin placed on TLC plates (Dickens and Whitaker, 1983). These two errors can cause underestimation of aflatoxin content by as much as 11%. Two-dimensional TLC procedures have now been developed which can reduce the interference and therefore the need for extensive clean-up procedures (Beljaars *et al.*, 1973). In a comparison with an ELISA estimation of aflatoxin, Chu *et al.* (1988) observed that the largest error in the determination of aflatoxin by the TLC method occurs during the silica gel clean-up and/or the TLC step itself.

(b) High performance liquid chromatography

HPLC provides a precise and sensitive method for determining aflatoxins in all types of groundnut products. Combined with flow injection analysis it also offers speed, simplicity and economy (Lazaro *et al.*, 1988). Both normal phase and reverse phase HPLC separations have been developed. For either method samples have to be extracted and cleaned up prior to analysis. Pons and Franz (1978) found an acidified aqueous methanol extraction yielded consistently higher results. Samples are purified on a small silica gel column before being eluted and quantified. Normal phase HPLC uses a silica-gel packed flow cell for either UV detection at 360–365 nm (AFB1/B2) or fluorometric detection (AFG1/G2). Using this method Pons and Franz (1978) were able to resolve completely the four aflatoxins in extracts of raw groundnuts, groundnut meal and peanut butter. The drawback with this method is the instability of the packed cell (Wilson, 1989).

Reverse phase HPLC separations are more widely used than normal phase (Hurst and Toomey, 1978) but the fluorescent intensities of AFB1 and AFG1 are reduced in reverse phase solvent mixtures. Derivatization with trifluoroacetic acid (TFA) prior to injection overcomes this by converting the weakly fluorescent AFB1 and AFG1 to their respective highly fluorescent hemiacetals AFB2a and AFG2a. Park *et al.* (1990) reported an average recovery of 88.1% and 63.8% for peanut butter and raw groundnuts respectively using this method. It has now been recommended as official first action for determination of AFB1, AFB2, AFG1 and AFG2 in peanut butter at concentrations $\geq 13 \mu g$ total aflatoxin/kg. Another alternative is post-column derivatization (PCD) with iodine in water to enhance AFB1 and AFG1 fluorescence after reversed phase chromatography. Dorner and Cole (1988) combined a minicolumn clean-up procedure with this method. It proved to be sensitive, efficient and extremely rapid.

13.9.4 Immunochemical methods

AFB1 in groundnuts can also be detected by immunochemical methods Two of these – the ELISA and the monoclonal antibody affinity column – use specific antibodies to detect aflatoxin in different foods and feed. They are comparatively quick, easy and potentially inexpensive. They have the ability to detect low aflatoxin levels and the capability for routine screening of large numbers of test samples.

The direct competitive ELISA is based on competition between an enzyme-conjugated AFB1 and (free) aflatoxins in the test sample. They compete for aflatoxin-specific antibodies, coated onto a solid phase, usually a microtitre plate. A comparison by Chu *et al.* (1988) between the ELISA method and each step of the TLC method showed a clean-up treatment is not necessary in the ELISA.

When a polyclonal antibody is used to coat the ELISA plate, the result is not confirmatory for AFB1. This is because a polyclonal antibody cross-reacts with AFB2 (70%), AFG1 (75%) and AFG2 (<10%) respectively (Trucksess *et al.*, 1989). Park *et al.* (1989) showed a 98% correlation between this method and TLC results for roasted groundnuts. As such the method has been approved as interim official first action as a screening method to determine the presence or absence of AFB1, AFB2 and AFG1 in groundnut products. The detection limit is set at 20 µg/kg – a positive result for total aflatoxins at >20 µg/kg or negative results at ≤20 µg/kg (Park *et al.*, 1989). Trucksess *et al.* (1989), investigating a modification of this method, recommended a detection limit of >30 µg/kg in raw groundnuts. Commercial ELISA kits are now available, providing a rapid screening method for large numbers of samples, but these kits vary in their specificity and sensitivity (Mortimer *et al.*, 1988; Dorner and Cole, 1989; Patey *et al.*, 1989).

The use of monoclonal antibodies enables the detection of very low levels of aflatoxin. The Aflatest has been developed, based on monoclonal antibodies adsorbed onto an affinity column, and is useful for the rapid isolation of aflatoxins from samples of food (including milk). As the methanol-extracted sample passes through the column, any aflatoxin in it adheres to the monoclonal antibody. Methanol is then used to elute the conjugated aflatoxin–antibody complex. This collects in a fluorosil tip. The fluorescence of the sample can be compared under UV light with a set of standards. This provides a test for total aflatoxin in minutes, using simple equipment with a minimum of expertise (Cole and Masuka, 1989).

The affinity column also provides a rapid one-step removal of interfering substances from diverse biological samples. The method is highly specific and sensitive (Groopman *et al.*, 1985) and has proved to be as effective as the CB extraction method (Trucksess *et al.*, 1990). The affinity column separation can then be coupled with solution fluorometry with bromine (SFB) to estimate total aflatoxins. Alternatively individual toxins can be determined by reverse phase liquid chromatography with post column derivatization with iodine (PCD). This method has now been recommended as an official first action for the determination of aflatoxins in raw groundnuts and peanut butter at total aflatoxin concentrations ≥10 ng/g (Trucksess *et al.*, 1991). It is interesting to note that a similar study by Patey *et al.* (1991) gave much lower levels of recovery (51–67%) from peanut butter samples compared with those reported by Trucksess *et al.* (1991) for the same PCD method (81% and 83% respectively for raw groundnuts and peanut butter).

13.9.5 *In vivo* detection of aflatoxin

The variability in dietary exposure to AFB1 makes it very difficult to determine the extent of cumulative aflatoxin exposure in an individual. Genetic susceptibility to adduct formation also makes it difficult to establish a dose–response relationship between the amount of aflatoxin ingested and the amount of adduct found in the tissues at a given time (Lee *et al.*, 1989).

Aflatoxin has a short half-life in serum. It is excreted in the urine and breast milk as a number of different metabolites which are excreted at different rates. An ELISA method developed to detect AFB1, AFG1 and AFQ1 in serum or plasma has proved to be a useful index of short-term exposure, provided a blood sample is taken within 24 hours (Denning, 1987). Wilkinson *et al.* (1988) reported that the pretreatment of sera with a 1:1 ratio (v/v) of methanol reduces the differing levels of interference in different sera, caused by serum proteins. Wild *et al.* (1987) have developed an ELISA for the detection of aflatoxin in human breast milk, with a sensitivity of 2 pg AFM1/ml milk. Eleven per cent of samples tested in rural Zimbabwe gave a positive result with levels up to 50 pg/ml.

The measurement of urinary metabolites is another alternative. Samples can be obtained easily and non-invasively. AFM1 is the major metabolite present in urine (Campbell *et al.*, 1970). It comprises 1–4% of ingested AFB1 and is mostly excreted within 48 hours; therefore, it can be used as a measure of recent aflatoxin ingestion. AFB1 and AFP1 (in rat urine) have also been recovered but in much smaller amounts. Measurement of AFB1-N7-Gua in urine reflects the proportion of AFB1 activated by exposed people (Groopman and Donahue, 1988).

The problem with measuring aflatoxin metabolites and/or nucleic adducts in body fluids is that they are rapidly excreted and therefore reflect a relatively short-term (few days) exposure. Another alternative is to measure the permanent adduct, AFB1-FAPyr. Conversion of the major AFB1-DNA adduct (AFB1-N7-Gua) to the more stable AFB1-FAPyr (imidazole ring-opened guanine) is known to occur in the liver. Inter-individual differences in levels of this permanent adduct suggest that susceptibility to aflatoxin may also be genetically restricted (Lee *et al.*, 1989). With the development of monoclonal antibodies to specifically quantify AFB1-FAPyr adducts, insight might be gained into the relationship between infection with HBV, dietary exposure to AFB1 and liver cancer. However the measurement of this adduct requires the taking of a liver tissue sample, an invasive technique (Hsieh *et al.*, 1988).

Ideally a laboratory assay is needed that will give assessment of long term aflatoxin exposure in a sample obtained non-invasively. A highly significant correlation between AFB1-N7-Gua excretion into urine and AFB1 intake has already been shown, but the measurement of this parameter only reflects relatively short-term exposure. Groopman *et al.* (1987)

have now shown a strong correlation between urinary excretion of AFB1-N7-Gua adduct and AF-albumin adducts in the peripheral blood. AFB1 binds in a dose-related manner to peripheral blood albumin in rats. Upon repeated exposure, accumulation of binding occurs and the level of albumin binding parallels the binding to liver DNA (Wild *et al.*, 1990). Because albumin has a half-life of about 20 days in humans, this adduct may be a useful marker of aflatoxin exposure over a period of weeks or months, with an accumulation of adducts to a level 30-fold higher than that induced by a single exposure.

A direct ELISA of the AF-albumin adducts has been shown to give false positive results. Hydrolysis of the sample gives an AF-lysine adduct. Measurement of this by the same assay is much more sensitive, and can detect 5 pg aflatoxin/mg albumin. This method can be used as an initial screening test, followed by HPLC to confirm positive samples (Wild *et al.*, 1990). Before blood AF-albumin levels can be extrapolated to liver DNA damage, they must be assessed under experimental conditions which reflect the human condition, especially when the liver is compromised by environmental variants, such as HBV, alcohol and/or nutritional deficiency. This is because studies with rats suggest that the two parameters may be affected by alcohol intake. Until then, the measurement of the albumin adduct will be useful only to identify populations at risk. Measurement of the excreted DNA adducts will continue to be a useful index for liver damage in exposed people (Wild *et al.*, 1990).

13.10 OVERCOMING AFLATOXIN CONTAMINATION

In the past, the fairly high levels of aflatoxin often found in groundnut meals restricted their use as a protein source in many animal diets. The alternative was to divert this meal to fertilizer, which grossly underutilized this valuable source of protein. Nowadays groundnuts are gaining even more importance as a source of low-cost protein in the human diet, especially in developing countries where aflatoxin contamination is most severe. This stresses the need to eliminate or remove aflatoxins from contaminated groundnuts.

Ideally the best control method is prevention – encouraging good farm management in cultivation and storage practices. However, damage and contamination still occur, which has led to various methods being employed in the attempt to eliminate or lower the contamination to acceptable levels for human and animal consumption. Aflatoxins have proved to be highly resistant to various physical and chemical treatments. A problem arises with the processing of the kernels into a variety of products, e.g. oil, press and expeller cake (containing some residual oil) and meal (solvent-extracted, therefore containing low oil levels). This means that a particular method may be severely restricted by the nature of the product to be detoxified (Feuell, 1966).

13.10.1 Removal of aflatoxin

Generally a relatively small number of kernels are contaminated and in many cases physical separation is the ideal method. This is based on the assumption that contaminated kernels are either discoloured or shrivelled. Methods include manual sorting (Dickens and Weltz, 1968), mechanical sorters or electronic scanning which examines each kernel separately, scanning with a photoelectric cell which accepts or rejects them on the basis of reflectance (Goldblatt, 1971).

Groundnut samples are currently graded as sound and mature if they contain less than 2% damaged kernels, including discoloured, broken, insect-infested or mould-damaged kernels. However, subsamples of a given lot of these groundnuts may vary greatly in aflatoxin content, due to the extreme variability in the degree of contamination in individual kernels (Hodges et al., 1964). Wrinkled nuts, thought to be caused by dampening during storage, can account for 5% of a sample previously graded as sound. While this is a relatively high level of contamination, the level of toxin within these kernels is usually comparatively low. Conversely, a second type of contamination is characterized by a low (0.24%) concentration of contaminated kernels but very high levels of toxin. Disconcertingly, the majority of these kernels appear normal on the exterior (Cucullu et al., 1966).

When aflatoxin-contaminated nuts are pressed for oil, most of the aflatoxin remains in the press-cake. The same occurs during oil extraction with commercial hexane, with most remaining in the extracted by-product meal. With the removal of the oil, the cake and meal have an enriched protein content, which make them cheap protein sources for compounding into animal feeds (Feuell, 1966). (The outbreak of Turkey X disease in the UK in 1961 was caused by prepressed, hexane-extracted groundnut meal containing a high level of aflatoxin.)

Several solvent extraction methods remove aflatoxins from contaminated oilseed meals. They have the advantage of being able to remove essentially all aflatoxins without affecting the nutritive value or forming adverse by-products. Polar solvents appear to be the most effective; acetone and methanol have been shown to extract aflatoxins completely from substrates. A binary solvent consisting of 90% acetone and 10% water (w/w), used for extracting aflatoxin from pre-extracted groundnut meal of low oil content, gave up to 88% removal of aflatoxin. With this system, an increased solvent temperature was shown to result in a higher percentage reduction of aflatoxin. A tertiary solvent of 54% acetone, 44% hexane and 2% water (w/w) was also developed for simultaneously extracting oil and aflatoxin from prepressed groundnut cake (Gardner et al.,

1968). The water level in each system is critical. While it facilitates extraction, too much removes excessive amounts of water-soluble meal constituents and leads to processing difficulties. Dollear *et al.* (1968) showed that a binary solvent (90% acetone, 10% water) completely extracted aflatoxin, leaving a meal with a chemical composition nearly identical to the original starting material.

Aqueous isopropanol has also been shown to be an effective solvent for the removal of aflatoxin from contaminated groundnut meals. Extraction with six passes of 80% aqueous isopropanol at a minimum of 60 °C resulted in complete removal. Extraction with the isopropanol–water azeotrope (87.7% isopropanol, w/w) was less effective, removing approximately 80% of the aflatoxin (Rayner and Dollear, 1968).

Crude oils may contain various amounts of aflatoxin, depending on the material used and the method of processing. Treatment with fuller's earth and a hot alkali wash (conventional refining) has been shown to remove essentially all aflatoxin present (Parker and Melnick, 1966). The aflatoxin in crude oil appears to be in suspension, associated with cell debris carried over mechanically, and not in true solution. This was confirmed by Basappa and Sreenivasamurthy (1977) who showed that sedimentation by centrifugation at 40 000 × g for 30 minutes removed a maximum of 65% of aflatoxin. (This compared with a maximum 50% removed by filtration.) The rest (35%) remained in solution in the supernatant. Studies with different adsorbants showed that activated fuller's earth (2%) increased sedimentation of aflatoxin to 92.5% at 10 000 × g for 30 minutes. Alternatively the use of casein or arecanut powder (35% total tannins), at 10% w/w, reduced the aflatoxin by 92% and 86% respectively but gave objectionable colour and flavour to the oil (Dwarakanath *et al.*, 1969).

Another recent study has shown hydrated sodium calcium aluminosilicate (HSCAS), used as an anti-caking agent in animal feed, to be a high affinity adsorbant for aflatoxin (Harvey *et al.*, 1989; Kubena *et al.*, 1990; Pettit *et al.*, 1990). Studies showed that more than 80% aflatoxin was bound to HSCAS. Less than 10% of this was subsequently extracted, indicative of the strong bonds that are formed. Kubena *et al.* (1990) found that the growth inhibitory effects of AFB1 were significantly diminished when 7.5 mg/kg AFB1 was added to poultry feed containing 0.5% HSCAS. Pigs fed a diet containing 3 mg/kg and 0.5% HSCAS showed no liver lesions (Harvey *et al.*, 1989). These findings suggest that HSCAS can reduce the bioavailability and therefore toxicity in growing animals. As such, it presents a novel approach to preventative management of the problem of aflatoxin contamination in such feeds.

13.10.2 Destruction of aflatoxin

Destruction of aflatoxin, whether by heat, chemical or biological methods, is another option. With such methods, however, there is always the possibility of diminished nutritive value or unwanted residues. Proteins are susceptible to denaturation; groundnut protein is also lysine-deficient, even before the terminal amine group of this amino acid is attacked by reactive chemicals.

Aflatoxin is very stable to heat at temperatures below 100 °C. Several studies have been carried out on the effects of dry- and oil-roasting on groundnut kernels. Lee *et al.* (1969) showed the average reduction of aflatoxin ranged from 43 to 83%, depending on time and temperature conditions and the level of toxin in the raw kernels. They reported an overall reduction of 65% in AFB1 for oil roasting compared with 69% for dry roasting, with the percentage reduction being highest at the higher aflatoxin contamination levels.

In a similar experiment, Pluyer *et al.* (1987) showed that oven-roasting at 150 °C for 30 minutes caused a 30–45% reduction of AFB1. They also showed that microwaving at 0.7 kW for 8.5 minutes gave a similar reduction in aflatoxin levels. (Another group reported a 95% reduction after 4 minutes at 6 kW, but to achieve this reduction at lower energy levels would lead to over-roasting the kernels.) Unlike Lee's group (1969), no correlation was found between the percentage destruction and the levels of aflatoxin in the samples.

Autoclaving wet meals reduces aflatoxin content but this can take up to 4 hours (Coomes *et al.*, 1966). Codifer *et al.* (1976) reduced this to 1 hour by the addition of 1.5% formaldehyde and 2% calcium hydroxide, but the nutritional value of the final product was in doubt.

Many chemicals have been screened for their potential to destroy aflatoxin (Feuell, 1966; Sreenivasamurthy *et al.*, 1967; Dollear *et al.*, 1968; Natarajan *et al.*, 1975b). Under alkaline conditions the lactone ring common to AFB1 and AFB2 hydrolyses to the hydroxy acid form, which is then susceptible to oxidation. Ammonia gas or ammonium hydroxide detoxify by alkaline hydrolysis of the lactone moiety. Vesonder *et al.* (1975) showed apparent detoxification with ammonium hydroxide at room temperature for 21 hours. Subsequent treatment with acid restored the AFB1 to its toxic form, which demonstrates the need for caution when assessing potential chemicals to process meals.

It is now known that several compounds are formed during the ammoniation process. If the process is carried out under mild conditions, then the reversibility demonstrated by Vesonder *et al.* (1975) can occur. However, when the reaction is allowed to proceed further, the compounds formed do not revert back to aflatoxin (Park *et al.*, 1988). Since ammoniation has been shown to reduce aflatoxin levels to less than 1%, it is becoming the method of choice. Ammonium hydroxide or gaseous ammonia can be used at either high temperature and pressure or low temperature and pressure (Dollear *et al.*, 1968; Gardner *et al.*, 1971; Viroben *et al.*, 1978; Delort-Laval *et al.*, 1980; Park *et al.*, 1988). Reaction products formed depend on the treatment and conditions used (Lee and Cucullu, 1978; Park *et al.*, 1988).

The ammoniation process reduces the amount of soluble nitrogen in the meal. Cystine undergoes partial destruction, which can be counterbalanced by the addition of synthetic methionine in animal diets. Protein efficiency ratios (PER) are generally lower, probably due to the decreased nitrogen solubility and available lysine levels (Gardner *et al.*, 1971; Viroben *et al.*, 1978; Delort-Laval *et al.*, 1980). However, Park *et al.* (1988) showed production parameters, e.g. milk and egg quality, to be significantly better or not adversely affected by the treatment. The total nitrogen content of the meals increased, resulting in positive weight gain and feed efficiency values in ruminants. Park *et al.* (1988) reported that either no effect or decreased weight gain or feed efficiency values were obtained with simple-stomached animals, which correlates with Dollear *et al.* (1968) who reported that treated meals gave lower weight gains with rats.

Frayssinet and Lafarge-Frayssinet (1990) showed the efficacy of ammoniation in a long-term feeding study with rats. The aflatoxin content decreased proportionately with the increased pressure of application. The degradation products of the reaction showed no carcinogenicity; the process did not produce any new toxin and no reversibility was observed.

There are conflicting reports as to the effectiveness of UV radiation in destroying the toxin (Feuell, 1966; Goldblatt, 1971). Shantha (1989) showed that the exposure of contaminated groundnut oil to bright sunlight for 15–30 minutes completely destroyed the toxin. Exposing it to UV light was far less effective. Gamma radiation at acceptable safety levels has been shown to be ineffective (Feuell, 1966; Chiou *et al.*, 1990). Van Dyck *et al.* (1982) had success with ionizing radiation in aqueous solutions, but the dose required to detoxify the aflatoxin (>10 kGy) was outside the permissible range for the irradiation of foodstuffs. Patel *et al.* (1989) expanded this work by demonstrating the synergistic effect of hydrogen peroxide and gamma radiation to degrade higher concentrations of aflatoxins at lower doses of radiation. This followed on from work by Sreenivasamurthy *et al.* (1967), who demonstrated the effectiveness of 6% hydrogen peroxide under alkaline conditions in destroying 85% of the toxin. Together, gamma radiation and hydrogen peroxide produce a higher amount of free radicals in aqueous solution than individually. This method is suited to groundnut kernels and has the advantage over ammoniation of being less time-consuming and a less expensive process. However, animal toxicity studies are required before commercial application is possible.

13.11 CONCLUSIONS

Aflatoxin contamination of groundnuts is the most significant economic problem facing the groundnut industry. Once groundnuts are contaminated, there are severe limitations placed on the removal of aflatoxin. For example, while solvent extraction may be applicable for the treatment of groundnut meals, it is not the method of choice for whole kernels. Apart from these individual limitations, there is a general one of cost; a detoxification step involving additional handling and processing would inevitably raise the final cost of the commodity. There is also the possible nutritive loss (Feuell, 1966; Goldblatt, 1971). Therefore it has to be stressed that prevention of contamination is still the best approach.

The marketable commodities produced from groundnuts, including edible oil, expeller meal, oil cake, confectionery products and peanut butter, each represent a different level in the trade chain. There is the potential for reduced export earnings and loss at each level if aflatoxin contamination is detected above certain limits.

An enormous range of aflatoxin concentrations in groundnut products has been reported between countries. It must be borne in mind that this could be due to variation in the sensitivity of the analytical methods employed rather than the true levels of contamination.

Many countries have now set limits for AFB1 and total aflatoxin in foodstuffs and feed, as well as allowable limits of AFM1 in dairy products (Van Egmond, 1989). Data collected over the last quarter of a century indicate that the risk from consumption of aflatoxin-contaminated groundnuts is likely to be more to animals than to humans (Bhat, 1989). Limits on animal feed not only protect the health and productivity of the animals consuming groundnut-containing feeds but also prevent secondary transmission to humans via the food chain.

The major problem lies with developing countries. The environmental conditions in these countries are often ideal for aflatoxin production. This, combined with poor management practices during production, harvest, handling and storage, must affect the degree of contamination. Farmers will surely sell their crops when heavier (and therefore insufficiently dry) if it means better financial returns. In many places up to 80% of the agricultural produce is consumed at home or at the village level, without entering the organized market chain (Bhat, 1989). This means that there is no dilution effect, and these people and livestock can be exposed to large and prolonged doses of aflatoxin. It is these same people (and their animals) who are often malnourished and therefore more susceptible to the toxin and its immunosuppressive effect.

ACKNOWLEDGEMENTS

The assistance of the staff of the Lincoln University Library in obtaining many of the references used in this work is gratefully acknowledged.

REFERENCES

Abalaka, J.A. and Elegbede, J.A. (1982) Aflatoxin distribution and total microbial counts in an edible oil extracting plant. I. Preliminary observations. *Food and Chemical Toxicology*, **20**, 43–46.

Aibara, K., Ichinoe, M, Maeda, K. *et al.* (1985) Storage conditions of imported raw shelled peanuts and production of aflatoxins. *Journal of the Food Hygienic Society of Japan*, **26**(3), 234–242.

Akano, D.A. and Atanda, O.O. (1990) The present level of aflatoxin in Nigerian groundnut cake ('Kulikuli'). *Letters in Applied Microbiology*, **10**, 187–189.

Allcroft, R., Carnaghan, R.B. and Sargeant, K. (1961) A toxic factor in Brazilian groundnut meal. *Veterinary Record*, **73**, 428–429.

Allcroft, R. and Carnaghan, R.B.A. (1962) Groundnut toxicity, *Aspergillus flavus* toxin in animal products. *Veterinary Record*, **74**, 863–864.

Allcroft, R. and Carnaghan, R.B.A. (1963) Toxic Products in Groundnuts: Biological effects. *Chemistry and Industry*, 12 January, 1963, pp. 50–53.

Allcroft, R. and Raymond, W.P. (1966) Toxic groundnut meal: biological and chemical assays of a large batch of a 'reference' meal used for experimental work. *Veterinary Record*, **79**, 122–123.

Allcroft, R. and Roberts, B.A. (1968) Toxic groundnut meal: the relationship between aflatoxin B_1 intake by cows and excretion of aflatoxin M_1 in milk. *Veterinary Record*, **82**, 116–118.

Amla, I., Shyamla, Y., Sreenivasamurthy, V. *et al.* (1970) Role of aflatoxin in Indian childhood cirrhosis. *Indian Paediatrics*, **7**, 262–270.

Amla, I., Kamala, C.S., Gopalakrishna, G.S. *et al.* (1971) Cirrhosis in children from peanut meal contaminated by aflatoxin. *American Journal of Clinical Nutrition*, **24**, 609–614.

Angsubhakorn, S., Get-Ngern, P., Miyamoto, M. and Bhamarapravati, N. (1990) A single dose-response effect of aflatoxin B_1 on rapid liver cancer induction in two strains of rats. *International Journal of Cancer*, **46**, 664–668.

Annau, E., Corner, A.H., Magwood, S.E. and Jericho, K. (1964) Electrophoretic and Chemical Studies on Sera of Swine Following the Feeding of Toxic Groundnut Meal. *Canadian Journal of Comparative Medicine and Veterinary Science*, **28**, 264–269.

Anon. (1984) Aflatoxins and kwashiorkor. *Lancet*, **2**, 1133–1134.

AOAC (1984) *Official Methods of Analysis*, 14th edn, Association of Official Analytical Chemists, Washington DC.

Asplin, F.D. and Carnaghan, R.B.A. (1961) The toxicity of certain groundnut meals for poultry with special reference to their effect on ducklings and chickens. *Veterinary Record*, **73**, 1215–1219.

Austwick, P.K.C. and Ayerst, G. (1963) Toxic Products in Groundnuts: Groundnut Microflora and Toxicity. *Chemistry and Industry*, January 12, 1963, 55–61.

Autrup, H., Seremet, T., Wakhisi, J. and Wasunna, A. (1987) Aflatoxin exposure measured by urinary excretion of aflatoxin B_1-guanine adduct and hepatitis B virus infection in areas with different liver cancer incidence in Kenya. *Cancer Research*, **47**, 3430–3433.

Azaizeh, H.A., Pettit, R.E., Sarr, B.A. and Phillips, T.D. (1990) Effect of peanut tannin extracts on growth of *Aspergillus parasiticus* and aflatoxin production. *Mycopathologia*, **110**(3), 125–132.

Basappa, S.C. and Sreenivasamurthy, V.S. (1977) State of aflatoxin in groundnut oil. *Journal of Food Science and Technology*, **14**(2), 57–60.

Baur, F.J. (1975) Effect of storage upon aflatoxin levels in peanut materials. *Journal of the American Oil Chemists' Society*, **52**(8), 263–265.

Becroft, D.M.O. and Webster, D.R. (1972) Aflatoxins and Reye's Disease. *British Medical Journal*, **4**, 117.

Beljaars, P.R., Fabry, F.H.M., Pickott, M.M.A. and Peeters, M.J. (1972) Quantitative fluorodensitometric measurement of aflatoxin B$_1$ with a flying-spot densitometer. (2) Comparative study of B$_1$ measurement in spiked and naturally contaminated peanut products. *Journal of Association of Official Analytical Chemists*, **55**(6), 1310–1315.

Beljaars, P.R., Verhulsdonk, C.A.H., Paulsch, W.E., and Liem, D.H. (1973) Collaborative study of two-dimensional thin layer chromatographic analysis of aflatoxin B$_1$ in peanut butter extracts, using the antidiagonal spot application technique. *Journal of the Association of Analytical Chemists*, **56**(6), 1444–1451.

Bennett, R.A., Essigmann, J.M. and Wogan, G.N. (1981) Excretion of an aflatoxin-guanine adduct in urine of aflatoxin B$_1$-treated rats. *Cancer Research*, **41**, 650–654.

Bhat, R.V. (1989) Risk to human health associated with consumption of groundnuts contaminated with aflatoxins, in *Aflatoxin contamination of groundnut: Proceedings of the International Workshop*, 6–9 Oct 1987, ICRISAT, Patancheru, pp. 19–30.

Blankenship, P.D., Cole, R.J., Sanders, T.H. and Hill, R.A. (1984) Effect of geocarpo-sphere temperature on pre-harvest colonization of drought-stressed peanuts by *Aspergillus flavus* and subsequent aflatoxin contamination. *Mycopathologia*, **85**, 69–74.

Blount, W.P. (1961) Turkey "X" Disease. *Journal of the British Turkey Federation*, **9**(2), 52–58, 61, 77.

Brown, G.H. (1984) The distribution of total aflatoxin levels in composited samples of peanuts. *Food Technology in Australia*, **36**(3),128–130.

Brudzynski, A., Pee, W. van and Kornaszewski, W. (1978) The occurrence of aflatoxin B$_1$ in peanuts, corn, and dried cassava sold at the local market in Kinshasa, Zaire; its coincidence with high hepatoma morbidity among the population. *Food Science and Technology Abstracts*, 10 12 C, 492.

Bushnell, D.G. (1965) The incidence of aflatoxin in the Rhodesian groundnut crop. *Rhodesia Agricultural Journal*, **62**, 94–96, 98.

Butler, W.H. and Barnes, J.M. (1968) Carcinogenic Action of Groundnut Meal Containing Aflatoxin in Rats. *Food and Cosmetics Toxicology*, **6**,135–141.

Campbell, T.C., Caedo, J.P.(Jr.), Bulatao-Jayne, J. *et al.* (1970) Aflatoxin M$_1$ in human urine. *Nature*, **227**, 403–404.

Campbell, T.C., Chen, J., Liu, C. *et al.* (1990) Nonassociation of aflatoxin with primary liver cancer in a cross-sectional ecological survey in the People's Republic of China. *Cancer Research*, **50**, 6882–6893.

Carnaghan, R.B.A. and Sargeant, K. (1961) The toxicity of certain groundnut meals to poultry. *Veterinary Record*, **73**, 726–727.

Carnaghan, R.B.A. (1967) Hepatic tumours and other chronic liver changes in rats following a single oral administration of aflatoxin. *British Journal of Cancer*, **21**, 811–814.

Chandra, R.K. (1970) Immunological picture in Indian childhood cirrhosis. *Lancet*, **1**, 537–540.

Chang-Yen, I and Felmine, J. (1987), Aflatoxin levels in selected bulk foods and feeds in Trinidad. *Tropical Agriculture (Trinidad)*, **64**(4), 283–286.

Chiou, R.Y.Y., Lin, C.M. and Shyu, S.L. (1990) Property characterization of peanut kernels subjected to gamma irradiation and its effect on the outgrowth and aflatoxin production by *Aspergillus parasiticus*. *Journal of Food Science*, **55**, 210–213.

Chu, J.S., Lee, R.C., Trucksess, M.W. and Park, D.L. (1988) Evaluation by enzyme-linked immunosorbent assay of cleanup for thin-layer chromatography of aflatoxin B$_1$ in corn, peanuts and peanut butter. *Journal of the Association of Official Analytical Chemists*, **71**(5), 953–956.

Coady, A. (1976) Tropical cirrhosis and hepatoma. *Journal of the Royal College of Physicians*, **10**, 133–144.

Codifer, L.P. Jr., Mann, G.E. and Dollear, F.G. (1976) Aflatoxin inactivation: treatment of peanut meal with formaldehyde and calcium hydroxide. *Journal of the American Oil Chemists' Society*, **53**(5), 204–206.

Coffey, M.T., Hagler, W.M. Jnr. and Cullen, J.M. (1989) Influence of Dietary Protein, Fat or Amino Acids on the Response of Weanling Swine to Aflatoxin B$_1$ *Journal of Animal Science*, **67**, 465–472.

Coker, R.D. (1979) Aflatoxin: past, present and future. *Tropical Science*, **21**(3),143–162.

Cole, R.J. (1989) Preharvest aflatoxin in peanuts. *International Biodeterioration*, **25**, 253–257.

Cole, D.L. and Masuka, A.J. (1989) Evaluation of New Rapid Methods for Aflatoxin Detection in Groundnuts in Zimbabwe. ICRISAT *Proceedings of the Third Regional Groundnut Workshop for Southern Africa*, 13–18 March 1988, Lilongwe, Malawi, pp. 185–189.

Cole, R.J., Sanders, T.H., Hill, R.A. and Blankenship, P.D. (1985) Mean geocarposphere temperatures that induce preharvest aflatoxin contamination of peanuts under drought stress. *Mycopathologia*, **91**, 41–46.

Cole, R.J., Hill, R.A., Blankenship, P.D. and Sanders, T.H. (1986a) Color mutants of *Aspergillus flavus* and *Aspergillus parasiticus* in a study of preharvest invasion of peanuts. *Applied and Environmental Microbiology*, **52**(5), 1128–1131.

Cole, R.J., Sanders, T.H., Blankenship, P.D. and Hill, R.A. (1986b) Environmental conditions required to induce preharvest aflatoxin contamination of peanuts. *Biodeterioration 6*. 6th International Biodeterioration Symposium, Washington, DC, August 1984 (eds S. Barry, D.R. Houghton) pp. 294–299.

Coomes, T.J., Crowther, P.C., Feuell, A.J. and Francis, B.J. (1966) Experimental detoxification of groundnut meals containing aflatoxin. *Nature*, **209**, 406–407.

Coon, F.B., Baur, F.J. and Symmes, L.R.L. (1972) International aflatoxin check sample program: 1971 study. *Journal of the Association of Official Analytical Chemists*, **55**(2), 315–327.

Cucullu, A.F., Lee, L.S., Mayne, R.Y. and Goldblatt, L.A. (1966) Determination of Aflatoxins in Individual Peanuts and Peanut Sections. *Journal of the American Oil Chemists' Society*, **43**, 89–92.

Cullen, J.M., Ruebner, B.H., Hsieh, L.S. *et al.* (1987) Carcinogenicity of dietary Aflatoxin M$_1$ in male Fischer rats compared to Aflatoxin B$_1$ *Cancer Research*, **47**, 1913–1917.

Davis, N.D., Guy, M.L. and Diener, U.L. (1980) A fluorometric rapid screen method for aflatoxin in peanuts. *Journal of the American Oil Chemists' Society*, **57**(3), 109–110.

de Iongh, H., Vles, R.O. and van Pelt, J.G. (1964) Milk of mammals fed on aflatoxin-containing diet. *Nature*, **202**, 466–467.

Delort-Laval, J., Viroben, G. and Borgida, L.P. (1980) Efficacité biologique pour le poulet de chair du tourteau d'arachide traité à l'ammoniac ou à la monométhylamine en vue de l'inactivation des aflatoxines. *Annales de Zootechnie*, **29**(4), 387–400.

Denning, D.W. (1987) Aflatoxin and human disease. *Adverse Drug Reaction and Acute Poisoning Reviews*, **4**, 175–209.

Denning, D.W., Allen, R., Wilkinson, A.P. and Morgan, M.R.A. (1990) Transplacental transfer of aflatoxin in humans. *Carcinogenesis*, **11**(6), 1033–1035.

de Vries, H.R., Maxwell, S.M. and Hendrickse, R.G. (1990) Aflatoxin excretion in children with kwashiorkor or marasmic kwashiorkor – a clinical investigation. *Mycopathologia*, **110**, 1–9.

Dichter, C.R. (1984) Risk estimates of liver cancer due to aflatoxin exposure from peanuts and peanut products. *Food and Chemical Toxicology*, **22**(6), 431–437.

Dickens, J.W. and Weltz, R.E. (1968) Detecting Farmers' Stock Peanuts containing Aflatoxin by Examination for Visible Growth of *Aspergillus flavus*. *Mycopathologia*, **37**, 65–69.

Dickens, J.W. and Whitaker, T.B. (1983) Dilution errors in aflatoxin determinations caused by compounds extracted from peanuts. *Journal of the Association of Official Analytical Chemists*, **66**(5), 1059–1062.

Diener, U.L. and Davis, N.D. (1968) Effect of environment on aflatoxin production in freshly dug peanuts. *Tropical Science*, **10**, 22–28.

DiProssimo, V.P. (1976) Distribution of aflatoxins in some samples of peanuts. *Journal of the Association of Official Analytical Chemists*, **59**(4), 941–944.

Dollear, F.G., Mann, G.E., Codifer, Jr., L.P. *et al.* (1968) Elimination of aflatoxins from peanut meal. *Journal of the American Oil Chemists' Society*, **45**, 862–865.

Dorner, J.W. and Cole, R.J. (1988) Rapid determination of aflatoxins in raw peanuts by liquid chromatography with postcolumn iodination and modified minicolumn cleanup. *Journal of the Association of Official Analytical Chemists*, **71**(1), 43–47.

Dorner, J.W., Cole, R.J., Sanders, T.H. and Blankenship, P.D. (1989) Interrelationship of kernel water activity, soil temperature, maturity, and phytoalexin production in preharvest aflatoxin contamination of drought-stressed peanuts. *Mycopathologia*, **105**(5), 117–128.

Dorner, J.E. and Cole, R.J. (1989) Comparison of two ELISA screening tests with liquid chromatography for determination of aflatoxins in raw peanuts. *Journal of the Association of Official Analytical Chemists*, **72**(6), 962–964.

Duthie, C.F., Lancaster, M.C., Taylor, J. *et al.* (1966) Toxic groundnut meal in feeds for pigs. (1) A trial made at two laboratories with pigs from about 40 to 200 lb. live weight fed to a restricted scale. *The Veterinary Record*, **79**(22), 621–625.

Duthie, I.F., Lancaster, M.C., Taylor, J. *et al.* (1968) Toxic groundnut meal in feeds for pigs. (2) The effect of consuming toxic groundnut meal during part of the growing period or during the finishing period. *Veterinary Record*, **82**, 427–430.

Dwarakanath, C.T., Sreenivusamurthy, V. and Parpia, A.B. (1969) Aflatoxin in Indian peanut oil. *Journal of Food Science and Technology*, **6**, 107–109.

Essigmann, J.M., Croy, R.G., Bennett, R.A. and Wogan, G.N. (1982) Metabolic activation of aflatoxin B_1: patterns of DNA Adduct Formation, Removal, and Excretion in Relation to Carcinogenesis. *Drug Metabolism Reviews*, **13**(4), 581–602.

Fehr, P.M., Bernage, L. and Vassilopoulos, V. (1968) Effet de la consommation de tourteau d'arachide pollué par *Aspergillus flavus* chez le ruminant en lactation. *Mémoires Originaux*, 377–390.

Ferrando, R., Palisse-Roussel, M. and Jacquot, L. (1984) Relay toxicology of aflatoxin M1 in dried milk. *Comptes Rendus Hebdomadaires des Séances Série III Sciences de la Vie*, **298**(13), 355–358.

Feuell, A.J. (1966) Aflatoxin in groundnuts. Part IX – Problems of detoxification. *Tropical Science*, **8**, 61–70.

Fong, L.Y.Y., Ton, C.C.T., Koonanuwatchaidet, P. and Huang, D.P. (1980) Mutagenicity of peanut oils and effect of repeated cooking. *Food and Cosmetic Toxicology*, **18**, 467–470.

Fong, L.Y.Y. and Chan, W.C. (1981) Long-Term effects of Feeding Aflatoxin-contaminated Market Peanut Oil to Sprague-Dawley Rats. *Food and Cosmetic Toxicology*, **19**, 179–183.

Frayssinet, C. and Lafarge-Frayssinet, C. (1990) Effect of ammoniation on the carcinogenicity of aflatoxin-contaminated groundnut oil cakes: long-term feeding study in the rat. *Food Additives and Contaminants*, **7**(1), 63–68.

Fukal, L., Prosek, J. and Sova, Z. (1987) The occurrence of aflatoxins in peanuts imported into Czechoslovakia for human consumption. *Food Additives and Contaminants*, **4**(3), 285–289.

Gardiner, E.E. (1962) A comparison of the toxicity to poults and chicks of a certain peanut oil meal. *Poultry Science*, **41**, 1348–1350.

Gardner, H.K. Jnr, Koltun, S.P. and Vix, H.L.E. (1968) Solvent extraction of aflatoxins from oilseed meals. *Journal of Agriculture and Food Chemistry*, **16**(6), 990–993.

Gardner, H.K. Jnr, Koltun, S.P., Dollear, F.G. and Rayner, E.T. (1971) Inactivation of Aflatoxins in Peanut and Cottonseed Meals by Ammoniation. *Journal of the American Oil Chemists' Society*, **41**, 70–73.

Gelda, C.S. and Luyt, L.J. (1977) Survey of total aflatoxin content in peanuts, peanut butter and other foodstuffs. *Annals of Nutrition and Metabolism*, **31**, 477–483.

Glen-Bott, A.M. (1987) Aspirin and Reye's Syndrome. A Reappraisal. *Medical Toxicology*, **2**, 161–165.

Goldblatt, A. (1971) Control and Removal of Aflatoxin. *Journal of the American Oil Chemists' Society*, **41**, 605–610.

Golden, M.H.N. and Golden, B.E. (1981) Trace elements. *British Medical Bulletin*, **37**, 32–34.

Groopman, J.D., Donahue, P.R., Zhu, J. *et al.* (1985) Aflatoxin metabolism in humans: detection of metabolites and nucleic acid adducts in urine by affinity chromatography. *Proceeding of the National Academy of Science*, **82**, 6492–6496.

Groopman, J.D., Donahue, P.R., Zhu, J. *et al.* (1987) Temporal patterns of aflatoxin metabolites in urine of people living in Guangxi Province, PRC. *Proceedings of the American Association of Cancer Research*, **28**, 130, Abstract no. 517.

Groopman, J.D. and Donahue, K.F. (1988) Aflatoxin, a human carcinogen: determination in foods and biological samples by monoclonal antibody affinity chromatography. *Journal of the Association of Official Analytical Chemists*, **71**(5), 861–867.

Groopman, J.D., Cain, L.G. and Kensler, T.W. (1988) Aflatoxin exposure in human populations: measurements and relationship to cancer. *CRC Critical Reviews in Toxicology*, **19**(2), 113–145.

Gumbmann, M.R. and Williams, S.N. (1969) Biochemical effects of aflatoxin in pigs. *Toxicology and Applied Pharmacology*, **15**, 393–404.

Gurtoo, H.L., Dahms, R.P. and Paigen, B. (1978) Metabolic activation of aflatoxins related to their mutagenicity. *Biochemical and Biophysical Research Communications*, **81**(3), 965–971.

Habish, H.A., Abdulla, M.H. and Broadbent, J.H. (1971) The incidence of aflatoxin in Sudanese groundnuts. *Tropical Science*, **13**(4), 279–287.

Harvey, R.B., Kubena, L.F., Phillips, T.D. *et al.* (1989) Prevention of Aflatoxicosis by addition of hydrated sodium calcium aluminosilicate to the diets of growing barrows. *American Journal Veterinary Research*, **50**(3), 416–420.

Hendrickse, R.G., Coulter, J.B.S., Lamplugh, S.M. *et al.* (1982) Aflatoxins and kwashiorkor: a study in Sudanese children. *British Medical Journal*, **285**, 843–846.

Hendrickse, R.G. (1988) Kwashiorkor and aflatoxins. *Journal of Paediatric Gastroenterology and Nutrition*, **7**, 633–636.

Hill, R.A., Blankenship, P.D., Cole, R.J. and Sanders, T.H. (1983) Effects of soil moisture and temperature on preharvest invasion of peanuts by the *Aspergillus flavus* group and subsequent aflatoxin development. *Applied and Environmental Microbiology*, **45**(2), 628–633.

Hintz, H.F., Booth, A.N., Cucullu, A.F. *et al.* (1967) Aflatoxin toxicity in swine. *Journal of PSEBM*, **124**, 266–268.

Hodges, R., Mortimer, P.H. and Drew Smith, J. (1964) Toxic Groundnuts offered for human consumption in New Zealand. *The New Zealand Veterinary Journal*, **12**, 19–20.

Holaday, C.E. (1968) Rapid method for detecting aflatoxins in peanuts. *Journal of the American Oil Chemists' Society*, **45**, 680–688.

Holaday, C.E. and Lansden, J. (1975) Rapid screening method for aflatoxin in a number of products. *Journal of Agriculture and Food Chemistry*, **23**(6), 1134–1136.

Horrocks, D., Burt, A.W.A., Thomas, D.C. and Lancaster, M.C. (1965) Effects of groundnut meal containing aflatoxin in cattle diets. *Animal Production*, **7**, 253–261.

Hsieh, L.-L., Hsu, S.-W., Chen, D.-S. and Santella, R.M. (1988) Immunological detection of aflatoxin B_1-DNA adducts formed *in vivo*. *Cancer Research*, **48**, 6328–6331.

Hurst, W.J. and Toomey, P.B. (1978) Determination of aflatoxins in peanut products using reverse phase HPLC. *Journal of Chromatographic Science*, **16**(8), 372–376.

Jarvis, B. (1971) Factors affecting the production of mycotoxins. *Journal of Applied Bacteriology*, **34**(1), 199–213.

Joffe, A.Z. (1969) Relationships between *Aspergillus flavus*, *A. niger* and some other fungi in the mycoflora of groundnut kernels. *Plant and Soil XXXI*, **1**, 57–63.

Kannaiyan, J., Sandhu, R.S. and Phiri, A.L. (1989) Aflatoxin and *Aspergillus flavus* contamination problems of groundnuts in Zambia, in *Aflatoxin contamination of groundnut: Proceedings of the International Workshop, 6–9 Oct. 1987*, ICRISAT, Patancheru, pp. 65–70.

Keen, P. and Martin, P. (1971) The toxicity and fungal infestation of foodstuffs in Swaziland in relation to harvesting and storage. *Tropical and Geographical Medicine*, **23**(1), 35–43.

Knutti, R. and Schlatter, C. (1982) Distribution of aflatoxin in whole peanut kernels, sampling plans for small samples. *Zeitschrift fur Lebensmittelchemie Unters Forsch*, **174**, 122–128.

Korobkin, M. and Williams, E.H. (1968) Hepatoma and groundnuts in the West Nile district of Uganda. *Yale Journal of Biology and Medicine*, **41**, 69–78.

Krishnamachari, K.A., Bhat, R.V., Nagaragan, V. and Tilak, T.B. (1975) Hepatitis due to aflatoxicosis – an outbreak in West India. *Lancet*, **1**, 1061–1063.

Kubena, L.F., Harvey, R.B., Phillips, T.D. *et al.* (1990) Diminution of Aflatoxicosis in growing Chickens by the dietary addition of a hydrated, sodium calcium aluminosilicate. *Poultry Science*, **69**, 727–735.

Lamplugh, S.M., Hendrickse, R.G., Apaegyei, F. and Mwanmut, D.D. (1988) Aflatoxins in breast milk, neonatal cord blood, and serum of pregnant women. *British Medical Journal*, **296**, 968.

Lancaster, M.C., Jenkins, F.P. and Philp, J.McL. (1961) Toxicity associated with certain samples of groundnuts. *Nature*, **192**, 1095–1096

Lazaro, F., de Castro, M.D.L. and Valcarcel, M. (1988) Fluorimetric determination of aflatoxins in foodstuffs by high-performance liquid chromatography with flow injection analysis. *Journal of Chromatography*, **448**, 173–181.

Lee, L.S., Cucullu, A.F., Franz, A.O. Jr. and Pons, W.A. Jr. (1969) Destruction of aflatoxins in peanuts during dry and oil roasting. *Journal of Agriculture and Food Chemistry*, **17**(3), 451–453.

Lee, L.S. and Cucullu, A.F. (1978) Conversion of aflatoxin B_1 to aflatoxin D_1 in ammoniated peanut and cottonseed meals. *Journal of Agriculture and Food Chemistry*, **26**(4), 881–884.

Lee, H.-S, Sarosi, I. and Vyas, G.N. (1989) Aflatoxin B_1 formamidopyrimidine adducts in human hepatocarcinogenesis: A preliminary report. *Gastroenterology*, **97**, 1281–1287.

Lim, Han Kuo and Yeap, Gim Sai. (1966) The occurrence of aflatoxin in Malayan imported oil cakes and groundnut kernels. *The Malaysian Agricultural Journal*, **45**(3), 232–244.

Lipsky, M.M., Cole, K.E., Hsu, I.-C. *et al.* (1990) Interspecies comparisons of *in vitro* hepatocarcinogenesis. *Progress in Clinical and Biological Research*, **331**, 395–408.

Llewellyn, G.C., Johnson, R.H. and O'Rear, C.E. (1983) A model for evaluating aflatoxin occurrence in shelled peanuts, in *Biodeterioration 5*, (ed. T.A. Oxley and S. Barry), 5th International Biodeterioration Symposium, Aberdeen, pp. 638–651.

Llewellyn, G.C., O'Rear, C.E., Sherertz, P.C. *et al.* (1988) Aflatoxin contamination of Virginia peanuts for the crop-years 1982–1986. *International Biodeterioration*, **24**(4 & 5), 399–407.

Loosmore, R.M. and Markson, L.M. (1961) Poisoning of cattle by Brazilian groundnut meal. *Veterinary Record*, **73**, 813–814.

Loosmore, R.M. and Harding, J.D.J. (1961) A toxic factor in Brazilian groundnuts causing liver damage in pigs. *Veterinary Record*, **73**, 1362–1364.

Lotlikar, P.D., Raj, H.G., Bohm, L.S. *et al.* (1989) A mechanism of inhibition of aflatoxin B_1-DNA binding in the liver by phenobarbital pretreatment of rats. *Cancer Research*, **49**, 951–957.

Lovelace, C.E.A. and Aalbersberg, W.G.L. (1989) Aflatoxin levels in foodstuffs in Fiji and Tonga Islands. *Plant Foods for Human Nutrition*, **39**, 393–399.

Lutwick, L.I. (1979) Relation between aflatoxin, hepatitis-B virus and hepatocellular carcinoma. *Lancet*, **1**, 755–775.

McLean. A.E.M. and Marshall. A. (1971) Reduced carcinogenic effects of aflatoxin in rats given phenobarbitone. *British Journal of Experimental Pathology*, **52**, 322–329.

Manzo, S.K. and Misari, S.M. (1989) Status and management of aflatoxin in groundnuts in Nigeria, in *Aflatoxin contamination of groundnut: Proceedings of the International Workshop, 6–9 Oct. 1987*, ICRISAT, Patancheru, pp. 77–90.

Mehan, V.K., Bhavanishankar, T.N.I. and Bedi, J. (1985) Comparison of different methods of extraction and estimation of aflatoxin B_1 in groundnut. *Journal of Food Science and Technology*, **22**(2), 123–125.

Mortimer, D.N., Shepherd, M.J., Gilbert, J. and Clark, C. (1988) Enzyme-linked immuno-sorbent (ELISA) determination of aflatoxin B_1 in peanut butter, collaborative trial. *Food Additives and Contaminants*, **5**(4), 601–608.

Moss, E. and Neal, G.E. (1985) The metabolism of aflatoxin B_1 by human liver. *Biochemical Pharmacology*, **34**(17), 3193–3197.

Murthy, T.R.K., Jammali, M., Henry, Y. and Frayssinet, C. (1975) Aflatoxin residues in tissues of growing swine: effect of separate and mixed feeding of protein and protein-free portions of the diet. *Journal of Animal Science*, **41**(5), 1339–1347.

Nakazato, M., Morozumi, S., Saito, K. *et al.* (1990) Interconversion of aflatoxin B_1 and aflatoxicol by several fungi. *Applied and Environmental Microbiology*, **56**(5), 1465–1470.

Natarajan, K.R., Rhee, K.C., Cater, C.M. and Mattil, K.F. (1975a) Distribution of aflatox-ins in various fractions separated from raw peanuts and defatted peanut meal. *Journal of the American Oil Chemists' Society*, **52**(2), 44–47.

Natarajan, K.R., Rhee, K.C., Cater, C.M. and Mattil, K.F. (1975b) Destruction of aflatoxins in peanut protein isolates by sodium hypochlorite. *Journal of the American Oil Chemists' Society*, **52**(5), 160–163.

Newberne, P.M. and Butler, W.H. (1969) Acute and chronic effects of aflatoxin on the liver of domestic and laboratory animals. A review. *Cancer Research*, **29**, 236–250.

Ngindu, A., Johnson, B.K., Kenya, P.R. *et al.* (1982) Outbreak of acute hepatitis caused by aflatoxin poisoning in Kenya. *Lancet*, **1**, 1346–1348.

O'Brien, K., Moss, E., Judah, D. and Neal, G. (1983) Metabolic basis of the species difference to aflatoxin B_1 induced hepatoxicity. *Biochemical and Biophysical Research Communications*, **114**(2), 813–821.

Paget, G.E. (1954) Exudative hepatitis in guinea-pigs. *Journal of Pathology and Bacteriology*, **67**, 393–400.

Park, D.L., Lee, L.S., Price, R.L. and Pohland, A.E. (1988) Review of the Decontamination of Aflatoxins by Ammoniation: Current Status and Regulation. *Journal of the Association of Official Analytical Chemists*, **71**(4), 685–703.

Park, D.L., Miller, B.M., Nesheim, S. *et al.* (1989) Visual and semiquantitative spectro-photometric ELISA screening method for aflatoxin B_1 in corn and peanut products: follow-up collaborative study. *Journal of the Association of Official Analytical Chemists*, **72**(4), 638–643.

Park, D.L., Nesheim, S., Trucksess, M.W. *et al.* (1990) Liquid Chromatographic Method for Determination of Aflatoxins B_1, B_2, G_1 and G_2 in Corn and Peanut Products: Collaborative Study. *Journal of the Association of Official Analytical Chemists*, **73**(2), 260–266.

Parker, W.A. and Melnick, D. (1966) Absence of Aflatoxin from Refined Vegetable Oils. *Journal of the American Oil Chemists' Society*, **42**, 635–638.

Patel, U.D., Govindarajan, P. and Dave, P.J. (1989) Inactivation of aflatoxin B_1 by using the synergistic effect of hydrogen peroxide and gamma radiation. *Applied and Environmental Microbiology*, **55**(2), 465–467.

Paterson, J.S., Crook, J.C., Shand, A. *et al.* (1962) Groundnut toxicity as the cause of exudative hepatitis (oedema disease) of guinea-pigs. *Veterinary Record*, **74**, 639–640.

Patey, A.L., Sharman, M., Wood, R. and Gilbert, J. (1989) Determination of aflatoxin concentrations in peanut butter by enzyme-linked immunosorbent assay (ELISA): Study of three commercial ELISA kits. *Journal of the Association of Official Analytical Chemists*, **72**(6), 965–969.

Patey, A.L., Sharman, M. and Gilbert, J. (1991) Liquid chromatographic determination of aflatoxin levels in peanut butters using an immunoaffinity column cleanup method: International Collaborative trial. *Journal of the Association of Official Analytical Chemists*, **74**(1), 76–81.

Pettit, R.E. (1986) Incidence of aflatoxin in groundnuts as influenced by seasonal changes in environmental conditions – A review, in *Agrometeorology of Groundnut*, Proceedings of an International Symposium, 21–26 August, 1985, ICRISAT Sahelian Center, Niamey, Niger. ICRISAT, Patancheru, pp. 163–174.

Pettit, R.E., Azaizeh, H.A., Taber, R.A. *et al.* (1989) Screening groundnut cultivars for resistance to *Aspergillus flavus, Aspergillus parasiticus*, and aflatoxin contamination, in *Aflatoxin contamination of groundnut: Proceedings of the International Workshop, 6–9 Oct. 1987*, ICRISAT, Patancheru, pp. 291–303.

Pettit, R.E., Sarr, B.A., Machen, M.D. and Phillips, T.D. (1990) Detection and detoxification of aflatoxin-contaminated groundnut products in West Africa, in *Summary Proceedings of the First ICRISAT Regional Groundnut Meeting for West Africa*, 13–16 September, 1988, pp. 97–98.

Platonow, N. and Beauregard, M. (1965) Feeding of ferrets with the raw meat and liver of chickens chronically poisoned with toxic groundnut meal. *Canadian Journal of Comparative Medicine and Veterinary Science*, **29**, 63–65.

Pluyer, H.R., Ahmed, E.M. and Wei, C.I. (1987) Destruction of aflatoxins on peanuts by oven and microwave-roasting. *Journal of Food Protection*, **50**(6), 504–508.

Pons, W.A., Jr. and Franz, A.O., Jr. (1978) High pressure liquid chromatographic determination of aflatoxins in peanut products. *Journal of the Association of Official Analytical Chemists*, **61**(4), 793–800.

Portman, R.S., Plowman, K.M. and Campbell, T.C. (1968) Aflatoxin metabolism by liver microsomal preparations of two different species. *Biochemical and Biophysical Research Communications*, **33**(5), 711–715.

Ramsdell, H.S. and Eaton, D.L. (1990) Species susceptibility of aflatoxin B$_1$ carcinogenesis: comparative kinetics of microsomal biotransformation. *Cancer Research*, **50**, 615–620.

Ray, A.C., Abbitt, B., Cotter, S.R. *et al.* (1986) Bovine abortion and death associated with consumption of aflatoxin-contaminated peanuts. *Journal of the American Veterinary Medical Association*, **188**(10), 1187–1188.

Rayner, E.T. and Dollear, F.G. (1968) Removal of aflatoxins from oilseed meals by extraction with aqueous isopropanol. *Journal of American Oil Chemists' Society*, **45**, 622–624.

Reed, J.D. and Kasali, O.B. (1989) Hazards to livestock of consuming aflatoxin-contaminated groundnut meal in Africa, in *Aflatoxin contamination of groundnut: Proceedings of the International Workshop, 6–9 Oct. 1987*, ICRISAT, Patancheru, pp. 31–38.

Robinson, P. (1967) Infantile cirrhosis of the liver in India with special reference to probable aflatoxin etiology. *Clinical Paediatrics*, **6**, 57–62.

Roebuck, B.D. and Wogan, G.N. (1977) Species comparison of *in vitro* metabolism of aflatoxin B$_1$. *Cancer Research*, **37**, 1649–1656.

Rogan, W.R., Yang, G.C. and Kimbrough, R.D. (1985) Aflatoxin and Reye's Syndrome: A Study of Livers from Deceased Cases. *Archives of Environmental Health*, **40**(2), 91–95.

Ryan, N.J., Hogan, G.R., Hayes, W. *et al.* (1979) Aflatoxin B$_1$: Its Role in the Etiology of Reye's Syndrome. *Paediatrics*, **64**(1), 71–75.

Salhab, A.S. and Edwards, G.S. (1977) Comparative *in vitro* metabolism of aflatoxicol by liver preparations from animals and humans. *Cancer Research*, **37**, 1016–1021.

Sanders, T.H., Davis, N.D. and Diener, U.L. (1968) Effect of carbon dioxide, temperature and relative humidity on production of aflatoxin in peanuts. *Journal of the American Oil Chemists' Society*, **45**, 683–685.

Sanders, T.H., McMeans, J.L. and Davidson, J.I. (1984) Aflatoxin content of peanut hulls. *Journal of the American Oil Chemists' Society*, **61**(12), 1839–1841.

Sanders, T.H., Cole, R.J., Blankenship, P.D. and Hill, R.A. (1985a) Relation of environmental stress duration to *Aspergillus flavus* invasion and aflatoxin production in preharvest peanuts. *Peanut Science*, **12**, 90–93.

Sanders, T.H., Blankenship, P.D., Cole, R.J. *et al.* (1985b) Conclusive test for aflatoxin resistance in peanuts. *Journal of the American Oil Chemists' Society*, **62**(4), 654–655.

Sargeant, K., O'Kelly, J., Carnaghan, R.B.A. and Allcroft, R. (1961a) The assay of a toxic principle in certain groundnut meals. *Veterinary Record*, **73**(46), 1219–1223.

Sargeant, K., Allcroft, R. and Carnaghan, R.B.A. (1961b) Groundnut toxicity. *Veterinary Record*, **73**, 865.

Sargeant, K., Sheridan, A. and O'Kelly, J. (1961c) Toxicity associated with certain samples of groundnuts. *Nature*, **192**, 1096–1097.

Sargeant, K., Carnaghan, R.B.A. and Allcroft, R. (1963) Toxic Products in Groundnuts: Chemistry and Origin. *Chemistry and Industry*, 12 January, 1963, 53–55.

Sarnaik, S., Godbole, S.H. and Kanekar, P. (1988) Incidence of toxigenic *Aspergillus flavus* in marketed edible vegetable oils. *Current Science*, **57**(24), 1336–1337.

Schindler, A.F., Palmer, J.G. and Eisenberg, W.V. (1967) Aflatoxin production by *Aspergillus flavus* as related to various temperatures. *Applied Microbiology*, **15**, 1006–1009.

Schroeder, H.W. and Boller, R.A. (1973) Aflatoxin production of species and strains of the *Aspergillus flavus* group isolated from field crops. *Applied Microbiology*, **25**(6), 885–889.

Serck-Hanssen, A. (1970) Aflatoxin-induced fatal hepatitis? A case report from Uganda. *Archives of Environmental Health*, **20**, 729–731.

Shantha, T. (1989) Detoxification of groundnut seed and products in India, in *Aflatoxin contamination of groundnut: Proceedings of the International Workshop, 6–9 Oct. 1987*, ICRISAT, Patancheru, pp. 153–160.

Shayiq, R.M. and Avadhani, N.G. (1989) Purification and characterization of a hepatic mitochondrial cytochrome P450 active in aflatoxin B_1 metabolism. *Biochemistry*, **28**, 7546–7554.

Shimada, T. and Guengerich, F.P. (1989) Evidence for cytochrome P-450$_{nf}$, the nifedipine oxidase, being the principal enzyme involved in the bioactivation of aflatoxins in human liver. *Proceedings of the National Academy of Science*, USA, **86**, 462–465.

Shotwell, O.L. and Holaday, C.E. (1981) Minicolumn detection methods for aflatoxin in raw peanuts: collaborative study. *Journal of the Association of Official Analytical Chemists*, **64**(3), 674–677.

Siller, W.G. and Ostler, D.C. (1961) The histopathology of an entero-hepatic syndrome of turkey poults. *The Veterinary Record*, **73**(7), 134–138.

Sisk, D.B., Carlton, W.W. and Curtin, T.M. (1968) Experimental aflatoxicosis in young swine. *American Journal of Veterinary Research*, **29**, 1591–1602.

Siwela, A.H. and Caley, A.D. (1989) Aflatoxin contamination of stored groundnuts in Zimbabwe, in *Aflatoxin contamination of groundnut: Proceedings of the International Workshop, 6–9 Oct. 1987*, ICRISAT, Patancheru, pp. 59–64.

Smith, J.W., Hill, C.H. and Hamilton, P.B. (1971) The effect of dietary modifications on aflatoxicosis in the broiler chicken. *Poultry Science*, **50**, 768–774.

Sreenivasamurthy, V., Parpia, H.A.B., Srikanta, S. and Shankar Murti, A. (1967) Detoxification of aflatoxin in peanut meal by hydrogen peroxide. *Journal of the Association of Official Analytical Chemists*, **50**(2), 350–354.

Stanton, D.W. (1977) A survey of some foods for aflatoxins. *Food Technology in New Zealand*, **12**(4), 25.

Stoloff, L. (1989) Aflatoxin is not a probable human carcinogen: the published evidence is sufficient. *Regulatory Toxicology and Pharmacology*, **10**, 272–283.

Strange, R.N. (1984) The possible role of phytoalexins in the resistance of groundnuts to *Puccinia arachidis* Speg. Groundnut Rust Disease, in *Groundnut Rust Disease: Proceedings of a Discussion Group Meeting 24–28 September, 1984*, ICRISAT, Patancheru, pp. 167–174.

Strzelecki, E.L. and Cader-Strzelecka, B. (1988) A survey of aflatoxin levels in peanut meal imported into Poland for animal feedingstuffs. *Food Additives and Contaminants*, **5**(4), 597–599.

Strzelecki, E.L., Gasiorowska, U., Gorazdowska, M. and Cader-Strzelecka, B. (1990) Contamination of groundnut meal by aflatoxin metabolites of *Aspergillus flavus*. *Microbios Letters*, **43**, 7–10.

Subrahmanyam, P., Mehan, V.K., Nevill, D.J. and McDonald, D. (1980) Research on

Fungal Diseases of Groundnut at ICRISAT, in *Proceedings of the International Workshop on Groundnuts*, 13–17 October, ICRISAT, Patancheru, pp. 197–198.

Subrahmanyam, P. (1990) Groundnut Diseases in West Africa and their management. Research problems, priorities and future strategies, in *Summary Proceedings of the First ICRISAT Regional Groundnut Meeting for West Africa*, 13–16 September, 1988, pp. 3–6.

Syamasundara Rao, P. (1970) Study of liver injury in mice fed with spoiled groundnuts. *Journal of the Indian Medical Association*, **54**(1), 6–11.

Sylos, C.M. and Rodriguez-Amaya, D.B. (1989) Inexpensive, rapid screening method for aflatoxins in peanuts and peanut products. *Journal of the Science of Food Agriculture*, **49**, 167–172.

Trucksess, M.W., Stack, M.E., Nesheim, S. *et al.* (1989) Enzyme-linked immunosorbent assay of aflatoxins B_1, B_2, and G_1 in corn, cottonseed, peanuts, peanut butter, and poultry feed: collaborative study. *Journal of the Association of Official Analytical Chemists*, **72**(6), 957–961.

Trucksess, M.W., Young, K., Donahue, K.F. *et al.* (1990) Comparison of two immunochemical methods with thin-layer chromatographic methods for determination of aflatoxins. *Journal of the Association of Official Analytical Chemists*, **73**(3), 425–429.

Trucksess, M.W., Stack, M.E., Nesheim, S. *et al.* (1991) Immunoaffinity column coupled with solution fluorometry or liquid chromatography postcolumn derivatization for determination of aflatoxins in corn, peanuts, and peanut butter: Collaborative Study. *Journal of the Association of Official Analytical Chemists*, **74**(1), 81–87.

Tsubouchi, H., Yamamoto, K., Hisada, K. and Sakabe, Y. (1983) Degradation of aflatoxins by *Aspergillus niger* and aflatoxin non-producing *Aspergillus flavus*. *Journal of the Food Hygienic Society of Japan*, **24**(2), 113–119.

Tupule, P.G. (1969) Aflatoxicosis. *Indian Journal of Medical Research*, **57**(8), 102–114.

van der Linde, J.A., Frens, A.M., de Iongh, H. and Vles, R.O. (1964) Oonderzoek van melk afkomstig van koejen gevoed met aflatoxinehoudend grondnotenmeel. (Inspection of milk from cows fed aflatoxin-containing groundnut meal). *Tijdschrift, Diergeneesk*, **89**(15), 1082–1088.

Van Dyck, P.J., Tobback, P., Feys, M. and van de Voorde, H. (1982) Sensitivity of aflatoxin B_1 to ionizing radiation. *Applied and Environmental Microbiology*, **43**(6), 1317–1319.

Van Egmond, H.P. (1989) Current situation on regulations for mycotoxins. Overview of tolerances and status of standard methods of sampling and analysis. *Food Additives and Contaminants*, **6**(2), 139–188.

Van Egmond, H.P. and Wagstaffe, P.J. (1989) Aflatoxin B_1 in peanut meal reference materials: intercomparisons of methods. *Food Additives and Contaminants*, **6**(3), 307–319.

Varsavsky, E. and Sommer, S.E. (1977) Determination of aflatoxins in peanuts. *Annals of Nutrition and Metabolism*, **31**(4/6), 539–544.

Vesonder, R.F., Beckwith, A.C., Ciegler, A. and Dimler, R.J. (1975) Ammonium hydroxide treatment of aflatoxin B_1. Some chemical characteristics and biological effects. *Journal of Agriculture and Food Chemistry*, **23**, 242–243.

Viroben, G., Delort-Laval, J., Colin, J. and Adrian, J. (1978) Inactivation des aflatoxines par traitement a l'ammoniac études *in vitro* de tourteaux d'arachide détoxiqués. *Annales de la Nutrition et de l'Alimentation*, **32**(1), 167–185.

Waltking, A.E. (1980) Sampling and preparation of samples of peanut butter for aflatoxin analysis. *Journal of the Association of Official Analytical Chemists*, **63**(1), 103–106.

Whitaker, T.B., Dickens, J.W., Monroe, R.J. and Wiser, E.H. (1972) Comparison of the observed distribution of aflatoxin in shelled peanuts to the negative binomial distribution. *Journal of the American Oil Chemists' Society*, **49**(10), 590–593.

Whitaker, T.B. and Dickens, J.W. (1983) Comparison of the amounts of aflatoxin extracted from raw peanuts using AOAC methods I and II. *Peanut Science*, **10**(2), 52–54.

Whitaker, T.B., Dickens, J.W. and Giesbrecht, F.G. (1984) Effects of methanol concentration and solvent: peanut ratio on extraction of aflatoxin from raw peanuts. *Journal of the Association of Official Analytical Chemists*, **67**(1), 35–36.

Whitaker, T.B., Dickens, J.W. and Chew, V. (1985) Development of statistical models to simulate the testing of farmers stock peanuts for aflatoxin using visual, thin layer chromatography, and minicolumn methods. *Peanut Science*, **12**, 94–98.

Whitaker, T.B. and Dickens, J.W. (1986) Efficacy of the visual minicolumn and thin layer chromatography methods to test farmers stock peanuts for aflatoxin. *Peanut Science*, **13**, 74–77.

Whitaker, T.B., Dickens, J.W. and Giesbrecht, F.G. (1986) Optimum methanol concentration and solvent/peanut ratio for extraction of aflatoxin from raw peanuts by modified AOAC Method II. *Journal of the Association of Official Analytical Chemists*, **69**(3), 508–512.

Whitaker, T.B. and Dickens, J.W. (1989) Simulation of aflatoxin testing plans for shelled peanuts in the United States and in the export market. *Journal of the Association of Official Analytical Chemists*, **72**(4), 644–648.

Wild, C.P., Pionneau, F., Montesano, R. *et al.* (1987). Aflatoxin detected in human breast milk by immunoassay. *International Journal of Cancer*, **40**(3), 328–333.

Wild, C.P., Jiang, Y.-Z., Sabbioni, G. *et al.* (1990) Evaluation of methods for quantitation of aflatoxin-albumin adducts and their application to human exposure assessment. *Cancer Research*, **50**, 245–251.

Wilkinson, A.P., Denning, D.W. and Morgan, M.R.A. (1988) An ELISA method for the rapid and simple determination of aflatoxin in human serum. *Food Additives and Contaminants*, **5**(4), 609–619.

Wilson, D.M., Jay, E. and Hill, R.A. (1985) Microflora changes in peanuts (groundnuts) stored under modified atmospheres. *Journal of Stored Product Research*, **21**(1), 47–52.

Wilson, D.M. (1989) Analytical methods for aflatoxins in corn and peanuts. *Archives of Environmental Contamination and Toxicology*, **18**, 308–314.

Wogan, G.N. and Newberne, P.M. (1967) Dose response characteristics of aflatoxin B_1 carcinogenesis in the rat. *Cancer Research*, **27**, 2370–2376.

Wogan, G.N. (1968) Aflatoxin risks and control measures. *Federation Proceedings*, **27**(3), 932–938.

Xiao, D.R. (1989) Research on aflatoxin contamination of groundnut in the People's Republic of China, in *Aflatoxin contamination of groundnut: Proceedings of the International Workshop, 6–9 Oct 1987*, ICRISAT, Patancheru, pp. 95–100.

Yabe, K., Ando, Y. and Hamasaki, T. (1988) Biosynthetic relationship among aflatoxins B_1, B_2, G_1 and G_2. *Applied and Environmental Microbiology*, **54**, 2101–2106.

Yadgiri, B., Reddy, V., Tulpule, P.G. *et al.* (1970) Aflatoxin and Indian childhood cirrhosis. *American Journal of Clinical Nutrition*, **23**, 94–98.

Yndestad, M. and Underdal, B. (1975) Aflatoxin in foods on the Norwegian market. *Nordic Veterinary Medicine*, **27**, 42–48.

CHAPTER 14

Groundnut breeding

T.G. Isleib, J.C. Wynne and S.N. Nigam

14.1 INTRODUCTION

Groundnuts (*Arachis hypogaea* L.) are grown throughout the tropical and warm temperate regions of the world, with commercial production principally between latitudes 40° N and 40° S. Leading producing nations are India (33.4% of global production), China (27.8%), USA (9.3%), Senegal (4.2%), Indonesia (4.2%), Nigeria (3.3%), Myanmar (3.0%), Sudan (2.7%) and Argentina (2.0%). Clearly, the crop is grown in several agroecological systems and under numerous socioeconomic environments. Yield of groundnuts is often low due to diseases and insects, unpredictable and variable rainfall, inability to apply improved agronomic practices and production technology, lack of cultivars adapted to local conditions, low financial inputs, lack of small-scale farm implements, and lack of the infrastructure required to supply quality seed of improved cultivars (Nigam *et al.*, 1991).

The primary objectives of groundnut breeders are to develop cultivars with high yield potential, adaptation to specific environments and production systems, resistance or tolerance to environmental stresses and resistance to diseases and insects. Because groundnuts are grown under many different cropping systems across a wide array of agroecological conditions, the specific objectives of breeding programmes vary considerably. Breeding is a continuing process as the crop is introduced to new environments and production systems, as market demands change, and as disease and insect pest populations shift in reaction to deployment of new cultivars. As the primary constraints to production are overcome by new cultivars and production practices, breeding for improved flavour and quality desired by processors and consumers becomes more important (Bunting *et al.*, 1985).

The Groundnut Crop: A scientific basis for improvement. Edited by J. Smartt. Published in 1994 by Chapman & Hall, London. ISBN 0 412 408201.

TABLE 14.1 *Botanical division of* A. hypogaea *L.*

A. *hypogaea* L.	Characteristics	Secondary centres of diversity
subsp. *hypogaea*	No flowering on mainstem Alternate branching	
var. *hypogaea*	Two seeds per pod	Bolivian Amazonian
var. *hirsuta* Kohler	Long mainstem Three or more seeds per pod	Peruvian
subsp. *fastigiata* Waldron	Flowering on mainstem Sequential branching	
var. *fastigiata*	Limited vegetative branching Three or more seeds per pod	Goiás and Minas Gerais Guaranian Goiás and Minas Gerais Peruvian North-east Brazil
var. *vulgaris* Harz	Short runs of reproductive branches Two seeds per pod	Goiás and Minas Gerais Guaranian Peruvian

14.2 VARIABILITY IN GROUNDNUT GERMPLASM

In order to develop cultivars with traits that overcome the constraints peculiar to a specific environment, there must be sufficient genetic variation to allow selection for desired traits. Assertions by American researchers regarding the paucity of genetic variation in groundnut referred to specific economically important characters of the extant cultivars and breeding stocks within market classes in the USA (Gregory *et al.*, 1973). More recently, molecular analyses have not detected significant amounts of variability in allozymes (Grieshammer and Wynne, 1990), restriction fragment length polymorphisms (Kochert *et al.*, 1991) or DNA fragments amplified by polymerase chain reaction (Halward *et al.*, 1992) in cultivated germplasm of broadly diverse origin.

The most commonly used botanical division of *A. hypogaea* into subspecies and varieties is that of Krapovickas (1968), based on patterns of reproductive and vegetative branching and on pod morphology as summarized in Table 14.1.

Because of strong local preferences for particular pod and seed characteristics, early breeders of groundnut often worked with limited numbers of parents possessing attributes acceptable to local consumers or processors. In the USA, market classes of groundnut roughly follow the botanical divisions of the cultivated species with the following exceptions. Firstly, the runner and virginia market classes are commonly equated with var. *hypogaea*, but do not have purely *hypogaea* ancestry. Both have had substantial introgression of fastigiate germplasm, primarily from spanish ancestors (var. *vulgaris*), in the course of plant improvement through breeding (Isleib and Wynne, 1992). Spanish parents were used to increase oil content, shorten maturity and increase the harvest index of the crop. The most common runner and virginia groundnut cultivars in the USA have 0–50% fastigiate ancestry (Table 14.2) and average 35%. Secondly, groundnuts of var. *hirsuta* are not represented in any market class, nor has *hirsuta* germplasm been used in the development of any released cultivars or registered germplasm to date. *Hirsuta* types are extremely rare in the USA national collection and in the global collection maintained at the International Crops Research Institute for the Semi-Arid Tropics (ICRISAT) in Andhra Pradesh, India. Chinese scientists report that 'peruvian-type' groundnuts are grown commonly in their central production region, but it is not clear whether the peruvian type mentioned is *hirsuta* or *fastigiata* with the typical peruvian pod configuration. *Hirsuta* groundnuts remain a garden crop in north-western South America and in Mexico (D. Williams, personal communication), and collection efforts should be focused there.

It was not until the late 1970s that the extent of natural genetic variability available to groundnut breeders was fully appreciated and widely recognized (Norden, 1980). Collections of cultivated groundnuts were considered extensive by the mid 1970s (Banks, 1976). Many of these and other accessions in germplasm collections in the USA or at ICRISAT since 1976 were obtained from expeditions made to South America, the centre of origin and diversity for groundnuts, under the sponsorship of the US Department of Agriculture (USDA) and the International Board for Plant Genetic Resources (IBPGR) in co-operation with state experiment stations in the USA and several other countries. The most important expeditions were those of Archer in 1936; Stephens and Hartley in 1947–48; Gregory, Krapovickas, Pietrarelli and others in 1959, 1961, and 1967; Hammons, Langford, Krapovickas, Pietrarelli and others in 1968; and Gregory, Banks, Simpson, Krapovickas, Pietrarelli and others in 1976, 1977, 1979 and 1980 (Wynne and Gregory, 1981) as well as those of IBPGR/CENARGEN/ICRISAT teams made during 1989 (Simpson, 1990). Collection of South American genetic resources continues today with particular emphasis on the wild species native to areas of Brazil undergoing rapid development. Much of the germplasm from Africa, an important centre of secondary variation (Gibbons *et al.*, 1972), was introduced into the United States by Smartt in 1959 (Wynne and Gregory, 1981). These accessions are maintained by the Southern Regional Plant Introduction Station at Experiment, Georgia. The most extensive collection of cultivated groundnut germplasm is now maintained by ICRISAT.

The Genetic Resources Unit there maintains a global collection of more than 12 000 accessions (Nigam *et al.*, 1991). In addition to the cultivated germplasm, there are more than 70 wild species of *Arachis* (Stalker and Moss, 1987). Some of the wild species have direct value as forages (Prine *et al.*, 1981) but for the most part they constitute a genetic reservoir of useful characteristics for the improvement of the cultivated groundnut, notably with respect to host-plant resistance to diseases and insects but perhaps also for agronomic traits (Guok *et al.*, 1986).

Although it was widely recognized that there was tremendous morphological variation among the accessions of cultivated groundnuts, it has only been during the last decade that the extent of the desirable variation has been demonstrated (Wynne *et al.*, 1991). As late as 1973, the widely held view was that there were many defects, with respect to the requirements of man, in the genetic composition of groundnuts (Gregory *et al.*, 1973). This view resulted from inadequate collection and evaluation of the germplasm of groundnuts (Wynne and Halward, 1989b). Systematic and extensive screening of the cultivated germplasm was not practised until ICRISAT adopted groundnut as a mandate crop in 1976 (Nigam *et al.*, 1991).

Several groundnut breeders had large collections of groundnuts by the 1970s but these were inadequately evaluated due to limited funds and personnel (Norden, 1980). Although many of the efforts to evaluate groundnut germplasm have not been systematic or exhaustive, a large number of accessions of cultivated groundnuts have been identified that contain desirable variation for yield, fruit size, morphological traits, tolerance to environmental stresses, disease and insect resistances, and seed composition.

In many developing nations, groundnut's primary use is as an oilseed either for domestic use or for export. Other countries view groundnut primarily as a food crop. In either case, the composition of groundnut has become an issue of increasing importance in the past ten years. For use as an oilseed, the market demands groundnuts with high oil content and good storability. Where groundnut is used as a food, whether as whole or processed seeds, attributes important to the consumer such as flavour, protein quality and shelf life have long been primary concerns of the groundnut marketing and processing industries. Shelf life is largely a function of the time required for auto-oxidation of linoleic fatty acid in the oil fraction of the seed to produce a characteristic rancid flavour. Increased use of oxygen-permeable packaging materials by groundnut processors has increased the need to extend shelf life through genetic improvement of the seed itself. Oleic acid, the 18-carbon mono-unsaturated (18:1) precursor to linoleic acid (18:2), is less reactive with oxygen and is therefore more desirable in the fatty acid profile of groundnut oil. A commonly used index of the storability of groundnut is the ratio of oleic to linoleic fatty acids (O/L ratio). This ratio ranges in value from under 1 to over 2 in cultivars used in the USA (Brown *et al.*, 1975; Ahmed and Young, 1982). O/L ratio

TABLE 14.2 *Fastigiate ancestry of runner- and virginia-type groundnut cultivars currently or formerly grown in the United States*

Cultivar	Market class	Share of US certified area[1]	Specific fastigiate ancestry	Total fastigiate ancestry
Dixie Runner	Runner	0%	50% Small White Spanish 3x-1	50%
Early Runner	Runner	0%	50% Small White Spanish 3x-2	50%
Florunner	Runner	47.4% (61.8%)	37.5% Small White Spanish 3x-2 12.5% Spanish 18–38	50%
GK 7	Runner	7.0% (9.1%)	15.625% Small White Spanish 3x-2 15.625% Spanish 18–38	31.25%
MARC I	Runner	0% (New release)	1.5625% McSpan Spanish 3.125% Pearl 6.25% Small White Spanish 3x-1 21.875% Small White Spanish 3x-2 7.8125% Spanish 18–38	40.625%
Okrun	Runner	3.7% (4.8%)	50% Argentine 18.75% Small White Spanish 3x-2 6.25% Spanish 18–38	75%
Southern Runner	Runner	2.2% (2.9%)	18.75% Small White Spanish 3x-2 6.25% Spanish 18–38	25%
Sunrunner	Runner	3.5% (4.5%)	25% Small White Spanish 3x-2 12.5% Spanish 18–38	37.5%

TABLE 14.2 *Cont.*

Cultivar	Market class	Share of US certified area[1]	Specific fastigiate ancestry	Total fastigiate ancestry
Florigiant	Virginia	0.3% (1.9%)	25% Small White Spanish 3x-2 12.5% Spanish 18-38	37.5%
NC 2	Virginia	0%	25% Spanish 18-38	25%
NC 5	Virginia	0%	25% Improved Spanish 2B	25%
NC 6	Virginia	1.3% (8.0%)		0%
NC 7	Virginia	5.2% (33.4%)	6.25% Improved Spanish 2B 25% Small White Spanish 3x-2 12.5% Spanish 18-38	43.75%
NC 9	Virginia	3.6% (23.0%)	12.5% Small White Spanish 3x-2 18.75% Spanish 18-38	31.25%
NC 10C	Virginia	2.2% (13.9%)	6.25% Improved Spanish 2B 18.75% Small White Spanish 3x-2 12.5% Spanish 18-38	37.5%
NC-V11	Virginia	3.0% (19.1%)	6.25% Improved Spanish 2B 12.5% Small White Spanish 3x-2 6.25% Spanish 18-38 25% PI 337396 (var. *fastigiata*)	50%
VA-C 92R	Virginia	0% (New release)	6.25% Improved Spanish 2B 12.5% Small White Spanish 3x-2 9.375% Spanish 18-38	28.125%

[1]Proportion of total area certified and (in parentheses) proportion of market class area certified.

is commonly used as a criterion for release of new cultivars in the USA with high values viewed as desirable by the groundnut industry. It is interesting to note that this demand for mono- rather than di-unsaturated fat in groundnut contradicts the general demand for less saturated oils in components of human diets. Norden *et al.* (1987) reported a groundnut variant with an extremely high O/L ratio greater than 30. Moore and Knauft (1989) determined that this trait was governed by two recessive genes and is therefore easily transferable to existing cultivars through backcrossing.

Recently, concern over the high contribution of fats to daily caloric intake by the populations of industrialized nations has created demand for low- or reduced-fat foods. The range of oil content in cultivars in the USA is 43.6–55.5% (Norden *et al.*, 1982). Among over 6000 groundnut accessions evaluated at ICRISAT, the range was 31.8–55.0% (ICRISAT, personal communication). This range is too high to permit reference to groundnut, even at the lower extreme of the distribution of fat contents, as a low-fat food, but it is clear that selection for reduced oil content should be effective.

14.2.1 Foliar fungal pathogens

Much of the screening of the groundnut germplasm for desirable variation during recent years has emphasized biotic stresses. This work was recently reviewed (Nigam *et al.*, 1991; Wynne *et al.*, 1991). Three foliar fungal diseases – late leaf spot [*Phaeoisariopsis personata* (*Mycosphaerella berke leyi*)], early leaf spot [*Cercospora arachidicola* (*Mycosphaerella arachidis*)] and rust (*Puccinia arachidis*) – are the most widely distributed and economically important diseases of groundnut. They are common wherever groundnuts are grown but they vary in incidence and severity among locations and years. Each disease alone can cause severe damage but yield losses are generally increased when they occur together. For example, rust and late leaf spot together can cause up to 70% yield loss in India (Subrahmanyam *et al.*, 1984). These diseases also affect seed grade adversely and they markedly reduce haulm yields – an effect that is of particular importance in those regions of the semi-arid tropics where small farmers maintain significant numbers of livestock.

Effective field screening methods have been developed for use in areas where natural disease pressure is high or where such pressure can be artificially induced. At the ICRISAT Center, field screening with infector rows is used to challenge host plants in a worst-case situation (Subrahmanyam *et al.*, 1982a). Genotypes and breeding populations to be screened are planted in a nursery together with rows of highly susceptible cultivars arranged systematically throughout the nursery. The ratio of test and infector rows varies from season to season and location to location. A mixture of short- and long-season susceptible cultivars is used to ensure inoculum supply for a longer period. Plants in infector rows are inoculated with spore suspensions to enhance disease development. This procedure is most successful if infector rows are inoculated in the evening immediately following overhead irrigation. Potted 'spreader' plants heavily infected with rust are also placed systematically throughout the field to provide another source of inoculum. The nursery may be irrigated by overhead sprinklers until harvest as required by climatic conditions.

Disease reaction on test plants is scored using a nine-point scale (Subrahmanyam et al., 1982a, b). Disease scores are recorded about 10 days before harvest in preliminary screening and at several growth stages in advanced screening and other studies. These techniques are useful for grouping lines into resistant and susceptible classes but not for identifying moderate levels of resistance. Germplasm and advanced breeding lines can also be screened in the glasshouse using potted plants or in the laboratory using detached leaves to measure components of disease resistance such as latent period, lesion number, lesion size and sporulation rate.

Screening methods similar to those used for rust are also used for late leaf spot using plots of test genotypes interspersed at regular intervals with susceptible infector rows inoculated with late leaf spot spores. Additional inoculum is provided by scattering on the infector rows leaf debris collected from infected plants in the previous season.

In the USA, field methods for identifying moderate resistance to leaf spots generally require isolation of test genotypes from one another to minimize the effect of adjacent plots. In North Carolina, isolation has been accomplished with border rows of non-host crop species such as maize (*Zea mays* L.), soybean (*Glycine max* [L.] Merr.) or cotton (*Gossypium hirsutum* L.) with eight border rows at 90 cm spacing between adjacent plots and 6–7 m of border between plots occupying the same rows in the field. Each plot is inoculated with spore suspension to ensure the presence of the pathogen, and disease progress is monitored after inoculation. Defoliation of the mainstem is the primary criterion of resistance and is expressed as a proportion or percentage of nodes defoliated. Isolation of the plots reduces the influence of neighbouring susceptible plants on accessions with partial resistance.

Sources of resistance to rust were reported by Bromfield and Cevario (1970). Hammons (1977) summarized the screening of groundnuts for rust resistance and concluded that resistant sources originated from three sources: Tarapoto (PIs 259747, 341879, 350680, 381622, and 405132), Israeli line 136 (PIs 298115 and 315608), and DHT 200 (PI 314817). Tarapoto and DHT 200 both originated in Peru. ICRISAT has screened more than 12 000 accessions of groundnut for rust resistance in the field, using infector rows to develop disease pressure (Subrahmanyam and McDonald, 1983),and 124 lines have been found with rust resistance (Nigam et al., 1991). These include 14 rust-resistant lines released jointly by USDA and ICRISAT (Subrahmanyam and McDonald, 1983). In addition, several wild *Arachis* species and their interspecific derivatives with cultivated groundnuts have been screened for resistance to rust under both field and laboratory environments (Subrahmanyam *et al.*, 1983c). The rust resistance identified in the cultigen is of a 'slow-rusting' type. Resistant lines exhibit increased incubation period, decreased infection frequency and reduced pustule size, spore production and spore germinability (Subrahmanyam *et al.*, 1983a, b). Many wild *Arachis* species, and lines derived from their hybridization with the cultigen, have been screened for resistance to rust under field and laboratory conditions. Accessions of several species were found to be immune to rust: *A. batizocoi* (PI 298639, PI 338312), *A. duranensis* (PI 219823), *A. cardenasii* (PI 262141), *A. chacoensis* (PI 276235), *A. pusilla* (PI 338449), *A. villosa* (PI 210554) and *A. correntina* (PI 331194) among others (Subrahmanyam *et al.*, 1983c). Most of the interspecific derivatives showed a high degree of resistance to rust. They had small and slightly depressed uredinia that did not rupture to release the comparatively few uredospores produced.

Screening for resistance to the leaf spots caused by *C. arachidicola* and *P. personata* has been extensive in recent years. Several sources of resistance to both early and late leaf spots have been reported (Foster *et al.*, 1980, 1981; Gorbet *et al.*, 1982; Hassan and Beute, 1977; Melouk *et al.*, 1984; Subrahmanyam *et al.*, 1985). Screening for late leaf spot resistance has been most extensive at ICRISAT where the 12 000 or more genotypes screened for rust have also been screened for late leaf spot. Fifty-three accessions of *A. hypogaea* have now been identified with documented resistance to late leaf spot (Nigam *et al.*, 1991) and 29 of these 53 lines are also resistant to rust (Table 14.3). Resistance to late leaf spot operates through much the same mechanisms as resistance to rust (Subrahmanyam *et al.*, 1982b).

Among the many accessions of wild *Arachis* species tested at ICRISAT Center, *A. chacoense* (PI 276325), *A. cardenasii* (PI 262141) and *A. stenosperma* (PI 338280) of section *Arachis* combined cross-compatibility with the cultigen and immunity or high resistance to the pathogen. Highly resistant wild species from other sections included *A. repens*, *A. appressipila*, *A. paraguariensis*, *A. villosulicarpa*, *A. hagenbeckii* and *A. glabrata* (Subrahmanyam *et al.*, 1985).

Several lines, including NC 3033, NC 5, PI 270806, GP-NC 343, PI 109839, PI 259747 and PI 350680, have been shown to possess epidemiological components of rate-reducing resistance to early leaf spot in the USA (Foster *et al.*, 1981; Green and Wynne, 1987; Hassan and Beute, 1977; Sowell *et al.*, 1976). Some of these lines (NC 3033, PI 270806, PI 259747 and PI 350680) did not show resistance in India or Malawi when infector-row inoculation techniques were used (Nigam, 1987; ICRISAT, 1984). Because early leaf spot does not usually occur readily in the field at Patancheru, screening has been less extensive at ICRISAT than for either rust or late leaf spot, but incidences of heavy disease at ICRISAT Center in

1983 and 1987 were utilized to screen germplasm already planted in the field. Also in 1987, screening for early leaf spot resistance on a limited scale was initiated by ICRISAT in Pantnagar, India, where *C. arachidicola* occurs more regularly. Of 3000 genotypes screened for early leaf spot, several showed moderate levels of field resistance at both locations (Waliyar *et al.*, 1989):

ICG 2711 (NC 5)
ICG 6709 (NC Ac 16163)
ICG 7291 (PI 262128)
ICG 7406 (PI 262121)
ICG 7630
ICG 7892 (PI 393527-B)
ICG 9990.

In Malawi, screening for early leaf spot resistance has not identified significant sources of resistance. More than 1000 selected germplasm lines of the cultigen have been screened individually but none showed any appreciable level of resistance to the disease. In 1986–87, 'bulk' testing was utilized to evaluate a large number of lines: 110 bulk populations were constructed by compositing five seeds from each of 100 lines. All lines in a given bulk population shared a common botanical variety. This method allowed representation of 11 000 lines in the screening although the identities of individual lines were lost. Only two bulks had a few plants which merited further testing. In the 1987–88 season, component lines of the two bulks were planted separately and scored for the disease. Only three germplasm lines – ICG 50, ICG 84 and ICG 11282 – were retained for further testing. Other lines that retained a higher than usual proportion of foliage despite heavy disease pressure (ICRISAT, 1986) were:

ICGM 189 (ICG 5216, PI 262087)
ICGM 197 (ICG 6012, NC Ac 16142, PI 262093)
ICGM 281 (ICG 8515)
ICGM 284 (ICG 8521)
ICGM 285 (ICG 8522)
ICGM 286 (ICG 8523)
ICGM 292 (ICG 8529)
ICGM 300 (ICG 8569, NC Ac 868, PI 119072)
ICGM 473 (ICG 3431)
ICGM 500 (ICG 3150)
ICGM 525 (ICG 6151).

Thirty-five lines reported to have resistance to early leaf spot at ICRISAT Center were not resistant in Malawi (ICRISAT, 1989).

Several wild species of *Arachis*, including *A. cardenasii*, have been reported to be resistant to early leaf spot; however, only *Arachis* species 30003 has shown consistent resistance when tested in Malawi using

TABLE 14.3 *Sources of resistance to both rust (Puccinia arachidis) and late leaf spot (Phaeoisariopsis personatum) available at ICRISAT Centre in 1989*

| | | | | | Disease score[2] | |
| | | | | | Rust | Late leaf spot |
ICG No.	Identity	Botanical variety	Seed colour	Origin		
1703	NC Ac 17127	*fastigiata*	Variegated	Peru	4.7	5.0
1707	NC Ac 17132	*fastigiata*	Purple	Peru	4.0	4.0
1710	NC Ac 17135	*fastigiata*	Dark purple	Peru	4.0	4.0
2716	EC 76446 (292)	*fastigiata*	Dark purple	Uganda	3.3	4.7
3527	USA 63	*fastigiata*	Purple	USA	4.7	4.7
4747	PI 259747	*fastigiata*	Purple	Peru	3.7	4.0
4995	NC Ac 17506	*fastigiata*	Purple	Peru	4.3	4.3
6022	NC Ac 927	*fastigiata*	Purple	Sudan	4.0	4.0
6330	PI 270806	*hypogaea*	Tan	Zimbabwe	2.1	3.3
6340	PI 350680	*fastigiata*	Dark purple	Honduras	3.0	4.0
7013	NC Ac 17133(RF)	*fastigiata*	Dark purple	India	3.3	4.0
7881	PI 215696	*fastigiata*	Dark purple	Peru	4.3	3.7
7884	PI 341879	*fastigiata*	Purple	Israel	3.0	3.7
7885	PI 381622	*fastigiata*	Purple	Honduras	3.0	4.3
7886	PI 390593	*fastigiata*	Tan	Peru	4.7	3.3
7894	PI 393641	*fastigiata*	Variegated	Peru	4.0	4.7
7897	PI 405132	*fastigiata*	Purple	Peru	2.7	4.0

TABLE 14.3 *Cont.*

| ICG No.[1] | Identity | Botanical variety | Seed colour | Origin | Disease score[2] | |
					Rust	Late leaf spot
10010	PI 476143	*fastigiata*	Variegated	Peru	4.0	5.0
10023	PI 476152	*fastigiata*	Tan	Peru	4.3	4.7
10028	PI 476163	*fastigiata*	Purple	Peru	4.7	5.0
10029	PI 476164	*fastigiata*	Tan	Peru	4.3	5.0
10035	PI 476172	*fastigiata*	Purple	Peru	4.0	3.7
10889	PI 476016	*fastigiata*	Red	Peru	3.3	4.3
10915	PI 476148	*fastigiata*	Variegated	Peru	2.3	5.0
10936	PI 476168	*fastigiata*	Dark purple	Peru	4.3	4.0
10940	PI 476173	*fastigiata*	Variegated	Peru	2.3	5.0
10941	PI 476174	*fastigiata*	Light purple	Peru	4.7	4.7
11182	PI 476174	*fastigiata*	Tan	Peru	2.7	5.0
11485	--	*fastigiata*	Light purple	Peru	5.0	3.7
Susceptible check cultivars:						
221	TMV 2	*vulgaris*	Tan	India	8.3	8.0
799	Robut 33–1	*hypogaea*	Tan	India	7.7	7.3

[1] ICRISAT groundnut accession number.
[2] Scored on a modified 9-point disease scale where 1 = 0%, 2 = 1 to 5%, 3 = 6 to 10%, 4 = 11 to 20%, 5 = 21 to 30%, 6 = 31 to 40%, 7 = 41 to 60%, 8 = 61 to 80% and 9 = 81 to 100% damage to foliage (ICRISAT Center, rainy season 1989).

TABLE 14.4 *Reaction of some groundnut germplasm lines with resistance to early (Cercospora arachidicola) and late leaf spot (*Phaeoisariopsis personatum*) and rust (*Puccinia arachidis*), ICRISAT Center, rainy season 1987*

Entry	Original name	Disease reaction[1]		
		Early leaf spot	Late leaf spot	Rust
ICG 1703	NC Ac 17127	4.7	5.0	4.7
ICG 6284	NC Ac 17500	5.0	7.0	3.3
ICG 7340	198/66 Coll 182	5.7	5.1	2.7
ICG 9294	58–295	5.1	6.0	2.7
ICG 10010	PI 476143	5.7	5.1	4.1
ICG 10040	PI 476176 (SPZ 451)	5.0	4.7	3.7
ICG 10900	PI 476033	5.3	4.7	4.1
ICG 10946	PI 476176	5.0	6.0	4.1
Susceptible controls				
ICG 799	Kadiri 3 (Robut 33–1)	8.0	7.0	7.0
ICG 221	TMV 2	8.0	8.0	8.0
Mean (n=500)		6.9	6.5	4.9
Standard error[2]		±0.48	±0.7	±1.1
CV (%)		7.0	11	22

[1] Mean of 3 plots, each 2 4-m rows, rated on a 1–9 scale where 1=no disease and 9=50–100% foliar destruction.
[2] Standard error and CV calculated on the basis of all 500 genotypes tested.

infector-row inoculation techniques (ICRISAT, 1989). Among other species, *A. chacoensis* and *A.* sp. 30085 showed high promise in the first year of screening but were susceptible in subsequent tests. *A. stenosperma* was found to be highly susceptible in Malawi (ICRISAT, 1988), contrary to reports from the USA. Several interspecific derivatives were found to retain more foliage than the susceptible control cultivar.

Table 14.4 shows that eight lines of groundnuts with moderate to high levels of resistance to all three foliar diseases – rust, early leaf spot and late leaf spot – have been identified (ICRISAT, 1988; Waliyar *et al.*, 1989). The rust and late leaf spot reactions of most accessions are stable over a wide range of geographic locations. Only for NC Ac 17090 and PI 298115 has variation in rust reaction been observed across locations. Reaction to early leaf spot has exhibited greater variation across locations. The eight lines in Table 14.4 are potentially the most useful parental lines available, since the foliar diseases generally occur in combination.

The genetics of resistance to these three diseases is not well understood. Bromfield and Bailey (1972) first reported that resistance to rust in the cultigen was controlled by two recessive genes. However, Nigam *et al.*

(1980) found continuous variation in the progeny of crosses among rust-resistant FESR lines (Bailey *et al.*, 1973) and suspected that rust resistance, though recessive in nature, might be governed by more than two genes. In generation means analysis of resistant-by-susceptible crosses, Reddy *et al.* (1987) found additive, additive-by-additive and additive-by-dominance effects for rust resistance. In some diploid wild *Arachis* species, resistance appeared to be partially dominant (Singh *et al.*, 1984).

Nevill (1982) studied five F_2 progenies from crosses between two resistant and three susceptible cultivars for components of resistance to late leaf spot in detached leaf tests. To account for the observed distribution of phenotypic values in the F_2, he postulated a five-locus polygenic system assuming resistance to be completely recessive. Non-additive gene action was concluded to be extremely important but its nature could not be elucidated due to the omission of the F_1 generation from the study.

14.2.2 Viral pathogens

Variation for resistance to several virus diseases has been reported in groundnut (Nigam *et al.*, 1991). The crop is host to several viruses but only a few are considered economically important. These include groundnut rosette (GRV) in Africa, bud necrosis (BNV) in India, tomato spotted wilt (TSWV) in the USA, peanut mottle (PMV) worldwide, peanut stripe (PStV) in east and south-east Asia, and peanut clump (PCV) in West Africa and India. Laboratory and field screening techniques have been developed for all these virus diseases. Resistance to rosette virus was discovered in local land races in Burkina Faso in the 1950s (de Berchoux, 1958, 1960). Of seven wild species of *Arachis* screened in an SADCC–ICRISAT regional groundnut project, two species (*A.* sp. 30003 and *A.* sp. 30017) remained symptom-free throughout the season. The apparent immunity of *A.* sp. 30003 to rosette and its high resistance to early leaf spot suggest that efforts to use this species should be emphasized (Bock, 1989).

Several groundnut accessions with consistently low symptoms of bud necrosis have been identified at ICRISAT, including:

C102
C121
C136
GP-NC 343
NC Ac 2232
NC Ac 2242
NC Ac 17888
ICGV 86029
ICGV 86031.

Only ICGV 86029 and 86031 showed tolerance to the virus (Nigam *et al.*, 1991). Southern Runner, a cultivar with resistance to late leaf spot, has shown fewer symptoms of tomato spotted wilt virus than other cultivars in the USA.

Peanut stripe, both aphid-transmitted and seed-borne, is composed of strains which can be distinguished on the basis of differential host reaction. Over 9000 lines of *A. hypogaea* were screened at two sites in Indonesia without any resistance being found (Nigam *et al.*, 1991). A few wild species have shown resistance with one species, *A. cardenasii*, being immune (Stalker and Moss, 1987).

Peanut mottle virus (PMV) disease of groundnut is widespread and generally present in varying intensity in all major groundnut-growing areas of the world. It can cause up to 30% loss in yield (Kuhn and Demski, 1975). Because PMV's foliar symptoms are inconspicuous, it has not received much attention in crop improvement programmes. Infected plants show mild mottling and vein clearing in newly formed leaves. Older leaves show upward curling and interveinal depression with occasional dark green islands. Infected plants are not severely stunted and older plants seldom show typical symptoms. The virus is sap-transmitted and its vectors are *Aphis craccivora*, *A. gossypii* and *Myzus persicae* among others. It is also seed-transmitted in a range from 0.1% to 3.5%, depending on the groundnut genotype (Ghanekar, 1980).

From a 5-year study on PMV epidemiology in Georgia, USA, Kuhn and Demski (1975) concluded that the initial inoculum of the disease in the field came from seedlings originating from infected seeds. Taking a lead from this observation, the groundnut group at ICRISAT adopted the approach of combining resistance/tolerance to PMV with absence of seed transmission in the disease resistance breeding. Limited breeding efforts are under way to achieve this objective.

A rapid method of field inoculation has been developed (Ghanekar, 1980), by means of which about 1000 plants can be inoculated in 1 hour with 80% infection frequency. The method involves the spraying of extracts from infected leaves, prepared in phosphate buffer containing celite and mercaptoethanol, onto test plants through a fine nozzle under pressure of 50 psi. More than 2500 germplasm lines of *A. hypogaea* have been screened in the field. No line has shown resistance to the virus; however, many germplasm lines suffered much lower yield loss than controls. Two germplasm lines, NC Ac 2240 and NC Ac 2243, have shown significantly low yield loss due to disease over years (ICRISAT, 1983). A few breeding lines have also shown tolerance to the disease.

Fifty wild *Arachis* species accessions have been screened for PMV resistance under glasshouse conditions using mechanical leaf rub and air brush inoculations. Of these, only two species, *A. chacoensis* (10602) and *A. pusilla* (12911), remained free from infection even after repeated graft inoculations (Subrahmanyam *et al.*, 1985).

Seeds of PMV-infected plants of several germplasm lines were screened in the laboratory for virus presence, using ELISA (Reddy, 1980). With this technique, 1000 seeds can be screened in 2 days. A small portion of cotyledon is adequate for the test. Two rust-resistant germplasm lines, EC 76446(292) and NC Ac 17133(RF), have failed to show any seed transmission in repeated tests over years on seeds totalling more than 13 000 (ICRISAT, 1988). A recently released Indian cultivar and many breeding lines with these rust-resistant parents in their ancestry also have shown no seed transmission. Lines with low yield loss and no seed transmission characteristics have been crossed and advanced generation lines are in field tests for measuring yield loss due to the disease. Promising lines from these tests will be studied for non-seed transmission in the laboratory.

Peanut clump virus (PCV) disease has been reported from West Africa (Trochain, 1931; Bouhot, 1967) and India (Sundararaman, 1926; Reddy *et al.*, 1979). The virus is soil-borne and seed-transmitted (ICRISAT, 1986). Infected plants are severely stunted with small, dark green leaves. The young tetrafoliolate leaves show mosaic mottling and chlorotic rings. Roots become dark in colour and the outer layers peel off easily. Most of the early-infected plants fail to produce pods. Even in case of late infection, losses of up to 60% are recorded. The virus has many serologically distinct isolates which produce varying severity of disease on groundnut varieties and different reactions on diagnostic hosts. A few soil fungi and nematode species have been suspected as possible vectors of the virus. Studies in India have shown that *Polymixa graminis*, a soil fungus, can transmit the virus (ICRISAT, 1988).

The disease occurs in both warm summer and rainy season crops. The extent of area infected with the disease is not well documented. Individual fields can become severely infected with the virus, forcing farmers to abandon groundnut cultivation in those fields. Chemicals such as Nemagone®, Temik®, and Carbofuran® can greatly reduce disease and increase yields. However, these chemicals are expensive for most farmers of the semi-arid tropics. Solarization treatment of the infected areas of the field greatly reduces the disease incidence (ICRISAT, 1987).

More than 7000 germplasm lines of the cultivated groundnut species *Arachis hypogaea* have been screened in farmers' diseased fields in the Indian states of Punjab and Andhra Pradesh. None of these lines showed resistance to the virus. A few lines showed tolerance to the disease as they did not suffer severely in growth and yield. Of 38 wild *Arachis* species and their 200 interspecific derivatives tested, only *Arachis* species 30036 did not become infected in the field (ICRISAT, 1985). Due to the genetic complexity of virus populations and lack of high-level tolerance in germplasm, no resistance breeding activity has been started for this disease.

14.2.3 Soil-borne pathogens

Screening of groundnut germplasm for resistance to soil-borne diseases has been less extensive than screening for resistance to foliar fungal pathogens because of the local prevalence of most soil-borne diseases. Nevertheless, variation for reaction to several soil-borne diseases has been found in groundnut. Resistance to bacterial wilt caused by *Pseudomonas solanacearum* was reported in the 1920s by Dutch scientists working in East Java (Indonesia) (Buddenhagen and Kelman, 1964). The disease occurs in several Asian and African countries but significant losses are reported only for Indonesia and China. Numerous resistant genotypes have been identified in those two countries (Nigam *et al.*, 1991).

From screening in North Carolina, USA, a few virginia and several spanish genotypes were reported to be resistant to *Cylindrocladium crotalariae*, which causes cylindrocladium black rot disease (CBR) (Green *et al.*, 1983). NC 3033, a line resistant to CBR, was also found to be resistant to *Sclerotium rolfsii*, the causal organism of southern stem rot (Beute *et al.*, 1976).

Toalson, PI 341885 and TxAG-3 (a selection from PI 365553) were found to be resistant to southern stem rot and pythium pod rot caused by *Pythium myriotylum* in Texas (Smith *et al.*, 1989). Resistance has also been found to *Sclerotinia minor* in screening studies in Oklahoma and Virginia (Coffelt and Porter, 1982). Sources of resistance include Chico, germplasm from Texas (TX 498731, TX 798736, TX 804475), germplasm from Virginia (TRC 02056-1), and seven accessions from China (PIs 467829, 476831, 476834, 476835, 476842, 467843, and 467844) (Wynne *et al.*, 1991).

14.2.4 Aflatoxin

Environment and cultural practices can make groundnut plants and seeds prone to invasion by toxigenic species of *Aspergillus* (*A. flavus* and *A. parasiticus*) – discussed also in Chapters 10 and 13. Seeds may be contaminated with aflatoxin before harvest, during post-harvest curing and drying, or during storage. In some regions the problem develops predominantly post-harvest while in others it is largely a preharvest phenomenon. Several recommendations have been made regarding cultural practices, curing and drying procedures, and storage conditions to minimize seed invasion by *A. flavus*. However, these recommendations have not been widely adopted in developing nations where groundnut production is subject to the vagaries of the weather.

Aflatoxin contamination was considered a post-harvest problem and received little attention in breeding programmes until it was reported by Mixon and Rogers (1973) that two germplasm lines, PI 337409 and PI 337394F, were resistant to seed invasion and colonization by *A. flavus*. Their screening method used rehydrated, sound, mature seeds inoculated artificially with *A. flavus* conidia in an environment favourable to fungal growth. They suggested that this resistance to invasion and colonization to *A. flavus*, associated with the seed coat, could be an effective means of preventing aflatoxin contamination. Varietal resistance to aflatoxin production in groundnut seed also was reported by others (Rao and Tulpule, 1967; Kulkarni et al., 1967). These findings stimulated further research on varietal resistance in several countries.

Resistance to A. flavus in groundnut may operate at three sites in the plant: the pod, the seed coat and the cotyledons. Genetic variation in pod resistance to A. flavus has been attributed to differences in pod-shell structure (Zambetakkis et al., 1981), presence of antagonistic microflora in the shell (Kushalappa et al., 1979; Mixon, 1980), and the presence of thick-walled parenchyma cells (Pettit et al., 1977). Field screening for pod resistance has been limited somewhat due to the problem of consistently reproducing the environmental conditions required to promote infection. Infection of seeds from the field may be assessed by surface sterilizing seeds from mature intact pods and then incubating them under conditions conducive to fungal growth. Disease reaction is typically expressed as the percentage of seeds exhibiting colonization.

Seed-coat resistance has also been associated with different character-istics such as the compact arrangement of testa cells and small hilum with little exposure of parenchyma cells (Taber et al., 1973), waxes deposited on the testa (LaPrade et al., 1973), 5,7-dimethoxyisoflavone (Turner et al., 1975), tannin (Sanders and Mixon, 1978; Lansden, 1982; Karchesy and Hemingway, 1986), and total soluble amino compounds and arabinose content (Amaya et al., 1980). However, Jambunathan et al. (1989) did not find significant correlation between seed colonization and polyphenol con-tent in seed coat. Procedures for assay of in vitro seed colonization by A. flavus (IVSCAF) utilize artificial inoculation to ensure uniform exposure of seeds to the pathogen. Sound mature seeds from intact, dried pods are surface sterilized, imbibed, and inoculated with a conidial suspension of a toxigenic strain of A. flavus or A. parasiticus, then incubated to promote mycelial growth (Mixon and Rogers, 1973; Mehan et al., 1981).

Many sources of resistance have now been reported for preharvest seed infection, in vitro seed colonization and aflatoxin production (Table 14.5). These include PI 337409, PI 337394F, UF 71513, J 11, Ah 7223, U-4-47-7, 55-437, and 73-30 for preharvest field infection and colonization and aflatoxin production. J 11 is grown commercially in India, as are 55-437 and 73-70 in Senegal and other West African nations. Three lines with resist-ance to IVSCAF (PI 337394F, PI 337409 and J 11) have been evaluated in more than one country. J 11 exhibited resistance to seed infection in India and the USA. PI 337409 was resistant in tests in Senegal and India, but was susceptible in the USA (Kisyombe et al., 1985). Mixon (1976) recorded percentage colonization of seeds in the F_1 and F_2 generations of crosses between PI 337409 and PI 331326, a susceptible line. Broad-sense heritabi-lity was estimated at 78.5%. Based on diallel and factorial matings con-ducted at ICRISAT Center, Vasudeva Rao et al. (1989) reported that UF 71513, Ah 7223, PI 337394F and PI 337409 had good combining abilities for seed-coat resistance. Resistance to IVSCAF in breeding lines

TABLE 14.5 *Sources of resistance to* Aspergillus flavus *or* A. parasiticus

Source of resistance	Type of resistance	Country where used	Reference
1–4	IVSCAF	India	Ghewande *et al.*, 1989
1–7	IVSCAF	India	Ghewande *et al.*, 1989
55–437	Field infection	Senegal	Waliyar and Bockelee-Morvan, 1989
			Zambetakkis *et al.*, 1981
	IVSCAF	Senegal	Zambetakkis *et al.*, 1981
73–30	Field infection	Senegal	Waliyar and Bockelee-Morvan, 1989
			Zambetakkis *et al.*, 1981
	IVSCAF	Senegal	Zambetakkis *et al.*, 1981
73–33	Field infection	Senegal	Waliyar and Bockelee-Morvan, 1989
			Zambetakkis *et al.*, 1981
	IVSCAF	Senegal	Zambetakkis *et al.*, 1981
A. cardenasii	Aflatoxin production	India	Ghewande *et al.*, 1989
	IVSCAF	India	Ghewande *et al.*, 1989
A. duranensis	Aflatoxin production	India	Ghewande *et al.*, 1989
	IVSCAF	India	Ghewande *et al.*, 1989
Acc 63	IVSCAF	Philippines	Pua and Medalla, 1986[*]
Ah 6487	IVSCAF	China	Tsai and Yeh, 1985
Ah 7223	Field infection	India	Mehan *et al.*, 1986b, 1987
	IVSCAF	India	Mehan and McDonald, 1980
			Ghewande *et al.*, 1989
AR-1	IVSCAF	USA	Mixon, 1983b
AR-2	IVSCAF	USA	Mixon, 1983b
AR-4	IVSCAF	USA	Mixon, 1983b
Basse	IVSCAF	China	Tsai and Yeh, 1985
C 116(R)	IVSCAF	China	Tsai and Yeh, 1985
C 184	IVSCAF	China	Tsai and Yeh, 1985
Celebes	IVSCAF	Philippines	Pua and Medalla, 1986
CES 48–30	IVSCAF	Philippines	Pua and Medalla, 1986
CGC 7	IVSCAF	India	Ghewande *et al.*, 1989
CGC-2	IVSCAF	India	Ghewande *et al.*, 1989
Darou IV	Pod infection	Senegal	Zambetakkis, 1975
F-7	IVSCAF	China	Tsai and Yeh, 1985
Faizpur	IVSCAF	India	Mehan and McDonald, 1980
GE 652	IVSCAF	China	Tsai and Yeh, 1985
GFA-1	IVSCAF	USA	Mixon, 1983a
GFA-2	IVSCAF	USA	Mixon, 1983a
J 11	Field infection	India	Mehan *et al.*, 1986b, 1987
	IVSCAF	India	Mehan and McDonald, 1980
			Ghewande *et al.*, 1989
		USA	Kisyombe *et al.*, 1985

TABLE 14.5 *Cont.*

Source of resistance	Type of resistance	Country where used	Reference
M395	IVSCAF	China	Tsai and Yeh, 1985
Maria-B	IVSCAF	China	Tsai and Yeh, 1985
Monir 240–30	IVSCAF	India	Mehan and McDonald, 1980
NC 449	IVSCAF	China	Tsai and Yeh, 1985
NC 482	IVSCAF	China	Tsai and Yeh, 1985
PI 196621	IVSCAF	China	Tsai and Yeh, 1985
PI 196626	IVSCAF	China	Tsai and Yeh, 1985
PI 337394F	Field infection	India	Mehan *et al.*, 1986b, 1987
		Senegal	Waliyar and Bockelee-Morvan, 1989
			Zambetakkis *et al.*, 1981
	IVSCAF	India	Mehan and McDonald, 1980
		Senegal	Zambetakkis *et al.*, 1981
		USA	Mixon and Rogers, 1973
PI 337409	Field infection	Senegal	Zambetakkis, *et al.*, 1981
	IVSCAF	India	Mehan and McDonald, 1980
		Senegal	Zambetakkis *et al.*, 1981
		USA	Kisyombe *et al.*, 1985
			Mixon and Rogers, 1973
PI 339407	Field infection	India	Mehan *et al.*, 1986b, 1987
		Senegal	Waliyar and Bockelee-Morvan, 1989
RMP 12	IVSCAF	China	Tsai and Yeh, 1985
Roxo (Sal)	IVSCAF	China	Tsai and Yeh, 1985
S 230	IVSCAF	India	Ghewande *et al.*, 1989
Shulamith	Pod infection	Senegal	Zambetakkis, 1975
Sp. 218	IVSCAF	China	Tsai and Yeh. 1985
Sp. 424	IVSCAF	China	Tsai and Yeh, 1985
U4–47–7	Field infection	India	Mehan *et al.*, 1986b, 1987
	IVSCAF	India	Mehan and McDonald, 1980
U4–7–5	Aflatoxin production	India	Mehan *et al.*, 1986a
UF 71513	Field infection	India	Mehan *et al.*, 1986b, 1987
	IVSCAF	India	Mehan and McDonald, 1980
	IVSCAF	USA	Bart *et al.*, 1978
UPL Pn4	IVSCAF	Philippines	Pua and Medalla, 1986
Var 27	IVSCAF	India	Mehan and McDonald, 1980
VRR 245	Aflatoxin production	India	Mehan *et al.*, 1986a

developed in India has remained stable over years and locations (Vasudeva Rao *et al.*, 1989).

In the United States, there is controversy as to the value of IVSCAF in practical control of *Aspergillus* contamination. Wilson *et al.* (1977) found production of aflatoxin in PI 339396F and PI 339407 to be similar to IVSCAF-susceptible genotypes PI 334360 and Florunner when seed lots were stored under high humidity. All lots exhibited 2–3% infection of seeds by *Aspergillus* spp. prior to storage. None of the lots was inoculated. Davidson *et al.* (1983) compared aflatoxin contamination in farm-grown samples of Florunner with Sunbelt Runner, a cultivar selected for resistance to IVSCAF. Seeds of Sunbelt Runner sampled prior to storage exhibited levels of natural infection and aflatoxin production comparable to Florunner. Seed-coat resistance is operative only in seeds with intact testae. The conditional nature of this resistance limits its utility under field conditions. Its effectiveness is reduced by mechanical operations causing pod and seed damage or by faulty curing, drying and storage conditions.

Genetic variation has been observed for the ability of groundnut cotyledons to support production of aflatoxins (Rao and Tulpule, 1967; Kulkarni *et al.*, 1967; Doupnik, 1969; Aujla *et al.*, 1978; Doupnik and Bell, 1969; Nagrajan and Bhat, 1973; Tulpule *et al.*, 1977). Very little is known about the mechanism of resistance to aflatoxin production. Several studies have reported effects of fungal nutrition on toxigenesis by *Aspergillus* spp. grown on defined media. Payne and Hagler (1983) observed differences in the growth of *Aspergillus* spp. on media containing different amino acids. Casein, proline, asparagine and ammonium sulphate supported fungal growth and toxin production better than did tryptophan or methionine. Venkitasubramanian (1977) found toxin production to be enhanced on defined media containing casamino acids rather than urea or ammonium nitrate as the nitrogen source. Maggon *et al.* (1973) studied the effects of micronutrients on aflatoxin biosynthesis, finding that toxin production was stimulated by copper but inhibited by cadmium, barium and vanadium. Screening methods for aflatoxin production are similar to those used for seed colonization. Some researchers have removed the testa of the seed prior to inoculation in order to remove any barrier to infection contained therein. Inoculated seeds are incubated and aflatoxin measured using thin layer chromatography (Mehan and McDonald, 1980). Mehan *et al.* (1986) identified U4-7-5 and VRR 245 as resistant to production of aflatoxin. U-4-7-5 and VRR 245 do not support high levels of aflatoxin production but are susceptible to colonization and seed invasion. A previous report of two wild species, *A. cardenasii* and *A. duranensis*, supporting production of only trace levels of aflatoxin (Ghewande *et al.*, 1989) was not confirmed in subsequent screening performed at ICRISAT (Mehan *et al.*, 1992).

A. flavus is a weak pathogen. Its ability to invade intact pods and seeds is strongly influenced by environmental conditions during pod and seed development. Developing pods must be predisposed to infection by the occurrence of water stress in the soil surrounding them and by high soil temperatures (38–40 °C) in the podding zone (Cole *et al.*, 1989). These conditions weaken the host plant and suppress the growth of soil microbes antagonistic to or competitive with *A. flavus*. At ICRISAT Center, field screening for resistance to preharvest infection is conducted in the post-rainy season; severe drought stress is imposed by withholding irrigation late in the growth cycle.

14.2.5 Insect pests

Groundnut is subject to reduction of yield and quality due to feeding by insects and arachnids on leaves, pegs, pods and seeds. In addition to causing damage directly, some insects serve as vectors of viral diseases. Insects of global importance include aphids, thrips, jassids and *Spodoptera*. Leaf miner, *Hilda*, *Helicoverpa* and other lepidopterous species present problems in specific regions. In Asia and Africa, white grub is the most economically important pod-feeding pest, but termites, millipedes and ants may also damage pods in specific regions. In the USA, lesser cornstalk borer (*Elasmopalpus lignosellus*), white-fringed beetle (*Graphognathus* spp.) and southern corn rootworm (SCR, *Diabrotica undecimpunctata howardii*) are the primary agents of damage to pegs and pods. Damage from pod feeders not only reduces yield but also permits entry into the pod of soil-borne pathogens such as *A. flavus*.

Sources of resistance to most insect pests have been identified (Lynch, 1990; Wightman *et al.*, 1990; Nigam *et al.*, 1991) although levels of resistance do not approach immunity. Some sources exhibit resistance to more than one pest:

NC 6
GP-NC 343
NC Ac 01705
NC Ac 02142
NC Ac 02214
NC Ac 02230
NC Ac 02232
NC Ac 02240
NC Ac 02242
NC Ac 02243
NC Ac 02460.

These sources of resistance trace ancestry to PI 121067 or to X-irradiated leaf mutants of NC 4 selected by W.C. Gregory and D.A. Emery at North Carolina State University in the 1950s. Several have dense, elongated or erect trichomes on leaflet surfaces.

Dwivedi *et al.* (1986) reported predominantly non-additive genetic variance for trichome characters. Additive genetic effects were important for trichome length and jassid damage. Holley *et al.* (1985) found additive genetic effects to predominate for resistance to a complex of insect pests (thrips, jassids and *Helicoverpa*) in North Carolina. Several breeding lines and cultivars resistant to foliar diseases (ICG [FDRS] 4, ICG [FDRS] 10, ICGV 86590, GP-NC 343 and NC 6) also exhibit tolerance to one or more insect species such as *Spodoptera*, leaf miner or jassids.

14.3 BREEDING METHODS

Early breeders of groundnut used mass selection to exploit natural variation in local cultivars. This method was commonly used in the USA during the 1950s but was gradually replaced by use of mutagenesis or hybridization as means of creating new genetic variation. It is interesting to note that mass selection is still used to some extent today, especially in conjunction with genetic stocks introduced from outside the USA. It is common for groundnut cultivars to exhibit some phenotypic variation in the field. This could be the result of segregation within the progeny of the last single plant selected in the course of cultivar development, of segregation and assortment following natural hybridization between pure-line components of a genetically heterogeneous but phenotypically homogeneous cultivar, or of duplication or deletion of chromosomal segments following the occasional formation of quadrivalents in the first meiotic division of the tetraploid groundnut. The most recently released American cultivar developed by mass selection was Avoca 11, a virginia cultivar selected from Florigiant and released in 1976.

The method most commonly used by groundnut breeders is the pedigree method. This allows the breeder to practise selection for highly heritable characters such as pod and seed size and shape, plant type and testa colour in early segregating generations. Because these traits determine market type and conformation to local standards, they are generally the focus of intensive early-generation selection. This practice serves to reduce quickly the size of individual segregating populations. Only when the desirable plant, pod and seed type have been recovered is emphasis placed on quantitative characters such as yield and seed composition.

Modified pedigree (single-seed descent) procedures and recurrent selection have been used in groundnut (Wynne, 1975; Hildebrand, 1985) but are not the methods of choice. Despite the recommendation of Brim (1966) that single-seed descent be used to allow segregating populations to resolve into collections of pure lines before selecting even for qualitative traits, groundnut breeders have continued to favour the pedigree method. The basis for this preference may lie in the space-intensive nature of plot work in groundnut. In modified pedigree procedures, the breeder must for several generations carry forward large populations of plants, selecting a single pod from each at random. This necessitates planting at population densities sufficiently low to allow identification of individual plants. Particularly in populations segregating for spreading growth habit, individual groundnut plants may occupy large areas relative to small grains or grain legumes bearing aerial fruit.

Backcross breeding has not been used extensively in groundnut due to the paucity of simply inherited disease and insect resistances. This methodology may find greater favour in the future as recently identified resistances to rust and late leaf spot or characters such as the Florida high O/L ratio are transferred into existing cultivars that meet exacting standards of processors and consumers. Backcrossing augmented by use of molecular techniques for identifying heterozygotes may be used for transfer of genes introduced into *A. hypogaea* through transformation procedures, although it is to be hoped that transformation protocols insensitive to the recipient genotype will be developed, allowing independent transformation of any existing cultivar.

Development of genetic maps utilizing allozymes, RFLPs or random amplified polymorphic DNAs (RAPDs) as markers has promised to resolve the poly- or oligogenically inherited, quantitative traits such as yield to essentially qualitative traits by allowing the breeder to identify chromosomal segments bearing genes with measurable effects on the quantitative traits. Such methodology requires the genomic map to be saturated with markers, i.e. that there be markers exhibiting polymorphism in the segregating population of interest at average intervals of 5–20 centimorgans. Unfortunately, cultivated groundnut exhibits very little polymorphism for allozymes or RFLPs, making this approach to groundnut improvement impractical at present. On the other hand, the diploid wild species of *Arachis* exhibit large amounts of polymorphism for allozymes and RFLPs. It may be possible to utilize these markers for construction of a genomic map in the diploid species and to monitor the incorporation of wild species' germplasm in populations of cultivated groundnut. The foremost potential obstacle to use of molecular markers in wild species is the possibility of abnormal recombination between homologous chromosomes of related species, especially if the specific genomes are differentiated by structural rearrangements such as inversions or translocations.

14.4 REGIONAL PROGRESS

14.4.1 Africa

An important cash and food crop in Africa, groundnuts have declined there in terms of area, yield and productivity over the past 20 years. Two epiphytotics of groundnut rosette virus in West Africa in 1975 and 1987 almost wiped out the crop in many countries, leaving not even enough seed for farmers to plant their next crop. The changing rainfall pattern in West Africa and other parts of the continent has resulted in reduction of the length of the rainy season and forced groundnut out of cultivation in desiccated areas where it once was a major crop.

Groundnut research in Africa began during the colonial period. Colonial governments made serious efforts to establish and increase groundnut production in their colonies to meet the increasing demands of home industries and population. This effort was strengthened during and after World War II, when shortages in Europe became acute. During that period, much research was conducted in Burkina Faso, Senegal, Nigeria, Uganda, Tanzania, Zambia, Zimbabwe, Malawi, Sudan and Zaire. After the decolonization of Africa the same impetus in research could not be maintained by newly independent nations. Civil strife, lack of physical resources, deteriorating infrastructure and lack of trained scientists and technicians resulted in the near-death of many national research programmes. Work was discontinued, valuable germplasm lost, records destroyed and cultivars mixed.

Over the past two decades many national programmes have been revitalized with the support of international organizations and donor agencies such as FAO, UNDP, ICRISAT, ODA, USAID, IDRC, IRHO, IRAT, GTZ and others. However, the revival process has been slow and many national programmes collapse as soon as financial support by donor agencies is withdrawn. Many countries have better trained and qualified scientists, but the lack of the resources necessary to conduct needed research continues to plague many national programmes. Lately the World Bank has taken interest in restructuring the national agricultural apparatus in Africa. IRHO, IRAT, and ICRISAT through its regional programmes in Malawi and Niger have made long-term commitments to the region and are making efforts to strengthen national programmes. USAID's Peanut Collaborative Research Support Program (CRSP) has been involved in development of West African peanut research for the past 10 years (Peanut CRSP, 1990).

Most results of research conducted in Africa are confined to annual reports of individual projects. Very little is published in international journals. Due to poorly developed seed production, distribution and extension programmes, most new cultivars and new cultural practices have not been adopted by producers at large. Uncertain tenure of land, lack of price support and unavailability of credit have discouraged farmers from increasing their groundnut production. Importing nations, particularly the European Community, have established extremely low tolerances for aflatoxin in imported groundnuts – levels difficult to meet for developing nations with generally poor storage and handling facilities. Export markets for African countries have declined due to poor quality and irregular supply of groundnuts.

From reports of 30 African nations published in proceedings of workshops conducted by ICRISAT and in reports of other organizations, the most important constraints on increased groundnut production in Africa (excluding socioeconomic factors) include important biotic stresses such as foliar fungal diseases (early leaf spot, late leaf spot, rust), viral diseases (groundnut rosette virus, peanut clump virus and peanut mottle virus), arthropod pests (aphids, thrips, leaf miner, *Spodoptera*, jassids, white grubs, *Hilda patruelis*, termites and millipedes) and other animal pests (nematodes, rats, squirrels and monkeys).

Abiotic stresses of primary importance are drought and poor soil fertility. Other stresses are restricted in distribution to one or two countries. They include bacterial wilt in Uganda; *Alectra* species; phanerogamic root parasitic weeds in Nigeria and Malawi; acid soils in Zaire, Zambia and Malawi; and *Phoma arachidicola* in Zimbabwe.

Breeding objectives of the national programmes in Africa can be summarized as development of high-yielding oil type and/or confectionery cultivars with adaptation to specific agroecological conditions and resistance to the stresses constraining yield. Resistance to leaf spots, rust, *A. flavus*, groundnut rosette virus, tolerance to drought and early maturity rate high in most breeding programmes. Very little effort has been expended on breeding for resistance to animal pests.

Breeding methods employed in Africa are similar to those used elsewhere in the world. Programmes with limited resources or technical expertise for hybridization and selection rely primarily on introduction and pure-line selection within local landraces. International institutes such as ICRISAT and bilateral programmes such as IRHO and USAID's Peanut CRSP continue to be major sources of new genetic material in African national programmes.

Hybridization has been used in only a few national programmes and only intermittently in those. Countries with stronger programmes distribute their cultivars to neighbouring nations and to nations sharing common linguistic or economic ties with a former colonial power. For example, Burkina Faso and Senegal have shared their cultivars with other countries in francophone West Africa while Zambia has provided cultivars to nations of southern Africa with ties to the UK. In programmes using hybridization, pedigree selection has been the most commonly used method of generation advance. The backcross method has been used in breeding for disease resistance. Zimbabwe's national programme used a modified pedigree method (single-seed descent) to develop two cultivars (Hildebrand, 1985). The Zambian national programme has also used single-seed descent.

Interspecific hybrids obtained from the University of Reading, North Carolina State University and ICRISAT have been evaluated for resistance to foliar diseases in Malawi and Zimbabwe and for resistance to foliar diseases and insect pests in Nigeria. A programme of mutation breeding was initiated in Uganda to create variability for selection because the breeder there found the time required for emasculation and pollination to be excessive (Busolo-Bulafu, 1990).

Increased desertification in sub-Saharan Africa has made breeding for drought resistance a primary objective in that region. The Senegalese programme at the Bambey centre of the Institut Sénégalais de Récherches Agricoles (ISRA) has developed many cultivars with improved resistance to drought, including 47-16, 50-127, 73-33 and 55-437 (Bockelee-Morvan *et al.*, 1974). Adaptation to dry climate was achieved by shortening the growing cycle of the breeding lines using 'Chico' as a source of early maturity and screening lines for tolerance to drought (Gautreau and De Pins, 1980). Recently, a joint programme between ISRA and the Sebele Research Station of the Botswana Department of Agriculture at Gaborone was initiated to improve drought tolerance in groundnut. Two crops are grown each year, one in Senegal and one in Botswana. Eight cultivars (virginia types 47-16, 57-422, 59-127 and 73-33 and spanish types 49-20, 55-437, 68-111 and TS 32-1) were used as parents in a convergent (pyramidal) crossing scheme (Mayeux, 1987). Drought-tolerant germplasm developed at ICRISAT Center near Hyderabad, India, has been introduced into southern and West Africa.

Breeding for resistance to rust and late leaf spot is ongoing in many national programmes including Burkina Faso, Malawi, Nigeria, Senegal, Zambia and Zimbabwe. These continue to emphasize the introduction of improved resistant germplasm from ICRISAT and the USA. 'RMP 91', a GRV-resistant cultivar developed in Burkina Faso, was found to be tolerant to leaf spots. A few programmes have crossed introduced sources of resistance with local cultivars. No cultivar with resistance to foliar fungal pathogens has been released in Africa to date.

African cultivars have been screened to identify resistance to early leaf spot, but no resistant cultivars have been found. At the SADCC-ICRISAT Regional Groundnut Program in Malawi, several germplasm lines and advanced breeding lines have been found to retain foliage longer than checks under intense disease pressure (Bock, 1987). These sources of resistance to defoliation are being intermated to improve the level of resistance. Of the *Arachis* species screened in Malawi, *A.* sp. 30003 exhibited a high level of resistance to early leaf spot. Unfortunately, this diploid species cannot be crossed directly with *A. hypogaea*.

Breeding for resistance to groundnut rosette virus has been remarkably successful in Africa. Resistance to GRV was identified in local landrace cultivars in Burkina Faso by de Berchoux (1958),who later (1960) showed that the resistance was governed by two independent recessive genes. The resistance operates equally against both chlorotic (de Berchoux, 1960) and green (Harkness, 1977) rosette. Nigam and Bock (1990) confirmed de Berchoux's observations and described an effective field screening technique for rosette. Utilizing resistance from landraces, IRHO breeding programmes in Burkina Faso and Senegal have developed several GRV-resistant cultivars including RMP 12, RMP 91, 69-101, KH-149A and KH-241D. The last two cultivars are spanish type; the others are virginia type. In southern Africa, the Malawi national programme developed a

GRV-resistant cultivar, RG1. For many regions in Africa, current emphasis in breeding for resistance to rosette is on transferring resistance into early-maturing cultivars. The SADCC–ICRISAT Regional Groundnut Program and the Nigerian national programme are actively involved in GRV-resistance breeding.

Other than local landraces, the genetic source that has contributed most to varietal development in Africa is Mani Pintar. The history of this line illustrates the powerful role of introduction in crop improvement. Mani Pintar was collected from a market place in La Paz, Bolivia, by Stephens and Hartley in 1947 (Hartley, 1949). The name is undoubtedly a corruption of 'maní pintado' or 'painted groundnut'. The characteristic features of the line are red-and-white variegated testa and spreading bunch growth habit (cultivar group Nambyquarae). The original seed sample was shared by the Queensland Department of Agriculture and Stock in Australia and the USDA. In the USA, the accession was assigned plant introduction number PI 162404. In 1955 the accession was introduced to the Mount Makulu Research Station in Zambia, where pure line selection was practised in subsequent years. A single-plant selection with solid red testae led to the release of 'Makulu Red' in 1961 (Smartt, 1978). Mani Pintar and Makulu Red were introduced into Zimbabwe in 1960. Sigaro Pink, a variant with pink testae, arose from Makulu Red, presumably as a result of natural hybridization, and was released in Zimbabwe in 1968–69. Further selection within Sigaro Pink gave rise to Apollo in 1972–73 and Egret in 1975. Mani Pintar is also one of the parents of GRV-resistant cultivars RMP 12 and RMP 91, which are very popular in West Africa.

There are more than 65 released cultivars reported in the literature from Africa. However, only a few are grown on a large scale and are pan-African in nature (Table 14.6). Most of the common cultivars of West Africa were developed by ISRA's Centre Nationale pour les Récherches Agricoles (CNRA) at Bambey, Senegal, and by IRHO in Burkina Faso.

14.4.2 East Asia

China, Japan and South Korea are the major groundnut-growing countries in east Asia. China is the leading groundnut producer in the world. In 1989, the groundnut area in the country was 2 946 000 ha and the total production was 5 362 000 t with an average yield of 1815 kg/ha. Compared with the 1970s, the groundnut area in China in the 1980s increased by 50%, the production by 124% and the yield by 48%. In this period, old cultivars were replaced by new improved cultivars in 95% of the groundnut area of the country. Groundnut cultivation in China is concentrated in the northern region, which accounts for 60% of the total groundnut area. Shandong Province in the northern region is the leading groundnut-producing province in China with an average pod yield of 2.7 t/ha. Other important areas are the southern (21%) and central (12%) region.

TABLE 14.6 *Cultivars released in Africa*

Cultivar	Type	Origin or pedigree	Year	Description
Burkina Faso (IRHO, Niangoloko Station)				
Te.3	Spanish	Selection from a local population from Upper Volta	1958	90-day cycle, erect growth habit, medium leaflet size, semi-compact fruiting habit, small (70–80 g/100) 2-seeded pods with crest and slight constriction, no beak, small (38–40 g/100) flattened seeds, salmon pink testa, 67–70% meat content, 47–48% oil content, no seed dormancy, 41–43% oleic acid content, 33–35% linoleic acid content, resistant to drought. Used in Benin, Burkina Faso.
RMP 12	Virginia	F₉ selection following hybridization, 1036 / Mani Pintar	1963	135–150-day cycle, semi-spreading growth habit, medium leaflet size, compact fruiting habit, medium (80–90 g/100) 2-seeded pods with no crest or constriction, marked reticulation, moderate beak, medium (50–55 g/100) oblong flattened seeds, pink variegated testa, 72% meat content, 49% oil content, 98% seed dormancy, 55–58% oleic acid content, 24–26% linoleic acid content, sensitive to drought, excellent resistance to GRV, susceptible to rust. Used in Benin, Burkina Faso, Mozambique, Nigeria.
RMP 91	Virginia	F₉ selection following hybridization, 48–37 / Mani Pintar	1963	135–150-day cycle, semi-spreading growth habit, medium leaflet size, compact fruiting habit, small (75–85 g/100) 2-seeded pods with no crest or constriction, marked reticulation, moderate beak, small (48–50 g/100) oblong seeds, pink testa, 68% meat content, 48% oil content, 98% seed dormancy, 55–58% oleic acid content, 24–26% linoleic acid content, sensitive to drought, excellent resistance to GRV, tolerant to early and late leaf spots. Used in Benin, Burkina Faso, Cameroon, Nigeria.
KH-149 A	Spanish	F₇ selection following hybridization, GH 119–7.III-III / 91 Saria	1964	90-day cycle, semi-spreading growth habit, medium leaflet size, semi-compact fruiting habit, small (65–75 g/100) 2-seeded pods with deep constriction, no crest, slight beak, small (30–35 g/100) oblong seeds, red testa, 67–70% meat content, 48–50% oil content, no seed dormancy, 37–39% oleic acid content, 34–36% linoleic acid content, low resistance to drought, resistant to GRV. Used in Benin, Burkina Faso, Niger.
KH-214 D	Spanish	F₇ selection following hybridization, GH 1185.2 II/ 91 Saria	1964	90-day cycle, semi-spreading growth habit, medium leaflet size, semi-compact fruiting habit, medium (80–90 g/100) 2-seeded pods with very slight constriction, no crest, moderate beak, small (35–40 g/100) flattened seeds, red testa, 70% meat content, 49–50% oil content, no seed dormancy, 38–40% oleic acid content, 35–37% linoleic acid content, resistant to drought, resistant to GRV. Used in Benin, Burkina Faso.

TABLE 14.6 *Cont.*

Cultivar	Type	Origin or pedigree	Year	Description
TS 32–1	Spanish	Selection following hybridization, Spanlex Te. 3	1966	90-day cycle, erect growth habit, medium leaflet size, semi-compact fruiting habit, small (70–80 g/100) 2-seeded pods with moderate constriction, no crest, slight beak, small (35–38 g/100) slightly flattened seeds, pink testa, 68–70% meat content, 50–51% oil content, no seed dormancy, 44–46% oleic acid content, 31–33% linoleic acid content, resistant to drought. Used in Benin, Burkina Faso, Chad, Niger.
Congo (IRHO, Loudima)				
A-124 B	Valencia	Selection from a local population, Loudima Red	1956	Long Manyema group, 90-day cycle, erect growth habit, large leaflet size, large (165 g/100) 3- or 4-seeded pods with deep dorsal constriction, no crest, marked reticulation, prominent beak, small (42 g/100) oblong seeds, red to purplish-blue testa, 69% meat content, 48–50% oil content, no seed dormancy, 45–47% oleic acid content, 31–33% linoleic acid content, low resistance to drought.
Malawi				
RG 1	Virginia (bunch)	Selection following hybridization, Makulu Red / 48-34	1976	Resistant to GRV.
Chitembana	Virginia (runner)	Selection following hybridization, Chalimbana / RJ5	1980	Confectionery type.
Mawanga	Virginia (bunch)	Introduced from Bolivia	1980	Oil type.
ICGM 42	Virginia (bunch)	Selection following hybridization, USA 20 / TMV 10	1990	Red testa.
Senegal (IRHO)				
756A	Virginia	Selection from a local population from the Casamance region of Senegal	1951	125-day cycle, erect growth habit, medium leaflet size, semi-compact fruiting habit, large (160–200 g/100) 2-seeded pods with no constriction or crest, no beak, medium (65–75 g/100) round distinctly flattened seeds, pink testa, 70% meat content, 48% oil content, complete seed dormancy, 55–58% oleic acid content, 18–20% linoleic acid content, sensitive to drought.

TABLE 14.6 *Cont.*

Cultivar	Type	Origin or pedigree	Year	Description
Senegal (ISRA, Bambey CNRA)				
28–206	Virginia	Selection from a population from Bamako, Mali	1928	Samaru group, 120-day cycle, erect growth habit, medium leaflet size, compact fruiting habit, medium (100–125 g/100) 2-seeded pods with very slight constriction, no crest, fine reticulation, no beak, small (45–49 g/100) round distinctly flattened seeds, pink testa, 73% meat content, 50% oil content, complete seed dormancy, 65–68% oleic acid content, 15–18% linoleic acid content, sensitive to drought. Used in Cameroon, Gambia, Mali, Niger, Senegal.
47–10	Spanish	Selection from a population received from Madagascar, Ambata B / Morovoay	1947	Manyema group, 90-day cycle, erect growth habit, large leaflet size, medium (105 g/100) 2-seeded pods with moderate dorsal constriction, prominent crest, prominent reticulation, very prominent beak, small (45 g/100) slightly flattened seeds, salmon pink testa, 71% meat content, 48% oil content, no seed dormancy, 43–45% oleic acid content, 32–33% linoleic acid content, moderate resistance to drought, low resistance to *Pythium myriotylum*. Used in Mali.
55–437	Spanish	Selection from a population of probable South American received from Hungary	1955	Natal Barberton group, 90-day cycle, erect growth habit, large leaflet size, compact fruiting habit, small (85–95 g/100) 2-seeded pods with slight constriction, prominent reticulation, almost no beak, small (35–38 g/100) slightly flattened seeds, pale pink testa, 75% meat content, 49% oil content, 30% seed dormancy, 46–49% oleic acid content, 27–30% linoleic acid content, resistant to drought. Used in Botswana, Cameroon, Chad, Gambia, Mali, Niger, Nigeria, Senegal, Uganda.
57–422	Virginia	Selection from a hybrid population imported from Tifton, Georgia, USA, F334–3–404	1957	105–110-day cycle, erect growth habit, large leaflet size, large (165–175 g/100) 2-seeded pods with very deep constriction, no crest, very slight reticulation prominent beak, medium (65–69 g/100) slightly flattened oblong bumpy seeds, yellowish pink testa, 78% meal content, 50% oil content, 95–100% seed dormancy, 50–53% oleic acid content, 27–30% linoleic acid content, moderate resistance to drought, susceptible to late leaf spot and *A. niger*, tolerant to PCV. Used in Mozambique, Niger, Senegal.
57–313	Virginia	Selection from a population from Ouagadougou, Burkina Faso	1957	Samaru group, 125-day cycle, erect growth habit, medium leaflet size, diffuse fruiting habit, medium (125–130 g/100) 2-seeded pods with slight constriction, no crest, fine reticulation, no beak, small (48–52 g/100) round distinctly flattened seeds, pink testa, 75% meat content, 50% oil content, complete seed dormancy, 64–67% oleic acid content, 14–17% linoleic acid content, sensitive to drought.

TABLE 14.6 *Cont.*

Cultivar	Type	Origin or pedigree	Year	Description
GH 119–20	Virginia	Introduced from Tifton, Georgia, USA, in 1960. F_4 selection from Southeastern Runner / Dixie Giant. 210–4 // Virginia Runner	1960	Jumbo group, 110-day cycle, erect growth habit, large leaflet size, fair fruiting habit, large (230–240 g/100) 2-seeded pods with moderate constriction, no crest, marked reticulation, large (85–90 g/100) oblong seeds, pink testa, 70% meat content, 43–46% oil content, medium seed dormancy, 63–66% oleic acid content, 14–17% linoleic acid content, sensitive to drought. Used in Ethiopia, Senegal.
69–101	Virginia	BC_3F_5 selection following hybridization, 55–455 / 4*28–206	1969	Saloum group, 125-day cycle, erect growth habit, medium leaflet size, compact fruiting habit, medium (130 g/100) 2-seeded pods with very slight constriction, no crest, fine reticulation, no beak, small (46–50 g/100) round distinctly flattened seeds, pink testa, 73% meat content, 50% oil content, complete seed dormancy, 65–68% oleic acid content, 14–17% linoleic acid content, sensitive to drought, resistant to GRV. Used in Benin, Burkina Faso, Senegal.
73–27	Virginia	F_8 selection following hybridization, 756A / GH 119–20, Line 252	1972	Jumbo group, 120–125-day cycle, erect growth habit, large leaflet size, fair fruiting habit, large (200, 210 g/100) 2-seeded pods with moderate constriction, no crest, slight reticulation, no beak, large (85–90 g/100) oblong seeds, salmon pink testa, 71% meat content, good seed dormancy, 58–61% oleic acid content, 20–22% linoleic acid content, sensitive to drought, used for confectionery purposes.
73–28	Virginia	F_8 selection following hybridization, 756A / GH 119–20, Line 255	1972	Jumbo group, 120–125-day cycle, erect growth habit, large leaflet size, fair fruiting habit, large (190 –200 g/100) 2-seeded pods with moderate constriction, no crest, slight reticulation, no beak, large (85–90 g/100) oblong seeds, salmon pink testa, 72% meat content, good seed dormancy, 55–58% oleic acid content, 21–23% linoleic acid content, sensitive to drought, used for confectionery purposes.
73–30	Spanish	F_8 selection following hybridization, 61–24 (spanish) / 59–127 (virginia type Saloum)	1973	95-day cycle, erect growth habit, medium to large leaflet size, compact fruiting habit, medium (100 g/100) 2-seeded pods with slight constriction, no crest, slight reticulation, no beak, small (40 g/100) oblong seeds, salmon pink testa, 73% meat content, 48% oil content, complete seed dormancy, 60–63% oleic acid content 18–21% linoleic acid content, resistant to drought.
73–33	Virginia	F_{12} selection following hybridization, 58–650 / 59–46	1973	Fung group, 105–110-day cycle, very erect growth habit, compact fruiting habit, medium (120–125 g/100) 2-seeded pods with deep constriction, no crest, marked reticulation, medium beak, small (50–52 g/100) oblong seeds, pink, 73% meat content, 50% oil content, 95% seed dormancy, 58–61% oleic acid content, 20–22% linoleic acid content, resistant to drought. Used in Gambia, Senegal.

TABLE 14.6 *Cont.*

Cultivar	Type	Origin or pedigree	Year	Description
South Africa				
Natal Common	Spanish			Erect growth habit, large leaflet size, 2 seeds per pod, no constriction or beak, pale tan testa. Used in Mozambique, South Africa, Tanzania, Zambia, Zaire.
Tanzania				
Nyota	Spanish	Introduced from USA in 1978 (Spancross)	1983	
Johari	Spanish	Introduced from India in 1980 (Robut 33–1)	1985	
Zaire				
A 65	Valencia	Introduced from Brazil	1958	90-day cycle, erect growth habit, rose-tan testa. Used in Burundi, Zaire.
G 17	Valencia	Selection from a local landrace following apparent natural hybridization	1975	105-day cycle, erect growth habit, rose-tan testa.
Zambia (Mount Makulu Research Station)				
Mani Pintar	Virginia	Collection from a market in La Paz, Bolivia, introduced to Mt. Makulu Station in 1955	1955	130–140-day cycle, spreading bunch growth habit, dark green leaves, large 2-seeded pods with no constriction, pronounced beak, medium large flattened seeds, red and white variegated testa. Used in Malawi, Uganda, Zambia.
Makulu Red	Virginia	Selection from Mani Pintar	1961	130–140-day cycle, spreading bunch growth habit, dark green leaves, large leaflet size, large 2-seeded pods with no constriction, pronounced beak, medium large flattened seeds, red testa, 67% meat content, 45% oil content, field resistance to early leaf spot. Used in Tanzania, Uganda, Zambia, Zimbabwe.

TABLE 14.6 *Cont.*

Cultivar	Type	Origin or pedigree	Year	Description
Chitembana	Virginia	Selection from a local population from eastern Zambia	1964	140–150-day cycle, runner growth habit, large leaflet size, thick stems, large coarse 2-seeded pods with slight constriction, no beak, flattened seeds, dark tan testa, used for confectionery purposes. Used in Malawi, Zambia.
Comet	Spanish	Introduced from USA (Comet)	1984	
MGS 2	Virginia (runner)	Introduced from India (M13)	1988	
Zimbabwe				
Egret	Virginia	Selection from naturally occurring pink variants in Makulu Red	1974	130–140-day cycle, spreading bunch growth habit, large leaflet size, large 2-seeded pods with no constriction, pronounced beak, medium large flattened seeds, pink testa, 67% meat content, 45% oil content, field resistance to early leaf spot.
Flamingo	Virginia (bunch)	PI 261911 / Natal Common	1982	
Plover	Spanish	Introduced from Brazil (PI 336954)	1982	
Swallo	Virginia (bunch)	PI 261911 / Makulu Pink selection	1982	

In the northern region, the main constraints to groundnut production are early and late leaf spots, viruses (peanut stripe, peanut stunt, cucumber mosaic, TSWV), aphids, *Helicoverpa*, *Spodoptera*, thrips, nematodes and drought. Surveys conducted in the 1970s indicated that more than 300 000 ha were infested with nematodes in nine provinces of China. *Meloidogyne hapla* is widespread in the north, whereas it is *M. arenaria* in the south of the country. These nematodes cause on average 20–30% yield loss in the country. Breeding began at the Peanut Research Institute at Laixi in Shandong Province in 1959. Since then 15 cultivars have been released for cultivation in the province and other parts of the country (Table 14.7). Following hybridization, the single-seed descent method has been used to advance breeding generations. Twelve of the 15 cultivars released by the institute are the result of hybridization and the remaining three are pure line selections among local landraces. Hua 37 and Luhua 4 are very popular among farmers and have good export quality. Hua 37 covers more than 100 000 ha in the country. With new production technology, which includes polyethylene mulching, these cultivars can produce 7.5 t pods/ha. The main emphasis in groundnut breeding in Shandong Province has been to increase pod yield and improve quality. Quality parameters that have received attention in breeding are large elongated seed, high oil (55%), O/L ratio (>1.4 for large-seeded virginia types, >1.2 for spanish types), high protein (>30%), high blanchability, pink testa colour, and flavour (by organoleptic test).

In the central region, early and late leaf spots, rust, bacterial wilt, viruses (peanut stripe, peanut stunt, cucumber mosaic, TSWV), aphids, *Helicoverpa*, *Spodoptera*, thrips, leafhoppers, white grub, drought, waterlogging and high temperature are serious constraints to production. The Oil Crops Research Institute at Wuhan is responsible for groundnut research in Hubei Province in this region. The Institute maintains a collection of 4350 accessions of groundnut, including 130 wild *Arachis* species. All accessions have been characterized for agronomic characters. In collaboration with the Peanut Research Institute in Shandong Province, 4029 lines have been screened for resistance to nematodes (*M. hapla*), to which two lines – Tian Fu No. 4 and Da Hua Cheng – have shown a high level of resistance. Three other lines with moderate resistance and five lines with tolerance also have been identified. Four thousand lines have been screened for bacterial wilt, rust, late leaf spot and early leaf spot. Seventy lines with resistance to bacterial wilt and many lines with resistance to rust and late leaf spot have been identified, but a satisfactory level of resistance to early leaf spot has not yet been located. Many of the lines resistant to foliar disease were obtained from ICRISAT. The germplasm has also been screened for biochemical factors. The protein content in the collection ranges from 14.0% to 36.8% and oil content from 36.0% to 60.21%. There are many lines with O/L ratios greater than 3.0.

Breeding objectives at the institute include high yield, early maturity, improved quality, and resistance to diseases and insect pests. Following hybridization, the pedigree method is followed to advance breeding generations. From 1986 to 90, the significant achievements of the breeding group at the institute included identification of sources of resistance to bacterial wilt and rust.

About 200 000 ha are infected with bacterial wilt in the central region of China. Yield loss to this disease averages 10–15% and may go up to 60%. Since 1970, more than 4000 germplasm accessions have been screened in the field and screenhouse; 70 resistant lines have been identified. The resistance in these lines is generally stable under field conditions but it can break down under heavy artificial inoculation and with a highly virulent strain. Inheritance studies involving spanish types indicated that resistance to bacterial wilt is partially dominant and is governed by three major genes with additive effects (Boshou *et al.*, 1990).

Peanut stripe virus (PStV), although widely distributed in the country, is mainly important in central and northern China. A 50% disease incidence is often found in these areas, reaching up to 100% in many fields. In southern China, the disease incidence is <1%. In laboratory and field studies, 20% yield loss was observed with early infection of the virus. More than 1300 germplasm lines have been screened without identifying any resistant accessions.

In the central region mostly spanish and peruvian types are grown. Four new groundnut cultivars have been released by the institute in the last five years (Table 14.7). Current breeding activities (1991–95) include development of cultivars with multiple resistance to diseases and pests, utilization of wild *Arachis* species to develop cultivars resistant to leaf spot, screening for resistance to virus diseases, screening for tolerance to acid soils and breeding for increased nitrogen fixation.

In China's southern region, the primary constraints to production are rust, bacterial wilt, waterlogging and soil acidity. Guangdong Province, where mainly spanish types are grown, leads the region in groundnut production and its Industrial Crops Research Institute, Guangzhou, carries out groundnut research in the region. The main objective of its present groundnut research programme is to develop high-yielding cultivars with resistance to bacterial wilt and rust and adaptation to different growing conditions in the province. The six sources of bacterial wilt resistance used in the breeding programme are Teishan Sanliyue (a valencia cultivar from China), Teishan Zhenzhu (a spanish cultivar from China), Xie Kong Chung (a spanish cultivar from China), Schwartz (a spanish cultivar from Indonesia), Yindu Huapi (a virginia cultivar from India) and Tianjin Don (a runner cultivar from China). Sources of rust resistance have been obtained from ICRISAT.

In Japan, groundnut is a minor crop. The main centre of production is the Kanto region in the central part of the country. The consumption of groundnut in Japan amounted to 85 000 t in 1989, of which 44% was

TABLE 14.7 *Groundnut cultivars released in East Asia*

Cultivar	Botanical type	Pedigree	Year of release	Characteristics
China (Northern Region)[1]				
Fuhuasheng	Spanish		1960	
Hua 27	Virginia		1967	
Hua 11	Spanish		1969	
Hua 19	Virginia		1975	
Hua 28	Intermed. between spanish and virginia (M)		1979	
Hua 31 (Hai Hua 1)	M		1984	
Hua 37	M		1985	Good quality
Hua 39 (Luhua 4)	Virginia		1986	Good quality
Hua 17	Virginia		1974	O/L ratio 1.45
Hua 98	Virginia		1974	Tolerant to drought
Luhua 3	Spanish		1982	High oil, resistant to bacterial wilt
Luhua 6	M		1986	
Luhua 8	M		1988	
Luhua 9	Virginia		1988	Good quality
China (Central Region)				
El Hua 4		Hongmei Zhao / El Hua 2		High yield, early maturity, high quality, tolerant to drought
Zhong Hua 1		El Hua 3 / Taishan Zenghou		High yield, tolerant to leaf spot
Zhong Hua 2		El Hua 4 / Taishan Sanlirou		High oil and protein, resistant to bacterial wilt
Zhong Hua 117				Resistant to rust, moderately resistant to bacterial wilt, high protein, high yield

TABLE 14.7 *Cont.*

Cultivar	Botanical type	Pedigree	Year of release	Characteristics
China Southern Region)				
Yue You 39		Yue You 116 / Yindu Huapi		Resistant to bacterial wilt and rust
Yue You 223		Shan You 26 / EC 76446 (292)		Tolerant to rust
Yue You 92		Yue You 116 / Xie Kang		Resistant to bacterial wilt, high in oil content (54%)
Yue You 256		Yue You 116 / Xie Kang		Resistant to bacterial wilt, high yield
Japan				
Wase-dairyu	Spanish			Early maturing, large seed
Tachi-masari	Spanish			Early maturing, large seed
Chiba-handachi				Medium maturing cultivar with large seed
Nakate-yutaka				Medium maturing, high yielding cultivar with good eating and external quality
Azuma-yutaka				Medium maturing, high yielding cultivar with good eating and external quality
Sayaka				Medium maturing (later than Nakate-Yutaka), high yielding, better suited for roasting due to its thicker shell than Nakate-yutaka
Yude-rakka				Early maturing, good eating quality, white pod colour with superior external appearance, suitable for unshelled whole pod or frozen boiled groundnut trade

TABLE 14.7 *Cont.*

Cultivar	Botanical type	Pedigree	Year of release	Characteristics
Korea				
Younghotangkong	Virginia		1980	Late maturing, large elongated seed, pods with deep constriction
Saedltangkong	Inter. between spanish and valencia		1983	Early maturing, erect plant type, an intermediate type between spanish and valencia
Jinpungtangkong (ICGS 35)			1986	Early maturing, small seeded, high yielding spanish type
Daekwangtangkong	Intermed. between spanish and valencia	Florigiant / Chiba-handachi // Chiba-handachi /3/ F393–6–3–2–3–1–2	1985	Early maturing, high yielding, high oil content, erect plant type with few branches and large seed.
Namdaettangkong	Virginia (bunch)	Virginia bunch Improved / Suwoen 30	1988	Large seeded, high yielding, high in oil content, tolerant to *Phoma arachidicola*

[1] Cultivars released by the Peanut Research Institute, Laixi, Shandong Province, China.

produced locally and the rest was imported. Since the end of World War II, groundnut breeding in Japan has pursued two main objectives: breeding early-maturing cultivars for warm and cool areas, and breeding medium and late-maturing cultivars. Because groundnut is a delicacy in Japan, both eating quality and external quality are important attributes (Gocho, 1991) and improvement in quality has received the most attention in groundnut breeding. Sucrose content and hardness of seed are closely related with eating quality and they decrease if harvesting is delayed. The seed hardness is measured when the moisture content in seed is in the range of 5–9%. (All cultivars under test should have the same moisture level within this moisture range.)

Groundnut is also a minor crop in Korea, where yields are affected by leaf spots, rust, and low temperature at the ripening stage. The main breeding objective at the Crop Experiment Station, Rural Development Administration, Suwan, is to develop cultivars with large seed and erect plant type (Lee *et al.*, 1986, 1989).

14.4.3 Southern Asia

Groundnut research and production in southern Asia are dominated by India. Other groundnut-growing countries in the region are Bangladesh, Bhutan, Myanmar, Nepal, Pakistan and Sri Lanka. Except for Myanmar, groundnut production in these countries is small. The crop in India and Myanmar is grown mainly for edible oil production and in other countries in the region for direct consumption or for use in confectionery. The region accounts for 43.4% of the area and 35.7% of the production of groundnut in the world. However, the average productivity in the region (0.94 t/ha) remains below the world average (FAO, 1990). The main biotic constraints to increased groundnut production in the region include diseases – late leaf spot, rust, early leaf spot, collar rot (*Aspergillus niger*), stem rot (*Sclerotium rolfsii*), *A. flavus*, bud necrosis disease (BNV) – and insects (thrips, jassids, aphids, leaf miner, *Spodoptera*, *Helicoverpa*, red hairy caterpillar, whitegrub and termites). Abiotic constraints include drought, lack of high-yielding cultivars adapted to local growing conditions, lack of availability of good quality seeds and lack of small-scale farm machinery for groundnut cultivation.

Introduction and reselection in introduced populations continue to be the main methods of crop improvement in the region – with the exception of India where, over the past decade, the majority of new cultivars have resulted from hybridization between parents selected for their desirable characteristics. However, in countries where the research programmes are small and the scientists are responsible for more than one crop, dependence on the introduction of improved germplasm from various sources is heavy. ICRISAT has played an important role in such introductions.

Prior to 1980, breeding efforts were directed mainly towards improving

TABLE 14.8 *Groundnut cultivars released in South Asia*

Cultivar	Botanical type	Pedigree	Year of release	Characteristics
India (old popular cultivars)				
Gangapuri	Valencia	–	–	Early maturing, 3–4-seeded, small–medium pods, preferred as table variety, good source for earliness, popular in central India
Spanish Improved	Spanish	Selection from spanish peanut	1905	Small-seeded, suitable for light soils (proposed for denotification)
Kopergaon I	Virginia (bunch)	Selection from local variety	1933	Medium sized pods (proposed for denotification)
TMV 2	Spanish	Selection from 'Gudhiatham Bunch', a local variety	1940	Widely adapted, well suited for rainy and summer season cultivation in southern India, a leading spanish variety in the past, still continues to be popular with farmers
AK 12–24	Spanish	Selection from local variety	1940	Widely adapted, suited for medium to heavy soils (proposed for denotification)
Punjab Groundnut I	Virginia (runner)	Selection from 'Samrala Local'	1953	Wide adaptability (proposed for denotification)
Karad 4–11	Virginia (runner)	Selection from local variety	1957	Late maturing, 1–3-seeded medium to long pods
RSB 87	Virginia (bunch)	Selection from a Brazilian collection	1961	3-seeded pods with dark red seeds
J 11 (SB XI)	Spanish	Ah 4218 / Ah 4354	1964	Widely adapted, resistant to collar rot and *A. flavus* seed colonization
S 206	Spanish	Selection from 'Manvi Local'	1969	Reticulated pods with slight beak and constriction
S 230	Virginia (runner)	Selection from 'Tandur Local'	1969	–
M 13	Virginia (runner)	Selection from NC 13	1972	Large-seeded variety with tolerance to leaf spots
JL 24	Spanish	Selection from EC 94943	1978	Large dark green leaves, smooth 2–3-seeded pods, early in maturity

TABLE 14.8 *Cont.*

India (cultivars released since 1980)

Cultivar	Botanical type	Pedigree	Year of release	Characteristics
Kisan	Spanish	Spanish Improved B 31	1980	Small pod with prominent reticulation, released for Orissa State
M 37	Virginia (runner)	A1 / C 6–4–7–2	1980	Two-seeded pods with small beak, light brown seed coat, released for Punjab State
KRG 1	Spanish	Selection from 'Argentine' variety	1981	Two-seeded medium sized pods, released for Karnataka State
TG 17	Spanish	'Dark Green' Mutant / TG 1	1982	Large-seeded, pinkish seed coat, high harvest index, fresh seed dormancy, released for Maharashtra State
M 197	Virginia (bunch)	C 501 / U 4–7–2	1982	Dark green leaves, large-seeded pods with smooth reticulation, released for Punjab State
GG 2	Spanish	J11 / EC 16659	1983	Two-seeded reticulated pods, early flowering with dark green leaves, released for Gujarat State
Jawan	Spanish	J11 / Asiriya Mwitunde	1983	Medium elongated pods with moderate beak. rose seed coat, released for Orissa State
CO 2	Spanish	EMS Mutant from Pollachi 1	1983	Two-seeded medium plumpy pods with rose colour testa, released for Tamil Nadu State
Dh 8	Spanish	Selection from RS 144	1984	Dark green leaves, compact plant, tolerant to late leaf spot, small pods with smooth rose seeds round at one end and sharply pointed at the other, released for Karnataka State
Chitra	Virginia (runner)	Spanish 5B–1 / EC 1688	1984	Dark green leaves, variegated testa with rose background, released for Uttar Pradesh State
Kaushal	Virginia (bunch)	Selection from T 28	1984	2–3–1-seeded pods, compact plant with dark green leaves early in maturity, released for whole of India
UF 70–103	Virginia (bunch)	Introduction from USA	1984	Suitable for summer cultivation in Maharashtra State
GG 11	Virginia (runner)	M 13 / Gaug 10	1984	Leaflets and pod bigger than Gaug 10, released for Gujarat State

TABLE 14.8 *Cont.*

Cultivar	Botanical type	Pedigree	Year of release	Characteristics
TG 3	Spanish	A mutant of Spanish Improved	1985	Spanish, medium–large pods, suitable for both rainy and summer seasons, tolerant to pod borer
MA 16	Virginia (bunch)	Selection from EC 16664	1986	Large seeded suitable for HPS trade
SG 84	Spanish	Selection from ICGS 1	1986	Mainly 2-seeded, medium sized pods, suitable for summer/spring cultivation in north India
M 335	Virginia (runner)	M 13 / F7	1986	2–1–3-seeded large pods with prominent reticulation and moderate constriction, seeds large with light brown testa, large dark green leaves with compact plant, released for Punjab State
ICGS 11 (ICGV 87123)	Spanish	Selection from natural hybrid population of Robut 33–1	1986	2-seeded smooth medium sized pods with no beak and slight to moderate constriction, seeds tan colour, 100-seed mass 60 g, oil 49%, protein 22%, above average tolerance of end-of-season drought, photoperiod insensitive. Field tolerance of bud necrosis disease, adapted to post-rainy season cultivation in India, performs well in West Africa also
VRI 1	Spanish	TMV 7/ FSB 7–2	1986	Large pods with deep constriction and prominent beaks
ALR 1	Spanish	Pollachi 2/ PPG 4	1987	Small dark green leaves, red testa, resistant to rust and late leaf spot
Girnar 1	Valencia	X14-4-B-19B / NC Ac 17090	1988	Early maturing, resistant to late leaf spot, rust, collar rot, and seed colonization by *A. flavus*, 2–3-seeded with reticulated, constricted and beaked pods
ICGV 87128 (ICGS 44)	Spanish	Selection from natural hybrid population of Robut 33–1	1988	2-seeded smooth small to medium sized pods with no or little beak, seeds tan in colour, 100-seed mass 60 g, oil 49%, protein 25%, field tolerance to bud necrosis disease, good recovery from midseason drought, relatively photoperiod insensitive, adapted to post-rainy season cultivation in India, performs well in Pakistan also
RG 141	Spanish	Robut 33–1/NC Ac 2821	1989	Spanish with dark green foliage suitable for black soils
VRI 2	Spanish	JL 24 / CO 2	1989	Mostly 2-seeded large pods with moderate beak, constriction and reticulation. Seeds light rose in colour, 100-seed mass 50 g, oil 48%

TABLE 14.8 Cont.

Cultivar	Botanical type	Pedigree	Year of release	Characteristics
ICGV 87141 (ICGS 76)	Virginia (bunch)	TMV 10 / Chico	1989	Mainly 2-seeded medium sized pods with moderate to prominent reticulation, slight to moderate constriction and beak, seeds tan in colour, 100-seed mass 44 g, oil 43%, protein 20%, good recovery for pod yield from midseason drought, field tolerance to bud necrosis, adapted to rainy season cultivation in India, performs well in Sudan also
ICGV 87121 (ICGS 5)	Virginia (bunch)	Robut 33–1 / NC Ac 316	1989	2-seeded medium sized pods with none to slight beak and reticulation, slight to moderate constriction, seeds tan in colour, seed mass 38 g/100, oil 58%, protein 22%, shows good recovery for pod yield from midseason drought, adapted to rainy season cultivation in India
ICGS 1 (ICGV 87119)	Spanish	Selection from natural hybrid population of Robut 33–1	1990	Mainly 2-seeded medium sized pods with slight to moderate constriction, none to slight beak, and smooth to slight reticulation, seeds tan in colour, oil 51.1%, protein 21%, 100-seed mass 35 g, shows good recovery for pod yield from midseason drought, field tolerance to bud necrosis
Birsa Groundnut-3	Virginia (bunch)	Early Runner / Asiriya Mwitunde	–	Early maturing
ICGV 87187 (ICGS 37)	Spanish	Selection from natural hybrid population of Robut 33–1	1990	Mainly 2-seeded medium sized pods with slight reticulation, slight to moderate constriction and none to slight beak, seeds tan in colour, 100-seed 53 g, oil 48%, protein 23%, tolerance to end-of-season drought, field tolerance of bud necrosis disease, photoperiod insensitive, tolerant to rust and late leaf spot, adapted to summer season cultivation in India, also performs well in Pakistan
ICGV 87160 [ICG(FDRS)10]	Spanish	Ah 63 / NC Ac 17090	1990	2-seeded stubby pods with moderate to prominent ridges, slight reticulation, beaks and constriction either absent or less conspicuous, seeds tan in colour, 100-seed mass 36 g, oil 48%, protein 27%, resistant to rust, tolerant to late leaf spot, field tolerance to bud necrosis disease, less susceptible to stem and pod rots caused by S. rolfsii, moderately resistant to leaf miner
VRI 3	Spanish	J 11 / Robut 33–1	1990	Small-seeded pods with moderate constriction and little or no beak
RSHY 1	Spanish	GDM / TMV 2	1990	Suitable for residual moisture situation
ICGV 86590	Spanish	X14–4–B–19–B / PI 259747	1991	3-seeded pods, resistant to rust, tolerant to late leaf spot, bud necrosis disease, stem and pod rots, and Spodoptera

TABLE 14.8 Cont.

Cultivar	Botanical type	Pedigree	Year of release	Characteristics
Pakistan				
Banki	Virginia (bunch)	Introduction in 1973	–	160–180 days to maturity
No. 334	Virginia (runner)	–	–	180–200 days to maturity
BARD 669	Spanish	A composite of ICGS 44 and ICGS 37	1989	150–160 days to maturity, high yielding, high in shelling turnover
Bangladesh				
Dhaka-1 (Maizchar Badam)	Spanish	–	1976	Oil 48–50%, shelling 75%, matures in 135–140 days, highly susceptible to leaf spots
DG 2 (Basanti Badam)	Virginia (bunch)	–	1979	Mainly 2-seeded large pods, 170–175 days in maturity, seed dormancy for 40–50 days, tolerant to leaf spots
DM 1	Valencia	Introduced from India	1987	Very dwarf, early in maturity, tolerant to leaf spots and rust
Acc 12	Valencia	–	1988	Tolerant to drought, leaf spots and rust
Sri Lanka				
Red Spanish	Valencia	–	1961	Semi-erect, large 3-seeded pods, dark pink seed colour, 100-seed mass 45 g, shelling 68%, 110–120 days maturity
MI 1	Spanish	–	–	2-seeded medium pods with pink colour seed, 100-seed mass 40 g, shelling 72%, 110–120 days maturity
No. 45	Spanish	Introduction from ICRISAT, India	1982	2-seeded medium pods with pink seed colour, 100-seed mass 45 g, shelling 75%, 110–120 days maturity
X14-4-1-6-19-6	Spanish	Introduction from India	1982	2-seeded medium pods with pink seed colour, 100-seed mass 48 g, 115–120 days maturity

TABLE 14.8 *Cont.*

Cultivar	Botanical type	Pedigree	Year of release	Characteristics
Nepal				
B 4	Virginia (bunch)	Introduction from Pakistan	1976	135–140 days to maturity, 3–2 seeded medium sized pods, tan colour seed with high oil content, oil purpose cultivar
Janak	Virginia (runner)	NC Ac 343	1987	140–145 days to maturity, 2-seeded large pods, tan colour seed with high oil content, moderately resistant to disease and insects, a dual purpose cultivar
Myanmar				
Sinpadetha 2	Spanish	JL 24	1984/85	–
Sinpadetha 3	Virginia (bunch)	Robut 33–1	1984/85	–
SP 121	Spanish	–	–	2-seeded small pods, early maturing
Magwe 10	Spanish	SP 121/070 / S 550–05	–	2-seeded small pods, high shelling (76%), high oil (54%), early maturing
Magwe 11	Spanish	Selection from Shawat 21/6	–	2-seeded small-medium pods with high oil content (55%)
Magwe 12	Spanish	–	–	2-1-3-seeded medium sized pods with high oil content (55%)
Magwe 15	Spanish	UPL Pn-2 / Kyaung Gone	–	2-1-3-seeded medium pods, high shelling (77%), high oil content (54%), seed dormancy for 2 weeks
Kyaung Gone	Virginia (bunch)	–	–	2-1-seeded, seed dormancy up to 2 months
MS 2	Virginia (runner)	–	–	2-3-1-seeded pods, seed dormancy up to 3 months
Bhutan				
				In Bhutan some undefined cultivars are grown in small pockets in the valleys for local consumption

yield potential. With the identification of sources resistant to major diseases and insect pests at ICRISAT and in the national programme in India, resistance breeding received a strong stimulus resulting in release of cultivars with multiple resistances in India. A genetic gain of 1.3–3.2% per annum was achieved during the 1980s under rainfed conditions in India (Nigam *et al.*, 1991). A large number of cultivars have been released in India, particularly since 1980 (Table 14.8). Notwithstanding the release of several improved cultivars, some very old ones are still grown extensively due to lack of availability of seed: only 20% of the seed requirement in improved cultivars is met at present in India. The situation is not very different in other countries of the region. Pakistan, Sri Lanka, Nepal, and Bangladesh have very small groundnut research programmes and rely mainly on introduction for improved germplasm. Although Myanmar has a sizeable area under groundnut, its research programme is hampered by lack of trained scientific manpower.

Approximately 80% of India's groundnuts are grown in the rainy season (July–October). The remaining 20% is grown with irrigation in the post-rainy season (October/November–March/April) and the summer (January/February–April/May). The groundnut area in the post-rainy/summer season has increased recently as pod yields are high at this time. Varietal requirements of rainy and post-rainy/summer seasons differ because of differing disease and insect pest complexes occurring in them. High pod yield, high shelling percentage and high oil content are requirements common to both growing seasons. Additional requirements of improved cultivars in the rainy season are: drought tolerance; adaptation to agroecological zones differing in rainfall pattern and length of growing season; fresh seed dormancy in spanish/valencia types; tolerance to insect pests such as aphids, jassids, thrips, leaf miner, *Spodoptera* and white grub; and tolerance to diseases such as early and late leaf spots, rust, collar rot, stem rot, *A. flavus* and bud necrosis. In the post-rainy/summer season, disease pressure is generally very low but tolerance/resistance to insect pests such as leaf miner and *Spodoptera*, tolerance of low temperature in the early stages of crop growth, early maturity, and responsiveness to fertilizers and irrigation are needed in new cultivars.

Much of the emphasis in the past in groundnut breeding in India was placed on the improvement of pod yield. The quality characteristics which received attention included shelling percentage and oil content. Oil quality itself received virtually no attention. During the VIII Plan (1990–95), India's most recent programme for the improvement of agricultural production, the following breeding activities have been accorded high priority:

- For dryland conditions, emphasis is on development of drought-tolerant, high-yielding, early-maturing spreading groundnut cultivars.
- For use in paddy fallows, early-maturing bunch cultivars able to extract residual soil moisture are being developed.

- For post-rainy/summer season irrigated conditions, the objective is to produce high-yielding spanish cultivars tolerant of iron chlorosis.
- For rainfed crops, resistance to foliar diseases is a high priority.

There is also demand for cold-tolerant, early-maturing cultivars possessed of fresh seed dormancy. High oil content is a primary objective for cultivars developed for use as oilseeds, while large seeds and less susceptibility to *Aspergillus* species are the objectives in cultivars for confectionery.

For each breeding activity, targets have been fixed and responsibilities have been assigned to main groundnut research centres under the aegis of the All-India Coordinated Research Project on Oilseeds (AICORPO) at the Indian Council of Agricultural Research (ICAR), New Delhi. Hybridization between adapted cultivars and donor parents of desirable characteristics, followed by selection for such traits combined with high yield in segregating populations, has been adopted to achieve the target of the breeding activities listed above. Wherever required, interspecific hybridization is also being pursued. Some of the sources of desirable characteristics in use in hybridization are:

For earliness:

Chico	JB(E)559
TG(E)1	ICGS 6
TG(E)2	ICGS 51
VG(E)55	ICGV 86309
91176	ICGV 86315
ICGS(E)21	ICG 11199
ICGS(E)22	CSMG 881
ICGS(E)52	CSMG 902
ICGS(E)217	CSMG 905
TG 7	CSMG 917
J(E)5	CSMG 918
J(E)6	CSMG 9102
JB(E)194	Kadiri 3.
JB(E)262	

For drought tolerance:

ICGV 86607	ICGV 87264
ICGV 86707	Gujarat Narrow Leaf Mutant, A 13
ICGV 87259	

For cold tolerance:

A. monticola	NRCG 9608
NGCG 1339	CGC 498

For seed dormancy:

Dh 8	TG 7
CGC 7	TG 9
ICGS 30	TG 17

ALG 56   C 390
Kadiri 3   CGC 3
TMV 10   RSHY 6

For high shelling percentage:
J 13   CSMG 916
Spancross   Kadiri 3

For bold seed:
ALG 62   CSMG 81-1
JSP(HPS)19   CSMG 83-1
Somnath   CSMG 9101
CSMG 33   M 13
CSMG 35

For high oil content:
NC Ac 17500   TMV 10
C 174   TG 7
TMV 3

For iron chlorosis tolerance:
NGS 7   GG 2
JL 24

For resistance to rust and late and early leaf spots:
PI 259747   ICG(FDRS)69
PI 270934   ICGV 86350
PI 393516   ICGV 86598
PI 393517   ICGV 86707
PI 393643   ICGV 87160
PI 393527   ICGV 87261
PI 414331   ICGV 87264
NC Ac 17090   ICG 1697
NC Ac 17129   ICG 7894
ICG(FDRS)43   CSMG 84-1
ICG(FDRS)68

For tolerance to bud necrosis disease:
ICGV 86031

For insect tolerance:
Leaf miner:
ICG 5240   ICG 11786
GBFDS 273   ICGV 86137
GBFDS 592
Spodoptera:
ICGV 86350   ALG 50
ICGV 87264
Jassids:
NC Ac 2663

Multiple insect resistance:
    ICG 2271   JL 116
    JL 83

For *A. flavus* tolerance:
    Monir 240-30   J 11
    UF 71513   PI 337409
    Ah 7223   PI 337394F

Indian scientists have attempted to access genetic variability in the wild relatives of groundnut. Interspecific hybridization between the tetraploid *A. hypogaea* and diploid wild species *A. cardenasii*, *A. stenosperma* and *A. chacoense* has been carried out in Tamil Nadu state in India. Derivatives of the interspecific hybridizations are currently under evaluation. Irradiation and chemical mutagens have been used frequently in India to create additional variability for use in breeding programmes. Cultivars such as MH 2, TG 1 (Trombay Groundnut 1), TG 3, BG 1 (Birsa Groundnut 1) and BG 2 were developed by mutation breeding using irradiation, and CO 2 (Coimbatore 2) from chemical mutagenesis.

14.4.4 South-east Asia

Groundnut is an important food legume and oil crop in south-east Asia. Indonesia, Vietnam, Thailand and the Philippines are the major producers in the region; other countries – Malaysia, Laos and Cambodia – have only small areas under groundnut. Thailand, Vietnam and Indonesia are able to meet their domestic demand but in the other countries there is a big gap between domestic production and demand. Consumption of groundnut pods and seeds in the boiled form is very popular in this region. Peanut butter is a popular groundnut product in Malaysia and the Philippines.

The region grows groundnut on about 0.92 million ha with a total production of 950 000 t. Average pod yields are low compared with China and the USA. Major production in the region comes from upland areas, where groundnut is generally grown as a monocrop. In plantation areas it is intercropped with young rubber, oil palm and coconut trees. A sizeable area of groundnut is grown in rice fallows under residual moisture conditions.

Several biotic and abiotic factors are responsible for low productivity in the region. The major constraints to increased groundnut production are late leaf spot, rust, sclerotium wilt, bacterial wilt, peanut stripe virus, leaf miner, leafhopper, *Spodoptera*, *Helicoverpa*, aphids, thrips, drought, acid soils, low soil fertility, shade under plantation crops, low price of groundnut and lack of seed availability of improved cultivars.

Groundnut research in Thailand, Vietnam, Indonesia and the Philippines is very active. Malaysia has a small groundnut research programme.

TABLE 14.9 *Improved groundnut cultivars released in South-east Asia*

Cultivar	Botanical type	Pedigree	Year of release	Characteristics
Indonesia				
Gajah	Spanish	Schwartz 21 / Spanish	1950	Adapted to upland cultivation, 95–100 days to maturity, seed size 45–50 g/100, pod yield 1.51 t/ha, resistant to bacterial wilt
Kidang	Spanish	Schwartz 21 / Small Japan	1950	Adapted to upland cultivation, 95–100 days to maturity, seed size 45–50 g/100, pod yield 1.5 t/ha, resistant to bacterial wilt
Macan	Spanish	Schwartz 21 / Spanish	1950	Adapted to upland cultivation, 95–100 days to maturity, seed size 45–50 g/100, pod yield 1.5 t/ha, resistant to bacterial wilt
Banteng	Spanish	Schwartz 21 / Spanish	1950	Adapted to upland cultivation, 95–100 days to maturity, seed size 45–50 g/100, pod yield 1.5 t/ha, resistant to bacterial wilt
Pelanduk	Spanish	Kidang / VB 1	1983	Adapted to upland cultivation, 95–100 days to maturity, seed size 45–50 g/100, pod yield 1.5 t/ha, resistant to bacterial wilt and *A. flavus*
Tapir	Spanish	Kidang / VB 1	1983	Adapted to upland cultivation, 95–100 days to maturity, seed size 45–50 g/100, pod yield 2.0 t/ha, resistant to bacterial wilt and *A. flavus*
Tupai	Spanish	US 26 / Kidang	1983	Adapted to upland cultivation, 95–100 days to maturity, seed size 45–50 g/100, pod yield 2.0 t/ha, resistant to bacterial wilt and *A. flavus*
Rusa	Spanish	Gajah / AH 223	1983	Adapted to upland cultivation, 100–110 days to maturity, seed size 35–40 g/100, pod yield 1.5 t/ha, resistant to bacterial wilt and rust
Anoa	Spanish	Gajah / AH 223	1983	Adapted to upland cultivation, 100–110 days to maturity, seed size 35–40 g/100, pod yield 1.5 t/ha, resistant to bacterial wilt and rust
Kelinci	Valencia	Acc 12	1987	Adapted to upland and lowland cultivation, 100–110 days to maturity, seed size 40–45 g/100, pod yield 2.0 t/ha, tolerant to bacterial wilt and resistant to leaf spot
Jepara	Spanish	–	1989	Adapted to lowland cultivation, 90–100 days to maturity, seed size 35–40 g/100, pod yield 1.2 t/ha, tolerant to bacterial wilt
Landak	Spanish	Schwartz 21 / Spanish	1989	Adapted to upland and lowland cultivation, 90–95 days to maturity, seed size 45–50 g/100, pod yield 1.8 t/ha, tolerant to bacterial wilt

TABLE 14.9 *Cont.*

Cultivar	Botanical type	Pedigree	Year of release	Characteristics
Mahesa	Spanish	PI 350680 / Kidang	1991	Adapted to upland and lowland cultivation, 95–100 days to maturity, seed size 46 g/100, pod yield 2.0 t/ha, resistant to bacterial wilt, tolerant to rust
Badak	Valencia	FESR 12 /Local Depok	1991	Adapted to upland and lowland cultivation, 95–100 days to maturity, small seed size, pod yield 2.0 t/ha, tolerant to bacterial wilt and leaf spot
Biawak	Spanish	–	1991	Adapted to upland cultivation, 90–95 days to maturity, medium seed size, pod yield 2.0 t/ha, resistant to bacterial wilt
Komodo	Spanish	–	1991	Adapted to upland cultivation, 95–100 days to maturity, medium seed size, pod yield 2.0 t/ha, resistant to bacterial wilt and rust
Vietnam				
Do Bac Giang	Spanish	Local cultivar	–	
Sen Nghe An	Spanish	Local cultivar	–	
Moket	Spanish	Local cultivar	–	
Ly	Spanish	Local cultivar	–	
Giay	Spanish	Local cultivar	–	
Cuc Nghe An	Spanish	Local cultivar	–	
Sutuyen	Spanish	Selection from China	1970	
Tram Xuyen	Spanish	Selection from China	1970	
V 79	Spanish	X-ray mutant of Bachsa 77	1980	
B 5000	Spanish	X-ray mutant of Bachsa 77	1983	
Sen Lai (75–23)	Spanish	Sen Nghe An / Tram Xuyen	1985	

TABLE 14.9 *Cont.*

Cultivar	Botanical type	Pedigree	Year of release	Characteristics
Philippines				
UPL Pn 6	–	CES 103 / PI 298115	1986	
UPL Pn 8	–	CES 101 / PI 298115	1989	
BPI Pn 2	–	–	–	
UPL Pn 2	Spanish	Moket	1976	
UPL Pn 4	Valencia	Acc 12 (PI 314817)	1978	
BPI P9	Spanish	E.G. Red / Fante 17	1973	
CES 101	Spanish	Pureline selection from unknown cultivar	1973	
Thailand				
Khon Kaen 60–1	Spanish	Moket	–	
Khon Kaen 60–2	Valencia	TMV 3	1988	
Khon Kaen 60–3	Virginia	Selection from NC 7	1988	
Lampang	Valencia	–		
SK 38	Valencia	Selection in local cultivar		
Tainan 9	Virginia (bunch)	Introduction		
Malaysia				
MKT 1			1990	

Not much is known about Laos and Cambodia. The Peanut CRSP of USAID in Thailand and the Philippines, ACIAR of Australia in Indonesia, and IDRC of Canada in Thailand have supported or continue to support groundnut research in the region. ICRISAT has played an important role in introducing improved germplasm in the region. In Thailand and the Philippines, the national groundnut programmes have strong multidisciplinary teams of scientists. In addition to introducing improved germplasm, hybridization has been commonly adopted to develop new cultivars in Indonesia, Thailand and the Philippines. Several cultivars have been released in the region (Table 14.9). In Indonesia, almost all improved cultivars are either resistant or tolerant to bacterial wilt; Schwartz 21, the first disease-resistant groundnut cultivar developed through hybridization, was released here as early as 1927.

Groundnut research activity in Malaysia is very limited. Improved germplasm introduced from ICRISAT and other sources is evaluated for local adaptation, including resistance to prevailing diseases and insect pests.

The research programme in Vietnam is in its infancy and suffers from lack of trained manpower, poor infrastructure and paucity of resources. However, in collaboration with ICRISAT, breeding activities covering resistance to foliar diseases (late leaf spot and rust) and bacterial wilt, earliness, high yield and improved seed quality have been initiated recently. ICRISAT is developing single-seed descent breeding populations derived from crosses between Vietnamese cultivars and other desirable donor parents at its centre in India: at the F_5 stage, these populations will be grown in Vietnam for *in situ* selection.

In Indonesia, the main objective of the groundnut improvement programme is to improve yield potential and adaptation to varying agroecology and cropping systems. The specific issues that receive attention are early maturity, tolerance to excess soil moisture, tolerance to drought, tolerance to soil acidity, tolerance to mineral toxicities, adaptation to inter- and mixed-cropping, tolerance/resistance to insect pests and diseases, and tolerance to heat. A massive field screening exercise was undertaken in Indonesia to evaluate *Arachis* germplasm for resistance to peanut stripe virus. No resistance was found among 9000 lines of *A. hypogaea*; among 54 accessions of wild *Arachis* species, only *A. cardenasii* was immune. A few others showed a resistant reaction.

The primary objective of groundnut breeding in the Philippines is to develop groundnut cultivars with desirable agronomic traits such as high yield, early maturity, acceptable quality and resistance to rust, late leaf spot, *A. flavus*, leafhopper and spider mites. In addition, the improved cultivars should have tolerance/adaptation to drought, partial shade and acidic soil conditions, and improved nitrogen-fixing ability. From the screening activities, several promising sources of desirable characters have been identified for use in the breeding programme (PCARRD, 1985).

They are:

Rust:
  PI 259653
  PI 109839
  ICGS 55

Sclerotium wilt:
  IPB Pn 82-71-27
  IPB Pn 82-68-16

Local factors:

Drought:
  Acc 847
  EG Pn 12

Acid soils:
  IPB Pn 24-2
  IPB Pn 24-3
  IPB Pn 26-4
  BPI P9
  UPL Pn 4

Multiple insect pests:
  NC Ac 343
  NC Ac 2214
  ICG(FDRS)11
  Bhairwa

Shade:
  UPL Pn 2
  IPB Pn 12-14
  ICGS(E)123
  ICGS(E)120

High nitrogenase activity:
  RLRS 5
  RLRS 7
  IPB Pn 49-12
  57-422

The most active groundnut breeding programme in the region is that of Thailand, the objectives of which include: high yield and earliness; adaptation to after-rice, unirrigated condition and before-rice growing conditions; resistance to foliar diseases (rust and late leaf spot); resistance to *A. flavus*, *A. niger* and *Sclerotium rolfsii*; and large-seeded confectionery and boiling-type cultivars. Significant progress is being made in achieving these objectives;. Two cultivars were released recently, and several breeding lines with good promise have been identified and are under evaluation.

14.4.5 Australasia

The Australasian region is not very important from the perspective of global groundnut production. Production in the region is dominated by Australia, which provides high quality groundnuts for world trade during the off-season for producing nations in the northern hemisphere. Major constraints to increased production in Australia include the foliar pathogens (early and late leaf spots and rust); soil-borne diseases (Cylindrocladium black rot, Sclerotinia blight, and *A. flavus*); and drought. Other countries in the region include Papua New Guinea, Solomon Islands, Vanuatu, Tonga, New Zealand and Fiji, all of which produce only limited amounts of groundnut for local consumption.

Australia has the most active research programme in the region. Prior to the programme of varietal improvement started in 1977–78, the primary source of cultivars in Australia was introduction. A high degree of mechanization permits widespread use of cultivars with spreading or runner growth habits. Large-seeded virginia-type cultivars such as Shulamith and NC 7 are preferred here. The spanish cultivar 'McCubbin' was released by the Australian national programme, the goals of which are yield improvement, quality maintenance (particularly shelf life), and resistance to foliar diseases.

Other countries in the region do not have breeding programmes but still rely exclusively on introduction for new cultivars. Recently, Papua New Guinea and Fiji obtained advanced breeding lines from ICRISAT for evaluation and *in situ* selection. Red-seeded spanish is the preferred type grown in the Solomon Islands and Papua New Guinea.

14.4.6 North America

The United States is the largest producer of groundnuts in North America and conducts the bulk of the groundnut research in the region. Collection, maintenance and evaluation of groundnut germplasm have been high priorities in the USA. Placement of a full-time groundnut curator for the national germplasm collection at Griffin, Georgia, has helped to organize efforts in this area. During the last decade, breeders have identified considerable germplasm that can be used to improve the groundnut (Wynne and Halward, 1989b). At the same time, collection expeditions have continued to add to the diversity available for improvement of the groundnut (Simpson, 1983, 1990).

Utilization of the wild species of *Arachis* to improve the cultigen has been investigated in the USA by research programmes in North Carolina, Oklahoma and Texas. Much of the research has been concerned with the crossing relationships among the various species and with cultivated groundnuts. Pathways for the transfer of genetic material from the species to cultivated groundnuts have been established (Simpson, 1991; Stalker and Moss, 1987). The progress of research in this area has been reviewed recently (Wynne and Halward, 1989b; Stalker and Moss, 1987).

Cultivar development programmes at state experiment stations in Florida, Georgia, Oklahoma, North Carolina, Texas and Virginia and at a private company (formerly Gold Kist; now Agratech) released numerous cultivars (Table 14.10). Over the past 10 years, these have broadened the genetic base of the groundnut crop in the USA and provided sources of pest resistance (Knauft and Gorbet, 1989). Knauft and Gorbet assessed the genetic diversity among cultivars released by 1988 and concluded that the genetic base had been broadened considerably since 1976. This broadening has continued through additional cultivar releases since this report (Isleib and Wynne, 1992).

Cultivars released for their pest resistance include NC 6 (southern corn rootworm), NC 8C and NC 10C (cylindrocladium black rot), Va 81B

TABLE 14.10 *Groundnut cultivars released in the United States*

Cultivar	Market type	Pedigree	Year of release
Florigraze*	Rhizoma	Selection from PI 118457 (*Arachis glabrata* Benth. cv. 'Arb', collected by W.A. Archer near Campo Grande, Brazil, in 1936)	1978
Arbrook	Rhizoma	PI 262817 (*Arachis glabrata* Benth. collected by W.C. Gregory (Col. No. 9569) near Trinidad, Itapua Department, Paraguay, in 1959)	1985
Dixie Runner	Runner	Small White Spanish 3x-1 / Dixie Giant	1943
Virginia Bunch 67*	Runner	Selection from 'Virginia Bunch' obtained in 1941 from East Georgia Peanut Co., Bulloch Co., GA	1945
Southeastern Runner 56–15*	Runner	Selection from 'Southeastern Runner'	1947
Early Runner	Runner	Small White Spanish 3x-2 / Dixie Giant	1952
Florispan Runner	Runner	Basse / Spanish 18–38, GA 207–3 // Early Runner	1953
Georgia 119–20	Runner	Southeastern Runner / Dixie Giant, 210–4 // Virginia Runner	1954
Florunner	Runner	F334A–3–14 (Florispan sib) / F230–118–B–8–1 (Early Runner sib)	1969
Altika	Runner	F393–7–1 (NC-FLA 14 sib) /3/ GA 119–20, Southeastern Runner / Dixie Giant, 210–14 // Virginia Runner	1972
GK19	Runner	F334–3–5–5–1 (Florispan derivative) / Jenkins Jumbo, F393–6 // F334–9 (Florispan sib)	1973
Tifrun	Runner	Florida Small Spanish / Dixie Giant, F231–51 /4/ F385–1–7–2, Pearl (F228) // F68–74 S_3–1–2, McSpan (F13, Small White Spanish) / Virginia Jumbo Runner (F14), F249–42–3–1 /3/ Jenkins Jumbo, T1645 (selection from F416) / T1861, selection made in 1966 from local virginia stock in Georgia, thought to have arisen from a virginia × spanish hybrid)	1977
GK7	Runner	F334–3–5–5–1 (Florispan derivative) / Jenkins Jumbo, F393–2 // GK 19	1982
Sunbelt Runner	Runner	F392–12–B–28 (Florigiant sib) / VA Bunch 67, A4–4 // Florunner	1982
Sunrunner	Runner	F439–16–10–1–1 (Florunner component) // UF393–7–1 (NC-FLA 14 sib), UF334A–3–5–5–1 (Florispan derivative) / Jenkins Jumbo	1982
Southern Runner	Runner	PI 203396 (resistant to *Cercospora arachidicola* and *Phaeoisariopsis personata*) / Florunner	1984
Langley	Runner	Florunner/PI 109839	1986

TABLE 14.10 *Cont.*

Cultivar	Market type	Pedigree	Year of release
Okrun	Runner	Florunner / Spanhoma	1986
Tamrun 88	Runner	Goldin I (Wilson County Peanut Co., Pleasanton, TX) / Florunner	1988
Georgia Runner	Runner	Krinkle-leaf (var. *vulgaris*) / PI 331334 ('Criollo', var. *hypogaea* from Bolivia)	1990
MARC I	Runner	Early Runner / Florispan, F439–17–2–1–1 (Florunner sib) // F459B–3–2–4–6–2–2–1 (Early Bunch component)	1990
Improved Spanish 2B*	Spanish	Selection from local Spanish made *c.* 1918 at Florence, SC	
Spanish 18–38*	Spanish	Selection from farmers' spanish stocks	
Spanish No. 146	Spanish	Spanish introduction (Coll. No. 146) obtained from India by Tom Huston Peanut Co.	
GFA Spanish*	Spanish	Selection from 'Small Spanish' obtained from a grower in 1930	1941
Spantex*	Spanish	Selection from farmers' spanish stocks	1948
Dixie Spanish*	Spanish	Selection from Spanish introduction (Coll. No. 146) obtained from India by Tom Huston Peanut Co.	1950
Argentine*	Spanish	Selection from PI 121070 (var. *vulgaris*)	1951
Spanette*	Spanish	Selection from Spanish 18–38	1959
Starr	Spanish	Spantex / PI 161317 (var. *vulgaris* obtained in 1947 from Salto, Uruguay)	1961
Spanhoma*	Spanish	Selection from 'Argentine'	1969
Comet*	Spanish	Selection from 'Starr'	1970
Spancross	Spanish	Argentine (PI 121070–1) / PI 405933 (*Arachis monticola*)	1970
Tifspan	Spanish	Argentine (PI 121070–1) / Spanette	1970
Tamnut 74	Spanish	Starr // TPL 647–2–5, Spantex / *Arachis monticola*	1974
Goldin I	Spanish	Obtained from E. Goldin, Faculty of Agriculture, Hebrew University of Jerusalem, Rehovot, Israel by the Wilson Co. Peanut Co., Pleasanton, TX	1976
Toalson	Spanish	PI 221057 (var. *vulgaris*) / Selection 26 (Spantex sib), TPL 673–A // Starr	1979
Pronto	Spanish	Chico / Comet	1980

TABLE 14.10 *Cont.*

Cultivar	Market type	Pedigree	Year of release
Spanco	Spanish	Chico / Comet	1981
Tennessee Red*	Valencia	Selection from farmers' valencia stocks	1971
New Mexico Valencia A*	Valencia	Selection from 'New Mexico Valencia'	1971
McRan*	Valencia	Selection from African plant introduction	1973
New Mexico Valencia C*	Valencia	Selection from PI 355987, irradiated 'Colorado Manfredi' obtained from the research station at Manfredi, Argentina	1979
Georgia Red	Valencia	UF439–16–10–3 (Florunner component) / New Mexico Valencia A	1986
NC 4*	Virginia	Selection from 100 plant isolations made in 1929 from NC farmers' cultivars by P.H. Kime, NCSU agronomist, Selection #4 deemed typical virginia bunch	1944
Holland Jumbo*	Virginia	Selection from farmers' virginia stocks	1945
Holland Virginia Runner*	Virginia	Selection from farmers' virginia stocks	1945
Virginia Bunch 46–2*	Virginia	Selection from 'Virginia Bunch Large'	1952
Virginia Bunch G2*	Virginia	Selection from 'Virginia Bunch' obtained in 1941 from East Georgia Peanut Co., Bulloch Co., GA	1952
Virginia Bunch G26*	Virginia	Selection from 'Virginia Bunch' obtained in 1941 from W.A. Groover, Bulloch Co., GA	1952
NC 1	Virginia	NC4 / Improved Spanish 2B	1952
NC 2	Virginia	Basse / Spanish 18–38, GA 207–2 // White's Runner	1952
Virginia 56R*	Virginia	Selection from 'Atkins Runner'	1956
NC 4X	Virginia	Selection from irradiated 'NC 4'	1959
Florigiant	Virginia	Basse / Spanish 18–38, GA 207–3 // F230–118–2–2 (same as F230), F334A–5–5–1 /3/ F359–1–3–14, Jenkins Jumbo // F230–118–5–1, Dixie Giant / Small White Spanish 3x-2	1961
Virginia 61R*	Virginia	Selection from 'Atkins Runner'	1962
NC 5	Virginia	NC 1 // C12, PI 121067 / NC Bunch	1964
Shulamith	Virginia	Florigiant / F334A–B–17–1 (Florispan derivative)	1968

TABLE 14.10 Cont.

Cultivar	Market type	Pedigree	Year of release
NC17	Virginia	F334A–3–5–5–1 (Florispan derivative) / Jenkins Jumbo	1969
Virginia 72R	Virginia	VA 61R / VA A89–15 (selection from farmers' stocks, perhaps Atkins Runner)	1971
NC-FLA 14	Virginia	Jenkins Jumbo / F334A–3–5–5–1 (Florispan derivative)	1973
Keel 29*	Virginia	Selection from 'Florigiant'	1974
Avoca 11*	Virginia	Selection from 'NC 2'	1976
GK 3	Virginia	Florida Small Spanish / Dixie Giant, F231–51 /4/ F385–1–7–2, Pearl (F228) // F68–74 S_3–1–2, McSpan (F13, Small White Spanish) / Virginia Jumbo Runner (F14), F249–42–3–1 /3/ Jenkins Jumbo, F416–2 /5/ F392 (Florigiant sib)	1976
NC 6	Virginia	NC Bunch / PI 121067, C12 // C37 (same as C12), GP-NC 343 (selection from NC Ac 4508) // VA 61R: Resistant to SCR	1976
Early Bunch	Virginia	Virginia Station Jumbo /4/ F385–1–7–4, Pearl (F228) // F68–74 S_3–1–2, McSpan (F13, Small White Spanish) / Virginia Jumbo Runner (F14), F249–42–3–1 /3/ Jenkins Jumbo, F406A /5/ F420, F231–51 (Dixie Runner sib) / F392–12–1–7 (Florigiant sib)	1977
NC 7	Virginia	NC 5 // F393, F334–3–5–5–1 (Florispan derivative) / Jenkins Jumbo	1978
VA 81B	Virginia	F392–8 (Florigiant sib) /3/ GA 119–20, Southeastern Runner / Dixie Giant, 210–14 // Virginia Runner	1981
NC 8C	Virginia	NC 2 // A48, NC 4 / Spanish 2B, NC Ac 3139 /3/ Florigiant	1982
NC 9	Virginia	NC 2 / Florigiant	1985
NC10C	Virginia	NC 8C / Florigiant	1988
NC-V 11	Virginia	Florigiant / NC 5 // Florigiant / PI 337396 (var. *fastigiata*)	1989
VA-C 92R	Virginia	Florigiant // F393, F334–3–5–5–1 (Florispan derivative) / Jenkins Jumbo, NC Ac 17213 /3/ NC 7	1992
VC-1	Virginia	F334–3–5–5–1 (Florispan derivative) / Jenkins Jumbo, F393 // F334 (Florispan derivative) / F393 /3/ F392 (Florigiant sib) / GA 186–28 // F439 (Florunner component)	1991

* Developed by selection within a plant introduction or existing cultivar.

(sclerotinia blight) and Southern Runner (late leaf spot). The cultivars that have been released primarily for their pest resistance have generally compromised one or more agronomic traits, making them less competitive in absence of the pest.

Considerable effort to develop pest-resistant groundnut cultivars began during the 1980s in the USA. Wynne *et al.* (1991) summarized progress in breeding for disease resistance. They concluded that although several USA breeding programmes had been initiated for resistance to diseases – *Aspergillus* spp. (aflatoxin), tomato-spotted wilt virus, nematodes, early and late leaf spots, sclerotinia blight, and cylindrocladium black rot – few cultivars had yet been released, due to the short duration of the efforts. However, many sources of disease resistance were identified by screening programmes during the 1980s and breeding for disease resistance is now a priority in most USA programmes. Much progress can be expected.

Considerable effort in the USA has also been devoted to the use of wild species of *Arachis* for sources of resistance to pests. Programmes to transfer the high levels of resistance or immunity to early and late leaf spots, rust, nematodes and viruses were active during the 1980s (Stalker and Moss, 1987; Wynne and Halward, 1989a). To date, no cultivar incorporating germplasm from diploid wild species has been released.

Recently the groundnut industry identified quality and aflatoxin resistance as two major issues that needed additional research and were considered of highest priority because of the effect they have on the export of groundnuts. Substantial funding from the National Peanut Foundation and USDA has increased conventional breeding and molecular genetic research to address these problems.

Several researchers in the USA are now investigating and developing methodologies to use molecular techniques for groundnut improvement. The use of RFLPs as molecular markers is being investigated by a University of Georgia researcher (Kochert and Branch, 1990) in cooperation with several others. Little variation has been reported among cultivars but abundant polymorphism has been found among the diploid species of *Arachis*. Similar results were found for isozymes (Grieshammer and Wynne, 1990; Stalker *et al.*, 1990).

Several USA researchers are investigating somatic embryogenesis and plant regeneration in groundnuts. At least two laboratories have developed a repetitive somatic embryogenesis system and have established plants in soil (Durham *et al.*, 1991; A. Weissinger, North Carolina State University, personal communication, 1991). These successes should expedite the use of gene transfer systems in the crop. The use of microprojectile bombardment as part of a gene transfer system in groundnut is being evaluated in at least two laboratories. The success of these systems will allow the movement of agronomically important genes into the groundnut.

Several laboratories are identifying and sequencing genes from viruses and from other plants that may be useful in improving the groundnut. This research is receiving funding support from the Peanut CRSP, private companies, the USDA and state experiment stations.

14.4.7 South America

The area under commercial groundnut production in South America is about 350 000 ha. Argentina ranks first in groundnut area in the region (180 000 ha), followed by Brazil (100 000 ha) and Paraguay (30 000 ha). The area in other countries, such as Bolivia, Chile, Ecuador, Peru, Uruguay and Venezuela, does not exceed 5000 ha. Although the region's area under groundnuts has been declining, the total production has not suffered significantly, due to increase in productivity: average yields in the 1980s were nearly 50% higher than those of the 1970s. New crop production technology and improved cultivars have contributed to increased yields.

Average seed yield in Argentina has increased from 0.79 t/ha in the 1970s to 1.20 t/ha in the 1980s. During the 1970s two valencia cultivars, Colorado Irradiado INTA and Blanco Rio Segundo, contributed 80% to the total groundnut production; the remaining 20% was contributed by Blanco Manfredi 68, a derivative of a cross between virginia and spanish types (Godoy and Giandana, 1992). Since then, the varietal picture in the country has changed completely as virginia runner types proved better adapted to the main groundnut-growing region of the country. By 1989, Florman INTA and Florunner accounted for 80% of the total groundnut area and production. This development is somewhat disturbing in view of the international community's expressed desire to maintain genetic diversity in food crops, particularly in the centres of diversity for those species. However, it is necessary to balance against this desire the needs of the individuals and nations in those regions. The good adaptation of runner types to the climate and soil in the country led Argentina to become the third largest exporter of edible groundnut in the world, after the USA and the People's Republic of China.

Average pod yields in Brazil in rainy seasons are 2.0–2.1 t/ha and in the dry season about 1.5 t/ha. The reduction in area and production has been dramatic: the cultivated area in 1972 was 759 000 ha and it declined to 100 000 ha in 1988, while production fell from 956 000 to 167 000 t in the same period. The main reason for such a sharp decline was establishment of soybean as the leading oilseed crop in the country. However, the average yield of groundnut has increased from 1.5 t/ha in the 1970s to 1.8 t/ha in the 1980s in São Paulo province (the main groundnut-growing area in the country). In the Ribeirao Preto region, the pod yield averages 2.5 t/ha but yields up to 4.0 t/ha can be obtained with the red valencia cultivar Tatu, which has a short growing cycle of 90–100 days and now occupies 80% of the groundnut-growing area in São Paulo. Another cultivar, Tatu Branco, which is similar to Tatu except for its seed colour, occupies 10% of the groundnut area; it has undefined tolerance to drought and is adapted to low fertility. Recently three new cultivars with 15–20% higher yield than Tatu have been released – Tupa, Oira and Poitara. All three are derived from valencia-by-spanish crosses; they mature in 110–120 days and have two-seeded medium sized pods.

Groundnut cultivation in Bolivia is manual and local cultivars are grown. The three main local cultivars are Coloradito Palmer, an erect type with 125 days maturity; Cuero Padilla, a semi-erect type with 135 days maturity; and Bayo Gigante (also called Colorado Grande), a runner type with 145 days maturity.

In Paraguay, groundnuts are grown in three regions which differ in soil type, climatic conditions and level of technology input. In the Chaco region, cultivation is mechanized and spanish cultivars are grown. In the central region, valencia and spanish types are grown by small farmers in less fertile soils with low levels of technology input. In the southern region, long-season virginia types are cultivated on fertile soils. The present yields in Paraguay (1.3 t/ha) are 50% higher than those of 20 years ago.

Pod yields in Uruguay range from 0.7–1.8 t/ha. Groundnuts are generally cultivated by small farmers on acidic sandy soils which are low in Ca content, with family labour and little technology. Predominantly valencia types are grown; spanish and virginia types are also cultivated to a limited extent.

In 1990, the southern nations of South America (Argentina, Brazil, Bolivia, Paraguay and Uruguay) initiated a co-operative research effort called PROMANI (pro = program; maní = groundnut). Its objective is to promote groundnut research and extension activities in the participating countries. Research in Argentina and Brazil has been intensified since the early 1980s. Both countries have their own active breeding programme, whereas other PROMANI countries rely mostly on introduction of improved cultivars and selection in local cultivars/landraces.

In Argentina, groundnut research is focused on studies of the taxonomy of the genus *Arachis* and on the development of new cultivars of medium duration (125–130 days) with tolerance to drought and leaf spots, resistance to *A. flavus* infection, high O/L ratio, low iodine value, improved content and quality of seed proteins, and improved flavour and aroma. In Brazil, research objectives include germplasm collection and taxonomic studies on genus *Arachis*; breeding for resistance to late leaf spots; selection of early-maturing, high-yielding, red-seeded valencia/spanish types with improved pod/seed appearance and shelling out-turn; development of high-yielding virginia runner cultivars with resistance to leaf spot and other diseases; resistance to *Aspergillus* infection; and resistance to thrips.

Development of high-yielding, leaf spot-resistant cultivars with acceptable agronomic and quality attributes will benefit the South American region most. Other diseases which could be potentially important in the region are rust, scab and *Sclerotium*. Except for some areas in Argentina, groundnuts in the region are generally grown under rainfed conditions. Drought is the most common abiotic stress in the region.

14.5 ACCOMPLISHMENTS AND FUTURE EFFORTS

On a worldwide basis, the most important results of groundnut breeding in the past 10–20 years have been the identification of sources of resistance to the three globally important foliar fungal pathogens and the transfer of resistance into breeding populations with the locally appropriate agronomic attributes. It remains to be seen whether release of cultivars resistant to rust and leaf spot will significantly affect patterns of groundnut production.

Closely following the foliar diseases in importance is the aflatoxin problem. Despite the identification of seed-coat resistance and the release of IVSCAF-resistant cultivars, aflatoxin contamination remains the largest single problem affecting international trade in groundnut. The recent adoption by the European Community of extremely low tolerances for aflatoxin may eliminate some nations from the array of groundnut exporters. Although this problem is certainly not confined to the realm of plant breeding, the international community looks primarily to a genetic solution.

Breeding for resistance to insect pests has not been emphasized to the same degree as breeding for resistance to foliar diseases. Common foliar diseases occur with great regularity in most parts of the world while many insect species require particular environmental conditions in order to reach the population densities necessary to cause economic damage. Under management systems with minimal or no application of insecticides and fungicides, insect pest populations may be curbed by the presence of predatory insects and animal or fungal parasites. Host, pests, predators and parasites exist in a balance sensitive to subtle changes in the ecological dynamic. In such production systems, pest resistance of low or intermediate level may be sufficient to shift the balance in favour of the host plant. In many developing nations, the microeconomics of the production system and the infrastructure for distribution and acquisition of pesticides prohibit widespread use of pesticides for control of insects. Host-plant resistance to insects will be the most effective means of reducing losses in yield and quality associated with insect depredation.

Under intensive management systems, insect pests are controlled by applications of pesticides that may also destroy beneficial species whose absence allows unchecked growth of pest species that develop later in the growing season, thereby necessitating further applications of pesticides. Recently, socioeconomic forces in developed nations have created the concept of LISA – low-input sustainable agriculture – as a paradigm for mechanized agricultural production systems. These forces include demand by consumers for agricultural products free from pesticide residues, public concern over the effects of pesticides on the environment, reduction of production costs, and the increasing difficulty encountered by manufacturers of pesticides in obtaining government approval for their registration and sale. Key concepts of LISA include minimal application of pesticides that have potentially harmful effects on consumers or environment; emphasis on soil conservation, including reduced tillage and use of green manure animal waste as sources of organic matter and incorporation of leguminous species into rotations to reduce use of mineral fertilizers that can contaminate groundwater supplies. In short, LISA comprises a set of production practices which by choice avoid extensive reliance on the products of the chemical revolution that has so dramatically changed the face of agriculture in developed nations in the last 50–60 years. While use of herbicides and fungicides are affected by these practices, insecticides are probably affected most because of their generally greater acute toxicity to mammals. This trend may provide impetus for increased efforts in breeding for insect resistance in developed nations. It remains to be seen whether the consuming public in developed nations is sufficiently desirous of pesticide-free produce to accept groundnuts bearing evidence of insect feeding. Assuming that it is not, countries supplying edible groundnuts will need to deploy cultivars with high levels of resistance to insects, a practice that will certainly place strong selective pressure on pest populations.

Until recently, the gene pool for cultivated groundnut comprised the global collection of the cultigen (some 12 000 accessions) and the smaller collection of *Arachis* species of which genes only from species of section *Arachis* were accessible through sexual transfer. In the summer of 1992, researchers from several public and private institutions reported success in transforming groundnut with exogenous DNA and regeneration of fertile plants from transformed tissues. Transformation has been effected through microprojectile bombardment of embryonic axes (Brar *et al.*, 1992), embryogenic immature cotyledonary tissue (Weissinger *et al.*, 1992) and callus derived from embryonic leaflets (Weissinger *et al.*, 1992), and through electroporation of protoplasts (Demski *et al.*, 1992). It would appear that the array of transformation techniques effective in soybean can be adapted to groundnut through modification of protocols. This effectively converts the gene pool from a portion of the genetic information in genus *Arachis* to virtually all genes in the planetary biosphere. The key problem in groundnut breeding is changing from location of sources of useful genes within the cultigen to identification of the genes *per se*, i.e. the DNA base sequences, of potential economic value in groundnut regardless of the source of those genes. Issues of the proprietary nature of such genes and the payment of royalties, particularly by groundnut producers in developing nations, will doubtless interest the groundnut breeding community for decades to come.

REFERENCES

Ahmed, E.M. and Young, C.T. (1982) Composition, quality, and flavour of peanuts, in *Peanut Science and Technology*, (eds C.T. Young and H.E. Pattee), American Peanut and Research Education Society, Yoakum, Texas, pp. 655–688.

Amaya, F.-J., Young, C.T., Norden, A.J. and Mixon, A.C. (1980) Chemical screening for *Aspergillus flavus* resistance in peanut. *Oléagineux*, **35**, 255–257.

Aujla, S.S., Chohan, J.S. and Mehan, V.K. (1978) The screening of peanut varieties for the accumulation of aflatoxin and their relative reaction to the toxigenic isolate of *Aspergillus flavus* Link ex Fries. *Journal of Research of the Punjab Agricultural University*, **15**, 400–403.

Bailey, W.K., Stone, E., Broomfield, K.R. and Garren, K.H. (1973) *Notice of release of peanut germplasm with resistance to rust*, Virginia Agricultural Experiment Station, Blacksburg, VA, and USDA Agricultural Research Service, Washington, DC, 3 pp.

Banks, D.J. (1976) Peanuts: Germplasm resources. *Crop Science*, **16**, 499–502.

Bartz, Z.A., Norden, A.J., LaPrade, J.C. and Demuynk, T.J. (1978) Seed tolerance in peanut (*Arachis hypogaea* L.) to members of the *Aspergillus flavus* group of fungi. *Peanut Science*, **5**, 53–56.

Beute, M.K., Wynne, J.C. and Emery, D.A. (1976) Registration of NC 3033 peanut germplasm. *Crop Science*, **16**, 887.

Blankenship, P.D., Cole, R.J., Sanders, T.H. and Hill, R.A. (1984) Effect of geocarposphere temperature on pre-harvest colonization of drought stressed peanuts by *Aspergillus flavus* and subsequent aflatoxin. *Mycopathologia*, **85**, 69–74.

Bock, K.R. (1987) Rosette and early leaf spot diseases: a review of research progress, 1984/85, in *Proceedings of the Second Regional Groundnut Workshop in Southern Africa, 10–14 February 1986, Harare, Zimbabwe*, ICRISAT, Patancheru, pp. 5–14.

Bock, K.R. (1989) ICRISAT Regional Groundnut Pathology Program: A review of research progress during 1985–87 with special reference to groundnut streak necrosis disease, in *Proceedings Third Regional Groundnut Workshop for Southern Africa, 13–18 March 1988, Lilongwe, Malawi*. ICRISAT, Patancheru, pp. 13–20.

Bockelee-Morvan, A. (1983) Les différentes variétés d'arachide: répartition géographique et climatique, disporobilité. *Oléagineux*, **38**, 73–116.

Bockelee-Morvan, A., Gautreau, J., Mortreuil, J.C. and Russel, O. (1974) Results obtained with drought-resistant groundnut varieties in West Africa. *Oléagineux*, **29**, 309–314.

Bouhot, D. (1967) Observations sur quelques affections des plantes cultivées au Senegal. *L'Agronomie Tropicale*, **22**, 888–890.

Boshou, L., Yuying, W., Xingming, X. *et al.* (1990) Genetic and breeding aspects of resistance to bacterial wilt in groundnut, in *Bacterial wilt of groundnut* (eds K.J. Middleton and A. C. Hayward), Proceedings of an ACIAR/ICRISAT Collaborative Research Planning Meeting held at Genting Highlands, Malaysia, 18–19 March 1990. ACIAR Proceedings No. 31, Australian Centre for International Agricultural Research, Canberra, pp. 39–43 of 58 pp.

Brar, G. (1992)

Brim, C.A. (1966) A modified pedigree method of selection. *Crop Science*, **6**, 220–221.

Bromfield, K.R. and Bailey, W.K. (1972) Inheritance of resistance to *Puccinia arachidis* in peanut. *Phytopathology*, **62**, 748 (Abstr.).

Bromfield, K.R. and Cevario, S.J. (1970) Greenhouse screening of peanut (*Arachis hypogaea*) for resistance to peanut rust (*Puccinia arachidis*). *Plant Disease Reporter*, **54**, 381–383.

Brown, D.F., Cater, C.M., Mattil, K.F. and Darroch, J.G. (1975) Effect of variety, growing location, and their interaction on the fatty acid composition of peanuts. *Journal of Food Science*, **40**, 1055–1060.

Buddenhagen, I.W. and Kelman, A. (1964) Biological and physiological aspects of bacterial wilt caused by *Pseudomonas solanacearum*. *Annual Review of Phytopathology*, **2**, 203–230.

Bunting, A.H., Wynne, J.C. and Gibbons, R.W. (1985) Groundnut (*Arachis hypogaea* L.), in *Grain Legume Crops*, (eds. R.J. Summerfield and E.H. Roberts), Collins Professional and Technical Books, London, pp. 747–800.

Busolo-Bufalu, C.M. (1990) Groundnut improvement program in Uganda, in *Proceedings of the Fourth Regional Groundnut Workshop for Southern Africa, 19–23 March, Arusha, Tanzania*, ICRISAT, Patancheru, pp. 55–59.

Coffelt, T.A. and Porter, D.M. (1982) Screening peanuts for resistance to sclerotinia blight. *Plant Disease*, **66**, 385–387.

Cole, R.J., Sanders, T.H., Dorner, J.W. and Blankenship, P.D. (1989) Environmental conditions required to produce preharvest aflatoxin contamination of groundnuts: summary of six years' research, in *Aflatoxin Contamination of Groundnuts: Proceedings of the International Workshop, 6–9 October 1987, ICRISAT Center*, (eds D. McDonald and V.K. Mehan), ICRISAT, Patancheru, pp. 279–287.

Davidson, J.I., Jr, Hill, R.A., Cole, R.J. *et al* (1983) Field performance of two peanut cultivars relative to aflatoxin contamination. *Peanut Science*, **10**, 43–47.

de Berchoux, C. (1958) Étude sur la résistance de l'arachide en Haute Volta. Premiers resultants. *Oléagineux*, **13**, 237–239.

de Berchoux, C. (1960) La rosette de l'arachide en Haute Volta. Comportement des lignes resistantes. *Oléagineux*, **15**, 237–239.

Doupnik, B., Jr (1969) Aflatoxins produced on peanut varieties previously reported to inhibit production. *Phytopathology*, **59**, 1554.

Doupnik, B., Jr and Bell, D.K. (1969) Screening peanut breeding lines for resistance to aflatoxin accumulation. *Journal of the American Peanut Research and Education Association*, **1**, 80–82.

Durham, R.E., Parrott, W.A., Baker, C.M. and Wetzstein, H.Y. (1991) Repetitive somatic embryogenesis and plant regeneration in peanut. *Agronomy Abstracts*, **83**, 194.

Dwivedi, S.L., Amin, 'P.W., Rasheedunisa, Nigam, S.N. *et al*. (1986) Genetic analysis of trichome characters associated with resistance to jassid (*Empoasca kerri* Pruthi) in peanut. *Peanut Science*, **13**, 15–18.

FAO (Food and Agriculture Organization of the United Nations) (1990) *FAO Year Book – Production 1989*, Statistics Series No. 94, Vol. 43, pp. 157–158.

Foster, D.J., Stalker, H.T., Wynne, J.C. and Beute, M.K. (1981) Resistance of *Arachis hypogaea* L. and wild relatives to *Cercospora arachidicola* Hori. *Oléagineux*, **36**, 139–143.

Foster, D.J., Wynne, J.C. and Beute, M.K. (1980) Evaluation of detached leaf culture for screening peanuts for leaf spot resistance. *Peanut Science*, **7**, 98–100.

Gautreau, J. and De Pins, O. (1980) Groundnut production and research in Senegal, in *Proceedings of the International Workshop on Groundnuts, 13–17 October*, ICRISAT, Patancheru, pp. 274–281.

Ghanekar, A.M. (1980) Groundnut virus research at ICRISAT, in *Proceedings of the International Workshop on Groundnuts, 13–17 October 1980*, ICRISAT, Patancheru, pp. 211–216.

Ghewande, M.P., Nagaraj, G. and Reddy, P.S. (1989) Aflatoxin research at the Indian National Research Center for Groundnut, in *Aflatoxin Contamination of Groundnuts: Proceedings of the International Workshop, 6–9 October 1987*, (eds D. McDonald and V.K. Mehan), ICRISAT, Patancheru, pp. 237–243.

Gibbons, R.W., Bunting, A.H. and Smartt, J. (1972) The classification of varieties of groundnut (*Arachis hypogaea* L.). *Euphytica*, **21**, 78–85.

Gocho, H. (1991) Breeding for eating quality in groundnut in Japan. Paper presented at the Second International Groundnut Workshop, 25–29 November 1991, ICRISAT Center, Patancheru.

Godoy, I.J. and Giandana, E. (1992) Groundnut production and research in South America, in *Proceedings of the Second International Groundnut Workshop, 25–29 November 1991*, ICRISAT, Patancheru.

Gorbet, D.W., Shokes, F.M. and Jackson, L.J. (1982) Control of peanut leafspot with a combination of resistance and fungicide treatment. *Peanut Science*, **9**, 87–90.

Green, C.C. and Wynne, J.C. (1987) Genetic variability and heritability for resistance to early leaf spot in four crosses of virginia-type peanut. *Crop Science*, **27**, 18–21.

Green, C.C., Beute, M.K. and Wynne, J.C. (1983) A comparison of methods of evaluating resistance to *Cylindrocladium crotalariae* in peanut field tests. *Peanut Science*, **10**, 66–69.

Gregory, W.C., Gregory, M.P., Krapovickas, A. *et al.* (1973) Structure and genetic resources of peanuts, in *Peanuts – Culture and Uses*, (ed. C.A. Wilson), American Peanut Research and Education Association, Inc., Stillwater, Oklahoma, pp. 47–133.

Grieshammer, U. and Wynne, J.C. (1990) Mendelian and non-Mendelian inheritance of three isozymes in peanut (*Arachis hypogaea* L.). *Peanut Science*, **17**, 101–105.

Guok, H.P., Wynne, J.C and Stalker, H.T. (1986) Recurrent selection within a population from an interspecific peanut cross. *Crop Science*, **26**, 249–253.

Halward, T.M., Stalker, T., LaRue, E. and Kochert, G. (1992) Use of single-primer DNA amplification in genetic studies of peanut (*Arachis hypogaea* L.). *Plant Molecular Biology*, **18**, 315–325.

Hammons, R.O. (1977) Groundnut rust in the United States and the Caribbean. *PANS*, **23**, 300–324.

Harkness, C. (1977) *The breeding and selection of groundnut varieties for resistance to rosette virus disease in Nigeria*, Institute for Agricultural Research Report, Ahmadu Bello University, Zaria, Nigeria, 45 pp.

Hartley, W. (1949) *Plant collecting expedition to sub-tropical South America 1947–48. Report.* DW. Plant Industry Australia No. 7.

Hassan, H.N. and Beute, M.K. (1977) Evaluation of resistance to cercospora leaf spot in peanut germplasm potentially useful in a breeding program. *Peanut Science*, **4**, 78–83.

Hildebrand, G. (1985) Use of the single-seed descent method of selection in groundnut breeding in Zimbabwe, in *Proceedings of the Regional Groundnut Research Workshop for Southern Africa, 26–29 March, 1984, Lilongwe, Malawi*, ICRISAT, Patancheru, pp. 137–140.

Holley, R.H., Wynne, J.C., Campbell, W.V. and Isleib, T.G. (1985) Combining ability for insect resistance in peanut. *Oléagineux*, **40**, 203–207.

ICRISAT (1983) *Annual report 1982*, ICRISAT, Patancheru.

ICRISAT (1984) *Annual report 1983*, ICRISAT, Patancheru, 186 pp.

ICRISAT (1985) *Annual report 1984*, ICRISAT, Patancheru, 212 pp.

ICRISAT (1986) *Annual report 1985*, ICRISAT, Patancheru, 250 pp.

ICRISAT (1987) *Annual report 1986*, ICRISAT, Patancheru, 226 pp.

ICRISAT (1988) *Annual report 1987*, ICRISAT, Patancheru, 235 pp.

ICRISAT (1989) *Annual report 1988*, ICRISAT, Patancheru.

Isleib, T.G. and Wynne, J.C. (1992) Use of plant introductions in peanut improvement, in *Use of Plant Introductions in Cultivar Development, Part 2*, (eds H.L. Shands and L.E. Weisner), CSSA Spec. Pub. No. 20, pp. 75–116.

Jambunathan, R., Mehan, V.K. and Gurtu, Santosh. (1989) Aflatoxin contamination of groundnut, in *Aflatoxin Contamination of Groundnuts: Proceedings of the International Workshop, 6–9 October 1987*, (eds D. McDonald and V.K. Mehan), ICRISAT, Patancheru, pp. 357–364.

Karchesy, J.J. and Hemingway, R.W. (1986) Condensed tannins (4B→8→2B→0→7)-linked procyanidins in *Arachis hypogaea* L. *Journal of Agricultural and Food Chemistry*, **34**, 966–970.

Kisyombe, C.T., Beute, M.K. and Payne, G.A. (1985) Field evaluation of peanut genotypes for resistance to infection by *Aspergillus parasiticus*. *Peanut Science*, **12**, 12–17.

Knauft, D.A. and Gorbet, D.W. (1989) Genetic diversity among peanut cultivars. *Crop Science*, **29**, 1417–1422.

Kochert, G. and Branch, W.D. (1990) RFLP analysis of peanut cultivars and wild species. *Proceedings American Peanut Research and Education Society*, **22**, 53 (Abstr.).

Kochert, G.D., Halward, T.M., Branch, W.D. and Simpson, C.E. (1991) RFLP variability in peanut (*Arachis hypogaea* L.) cultivars and wild species. *Theoretical and Applied Genetics*, **81**, 565–570.

Krapovickas, A. (1968) Origen, variabilidad, y difusión del maní (*Arachis hypogaea* L.). Actas y Memorias del XXXVII Congreso Internacional de Americanistas. English translation (1969) The origin, variability, and spread of groundnut (*Arachis hypogaea*), in *The domestication and exploitation of plants and animals* (eds P.J. Ucko and G.W. Dimbleby), Gerald Duckworth Co. Ltd., London, pp. 427–441.

Kuhn, C.W. and Demski, J.W. (1975) *The relationship of peanut mottle virus to peanut production*. Georgia Agricultural Experiment Station Research Report No. 213.

Kulkarni, L.G., Sharief, Y. and Sarma, V.S. (1967) Asirya Mwitunde groundnut gives good results in Hyderabad. *Indian Farming*, **17**, 11–12.

Kushalappa, A.C., Bartz, J.A. and Norden, A.J. (1979) Susceptibility of pods of different peanut genotypes to Aspergillus group of fungi. *Phytopathology*, **69**, 159–162.

Lansden, J.A. (1982) Aflatoxin inhibition and fungistasis by peanut tannins. *Peanut Science*, **9**, 17–20.

LaPrade, J.C., Bartz, J.A., Norden, A.J. and Demuynk, T.J. (1973) Correlation of peanut seed-coat surface wax accumulations with tolerance to colonization by *Aspergillus flavus. Journal of the American Peanut Research and Education Association*, **5**, 89–94.

Lee, J.I., Han, E.D., Park, H.N. and Park, R.K. (1989) '*Namdaettangkong*' *a new large-seed and high-yielding virginia bunch type peanut variety*. Korea Rural Development Administration Research Reports, Vol. 21, No. 4 (U&I), 1989:20–25 (Korean with English abstract).

Lee, J.I., Han, E.D., Park, H.W. *et al.* (1968) *An early, erect type, large grain and high-yielding peanut variety 'Daekwangtangkong'*. Korea Rural Development Administration Research Report 28, No. 2 (Crops), Dec. 1986: 197–202 (Korean with English abstract).

Lynch, R.E. (1990) Resistance in peanut to major arthropod pests. *Florida Entomologist*, **73**, 422–445.

Maggon, K.K., Gopal, S. and Venkitasubramanian, T.A. (1973) Effect of trace metals on aflatoxin production by *Aspergillus flavus. Biochem. Physiol. Pflanzen*. **164**, 523.

Mayeux, A. (1987) Groundnut research program in Botswana, in *Proceedings of the Second Regional Groundnut Workshop in Southern Africa, 10–14 February 1986, Harare, Zimbabwe*, ICRISAT, Patancheru, pp. 65–71.

Mehan, V.K. and McDonald, D. (1980) *Screening for resistance to Aspergillus flavus invasion and aflatoxin production in groundnuts*, ICRISAT, Groundnut Improvement Program Occasional Paper No. 2, ICRISAT, Patancheru (limited distribution).

Mehan, V.K., McDonald, D. and Rajagopalan, K. (1987) Resistance of peanut genotypes to seed infection by *Aspergillus flavus* in field trials in India. *Peanut Science*, **14**, 17–21.

Mehan, V.K., McDonald, D. and Ramakrishna, N. (1986a) Varietal resistance in peanut to aflatoxin production. *Peanut Science*, **13**, 7–10.

Mehan, V.K., McDonald, D., Ramakrishna, N. and Williams, J.H. (1986b) Effects of genotype and date of harvest on infection of peanut seed by *Aspergillus* and subsequent contamination with aflatoxin. *Peanut Science*, **13**, 46–50.

Mehan, V.K., McDonald, D., Nigam, S.N. and Lalitha, B. (1981) Groundnut cultivars with seed resistant to invasion by *Aspergillus flavus. Oléagineux*, **30**, 501–507.

Melouk, H.A., Banks, D.J. and Fanous, M.A. (1984) Assessment of resistance to *Cercospora arachidicola* in peanut genotypes in field plots. *Plant Disease*, **68**, 395–397.

Mixon, A.C. (1976) Peanut breeding strategy to minimize aflatoxin contamination. *Journal of the American Peanut Research and Education Association*, **8**, 54–58.

Mixon, A.C. (1980) Comparison of pod and seed screening methods on *Aspergillus* spp. infection of peanut genotypes. *Peanut Science*, **7**, 1–3.

Mixon, A.C. (1983a) Peanut germplasm lines, AR-1, -2, -3, and -4. *Crop Science*, **23**, 1021.

Mixon, A.C. (1983b) Two peanut germplasm lines, GFA-1 and GFA-2. *Crop Science*, **23**, 1020–1021.

Mixon, A.C. and Rogers, K.M. (1973a) Peanut accessions resistant to seed infection by *Aspergillus flavus*. *Agronomy Journal*, **65**, 560–562.

Mixon, A.C. and Rogers, K.M. (1973b) Peanuts resistant to seed infection by *Aspergillus flavus*. *Oléagineux*, **28**, 85–86.

Moore, K. and Knauft, D.A. (1989) The inheritance of high oleic acid in peanut. *Journal of Heredity*, **80**, 252–253.

Nagrajan, V. and Bhat, R.V. (1973) Aflatoxin production in peanut varieties by *Aspergillus flavus* Link and *A. parasiticus* Speare. *Applied Microbiology*, **25**, 319–321.

Nevill, D.J. (1982) Inheritance of resistance to *Cercosporidium personatum* in groundnuts: a genetic model and its implications for selection. *Oléagineux*, **37**, 355–362.

Nigam, S.N. (1987) A review of the present status of the genetic resources of the ICRISAT Regional Groundnut Improvement Program of the Southern African Cooperative Regional Yield and of rosette virus resistance breeding, in *Proceedings of the Regional Groundnut Workshop South Africa, 2nd, 10–14 February 1986, Harare, Zimbabwe*, ICRISAT, Patancheru, pp. 15–30.

Nigam, S.N. and Bock, K.R. (1990) Inheritance of resistance to groundnut rosette virus in groundnut (*Arachis hypogaea*). *Annals of Applied Biology*, **117**, 553–560.

Nigam, S.N., Dwivedi, S.L. and Gibbons, R.W. (1980) Groundnut breeding at ICRISAT, in *Proceedings of the International Workshop on Groundnuts, 13–17 October 1980*, ICRISAT, Patancheru, pp. 62–68.

Nigam, S.N., Dwivedi, S.L. and Gibbons, R.W. (1991) Groundnut breeding: constraints, achievements, and future possibilities. *Plant Breeding Abstracts*, **61**, 1127–1136.

Norden, A.J. (1980) Crop improvement and genetic resources in groundnuts, in *Advances in Legume Science*, (eds R.J. Summerfield and A.H. Bunting), Royal Botanic Gardens, Kew, UK, pp. 515–523.

Norden, A.J., Gorbet, D.W., Knauft, D.A. and Young, C.T. (1987) Variability in oil quality among peanut genotypes in the Florida breeding program. *Peanut Science*, **14**, 7–11.

Norden, A.J., Smith, O.D and Gorbet, D.W. (1982) Breeding of the cultivated peanut, in *Peanut Science and Technology*, (eds H.E. Pattee and C.T. Young), American Peanut Research and Education Society, Inc., Yoakum, TX, pp. 95–122.

Payne, G.A. and Hagler, W.M., Jr (1983) Effect of specific amino acids on growth and aflatoxin production by *Aspergillus flavus* in defined media. *Applied Environmental Microbiology*, **46**, 805–812.

Peanut C.R.S.P., United States Agency for International Development (1990) *1989/90 Annual Report. Peanut Collaborative Research Support Program*, The University of Georgia, Georgia Experiment Station, Griffin, GA.

Pettit, R.E., Taber, R.A., Smith, O.D. and Jones, B.L. (1977) Reduction of mycotoxin contamination in peanuts through resistant variety development. *Ann Tech Agric* **27**, 343–351.

Philippine Council for Agriculture and Resources Research and Development (1985) *Peanut Proceedings. PCARRD Book Series No. 39*, PCARRD, Los Baños, Laguna, Philippines, 116 pp.

Prine, G.M., Dunavin, L.S., Moore, J.E. and Roush, R.D. (1981) 'Florigraze' rhizoma peanut – a perennial forage legume. *Florida Agricultural Experiment Station Bulletin*, 275.

Pua, A.R. and Medalla, E.C. (1986) Screening for resistance to *Aspergillus flavus* invasion in peanut. *17th Anniversary and Annual Convention*, Pest Control Council of the Philippines, 8–10 May, Iloila City, Philippines. (Abstract.)

Rao, K.S. and Tulpule, P.G. (1967) Varietal differences of groundnut in the production of aflatoxin. *Nature*, **214**, 738–739.

Reddy, D.V.R. (1980) International aspects of groundnut virus research, in *Proceedings of the International Workshop on Groundnuts, 13–17 October 1980*, ICRISAT, Patancheru, pp. 203–210.

Reddy, D.V.R., Iizuka, N., Subrahmanyam, P. *et al.* (1979) A soil borne virus disease of peanuts in India. *Proceedings of the American Peanut Research and Education Society*, **11**, 49 (Abstract).

Reddy, L.J., Nigam, S.N., Dwivedi, S.L. and Gibbons, R.W. (1987) Breeding groundnut cultivars resistant to rust (*Puccinia arachidis* Speg.), in *Groundnut Rust Disease: Proceedings of a Discussion Group Meeting, 24–28 September 1984*, ICRISAT, Patancheru, pp. 17–25.

Sanders, T.H. and Mixon, A.C. (1978) Effect of peanut tannins on percent seed colonization and *in vitro* growth by *Aspergillus parasiticus*. *Mycopathologia*, **66**, 169–173.

Simpson, C.E. (1983) Plant exploration: planning, organization and implementation with special emphasis on *Arachis*, in *Conservation of Crop Germplasm: an International Perspective*, (eds W.L. Brown, T.T. Chang, M.M. Goodman and Q. Jones), Crop Science Society of America Special Publication, No. 8, CSSA, Madison, Wis., pp. 1–20.

Simpson, C.E. (1990) Collecting wild *Arachis* in South America past and future, in IBPGR International Crop Network Series. 2. *Report of a Workshop on the Genetic Resources of Wild Arachis Species Including Preliminary Descriptors for Arachis*. IBPGR/ICRISAT, Rome, pp. 10–17.

Simpson, C.E. (1991) Pathways for introgression of pest resistance into *Arachis hypogaea* L. *Peanut Science*, **18**, 22–25.

Singh, A.K., Subrahmanyam, P. and Moss, J.P. (1984) The dominant nature of resistance to *Puccinia arachidis* in certain wild *Arachis* species. *Oléagineux*, **39**, 535–538.

Smartt, J. (1978) Makulu Red – a 'Green Revolution' variety? *Euphytica*, **27**, 605–608.

Smith, O.D., Boswell, T.E., Gricher, W.J. and Simpson, C.E. (1989) Reaction of select peanut (*Arachis hypogaea* L.) lines to southern stem rot and *Pythium* pod rot under varied disease pressure. *Peanut Science*, **16**, 9–13.

Sowell, G., Smith, D.H. and Hammons, R.O. (1976) Resistance of peanut plant introductions to *Cercospora arachidicola*. *Plant Disease Reporter*, **60**, 494–498.

Stalker, H.T. and Moss, J.P. (1987) Speciation, cytogenetics, and utilization of *Arachis* species. *Advances in Agronomy*, **41**, 1–40.

Stalker, H.T., Jones, T.M. and Murphy, J.P. (1990) Isozyme variability among *Arachis* species. *Proceedings of the American Peanut Research and Education Society*, **22**, 50 (Abstract).

Subrahmanyam, P., Ghanekar, A.M., Nolt, B.L. *et al.* (1985) Resistance to groundnut diseases in wild *Arachis* species, in *Proceedings of the International Workshop on Cytogenetics of Arachis, 31 October–2 November 1983*, ICRISAT, Patancheru, pp. 49–55.

Subrahmanyam, P. and McDonald, D. (1983) *Rust disease of groundnut*, ICRISAT Information Bulletin No. 13, ICRISAT, Patancheru.

Subrahmanyam, P., Gibbons, R.W., Nigam, S.N. and Rao, V.R. (1982a) Screening methods and further sources of resistance to peanut rust. *Peanut Science*, **7**, 10–12.

Subrahmanyam, P., McDonald, D., Gibbons, R.W. *et al.* (1982b) Resistance to rust and late leaf spot diseases in some genotypes of *Arachis hypogaea*. *Peanut Science*, **9**, 6–10.

Subrahmanyam, P., McDonald, D. and Subba Rao, P.V. (1983a) Influence of host genotype on uredospore production and germinability in *Puccinia arachidis*. *Phytopathology*, **73**, 726–729.

Subrahmanyam, P., McDonald, D., Gibbons, R.W. and Subba Rao, P.V. (1983b) Components of resistance to *Puccinia arachidis* in peanuts. *Phytopathology*, **73**, 253–256.

Subrahmanyam, P., Moss, J.P. and Rao, V.R. (1983c) Resistance to peanut rust in wild *Arachis* species. *Plant Disease*, **67**, 209–212.

Subrahmanyam, P., Williams, J.H., McDonald, D. and Gibbons, R.W. (1984) The influence of foliar diseases and their control by selective fungicides on a range of groundnut (*Arachis hypogaea* L.) genotypes. *Annals of Applied Biology*, **104**, 467–476.

Sundararaman, S. (1926) The clump disease of groundnuts. *Madras Agricultural Yearbook*, **1926**, 13–14.

Taber, R.A., Pettit, R.E., Benedict, C.R. *et al.* (1973) Comparison of *Aspergillus flavus* tolerant and susceptible lines. I. Light microscopic investigation. *Proceedings of the American Peanut Research and Education Association*, **5**, 206–207.

Trochain, J. (1931) La lepre de l'arachide. *Revue de Botanique Appliquée et d'Agriculture Tropicale*, **11**, 330–334.

Tsai, A.H. and Yeh, C.C. (1985) Studies on aflatoxin contamination and screening for disease resistance in groundnuts. *Journal of Agricultural Research of China*, **34**, 79–86.

Tulpule, P.G., Bhat, R.V. and Nagraj, V. (1977) Variations in aflatoxin production due to fungal isolates and crop genotypes and their scope in prevention of aflatoxin production. *Archives d'Institut Pasteur, Tunis*, **54**, 487–493.

Turner, R.B., Lindsey, D.L., Davis, D.D. and Bishop, R.D. (1975) Isolation and identification of 5,7-dimethoxyisoflavone, an inhibitor of *Aspergillus flavus* from peanut. *Mycopathologia*, **57**, 39–40.

Vasudeva Rao, M.J., Nigam, S.N., Mehan, V.K. and McDonald, D. (1989) *Aspergillus flavus* resistance breeding in groundnut: progress made at ICRISAT Center, in *Aflatoxin Contamination of Groundnuts*, (eds D. McDonald and V.K. Mehan), Proceedings of the International Workshop, 6–9 October 1987, ICRISAT, Patancheru, pp. 345–355.

Venkitasubramanian, T.A. (1977) Biosynthesis of aflatoxin and its control, in *Mycotoxins in Human and Animal Health*, (eds J.V. Rodricks, C.W. Hesseltine and M.A. Mehlman), Pathotox Publications, Inc., Park Forest South, FL, pp. 81–98.

Waliyar, F. and Bockelee-Morvan, A. (1989) Resistance of groundnut varieties to *Aspergillus flavus* in Senegal, in *Aflatoxin Contamination of Groundnuts*, (eds D. McDonald and V.K. Mehan), Proceedings of the International Workshop, 6–9 October 1987, ICRISAT, Patancheru, pp. 305–310.

Waliyar, F., McDonald, D., Nigam, S.N. and Subba Rao, P.V. (1989) Resistance to early leafspot of groundnut, in *Proceedings of the Third Regional Groundnut Workshop, 13–18 March 1988, Lilongwe, Malawi*, ICRISAT, Patancheru, pp. 49–54.

Wightman, J.W., Dick, K.M., Ranga Rao, G.V. *et al.* (1990) Pests of groundnut in the semi-arid tropics, in *Insect Pests of Food Legumes*, John Wiley and Sons, New York, pp. 243–322.

Wilson, D.M., Mixon, A.C. and Troeger, J.M. 1977. Aflatoxin contamination of peanuts resistant to seed invasion by *Aspergillus flavus*. *Phytopathology*, **67**, 922–924.

Wynne, J.C. and Gregory, W.C. (1981) Peanut breeding, in *Advances in Agronomy*, (ed. N.C. Brady), Vol. 34. Academic Press, New York, pp. 39–72.

Wynne, J.C. and Halward, T. (1989a) Cytogenetics and genetics of *Arachis*, in *Critical reviews in plant science*, (ed. B.V. Conger), CRC Press, Boca Raton, Florida, pp. 189–220.

Wynne, J.C. and Halward, T.M. (1989b) Germplasm enhancement in peanut, in *IBPGR Training Courses: Lecture Series 2. Scientific Management of Germplasm: Characterization, Evaluation and Enhancement*, (eds H.T. Stalker and C. Chapman), International Board for Plant Genetic Resources, Rome, pp. 155–174.

Wynne, J.C., Beute, M.K. and Nigam, S.N. (1991) Breeding for disease resistance in peanut (*Arachis hypogaea* L.). *Annual Reviews of Phytopathology*, **29**, 279–303.

Xeyong, X. (1991) Groundnut production and research in East Asia in the 1980s. *Proceedings of the Second International Groundnut Workshop, 25–29 November 1991*, ICRISAT, Patancheru.

Zambetakkis, C. (1975) Étude de la contamination de quelques variétés d'arachide par l'*Aspergillus flavus*. *Oléagineux*, **30**, 161–167.

Zambetakkis, C., Waliyar, F., Bockelee-Morvan, A. and dePins, O. (1981) Results of four years of research on resistance of groundnut varieties to *Aspergillus flavus*. *Oléagineux*, **36**, 377–385.

CHAPTER 15

Utilization of *Arachis* species as forage

B.G. Cook and I.C. Crosthwaite

15.1 INTRODUCTION

Legumes are no less important in the nutrition of grazing animals than they are in human nutrition. They serve a dual role: firstly as a source of high protein fodder in the animal diet, and secondly to inject nitrogen into the forage system by virtue of the symbiotic relationship between many leguminous species and the bacterial genera, *Rhizobium* and *Bradyrhizobium*. Nitrogen input from the legume is one of the prime determinants of the productivity of a mixed grass/forb pasture, and high levels can only be achieved through maintaining a healthy, productive legume component in the pasture. While nitrogen can also be supplied as a constituent of chemical fertilizers, there is often farmer resistance to such practice based on economic and environmental considerations. Legumes vary in their capacity to fix nitrogen, as well as in their general suitability as forages in terms of productivity, palatability, toxicity, growth habit and adaptation to environment. Further, the legume flora varies considerably from country to country and region to region. Where there is a natural dearth of suitable forage legumes, researchers draw germplasm from similar environments elsewhere in an endeavour to provide more productive, stable pasture systems. Thus, traditional temperate legume genera such as *Trifolium*, *Medicago* and *Lotus* have proved to be of less value in the tropics and subtropics than they are in temperate and mediterranean regions, necessitating a search for, and evaluation of, a range of tropical forage species.

The science of tropical forages is in its infancy in comparison with that of temperate forages. It spans a period of little more than 50 years, during which commercial forage legumes have been drawn from many genera, the most widely planted probably coming from *Centrosema*, *Macroptilium*, *Stylosanthes* and *Desmodium*. However, the genetic potential of these and

The Groundnut Crop: A scientific basis for improvement. Edited by J. Smartt. Published in 1994 by Chapman & Hall, London. ISBN 0 412 408201.

other genera, including *Arachis*, has been only partly exploited and evaluation continues. The considerable contribution of *Arachis* to the pasture lands of South America is well recognized (Hartley, 1949; Higgins, 1951b), leading pasture plant evaluators to assess *Arachis* for pasture improvement in Australia (Miles, 1949), India (Aiyadurai, 1959) and the USA (Prine, 1964; Blickensderfer *et al.*, 1964; Beaty *et al.*, 1968), resulting in the release of a number of cultivars.

15.2 *ARACHIS HYPOGAEA*

While the commercial groundnut or peanut (*Arachis hypogaea*) has gained greatest prominence by virtue of the quality and quantity of its seed, the value of the vegetative portion of the plant, both fresh and as hay, has long been recognized (Burke, 1850; Easby, 1851). The crop is grown in many parts of the world, and in various ways makes a significant contribution to animal production.

15.2.1 'Hogging off'

Whole groundnut plants have been grazed and rooted up by pigs for generations. 'Hogging off' or turning pigs into the fields to root out the nuts has been very popular in the United States. During the 10-year period 1925–34, approximately 35% of the total groundnut crop of 785 000 ha was hogged off rather than being harvested (Morrison, 1938). As late as 1973, 3% of groundnuts were used in the field for fattening pigs (Bogdan, 1977).

Groundnut crops which are hogged off have a beneficial effect on the soil compared with those harvested conventionally (York and Colwell, 1951). Corn and cotton yields increased after hogged off groundnuts, probably due to the nitrogen returned to the soil.

15.2.2 Use of haulms

Haulms can be defined as that part of the plant remaining after the pods have been removed. This includes leaves, stems and pegs. The amount of root material remaining depends on the harvesting method.

(a) Asia

In northern China, up to 80% of the haulm may be stored for use as a fodder for horses, pigs and sheep during winter, whereas only 50% is stored in southern China where it is warmer and there is a lower need for stored hay (Liao Boshu, personal communication).

South-east Asian farmers make significant use of groundnut haulms, which are sold in bundles and cartloads during harvest time and are often fed green. In Indonesia, horses carry fresh haulms for consumption while away from home. In the Philippines, groundnut haulms are transported to animal sheds away from the riverine production areas, which are subject to inundation during the wet season (R.B. Santos, personal communication). Dry haulms are fed to pigs and cattle in Vietnam, where the crop is windrowed. However, few Thai farmers use groundnut haulms to feed stock (V. Benjasil, personal communication).

(b) Africa

Groundnut haulms are an important feed source in many parts of Africa. Haulms are commonly used in the northern part of Cameroon as cattle feed and are actively traded during harvest (Essomba *et al.*, 1990). They are also used in Gambia (Drammeh, 1990), Nigeria (Misari *et al.*, 1990) and Ethiopia (A. Wakjira, personal communication), but feeding of haulms to cattle is not yet widespread in the Republic of Guinea (Tounkara, 1990). While all of the haulm crop is fed to livestock in South Africa (C.J. Swanevelder, personal communication), they are not used as stock feed to any great extent by smallholder farmers in other southern African countries because of spoilage by foliar disease and since the majority of these smallholder farmers do not own livestock (G. Hildebrand, personal communication).

(c) India

Groundnut haulms are widely used in India as cattle feed (Nagaraj, 1988). They are fed green or stored in 'hay stacks' and are often mixed with other fodder materials. Farmers are happy with good haulm yields, even when pod yields are poor, because their economy depends on milk animals.

(d) Australia

The sale of groundnut hay in southern Queensland can add significantly to the income generated from the crop. Typically the value of the total crop could be increased by 10–20% by selling the haulms. In some drought years, crops have been baled and sold without threshing.

(e) USA

Widespread use was made of groundnut haulms in the United States. Bogdan (1977) estimated that haulms from 20% of the peanut area or 100 000 ha was utilized as hay. The value of goober pea (groundnut) vines as stockfeed was recognized in the USA as early as 1851 (Higgins, 1951).

15.2.3 Haulm quality

Groundnut haulms are valued for their nutrient content, particularly their high digestibility and protein. However, as the haulm component is only secondary to the production of kernels, its quality varies considerably (Table 15.1). In most situations the optimum time for haulm production and quality would be well before the crop is harvested for maximum kernel yields. Halevy and Hartzook (1988) found that the decline in nitrogen and phosphorus levels in the stems and leaves began 64 days before harvest and potassium levels 37 days later. The number of pods remaining in the hay also influences quality (Table 15.2), particularly in relation to digestible protein and metabolizable energy.

Weather conditions may affect haulm quality. A prolonged drought or a long period of wet weather prior to harvest leads to leaf and stem deterioration. The time that is most critical to haulm quality is between lifting and removal of the haulms from the field: wet weather when the plants are drying on the soil surface causes a rapid fall in quality.

Where mechanized threshing is used, the weather and bush condition at threshing are important to haulm quality. If the humidity is very low, much of the valuable leaf material is broken up into fine particles and lost when passing through the thresher, leading to very stalky hay. Groundnut leaf can contain more than twice the nitrogen or crude protein content of either stem or root fractions (Powell, 1986). Some threshers in Australia have been modified to leave the haulms in windrows for easy pick-up by haybalers instead of spreading the discarded haulms evenly over the paddock.

Disease is also a vital factor in haulm quality and quantity. Foliar diseases such as leaf spots (*Cercospora arachidicola* and *Phaeoisariopsis personata*), rust (*Puccinia arachidis*) and some virus diseases can reduce the leaf component significantly.

Quality can be influenced by plant population, through its influence on leaf to stem ratios. Increases in plant population significantly increase the percentage of total dry matter partitioned to stems and decrease the percentage partitioned to pods, while the partitioning to leaves does not change (Kvien and Bergmark, 1987). Similar effects have been recorded with planting dates, where a greater proportion of vegetative dry matter was fixed in stems rather than leaves of early planted crops compared with later plantings (Bell, 1986).

Significant differences among cultivars in leaf concentrations of potassium, calcium and magnesium (Hallock and Martens, 1974) are of little consequence in terms of feed quality.

15.2.4 Improving nutritive value

The quality of groundnut haulms can be changed by treating with an alkali or urea. *In vitro* dry matter (DM) and organic matter digestibility were

TABLE 15.1 *Nutritional value of post-harvest groundnut haulms*

	Source								
	1	2	3	4	5	6	7	8	9
Dry Matter (%)	90	89–91		90–93		85	87–89		92.2
Crude protein (%)	6–20	13–14	19.9	8.3–15	12	9.6	9.6–12.1	14.6–15	9.3
Digest. protein (%)	2–6			7.3–13.2	5.7	6.5	4.9–6.1	7.6–7.9	
TDN (%)	50			53	56			46.8–56.4	
ME (MJ/kg)	8–11			9.8					
Fat (%)		2	4.8	1.4–2.9	4.6	1.1			
Ether extract (%)								0.8–2	2.3
NFE (%)					40.7	48.4	38.9–45.9	33.8	
Carbohydrate (%)		42–46		38–47					
Carbon hydrolysis (%)			30.9						
A-D-F (%)			38.4				40.3–43.1		
N-D-F (%)			42						
Crude fibre (%)		22–24	34.5	22–35	30.0	19		28–30.5	37.6
Ash (%)		5.7–6	9.9	9–17	11.9			9.1–15.3	9.1
Minerals (%)									
Phosphorus (%)		0.7			0.42				0.3
Calcium		1.7			2.4				1.5

Sources: 1. W.J. Edwards, pers. comm. 2. Anon. (1982) 3. V. Benjasil, pers. comm. 4. Nagaraj (1988) 5. Shukla *et al.* (1985) 6. Ikhatua and Adu (1984) 7. Combellas *et al.* (1972) 8. Velasquez and Gonzalez (1972) 9. Leche *et al.* (1982)

TABLE 15.2 *Nutrient value of groundnut hay of differing quality (after Bredon et al., 1987)*

	Good	Average	With some pods left	With pods
Good dry matter (%)	90	92	91	92
Crude protein (%)	12.3	9.2	10.8	13.4
Digestible protein (%)	7.9	6.0	6.5	10.2
TDN (%)	50.0	49.0	62.6	71.6
ME (MJ/kg)	7.61	7.36	9.40	10.74
Ether extract (%)	2.52	1.85	5.0	12.6
NFE (%)	44.8	45.9	41.8	34.9
Crude fibre (%)	21.0	24.1	24.5	23.0
Phosphorus (%)	0.13	–	0.16	0.15
Calcium (%)	1.12	–	1.14	1.13
Potassium (%)	1.60	–	1.22	2.14

increased by addition of sodium hydroxide and further improved by storing the treated hay for up to three days (Myung *et al.*, 1986).

The beneficial effect of alkali treatment was confirmed by Abou-Raya *et al.* (1971) by feeding groundnut hay to adult sheep. They found that the digestibility of all components, except N-free extract, was increased by treatment with calcium hydroxide, sodium hydroxide, and sodium hydroxide plus urea. However, the total dry matter was reduced by the hydroxide treatments.

Chopping or grinding had no effect on the voluntary DM consumption and digestibility coefficients of crude nutrients of groundnut haulms fed to goats (Ayoade and Njewa, 1983/1984). Even so, in South Africa the haulms are usually hammer-milled before being used in the ration (C.J. Swanevelder, personal communication).

15.2.5 Haulm yields

Haulm yields vary markedly and are affected by variety, pre-harvest conditions, disease level, time of digging, conditions after digging and method of retrieval and storage.

Maximum dry matter yield of groundnut tops is reached before the optimum lifting time for kernel yield (Wright *et al.*, 1991). Under dryland conditions, vegetative dry matter yield peaked at 90 days after planting and, for most cultivars, declined steadily to digging at 130 days. Under irrigation, this decline did not begin until 100–130 days after planting, with different varieties showing different responses. Halevy and Hartzook (1988) showed that the shedding of leaves and stems caused a reduction of 1386 kg from a potential haulm yield of 6006 kg/ha. This decline was also recorded by Bell (1986), who observed the decline was accelerated by foliar diseases.

Powell (1986), working with spanish varieties in central Nigeria, noted a highly significant positive correlation between pod yield and yield of vegetative dry matter. In the USA, spanish types yield 1–1.5 tonnes of hay per tonne of pods, while runner types yield 1.5–2 tonnes per tonne of pods (York and Colwell, 1951). This compares with results from southern Queensland, Australia, where haulm:pod ratios are generally in the range of 1 to 2:1 for virginia bunch under dryland conditions and 0.6 to 1.2:1 with irrigation (G.R. Harch, personal communication). Ratios for spanish varieties in the same dryland environment range from 1.4 to 1.7:1 and for irrigated crops from 0.7 to 1:1.

15.2.6 Animal performance

Groundnut haulms have been fed to a range of animals including cattle (Shukla *et al.*, 1985), goats (Jamarun, 1984), pigs, rabbits (Aduku *et al.*, 1986), sheep and fish (Liao Boshu, personal communication), with varying success in terms of animal performance and economic returns.

Goats fed groundnut haulms (6.5% digestible protein) gained weight 60% faster than on *Digitaria* hay (3.5% digestible protein) but at almost double the cost per unit weight gain. Up to 75% groundnut hay can be substituted for fresh ipil-ipil (*Leucaena leucocephala*) as a feed for goats without reducing growth rates (Robles and Capitan, 1985). Similarly groundnut haulms, when replacing maize stover, can almost double the growth rate and increase the feed conversion efficiency of stall fed bullocks, although this advantage was lost through feeding *ad lib.* maize bran with either feed (Addy and Thomas, 1977).

Groundnut straw can comprise 60% of the ration of growing lambs without adversely affecting their growth rate and feed efficiency (Krishna Mohan *et al.*, 1985). However, as groundnut straw increases in the ration, the cost per unit weight gain increases and the digestibility of all nutrients tends to decrease. The most economical ration for growing lambs is 40% groundnut hay and 60% protein concentrate (Durga Prasad *et al.*, 1986). Increasing the protein content of the concentrate can increase the intake of groundnut hay by calves (Ahmed and Pollett, 1979).

15.2.7 Haymaking

Mechanical haymaking is feasible with groundnut herbage but losses are high (Combellas *et al.*, 1972). Groundnut hay in Australia is baled into small rectangular bales weighing around 25 kg or larger round bales of 300–500 kg. The cost of baling groundnut haulms is higher than for other hay because soil contamination leads to greater machinery maintenance costs. Soil adhering to roots can make up 10% of the weight of groundnut residues (Powell, 1986). In South Africa, sand contamination of baled haulms can have a detrimental effect on animals' teeth (C.J. Swanevelder, personal communication).

15.2.8 Storage

Long-term storage of groundnut hay in Australia is difficult. Mice are attracted by unthreshed kernels and destroy hay bales that have been put in sheds to reduce weather damage.

15.2.9 Aflatoxin

Aflatoxin poisoning may result from feeding groundnut haulms. Kernels and shells present the greatest risk. Poisoning can result from animals consuming groundnuts in the field, eating hay containing nuts or being fed shells remaining after the kernels have been removed. Hay made from groundnuts severely affected by drought, where threshing is uneconomic, may have more nuts attached and thus present a higher risk, particularly for young or weak animals with little or no other feed source.

Calves have been killed from eating groundnut hay containing pods with high aflatoxin levels (McKenzie *et al.*, 1981). Jaundice, photosensitization, diarrhoea, anorexia and depression were observed prior to death, while post mortems showed haemorrhage, hepatocyte damage and bile duct proliferation.

15.2.10 Chemical residues

Agricultural chemicals can improve the yield and quality of groundnut haulms but their misuse can lead to haulm contamination, rendering them unfit for use as animal fodder. Farmers in South Africa will even sacrifice a little pod yield to ensure a good forage free from chemical residue (C.J. Swanevelder, personal communication).

Haulm quality will generally be much higher from groundnuts treated with a fungicide. By controlling leaf diseases, the leaf-to-stem ratio of the haulms will be higher. However, Australian label requirements of some fungicides (e.g. chlorothalonil) state that groundnut haulms should not be fed to livestock. Other fungicides have a withholding period after spraying before haulms can be fed; for example, groundnuts treated with cyproconazole should not be grazed or cut for stock food for 14 days after application. Sulphur-treated groundnuts have no withholding period.

Livestock should not be fed haulms and hulls from crops treated with some nematicides, such as aldicarb, fenamiphos and ethoprophos. Similarly the haulms from crops treated with herbicides, such as acifluorfen, sethoxydim and 2,4-DB, should not be fed to livestock in Virginia, USA, yet haulms from crops sprayed with 2,4-DB in Queensland, Australia, can be fed to livestock seven days after spraying.

TABLE 15.3   *Nutrients removed from the soil by both groundnuts and haulms (after Collins and Morris, 1913)*

| Part of the plant | Yield (kg/ha) | Nutrient content (kg) | | | | |
		N	P	K	Ca	Mg
Haulms	4540	89.8	5.2	77.8	44.5	17.2
Nuts	2250	69.4	6.7	19.7	3.2	3.5
Percent of nutrients in hay	56.4	43.6	79.8	93.4	83.2	

Organochlorine pesticides (DDT, dieldrin, endrin and BHC) have been banned for crop use in many countries because of their long residual life in soil and subsequent contamination of crops. To reduce such a risk in Australia, soils in districts where organochlorines have been used are tested prior to growing groundnuts.

15.2.11 Soil depletion

Removal of groundnut haulms affects the soil in two ways: firstly the direct removal of nutrients from the soil, and secondly the depletion of organic matter which is essential to maintain soil structure. Large quantities of nitrogen and potassium are removed when haulms, as well as pods, are taken from the soil (Collins and Morris, 1941) (Table 15.3).

In assessing the monetary value of groundnut hay the cost of replacing these nutrients must be taken into account. At current prices in Australia, to replace major and minor nutrients (e.g. N, P, K, Ca, Mg, S, Cu, B, Zn, Mn) would cost approximately 35% of the market value of the hay.

A more difficult factor to measure is the long-term effect on organic matter depletion. Most groundnut-growing soils are of a light texture and organic matter is vital to maintain their structure and soil moisture-holding capacity. Groundnuts grown in soils low in organic matter are more susceptible to attack by *Sclerotium rolfsii* (K.J. Middleton, personal communication) and diplodia blight (*Diplodia natalensis*) (Queensland Department of Primary Industries, 1985).

Removing groundnut haulms may, therefore, provide a short-term gain with serious long-term effects unless steps are taken to counter the effect of nutrient and organic matter removal.

15.2.12 Disease reduction

Alternatively, removal of disease-affected parts of groundnut plants may have an effect in reducing subsequent crop infection by lowering the level of inoculum (K.J. Middleton, personal communication). Diseases that could be controlled partially in this way include leaf spots (*Cercospora arachidicola* and *Phaeoisariopsis personata*) and verticillium wilt (*Verticillium dahliae*).

15.3 WILD SPECIES

Resslar (1980) points out that many of the subgeneric epithets used in current classifications are not valid according to the International Code of Botanical Nomenclature (Stafleu, 1978). However, by virtue of common usage, the classificatory system proposed by Gregory *et al.* (1973) and modified by Krapovickas (1990) will be followed. *Arachis* wild species have largely been collected to provide genetic material for groundnut improvement programmes, particularly in relation to disease resistance (Stalker and Moss, 1987). Simpson (1990) notes that some 1500 accessions of wild species have been collected since the early 1950s, but less than half survive in world collections. Krapovickas (1990) recognizes 77 wild species, and most are represented by at least one accession in current collections (Simpson, 1990). Only a small proportion of these has been widely evaluated for forage potential.

One of the most valuable attributes displayed by many members of the genus is persistence under grazing. Several have persisted in grazed pastures for 30 to 50 years after planting, and others show clear indications of being able to do so. Many of the tropical legumes released to date decline over a period of 5 to 7 years, after which the pasture becomes unproductive. To date, most attention has been given to section *Rhizomatosae*, and more recently to section *Caulorhizae*. However, a limited range of material from other sections has been evaluated and shows considerable promise.

15.3.1 *Rhizomatosae*

(a) Description

Members of section *Rhizomatosae* are perennial and are characterized by the presence of rhizomes. The section comprises two series: *Prorhizomatosae*, a diploid group of somewhat delicate types, and *Eurhizomatosae*, a more robust tetraploid group (Gregory *et al.*, 1973) containing *A. glabrata* and *A. hagenbeckii*, and two or three unnamed species (Valls *et al.*, 1985). Forage cultivars selected so far have been drawn from *Eurhizomatosae* and are commonly referred to as rhizoma peanuts in the USA.

The rhizomes, which form a dense tangled mat (mostly in the top 5 cm of soil), vary in thickness from a few millimetres to about 1 cm in the sward of a variety, with average thickness varying from variety to variety. They are woody structures surrounded by a papery, brown, deciduous bark. Contributing to the tangled effect of the rhizomes are the often heavy tap

Figure 15.1 Typical *Eurhizomatosae* growth habit, with horizontal rhizomes, erect stems and vertical taproots (scale two-thirds actual size).

roots and a multitude of fine, fibrous roots, both of which bear the characteristic small, oblate, aeschynomenoid nodules described by Corby (1981). Herbaceous vegetative shoots arise from the rhizomes (Figure 15.1), producing a dense leafy sward in pure stands, which, when mature, may vary from about 5 cm to 35 cm in depth, depending on variety. This variation is partly attributable to growth habit, since in some varieties stems tend to be decumbent and in others more erect, but also to inherent differences in robustness.

Leaves are tetrafoliolate, with leaflets varying from accession to accession in overall leaflet shape from linear lanceolate to oblanceolate and obovate, and in shape of leaflet apex from acute to obtuse mucronate (Figure 15.2). Leaf size depends largely on plant vigour and frequency of defoliation but also differs markedly among accessions. In their key in Gregory *et al.* (1973), Krapovickas and Gregory describe flowers as 'large,

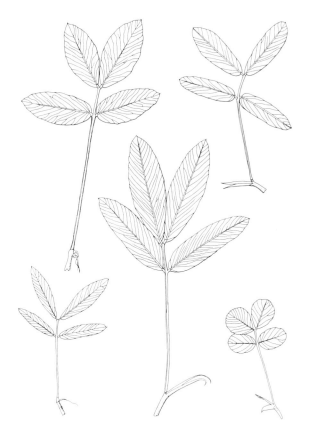

Figure 15.2 Variety of leaf shapes in *Eurhizomatosae*.

standard (22–25 mm wide, 18–22 mm high), yellow, soft orange to brilliant orange, no red veins on back of standard, wings usually not flared'. However, flowers of some accessions may be somewhat smaller than described, attaining a standard width of only 15 mm (B.G. Cook, unpublished data). Krapovickas (1973) notes that it is difficult to find fruit of the *Rhizomatosae* in the wild, although fruit can be set under cultivated conditions. Personal observation is that some accessions fruit freely in an immature state in cultivation but seedset declines rapidly with sward development, despite profuse flowering. Niles (1990) ascribes this effect in the American cultivar, Florigraze, to low pollen germination due to a dry stigma and a high rate of embryo abortion due to competition in dense canopies. The work of Lu *et al.* (1990) suggests stigma morphology may also be a contributing factor. Rhizomatous species studied had small stigmas capable of carrying few pollen grains, in contrast with the large stigmas of annual species which could accommodate more than 15 pollen grains.

(b) Origin

Current theory holds that *Arachis* evolution follows movement and adaptation from an uplifted and dissected Tertiary peneplane in the region of Mato Grosso in Brazil (Gregory *et al.*, 1980). With geocarpy a characteristic of the genus, dispersal could only occur through vegetative advance, peg elongation and erosive water movement. The other common dispersal mechanism for legumes, voiding ingested seed, is unlikely since seed is readily digested (Krapovickas, 1973). It is therefore not surprising that geographical distribution of the various sections of *Arachis* largely follows ancient and current drainage patterns. The *Eurhizomatosae* have been collected between 13° and 28° S, in a triangular area extending south of Corrientes (Argentina) to north of Rosario Oeste and east into São Paulo state (both Brazil) (Gregory *et al.*, 1973). While they have mostly occurred on red soils (Krapovickas, 1973), passport data in Gregory *et al.* (1973) show that *Eurhizomatosae* have also been found on soils ranging from light-coloured sands to fertile clays. Of the 186 accessions collected between 1936 and 1983, only 110 are currently conserved at the main germplasm centres (Valls *et al.*, 1985).

(c) Cultivars

Rhizomatous cultivars have been released only in Florida. The first, Arb, was registered by the Soil Conservation Service in Arcadia, Florida, following an evaluation programme commencing in 1937 (Prine, 1964). A variety of *A. glabrata*, it was initially collected from 'a black rich earth in the streets of Campo Grande, Mato Grosso do Sul' in Brazil (20° 27 ' S, 54° 37' W), and catalogued as PI 118457. The next cultivar, Arblick, was a lower growing and more rapidly spreading type of *A. glabrata*, derived from PI 262839. This accession was collected about 300 km south-west of Campo Grande, near Bela Vista in Paraguay (22° 8' S, 56° 16' W). The release of Florigraze by the University of Florida Institute of Food and Agricultural Sciences and the USDA-SCS gave greater impetus to the development of forage groundnuts in the USA. This variety arose as a seedling on the border between year-old plots of the morphologically distinct Arb and PI 151982, on the Agronomy Farm at Gainesville (Prine *et al.*, 1981). The new variety, initially catalogued GS-1, differed from the other two in being fast spreading. It had more shoots per unit of soil surface and a greater number of finer rhizomes than Arb – characteristics more in common with PI 151982. Leaf shape was intermediate. In 1985, IFAS and SCS released a further cultivar, Arbrook, a more robust variety with coarser rhizomes than Florigraze, to improve the drought tolerance of the species (Prine *et al.*, 1990). Arbrook, introduced as PI 262817, was collected on the top of a hill near Trinidad in Paraguay (27° 10' S, 55° 40' W).

Research in eastern Australia has shown that another Paraguayan variety, GKP 9618 (introduced into Australia as CPI 93469), an Argentinian variety, PI 231318/CPI 93483 and a third variety, PI 151982/ CPI 22762, whose origins are unknown, have sufficient promise to warrant release in the near future. The first was collected near the Rio Peribebuy (25° 12′ S, 57° 5′ W) (Gregory *et al.*, 1973) but precise origins of the second are not available (Anon., 1966).

(d) Adaptation

Forage research on *Eurhizomatosae* has been carried out in various South American countries, USA, Australia, south-east Asia and India. Results to date have shown various members of the series to have potential over a latitudinal range from near the equator in Sulawesi (W.W. Stür, personal communication) to 30° N in Louisiana (Caldwell *et al.*, 1991) and 30° S in New South Wales (G.P.M. Wilson, personal communication). Soil textures at evaluation sites range from the sands of Florida to the clay soils of India and Australia. Soil fertility also varies markedly. The main prerequisite appears to be good drainage. Prine *et al.* (1981) make the point that Florigraze 'is adapted to well-drained soils, but not to "flatwoods soils" or to any soil which is subject to high water tables.' Experience in Queensland supports this statement and shows that, although the suite of rhizomatous accessions can survive temporary waterlogging, most vigorous stands are obtained on well-drained flats and those parts of the slope not subject to seepage. At the other extreme, various members of *Eurhizomatosae* have demonstrated a strong degree of drought tolerance, one particular set surviving under non-irrigated conditions at the ICRISAT Center near Hyderabad, India, where the average annual rainfall of 750 mm is largely concentrated over a 4-month period. PI 163452 has persisted for some years in the Burnett region of southern Queensland (Jones and Rees, 1972), while Florigraze, not noted for its drought tolerance (Prine *et al.*, 1990), established and proved productive in semi-arid southern Texas (Ocumpaugh, 1990) – both areas with similar annual rainfall totals to those of Hyderabad but better distributed. The more drought tolerant Arbrook is often more productive than Florigraze during dry periods, although yields are similar under good rainfall conditions (Prine *et al.*, 1990). In severe drought, Prine *et al.* (1981) observed in Florigraze that top growth may die off completely, the stand regenerating from rhizomes with the onset of moist conditions. Vos and Jones (1986), assessing the importance of rhizomes and stolons in legume persistence under grazing and using *A. glabrata* PI 163452 and *Lotus pedunculatus* cv. Grasslands Maku as test varieties, measured a small increase in *Arachis* rhizome yield per unit area between an average rainfall year and an extremely dry year. In the more moisture sensitive *Lotus*, rhizome yield was reduced by 96%.

A. glabrata and *Eurhizomatosae* generally are adapted to a wide range of soil reaction. For Florigraze, Prine *et al.* (1981) suggest a soil pH of

5.8–6.5, although good yields have been obtained at pH 7.4 (Ocumpaugh, 1990), and no response to lime was obtained in an unidentified variety of *A. glabrata* in any of three Florida soils with pH (in water) as low as 4.5 (Martinez and Blue, 1978). Niles *et al.* (1990), working with soils over a pH range of 5.2–7.9, found that Florigraze showed a tendency to establish better in soils with lower pH and higher aluminium. Contrasting somewhat with this finding, Caldwell *et al.* (1991) obtained a response to dolomite in Florigraze on a soil of pH 5.7. This is an enigmatic result, in that there was no response to the addition of 2.5 t dolomite/ha producing a pH of 6.2, but there was a response to 5 t dolomite/ha which raised the pH to 6.7. In Australia, *A. glabrata* PI 163452/CPI 12121, perhaps the most widely planted accession of the series, has been grown successfully on soils of widely divergent reaction. This variety, along with several others, has shown potential in a series of evaluation experiments in the wet tropics of Malaysia and Indonesia on soils of pH 4.5–7 (W.W. Stür, personal communication). Results from more recent plantings of a wider range of *Eurhizomatosae* on a variety of soil types in Australia suggest that the adaptability to soil reaction displayed by selected members of the series is common in the series. However, Reed and Ocumpaugh (1991) have demonstrated variable manifestation of high pH-induced iron deficiency symptoms among a range of rhizomatous accessions.

Since members of this series form an effective association with *Bradyrhizobium* in the root nodules, nitrogen fertility of the soil is not important to their long-term productivity. Nitrogen fertilizer applied to pure swards of Florigraze had no effect on yield or persistence (Valentim *et al.*, 1988). However, when nitrogen was applied to a mixture of grass and Florigraze, legume yields declined linearly with increasing nitrogen application as grass yields increased (Valentim *et al.*, 1986, 1988). It has also been shown that nitrogen fertilizer applied at planting retards establishment, probably due to stimulated weed competition and depressed peanut nodulation (Adjei and Prine, 1976).

Perhaps the most important nutrient affecting pasture productivity, after nitrogen, is phosphorus. Niles *et al.* (1990) were unable to demonstrate an establishment response in Florigraze to applied phosphorus at 12 sites in Florida soils with extractable phosphorus levels (Mehlich, 1953) ranging from 11 to 317 ppm. It is worth noting that applications of potassium, magnesium and sulphur also produced no measurable response. In pot experiments, Blue *et al.* (1989), working on a Cameroon soil with Florigraze as the test variety, and G.E. Rayment (personal communication), working on a dark clay loam from south-east Queensland with GKP 9618/CPI 93469 as the test variety, demonstrated little or no response in either variety to applied phosphorus. The Cameroon soil had a Mehlich P level of 4.4 ppm and the Queensland soil a bicarbonate extractable P level (Colwell, 1963) of 12 ppm. In both cases, test varieties from other genera had proved responsive. From these experiments, and in view of the long-term persistence of a range of rhizomatous varieties at discontinued unfertilized evaluation sites in southern Queensland, it appears that the *Eurhizomatosae* share the characteristic common in the closely related genus, *Stylosanthes* (Krapovickas, 1973) of being tolerant of low phosphorus levels in soils.

Eurhizomatosae have been successfully cultivated from about 0° to 30° latitudinal range, although low temperatures in the higher latitudes can cause significant damage. This may vary from death of plant tops, which occurs even with light frosts, through to rhizome mortality under more extreme conditions. Prine (1983) suggests rhizoma peanuts should be grown in areas where temperatures do not fall below −10 °C. There appear to be varietal differences in cold tolerance: Florigraze survived low temperatures better than Arbrook (Prine *et al.* 1990); Arbrook was damaged severely by freezes of lower than −12 °C at Jay, Florida, and Americus, Georgia (Prine *et al.*, 1990); while Florigraze in Louisana, although markedly reduced by a December temperature of −15 °C had recovered to about 70% coverage by November the following year. Such temperatures are not experienced in those parts of eastern Australia where this series is being evaluated.

(e) Yield

Rhizoma peanuts are best adapted to warm environments, preferring a climate where temperatures are greater than 16 °C for 5 months or longer (Prine 1983). In southern Queensland, where monthly daily means range from 14–15 °C during June–August to about 25 °C from December to February, the average 6-weekly dry matter yield of 7 rhizomatous varieties was negligible in the cool season but over 2 t/ha in the warm season. In the transitional period, yields of 0.1–1.0 t/ha were recorded (B.G. Cook, unpublished). Likewise, dry matter yield/ha of Florigraze declined from 5.4 t in the summer to 2.9 t in the fall in Florida (Romero, 1986). This apparent lower growth rate may be attributable to a partitioning shift from shoots to rhizomes when daylength falls to 11 hours and temperatures to 20–22 °C as demonstrated in Florigraze by Niles (1990). In the southern Queensland experiment, the highest dry matter yield for a growing season was 12 t/ha measured over six harvests in PI 231318/CPI 93483, and the lowest was 3 t/ha in PI 338316/CPI 93464. Similar ranges in comparable environments are reported for a multivarietal set in Florida by Prine (1973), and for Florigraze in Louisiana (Caldwell *et al.*, 1991) and in a semi-arid Texas environment (Ocumpaugh 1990). Both Florigraze and Arbrook have produced yields in excess of 16 t/ha in Florida (Prine *et al.*, 1986). Maximum yields obtained in the wet tropics of Malaysia and Indonesia in a limited range of material were mostly below 5 t/ha/year (W.W. Stür, personal communication), although these may have been influenced by the shading effects of rubber trees and coconut palms.

However, comparisons between experiments should be viewed with some caution, since yields of a particular variety may also be affected by cutting height and frequency. Beltranena *et al.* (1981) obtained a 75% increase in yield of Florigraze by taking 4×6-week cuts rather than 12×2-week cuts, and a 17% increase through cutting at 3.8 cm rather than 7.6 cm. Ortega *et al* (1992) measured similar responses in grazed mixed swards of Florigraze and bermuda grass, where grazing intensity varied from 500 to 2500 kg/ha of residual dry matter and grazing frequency from 7 to 63 days. They concluded that to maintain 80%, groundnut combinations of at least 42 days between grazings and 1700 kg/ha residual dry matter, or 21 days between grazings and 2300 kg/ha residual dry matter, are necessary.

In areas with a pronounced cool season, annual forage yields may be improved by sowing cool-season species into the groundnut stand. Dunavin (1990), working in Florida, obtained yields of 3–4.5 t DM/ha from broadcasting seed of rye (*Secale cereale*) or italian ryegrass (*Lolium multiflorum*) into a 5 cm stubble of Florigraze. Yields of Florigraze in the following warm season were not affected. Cultivation to introduce the winter crop of ryegrass or oats (*Avena sativa*) in southern Queensland had no detrimental effect on the groundnut stand, and actually enhanced its spread.

(f) Feeding value

Eurhizomatosae produce unusually high quality feed for tropical pasture legumes, often approaching the high levels found in temperate species such as white clover (*Trifolium repens*) and lucerne (*Medicago sativa*). The most commonly published estimates of forage quality are nitrogen percentage or crude protein (N% $\times$ 6.25), *in vitro* organic matter digestibility (IVOMD) and mineral levels, particularly phosphorus. In plants of Arb, Arblick and Florigraze cut twice per season over 5 seasons, Prine (1973) measured N levels ranging from 1.6% to 2.86%, P levels from 0.24% to 0.38%, and IVOMD from 45.3% to 67.7%. With different accessions in south-eastern Queensland, and plants cut at ground level every 6 weeks, N levels in the dry matter ranged from 2.39%, to 3.56%, P levels from 0.3% to 0.54% and IVOMD values from 69.9% to 77.7% (B.G. Cook, unpublished data). Like yield, feed quality is influenced by height and frequency of cut. Beltranena *et al.* (1981) increased leaf content from 72% to 93%, N levels in the dry matter from 2.35% to 3.50%, and IVOMD from 64.8% to 74.3% by increasing cutting frequency from every 12 weeks to every 2 weeks. This improvement in quality is undoubtedly related to the higher crude protein and IVOMD in leaves than in stems (Romero *et al.*, 1987; Saldivar *et al.*, 1990). Ocumpaugh (1990) demonstrated that nutritive value is also affected by rainfall through its effect on dry matter production patterns, leafiness, nitrogen levels in leaf and stem, and stem IVOMD.

All the above aspects should be considered in order to achieve maximum benefit from the crop. Beltranena *et al.* (1981) suggest from their data that the crop be grazed every 4 weeks, green chopped or dehydrated as hay, leaf meal or pellets every 4–6 weeks, or cut for hay every 8 weeks, though Ocumpaugh (1990) suggests this should be less frequent in a semi-arid environment. These qualities of rhizoma peanut have been shown to translate into animal product using various classes of grazing animal. Sollenberger *et al.* (1987) at Gainesville, Florida, achieved an average daily liveweight gain (LWG) of 0.98 kg/head in yearling steers grazing Florigraze compared with 0.37 kg/head in animals grazing Pensacola bahiagrass (*Paspalum notatum*), reflecting the marked difference in quality between the two forages. A similar pattern emerged at Brooksville (also in Florida) where steers on Florigraze/grass pasture gained 0.68 and 0.90 kg/head in consecutive years, compared with 0.52 and 0.50 kg/head on grass alone (Williams *et al.*, 1991). Although stocking rate in the first study was 43% higher in the Pensacola than in Florigraze treatments, due to higher plant growth rates, it is worth noting that LWG/ha was still 27% greater in the groundnut than in the grass treatment. While it is common for LWG/head to be greater on legume compared with grass pastures, it is unusual for LWG/ha to be superior. However, rhizoma peanut compares favourably with the 'king of legumes', lucerne or alfalfa (*Medicago sativa*). Gelaye *et al.* (1990), feeding Florigraze and lucerne hay *ad lib.* to 4–5-month-old 20 kg goats, achieved daily LWGs of 63.8 g/head from the groundnut and 46.2 g/head from the lucerne. Feeds had similar analyses and were eaten in similar quantities. This favourable comparison was borne out by Romero's (1986) work with dairy cows, which gave similar milk yields from both legumes, despite more favourable analysis of the lucerne due to its being cut at an earlier stage than the groundnut. Romero *et al.* (1987) interpret this in terms of slower ruminal breakdown of the protein in the groundnut allowing a greater proportion of original protein to be utilized beyond the rumen.

(g) Establishment

Since *Eurhizomatosae* generally produce poor seed yields, they are mostly propagated from rhizomes. A commercial procedure has been developed in Florida whereby rhizomes which have been produced in sandy soils are lifted using a bermudagrass sprig harvester or a modified potato digger (Prine *et al.*, 1981). Where the potato digger is used, the rhizome mat is cut into squares or rectangles with gangs of rolling coulters before lifting. These pieces, normally about 30 cm square, can be planted in mounds no more than 1.8 m apart, or preferably divided into four and planted 0.9 m apart (Adjei and Prine, 1976). The recommended planting rate for both Florigraze and Arbrook is at least 3.5 m^3 of rhizomes/ha (40 bushels/acre) (Prine *et al.*, 1981, 1986). Doubling this planting rate led to a first year increment of 59% in one experiment involving Florigraze, but by the second year differences were not significant (Valentim *et al.*, 1988). A similar trend was observed by Canudas *et al.* (1989) using lower planting rates for Florigraze and Arbrook. The easiest way to establish rhizoma peanut is to broadcast the rhizome pieces over the surface and cover them using a disc harrow, then roll the soil to achieve rhizome/soil contact and to retain moisture (Prine *et al.*, 1986). Alternatively the rhizome pieces can be planted in furrows from 3 cm deep in clay soils to 6.5 cm deep in coarse sands. As a precaution, Prine *et al.* (1981) recommend inoculation of rhizomes with an effective *Bradyrhizobium* strain at planting. However, experience in Australia suggests that this is rarely necessary, due to the presence of effective native bacteria either in the soil or already on the rhizome surface.

In Florida, the preferred planting time is January or February, though this could be as early as mid-December or as late as mid-March, allowing for a slightly greater risk factor. At this stage, rhizomes are dormant. In Australia, the slightly later southern hemisphere equivalent of August to October has given good results, with January or February plantings resulting in very poor establishment. Saldivar *et al.* (1992a) showed that levels of nitrogen and total non-structural carbohydrate in the rhizomes were highest in the coolest part of the year and that levels of both were reduced by defoliation. These data support the preferred winter/spring planting time and suggest that rhizome nurseries should not be defoliated until immediately before planting. Defoliation also reduced rhizome production (Saldivar *et al.*, 1992b).

Effective weed control is important during the relatively slow establishment period. Canudas *et al.* (1989) demonstrated this for both Florigraze and Arbrook, achieving about double their ground cover and yield at the end of the establishment year through controlling grasses and broadleaf weeds. The broad-spectrum tolerance of rhizoma groundnut to herbicide has facilitated preparation of a weed control programme incorporating the use of pre-emergence applications of benefin, trifluralin or vernolate, post-emergence applications of alachlor and dinoseb, and routine applications of bentazon and 2,4-DB for broadleaf weed control, and sethoxydim and fluazifopbutyl for grass control as required (Prine *et al.*, 1986). Sethoxydim, but not dalapon, can be used to kill bermudagrass (*Cynodon dactylon*) rhizomes contaminating groundnut rhizome planting material, thus improving initial performance significantly (Canudas-Lara *et al.*, 1984). Mowing just above groundnut top growth may also reduce weed competition and enhance establishment (Prine *et al.*, 1986).

(h) Companion species

Although it might be important to suppress or control grasses during establishment, it may well be desirable in the interest of long-term stability to introduce a grass into the groundnut sward to absorb nitrogen from the legume system, thus reducing the likelihood of invasion by unpalatable nitrophilous weed species. Valentim *et al.* (1986, 1988), using Florigraze in association with a range of grasses, have demonstrated that, although groundnut yields might be reduced due to the presence of a companion grass, total pasture yields are at least the same and possibly enhanced.

By virtue of their unusual growth habit, *Eurhizomatosae* have been able to grow effectively with grass species whose compact growth habit and competitive ability may have precluded a successful association with other legumes. Perhaps the most notorious of these grasses is Pensacola bahia-grass (*Paspalum notatum*) although pangolagrass (*Digitaria eriantha*) and bermudagrass are also recognized as being extremely competitive. Prine (1964) records that Arb, once established, comprised 44%, 49% and 73% of the top growth when grown in association with these respective grasses. Prine *et al.* (1981) estimated that Florigraze, Arb and Arblick comprised, on average over five seasons, 90%, 60% and 60% of the total forage yield when grown with Pensacola bahiagrass. Valentim *et al.* (1986) found that Florigraze comprised 86% of the forage yield in association with Pensacola, 73% with bermudagrass and 81% with Survenola digitgrass (*Digitaria × umfolozi*) in the first year after establishment. In a later experiment (Valentim *et al.*, 1988), this time with Mott dwarf elephant grass (*Pennisetum purpureum*), Florigraze comprised 53% and 54% of the forage yield in the first two years. All comparisons were in the absence of added nitrogen. In Australia, the various accessions tested have been grown successfully with similar grasses as well as with other sward-forming species, rhodes grass (*Chloris gayana*), paspalum or dallis grass (*Paspalum dilatatum*), brunswickgrass (*P. nicorae*) narrowleaf carpet grass (*Axonopus affinis*) and blue couch grass (*Digitaria didactyla*) and tussock grasses such as setaria (*Setaria sphacelata* var *sericea*) and the native black speargrass (*Heteropogon contortus*). While there are many instances of long-term persistence of rhizoma peanut in grass swards, it should be noted that at Jay, in north-western Florida, the proportion of Florigraze in association with a range of sward-forming grasses declined in the latter part of an 8-year period (Dunavin, 1992).

The ability to compete with grass associates varies from accession to accession. In an experiment near Gympie in south-east Queensland comparing spread over 5 years of 31 *Eurhizomatosae* accessions through an established grass sward under grazing, radial spread varied from a total of 3 cm in PI 338284/CPI 93473 to 143 cm in PI 262814/CPI 93475 (B.G. Cook, unpublished). The American cultivars, Arb and Arblick, performed poorly. Rate of spread in the absence of grass was much greater. Calculating from the data of Adjei and Prine (1976), Florigraze spread at 2 m per year. A similar rate was observed with PI 338316/CPI 93464 in south-east Queensland.

Competitive or spreading ability is not necessarily related to yielding ability. In the Gympie study, those accessions with the better spreading ability tended to be low growing, and those that had produced high dry matter yields in other experiments performed relatively poorly in the spread experiment. The exception was PI 231318/CPI 93483 which produced the highest yield and spread through grass at 14 cm per year, ranking 8 for rate of spread. The implication is that there is sufficient diversity within the series to select hay types which are taller varieties mostly grown in pure stands, pasture types which are capable of reasonable production when grown with grasses, and the lower growing fast-spreading types which may be more appropriate for soil conservation and amenity use.

(i) Alternative uses

Eurhizomatosae have potential beyond the provision of hay and forage by virtue of their strong perenniality and competitiveness, their tolerance of mowing and of herbicides, associated with their dense generally low growth habit, and mid to deep green foliage with periodic flushes of brightly coloured flowers. Prine *et al.* (1981, 1986) suggest that the American cultivars could be used as ornamental ground covers and for stabilizing road verges and embankments. These cultivars, in descending order of suitability, are Arblick, Florigraze, Arb and Arbrook. They also note that rhizoma peanuts have been planted as cover crop between rows of orchard trees and vines but the long-term implications of such a practice have not been determined. Among the current Australian collection, the most suitable accessions for amenity use are CPI 19898, PI 151982/CPI 22762, PI 338316/CPI 93464, GKP 9618/CPI 93469, PI 262814/CPI 93475 and PI 262841/CPI 93476. These are dense low-growing types and most of them spread rapidly.

(j) Pests and diseases

High levels of resistance to insect and disease attack have been identified in the *Eurhizomatosae*, although such resistance to any one organism is not necessarily universal throughout the series. Herbert and Stalker (1981) found that, out of 55 rhizomatous accessions screened, 39 displayed a high level of resistance to peanut stunt virus. Prine *et al.* (1986) noted that there was also resistance to peanut stripe and mosaic viruses within the series. Varying resistance to corn earworm (*Heliothis zea*) has been demonstrated by Campbell *et al.* (1982); and to this and other insects – tobacco thrips (*Frankliniella fusca*) and potato leafhopper (*Empoasca fabae*) – by Stalker and Campbell (1983). All members of the Australian collection of the series are resistant to races of organisms causing early leaf spot (*Cercospora arachidicola*), late leaf spot (*Phaeoisariopsis personata*) and peanut rust (*Puccinia arachidis*) in Australia (K.J. Middleton, personal communication). Although immunity to rust appears common in the *Rhizomatosae*, susceptible accessions exist (Ramanatha Rao, 1987). Stands of a number of cultivated *Eurhizomatosae* were reduced or eliminated by rust near Brasilia (E.A. Pizarro, personal communication). White mould (*Sclerotium rolfsii*) and lepto spot (*Leptosphaerulina brisianna*) have been identified on Arbrook, and leaf spots caused by *Phyllosticta* and *Stemphylium* on both Florigraze and Arbrook, but none has caused long-term or serious damage (Prine *et al.*, 1981, 1986). Cotton root rot caused by the fungus *Phymatotrichum omnivorum* has produced temporary reduction of stands of various accessions in southern Texas (W.R. Ocumpaugh, personal communication).

Root-knot nematode *Meloidogyne* sp. can weaken the legume component in pasture, so that level of resistance to these nematodes is often an important determinant of the utility value of various legumes when grown in association with other crops. The American cultivars Florigraze and Arbrook proved highly resistant to the root-knot nematodes, *M. arenaria* race I, *M. javanica* and *M. incognita* races I and II (Baltensperger *et al*, 1986). Resistance to *M. arenaria* (Nelson *et al.*, 1989) and *M. hapla* (Castillo *et al.*, 1973) has been identified in other rhizomatous accessions.

15.3.2 Caulorhizae

(a) Description

Section *Caulorhizae* is characterized by the presence of stolons. While many *Arachis* have prostrate stems and may appear to be stoloniferous by the rooting effect of the fruiting pegs, only the *Caulorhizae* actually form regular nodal root systems (Gregory *et al.* 1973). The original accessions of the two species within the section, *A. pintoi* and *A. repens* (Figure 15.3), are morphologically distinct. However, more recent collections tend to be intermediate between the two, making it difficult to ascribe them to one or the other species (Valls, 1983). Descriptions will therefore largely apply to the original types. Both species are strongly perennial, developing roots along the stolons on those nodes touching the ground, or those in the high humidity of a dense canopy. These roots, fine at first, ultimately develop into large woody taproot systems, 1–2 cm in diameter near the crown. Roots carry large numbers of small aeschynomenoid nodules. In southern Queensland, *A. pintoi* may form a canopy up to 20 cm deep in a pure sward, while *A. repens* rarely exceeds 10 cm. These depths may be exceeded when either is grown with a taller growing grass, when the groundnut stolons adopt a more ascendant growth habit. *A. repens* tends to produce a denser, more closely appressed stolon mat than does *A. pintoi*. Leaves of both are tetrafoliolate, with distal (apical) leaflets obovate and the proximal (basal) leaflets oblong–obovate. Leaflet size and proportion vary between species. *A. pintoi* leaflets grow to about 45 mm long and 35 mm wide, while *A. repens* leaflets rarely exceed about 30 × 15 mm. Leaflet

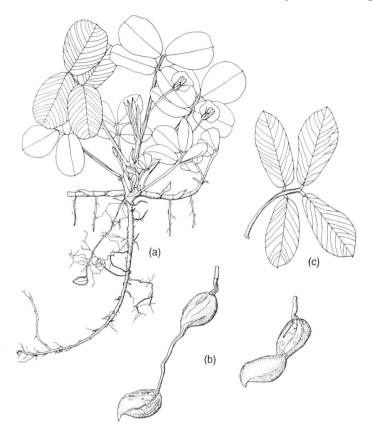

Figure 15.3 (a) Part of *Arachis pintoi* plant with stolon, taproot, foliage, flower and pods; (b) variation in pod types of *A. pintoi*; (c) leaf of *A. repens*. (Scales: (a) 2/3 actual size, (b) 1.5 × actual size, (c) 1–1.5 × actual size.)

apices are mostly obtuse and bases slightly cordate in *A. pintoi*, the apices of *A. repens* being mostly mucronate acute, and the bases cuneate to obtuse. *A. repens* also lacks the prominent spreading tuberculate hairs that are common on the proximal leaflet of *A. pintoi*.

Both species have relatively small yellow flowers, although one of the recent collections which is closer to *A. repens* has orange flowers (Valls, 1983). One of the more significant differences between the two species is that *A. pintoi* is capable of producing high yields of seed (Cook and Franklin, 1988) and the original *A. repens* rarely produces seed at all. *A. pintoi* pods are mostly single-seeded, about 6 mm in diameter and 11 mm long, and borne usually singly, on pegs 1–27 cm long with the terminal pod being prominently beaked. There are about 6000–8000 seed-in-pod/kg (Oram, 1990).

(b) Origin

A. pintoi was first collected in April 1954 by Professor Geraldo C.P. Pinto near the mouth of the Jequitinhonha River in Brazil at 15° 52′ S, 39° 6′ W, and 50 m A.S.L. (Gregory *et al.*, 1973). Valls (1983) and C.E. Simpson (personal communication), after visiting the collection site and many sites upriver, concluded that this material had been transported downriver from around Araçuai, that the collection site no longer exists due to southward displacement of the river mouth, and that the altitude cited above would more likely have been 5 m A.S.L.. It has travelled a circuitous route around the world – from the collection site, in turn to Cruz das Almas (Brazil), Corrientes (Argentina), Experiment (USA), Brisbane (Australia) and Cali (Colombia), ultimately returning to Brazil at Belém in the north. It has now been catalogued under a variety of collection numbers and accession numbers: GK 12787 = PI 338447 (US), HL 323 = PI 338314 = CPI 58113 (Australia) = CIAT 17434 (Colombia).

The site where Dr Thomas Dalton first discovered *A. repens* was at 600–700 m A.S.L. along the Jequitai River (18° 10′ S, 43° 26′ W). Gregory *et al.* (1980), believing this to be a tributary of the Jequitinhonha rather than of the São Francisco, recorded that *Caulorhizae* has been found only in the valley of the Jequitinhonha. Subsequent collecting expeditions have extended the known limits of the section to include the valleys of the São Francisco (Valls, 1983) and Tocantins (C.E. Simpson, personal communication). The original *A. repens* collection is now catalogued under seven different collection numbers, seven USA accession numbers and a variety of other accession numbers throughout the world.

A. pintoi has generally been collected in infertile red, sandy loam river-bottom soils of high aluminium saturation, appearing to do best in low areas which are wet to flooded during the wet season. Rainfall in the region during the wet season (October to May) is about 2000 mm, a further 200 mm falling during the remainder of the year. The native vegetation is *mata baixo* or low forest, with trees to 5 m and very little sunlight reaching the forest floor (C.E. Simpson, personal communication).

A. repens originates from similar environments. The only published noting about the first collection site was 'from an area sometimes inundated by water' (Anon., 1967). One recent collection was from a grey, silty loam soil, where the plants were growing among grasses in standing water, and another from a red silty clay where plants were growing on the roadside and in low areas (C.E. Simpson, personal communication).

(c) Cultivars

The only cultivar from *Caulorhizae* is *A. pintoi* PI 338314, which was introduced into Australia as CPI 58113 in 1972, released commercially in 1987 and registered as cv. Amarillo (mistakenly believing this to be

Portuguese for 'yellow', referring to the flower colour) in 1990 (Oram, 1990). Amarillo has been exported to Colombia, Indonesia, Malaysia, Thailand, Sri Lanka and various Pacific countries including Fiji, Western Samoa and Tonga.

(d) Adaptation

In general, *A. pintoi* and *A. repens* are well adapted to moderately drained, or seasonally inundated or poorly drained soils in the humid tropics and subtropics. They have performed well between latitudes of 1° 30′ near Manado in North Sulawesi, Indonesia (Kaligis and Sumolang, 1991), to 29° 30′ at Grafton in New South Wales, Australia (G.P.M. Wilson, personal communication), in areas receiving average annual rainfall from about 1100 mm to more than 3000 mm. Soil textures at successful evaluation and development sites have ranged from sands to well structured clays, and soil reaction mostly from very acid to neutral. Due to more intensive testing of *A. pintoi*, there is a more detailed understanding of its adaptational range than for *A. repens*. The only soil texture limitation for *A. pintoi* appears to be on poorly structured, medium to heavy clays, often referred to as 'puggy' soils. In two instances in southern Queensland, it has grown poorly or failed to persist on such soils (R.M. Jones, B.G. Cook, unpublished); this does not appear to be related to drainage, since experience in Brazil shows it to be well adapted to seasonally waterlogged conditions (Anon., 1989b). It appears to be less sensitive than many species to pH, having grown well on soils with pH values ranging from 4.3 (in water) on the Llanos Orientales of Colombia (Grof, 1985) to 7.2 (1:5 water) on the Darling Downs of south-east Queensland (J.N. Coote, personal communication). As might be expected, the adaptation to extreme acidity is accompanied by a tolerance of soluble aluminium and manganese. G.R. Rayment (unpublished data) demonstrated tolerance of manganese levels which would have been detrimental to many other legumes. Aluminium tolerance is indicated by soil analysis data from Grof's (1985) experiment, which showed an 86% aluminium saturation of the exchange complex.

A. pintoi may have a greater requirement for soil phosphorus than the rhizomatous species (section 15.3.1(d)). In a pot trial comparison between the rhizomatous GKP 9618/CPI 93469 and *A. pintoi*, using a soil of 12 ppm available P (bicarbonate extract; Colwell, 1963), the former gave little or no response, while the latter responded up to the equivalent of 50 kg/ha of applied phosphorus. Dextre (1984), Turnour (1987) and J.X. Kui (unpublished data) also obtained significant responses in *A. pintoi* to single rate applications of 20, 100 and 40 kg/ha of phosphorus to soils of 2.7–4.9 ppm (extractant not reported), 13 ppm and 19 ppm (both bicarbonate extract) respectively.

While *A. pintoi* and *A. repens* are adapted over a wide latitudinal range, best performance is obtained in areas not subject to frost. Top growth is 'burnt' by frost but stands recover from unaffected stolons or crowns.

One of the most significant characteristics of *A. pintoi* is its high level of shade tolerance. In a study by the University of Queensland comparing 84 legume accessions from many genera, *A. pintoi* was the most shade tolerant; at 20% light it produced 48% of the mean of yields produced at 100% and 70% light (W.W. Stür, 1991). This has far-reaching implications in its use as a forage and ground cover. *A. repens* was not as shade tolerant as *A. pintoi*; at 20% light it produced only 28% of the mean of yields produced at 100% and 70% light.

(e) Yield

The magnitude of dry matter yields reported for a particular species is often strongly dependent on height and frequency of cut. It is not surprising, then, to find that yields from experiments in various countries differ markedly. In Colombia, harvesting every 4 weeks at 3 cm, Grof (1985) obtained dry matter yields from 5.2 to 9.6 t/ha, depending on the companion grass. Yields in Costa Rica varied from 2.6 t/ha (11 cuts/yr) at San Isidro to 10.9 t/ha (9 cuts) at Guápiles (Anon., 1990b). Working in maturing rubber plantation in Malaysia, Ng (1991) recorded only 2.4 t DM/ha. Kaligis and Sumolang (1991) and Rika *et al.* (1991), working in coconut plantations in north Sulawesi and Bali in Indonesia, obtained yields of 4.8 and 2.7 t DM/ha for *A. pintoi* and 4.4 and 3.2 t DM/ha for *A. repens*. In all south-east Asian situations, *Arachis* was grown in pure stands and cut at 8-weekly intervals. In the humid subtropics of south-eastern Queensland, harvesting pure stands at 2–3 cm every 6 weeks, irrigated and unirrigated yields of 7.3 and 6.5 t DM/ha were obtained from cv. Amarillo (B.G. Cook, unpublished).

In the Guápiles experiment, two more recently collected accessions of *A. pintoi* (CIAT 18744, 18747) produced similar yields to the original collection (CIAT 17434) while others (CIAT 18745, 18746) produced only 3.7 and 3.8 t DM/ha.

(f) Feeding value

Both *A. pintoi* and *A. repens* are readily eaten by grazing livestock but information on feeding value is restricted to the former. Animals may select *A. pintoi* over other components of a mixed pasture. In pastures comprising *A. pintoi* and different species of the grass genus, *Brachiaria*, selection indices of 0.57 to 0.61 were calculated for three of the four mixtures, the other being 0.48 (Lascano and Thomas, 1988). Selection index relates the proportion of a species in the diet to the proportion in the pasture, a level <0.5 indicating selection against and >0.5 selection in favour (Chesson, 1983). Analysis of the feed on offer revealed leaf:stem ratios of about 0.24 in the dry season and 0.97 in the wet season but, unlike most forage legume species, there was little difference between IVOMD of leaf and stem, the greatest difference being between IVOMD of leaf in the dry (67%) and the wet season (60%). Mean value of crude protein (CP) in the leaf was 13.9% in the dry season and 18.0% in the wet season, while stem values were about 11% in both seasons. In earlier measurements from the same pastures, Grof (1985) measured average crude protein levels of 14.8% in the first year and 16.6% in the second, and phosphorus levels in the dry matter of 0.18–0.20%. With an average stocking rate of 2.4 animal units/ha (1 a.u. = 420 kg), these pastures proved capable of producing 515 g/head/day over a 594-day grazing period. In south-east Queensland, a small plot cutting trial produced dry matter of similar high quality to that obtained in the above experiment in Colombia – 20% CP, 73% IVOMD and 0.35% P (B.G. Cook, unpublished data), the latter figure probably reflecting the high levels of phosphatic fertilizer applied.

(g) Establishment

Due to poor seed set, *A. repens* is propagated only vegetatively. It strikes readily from lengths of stolon with two to three nodes, providing the soil is kept moist.

A. pintoi produces high seed yields and may therefore be propagated from either seed or cuttings. A laboratory technique for *in vitro* culture has also been described (Burtnik and Mroginski, 1985) but has no immediate application in commercial pasture establishment. The simplest method is to establish from seed. Freshly harvested seed should be dried at 35–40 °C for 10 days to improve germination (J. Butler, unpublished data). Cook and Franklin (1988) describe a method of precision planting to establish a seed crop. Seed should be inoculated with a specific strain of *Bradyrhizobium* prior to planting. Early evaluation of the species was hampered by lack of a highly effective strain of root nodule bacteria, but the strain CIAT 3101, initially isolated from *Centrosema macrocarpum*, was selected after a screening programme in Colombia (Sylvester-Bradley *et al.*, 1988). This strain, different from that used with *A. hypogaea*, is now used in all commercial plantings. For the seed crop, a sowing rate of 25 kg seed-in-pod/ha was used, but 10 kg/ha is adequate for pasture establishment providing seed is of good quality. Seed is best sown 2–6 cm deep. Specialized planting machinery is most economical of seed but other methods are available. Perhaps the most practical method is to broadcast seed onto a fairly well prepared seed-bed, cultivate to cover the seed, and roll to achieve good soil/seed contact. Some coverage is essential as surface-sown seed germinates poorly (Ogawa *et al.*, 1990) and is attractive to birds and rodents (B.G. Cook, unpublished data). Ogawa *et al.* (1990) noted that some seeds that did germinate failed to develop because the thick radicle could not penetrate the hard surface layer of uncultivated soil.

A. pintoi is tolerant of a range of pre- and post-emergence herbicides facilitating effective chemical weed control in an establishing stand. Its low growth habit in the early stages permits use of rope-wick herbicide applicators (Cook and Franklin, 1988). Sowing time depends on seasonal constraints such as probability of frost in the higher latitudes, and onset of wet and dry seasons in monsoonal climates. Seedlings establish quickly and complete groundcover can be achieved in 5 months from the higher sowing rate given above. Seed is not always available, due to difficulties in seed production and storage, but established stands may be available as a source of planting material. Asakawa and Ramírez (1989) describe an effective method of manual or mechanical vegetative establishment used in Colombia. They suggest planting 20 cm lengths of stolon about 15 cm deep in rows 35 cm apart or, if planted with a grass, in rows 1 m apart. Cuttings should be inoculated with *Bradyrhizobium* using an adhesive such as molasses.

(h) Seed production

A. pintoi seedlings usually commence flowering 3–4 weeks after emergence. The flowering continues through much of the growing season, with periodic flushes apparently in response to improved soil moisture conditions (Cook and Franklin, 1988). In the subtropics, flowering declines with the onset of cooler conditions in May, but in the tropics it continues through most of the year (D.G. Cooksley, personal communication). Initial flowers usually produce few pegs, and pegs produced during dry conditions fail to develop fruit. However, with these exceptions, seed set spans at least 5 months in subtropical Queensland, where all Australian seed is currently produced. By the time end-of-season fruit have matured, pegs of early-set fruit have decayed, giving low seed yields using conventional groundnut harvesting techniques. Cook and Franklin (1988) describe a more efficient harvesting principle that involves sieving the whole soil volume through rotating, coaxial, cylindrical screens once the seed is mature. Seed crops are grown in sandy soil to facilitate separation of seed and soil. This system, which has now been commercialized, has produced yields of about 1 t/ha of seed-in-pod per year. It is necessary to rotary-hoe prior to harvest but only a small proportion of seed is damaged in the process. No yield response or reduction in the proportion of 'pops' has been obtained through applications of calcium as gypsum or as lime in experiments at Gympie (D.S. Loch, unpublished data). There is no need to resow seed crops, since stands recover from unharvested seed and from new crowns formed at depth on taproots severed by the rotary hoe. In Costa Rica, Diulgheroff *et al.* (1990), also using a sieving technique, obtained seed yields ranging from 0.6 to 2.0 t/ha from experimental plots. Smallholder farmers at Yapacani, near Santa Cruz in Bolivia, have produced large amounts of seed, at times reaching 2.6 t/ha, using simple cradle-mounted flat sieves and digging hoes (E. Ramírez, personal communication). Some of the more recently collected *A. pintoi* varieties have proved capable of seed yields of about 1.5 t/ha in Peru in an experiment in which CIAT 17434 (the original variety) inexplicably produced no seed (Anon., 1990a).

(i) Companion species

These stoloniferous *Arachis* species are capable of forming stable associations with a variety of grasses. In the native habitat, *A. pintoi* has been collected growing with *Panicum maximum*, *Eriochloa polystachia* and *Paspalum* sp. and *A. repens* with *Paspalum* and *Cynodon dactylon* (C.E. Simpson, personal communication). In Colombia, Grof (1985) showed that *A. pintoi* could also grow successfully with the competitive *Brachiaria* species (*dictyoneura*, *brizantha*, *humidicola* and *ruziziensis*) comprising 20–45% of the pasture. In another wet tropical environment at South Johnstone in north Queensland, *A. pintoi* has proved capable of forming a balanced pasture with *B. decumbens*. In the subtropics, it has been grown with a wide range of native and introduced species (Oram, 1990). *A. repens* has shown similar persistence in a variety of grasses in tropical and subtropical Queensland. In an experiment in northern Queensland, both *A. pintoi* and *A. repens* have spread about 1.5 m in 15 months, when grass competition was initially low (D.G. Cooksley, unpublished data). Spread with vigorous grass competition has been slower in the subtropics. *A. pintoi* spread at 5 cm/yr over 5 years, and *A. repens* at up to 15 cm/yr (B.G. Cook, unpublished data). R.M. Jones (unpublished data) measured a similar rate of spread in *A. repens* over 15 years at Beerwah in southeastern Queensland, under competition from *Axonopus affinis* and *Digitaria eriantha* subsp. *pentzii*.

(j) Alternative uses

Both *Caulorhizae* species have application other than as forages, by virtue of sward-forming habit, shade tolerance or periodic flushes of flowering. Soil loss is a serious problem in orchards and plantations. Ground-cover is often destroyed intentionally to minimize competition with the developing crop, or to facilitate ease of movement through the crop, or it may simply succumb to increasing shade from the developing crop. Once ground-cover is lost, erosion becomes a problem. Dwyer *et al.* (1989) suggest *A. pintoi* as a living mulch, that should not compete with the crop to any great extent because of its low nutrient demand and its ability to fix its own nitrogen. However, G. Johns (Anon., 1991) recommends caution with this practice in bananas: despite an improvement in soil fertility under *A. pintoi* at Alstonville in northern New South Wales, fruit production over 3 years was reduced by about 13% compared with a crop on bare soil. He suggests that this may have been due to reduced soil temperatures rather than competition. The benefit is that there is no soil loss, which in turn implies a more sustainable practice.

A. pintoi is also being used in north Queensland for bank stabilization and as an ornamental ground-cover in shaded or erosion-prone areas (P. Bolton, personal communication). Although *A. repens* has not been similarly exploited in Queensland, it is used as an ornamental in its native environment. C.E. Simpson (personal communication) observed *A. repens* growing in domestic and commercial gardens during a collection trip in Brazil, where it is locally known as 'grama amendoim' or groundnut grass. PI 338277 was used principally for lawn planting at the collection site (Anon., 1970).

(k) Pests and diseases

Neither of the original collections of the two species is severely affected by disease. Both appear to be resistant to the common groundnut diseases, rust (*Puccinia arachidis*) and leaf spots (*Cercospora arachidicola*) and *Phaeoisariopsis personata*) (K.J. Middleton, unpublished data), although C.E. Simpson (personal communication) has observed late leaf spot on leaves of one of the more recent *A. repens* collections. *A. pintoi* CIAT 17434 has proved susceptible to two fungal diseases of the commercial peanut – peanut scab (*Sphaceloma arachidis*) in Colombia, and pepper leaf spot (*Leptosphaerulina arachidicola*) in Costa Rica (Anon., 1989a,c) – both having only a limited effect on productivity. Two other lines, CIAT 18752 and 18746, were resistant to *Sphaceloma* in a field screening. Other fungi – *Phomopsis* sp., *Cylindrocladium* sp. and *Colletotrichum gloeosporioides* – have been isolated from leaf spots on *A. pintoi* in Australia but none is serious or widespread. *A. pintoi* and *A. repens* are both subject to often extensive black stem lesions, which apparently have little effect on the overall vigour of the plant. *Colletotrichum gloeosporioides* has been isolated from these lesions in Australia (Oram, 1990) and *C. truncatum* (*C. dematium*) in Colombia (Grof, 1985). Inoculation of *A. pintoi* with strains of *C. gloeosporioides* that cause severe damage to stands of the related genera, *Stylosanthes* and *Aeschynomene*, failed to produce infection (Vinijisanum *et al.*, 1987), suggesting that these strains are different from those mentioned above.

Although a small proportion of leaves of *A. pintoi* cv. Amarillo in Australia are variegated, no virus particles have been observed in the sap under electron microscopic examination. However, peanut mottle virus has been isolated from material showing virus-like symptoms in Colombia (Anon., 1990c). Resistance to peanut stunt virus has been recognized in *A. repens* (Herbert and Stalker, 1981).

Nematodes and insects have not proved to be a field problem with either species. *A. pintoi* has a high level of resistance to *Meloidogyne incognita* and *M. arenaria*, and moderate resistance to *M. hapla*, but is susceptible to root lesion nematode (*Pratylenchus brachyurus*) (R.W McLeod and P.C. O'Brien, personal communication). Stalker and Campbell (1983) demonstrated resistance to tobacco thrips, potato leafhopper and corn earworm in *A. repens*.

Perhaps the greatest potential problem lies not with the usual microorganisms but with rodents. In several situations, rat and mouse populations have increased in response to build-up of seed under *A. pintoi*. While this has had no great effect on the legume stand, it can have undesirable consequences with the associated tree crop, through excessive soil aeration and drying from the burrowing animals, and loss of fruit from the trees as the animals seek an alternative diet.

15.3.3 Other species

Various other species have been identified as having forage potential. Following a plant collecting trip in 1947–48 through parts of Argentina, Uruguay, Paraguay and Brazil, Hartley (1949) nominated a number of species he considered to have merit by virtue of performance under grazing in their natural habitat. Unfortunately, as intimated by Gregory *et al.* (1973), specific affiliations at that time are different from those that would be applied today. Although Hartley provides reasonable descriptions of material collected, in many cases it is insufficient to identify with any degree of certainty to which current species he is alluding. His selection of highest potential is clear: through reference to rhizomes, his *A. prostrata* is obviously the current *A. glabrata* or, at least, section *Rhizomatosae*. The identity of his second choice *A. glabrata* is somewhat clouded. It is possible, from the plant descriptions and the distributions given, that these accessions and those he assigns to *A. diogoi* ('one of the most promising forage species') may all, or in part, belong to the sections *Erectoides* and *Procumbensae*. Hartley's judgement has been borne out in that members of these sections have come to prominence in Queensland and in Florida.

The Queensland accession is noteworthy by its record of persistence. Now simply accessed as CQ 1780, this line probably derives from material initially identified as *A. prostrata* CPI 6930 or *A. diogoi* CPI 8987-8 (Miles, 1949). These lines were planted in the late 1930s or early 1940s at Fitzroyvale near Rockhampton in central Queensland. Despite extremely heavy grazing and lack of fertilizer, one type has persisted and spread at this site. This line has recently been evaluated at other sites in coastal and subcoastal Queensland, and appears to be well adapted to subhumid environments (B.G. Cook, unpublished data).

The variety showing early promise in Florida, PI 446898, was collected by Professor A.E. Kretschmer, Jr, on the Pantanal of Brazil (20° 7′ S, 56° 43′ W), where it was growing in 10–20 cm of standing water, on a clay soil of pH 6.0–7.0 (Kretschmer and Wilson, 1988). This accession has a number of desirable attributes including fair yields of high quality forage (Kretschmer and Wilson, 1988; B.G. Cook, unpublished data), resistance to early and late leaf spot and rust (K.J. Middleton, unpublished data) and high seed yields (D.S. Loch, unpublished data). It has performed well under grazing in Florida but has not persisted well under grazing in coastal Queensland, possibly due to attack by various root-knot nematodes and several fungal diseases. Further evaluation is required.

All material considered thus far has been perennial. Beaty *et al.* (1968) assessed the annual, *A. monticola* PI 263393, for its persistence in swards of Pensacola bahia grass and Coastal bermudagrass. Results were encouraging but this accession, when screened for disease susceptibility in Queensland, proved susceptible to rust (K.J. Middleton, unpublished data). It is important that forage varieties are not susceptible to the more serious peanut diseases, not only because of the disease effect on forage productivity but also because diseased plant parts are a source of inoculum for nearby commercial crops.

15.4 BREEDING

Considering that the genus *Arachis* possesses many desirable forage attributes among its various sections and species, it is remarkable that so little breeding work appears to have been carried out with the specific aim of producing improved forage types. Gregory and Gregory (1979) demonstrated intra- and intersectional compatibility from a programme involving a total of 1075 crosses among 91 parental forms representing at least 50 species. Although the success rate was not high, the work does demonstrate some potential for improvement through breeding. Such improvement has already been achieved at Coimbatore, India, from a cross between the diploid *A. diogoi* and tetraploid *A. glabrata* (Chandrasekhar, 1980). The hybrid, which it is suggested be called cultivar Forachis, produces higher yields than either parent, yet retains the desirable rhizomatous growth habit of *A. glabrata*. Gregory and Gregory (1979) showed that *A. glabrata* crossed successfully with series *Annuae* of section *Arachis* and series *Tetrafoliolatae* and *Procumbensae* of section *Erectoides*, but not with series *Perennes* of section *Arachis*, of which *A. diogoi* is a member. This places doubt on the identity of the non-rhizomatous parent of Forachis.

Natural hybrids are common in pasture nurseries where representatives of several sections are grown in close proximity. The American cultivar Florigraze arose as an intrasectional cross under such circumstances at Gainesville, Florida (Prine *et al.*, 1981). A probable intrasectional cross occurred at Grafton in New South Wales, where a type, intermediate between *A. repens* and *A. pintoi*, appeared in a plot of *A. pintoi* growing close to a plot of *A. repens* (G.P.M. Wilson, personal communication). No other compatible types were growing nearby. Intersectional crosses between *Erectoides/Procumbensae* and *Rhizomatosae* have also developed at Grafton, and at Gympie, in south-east Queensland. These have appeared in plots of the freely seeding *Erectoides* accession, CQ 1780, and *Procumbensae* accession, PI 446898. The hybrids do not flower; they have different leaf characteristics from the female parent and produce a multitude of shoots arising from short rhizomes. Similar types have been observed at Gainesville, Florida (K.H. Quesenberry, personal communication). None of these appears to have an agronomic advantage over either parent. Other possible spontaneous intersectional hybrids have been observed in Costa Rica, with *A. pintoi* (*Caulorhizae*) as the female parent and PI 446898 (*Procumbensae*) as the male (P. Argel, personal communication).

However, once the limitations of the wild types have been more completely identified, there seems real potential to incorporate valuable attributes into more promising types – say nematode resistance into PI 446898 – or to produce a completely different type combining desirable qualities from several sources, using the various breeding techniques available.

15.5 CONCLUSION

The genus *Arachis* originates mainly from Brazil but is also found in Argentina, Bolivia, Paraguay and Uruguay, in an area extending from near the Equator to 34° S, and from the Atlantic coast to the foothills of the Andes. The altitude range is from sea level to 1450 m. Gregory *et al.* (1980) describe the diversity of habitat:

> It is found among vegetation types from broken forest to the open grasslands of 'Pantanal' and 'Campo', in 'Caatinga', 'Parque', 'Chaco', and 'Cerrado'. Its species grow submerged, among stones bathed with water, in dry gravel, and in flood-plain alluvium. They are found from semi-arid locations to regions that receive 2000 mm or more of rain per year and in regions subject to great floods and intense droughts. They grow most commonly in friable or somewhat sandy soils, but also in soils that vary from almost pure sand to reduced, humic clays and sands, various alluvia, dark red humic latosols, to pure lateritic caps and gravel. Underlying parent rock may be sandstone, limestone, basic gneiss, granite, basaltic lava, and less often mica and mica schists.

There are annual, biennial and perennial types, with procumbent, decumbent or erect growth habits. Prostrate stems develop nodal roots in some species; in other species, rhizomes develop. Plants may have simple taproots or varying degrees of tuberous development. Seed is produced underground on vertical or horizontal pegs. The genus has been divided into sections based on these characters. To date, forage potential has been recognized and commercialized in only three of these sections – *Arachis*,

Rhizomatosae and *Caulorhizae*. Work with material from these sections has demonstrated a number of particularly desirable characteristics within the genus, though not all necessarily possessed by any one variety – high yields and high quality of forage, high palatability, excellent haymaking qualities, persistence under intensive grazing and among competitive grasses, tolerance of low fertility and high aluminium and manganese, good drought tolerance, and minimal losses due to pests and disease. These characteristics have translated successfully to animal production.

Outside the native habitat, the bulk of forage derived from *Arachis* has come from haulms from the cultivated groundnut, *A. hypogaea*. At this stage, only a relatively small proportion of the world collection of wild types has been evaluated for forage potential in the tropics and subtropics. With the diversity of growth forms and adaptation within *Arachis*, there is little doubt that, as more of the collection finds its way into forage germplasm banks, the contribution of the genus to world livestock production will increase markedly.

ACKNOWLEDGEMENTS

The authors gratefully acknowledge the efforts of Mr Will Smith, botanical illustrator, Botany Branch, QDPI, in preparing the splendid illustrations used in this chapter.

REFERENCES

Abou-Raya, A.K., Abou-El-Hassanand, A. and Al-Rehab, F.A. (1971) The feeding value of peanut hay and its response to alkali treatment and addition of urea. *United Arab Republic Journal of Animal Production*, **11(2)**, 291–93.

Addy, B.L. and Thomas, D. (1977) Intensive fattening of beef cattle by stall feeding on the Lilongwe Plain, Malawi. 2. Utilisation of crop residues, crop by-products and leucaena. *Tropical Animal Health and Production* **9(4)**, 191–96.

Adjei, M.B. and Prine, G.M. (1976) Establishment of perennial peanuts (*Arachis glabrata* Benth.). *Soil and Crop Science Society of Florida Proceedings*, **35**, 50–53.

Aduku, A.O., Okoh, P.N., Njoku, P.C. *et al.* (1986) Evaluation of cowpea (*Vigna unguiculata*) and peanut (*Arachis hypogaea*) haulms as feedstuffs for weanling rabbits in a tropical environment (Nigeria). *Journal of Applied Rabbit Research*, **9(4)**, 178–180.

Ahmed, F.A. and Pollett, G.E. (1979) The performance of yearling Kenana (Sudan Zebu) calves given three levels of crude protein as a concentrate supplement to ad libitum groundnut hay. *Tropical Animal Production*, **4(1)**, 65–72.

Aiyadurai, S.G. (1959) Wild species of *Arachis* as forage and soil binding plants. *Indian Oilseeds Journal*, **3(3)**, 149–51.

Anon. (1966) *Plant Inventory No. 164*, January 1 to December 31, 1956, United States Department of Agriculture, Beltsville, Maryland.

Anon. (1967) *Plant Inventory No. 169*, January 1 to December 31 1961, United States Department of Agriculture, Beltsville, Maryland.

Anon. (1970) *Plant Inventory No. 176*, January 1 to December 31, 1968, United States Department of Agriculture, Beltsville, Maryland.

Anon. (1982) *Groundnut Cultivation in China*. Chap 19, Utilisation of Groundnut.

Anon. (1989a) Plant pathology, in *Annual Report 1988, Tropical Pastures*, Working Document No. 58, Centro Internacional de Agricultura Tropical, Cali, Colombia, p. 4.34.

Anon. (1989b) Agronomy Cerrados, in *Annual Report 1988, Tropical Pastures*, Working Document No. 58, Centro Internacional de Agricultura Tropical, Cali, Colombia, p. 7.20.

Anon. (1989c) Agronomy Central America and the Caribbean, in *Annual Report 1988, Tropical Pastures*, Working Document No. 58, Centro Internacional de Agricultura Tropical, Cali, Colombia, p. 9.17.

Anon. (1990a) Agronomy humid tropics, in *Annual Report 1989, Tropical Pastures*, Working Document No. 70, Centro Internacional de Agricultura Tropical, Cali, Colombia, pp. 9.21–23.

Anon. (1990b) Agronomy Central America and the Caribbean, in *Annual Report 1989, Tropical Pastures*, Working Document No. 70, Centro Internacional de Agricultura, Cali, Colombia, p. 8.3.

Anon. (1990c) Virus detection in forage legumes, in *CIAT Report 1990 – Report on 1989 Highlights*, CIAT Publication No. 162, Centro Internacional de Agricultura Tropical, Cali, Colombia, pp. 129–33.

Anon. (1991) Banana cover crop shows promise, but has its problems. *Commercial Horticulture*, **Autumn**, 24–25.

Asakawa, N.M. and Ramírez R., C.A. (1989) Metodología para la inoculación y siembra de *Arachis pintoi*. *Pasturas tropicales*, **11(1)**, 24–26.

Ayoade, J.A. and Njewa, B.J. (1983/84) Utilisation of agricultural residues by livestock. (I) Physical forms as a factor affecting the utilisation of bean and groundnut haulms by goats. *Research Bulletin, Bunda College of Agriculture, University of Malawi*, **12**, 61–68.

Baltensperger, D.D., Prine, G.M. and Dunn, R.A. (1986) Root-knot nematode resistance in *Arachis glabrata*. *Peanut Science*, **13**, 78–80.

Beaty, E.R., Powell, J.D. and Stanley, R.L. (1968) Production and persistence of wild annual peanuts in bahia and bermudagrass sods. *Journal of Range Management*, **21(5)**, 331–33.

Bell, M. (1986), Effect of sowing date on growth and development of irrigated peanuts, *Arachis hypogaea* L. cv. Early Bunch, in a monsoonal tropical environment. *Australian Journal of Agricultural Research*, **37 (4)**, 361–373.

Beltranena, R., Breman, J. and Prine, G.M. (1981) Yield and quality of Florigraze rhizoma peanut (*Arachis glabrata* Benth.) as affected by cutting height and frequency. *Soil and Crop Science Society of Florida Proceedings*, **40**, 153–156.

Blickensderfer, C.B., Haynsworth, H.J. and Roush, R.D. (1964) Wild peanut is promising forage legume for Florida. *Crops and Soils*, **17(2)**, 19–20.

Blue, W.G., Njwe, R.M., Nair, K.P.P. *et al.* (1989) Forage legume response to lime and phosphorus on the highland soils of northwest Cameroon. *Soil and Crop Science Society of Florida Proceedings*, **48**, 68–71.

Bogdan, A.V. (1977) *Tropical Pasture and Fodder Plants (Grasses and Legumes)*, Longman, London and New York, p. 323.

Bredon, R.M., Steward, P.G. and Dugmore, T.J. (1987) *A manual on the nutritive value and chemical composition of commonly used South African Farm Feeds*. Natal Region, Department of Agriculture and Water Supply, South Africa.

Burke, E.P. (1850) *Reminiscences of Georgia*, James M. Fitch, Ohio, cited by Higgins (1951a).

Burtnik, O.J. and Mroginski, K.A. (1985) Regeneración de plantas de *Arachis pintoi* (*Leguminosae*) por cultivo *in vitro* de tejidos foliares. *Oléagineux*, **40(12)**, 609–611.

Caldwell, A.G., Morris, D.R., Joost, R.E. *et al.* (1991) Perennial peanut, a summer legume for Louisiana. *Louisiana Agriculture*, **43(2)**, 14–18.

Campbell, W.V., Wynne, J.C. and Stalker, H.T. (1982) Screening groundnut for *Heliothis* resistance, in *Proceedings of the International Workshop on Heliothis Management, 15–20 November 1981*, ICRISAT, Patancheru, A.P., India, pp. 267–75.

Canudas-Lara, E.G., Quesenberry, K.H., Teem, D.H. and Prine, G.M. (1984) Sethoxydim and dalapon application to rhizomes for common bermudagrass control in rhizoma peanut. *Soil and Crop Science Society of Florida Proceedings*, **43**, 174–77.

Canudas, E.G., Quesenberry, K.H., Sollenberger, L.E. and Prine, G.M. (1989) Establishment of two cultivars of rhizoma peanut as affected by weed control and planting rate. *Tropical Grasslands*, **23(3)**, 162–169.

Castillo, M.B., Morrison, L.S., Russell, C.C. and Banks, D.J. (1973) Resistance to *Meloidogyne hapla* in peanut. *Journal of Nematology*, **5**, 281–85.

Chandrasekhar, N.R. (1980) Forachis – a perennial forage groundnut, in *National seminar on the application of genetics to improvement of groundnut, 16–17 July, 1980*, Tamil Nadu Agricultural University, Coimbatore, India.

Chesson, J. (1983) The estimation and analysis of preference and its relationship to foraging models. *Ecology*, **64**, 1297–1304.

Collins, E.R. and Morris, H.D. (1941) *Soil fertility studies with peanuts*, North Carolina Agricultural Experiment Stations Bulletin 330.

Colwell, J.D. (1963) The estimation of phosphorus fertilizer requirements of wheat in southern New South Wales by soil analysis. *Australian Journal of Experimental Agriculture and Animal Husbandry*, **3**, 190–97.

Combellas, J., Centeno, A., Mazzini, B. and Combellas, J. (1972) Use of the aerial parts of groundnut. 2. Making into hay, intake and digestibility in vivo. *Agronomia Tropical*, **22(3)**, 281–285.

Cook, B.G. and Franklin, T.G. (1988) Crop management and seed harvesting of *Arachis pintoi*, Krap et Greg. *nom. nud. Journal of Applied Seed Production*, **6**, 26–30.

Corby, H.D.L. (1981) The systematic value of leguminous root nodules, in *Advances in Legume Systematics* (eds R.M. Polhill and P.H. Raven), Royal Botanic Gardens, Kew, Richmond, Surrey, pp. 657–69.

Dextre, R. (1984) *Requerimientos nutricionales de Arachis pintoi CIAT 17434*). Centro Internacional de Agricultura Tropical (CIAT), Cali, Colombia (Mimeograph).

Diulgheroff, S., Pizarro, E.A., Ferguson, J.E. and Argel, P.J. (1990) Multiplicación de semillas de especies forrajeras tropicales en Costa Rica. *Pasturas tropicales*, **12(2)**, 15–23.

Drammeh, S. (1990). Groundnut improvement, production, management, and utilisation in the Gambia, in *Summary Proceedings of the First ICRISAT Regional Groundnut Meeting for West Africa*, International Crops Research Institute for the Semi-Arid Tropics, Patancheru, Andhra Pradesh, India, pp. 41–44.

Dunavin, L.S. (1990) Cool season forage crops seeded over dormant rhizoma peanut. *Journal of Production Agriculture*, **3(1)**, 112–14.

Dunavin, L.S. (1992) Florigraze rhizoma peanut in association with warm-season perennial grasses. *Agronomy Journal*, **84(2)**, 148–151.

Durga Prasad, R.D., Prasad, D.A. and Ramachandra Reddy, R. (1986) Evaluation of complete rations containing groundnut haulms for lambs. *Indian Journal of Animal Sciences*, **56(2)**, 258–261.

Dwyer, G.T., O'Hare, P.J. and Cook, B.G. (1989) Pinto's peanut: a ground cover for orchards. *Queensland Agricultural Journal*, **115(3)**, 153–54.

Easby, W.B. (1851) Letter to Commissioner of Patents. *Report Commissioner of Patents 1851* (21), 353–55, cited by Higgins (1951a).

Essomba, N.B., Mekontchou, T. and Iroume, R.N. (1990) Groundnut production and research in Cameroon, in *Summary Proceedings of the First ICRISAT Regional Groundnut Meeting for West Africa*, International Crops Research Institute for the Semi-Arid Tropics, Patancheru, Andhra Pradesh, India, pp 37–39.

Gelaye, S., Amoah, E.A. and Guthrie, P. (1990) Performance of yearling goats fed alfalfa and florigraze rhizoma peanut hay. *Small Ruminant Research*, **3**, 353–61.

Gregory, M.P. and Gregory, W.C. (1979) Exotic germ plasm of *Arachis* L. interspecific hybrids. *The Journal of Heredity*, **70**, 185–93.

Gregory, W.C., Gregory, M.P., Krapovickas, A. *et al.* (1973) Structure and genetic resources of peanuts, in *Peanuts – Culture and Uses*, American Peanut Research and Education Association, Inc., Stillwater, Oklahoma.

Gregory, W.C., Krapovickas, A. and Gregory, M.P. (1980) Structure, variation, evolution, and classification in *Arachis*, in *Advances in Legume Science* (eds R.J. Summerfield and

A.H. Bunting), University of Reading, England, pp. 469–81.

Grof, B. (1985) Forage attributes of the perennial groundnut *Arachis pintoi* in a tropical savanna environment in Colombia, in *Proceedings of the XV International Grasslands Congress, Kyoto, Japan,* Science Council of Japan and Japanese Society of Grassland Science, Nishi-nasuno, Japan, pp. 168–70.

Halevy, J. and Hartzook, A. (1988) Dry matter accumulation and nutrient uptake of high-yielding peanuts (*Arachis hypogaea* L.) grown in a sandy soil. *Peanut Science,* **15**, 5–8.

Hallock, D.L. and Martens, D.C. (1974) Content of eight nutrients in central stem leaf segments of ten peanut cultivars and lines. *Peanut Science,* **1 (2)**, 53–56.

Hartley, W. (1949) *Plant collecting expedition to subtropical South America 1947–8,* Divisional Report No. 7, CSIRO Division of Plant Industry.

Herbert, T.T. and Stalker, H.T. (1981) Resistance to peanut stunt virus in cultivated and wild *Arachis* species. *Peanut Science,* **8**, 45–47.

Higgins, B.B. (1951a) Economic Importance of Peanuts, in *The Peanut – the Unpredictable Legume,* The National Fertiliser Association, Washington, DC, pp. 3–17.

Higgins, B.B. (1951b) Origin and early history of the peanut, in *The Peanut – the Unpredictable Legume,* The National Fertilizer Association, Washington, DC, pp. 18–27.

Ikhatua, U.J. and Adu, I.F. (1984) A comparative evaluation of the utilisation of groundnut haulms and *Digitaria smutsii* hay by Red Sokoto goats. *Journal of Animal Production Research,* **4(2)**, 145–152.

Jamarun, N. (1984) The voluntary intake and digestibility of sugarcane (*Saccharum officinarum* L.) top hay and peanut (*Arachis hypogaea* L.) hay in combination with ipil-ipil (*Leucaena leucocephala* (*Lam*) *de Wit*) fed to Philippine native goat. Thesis (MS in Animal Science), University of the Philippines at Los Banos College, Laguna, Philippines.

Jones, R.M. and Rees, M.C. (1972) Persistence and productivity of pasture species at three localities in subcoastal south east Queensland. *Tropical Grasslands,* **6(2)**, 119–34.

Kaligis, D.A. and Sumolang, C. (1991) Forage species for coconut plantations in North Sulawesi, in *Forages for Plantation Crops* (eds H.M. Shelton and W.W. Stür), Proceedings of a Workshop, Sanur Beach, Bali, Indonesia 27-29.6.1990, ACIAR Proceedings No. 32 (in press).

Krapovickas, A. (1973) Evolution of the genus *Arachis* in *Agricultural Genetics – Selected Topics* (ed. R. Moav), John Wiley and Sons, New York, pp. 135–51.

Krapovickas, A. (1990) A taxonomic summary of the genus *Arachis,* in *Report of a Workshop on the Genetic Resources of Wild Arachis Species,* held at CIAT, Cali, Colombia, 29 February to 2 March 1989, International Crop Network Series 2, IBPGR, Rome, p. 9.

Kretschmer, A.E., Jr. and Wilson, T.C. (1988) A new seed-producing *Arachis* sp. with potential as forage in Florida. *Soil and Crop Science Society of Florida Proceedings,* **44**, 229–33.

Krishna Mohan, D.V.G., Murthy, P.R.S., Naidu, C.M. *et al.* (1985) Performance of weaner lambs fed rations containing different proportions of groundnut straw. *Indian Journal of Animal Sciences,* **55(6)**, 464–467.

Kvien, C.S. and Bergmark, C.L. (1987) Growth and development of the Florunner peanut cultivar as influenced by population, planting date and water availability. *Peanut Science,* **14**, 11–16.

Lascano, C.E. and Thomas, D. (1988) Forage quality and animal selection of *Arachis pintoi* in association with tropical grasses in eastern plains of Colombia. *Grass and Forage Science,* **43**, 433–39.

Leche, T.F., Groenendyk, G.M., Westwood, N.H. and Jones, M.W. (1982) *Composition of Animal Feedstuffs in Australia,* Australian Feeds Information Centre, Commonwealth Scientific and Industrial Research Organisation, Australia, p. 124.

Lu, J., Mayer, A. and Pickersgill, B. (1990) Stigma morphology and pollination in *Arachis* L. (Leguminosae). *Annals of Botany,* **66**, 73–82.

Martinez, B.F. de and Blue, W.G. (1978) Effects of calcium carbonate on chemical characteristics of three Florida soils and response of some agronomic plants. *Soil and Crop Science Society of Florida Proceedings,* **37**, 188–92.

Canudas, E.G., Quesenberry, K.H., Sollenberger, L.E. and Prine, G.M. (1989) Establishment of two cultivars of rhizoma peanut as affected by weed control and planting rate. *Tropical Grasslands*, **23(3)**, 162–169.

Castillo, M.B., Morrison, L.S., Russell, C.C. and Banks, D.J. (1973) Resistance to *Meloidogyne hapla* in peanut. *Journal of Nematology*, **5**, 281–85.

Chandrasekhar, N.R. (1980) Forachis – a perennial forage groundnut, in *National seminar on the application of genetics to improvement of groundnut, 16–17 July, 1980*, Tamil Nadu Agricultural University, Coimbatore, India.

Chesson, J. (1983) The estimation and analysis of preference and its relationship to foraging models. *Ecology*, **64**, 1297–1304.

Collins, E.R. and Morris, H.D. (1941) *Soil fertility studies with peanuts*, North Carolina Agricultural Experiment Stations Bulletin 330.

Colwell, J.D. (1963) The estimation of phosphorus fertilizer requirements of wheat in southern New South Wales by soil analysis. *Australian Journal of Experimental Agriculture and Animal Husbandry*, **3**, 190–97.

Combellas, J., Centeno, A., Mazzini, B. and Combellas, J. (1972) Use of the aerial parts of groundnut. 2. Making into hay, intake and digestibility in vivo. *Agronomia Tropical*, **22(3)**, 281–285.

Cook, B.G. and Franklin, T.G. (1988) Crop management and seed harvesting of *Arachis pintoi*. Krap et Greg. nom. nud. *Journal of Applied Seed Production*, **6**, 26–30.

Corby, H.D.L. (1981) The systematic value of leguminous root nodules, in *Advances in Legume Systematics* (eds R.M. Polhill and P.H. Raven), Royal Botanic Gardens, Kew, Richmond, Surrey, pp. 657–69.

Dextre, R. (1984) *Requerimientos nutricionales de Arachis pintoi CIAT 17434)*. Centro Internacional de Agricultura Tropical (CIAT), Cali, Colombia (Mimeograph).

Diulgheroff, S., Pizarro, E.A., Ferguson, J.E. and Argel, P.J. (1990) Multiplicación de semillas de especies forrajeras tropicales en Costa Rica. *Pasturas tropicales*, **12(2)**, 15–23.

Drammeh, S. (1990). Groundnut improvement, production, management, and utilisation in the Gambia, in *Summary Proceedings of the First ICRISAT Regional Groundnut Meeting for West Africa*, International Crops Research Institute for the Semi-Arid Tropics, Patancheru, Andhra Pradesh, India, pp. 41–44.

Dunavin, L.S. (1990) Cool season forage crops seeded over dormant rhizoma peanut. *Journal of Production Agriculture*, **3(1)**, 112–14.

Dunavin, L.S. (1992) Florigraze rhizoma peanut in association with warm-season perennial grasses. *Agronomy Journal*, **84(2)**, 148–151.

Durga Prasad, R.D., Prasad, D.A. and Ramachandra Reddy, R. (1986) Evaluation of complete rations containing groundnut haulms for lambs. *Indian Journal of Animal Sciences*, **56(2)**, 258–261.

Dwyer, G.T., O'Hare, P.J. and Cook, B.G. (1989) Pinto's peanut: a ground cover for orchards. *Queensland Agricultural Journal*, **115(3)**, 153–54.

Easby, W.B. (1851) Letter to Commissioner of Patents. *Report Commissioner of Patents 1851* (21), 353–55, cited by Higgins (1951a).

Essomba, N.B., Mekontchou, T. and Iroume, R.N. (1990) Groundnut production and research in Cameroon, in *Summary Proceedings of the First ICRISAT Regional Groundnut Meeting for West Africa*, International Crops Research Institute for the Semi-Arid Tropics, Patancheru, Andhra Pradesh, India, pp 37–39.

Gelaye, S., Amoah, E.A. and Guthrie, P. (1990) Performance of yearling goats fed alfalfa and florigraze rhizoma peanut hay. *Small Ruminant Research*, **3**, 353–61.

Gregory, M.P. and Gregory, W.C. (1979) Exotic germ plasm of *Arachis* L. interspecific hybrids. *The Journal of Heredity*, **70**, 185–93.

Gregory, W.C., Gregory, M.P., Krapovickas, A. *et al.* (1973) Structure and genetic resources of peanuts, in *Peanuts – Culture and Uses*, American Peanut Research and Education Association, Inc., Stillwater, Oklahoma.

Gregory, W.C., Krapovickas, A. and Gregory, M.P. (1980) Structure, variation, evolution, and classification in *Arachis*, in *Advances in Legume Science* (eds R.J. Summerfield and

Prine, G.M., Dunavin, L.S., Moore, J.E. and Roush, R.D.(1981) 'Florigraze' rhizoma peanut, a perennial forage legume, (Agricultural Experiment Station Circular S-275, University of Florida, Gainesville.

Queensland Department of Primary Industries (1985) A Handbook of Plant Diseases in Colour, Vol. 2, Field Crops, 2nd edn Information Series Q185016, Queensland Department of Primary Industries, Brisbane, Australia.

Ramanatha Rao, V. (1987) Origin, distribution, and taxonomy of *Arachis* and sources of resistance to groundnut rust (*Puccinia arachidis* Speg.), in *Groundnut rust disease: Proceedings of a Discussion Group Meeting, 24–28 September 1984*, ICRISAT, Patancheru, A.P., India, pp. 3–15.

Reed, R.L. and Ocumpaugh, W.R. (1991) Screening rhizome peanut for adaptation to calcareous soils. *Journal of Plant Nutrition* **14(2)**, 163–74.

Resslar, P.M. (1980) A review of the nomenclature of the genus *Arachis* L.*Euphytica*, **29**, 813–17.

Rika, I.K., Mendra, I.K., Gusti Oka, M. and Oka Nurjaya, M.G. (1991) New forage species for coconut plantations in Bali, in *Forages for Plantation Crops* (eds H.M. Shelton and W.W. Stür), Proceedings of a Workshop, Sanur Beach, Bali, Indonesia 27–29.6.1990, ACIAR Proceedings No. 32 (in press).

Robles, A.Y. and Capitan, F.H. (1985) Growth performance of peanut hayfed goats in combination with ipil-ipil, in *Management, nutrition and reproduction studies for improved goat production in the Philippines* (ed. A.Y. Robles), Philippine Council for Agriculture and Resources Research and Development, Los Banos, Laguna, Philippines, pp. 231–235.

Romero, F. (1986) Nutritional evaluation of Florida 77 alfalfa and 'Florigraze' rhizoma peanut as forages for dairy cattle. *Dissertation Abstracts International*, **47(4)**, 1343-B.

Romero, F., Van Horn, H.H., Prine, G.M. and French, E.C. (1987) Effect of cutting interval upon yield, composition and digestibility of Florida 77 alfalfa and Florigraze rhizoma peanut. *Journal of Animal Science*, **65**, 786–96.

Saldivar, A.J., Ocumpaugh, W.R., Gildersleeve, R.R. and Moore, J.E. (1990) Growth analysis of 'Florigraze' rhizoma peanut: forage nutritive value. *Agronomy Journal*, **82**, 473–77.

Saldivar, A.J., Ocumpaugh, W.R., Gildersleeve, R.R. and Prine, G.M. (1992a) Total nonstructural carbohydrates and nitrogen of 'Florigraze' rhizoma peanut. *Agronomy Journal*, **84(3)**, 439–444.

Saldivar, A.J., Ocumpaugh, W.R., Gildersleeve, R.R. and Prine, G.M. (1992b) Growth analysis of 'Florigraze' rhizoma peanut: shoot and rhizome dry matter production. *Agronomy Journal*, **84(3)**, 444–449.

Shukla, P.C., Talpada, P. M., Desai, M. C. *et al.* (1985) Composition and nutritive value of groundnut haulm as an industrial by-product. *Indian Journal of Animal Nutrition*, **2(2)**, 89–90.

Simpson, C.E. (1990) Collecting wild *Arachis* in South America past and future, in *Report of a Workshop on the Genetic Resources of Wild Arachis Species*, held at CIAT, Cali, Colombia, 29 February to 2 March 1989, International Crop Network Series 2, IBPGR, Rome, pp. 10–17.

Sollenberger, L.E., Prine, G.M. and Jones, C.S. (1987) Animal performance on perennial peanut pastures. *Agronomy Abstracts*, 145–46.

Stafleu, F.A. (Chm., Ed. Comm.) (1978) International code of botanical nomenclature. Adopted by the 12 International Botanical Congress, Leningrad, July 1975. Bohn, Scheltema, and Holkema, Utrecht, The Netherlands.

Stalker, H.T. and Moss, J.P. (1987) Speciation, cytogenetics, and utilization of *Arachis* species. *Advances in Agronomy*, **41**, 1–40.

Stalker, H.T. and Campbell, W.V. (1983) Resistance of wild species of peanut to an insect complex. *Peanut Science*, **10**, 30–33.

Stür W.W. (1991) Screening forage species for shade tolerance – a preliminary report, in *Forages for Plantation Crops* (eds H.M. Shelton and W.W. Stür), Proceedings of a Workshop, Sanur Beach, Bali, Indonesia 27–29.6.1990, ACIAR Proceedings No. 32 (in press).

Sylvester-Bradley, R., Mosquera, D. and Méndez, J.E. (1988) Selection of rhizobia for inoculation of forage legumes in savanna and rainforest soils of tropical America, in *Nitrogen fixation by legumes in mediterranean agriculture*, (eds D. P. Beck and L. A. Materon), Martinus Nijhoff, Dordrech, The Netherlands, pp. 225–34.

Tounkara, N.B. (1990) Groundnut production in the Republic of Guinea, in *Summary Proceedings of the First ICRISAT Regional Groundnut Meeting for West Africa*, International Crops Research Institute for the Semi-Arid Tropics, Patancheru, Andhra Pradesh, India, pp. 49–51.

Turnour, J.P. (1987) A glasshouse assessment of the effect of a number of different elements and shading on chlorosis in *Arachis pintoi* grown on a Mt Cotton red-yellow podzolic using *Macroptilium atropurpureum* cv. Siratro as a test plant. Department of Agriculture, University of Queensland, St. Lucia, Queensland (Mimeograph).

Valentim, J.F., Ruelke, O.C. and Prine, G.M. (1986) Yield and quality responses of tropical grasses, a legume and grass–legume associations as affected by fertilizer nitrogen. *Soil and Crop Science Society of Florida Proceedings*, **45**, 138–43.

Valentim, J.F., Ruelke, O.C. and Prine, G.M. (1988) Evaluation of forage yield, quality, and botanical composition of a dwarf elephantgrass–rhizoma peanut association as affected by nitrogen fertilization. *Soil and Crop Science Society of Florida Proceedings*, **47**, 237–42.

Valls, J.F.M., Ramanatha Rao, V., Simpson, C.E. and Krapovickas, A. (1985) Current status of collection and conservation of South American groundnut germplasm with emphasis on wild species of *Arachis* in *Proceedings of an International Workshop on Cytogenetics of Arachis, 31 October–2 November 1983*, ICRISAT, Patancheru, A.P., India, pp. 15–35.

Valls, J.F.M. (1983) Collection of *Arachis* germplasm in Brazil. *Plant Genetic Resources-Newsletter*, **53**, 9–14.

Velasquez, G. and Gonzalez, E. (1972) The nutritive value of groundnut (*Arachis hypogaea*) straw. *Agronomia Tropical*, **22(3)**, 287–290.

Vinijisanum, T., Irwin, J.A.G., and Cameron, D.F. (1987) Host range of three strains of *Colletotrichum gloeosporioides* from tropical pasture legumes, and comparative histological studies of interactions between type B disease-producing strains and *Stylosanthes scabra* (non-host) and *S. guianensis* (host). *Australian Journal of Botany*, **35(6)**, 665–77.

Vos, G. and Jones, R.M. (1986) The role of stolons and rhizomes in legume persistence. CSIRO Division of Tropical Crops and Pastures Annual Report, Brisbane, Australia, pp. 70–71.

Williams, M.J., Hammond, A.C., Kunkle, W.E. and Spreen, T.H. (1991) Stocker performance on continuously grazed mixed grass – rhizoma peanut and bahiagrass pastures. *Journal of Production Agriculture*, **4(1)**, 19–24.

Wright, G.C., Hubick, K.T. and Farquhar, G.C. (1991) Physiological analysis of peanut cultivar response to timing and duration of drought stress. *Australian Journal of Agricultural Research*, **42**, 453–470.

York, E.T., Jr and Colwell, W.E. (1951) Soil Properties, Fertilisation and Maintenance of Soil Fertility, in *The Peanut – the Unpredictable Legume*, The National Fertilizer Association, Washington, DC, pp. 122–172.

CHAPTER 16

The groundnut in farming systems and the rural economy – a global view

J. Smartt

It is perhaps axiomatic to state that the groundnut crop is produced in farming systems ranging from the most primitive to the most advanced. The role it performs in these systems varies considerably, from labour-intensive shifting cultivation to highly capital-intensive systems in North America and Australia.

The groundnut has an enormous advantage vis-à-vis many other crops in economies at or above subsistence level in that it has a dual-purpose role: it is both a subsistence and a cash crop, in contrast to beverage and industrial crops. Although it is a highly input-intensive crop, the nature of the inputs can be varied according to local circumstances; it is energy demanding and how these demands are met is open to alternatives. In areas where labour is abundant, all operations can be carried out manually; where it is in short supply, most can be mechanized. In intermediate situations it is possible, by appropriate injection of capital and energy resources, to optimize production efficiency for any particular set of circumstances. The development of intermediate technologies is an important issue in developing (as against stagnant) agricultural economies.

16.1 FARMING SYSTEMS

To consider the varied roles that groundnuts can perform, it is appropriate to appreciate the actual diversity of farming systems in which it can be a component. Farming systems can be classified according to a number of different criteria. Ruthenberg (1980) classifies them on the following bases:

1. Type of land management
2. Intensity of cropping

The Groundnut Crop: A scientific basis for improvement. Edited by J. Smartt. Published in 1994 by Chapman & Hall, London. ISBN 0 412 408201.

3. The source of water
4. The type of crops produced
5. Levels of production technology
6. Intensity of commercial production.

16.1.1 Land management systems

Five systems are generally recognized:

- Cropping plus fallowing
- Ley farming – alternate crop and livestock husbandry
- Field farming – permanent arable cultivation
- Perennial cropping – plantation style
- Agroforestry systems.

This first criterion in Ruthenberg's list is the most important basis for classification of farming systems but the remaining elements also need to be considered.

16.1.2 Intensity of cropping

This element is very closely related to section 16.1.1. Intensity of cultivation can conveniently be expressed in terms of an index, R, calculated in terms of three variables:

$$R = \frac{yc}{d} \times 100$$

where y = number of years under cultivation, c = number of crops per annum, and d = duration of land-use cycle (crops + fallow).

Ruthenberg makes extensive use of this index and uses it to characterize farming systems:

- Shifting cultivation: R <30 – fallows are long.
- Semi-permanent cultivation: R = 30–70 – fallows relatively short.
- Permanent cultivation: R >70 – fallows very short or occasional.

16.1.3 Water supply

The water needs for cropping can be supplied by:

- natural rainfall; or
- irrigation; or
- a combination of both.

16.1.4 Type of crops produced

Crops are produced broadly for four purposes:

- Food
- Beverages
- Industrial use
- Medicinal.

Individual crops may fall into more than one of these categories.

16.1.5 Level of production technology

Technological levels of farming systems can be considered in terms of increasing levels of technological input and sophistication.

- Planting or digging-stick systems – the basic level and most primitive.
- Hoe or spade farming – representing a simple technological advance with human labour being the major source of energy input.
- Draught implements with animal power – the development of more sophisticated equipment and technology with the supplementation of human with animal power in energy input.
- Draught implements with mechanical power – producing a further reduction in human energy input and replacement of animal energy by that derived from fossil fuels.

16.1.6 Degree of commercialization

This is a pragmatic breakdown of what is clearly a continuum, compared with 16.1.2 (intensity of cropping):

- Subsistence farming: sales <25% production.
- Partially commercialized farming: sales 25–50% of production.
- Semi-commercial farming: sales 50–75% of production.
- Commercial farming: sales >75% of production.

16.2 CULTIVATION SYSTEMS

The criteria in sections 16.1.2–6 superimposed on 16.1.1 give an adequate basis for describing and interrelating different farming systems and the role that the groundnut actually performs in them. For purposes of discussion, systems of cultivation can be considered under seven different headings, the first five of which will be discussed in detail:

1. Shifting cultivation
2. Semi-permanent cultivation
3. Ley farming
4. Permanent cultivation on natural rainfall
5. Permanent cultivation dependent in part or whole on irrigation
6. Plantations
7. Agroforestry systems.

The complexity of these systems is considerable and there are environmental and other constraints which determine whether or not they are practised in a given area and the precise form they take.

16.2.1 Shifting cultivation

This system is profoundly misunderstood in the western world, where land is generally regarded as the most precious environmental resource and anything but its most intensive use is considered profligate. This point of view, though entirely understandable in those brought up in the developed world, is certainly not tenable in the developing world and it would be misguided in the extreme to maintain such a standpoint in these areas.

At first sight, shifting cultivation is aesthetically and intellectually unappealing, but on further examination it is revealed as astonishingly effective. It has been dismissed in a somewhat pejorative fashion as a 'technology of expediency'. There is undoubtedly an element of truth in such a label but what practical alternative is there in vast areas of Africa, Asia and South America? It is also a viable agricultural system over a wide climatic range, from tropical rain forest to semi-arid environments. In these areas the system works well provided that the alternating cropping and fallowing phases are in balance. The essential requirement is that the fallow period is of sufficient duration to restore soils and vegetation to the state in which a further cycle of cropping can be supported. Depending on soil type and present and past climatic conditions, the fallow period might be quite short or extremely protracted. The practice of shifting cultivation in the rainforest regions or of 'chitemene' in the woodland areas of Central Africa may require something approaching the re-establishment of climax vegetation before the cultivation cycle can be repeated (Trapnell, 1953; Trapnell and Clothier, 1957; Allan, 1967). A great depth of ecological perception is often apparent in the operation of these systems – in knowing when the succession has reached the stage at which cropping can take place.

The system works very well when populations are relatively low and within the capacity of the environment to support. When population pressures mount, there is an irresistible tendency to attempt cropping before the soil has recovered fully from the previous cycle. Recovery of the soil has physical and chemical components. In the fallowing phase, the growth of grasses – especially their fibrous root systems – rebuilds the physical structure of the soil, improving aeration and water acceptance, so that field capacity is enhanced while allowing drainage of excess. Nutrients released by weathering of soil minerals, and through activities of mycorrhiza and nitrifying bacteria, are recycled *in situ* and accumulate until an equilibrium is established, at which point the cropping sequence can be taken.

Fallowing may also be resorted to when problems of weed infestation occur together with pest and disease build-up. This can happen on inherently fertile soils of good physical structure and high intrinsic nutrient levels. There may come a time when the energy expenditure in maintaining clean cultivation on an established plot is greater than that involved in clearing a new one. It is primarily a question of labour input in relation to the crop produced. In the case of the most primitive farming systems (using a digging or planting stick, for example), the really significant inputs are labour and seed or other propagating material which represents the capital element. In the technologically more advanced systems of hoe and spade culture, the capital element is somewhat higher. The substitution of animal or mechanical power for human labour greatly increases the level of capital commitment and associated risk.

Primitive agricultural systems (in the technological sense) using shifting cultivation are in fact very much more efficient in agronomic terms and in the use of environmental resources than appears at first sight to those conditioned to different and more intensive agricultural practices. They are also efficient in economic terms – that is, in the return on the labour investment. Furthermore, the lifestyle which can be followed offers considerable leisure time: the working day may be no longer than five hours on average and peak demands for labour may be relatively few, apart from sowing and harvesting. An important point is that more intensive systems may demand a greater labour input for no greater return. The most important point to appreciate in attempting to increase the intensity of land exploitation is that increased levels of inputs are required, initially in the form of plant nutrients, in order to increase levels of production. These inputs must result in a clear surplus over subsistence needs which must be adequate to cover their cost when marketed. Encouragement of increased input levels without a reasonably assured saleable surplus can lead to extremely serious problems of indebtedness, since farmers are essentially living off capital which they do not actually possess. If they have title to the land they cultivate, then they may well lose it in order to pay their debts.

16.2.2 Semi-permanent cultivation

This system, in which intensity of land use ranges from R values of 30 to 70, is found in the seasonally dry to semi-arid tropics rather than in the humid areas. Holdings tend to be more clearly defined; land occupancy by the farmer and family may be of an indefinite term. The actual cropping patterns and practices may be similar to those of shifting cultivation practised in the same geographic area.

Semi-permanent cultivation is efficient and effective when a relatively short fallow period is sufficient to restore fertility and reduce weed problems. However, where such a system of land-use is enforced by population pressures, the fertility and general condition of the land may become impoverished, with continuing depression of yields and increased erosion hazards. In such situations, research and extension operations should focus on income generation once subsistence needs have been met so that necessary inputs to maintain fertility and hence productivity can be afforded. In these conditions groundnut could be highly desirable as a valuable subsistence-cum-cash crop. The labour demands tend to be higher under this system; the weed problems are usually more severe and little in the way of land preparation may be possible during the dry season (as would be the case with shifting cultivation). Peaks of labour demand tend to be higher.

16.2.3 Ley-farming systems

After a period of popularity in the 1950s and of eclipse in the 1960s, ley-farming or alternate husbandry systems are poised to stage something of a comeback. Part of the reason is the increasing popularity of the ideas, at any rate, of organic farming. Ley-farming systems are not of necessity wholly organic – there is no reason why artificial fertilizers should not be applied or agrichemicals used. However, there is invariably a substantial cycling of organic matter within the system. Nutrients removed from the farm by sale of crops and livestock must be replaced in order that the system is sustainable at a reasonable level of productivity.

Ley-farming or alternate husbandry can be considered as being derived from the semi-permanent system by the exploitation of the fallow for grazing. In the primitive or unregulated form of ley-farming, the grazing of livestock (where kept) is almost inevitable if grasses predominate in the fallow vegetation when cultivation is abandoned. This system can be developed or regulated to involve the deliberate sowing of grasses or a grass–legume mixture in the ley. In the unregulated system, there need be little capital investment over and above that required for semi-permanent cultivation. The regulated system, however, can become fairly capital intensive. In order to control grazing effectively, some kind of fencing is required. At the time when the ley's useful life is at an end and it has to be ploughed, investment in an effectively designed plough for the purpose might well be necessary. Draught power could be provided by beasts produced in the system.

The cropping system in the arable phase could take into account not only the subsistence needs of the farmer (and the local community) but also suitable crops for marketing in the immediate hinterland and those which can be produced for export markets. On the livestock side, apart from the draught animals already mentioned, the produce of a ley-farming system would include meat or dairy products or both. In the tropics, fresh milk production would imply a need for a site close to an urban population which would provide a market. Meat production would not be subject to this particular constraint if live animals could be marketed; otherwise there would be a need for an abattoir, cold storage and refrigerated transport in the vicinity.

Groundnuts could play a useful role in such a system. After harvest, the haulm could be used as a very acceptable livestock feed and, if the seed produced was used for oil extraction, the cake could be returned to feed cattle and its nutrient content restored in part to the soil.

In theory this system seems to be well nigh perfect – a model of sustainability, in fact. In the cropping phase soil nutrients are depleted, physical condition of the soil deteriorates and weed populations may well build up. Under the ley phase, arable weeds will be suppressed, soil structure improved by grass root growth and the nutrient status enhanced by the activity of the root-nodule bacteria of pasture legumes as well as the dung of livestock, particularly if the animals are fed on imported concentrates. There is intrinsically a very good cycling of nutrients and organic matter.

Given the right economic conditions, this system can be very attractive – particularly in temperate climates, for which excellent grass–clover mixtures are available. The system has been viable in New Zealand and contributes largely to the country's exports of meat and dairy products. There are problems, however, in non-temperate zones. At lower elevations in the tropics, it is difficult to establish and maintain stable grass–legume mixtures in pastures. A further problem is that the nutritive value of tropical grass species is appreciably lower than that of the temperate rye grasses. Tropical grasses tend to be lower in dry matter when they are young and succulent and also tend to become fibrous more quickly with age.

Although in ley-farming the land is being used productively all the time, when pressures mount for maximization of production or income a greater return can be obtained from continuous arable cultivation. For a ley-farming system to be successful, population pressures in actual or potential ley-farming areas have to be such as to encourage a higher level of productivity but not so high that only continuous cultivation will meet subsistence and other needs. The system also demands a high level of farming skills and versatility: the farmer must be equally good with animals as with crops and must achieve self-sufficiency of forage production. In addition there must be no serious endemic disease problems and good veterinary services should be available when needed.

These constraints are very severe in many if not most parts of the tropics and preclude extensive development of the ley-farming system. Its intrinsic intellectual and aesthetic appeal is considerable and it is well worth bearing in mind under appropriate economic conditions.

16.2.4 Permanent cultivation on natural rainfall

This system can arise when arable cultivation is extended at the expense of fallow or grass ley but it does include systems in which short-term fallows or leys occur from time to time. A distinguishing feature of permanent agriculture is the dedication of land on a permanent basis for arable or grazing use. The arable land tends to be that which is most productive and easiest to manage, while the grazing lands are most often those with difficult topography, poor drainage or rocky outcrops which make them unsuitable for arable crop production. In the seasonally dry tropics, annual (sometimes biennial) crops are taken under arable systems; in the wet tropics, plantation tree crops are customary.

The most satisfactory systems of permanent cropping develop in areas of high intrinsic soil fertility, where this does not decline rapidly or drastically under cultivation. Such areas are those with young soils of volcanic origin or those developed in Rift Valley areas following the causative earth movements. They are characterized by seasonal rainfall or well-distributed rainfall in which leaching is not severe. This system is most readily maintained where fertilizer use is established on an intensive basis – a practice which maintains permanent cultivation in advanced agricultural economies, where intensive fertilizer use is paid for as a charge on the crop. Without this high level of input, it is questionable whether intrinsic fertility would sustain cropping at economic levels for any great length of time. In western Europe over the past two centuries, fertility has been imported not only in the form of N P K fertilizers but also in livestock feeds, often imported from the developing world – such as groundnut cake produced from oil extraction.

Permanent agriculture in the tropics is quite commonly poles apart from that found in the developed agricultural economies. In the Indian subcontinent, for example, population pressures may maintain land in permanent cultivation which would be better managed under a system with fallows. Inputs of fertilizer and manure are commonly inadequate and often negligible. Cow dung in India is too valuable as a fuel to be used as manure, while in Africa there is frequently no established tradition of using cattle manure. Farmers, particularly in tsetse-infested areas, have no experience of cattle management; on the other hand, some herdsmen have never produced crops. Agricultural development has often brought tsetse under control, enabling farmers to keep cattle, and settlement schemes have introduced pastoralists to agriculture. Both groups have to acquire the necessary skill in the handling of manure and this requires investment in carts if manure has to be transported for any substantial distance.

A general characteristic of permanent cultivation in the tropics is that demand for labour per unit area is high and that output may be relatively low in impoverished or degenerate systems. Ruthenberg (1976) described a system of agriculture developed by the Wakara people on an island in Lake Victoria. As population grew, they were unable to expand the area under cultivation and had perforce to intensify their agriculture. They developed a system in which fertility resources of all kinds, notably animal and green manures, are used to the full to maintain their land at an adequate level of productivity. The labour demand for this system is very high and those who have emigrated from the island and who continue to farm readily abandon their remarkable system of cultivation: they are very happy to adopt semi-permanent or shifting agricultural systems, which are much less laborious. The farming of the Wakara is particularly interesting because it shows the development of intensive land use comparable to that of many peoples of the Far East.

The role of the groundnut in such systems of permanent cultivation is basically similar to the role that it performs in fallowing or ley-farming systems. There is, however, the constraint that the high labour demand of this crop may sharply limit the extent to which it can be grown. Since it is a highly regarded food crop, it is likely that it would be grown anyway in suitable climatic areas on appropriate soil, even if only to a small extent; on the other hand, if a good local market exists it could be useful as a cash-earner, again even if only to a limited extent. In such areas it would obviously not be easy to expand production without some additional technological input.

16.2.5 Permanent agriculture under irrigation

In many ways this system represents an ideal. By implication it is practised in arid areas where leaching is minimal and fertility is high and the crop's needs for water can be met exactly, assuming that water supplies are constantly available. But for irrigation, some highly productive agricultural areas such as Egypt and California would be desert. In these areas temperatures are not limiting, although the diurnal range may be quite wide. The important considerations are availability of water when needed and its quality. Without appropriate care and attention, solute accumulation can occur to the detriment of future production. The maintenance of irrigation agriculture in the Nile Valley shows that such systems can be sustainable indefinitely. Problems can arise when attempts are made to expand cultivation with a limited water supply, which can bring disaster in their wake, or where there are drainage problems due to local topography.

Groundnuts have been produced successfully in the irrigation agriculture of Egypt for many years. The crop lends itself well to such a system. It has clearly defined critical periods when adequate water supplies are necessary for peak yield, the most important (after germination and establishment of the crop) being from flowering until pod-filling is complete. There are other situations where irrigation can supplement natural rainfall to ensure an adequate length of growing season. In semi-arid tropical areas, irrigation may be used to prolong the growing season by establishing crops prior to the onset of the rainy season, a practice which also gives the crop an extended period of freedom from the normal pests and diseases of the growing season. On both counts, yields can be increased without the cost of agrichemicals and their undesirable environmental effects. Irrigation can also be used at the end of the season to facilitate crop maturation when soil water has become limiting.

In certain areas the groundnut crop can be linked with production of rice, particularly in the humid tropics. In many parts of the wet tropics, rainfall can be expected at any time of the year but distribution is not uniform and it may be possible to combine production of paddy rice at the wetter times of the year with groundnut production in the 'dry' season. During this period less difficulty will be experienced with the drying and curing of the crop than in periods of the heaviest rainfall.

Groundnuts can be grown on quite heavy soils, provided that they are ridged and other suitable practices are adopted. They have been grown on the Gezira scheme in the Sudan on heavy cotton soils. In fact more serious problems have arisen in Africa from attempts to produce the crop on light soils (the East African Groundnut Scheme) than on heavy soils.

16.2.6 Mixed cropping versus sole cropping

In traditional systems, groundnut has often been produced in some kind of a mixture with other crops. Neither the components of the mixture nor their proportions are fixed. Such systems have considerable advantages, the chief of which is flexibility. In a subsistence economy, the balance between the components can be determined on the basis of need. The staple crop will naturally occupy the greater proportion of the land and the amounts grown of subordinate crops can be varied according to need. The mixture tends to be sown in a succession. In the mound culture systems of Central Africa (Trapnell 1953), the mounds are often set up by burying materials from land clearance or previous crop residues in the late dry season or early rains, followed by sowing maize and sorghum centrally on the mound and then a succession of other crops such as groundnuts, cowpeas, cucurbits, sweet potatoes and *Phaseolus* beans. At the peak of the season the ground cover can be excellent and erosion minimized.

This simple succession can lead to sowing of crops in a series of more or less concentric circles, which is satisfactory in a hand tillage system. It can be adapted to animal-powered tillage by producing a linear version. The staple crop can be sown in widely spaced lines and, following each cultivation, additional components can be added to the system in amounts determined by need. This 'relay cropping' (as it is commonly called) can be related to traditional farming practices adapted and modified to suit more advanced technology.

There are additional advantages of traditional and modern systems of mixed cropping in that they tend to reduce the incidence and severity of pest and disease attack. In a pure stand the pest needs to expend little energy in finding a host plant; similarly with plant pathogens the dispersed inoculum has nothing to alight on other than host plants. In mixed crops the effectiveness of pests and disease organisms in finding suitable host plants is substantially reduced. Development of infestations and infections is slower and epidemics are much less likely to occur. Also, the rooting zones of the component species in the mixture are unlikely to coincide, so that more effective exploitation of the whole soil profile is possible. Appropriate plant spacing could minimize competition between the root systems and allow the effective interception of light by the canopies. Light shading may even benefit groundnuts at pegging. These factors could have a cumulative effect and enhance overall yield relative to what the separate components might have achieved in pure stands from an equal area of land.

In traditional systems where mixed cropping is standard practice, some crops – such as finger millet (*Eleusine coracana*) – are invariably produced in pure stands. This is usually related to ease of harvesting of the small grains. When cash crops are introduced to such agricultural economies, they are frequently grown as pure stands even when they are part of a traditional system (as groundnuts are). Several considerations lead to the adoption of this practice in such circumstances. In the case of tobacco and cotton, ease of harvesting is an important consideration; in addition, in cotton, economic use of pesticides would be difficult in mixed crops even though the pests themselves might be less of a problem in such circumstances.

The change from mixed to sole cropping has been hastened by use of animal and mechanized power in crop production. It has without doubt increased the biotic pressure of pests and pathogens on the crops, so much so that interest has revived in mixed cropping systems such as relay planting and agroforestry. They provide alternatives which seem to be more environmentally friendly as well as meeting the subsistence needs of the farmer more effectively. There is no reason why cash crops could not be produced by relay planting and from agroforestry systems where, for example, no restrictive considerations with respect to harvesting were involved.

The theoretical basis for agroforestry systems is similar to that of relay cropping. The principal difference is in the time scale: relay cropping typically involves cropping one season at a time while in agroforestry a long-term perennial element is combined with production of annual crops. Agroforestry could also be regarded as an indefinite extension of the system used in establishing a plantation of tree crops when annual or other short-term crops are taken before the plantation comes into production.

16.3 ADAPTIVE RESEARCH

The subject of mixed cropping, intercropping, multiple cropping or poly-culture has received some attention in the past two decades in the developed world. The American Society of Agronomy published its symposium on multiple cropping in 1976. Among numerous other publications in the area, one can mention Vandermeer (1989), who develops a detailed theoretical consideration of the subject. This is also considered in the broader ecological context by authors such as Tivey (1990) and Loomis and Connor (1992). The subject is one of some complexity, as Vandermeer's (1989) work shows. In practice, it is necessary to develop an understanding of the needs of individual crop species for optimal production before fully effective manipulation of multiple cropping can be achieved.

In the context of improving production of the crop in the developing world, or in its introduction to new areas, it makes good sense to use existing knowledge of the crop's production to the full, and to draw upon experience with the crop in technologically advanced areas. However, it is foolhardy to proceed without preliminary exploratory trials – in other words, to undertake what is now called adaptive research. There is a real necessity to carry out locally-based programmes to determine how best to apply existing knowledge to other environments and how to fine-tune and adapt them to different locations.

The East African Groundnut Scheme provides a perfect illustration of what can go wrong when no prior adaptive research is carried out. From experience of the crop in North America and South Africa, it was believed that certain short cuts could be taken; it was assumed that lighter soils would suit the crop and that the short season cultivar Natal Common was a sensible choice. As it transpired, the light soils were a very different proposition from those of North America – instead of being loose and friable when dry, they set almost like concrete. It was most difficult to harvest the crop and the wear and tear on harvesting machinery was totally unacceptable. On the face of it, Natal Common was a reasonable choice – as a spanish variety it is quite early and capable of maturing in a relatively short growing season – but no pilot trials were conducted to determine whether in fact any groundnut genotype, no matter how early maturing, could succeed in the conditions at Kongwa.

16.3.1 Adaptive research in Zambia

In contrast to the East Africa scheme, an adaptive research programme did evolve in Zambia in the period 1954–61. It began as a breeding and selection programme. After the third season, when promising germplasm had been identified, it was apparent that available local knowledge was not such that the extension service could advise as to how to make the best use of any new material. The decision was made to undertake what would now be called a programme of adaptive research, the results of which are given by Smartt (1967). The outcome was a sound crop production package for the grower.

Three or four decades later, that approach is still relevant to any area where new productive varieties are being sought and where it is necessary to determine how to make the best use of any improved varieties that have been produced or identified. The following is an extended summary of the Zambia programme.

The groundnut crop in Zambia had been produced with varying success in different parts of the country. In the Eastern Province there was an export of high quality confectionery kernels, mainly from the variety Chalimbana (a Jumbo type). Production standards were high and yield levels were quite satisfactory. It was not clear why yields were so much lower outside the Eastern Province or why the crop was not a very attractive proposition as a generator of income elsewhere.

(a) Variety selection

A series of variety trials, run for four seasons from 1954 to 1958, included four local landrace varieties (Ngoni, Nsenga, Chewa and Chalimbana), the standard South African cultivar Natal Common and an introduction from Tanganyika (Asiriya Mwitunde). In the first season, 1954–55, the trial was sown at only one location – Mukulaikwa, the site of an abandoned Overseas Food Corporation pilot groundnut scheme. Yields in this first season were abysmal (Table 16.1); in the second season they were worse and no further groundnut trials were carried.out on this site after 1956. In the season 1955–56, experiments were carried out at two sites with the same varieties, one in the Eastern Province the other in the Southern Province. While the yields produced in the Eastern Province (Kalichero) were marginally better than those of the Southern Province (Monze), they were encouragingly close and gave an indication that acceptable, economic yield levels were possible elsewhere than in the Eastern Province. In 1956–57 the trials were again repeated in the Southern and Eastern Provinces and in this season the performance at Monze was actually considerably better than that at Petauke (Eastern Province). In the final season of these trials, the tables were turned and yields at Petauke were highest while those at Monze were among the lowest. In this last season, the trial was carried out at six locations with indications that at Kabompo, for example, there might be useful production potential.

An introduction nursery had been established at the Central Agricultural Research Station in 1954 and it became apparent in its early

TABLE 16.1 *Mean yields (kernels lb/acre) of groundnut varieties, 1954–8*

Season	Station	Province	Ngoni	Nsenga	Chewa	Chalim-bana	Natal Common	Asiriya Mwitunde	SE	LSD P = 0.05
1954–5	Mukulaikwa	Central	110	140	225	200	50	240	±22	69
1955–6	Mukulaikwa	Central	51	164	82	193	38	216	±33	99
	Kalichero	Eastern	795	870	906	872	710	900	±39	120
	Monze	Southern	813	783	776	872	588	680	±38	116
1956–7	Monze	Southern	1375	1188	1243	1254	957	1111	±32	97
	Petauke	Eastern	363	594	583	594	286	- 539	±24	75
1957–8	Petauke	Eastern	1232	1243	1199	1122	1078	704	±34	104
	Katapola	Eastern	693	836	792	748	429	616	±68	205
	Monze	Southern	429	484	418	429	264	429	±41	125
	Misamfu	Northern	660	693	649	726	528	660	±69	207
	Balovale	N. Western	407	396	429	275	33	374	±41	124
	Kabompo	N. Western	957	957	671	880	143	792	±72	219

TABLE 16.2 *Mean yields (kernels lb/acre) of groundnut varieties, Mount Makulu, 1956–9*

Variety	Habit	Season			Variety mean yields 1956–9
		1956–7	1957–8	1958–9	
Mani Pintar	Bunch	2192	1011	1472	1558
S183	Runner	1841	771	995	1202
GB1	Bunch	1738	846	1095	1168
Dixie Runner	Runner	1621	938	404	988
BS1	Runner	1603	696	741	1013
S731	Runner	1589	950	736	1092
Chitedze Mwitunde	Bunch	1512	631	630	924
TMV3	Bunch	1509	743	609	954
Asiriya Mwitunde	Bunch	1441	602	691	911
Seasonal mean yields		1672	799	819	
Standard error of variety means		±102	±49	±58	
LSD P = 0.05		299	158	190	

years that some introductions had a considerably greater yield potential than either the local landraces or introductions received from South Africa or Tanganyika. In the season 1956–57, sufficient bulk of material had been produced to undertake a new series of variety trials which were carried out from 1956 to 1959. These had only one variety (Asiriya Mwitunde) in common with the earlier series of trials. They showed that the introduced Bolivian variety Mani Pintar (undoubtedly a corruption of *maní pintado*) was the highest yielder in three very different and successive seasons. The other conclusion which could be drawn from the yield data (Table 16.2) was that, although many varieties performed well in the excellent 1956–57 season, there was little consistency of performance (apart from Mani Pintar) over all three seasons. It was possible to identify at a relatively early stage germplasm which showed promise and consistency in performance.

A consideration of the results of the earlier trials (Table 16.1) revealed a lack of consistently good performers but Natal Common was consistently poor. Of the six varieties tested, Chalimbana was generally preferred by growers and consumers on the grounds of its excellent quality and because in yielding ability it was marginally higher and more consistent than the others. In the planting of these trials (at the provincial sites) it was requested that they be sown at about the same time as the local crop. From nursery experience at Mount Makulu and anecdotal evidence from the Eastern Province, it became apparent that the crop benefited greatly from early sowing. When promising lines had been identified the need was felt to carry out agronomic investigations to determine how best to realize the potential for high consistent yield shown by Mani Pintar.

The variety Mani Pintar has an interesting and chequered history. It was collected in 1947–8 (Hartley, 1949) at a local market in La Paz, Bolivia. It was sent to the United States and to Australia, where it was grown experimentally in Queensland. A small sample was received in Zambia in 1955 and grown in nursery plots. Single plant selections were made and the remaining bulk was used for sowing in variety trials from 1956 onwards. Small samples were sent elsewhere in Africa and promising results were reported from Ghana initially (McEwen, 1961), Zimbabwe (Smartt, 1978; Hildebrand and Smartt, 1980), East Africa and as far north as the Sudan. Extensive trial sowings were made throughout Zambia and promising results and commercial bulking commenced in 1960.

The original material of Mani Pintar showed very striking variegation in pigmentation of the testa. This was thought to be a possible disadvantage in terms of marketing. Single plant selections were made of plants which produced seeds with self-coloured testa, both red and brown. The red-seeded selection bred reasonably true and was named Makulu Red; the brown-seeded material segregated and was clearly hybrid in nature. The behaviour of both Makulu Red and Mani Pintar when produced in bulk was interesting in that it was usually possible to select some self-coloured seed in a variegated batch and some variegated seed from a self-coloured batch. Since the two forms were otherwise identical, it was suggested that a position effect might be responsible (Smartt, 1960). Hammons (1973) showed that variegation behaves as a single locus effect but an alternative explanation is that the difference is due to a jumping gene or transposon. This possibility might be worthy of further study and it could, if established, have interesting implications.

Further selection in Zimbabwe (Hildebrand and Smartt, 1980) has shown that stable pink and brown testa forms can be selected from Mani Pintar and Makulu Red stock. The potential value of Bolivian germplasm in Central Africa (and elsewhere in the continent) has been suggested. This value derives from the fact that it evolved in a montane environment, probably at about 1500 m a.s.l., where day temperatures may be high with cool conditions at night. These conditions are in marked contrast to those which occur in many of the major growing areas closer to sea level, where temperature range is much less, and it seems likely that rather different physiologies have become established. Conditions in plateau Africa probably have a great deal in common with those in which Bolivian groundnuts have evolved. What is more certain is that they show remarkable pre-adaptation to conditions over much of Africa.

From a commercial standpoint, Mani Pintar (and Makulu Red) have certain good features and some shortcomings. The size and shape is relatively uniform but somewhat shorter in relation to diameter than in typical virginia varieties. The oil content is about 50% of dry matter; the kernel flavour is somewhat bland and thus it is better used where a strong flavour is not required, rather than for the preparation of salted or roasted peanuts. However, their potential value in breeding programmes is high. They have a further advantage in that their maturation period is intermediate between that of a typical virginia and a typical spanish/valencia type. This has proved to be particularly valuable in Central Africa, where virginia types require a somewhat longer maturation period than is available in many seasons, while the fastigiate valencia/spanish cultivars mature before the rains finish and losses may occur from sprouting. In Zambia the maturation period was found to be about 10 days less than the typical virginia and 20 days longer than spanish/valencia. A crossing programme between Mani Pintar and Chalimbana, although it was never launched, could have had exceptionally interesting results if it had proved possible to combine the confectionery quality of Chalimbana and the yielding capacity of the Bolivian germplasm.

(b) Agronomic practices

The identification of high-yielding and well-adapted genotypes is essential but it is only the first part of the task of the crop improver. A good genotype grown in poor conditions will not perform well, neither will an unproductive genotype grown in the best conditions that can be provided. The breeder initiating a selection programme in a new area will naturally look to local landrace material as well as introductions to provide the pool of variation from which to select. Local landraces are usually considered to be adapted to local conditions and this is frequently so in ways that are not appreciated. One would expect them to be attuned to local climatic conditions and those of the local environment generally. This might well include low levels of fertility: they may be able to produce a modest crop annually with negligible inputs of nutrient but they are commonly unable to respond effectively to substantially improved conditions. As local landraces they have never previously experienced such conditions and have been subjected to no previous selection for such responses, so that it should be no surprise if they do not respond markedly to improved cultural conditions.

In the course of nursery operations and carrying out the initial series of variety trials, the following were identified as the most important areas for agronomic investigation:

1. Time of sowing
2. Time of lifting
3. Plant population and spacing
4. Cultivation and weed control
5. Fertilizer use.

In addition, the effects of disease and (to a much lesser extent) pests were investigated.

1. Time of sowing

This was identified in the very first season as an important factor influencing yield. The initial nursery collection was sown in duplicate, the first replicate as soon as was practicable after the first rains in mid–late November and the second replicate approximately 4 weeks later in mid December. At harvest it was quite obvious that the early sowing had outperformed the later by a considerable margin and it was thought that an attempt to quantify the effects of delayed sowing would be a useful exercise. It soon became apparent that general practice among local traditional farmers was to sow their most important crops as early as possible. In the Eastern Province, where groundnuts were an important source of income, this category included groundnuts which could well be sown before 1 December. In other areas,where they were solely a subsistence crop, planting tended to be later after the sowing of staple crops – maize, sorghum or finger millet – and this could be as late as mid December. Sometimes, if onset of the rains was delayed, the earliest sowing might well be possible only as late as mid December.

Results from trials conducted in two seasons are presented in Tables 16.3 and 16.4. These show consistent trends, with delayed sowing producing substantial yield reductions. In the case of the Mount Makulu trials, where a kernel quality test was also carried out, reduction in quality follows a very similar trend to yield. A general recommendation can therefore be given, on the basis of these (and other similar) results, that advantage be taken of the earliest possible opportunity for sowing. Sometimes a problem arises in connection with the earliest sowings when moisture distribution is not uniform in the field and uneven germination results, but the consequences of delayed sowing are likely to be even more detrimental.

Another important effect of early sowing is to reduce the impact of some major pests and diseases on yield. This will be considered in more detail later in section 16.3.1(c).

2. Time of lifting

The question of harvesting a geocarpic crop with a fundamentally indeterminate flowering sequence is a vexed one. It is a matter of determining the optimum time when the highest proportion of pods are mature. Premature harvesting will obviously increase the proportion of immature fruits harvested as well as giving a yield reduction, while delayed harvesting produces loss of early-set pods by breakage or disintegration of pegs, severing the connection between plant and pods. Although all groundnuts are botanically indeterminate in flowering and fruiting, in practice the fastigiate varieties and the spreading bunch forms tend to have a defined period of effective flowering and fruiting. Only those pegs fairly close to the

TABLE 16.3 *Effect of sowing times on kernel yield and quality of three groundnut varieties at Mount Makulu, 1959–60*

Variety	GBI		Makulu Red		Local spreading 2	
Sowing date	Mean yields kernels lb/acre	% retained by 19/64″ × 3/4″ screen	Mean yields kernels lb/acre	% retained by 19/64″ × 3/4″ screen	Mean yields kernels lb/acre	% retained by 19/64″ × 3/4″ screen
SE	±52		±46		±52	
21 Nov.	690	17.00	1115	63.27	840	78.34
10 Dec.	761	26.44	828	75.90	707	80.02
23 Dec.	394	20.44	411	85.19	340	70.62
8 Jan.	109	17.74	154	67.07	88	56.89
21 Jan.	21	8.33	28	30.48	25	14.28
LSD P = 0.05	158		140		158	

TABLE 16.4 Effect of sowing time on kernel yield at several stations in Zambia (mean yields, kernels lb/acre)

Season:	1959–60		1960–1			
Station:	Petauke	Balovale	Monze 1	Monze 2	Monze 3	Monze 4
Variety:	Mani Pintar	Local Runner		Makulu Red		
Time of 1st sowing:	mid-December	mid-November	mid-November	mid-November	mid-November	mid-November
SE	±49	±39	±82	±73	±117	±73
1st sowing	2513	729	1298	1528	2279	1611
2nd sowing	1531	489	1133	1288	1864	1324
3rd sowing	104	358	524	705	1467	827
4th sowing	82	228	559	541	837	661
5th sowing	17	120	245	416	562	358
LSD P = 0.05	146	110	251	228	360	230

taproot borne on the lower nodes of the stem actually mature fruit. Pegs produced higher up the stem fail to make soil contact and develop into mature fruit. Maturity of individual pods is indicated by a darkening of the tissues on the inner pod wall. Since pods at varying stages of maturity are to be found on a single plant, it is not always easy to gauge maturity of a whole field crop from a few randomly selected plants.

It is important to know not only the optimal time for lifting but also the latitude that exists and the effects of premature or delayed harvesting on yield and quality. The question of quality these days embraces not only that of the kernel but also levels of infection by *Aspergillus flavus* and *A. parasiticus* and the resulting aflatoxin contamination. The most effective way of elucidating the effects of time of lifting on yield and quality is to carry out a quality test in addition to yield measurements soon after harvest. This involves screening the crop and recording the proportion of large kernels produced, a simple and effective quality assessment.

The results of a series of trials are presented in Tables 16.5, 16.6 and 16.7, which clearly show that the optimal lifting time for Mani Pintar/ Makulu Red is 150 days. There is a window between 150 and 160 days, after which serious yield loss can occur. For earlier maturing cultivars (Spanish GBI) the optimum appears to be 130 days; for the later maturing, spreading varieties it is about 160 days. While these indications are reliable on the whole, they can be affected by seasonal conditions. In particular, dry conditions post-sowing could cause delayed emergence and if this is longer than a few days due allowance has to be made in deciding when to lift.

3. Plant populations and spacing

The initial study of the effect of plant populations included a number of ridging treatments. These were recommended as good soil conservation practices and the opportunity was taken to assess possible effects on yield in a single season. The results presented in Table 16.8 show no significant effects of either cultural practices or plant populations. The latter were changed by varying density within rows; the rows themselves were on average 36 inches (91 cm) apart.

In order to obtain a satisfactory understanding of the question, it was clear that spacing between and within rows would have to be varied, subject only to the constraint of what could be accommodated in a standard size plot (25 ft × 10 ft). Experimental results are presented in Tables 16.9 and 16.10). The optimal plant population under the rain-fed conditions, in which these experiments were conducted was the intended density of 43 560 plants per acre as far as both yield and quality were concerned. It was interesting to note that yield fell less sharply than quality at higher than optimal plant populations. The higher proportion of shrivelled kernels suggested that, at the higher population densities, more pods were set but that water availability was inadequate to take the kernels through to

TABLE 16.5 *Effect of lifting on kernel yield and quality of three groundnut varieties at Mount Makulu, 1959–60*

Variety:			(a) GB1		(b) Makulu Red		(c) Local Spreading 2	
Time to lifting (days) for each variety (a) (b) (c)			Mean yields kernels lb/acre (SE ±27)	% retained by 19/64″ × 3/4″ screen	Mean yields kernels lb/acre (SE ±44)	% retained by 19/64″ × 3/4″ screen	Mean yields kernels lb/acre (SE ±79)	% retained by 19/64″ × 3/4″ screen
100	120	130	287	3.66	506	12.80	732	49.28
110	130	140	424	6.61	674	43.90	957	71.11
120	140	150	436	12.40	1183	69.18	817	79.44
130	150	160	495	19.43	1211	80.34	1120	84.08
140	160	170	518	15.88	1229	87.60	550	82.80
LSD P = 0.05			83		135		242	

TABLE 16.6 *Effect of time of lifting on kernel yield and quality of two groundnut varieties at Mount Makulu, 1960–61*

Variety:		(a) GB1		(b) Makulu Red	
Time to lifting (days) for each variety		Mean yields kernels lb/acre	% retained by 19/64″ × 3/4″	Mean yields kernels lb/acre	% retained by 19/64″ × 3/4″
(a)	(b)	(SE ±86)	screen	(SE ±74)	screen
100	130	318	4.56	1088	59.28
120	150	619	24.88	1322	83.88
130	160	700	18.25	1177	80.13
140	170	503	24.16	837	80.65
150	180	348	33.05	721	82.06
LSD P = 0.05		264		226	

TABLE 16.7 *Effect of time of lifting on kernel yield of groundnuts at two stations in Zambia, 1959–61 (mean yields, kernels lb/acre)*

Variety: Station:	Mani Pintar Petauke, 1959–60	Makulu Red Magoye, 1960–61
Days to lifting	(SE ±99)	(SE ± 127)
120	2046	1181
140	2151	1590
150	2152	2010
160	1888	1971
180	1877	1277
LSD P = 0.05	301	390

maturity. The inference could also be drawn that additional water supplied by irrigation could have taken more of the kernels to full maturity and improved both yield and quality from the higher populations.

4. Cultivations and weed control

The effects of the number of cultivations on yield were studied in season 1958–61. Initially up to five cultivations were compared with an unweeded control at three sites (Table 16.11). The unweeded control plots gave very low yields and at least two cultivations were necessary at some sites (and

TABLE 16.8 *Effects of ridging methods and plant populations on the yield of groundnuts, 1958–9 (kernels lb/acre)*

Station: Province: Variety:	Mount Makulu Central Mani Pintar	Ngoni Farm Eastern Local Runner	Petauke Eastern Local Runner	
Ridging treatments	(SE ±74)[a]	(±54)[c]	(±40)[e]	Mean
Flat	1061	483	552	699
Flat and ridged	1089	436	534	686
Narrow ridge	1067	536	499	701
Broad ridge	1094	576	525	732
Populations thousands/acre				
	(SE ±64)[b]	(±46)[d]	(±35)[f]	
20	955	461	491	636
40	1103	505	515	708
80	1175	560	576	770
Mean	1078	509	527	

Least significant differences P = 0.05

a	b	c	d	e	f
230	199	169	149	107	126

TABLE 16.9 *Effect of plant spacing on yield and quality of two varieties of groundnuts at Mount Makulu, 1959–60*

Variety:		GB1		Makulu Red	
Intended population plants/acre	Plant spacing	Mean yields kernels lb/acre (SE ±88)	% retained by 19/64″ × 3/4″ screen	Mean yields kernels lb/acre (SE ±117)	% retained by 19/64″ × 3/4″ screen
14 520	36″ × 12″	624	27.49	623	93.98
29 040	36″ × 6″	781	28.09	963	86.09
58 080	36″ × 3″	1030	20.68	1256	84.19
29 040	18″ × 12″	1116	26.86	1139	86.66
58 080	18″ × 6″	1329	17.64	1065	76.63
116 160	18″ × 3″	1333	8.95	1270	68.99
43 560	12″ × 12″	1443	21.56	1364	86.56
87 120	12″ × 6″	1290	6.81	1132	69.63
174 240	12″ × 3″	1268	6.64	1045	52.95
LSD P = 0.05		262		351	

TABLE 16.10 *Effect of plant spacing on kernel yield of groundnuts at Monze, 1960–1 (mean yields, (kernels lb/acre), variety Makulu Red*

Station:		Monze 2	Monze 3	Monze 4
Intended population plants/acre	Plant spacing	(SE ±42)	(SE ±163)	(SE ±102)
21 780	24″ × 12″	1341	2119	798
43 560	24″ × 6″	1400	2226	1197
43 560	12″ × 12″	1524	2562	1391
87 120	12″ × 6″	1493	2579	1352
LSD P = 0.05		125	492	310

TABLE 16.11 *Effect of number of cultivations on the yield of groundnuts, 1958–9 (kernels lb/acre)*

Station:	Mount Makulu	Ngoni Farm	Petauke
Province:	Central	Eastern	Eastern
Variety:	Makulu Red	Local Runner	Local Runner
Number of cultivations	±104	±61	±69
0	137	228	63
1	278	468	489
2	1053	762	440
3	1162	936	599
4	1489	980	651
5	1467	893	600
LSD P = 0.05	313	189	214

even more in some seasons) to approach maximum yield levels. There were additional benefits for up to four cultivations, even though the amount of weed growth removed was relatively small. Late in the season, competition between weeds and crop for water may produce some loss of yield and quality when water supply becomes limiting. When the trial was repeated in the following season (1959–60) at Mount Makulu (Table 16.12) there was a progressive increase in both yield and quality for every cultivation given. In provincial trials (Table 16.13) the results of Petauke and Balovale show an interesting contrast. Yield at Balovale is unresponsive to cultivation, reflecting very low weed populations, while the pattern of response at Petauke closely parallels that observed at Mount Makulu.

TABLE 16.12 *Effect of cultivations on groundnut yield and quality at Mount Makulu, 1959–60, variety Makulu Red*

Number of cultivations	Mean yields kernels lb/acre (SE ±65)	% retained by 19/64″ × 3/4″ screen
0	481	56.72
1	620	58.91
2	821	65.03
3	914	68.59
4	1139	71.43
LSD P = 0.05	195	

TABLE 16.13 *Effect on cultivations on yield at two stations in Zambia, 1959–60 (mean yields kernels lb/acre)*

Variety:	Mani Pintar	Local Runner
Station:	Petauke	Balovale
Number of cultivations	(SE ±77)	(SE ±59)
0	828	827
1	1683	849
2	2090	914
3	2132	870
4	2410	837
LSD P = 0.05	233	180

In the final season of experimentation at Mount Makulu a rather different approach was adopted, although an uncultivated control was also used. The crop was allowed to compete with weeds for varying periods and then clean weeded thereafter (Table 16.14). The results show that three weeks of weed competition between 6 and 27 January had little effect on yield and none on quality. Longer delays produced very substantial yield losses but very little reduction in quality. This result is particularly interesting because peak rainfall in Zambia usually occurs in January and there is a temptation to cultivate even though the soil may be in an unsuitable

TABLE 16.14 *Effect of withholding cultivations on yield quality of groundnuts at Mount Makulu, 1960–1, variety Makulu Red*

Date of first cultivation	Mean yields kernels lb/acre (SE ±61)	% retained by 19/64″ × 3/4″ screen
6 January	1011	79.35
27 January	933	80.47
10 February	440	78.15
10 March	172	71.19
Uncultivated	92	59.10
LSD P = 0.05	190	

condition. This result indicates that weeding might possibly be deferred without significant loss of yield until soils are less waterlogged and cultivation is likely to be more effective.

5. Fertilizer use

In what was current American practice during the 1950s and 1960s, the only direct fertilizer applied to the groundnut was gypsum to ensure supplies of calcium adequate for full normal pod development. While there is no doubt that groundnuts are responsive to soil fertility as generally perceived, responses to fertilizers in the season of application are inconsistent and erratic. This conclusion is supported by trials carried out in Zambia in 1958–59 (Table 16.15). It can be argued that the crop is in fact not responding to the fertilizers applied, since these are in all probability not in the same soil horizon as the feeding roots of the crop. This supports the wisdom of the old American guideline to fertilize the rotation as a whole rather than the groundnut break. Nutrients which have passed beyond the reach of the feeding roots of maize, for example, may be readily accessible to the groundnut root system. It would have been interesting to follow the effects of deep fertilizer placement.

(c) Pest and disease control

For local farmers to use pesticides and fungicides was not a practical proposition. However, the possibilities of cultural control were thought to be worth exploring. The two major disease problems studied were rosette virus and the cercospora leaf spots which, although produced by two different pathogens (*Cercospora arachidicola* and *Phaeoisariopsis personata*), can conveniently be considered as a single disease.

The rosette virus is transmitted by the aphid (*Aphis craccivora*) (Chapter

TABLE 16.15 *Responses of groundnuts to fertilizers, 1958–9 (kernels lb/acre)*

Station:	Mount Makulu	Magoye	Masaiti	Petauke	Misamfu
Province:	Central	Southern	Western	Eastern	Northern
Variety:	Mani Pintar	Mwitunde	Local Runner	Local Runner	Mwitunde
Effects	±37	±117	±139	±20	±26
N*	142	−104	135	−87	200
P†	−133	27	85	−52	176
K++	−78	−12	−50	−17	54
Mean yield	1760	1351	602	862	804
LSD P = 0.05	114	360	424	61	80

* Response to 100 lb ammonium sulphate per acre
† Response to 200 lb single superphosphate per acre
++ Response to 100 lb muriate of potash per acre

10) and there are two avenues of attacking the problem: controlling the insect vector by host-plant resistance and/or by resistance of the host plant to the virus itself. At the time of this experimental work, host resistance to the aphid had been found in East Africa in the variety Asiriya Mwitunde. Field tests at Mukulaikwa in Zambia showed that, under field conditions in which 100% infection was observed in Natal Common, only 2% occurred in Asiriya Mwitunde. Clearly antixenosis was operating, the basis of which has not been explored. Evans (1954) pointed out the difficulty of carrying out experimental studies on rosette using conventional small plot techniques. This difficulty was also experienced in Zambia but, very occasionally, analysable observations could be obtained. Such a dataset was collected at Petauke in 1954–60 on the incidence of rosette in relation to time of sowing. This clearly showed that establishment percentages decrease with delay in sowing, while percentage infection increased enormously (Table 16.16). These two effects are confounded in this analysis but it is known that establishment is less successful in delayed sowings, quite independently of rosette infection. Observation also suggests that poor gappy stands are more attractive to aphids, as are chlorotic plants resulting either from virus infection or through waterlogging. The suggestion has been made that, in closed stands, biological control of the insect vector through entomophagous fungi controls virus spread (which is less effective in poor stands). The attraction of aphids to chlorotic plants draws the vector to the virus source in many instances, from which it can spread rapidly. The rosette virus produces disease as a rule in crops grown in poor conditions: the commonest factors producing outbreaks are late sowing and drought. With good husbandry and in seasons of good, well-distributed rainfall, it is not serious as a rule.

TABLE 16.16 *Effect of time of sowing incidence of rosette, Petauke, 1959–60, variety Mani Pintar*

Sowing dates	Total plants infected	% plants infected	No. plants harvested	% plants harvested
17 December	28	2.3	1218	81.2
31 December	76	6.3	1197	79.8
14 January	327	35.5	920	61.3
28 January	358	46.7	767	51.1
8 February	325	85.1	382	25.5
	SE ±35	LSD P = 0.05	105	

This is not the case, however, with leaf spot diseases. These are universal and occur in good seasons and bad. They can produce very high levels of yield loss in conditions particularly favourable for disease development. The spanish/valencia group of varieties (subsp. *fastigiata*) are much more susceptible than are the virginia cultivars (subsp. *hypogaea*). Leafspot control trials were carried out at Mount Makulu between 1957 and 1960 and results are presented in Tables 16.17 and 16.18. In the susceptible Spanish 809 variety, yield reductions of >50% were produced in the first season (1957–58) and *c.* 30% in the second (1958–59), in which the overall yields of all treatments were higher. In the final season a comparative pair of trials using Spanish 809 and the more resistant/tolerant Makulu Red showed yield levels which were comparable in the fungicide treatments of both trials; however, the control treatment in Makulu Red yielded approximately 50% higher and still retained leaf at harvest, whereas Spanish 809 was completely defoliated. There was also an interesting effect on quality. In Makulu Red, kernel size was not significantly affected by the disease; in Spanish 809, on the other hand, the proportion of larger kernels doubled in the spray treatments. These experiments were conducted on sites adjacent to areas in which the crop had been cultivated in the previous season and a very high inoculum potential was ensured. Although no investigations were carried out on possible means of delaying the onset of leaf spot attack by the use of windbreaks and maximizing distances between sites of successive crops, these were thought to be areas worth exploring as possible methods of cultural control.

There were other pests and diseases of sporadic occurrence which were noted but not investigated in detail. Insects which caused some damage included thrips, leafhoppers, caterpillars, white grubs and termites. No control was attempted and problems never got seriously out of hand. In retrospect it is perhaps as well that control by pesticides was not attempted – natural control of these pests seemed quite effective with reasonably stable pest populations

TABLE 16.17 *Leaf spot control trials with fungicides, 1957–9 (mean yields, kernels lb/acre), variety Spanish 809*

Variety:	Spanish 809		
Season:	1957–8	1958–9	mean 1957–9
Fungicide treatment	(SE ±94)	(±103)	
Control	657	1486	1072
Sulphur dust	961	2105	1533
Bordeaux mixture	1566	2235	1901
Zineb	1391	2248	1820
Maneb	1399	2227	1813
LSD P = 0.05	286	311	

TABLE 16.18 *Effect of controlling leaf spot by fungicides on yield and quality of two groundnut varieties at Mount Makulu, 1959–60*

Variety:	Spanish 809		Makulu Red	
Fungicide treatment	Mean yield kernels lb/acre (SE ±177)	% retained by 19/64″ × 3/4″ screen	Mean yield kernels lb/acre (SE ±182)	% retained by 19/64″ × 3/4″ screen
Control	1202	30.66	1822	91.55
Dispersible sulphur	2273	65.62	2329	93.04
Copper oxychloride	2193	61.73	2062	93.90
Zineb	2261	65.28	1923	92.53
Maneb	2070	56.98	2086	92.04
LSD P = 0.05	546		559	

Diseases other than cercospora leaf spots and rosette were never serious. *Phyllosticta*, *Pleospora* and *Pseudoplea* were sometimes isolated from leaf lesions but attacks by these fungi never reached epidemic proportions. Bacterial wilt (*Pseudomonas solanacearum*) was sporadic in occurrence, often most severe in the first years after ploughing up resting land and declining thereafter. *Sclerotium rolfsii*, although destructive where infection occurred, was only sporadic and effective disposal of crop residues by turning in prevented this potentially serious pathogen becoming a problem. Seedling rots produced by fungi such as *Aspergillus niger* were effectively controlled by seed dressings. Concealed damage of kernels could be incited by a variety of species, including *Corticium solani* in stored pods, but the local practice of shelling groundnuts as soon as practicable after harvest reduced this problem.

TABLE 16.19   *Groundut yields and cultural practices used by the winning competitor in the Rhodesian National Groundnut Competition, 1972–74*

	1972/73	1973/74
Area of groundnuts grown:	6 ha	40 ha
Yield of 4 ha competition plot (kgs/ha unshelled):	9400 kg/ha	9600 kg/ha
Other crops grown:	Maize 94 ha	Maize 50 ha Winter wheat 25 ha
Variety grown:	Makulu Red	Makulu Red
Seeding rate:	150–160 kg/ha	140 kg/ha
Row spacing:	4 rows × 35 cm on 1.8 m bed	6 rows × 23 cm on 1.8 m bed
Fertilizer application:	400 kg/ha compound S 6%:N 18%:P_2O_5 6%:K_2O	500 kg/ha compound S
	220 kg/ha gypsum broadcast 8 weeks after sowing	150 kgs/ha gypsum at 6 and 12 weeks after sowing
Herbicide	Dyanap	Dyanap, alachlor
Foliage insecticide:	None	Nuvacron (20/11/73)
Leaf spot control:	None	Agricura special fungicide plus dithane applied 15/11/73, 15/12/73 and 17/1/74
Sowing date:	7 November 1972	1 October 1973
Digging date:	15 April 1973	1 March 1974
Growing season length:	160 days	150 days

Winner 1972/73 and 1973/74: B. Huntsman Williams   Altitude: 1233 m
Farm: Portwe Estates   Soil type: sandy clay loam

(d) Adaptive research conclusions

As a result of this experimental programme carried out between 1954 and 1961 (Smartt, 1967)) a package was produced for the grower which included a high-yielding variety and a set of recommended cultural practices to produce considerably improved levels of yield without involving additional inputs in the way of fertilizers, herbicides, insecticides or fungicides. It was interesting to note some 11 or 12 years later in Zimbabwe that this package, produced with the indigenous farmer in mind, was elaborated very successfully into a highly productive system which involved high levels of inputs. Details of the system are presented in Table 16.19.

The system itself has a number of interesting features. The crop was sown early and irrigated until the rainy season set in. In the first season only fertilizers and herbicides were used; in the second, three fungicide sprays were given at monthly intervals and a single foliage protectant spray. The result was pod yields in two successive seasons of 9.4 and 9.6 t/pods/ha (6–7 t/ha shelled). The factors which produced this remarkable yield were the availability of irrigation water and substantial avoidance and/or control of leaf spot disease. Very high plant population densities were established, with adequate water supplies to produce a fully mature crop. The leaf canopy which produced this crop had been protected from insect and fungus attack when necessary. It is interesting to note that this was necessary in the second season but not the first. Data from further crops produced under similar conditions in the same area would have been of interest. With extension of the cultivated area in the two seasons, it might be expected that leaf spot control would be necessary to maintain yield level; if control of foliage-feeding insects proved to be necessary again, it would suggest that perhaps natural population control mechanisms of the pest had been adversely affected. In this development of the initial 1961 package, the two major constraints identified had been removed: water availability limiting plant populations and the yield reduction due to leaf spot.

This indicates the need for clearly identifying the major yield constraints so that they may be addressed progressively as and when this is practicable. In the original 1961 package, the constraints were imposed by limiting rainfall and the impracticability of leaf spot control by fungicides and of weed control other than by mechanical means. In the 1972–74 development, all these constraints had been removed by supplementary irrigation and use of fungicides and herbicides. Experimental programmes should be designed to answer basic questions and, in applying the findings, due regard must be paid to practical considerations.

In designing such an experimental programme, it is desirable to make the experiments as simple and straightforward as possible and to design them to answer clear and well-defined questions. This is especially so with experiments that are being conducted by research stations in farmers' fields. These can be remarkably successful if the willing co-operation and interest of the farmer is secured – and not only of the farmer but also that of the extension service. Some of the most successful experiments carried out in Zambia in the period 1954–61 were those designed at the Mount Makulu Central Agricultural Research Station planted by extension staff of the Southern Province on the family holding of a local Tonga farmer. All concerned derived the greatest satisfaction from their successful outcome.

16.3.2 Adaptive research and agricultural development strategies

It has come to be appreciated that there is a very considerable gap between the generation of research findings and their application to practical agriculture. This has led to the formulation of an approach to research aimed at bridging the gap under the name of adaptive research – a new name but not a new activity, as has been shown. It was practised in the later phases of British colonial rule with some success (Smartt, 1967) but it has been in a state of eclipse because of the success of the Green Revolution. Many of us who had been working in agriculture in the developing world were heartened by the great leap forward of this revolutionary change. Although its shortcomings were recognized at a surprisingly early date (Brown, 1970), to some it showed that the agriculture of the Third World could be developed in a positive and constructive way and that wise and judicious investment could be justified in strict economic terms. Prior to this, agricultural development had been regarded more or less as an economic black hole.

Possibly the most unfortunate consequence of the Green Revolution approach was that it led the 'top down' strategy of development. The assumption on which it was based was that improved practices adopted by the more successful and wealthier farmers who could afford the new technology would tend to trickle down to the lower levels of agricultural society. This arose from the attempt to economize on the Green Revolution package. Recommended agricultural practices, miracle seeds and appropriate technology were transferred without the economic underpinning. Circulating loan funds were not made available,with the result that those poorer farmers who attempted to invest in the new style farming fell prey to money-lenders. From being small-scale peasant farmers, they often became landless labourers.

In agricultural terms, the Green Revolution concept did not take any account of traditional agricultural practices. It was a system of agriculture designed on purely logical, scientific lines – in fact rather narrowly based, with the emphasis on an area's staple crop (actual or potential). The narrowness of the basic concept had a number of unfortunate consequences in that improved high-yielding new cultivars made landraces of not only the staple crop species but also those of other crops uncompetitive. This tended to produce large areas of monocrops – conditions favourable for the generation of pest and disease epidemics. A reaction to this, which has set the clock back approximately 40 years, has been a return to systems closer to traditional farming which tend to maintain stability of production rather than the boom-and-bust cycles of monocrop agriculture that involve large areas of genetically uniform populations of a single species. This reaction has led to a reversal of strategy in favour of a grassroots-upwards approach, involving close collaboration between research and extension workers on the one hand and farmers on the other. It necessitates building up knowledge of local farming systems and agricultural practices so that the constraints in the system can be identified and defined and research directed to their relief. In this way it is to be hoped that much (if not all) of the inherent stability of the traditional system can be preserved while productivity is improved.

It is obvious that the personnel involved in institutionalized basic research do not include those who would engage most effectively in adaptive research. Quite properly, the scientific staff at an international agricultural research institute includes highly trained and specialized scientists, who would not be expected to have agronomic training to any great extent. The scientist engaged in adaptive research would need to have a good, broad agronomic training and be able to collaborate effectively with specialists in plant breeding, plant pathology, economic entomology and soil science. In order to do so, such a scientist must have some training in all these fields, even if only at a rather basic level. It is interesting to consider how, in the past, such training was given. In the days of the British Colonial Agricultural Service, there were two types of graduate who were recruited for training: those with a training in agriculture on the one hand (who were largely destined for extension or agronomic investigation) and, on the other, graduates in the natural sciences who were trained as plant breeders, pathologists, entomologists and soil scientists. The latter also received a broad general training in agriculture and were able to appreciate their own research field in its proper context. In addition, such individuals were quite well placed to carry out experimentation in fields related to their own specialization in the absence (all too frequent!) of the appropriate specialist. In the dependent territories 40 years ago, it was not uncommon for pathologists to be instigating and executing variety trials or for plant breeders to be carrying out agronomic investigations. With a broad agricultural background and an adequate training in field experimentation, this often worked quite well.

Another interesting approach has been developed at Wageningen, in the Netherlands, where specialist training in plant taxonomy has been combined with agronomy. This has produced a number of excellent agricultural botanists who have made very significant contributions in the general field of the botany and taxonomy of crop plants and especially in the specialist field of crop plant genetic resources. At first sight this combination of disciplines seems unusual, if not bizarre, but it shows a deep appreciation of the kind of scientific personnel needed to maintain progress in research and development for the Third World. There is a very real need for versatility and adaptability in research personnel engaged in crop improvement and agricultural development. In addition to the training and qualities already mentioned, they need to be open-minded, receptive to new ideas, very resourceful and able to carry out sound experimentation with the simplest equipment. Fortunately simple experiments are often all that are necessary. Complex experiments have a nasty habit of generating too many missing plots for effective analysis, while simple randomized block and Latin Square trials can prove to be remarkably resilient.

In colonial times, when agricultural research was undertaken by the departments of agriculture in the various dependent territories, the quality of research management was variable. In some instances it was so poor as to inhibit any useful work being done; in others it was surprisingly good due largely to commonsense combined with a basic appreciation of what research was about and some inkling as to how it might be undertaken. All too often the attempt was made with inadequate personnel. For example, a single plant breeder might be made responsible for the improvement of the full range of crops in the local agricultural calendar. Needless to say in such conditions the effort was so widely dispersed and thinly spread as to be totally ineffective. A concentrated but well focused effort on a narrow, well-defined range of crops or even a single crop was much more effective, provided of course that clear goals were set and that adequate logistical support was available. Such efforts can be extremely productive and highly cost-effective as long as appropriate exploratory work has been carried out, the problems clearly defined and adequate procedures devised to implement the research programme designed to provide solutions.

Research management implies more than the mere management of research personnel. The current idea that a 'good manager' can manage anything from a whelk stall to a giant multinational enterprise can be very far indeed from the truth. The manager must know and understand what it is that is being managed. The ideal manager would be a polymath with a good background in and understanding of the earth and physical sciences, in addition to considerable experience in the biological sciences as applied in the agricultural context. An ability to communicate with agricultural economists and sociologists would be advantageous. Such paragons do not exist in any great numbers! Perhaps it is more reasonable to expect to find a broad-minded agricultural scientist who can look inwards to the work of his research organization, organize it in a coherent and integrated fashion and not only package, present and 'sell' its product but also be fully cognizant of the needs and aspiration of those outside the organization, especially those whose activities have interfaces in common.

16.4 CONCLUSIONS

In any research endeavour, it is important that both the immediate and the more remote contexts of the work be fully appreciated. That is why farming systems and adaptive research have been considered at some length in this chapter. In specialist crop circles it is very easy indeed to lose sight of such considerations. The awareness of the relationship between one's own work and that of one's colleagues greatly facilitates the flow of information. Information can be one of the most underused of our resources – not least that which, although gleaned decades ago, remains of considerable current relevance. Lack of awareness of previous work leads to attempts to re-invent the metaphorical wheel or complete failure to learn the lessons of the past. We in the groundnut world have the spectre of the East African Groundnut Scheme to remind us of the consequences of ill advised and poorly considered development projects. The present era is that of the 'instant fix' the politician is constitutionally impatient and tends to favour the short-term 'solution' to development problems produced by a consultant, notwithstanding the totally inadequate factual base on which developmental initiatives might be founded. Quite obviously resources would be better directed into a programme of on-the-spot adaptive research, which could ensure that development could be both compatible with the local farming systems and sustainable in the long term.

REFERENCES

Allan, W. (1967) *The African husbandman*, Oliver and Boyd, London.

American Society of Agronomy (1976, reprinted 1977) *Multiple Cropping*, ASA Special Publication no. 27, Madison, Wisconsin.

Brown, L.R. (1970) *Seeds of change*, Praeger, New York.

Evans, A.C. (1954) A study of a rosette resistant variety Asiriya Mwitunde. *East African Agricultural Journal*, **22**, 27–31.

Hammons, R.O. (1973) Genetics of *Arachis hypogaea*, in *Peanuts – culture and uses*, American Peanut Research and Education Association, Inc., Yoakum, Texas.

Hartley, W. (1949) *Plant collecting expedition to sub-tropical South America*. Report Division of Plant Industry Australia no. 7 pp. 96.

Hildebrand, G.L. and Smartt, J. (1980) The utilization of Bolivian groundnut (*Arachis hypogaea* L.) germplasm in Central Africa. *Zimbabwe Journal of Agricultural Research*, **18**, 39–48.

Loomis, R.S. and Connor, D.J. (1992) *Crop Ecology*, Cambridge University Press.

McEwen, J. (1961) Groundnut variety Maní Pintar. *Nature*, **192**, 92.

Ruthenberg, H. (1976 and 1980) *Farming systems in the tropics*, 2nd and 3rd edns, Oxford University Press.

Smartt, J. (1960) Genetic instability and outcrossing in the groundnut variety Maní Pintar. *Nature*, **186**, 1070–1071.

Smartt, J. (1967) *Groundnut production in Zambia*, Government Printer, Lusaka.

Smartt, J. (1978) Makulu Red – a 'Green Revolution' groundnut variety. *Euphytica*, **27**, 665–675.

Tivey, J. (1990) *Agricultural Ecology*, Longman, London.

Trapnell, C.G. (1953) *The soils, vegetation and agriculture of North-eastern Rhodesia*, Government Printer, Lusaka.

Trapnell, C.G. and Clothier, J.N. (1957) *The soils, vegetation and agricultural systems of North-western Rhodesia*, Government Printer, Lusaka.

Vandermeer, J. C. (1989) *The ecology of intercropping*, Cambridge University Press.

CHAPTER 17

The future of the groundnut crop

J. Smartt

In assessing the future prospects of any crop, it is sensible to consider what factors contribute to the crop's present position and how these are likely to change, if at all, in the immediate and near future. The considerations tend predominantly to be economic in that the output from producing the crop must be commensurate with the input, though the input may not always be easy to quantify in purely monetary terms. Economic viability needs to be assessed by a broad and open-minded economist, not by an accountant who will tend to ignore what cannot be expressed in cash terms.

17.1 PRODUCT DIVERSITY

There are crops such as coconuts and bananas for which the level of inputs (in some circumstances at least) may be quite low. After the crop is established the farmer has little to do but to watch it grow and harvest it. A little gentle weeding and cultivation may be all that is needed. This is never the case with groundnuts: the returns, good though they may be, are not won easily nor without considerable effort. One might ask at this stage why the effort still continues to be forthcoming and the reason lies in the very high regard in which the product is held and its great versatility in use. The uses vary quite remarkably in the various areas of production. In subsistence economies the seed is used directly for human consumption while the crop residue is a useful livestock food. It is interesting to note the use that was made of the crop in the United States for 'hogging off', in which groundnut crops at or near maturity were in essence harvested by pigs. It is not beyond the bounds of possibility that such usage, given favourable circumstances, might return.

The utilization of the groundnut as food by subsistence cultivators is varied – limited, one might say, only by human ingenuity. The crop can be

The Groundnut Crop: A scientific basis for improvement. Edited by J. Smartt. Published in 1994 by Chapman & Hall, London. ISBN 0 412 408201.

used in similar ways to peas and beans in what might be called the 'green mature' stage before the fruit dries out, or even earlier when the pod is soft and juicy (Krapovickas, 1969). According to the same author, it can be used as the source of a refreshing drink, *chicha de maní*, and also a soap can be made from the seed by saponification. Since the seed may contain 50% of its dry matter as oil, it is a relatively simple matter to extract the oil. The mature seed can be parched or ground to a paste and used in the preparation of soups and relishes. The seed is also amenable to use in the traditional fermentation technologies practised at village level in south-east Asia and the Far East.

By the nature of things, subsistence cultivators cannot be totally self-sufficient: they usually produce some surplus which can be sold or exchanged locally for other essential goods. There is usually some small local trade in the crop since it is such a popular article of consumption. This surplus provides the base from which (in favourable areas) an export trade might develop. Historically, with urbanization there was no doubt a demand from town and city markets for this popular and well-regarded food material. The overseas export market was initially for the oil: Archer (1853) reports that the crop was sold by dry-fruiterers but the bulk of imports went to the oil mills. He noted, however, that the supply was irregular – 'sometimes several hundred quarters in the year, and at other times none at all'. He went on to say: 'It is never likely to be much used as an edible fruit in this country [Britain], being in no respect superior to the common grey pea of our fields.' Clearly the development of export markets depended on the reliability of production and relative stability of markets which has developed in the past 100–150 years. Johnson (1964) presents an extensive historical review of the history and past utilization of the crop and shows how, in the United States, slow progress was achieved towards general acceptance of the groundnut by the apostle of the peanut, George Washington Carver, in the early part of the present century.

The bulk of initial exports found an industrial market in the edible oil expression industry. The press cake became a valuable and lucrative by-product and a well-regarded, high-protein feeding stuff, particularly for dairy cattle, in the first half of the present century. In North America, the edible oil industry was eclipsed as a market with the development of confectionery uses in candies and sweets in addition to traditional roasted and salted peanuts, but most of all by the development of the peanut butter market. Peanut butter, together with cola beverages, has become some-thing of a North American addiction.

The result of all this is that the groundnut has an enormous diversity of use at present, which tends to promote stability of markets in the long term. It does mean that the quality prescriptions of the market must be met. Then there is the question of adjustment to changes in market demand. That for groundnut as an oilseed is slackening for a number of reasons, principally because of competition from other oilseeds (maize, sunflower, soya and rape) but also because of a change in the nature of the material exported. Until comparatively recently, oil expression depended on imported stock but there has been a tendency for this process to be carried out in the country of production. Unfortunately this has led to the problem of adulteration of groundnut oil with cheaper oils and, naturally, the reputation and standing of the product has suffered.

In contrast, the demand at the upper market level has remained good and is buoyant. This area is particularly attractive to food and confectionery manufacturers because the added value potential is enormous. The retail value of a tonne of groundnuts sold as roasted peanuts (approximately 40 000 packets) compares very well indeed with the import cost of the raw material. Just as there has been an attempt by producers to reap some benefit from groundnut processing in production of oil and press-cake, so it is reasonable to expect that they may well wish to benefit from the added value arising from peanut butter and confectionery manufacture. This could be attractive if the groundnut hulls could be used as an energy source for manufacture. The groundnut confectionery sector might well seek to emulate the processing of sugarcane as an eminently energy efficient industry.

Meeting the market demands for such an industry poses a challenge and imposes some constraints. Quality standards for oil stock have not been high – in the United States only reject nuts are used and the problem of aflatoxin is less important as it remains in the cake and is virtually absent from the oil. For confectionery use, quality standards are very exacting indeed: aflatoxin levels must be low and demands for even lower levels are likely to be imposed. The physical properties of the kernels are important; size, shape and even testa colour (although removed in blanching) must conform to standards. Chemical and organoleptic properties are similarly important.

The multiplicity of outlets and the diversity in utilization provides stability in the market, which is a great advantage in planning future production initiatives since these can be based on a firmer foundation than might otherwise be the case. However, the market's exacting standards must be met and fully taken into account in the course of planning.

17.2 FARMING GROUNDNUT

The groundnut is a high-input crop at all levels of production. It is basically demanding in terms of labour and time for land preparation, sowing, cultivation, harvesting, stripping and shelling. The time and labour commitments can be reduced by mechanization and the use of chemicals to reduce cultivation. In more advanced production systems, applications of gypsum or land plaster and insecticides and fungicides are commonplace and routine. These inputs are additional to those used by the less sophisticated cultivator.

17.2.1 Levels of input and productivity

In Central Africa it used to be a rule of thumb that a yield level of about 200 kg/ha was critical. Below this level the subsistence cultivator would not find it worthwhile to produce the crop; above, it was worth considering. The level of input necessary for a shifting cultivator was the effort required to clear and prepare the land, sow and cultivate the crop, harvest it and process it post-harvest. The inputs could be considered as labour, time, seed and the use of simple tools such as the hoe. With increasing sophistication the farmer in the developing world, operating above the subsistence level, will be able to extend the cropping area if animal power is available. The use of animal power incurs the need to invest additionally in at least a plough, and probably a cart and cultivating machinery. While the farmer may be relatively self-sufficient for animal power, clearly the operation must be capable of generating a cash surplus to pay for the capital cost of equipment. This is even more the case when mechanization takes place: a further reduction in self-sufficiency is incurred as well as expenditure on fuels, lubricants and maintenance costs in addition to capital and interest payments. This generates further pressure to increase either the land area cultivated or its productivity, or both. At the limit of available land area, productivity increases are necessary in order to meet these additional costs. Productivity can be improved by the application of fertilizers, and by the use of pesticides (herbicides, insecticides and fungicides) to control losses due to pest depredations. The increased costs must be scrupulously balanced against additional income produced and the rising curve of increasing expenditure must be less steep than that of income. Obvious though this is, it can be overlooked – especially when the farmer, as in the developing world, does not have the services of an accountant giving advice about what is producing a sufficient return and what is not.

17.2.2 Energy budgets

An area of increasing concern is that of the energy budgets of different production systems. In terms of productivity per unit of non-solar energy input, it has been claimed that the least sophisticated systems are the most efficient and the more sophisticated the least so. The energy investment from outside the farming system begins with the use of the hoe, the manufacture of which involves comparatively small amounts of energy. This increases with the use of ploughs, cultivators and other tools that require animal power. Additional increases are incurred with mechanization, the consumption of oil fuels, the manufacture of agri-chemicals and the energy cost of their distribution. In the most advanced systems of groundnut production, where field-curing is abandoned, there is considerable energy expenditure in drying the crop. The sum total of energy expenditure in producing the final, saleable crop is substantial.

The point can be made, with some justification, that the energy which is expended in the production of the crop may, if the trend continues, be greater than the intrinsic useful harvested energy in the crop itself. Such an energy deficit would clearly be made up by using fossil fuels. This poten-tially negative energy balance is a question of which we would be aware and attempt to address with circumspection. The question can be asked: how are we to set about improving the energy balance equation? We need in effect to be energy misers, using materials such as fertilizers only insofar as they give a sensible return and applying them at the most opportune times. In addition we should be seeking to reduce applications of pesticides by increasing levels of pest and disease resistance, seeking energy efficient use of herbicides (or alternatives to them) and the use of natural energy as far as possible in drying the crop. The energy balance may, in spite of improvements, remain adverse. Another pressure to which the producer is subjected arises from the pest residue problem. Economy in the use of agrichemicals may thus be doubly friendly to the environment.

17.2.3 Aflatoxin

The aflatoxin problem, which has been recognized for 30 years and more (Chapter 13), has not been finally resolved, although there has been progress in knowledge of the causal organisms and the way they produce their effects. Possibly the most productive approach is first of all to attempt known environmental approaches to minimize invasion, then to select host genotypes in which penetration of the pod wall is slow and finally to select host genotypes which do not induce mycotoxin production by the fungi. The prospects of more rigorous standards and lower tolerance levels is a very real one. Conditions of storage in producing countries also need to be of a high standard so that produce is not denied entry into importing countries. Continued mycological and breeding studies are required.

17.2.4 Climate changes

Climate change and the spread of cultivation to new areas, most notably south-east Asia, both raise problems. With shorter and less reliable wet seasons in some areas, particularly in sub-Saharan Africa, there is con-siderable interest in the breeding of ever more early-maturing cultivars. The questions are: what is the shortest period to maturity in which a useful and worthwhile yield can be achieved, and what is the minimum canopy size with which this end can be achieved? Success has been achieved with ultra-early maturing cowpeas and this could be repeated in the groundnut. Something like the valencia growth form might be appropriate with a pattern of assimilate partitioning producing a high harvest index.

17.2.5 Cultivation in new areas: adaptive research

In south-east Asia the problem is quite different. The groundnut in the tropics has been a feature of semi-arid environments where a single crop is taken per year. In the more humid tropics, there is a problem which arises from the fact that it may be possible to grow the crop all year round: the question then is to determine how best to go about this. A difficulty can arise in drying seed post-harvest, which is important for maintaining seed stocks. No problem arises for that part of the crop which is boiled for eating but for the part of the crop which is to be dried the drying process must be continuously maintained. Similar procedures to those developed for drying cocoa beans might be suitable. In areas where convectional rainfall is characteristic, the drying crop could simply be protected at the appropriate time of day. Periods of unremitting rainfall, however, could cause difficulties with moulding and mycotoxin generation.

The problems which could arise from effectively continuous groundnut cultivation in the humid tropics need to be evaluated very carefully. Certainly in the semi-arid tropics the occurrence of a close season has been regarded as an advantage. If cultivars with short-term seed dormancy are grown, then there are few (if any) volunteer plants to carry over viruses, vectors and fungal pathogens directly between successive crops. There would be no such check with continuous cultivation and the consequences of this need to be studied. These are not necessarily what might be anticipated. In areas where three and perhaps even four crops might be taken in a single calendar year, it is quite possible that one particular time of year might have particular advantages. This could be advantageous *vis à vis* the position in the semi-arid tropics (S.A.T.) where debris from one season's crop persists through the dry season and produces inoculum for the cercospora leaf spots at the beginning of the following rainy season. In the humid tropics were there to be a close season the debris would have decomposed before the next crop was sown.

While the agronomic position is well explored in traditional areas of groundnut cultivation, current studies are regarded as fine tuning. There is scope for and a need to explore the situation thoroughly in areas where the crop has been introduced comparatively recently. The question of cultivar selection is obviously an open one, too. For example, is it more productive to produce four crops per annum of short-season cultivars than three of long-season varieties? It would be worthwhile to explore the performance characteristics of the very longest season varieties and it would be useful if comparable results could be obtained from a single long-season crop as from two crops from an early maturing cultivar. There is a supposition that late-maturing forms tend to have low harvest indexes and would not be viewed favourably if seed production was the major concern. However, if the haulm was produced in a useful amount and was of good quality, such varieties could have a useful role. In such novel situations it is essential to be both objective and open-minded: received ideas from elsewhere may be of very limited application. The basic knowledge we have of the crop should be the starting point for a programme of adaptive research. Such a programme should take into account the objectives which are practicable in the light of present constraints but one should also bear in mind other objectives which might be explored when present constraints no longer operate. While we must obviously take into account present realities, we should keep a weather eye open for future possibilities.

17.2.6 Nitrogen fixation

The great advantage claimed for legumes in general (and not always realized) is that they are virtually self-sufficient for nitrogen. In the semi-arid tropics, this is very largely true. At the end of the dry season the first rains initiate a flush of nitrification by free-living soil bacteria which early sown crops are in a position to exploit. An early-sown groundnut crop in such circumstances can make use of this available soil nitrogen while its own nodular system is built up. With further rainfall, soil N levels can be expected to decline as the nodules on groundnut roots become effective. Later-sown crops do not derive any such benefit and may suffer not only the disadvantage of less soil nitrogen but also soils which are cooler and less well aerated on account of the rain. The crops establish more poorly as a result and they are also more prone to infection by aphid-borne viruses and to receiving leaf spot inoculum very early in their growth.

Root nodules in the groundnut are formed by *Bradyrhizobium* (Chapter 8) and there has been considerable research into improvement of the efficiency of N_2 fixation in the root nodules. Much attention has been focused on selection of strains which are especially efficient in N_2 fixation, in the hope that these might be released and establish themselves, infect roots and produce more effective nodules. This strategy appears to fail on two counts: that laboratory selected strains are often less competitive in the wild than the native bacteria, and that there is no possibility of establishing super-efficient *Arachis*-specific *Bradyrhizobium* populations because the cowpea cross inoculation group is all but universal in the tropics and subtropics. The best approach (Chapter 8) seems to be to select host genotypes with a propensity to produce effective nitrogen-fixing nodules with a very broad cross-section of the native *Bradyrhizobium* population. It is possible that the nitrification system might be manipulated in this way to advantage. Selection and evaluation would probably not be straightforward, however.

17.2.7 Fertilization

Fertilizer studies in the groundnut are fraught with difficulty. There is no question that the crop, judged by the yields it produces, can respond handsomely to high fertility conditions. The experience of agronomists who have attempted to study fertilizer responses has frequently been one of frustration, the exception being calcium response. On calcium-deficient soils there are clear-cut responses to applied gypsum or lime in terms of improved seed development and reduction of 'pops'. Apparent response to application of other nutrients is erratic and the reason seems to be that the nutrients are not in fact reaching the soil horizon in which groundnut roots are actually taking them up. Deep placement of nutrients is not practicable and the received wisdom of 40 years ago (York and Colwell, 1951) is probably as valid today as then: fertilize the rotation rather than the groundnuts. They will pick up nutrients which have passed beneath the root range of other crops and will not be wasted. This is consistent with the observed calcium response: the pegs and pods which take up this element are in the top 5 cm of soil and top dressings are readily accessible. The problem of apparent lack of response is actually one of accessibility, and incidentally probably led to the soubriquet of 'the unpredictable legume' being applied to the groundnut. It is less unpredictable now than it was, perhaps, but it is still by no means an open book.

17.2.8 Mixed cropping

The role of groundnuts in farming systems, particularly those appropriate to the developing world, would probably repay further study. The relatively low incidence of pests and diseases in traditional farming practices has often been remarked. In Mexico (Miranda, personal communication), the Mexican leaf beetle became a serious pest when the *Phaseolus* bean ceased to be grown as a mixed crop with maize and was cultivated as a sole crop. Mixtures of crops and mixtures of genotypes within a crop stand can both exert a controlling effect on pest and disease incidence. The challenge is how to gain advantage from the lower pest and disease incidence in mixed crops and yet benefit from mechanization and advanced technology. Basically similar approaches are adopted in relay sowing and in agroforestry. The effects of combining species of high and low growth can have beneficial effects in delaying arrival of inoculum and reducing spread by wind. I have observed instances of groundnut crops grown in relatively small woodland clearings which had markedly lower incidence of leaf spot diseases. It seems that anything which breaks up the uniformity of the potential host substratum can have a beneficial effect in reducing pest and disease incidence.

Such systems of mixed cropping have other beneficial effects in that the soil becomes comprehensively permeated by root systems of different species exploiting different soil horizons. This results in more efficient exploitation of the soil's bank of nutrients and recycling them and at the same time more effectively binding the soil and providing protection against erosion. There is wide scope here for the study of relay cropping systems and agroforestry in conjunction with local traditional farming systems, with a view to identifying the constraints which inhibit their further development and removing them to enable development to proceed.

17.2.9 Water resources

The question of water resources and their effective use is another import-ant issue which has to be addressed. Groundnuts have a reputation as a drought tolerant crop. They owe this mainly to their deep-rooting habit. While they may survive drought conditions if these are not too prolonged, they may not produce a satisfactory yield. However, their tolerance of drought can be put to good use in the semi-arid tropics. At the end of the dry season it often happens that there is a false start to the rains and there may be sufficient rainfall to allow preparation of land for sowing to be completed. The natural sequence of events is to follow this with the sowing of the staple cereal crop. In the case of maize, when no further rain falls for a few weeks the crop may emerge and be lost from drought. Groundnuts sown under such conditions will germinate, send down quite a long tap-root, produce a small tuft of leaves and sit out the drought. Further growth and development occur when the rains resume.

Groundnuts do require a good soil moisture status from flowering through pod-filling. Shortage of water during this period results in poorly filled pods and low quality kernels (Chapter 9). In terms of pod-filling, plant population density must be related to expected soil moisture contents at this critical stage. If it is inadequate, the level of yield may be maintained but the quality of the product is drastically affected. Supplementary irriga-tion (if water and the means to distribute it are available) could be used quite effectively to increase yields of good quality. This would enable the larger pod set from a high plant population to mature a larger crop more fully. If the length of rainy season is limiting but irrigation water is available at the end of it, groundnut crops could be started off under irrigation and maintain themselves with relatively low water use until the major rains come. This would give the benefits of a longer growing season combined with the advantage of the crop having a longer period free from pests and disease before pest and pathogen population densities build up. This is an interesting area for further exploration by both the water-relations physiologist and the irrigation specialist. In addition there is scope for detailed study of growth and development of root systems, with a number of ends in view, providing useful information for those investigat-ing rhizobial activity and the uptake of both water and nutrients.

Comparative studies of the patterns of root growth in different genotypes and the morphological types of the groundnut would be particularly valuable, though not an easy area for the breeder to explore. The root system of the groundnut underpins and supports so many of its useful attributes that we really ought to know more about it.

17.3 IMPROVING THE GROUNDNUT

Since the pioneering days of Gregory, Smith and Yarborough (1951), there has been a continuing development of botanical knowledge which provides the foundation for much of the work of groundnut improvement. In many areas the situation is highly satisfactory but in others there are large lacunae which have to be filled. Our basic morphological and anatomical knowledge of the cultigen and its wild relatives is reasonably satisfactory (Chapter 3). Our knowledge of its reproductive biology and development is also quite satisfactory (Chapter 5). This is not to imply that nothing further remains to be done in these areas, but there are others where lack of information and deficiencies of published knowledge are a decided handicap, especially in biosystematics related to genetic resources (Chapter 4). This has repercussions affecting the use of genetic resources in breeding (Chapter 14) as well as the development of new uses for members of the genus as forage in addition to their use as a genetic resource for improvement of the cultigen (Chapter 15).

17.3.1 Taxonomy and biosystematics

Since the 1950s there has been an active interest in the collection of the genetic resources not only of the cultigen *Arachis hypogaea* but also its wild relatives. The volume of material collected by Krapovickas and Gregory has been most impressive. Collection has been carried on in all areas of South America where the genus is known to occur wild, and also of landrace materials, particularly in the Americas and Africa. Unfortunately the substantial volume of wild material collected (including many species new to science) remains undescribed. Some of it has been given informal names and an informal taxonomy has been put forward (Resslar, 1980). This has been useful as far as it goes but it does not go far enough. The users of wild species as genetic resources need to know the status of their material to guide them in planning strategies for its use as a germplasm resource and in putting their plans into effect. The reasons for the delay are not clear but it is a matter of some urgency that this particular nettle is grasped and that a scientifically sound monograph be published of the genus as it is known at present. It is unrealistic to hope that a definitive monograph can be produced now which will stand for ever without revision – after all we must provide some useful future employment for taxonomists! The field of taxonomy and biosystematics is a vital one and essential to the development of sound and sensible strategies of genetic resource utilization. It is possible that this particular programme of work is under-resourced and that the first priority has been that the material in the wild be collected before it is made extinct by encroaching development and habitat destruction. The taxonomic synthesis may have had to take second place. We need to know what the situation is so that any necessary representations can be made and appropriate pressure applied where necessary to secure funding for the conclusion of this very necessary work.

17.3.2 Evolutionary history

In order to make effective use of wild *Arachis* species as a genetic resource we need to understand the evolutionary history of the groundnut itself (Chapters 2 and 4). Interesting problems are raised by the fact that the cultigen is a tetraploid which, together with its wild prototype *Arachis monticola*, constitutes a single biological species and that it is placed in an otherwise diploid section of the genus. The wild prototype material that is available amounts to a very slender genetic resource but the diploid species (most of which can be crossed with the cultigen) together constitute a very valuable resource. In spite of the ploidy level difference, it is possible to induce gene flow from them into the cultigen. Some species have been identified as sources of pest and disease resistance, among them *Arachis cardenasii* – one of the first species to produce viable and partially fertile triploid hybrids with the cultigen. It has resistance to late leaf spot and nematodes (Stalker, 1992) and is being used in breeding programmes at present. Studies of interspecific hybrids between diploid species have shown that three distinct genomes probably exist: A and B (Smartt *et al.*, 1978) and D (Stalker, 1991). It has been suggested that A and B genomes are present in the cultigen. While similar in genetic complement, these genomes are differentiated from each other by differences in chromosome structure and this confers a high level of genetic stability in the amphidiploid (allotetraploid) cultigen. There have been attempts to identify genome donor species and the leading candidates at the present time are *A. duranensis* and *A. batizocoi*. This hypothesis is supported by the results of experimental crosses involving the cultigen, wild species and experimentally produced amphidiploids. The suggestion has also been advanced (Singh, 1988) that the two groups virginia and spanish/valencia might have arisen from separate amphidiploidizations. This idea is supported by the fact that there does seem to be some chromosome structural differentiation between the two subspecies.

While the hypothetical ancestry of *A. hypogaea* is supported by the studies of Klozová *et al.* (1983a,b) it has not been supported by the RFLP studies of Kochert and co-workers (Kochert *et al.*, 1991). RFLP evidence suggests that *A. hyspogaea* is monophyletic: results from all samples tested show very little variation. Diploid species of section *Arachis* do show variation. Assuming that RFLP is an additive character it is possible, by a summation of the separate RFLP patterns of *A. duranensis* and *A. ipaenis*, to produce a combined pattern similar to that of *A. hypogaea*. However, both *A duranensis* and *A. ipaensis* are 'A' genome species and an amphidiploid produced from this pair of species would not be cytogenetically stable. There are suggestions (Kochert, unpublished) that more than a single RFLP pattern may occur in a species such as *A. duranensis* and it may well be that further search is necessary to identify complementary morphs from two species which would produce the *A. hypogaea* pattern and be cytogenetically credible. It is important to determine the extent to which gene exchange or recombination is possible between the A and B genomes. Some quadrivalent formation is observed from time to time in the meiosis of *A. hypogaea*, which could possibly bring about some intergenomic gene transfer. This could facilitate expression of desirable recessive characters and enhance expression of additive genes.

17.3.3 Role of the plant breeder

Sooner or later, and inevitably, crop improvement involves the breeder. We have already considered that the work of the breeder would be appropriate in the reduction of susceptibility to aflatoxin contamination, the improvement of the efficiency of rhizobial activity by host selection, and the improvement of root growth and function, to which also can be added that of the shoot in physiological terms. The range of breeding tasks which can be assigned to the groundnut breeder is enormous.

(a) Genetic resistance

One of the most attractive areas in which the plant breeder can operate is in the improvement of resistance to pests and disease attack. The high levels of mammalian toxicity in many insecticides, and their persistence, have aroused particular concern. There is also the fact that many have lost their efficacy in a very short time and that they are expensive and are costly to apply. Similar considerations apply to fungicides.

Successful resistance breeding provides both a cost-effective and a 'green' solution to this kind of problem. Even a partial solution has value: if application of agrichemicals can be delayed and their frequency of application reduced, this is of positive value. Pest and disease control by this means could enhance profitability considerably. But the breeder must not be complacent – there are instances where breeders have exacerbated the problem they aimed to solve, as in the loss of horizontal resistance in cereal rust breeding. The entomologist faces a similar situation in that the use of insecticides on the cotton crop has exacerbated the insect pest problems there.

The increasing levels of mammalian toxicity in insecticides and their decreasing periods of effective use have given urgency to the search for alternatives. This has led to the rise of integrated pest management (IPM), which takes an ecological approach in seeking how best to manipulate pest problems by maintaining their populations below damaging levels. Genetically controlled resistance or tolerance can be a component of such systems in combination with cultural control, biological control and minimal use of agrichemicals.

Control of virus diseases by breeding is interesting. The East African groundnut variety Asiriya Mwitunde was shown in the early 1950s to be a very poor aphid host and although it could be readily infected with the chlorotic rosette virus was field resistant (Smartt, unpublished). Genetic resistance to the virus itself has been found in West Africa and has been incorporated in many new varieties. Cultural control of rosette in good growing conditions is very effective but yields can be reduced to almost zero under conditions of drought and poor stands.

The leaf spots are the universal pathogens of the groundnut and are the cause of American growers undertaking costly spraying programmes. High levels of resistance to both early and late leaf spots would save a great deal of money where spraying is practised and bring about considerable and virtually cost-free yield increases where it is not. Wild *Arachis* species such as *A. cardenasii* and *A. chacoensis* have been shown to have leaf spot resistance and breeding efforts have been concentrated on these. However, as anyone who has worked with large numbers of accessions in a germplasm nursery knows, there are very considerable differences in leaf spot susceptibility between accessions of the cultigen. There is a distinct possibility that we may be overlooking useful sources of resistance here. Hemingway (1957) showed that there are morphological resistance mechanisms, and a study of a possible range of morphological and physiological mechanisms in South American landrace material might be productive. If such resistances exist, they might be combined in a single variety which might very well have a considerably higher level of resistance than any current variety.

Nematodes are a problem in many areas (Chapter 10) and there is considerable interest in controlling them by genetic resistance. Chemical control is difficult and expensive and thus impracticable on a large scale. Genetic resistance has been identified in wild species and active attempts are underway to transfer this to commercial cultivars. Pest and disease controls by incorporating genetic resistance are in essence the means of protecting potential crop yield in the field.

(b) Efficiency of physiology

There is a continuing study of the physiology underpinning the production of high yields and the improvement of harvest index. Manipulation of the type of canopy produced is possible by genetic means, as is the partitioning of assimilates between the various sinks. Selection for an efficient physiology is clearly a complex matter. It is unlikely that there will be a single optimal idiotype but it may be that a series of 'horses for courses' idiotypes might be devised which would enormously enhance the efficiency and effectiveness of the exchange of breeding materials internationally. It is quite possible that certain materials may prove to be in effect pre-adapted to a number of regions other than that in which they were produced. A good illustration of this is furnished by the strong and clear pre-adaptation of Bolivian landraces to conditions in Central Africa (Smartt, 1978; Hildebrand and Smartt, 1980).

(c) Nutritional value and chemical composition

There is potential for breeding and selection work in the improvement of chemical composition and nutritional value, on which little has as yet been done. For example, contents of polyunsaturates and susceptibility to oxidative rancidity are of concern to processors. The high concentration of polyunsaturates held to be desirable by nutritionists can be combined with a high content of the antioxidant tocopherols, a means by which this particular circle can be squared. There is some interest in the confectionary trade in kernels with intrinsically lower oil contents and these could in all probability be achieved by selection. However, the same genotype (such as Chalimbana) may produce mature kernels of very different oil contents when grown at different elevations in Central Africa – those at lower elevations having higher oil contents than those from higher altitudes.

Protein content, protein quality and organoleptic properties are also heritable and could be bred and selected should the need arise. Other commercial specifications set by the trade concerning heritable characters (such as size and shape of seed and pod, and testa colour) cannot be ignored but often have rather unfortunate repercussions on the grower and engender great frustrations for the breeder. This is well exemplified by the case of the cultivar NC2 when it was nearing the end of its useful life as a major cultivar in the peanut belt of Virginia and North Carolina. In the 1950s and 1960s it met the millers' and processors' needs perfectly; they had no interest in changing varieties and had apparently set their faces against such a move. Eventually the supply situation deteriorated so drastically that they were forced to change. Market forces will prevail ultimately but their operation should be tempered by intelligent anticipation and vision and a consensus between all sections of the industry. Quite justifiably, the breeders felt aggrieved that new varieties were rejected because the commercial specification was not met by only the narrowest margins. The industry appears to have learned its lesson but at considerable cost and damage to itself.

17.4 THE ROLE OF BIOTECHNOLOGY

In the opinion of many, the future of crop improvement seems to lie in the hands of the biotechnologist. To many in the field, this sounds like an enormous oversimplification. It would be equally ill-advised to deny the possibility of potential positive value as it would be to entertain totally unrealistic ideas about what can actually be accomplished that is of practical value – and the cost-effectiveness of attempting to do so (Nigam, 1992). Wild extrapolations from present accomplishments can lead to the setting of over-ambitious targets in the future and almost inevitable disappointment. In biotechnology, if we do not manage to keep our feet firmly on the ground, we are likely to dress up flights of fancy as realistic expectations. In other words,we shall re-run the time-honoured bandwagon saga yet again.

In this connection it is interesting to consider a bandwagon which was in motion 40 years ago – that of mutation breeding, which involved one of the giants of peanut research, Walton C. Gregory of North Carolina State University. His studies of mutation breeding were classic and well worth carrying out for the light they shed on the operation of the genetic system of the groundnut in particular and crop plants in general. In terms of useful output of cultivars released, NC4X resulted but not a great deal besides. New cultivars could be produced but less efficiently than by conventional breeding programmes. There is no reason why useful mutations artificially induced should not be fed into conventional breeding programmes. Where mutation breeding has made its mark and continues to be of high value and significance is in the improvement of vegetatively propagated crops. Artificial mutagenesis plus selection of somaclonal variants have revolutionized the breeding of chrysanthemums. It is a highly cost-effective breeding technique for such decorative species and produces results in a much shorter time than traditional breeding methods.

There are a number of reasons why the potential value of biotechnology in groundnut improvement should be assessed with an appropriate measure of caution. There are undoubted successes in certain crop species, such as tomato (Rick, in press), but the state of genetical knowledge of these species is vastly greater than that of the groundnut. Genetic maps of morphological traits in tomato, are well developed, while little progress, if any, has been achieved in this area for the groundnut. There has been some progress in producing RFLP maps in *Arachis* until these can be related to important morphological (and other) characters they are likely to remain of purely theoretical interest. Another problem, which does now seem to be approaching a solution, is that legumes as compared with members of the Solanaceae are not the most responsive to tissue culture and *in vitro* manipulation generally. Until comparatively recently it has not been possible to obtain regeneration from callus tissue cultures. The achievement of this is a considerable breakthrough. Nevertheless legumes in general, and groundnuts in particular, tend to provide highly recalcitrant and intractable material when it comes to *in vitro* manipulation (Nigam, 1992).

The techniques of achieving transformation of groundnuts are still in the development phase but there are indications that effective procedures will be devised. Whether transformed plants will behave as expected is an open question. Many of us are aware of the genetically transformed arthritic pig of some years ago. Quite clearly the first stage is to develop transformation techniques that work; the next task is to identify, extract and clone suitable genes for introduction. Some genes are already available (the *Bacillus thuringiensis* factor might have value in protecting the crop against foliar feeders) but they are few in number and the great bulk of potentially useful genes will have to come from other sources. The genus *Arachis* itself is the repository for most of the genes whose effects would be valuable if they could be used to transform *A. hypogaea* – many pest and disease resistance genes are in this category. Unfortunately the genetic systems of wild *Arachis* species are even less well understood than that of the cultigen; the identification of useful genes and their attraction for cloning are well beyond our present capability. To achieve the state of knowledge which would enable this to be done would require an enormous investment and the question remains as to whether such an investment would be cost-effective. There is no guarantee that, if any major resistance genes were incorporated through transformation, the magnitude of the effect would be comparable in the donor and recipient species. In practice, the most effective approach would be to exhaust the genetic resources of section *Arachis* species before attempting biotechnological manipulations, in any but an exploratory way, involving species in the higher order gene pools of Harlan and de Wet (1971).

The idea that the laboratory will be the source of cultivars in the future is one which has become lodged in the minds of many, including some of those involved in making decisions about setting research priorities and allocating funds. To the politicians, biotechnology has the great attraction of offering the quick fix. Their time frames seem to coincide with intervals between elections and there are those in the scientific community who are prepared to exploit such myopia. However, the short-term advantage is likely to be counter-productive when eventually the time comes for the return on investment to be measured and evaluated. The reaction is likely to be the discrediting of biotechnology as an instrument of crop improvement. A grave error is being made in the implicit promotion of biotechnology cum genetic engineering as a substitute for plant breeding rather than as a potentially valuable supplement to it.

It is becoming increasingly difficult to maintain a spirit of scientific objectivity and balance in the present age. Although academic freedom could be regarded by some as a meal ticket for life and justification for a long-term ego trip exploring esoteric and arcane avenues of research, it did mean that scientific exploration in the medium and long term could take place. Nowadays the short term rules so that results are expected within three years or so, which can lead to sleight of hand and slick footwork in what is becoming the sordid business of producing grant applications. There is, for example, the project-in-hand scenario, where a smooth and credible proposal can be submitted because the work is already virtually complete. This type of proposal fares better before a granting committee than one which seeks to explore difficult areas where problems are ill-defined and solutions uncertain. The honest and straightforward scientist is likely to be ever more frequently sidelined by the opportunistic (entrepreneurial) operator who may have very little of real value to offer. The field is open to the plausible charlatan with a superabundant vocabulary of buzz words who manages to keep ahead of the field and avoids ever being called to account.

17.5 AGRICULTURAL CONSTRAINTS, RESEARCH AND THE FUTURE

The International Groundnut Workshop at ICRISAT in late 1991 was an extremely important occasion for groundnut research workers. The publication of its proceedings in the following year enabled those who were not fortunate enough to attend to enjoy the fruits of the presentations and discussion to the full. In particular, the workshop recommendations (Nigam, 1992) provide a succinct résumé of the general tenor of the workshop's deliberations. Those of the greatest relevance to the content of this chapter relate to biotic constraints, current agronomic problems, genetic resources and germplasm enhancement and utilization.

17.5.1 Biotic constraints

The identification of constraints is a prime prerequisite for implementation of integrated pest management, the ideals and objectives of which are now better formulated than in the days of the pioneer adaptive researchers in the 1950s and 1960s. Emphasis was placed on broadening the deepening biological understanding of the pests and their epidemiology and on close monitoring of the effectiveness of such management programmes in the field and the possible development of forecasting systems. The use of modelling systems in deciding the relative effectiveness of genetic and management solutions in relief of biotic constraints was advocated. In practice these would most often be viewed as complementary to each other rather than as alternatives. The production and distribution of early-generation hybrids of short-season materials involving rosette-resistant parentage was recommended, to be distributed over a wide area for selection to be carried out. This could result in a range of genotypes carrying genetic resistance with a range of environmental adaptation.

17.5.2 Current agronomic problems and objectives

Drought resistance and water use economy in the crop are important aspects (see also Chapter 9) in the context of desertification and increased frequencies of drought in the African Sahel. A partial solution to the problem is to change to shorter season varieties. This could be reinforced by selection for selectable traits of the root system (which could improve water uptake) and characters of the shoot system (thicker cuticles, for example, which could reduce water loss). In addition, physiological drought resistance and ability to recover from desiccation would be desirable features to incorporate. These questions are closely related to high-temperature tolerance and could perhaps form a cluster of selection objectives. In some areas low-temperature tolerance could be important, especially in the seedling stage.

Partitioning patterns of assimilates could be manipulated to maximize seed production, for example, and equally to provide bulk of haulm if this was desired. With the development of mixed cropping and agroforestry systems, shade tolerance could become of increasing interest. Soil constraints relating to both nutritional deficiencies and acid soil problems could be the subject of potentially useful study, such as the search for rhizobial bacteria tolerant to acid soil conditions. The iron deficiency problem might possibly be relieved by a selection programme on a similar basis to that suggested for rosette resistance selection.

The plastic mulch technology developed in the Far East (China and Japan) could be investigated for use elsewhere. It has the advantage in temperate climates of speeding up soil warming after spring sowing and more generally of promoting water economy and suppressing weeds. The plastic residue generates disposal problems but these could be eased, if not solved, by use of thin films of biodegradable plastics which disintegrate readily when the crop is mature. The role of the groundnut as an important component of sustainable farming systems through the incorporation of the resistance and tolerances mentioned should be explored. The necessity for appropriate technical training for carrying out these recommendations was stressed.

17.5.3 Genetic resources and germplasm enhancement

The future of the groundnut crop depends directly on the nature and extent of genetic resources available for its improvement and adaption to future contingencies, only some of which can be foreseen. These include adaptation to novel systems of production which could evolve in new areas of cultivation. There can be little doubt that there are extensive areas in the Americas, for example, where groundnuts could be grown but are not. The same is true of North Africa and southern Europe: should circumstances change and it became more attractive to grow the crop under irrigation, there is little doubt that production could be enormously increased in the Mediterranean basin. The guided evolution and development of the traditional farming systems of the developing world could make further demands on the breeder initially, and ultimately on the pool of genetic resources.

Resistance to pests and diseases is, as ever, a vital area for further breeding effort. The expansion of selection at the local level of early generation hybrid progenies is an interesting and key area in bringing together the work of breeders at the international agricultural research institutes and those engaged in adaptive research at the grass-roots level. This could greatly increase the efficiency of breeding operations in selecting for a considerably enhanced range of environmental adaptation and resistance to or tolerance of local biotic and abiotic problems and hazards. Such selection programmes would not require high levels of sophisticated training in staff but would rely upon intelligent people endowed with good commonsense and appropriately trained.

Genetic resources need to be identified which could serve to relieve problems relating to acid soils and those deficient in micronutrients. Problems of drought resistance and tolerance breeding could probably be eased by appropriate use of available germplasm resources, as could questions relating to high and low temperature tolerances and aflatoxin contamination. The extent and accessibility of the available genetic resources in the genus *Arachis* are considered to be such that biotechnological solutions are the last resort. *Arachis hypogaea* is really quite exceptional among legumes in that barriers to interspecific gene exchange are not as well developed as in many other legume genera, such as *Vigna*, at least within section *Arachis*. Outside this section, but still within the genus, biotechnology would probably have to be invoked in making intersectional gene transfer. In the current absence of any practically useful genetic maps, such manipulations are definitely for the future and not of immediate relevance.

There is a very real problem to be faced in the dissemination of new and improved materials and technology. This would clearly be facilitated by development of a cadre of adaptive researchers acting as an interface between extension services and the research services of national and international agricultural research institutes. The transfer of information as well as materials could be affected by such a chain of communication.

17.5.4 Utilization

Those at the ICRISAT workshop were well aware of the changing patterns of groundnut utilization. Clearly trends need to be identified and if possible anticipated, so that appropriate materials can be fed into the farming systems to maintain their economic viability. This will necessitate market monitoring and research combined with development of new markets in order to maintain demand for the crop and keep the producers in business. With the present slackening demand for groundnut oil, maintenance of current levels of production will involve development or expansion of new and existing outlets – markets with marked potential for adding value to the basic product. The marketing and processing needs to be arranged so that the countries of production (and especially the producers) derive a fairer share of this added value than has been the case in the past.

17.6 A NEW PHILOSOPHY

Research developments to the groundnut since the establishment of the International Crops Research Institute for the Semi-Arid Tropics have been particularly exciting to those with prior involvement in groundnut research in both the developing and developed worlds. At the time of ICRISAT's inception, the concepts and ideas of the 'Green Revolution' were very much to the fore. However, 20 years later our approach is much more broadly based. We have perhaps embraced a philosophy and policy of 'Green **Evolution**' based on stable traditional farming systems rather than the numerous imponderables of the Green Revolution and its aftermath.

Fortunately groundnut research has definable goals and progress towards them can be monitored and measured. There are many able and gifted researchers operating in the field and it is to be hoped that the ultimate objectives of groundnut research, in helping the world's farming communities which produce the crops, can be reached: to achieve, maintain and where possible, improve the level of prosperity and well-being of the farmers as well as that of the communities they serve.

REFERENCES

Archer, T.C. (1853) *Popular Economic Botany*, Reeve, London.

Gregory, W.C., Smith, B.W. and Yarbrough, J.A. (1951) Morphology, genetics and breeding, in *The Peanut – the Unpredictable Legume*, National Fertilizer Association, Washington DC, pp. 28–88.

Harlan, J.R. and de Wet J.M.J. (1971) Toward a rational classification of cultivated plants. *Taxon*, **20**, 509–517.

Hemingway, J.S. (1957) The resistance of groundnuts to *Cercospora* leaf-spots. *Empire Journal of Experimental Agriculture*, **25**, 60–68.

Hildebrand, G.L. and Smartt, J. (1980) The utilization of Bolivian groundnut (*Arachis hypogaea* L.) germplasm in Central Africa. *Zimbabwe Journal of Agricultural Research*, **18**, 39–48.

Johnson, F.R. (1964) *The Peanut Story*, Johnson Publishing Co., Murfreesboro, North Carolina.

Klozová, E., Svachulová, Smartt, J. *et al.* (1983a) The comparison of seed protein patterns within the genus *Arachis* by polyacrylamide gel electrophoresis. *Biologia Plantarum*, **25**, 266–272.

Klozová, E., Turková, V., Smartt, J. *et al.* (1983b) Immunochemical characterization of seed proteins of some species of the genus *Arachis* L. *Biologia Plantarum*, **25**, 201–208.

Kochert, G., Halward, T., Branch, W.D. and Simpson, C.E. (1991) RFLP variability in peanut (*Arachis hypogaea* L.) cultivars and wild species. *Theoretical and Applied Genetics*, **81**, 565–570.

Krapovicakas, A. (1969) The origin, variability and spread of the groundnut (*Arachis hypogaea*), in *The Domestication and Evolution of Plants and Animals*, (eds P.J. Ucko and G.W. Dimbleby), Duckworth, London, pp. 427–441.

Nigam, S.N. (ed.) (1992) *Groundnut – a global perspective*. Proceedings of an International Workshop 25–29 November 1991 ICRISAT Centre. ICRISAT, Pantancheru, Andra Pradesh, India, pp. 375–378.

Resslar, P.M. (1980) A review of the nomenclature of the genus *Arachis* L. *Euphytica*, **29**, 813–817.

Rick (in press) in Smartt, J. and Simmonds, N.W. (eds) *Evolution of Crop Plants* (2nd edn) Longman, London.

Singh, A.K. (1988) Putative genome donors of *Arachis hypogaea* (Fabaceae), evidence from crosses with synthetic amphidiploids. *Plant Systematics and Evolution*, **160**, 143–151.

Smartt, J. (1978) Makulu Red – a 'Green Revolution' groundnut variety. *Euphytica*, **27**, 605–608.

Smartt, J., Gregory, W.C. and Gregory, M.P. (1978) The genomes of *Arachis hypogaea* L. 1. Cytogenetic studies and putative gene donors, *Euphytica*, **27**, 665–675.

Stalker, H.T. (1991) A new species in section *Arachis* of peanuts with a D genome. *American Journal of Botany*, **78**, 630–637.

Stalker, H.T. (1992) Utilizing *Arachis* germplasm resources in *Groundnut – a Global Perspective*, (ed. S.N. Nigam), ICRISAT Patancheru, A.P., India, 281–295.

York, E.T. and Colwell, W.E. (1951) Soil properties, fertilization and maintenance of soil fertility, in *The Peanut – the Unpredictable Legume*. National Fertilizer Association, Washington DC, pp. 122–172.

Index

DATE DUE

MAY 0 9 1995

FEB 2 3 1996

FEB 2 1 1996

JUN 0 4 1996

MAY 0 2 1996

RET'D SEP 0 5 2007

SEP 0 6 2007